Lignans

Lignans

Special Issue Editor

David Barker

MDPI • Basel • Beijing • Wuhan • Barcelona • Belgrade

MDPI

Special Issue Editor
David Barker
University of Auckland
New Zealand

Editorial Office
MDPI
St. Alban-Anlage 66
4052 Basel, Switzerland

This is a reprint of articles from the Special Issue published online in the open access journal *Molecules* (ISSN 1420-3049) from 2018 to 2019 (available at: https://www.mdpi.com/journal/molecules/special_issues/lignans).

For citation purposes, cite each article independently as indicated on the article page online and as indicated below:

LastName, A.A.; LastName, B.B.; LastName, C.C. Article Title. *Journal Name* **Year**, *Article Number, Page Range.*

ISBN 978-3-03897-908-1 (Pbk)
ISBN 978-3-03897-909-8 (PDF)

Contents

About the Special Issue Editor   ix

Preface to "Lignans"   xi

David Barker
Lignans
Reprinted from: *Molecules* **2019**, *24*, 1424, doi:10.3390/molecules24071424   1

Yongbei Liu, Yupei Yang, Shumaila Tasneem, Nusrat Hussain, Muhammad Daniyal, Hanwen Yuan, Qingling Xie, Bin Liu, Jing Sun, Yuqing Jian, Bin Li, Shenghuang Chen and Wei Wang
Lignans from Tujia Ethnomedicine Heilaohu: Chemical Characterization and Evaluation of Their Cytotoxicity and Antioxidant Activities
Reprinted from: *Molecules* **2018**, *23*, 2147, doi:10.3390/molecules23092147   5

Jiwon Baek, Tae Kyoung Lee, Jae-Hyoung Song, Eunyong Choi, Hyun-Jeong Ko, Sanghyun Lee, Sang Un Choi, Seong Lee, Sang-Woo Yoo, Seon-Hee Kim and Ki Hyun Kim
Lignan Glycosides and Flavonoid Glycosides from the Aerial Portion of *Lespedeza cuneata* and Their Biological Evaluations
Reprinted from: *Molecules* **2018**, *23*, 1920, doi:10.3390/molecules23081920   16

Ya Li, Shuhan Xie, Jinchuan Ying, Wenjun Wei and Kun Gao
Chemical Structures of Lignans and Neolignans Isolated from Lauraceae
Reprinted from: *Molecules* **2018**, *23*, 3164, doi:10.3390/molecules23123164   27

Maria Carla Marcotullio, Massimo Curini and Judith X. Becerra
An Ethnopharmacological, Phytochemical and Pharmacological Review on Lignans from Mexican *Bursera* spp.
Reprinted from: *Molecules* **2018**, *23*, 1976, doi:10.3390/molecules23081976   45

Patrik Eklund and Jan-Erik Raitanen
9-Norlignans: Occurrence, Properties and Their Semisynthetic Preparation from Hydroxymatairesinol
Reprinted from: *Molecules* **2019**, *24*, 220, doi:10.3390/molecules24020220   65

Xianhe Fang and Xiangdong Hu
Advances in the Synthesis of Lignan Natural Products
Reprinted from: *Molecules* **2018**, *23*, 3385, doi:10.3390/molecules23123385   77

Samuel J. Davidson, Lisa I. Pilkington, Nina C. Dempsey-Hibbert, Mohamed El-Mohtadi, Shiying Tang, Thomas Wainwright, Kathryn A. Whitehead and David Barker
Modular Synthesis and Biological Investigation of 5-Hydroxymethyl Dibenzyl Butyrolactones and Related Lignans
Reprinted from: *Molecules* **2018**, *23*, 3057, doi:10.3390/molecules23123057   99

Jian Xiao, Guangming Nan, Ya-Wen Wang and Yu Peng
Concise Synthesis of (+)-β- and γ-Apopicropodophyllins, and Dehydrodesoxypodophyllotoxin
Reprinted from: *Molecules* **2018**, *23*, 3037, doi:10.3390/molecules23113037 125

Patrik A. Runeberg, Yury Brusentsev, Sabine M. K. Rendon and Patrik C. Eklund
Oxidative Transformations of Lignans
Reprinted from: *Molecules* **2019**, *24*, 300, doi:10.3390/molecules24020300 132

Mayra Antúnez-Mojica, Andrés M. Rojas-Sepúlveda, Mario A. Mendieta-Serrano,
Leticia Gonzalez-Maya, Silvia Marquina, Enrique Salas-Vidal and Laura Alvarez
Lignans from *Bursera fagaroides* Affect In Vivo Cell Behavior by Disturbing the Tubulin
Cytoskeleton in Zebrafish Embryos
Reprinted from: *Molecules* **2019**, *24*, 8, doi:10.3390/molecules24010008 . . . . **170**

Marina Pereira Rocha, Priscilla Rodrigues Valadares Campana, Denise de Oliveira Scoaris,
Vera Lucia de Almeida, Julio Cesar Dias Lopes, Julian Mark Hugh Shaw and
Claudia Gontijo Silva
Combined In Vitro Studies and in Silico Target Fishing for the Evaluation of the Biological
Activities of *Diphylleia cymosa* and *Podophyllum hexandrum*
Reprinted from: *Molecules* **2018**, *23*, 3303, doi:10.3390/molecules23123303 . . . . **183**

Agnieszka Szopa, Michał Dziurka, Angelika Warzecha, Paweł Kubica,
Marta Klimek-Szczykutowicz and Halina Ekiert
Targeted Lignan Profiling and Anti-Inflammatory Properties of *Schisandra rubriflora* and
Schisandra chinensis Extracts
Reprinted from: *Molecules* **2018**, *23*, 3103, doi:10.3390/molecules23123103 . . . . **214**

Shuyu Chen, Jingjing Shi, Lisi Zou, Xunhong Liu, Renmao Tang, Jimei Ma,
Chengcheng Wang, Mengxia Tan and Jiali Chen
Quality Evaluation of Wild and Cultivated *Schisandrae Chinensis* Fructus Based on Simultaneous
Determination of Multiple Bioactive Constituents Combined with Multivariate Statistical
Analysis
Reprinted from: *Molecules* **2019**, *24*, 1335, doi:10.3390/molecules24071335 . . . . **230**

Lisa I. Pilkington
Lignans: A Chemometric Analysis
Reprinted from: *Molecules* **2018**, *23*, 1666, doi:10.3390/molecules23071666 . . . . **245**

Carmen Rodríguez-García, Cristina Sánchez-Quesada, Estefanía Toledo,
Miguel Delgado-Rodríguez and José J. Gaforio
Naturally Lignan-Rich Foods: A Dietary Tool for Health Promotion?
Reprinted from: *Molecules* **2019**, *24*, 917, doi:10.3390/molecules24050917 . . . . **269**

Laurine Garros, Samantha Drouet, Cyrielle Corbin, Cédric Decourtil, Thibaud Fidel,
Julie Lebas de Lacour, Emilie A. Leclerc, Sullivan Renouard, Duangjai Tungmunnithum,
Joël Doussot, et al.
Insight into the Influence of Cultivar Type, Cultivation Year, and Site on the Lignans
and Related Phenolic Profiles, and the Health-Promoting Antioxidant Potential of Flax
(*Linum usitatissimum* L.) Seeds
Reprinted from: *Molecules* **2018**, *23*, 2636, doi:10.3390/molecules23102636 . . . . **294**

André F. Brito and Yu Zang
A Review of Lignan Metabolism, Milk Enterolactone Concentration, and Antioxidant Status of
Dairy Cows Fed Flaxseed
Reprinted from: *Molecules* **2019**, *24*, 41, doi:10.3390/molecules24010041 . . . . **309**

Alessandra Durazzo, Massimo Lucarini, Emanuela Camilli, Stefania Marconi,
Paolo Gabrielli, Silvia Lisciani, Loretta Gambelli, Altero Aguzzi, Ettore Novellino,
Antonello Santini, Aida Turrini and Luisa Marletta
Dietary Lignans: Definition, Description and Research Trends in Databases Development
Reprinted from: *Molecules* **2018**, *23*, 3251, doi:10.3390/molecules23123251 . . . . **330**

Delphine Winstel and Axel Marchal
Lignans in Spirits: Chemical Diversity, Quantification, and Sensory Impact of (±)-Lyoniresinol
Reprinted from: *Molecules* **2019**, *24*, 117, doi:10.3390/molecules24010117 . . . . **344**

Thomas Olof Sandberg, Christian Weinberger and Jan-Henrik Smått
Molecular Dynamics on Wood-Derived Lignans Analyzed by Intermolecular Network Theory
Reprinted from: *Molecules* **2018**, *23*, 1990, doi:10.3390/molecules23081990 . . . . **359**

About the Special Issue Editor

David Barker, Associate Professor in Organic and Medicinal Chemistry. David Barker was born in Altrincham, UK. After moving to Australia, he graduated from the University of Sydney with a BSc degree (Honours, First Class) and then completed his PhD in 2002 at the same university, under the supervision of Prof. Margaret Brimble and Associate Professor Malcolm McLeod. After post-doctoral research at the School of Medical Sciences at the University of New South Wales working with Prof. Larry Wakelin, in 2004 he joined the University of Auckland as a lecturer. He is currently Associate Professor in Organic and Medicinal Chemistry and he has a diverse range of synthetic interests, including biologically active natural products, especially lignans and molecules of a marine origin. He also works on a range of drug discovery projects, particularly targeting cancer, and on the development of novel polymeric scaffolds.

Preface to "Lignans"

Lignans are traditionally defined as a class of secondary metabolites that are derived from the dimersation of two or more phenylpropanoid units. Despite their common biosynthetic origins, they boast a vast structural diversity. It is also well-established that this class of compounds exhibits a range of potent biological activities. Owing to these factors, lignans have proven to be a challenging and desirable synthetic target and have instigated the development of a number of different synthetic methods, advancing our collective knowledge towards the synthesis of complex and unique structures. Lignans are also well-known components of a number of widely eaten foods and are frequently studied for their dietary impact. This book is based on the Special Issue of the journal *Molecules* on 'Lignans'. This collection of research and review articles describe topics ranging in scope from recent isolation and structural elucidation of novel lignans, total syntheses and strategies towards lignan synthesis, assessment of their biological activities and potential for further therapeutic development. Research showing the impact of lignans in the food and agricultural industries is also presented.

David Barker
Special Issue Editor

molecules

MDPI

Editorial

Lignans

David Barker

School of Chemical Sciences, University of Auckland, Private Bag, Auckland 92019, New Zealand;
d.barker@auckland.ac.nz

Received: 8 April 2019; Accepted: 10 April 2019; Published: 11 April 2019

The 13 research articles/communications, six reviews, and one perspective that comprise this Special Issue on Lignans, highlight the most recent research and investigations into this diverse and important class of bioactive natural products.

Lignans are traditionally defined as a class of secondary metabolites that are derived from the oxidative dimerization of two or more phenylpropanoid units. Despite their common biosynthetic origins, they boast a vast structural diversity. It is also well-established that this class of compounds exhibit a range of potent biological activities. Owing to these factors, lignans have proven to be a challenging and desirable synthetic target that has instigated the development of a number of different synthetic methods, advancing our collective knowledge towards the synthesis of complex and unique structures.

New lignans are constantly being found and this Special Issue details some of the most recently discovered novel lignans—Liu et al. isolated three new dibenzocyclooctadiene lignans, heilaohulignans A–C from Heilaohu, the roots of Kadsura coccinea, which have a long history of use in Tujia ethnomedicine for the treatment of rheumatoid arthritis and gastroenteric disorders [1]. Heilaohulignan C, in particular, demonstrated cytotoxic activity in a number of human cancer cell lines. Two new lignan glycosides have also been found in the aerial portion of Lespedeza cuneata (Fabaceae), known as Chinese bushclover, a plant that has been used in traditional medicine for the treatment of diseases including diabetes, hematuria, and insomnia [2]. These newly-discovered compounds were tested for their biological activities against human breast cancer cell lines, showing some cytotoxic activity. A review detailing over 270 lignans isolated from Lauraceae, a valuable source of lignans and neolignans is also presented, compiled by Li et al. [3]. Furthermore, Mexican Bursera plants have been used in traditional medicine for treating various pathophysiological disorders and are a rich source of lignans. An Italian research group have summarized the biological activities of lignans isolated from selected Mexican Bursera plants in their review [4].

A subclass of lignans, norlignans lack a carbon present in the parent lignan structure, with 9-norlignans lacking a terminal carbon (C-9). An overview of the occurrence and biological activity of all the 9-norlignans reported to date are given in the article by Eklund and Raitanen, which also reports the semisynthetic preparation of a number of 9-norlignans using the natural lignan hydroxymatairesinol, obtained from spruce knots, as the starting material [5].

As stated above, owing to their potent biological activities, lignans are a popular synthetic target. A summary of the advances in lignan natural product synthesis over the last decade is outlined in the review by Fang and Hu [6].

Davidson et al. have presented their work on their novel, efficient, convergent, and modular synthesis of the well-known dibenzyl butyrolactone lignans through the use of the acyl-Claisen rearrangement to stereoselectively prepare a key intermediate [7]. Not only were the natural products able to be obtained, but the reported synthetic route also enabled the modification of these lignans to give rise to 5-hydroxymethyl derivatives, which were then shown to have an excellent cytotoxic profile which resulted in programmed cell death of Jurkat T-leukemia cells with less than 2% of the incubated cells entering a necrotic cell death pathway.

Advances in the synthesis of aryldihydronaphthalene and arylnapthalene lignans are also detailed in this Special Issue through the concise synthesis of (+)-β- and γ-apopicropodophyllins and dehydrodesoxypodophyllotoxin [8]. This was achieved using the key reaction involving regiocontrolled oxidations of stereodivergent aryltetralin lactones, which were easily accessed from a nickel-catalyzed reductive cascade approach.

As stated, lignans are formed from the oxidative dimerization of two or more phenyl propanoid units. However, numerous oxidative transformations of lignans themselves have been reported in the literature. Runeberg et al. provide an overview on the current findings in this field, focusing on transformations targeting a specific structure, reaction, or an interconversion of the lignan skeleton [9].

The extensive analysis of the potent biological activities of lignans remains a popular avenue of investigation. Antunez-Mojica et al. used a zebrafish embryo model to guide the chromatographic fractionation of antimitotic secondary metabolites, ultimately leading to the isolation of several podophyllotoxin-type lignans from the steam bark of *Bursera fagaroides* [10]. Subsequent to their isolation, the biological effects on mitosis, cell migration, and microtubule cytoskeleton remodeling of the isolated lignans were then further evaluated in zebrafish embryos through various methods. Ultimately, it was demonstrated that the zebrafish model can be a fast and inexpensive in vivo model to identify antimitotic natural products through bioassay-guided fractionation.

Pereira Rocha et al. combined the in silico prediction of biological activities of lignans from *Diphylleia cymosa* and *Podophyllum hexandrum* with in vitro bioassays testing the antibacterial, anticholinesterasic, antioxidant, and cytotoxic activities of these lignans [11]. In this study, the in silico approach was validated and several ethnopharmacological uses and known biological activities of lignans were confirmed, whilst it was shown that others should be investigated for new drugs with potential clinical use.

To explore the differences in lignan composition profiles between various parts and genders of *Schisandra rubriflora* and *Schisandra chinesis* (wuweizi), Szopa et al. used UHPLC-MS/MS [12]. Additionally, the anti-inflammatory activity of plant extracts and individual lignans was tested in vitro for the inhibition of 15-lipooxygenase (15-LOX), phospholipases A2 (sPLA2), cyclooxygenase 1 and 2 (COX-1; COX-2) enzyme activities. The results of anti-inflammatory assays revealed higher activity of *S. rubriflora* extracts, while individual lignans showed significant inhibitory activity against 15-LOX, COX-1 and COX-2 enzymes. Closely related, Chen et al. evaluated the quality and effect of cultivated and wild growing methods on the lignan composition of *Schisandra chinesis* through the use of UFLC-QTRAP-MS/MS in combination with multivariate statistical analysis, demonstrating that the composition differs between plants grown in these conditions and the quality of cultivated wuweizi was not as good as wild wuweizi [13].

While lignans have been shown to exhibit extensive potent biological activities, other factors need to be considered for them to be potential drugs. The physicochemical properties of various lignans subclasses were analyzed by Dr Lisa Pilkington to assess their Absorption, Distribution, Metabolism, Excretion and Toxicity (ADMET) profiles and establish if these compounds are lead-like/drug-like and thus have potential to be or act as leads in the development of future therapeutics [14]. Overall, she established that lignans show a particularly high level of drug-likeness, an observation that, coupled with their potent biological activities, demands future pursuit into their potential for use as therapeutics.

Traditionally, health benefits attributed to lignans have included a lowered risk of heart disease, menopausal symptoms, osteoporosis, and breast cancer. Rodriguez-Garcia et al. present a review that focuses on the potential health benefits attributable to the consumption of different diets containing naturally lignan-rich foods [15]. Current evidence endorses lignans as human health-promoting molecules and, therefore, dietary intake of lignan-rich foods could be a useful way to bolster the prevention of chronic illness, such as certain types of cancers and cardiovascular disease.

Lignan composition profiles of flaxseed, the richest grain source of lignans, was also studied, assessing the relative impact of genetic and geographic parameters on the phytochemical yield and composition [16]. It was found that cultivar is more influential than geographic parameters on the flaxseed phytochemical accumulation yield and composition. In addition, the corresponding antioxidant activity of these flaxseed extracts was evaluated using both in vitro, and in vivo methods, which confirmed that flaxseed extracts are an effective protector against oxidative stress and that secoisolariciresinol diglucoside, caffeic acid glucoside, and *p*-coumaric acid glucoside are the main contributors to the antioxidant capacity. A review of the use and effect of flaxseed as a food source for dairy cows has also been presented [17], covering the gastrointestinal tract metabolism of lignans in humans and animals. The review also provided an in-depth assessment of research towards the impacts of flaxseed products on milk enterolactone concentration and animal health, and the pharmacokinetics of enterolactone consumed through milk, which may have implications to both ruminants and humans' health.

With the rise in exploration of dietary lignans and their various effects, exemplified by the aforementioned studies, the study by Durazzo et al. provides assessment and analysis of the development and management of databases on dietary lignans, which includes a description of the occurrence of lignans in food groups, the initial construction of the first lignan databases, and their inclusion in harmonized databases at national and/or European level [18].

In addition to work into their notable biological activities, there has been a recent increase in investigations exploring lignans in other roles. This includes gaining insight into the effects of barrel-aging on spirits, whereby lignans present in the wooden barrels are released into the aging spirit. To evaluate the impact of lignans in spirits, screening of a number of lignans was set up and used to validate their presence in the spirit and release by oak wood during aging [19]. The most abundant, and also the bitterest, lignan, (±)-lyoniresinol was detected and quantified in a large number of samples to be above the gustatory threshold, suggesting its effect of increased bitterness in spirit taste. Related to this, the molecular dynamics on wood-derived lignans were analyzed by intramolecular network theory by Sandberg et al. [20]. These wood-derived lignan-based ligands called LIGNOLs were studied, where it was found in the hydration studies that tetramethyl 1,4-diol is the LIGNOL which was most likely to form hydrogen bonds to TIP4P solvent.

In summary, it can be seen in this Special Issue that research in natural lignans and lignin-derived compounds continues to be a fruitful area of research. Scientists working across a large number of disciplines continue to be attracted to work on lignans due to their relatively high natural abundance, coupled with their highly potent and diverse range of biological activities.

Conflicts of Interest: The author declares no conflict of interest.

References

1. Liu, Y.; Yang, Y.; Tasneem, S.; Hussain, N.; Daniyal, M.; Yuan, H.; Xie, Q.; Liu, B.; Sun, J.; Jian, Y.; et al. Lignans from Tujia Ethnomedicine Heilaohu: Chemical Characterization and Evaluation of Their Cytotoxicity and Antioxidant Activities. *Molecules* **2018**, *23*, 2147. [CrossRef] [PubMed]
2. Baek, J.; Lee, T.K.; Song, J.-H.; Choi, E.; Ko, H.-J.; Lee, S.; Choi, S.U.; Lee, S.; Yoo, S.-W.; Kim, S.-H.; et al. Lignan Glycosides and Flavonoid Glycosides from the Aerial Portion of Lespedeza cuneata and Their Biological Evaluations. *Molecules* **2018**, *23*, 1920. [CrossRef] [PubMed]
3. Li, Y.; Xie, S.; Ying, J.; Wei, W.; Gao, K. Chemical Structures of Lignans and Neolignans Isolated from Lauraceae. *Molecules* **2018**, *23*, 3164. [CrossRef] [PubMed]
4. Marcotullio, M.C.; Curini, M.; Becerra, J.X. An Ethnopharmacological, Phytochemical and Pharmacological Review on Lignans from Mexican *Bursera* spp. *Molecules* **2018**, *23*, 1976. [CrossRef] [PubMed]
5. Eklund, P.; Raitanen, J.-E. 9-Norlignans: Occurrence, Properties and Their Semisynthetic Preparation from Hydroxymatairesinol. *Molecules* **2019**, *24*, 220. [CrossRef] [PubMed]
6. Fang, X.; Hu, X. Advances in the Synthesis of Lignan Natural Products. *Molecules* **2018**, *23*, 3385. [CrossRef] [PubMed]

7. Davidson, S.J.; Pilkington, L.I.; Dempsey-Hibbert, N.C.; El-Mohtadi, M.; Tang, S.; Wainwright, T.; Whitehead, K.A.; Barker, D. Modular Synthesis and Biological Investigation of 5-Hydroxymethyl Dibenzyl Butyrolactones and Related Lignans. *Molecules* **2018**, *23*, 3057. [CrossRef] [PubMed]

8. Xiao, J.; Nan, G.; Wang, Y.-W.; Peng, Y. Concise Synthesis of (+)-β- and γ-Apopicropodophyllins, and Dehydrodesoxypodophyllotoxin. *Molecules* **2018**, *23*, 3037. [CrossRef] [PubMed]

9. Runeberg, P.A.; Brusentsev, Y.; Rendon, S.M.K.; Eklund, P.C. Oxidative Transformations of Lignans. *Molecules* **2019**, *24*, 300. [CrossRef] [PubMed]

10. Antúnez-Mojica, M.; Rojas-Sepúlveda, A.M.; Mendieta-Serrano, M.A.; Gonzalez-Maya, L.; Marquina, S.; Salas-Vidal, E.; Alvarez, L. Lignans from Bursera fagaroides Affect in Vivo Cell Behavior by Disturbing the Tubulin Cytoskeleton in Zebrafish Embryos. *Molecules* **2019**, *24*, 8. [CrossRef] [PubMed]

11. Pereira Rocha, M.; Valadares Campana, P.R.; de Oliveira Scoaris, D.; de Almeida, V.L.; Dias Lopes, J.C.; Shaw, J.M.H.; Gontijo Silva, C. Combined in Vitro Studies and in Silico Target Fishing for the Evaluation of the Biological Activities of Diphylleia cymosa and Podophyllum hexandrum. *Molecules* **2018**, *23*, 3303. [CrossRef] [PubMed]

12. Szopa, A.; Dziurka, M.; Warzecha, A.; Kubica, P.; Klimek-Szczykutowicz, M.; Ekiert, H. Targeted Lignan Profiling and Anti-Inflammatory Properties of Schisandra rubriflora and Schisandra chinensis Extracts. *Molecules* **2018**, *23*, 3103. [CrossRef]

13. Chen, S.; Shi, J.; Zou, L.; Liu, X.; Tang, R.; Ma, J.; Wang, C.; Tan, M.; Chen, J. Quality Evaluation of Wild and Cultivated Schisandrae Chinensis Fructus Based on Simultaneous Determination of Multiple Bioactive Constituents Combined with Multivariate Statistical Analysis. *Molecules* **2019**, *24*, 1335. [CrossRef]

14. Pilkington, L.I. Lignans: A Chemometric Analysis. *Molecules* **2018**, *23*, 1666. [CrossRef] [PubMed]

15. Rodríguez-García, C.; Sánchez-Quesada, C.; Toledo, E.; Delgado-Rodríguez, M.; Gaforio, J.J. Naturally Lignan-Rich Foods: A Dietary Tool for Health Promotion? *Molecules* **2019**, *24*, 917. [CrossRef] [PubMed]

16. Garros, L.; Drouet, S.; Corbin, C.; Decourtil, C.; Fidel, T.; Lebas de Lacour, J.; Leclerc, E.A.; Renouard, S.; Tungmunnithum, D.; Doussot, J.; et al. Insight into the Influence of Cultivar Type, Cultivation Year, and Site on the Lignans and Related Phenolic Profiles, and the Health-Promoting Antioxidant Potential of Flax (Linum usitatissimum L.) Seeds. *Molecules* **2018**, *23*, 2636. [CrossRef] [PubMed]

17. Brito, A.F.; Zang, Y. A Review of Lignan Metabolism, Milk Enterolactone Concentration, and Antioxidant Status of Dairy Cows Fed Flaxseed. *Molecules* **2019**, *24*, 41. [CrossRef] [PubMed]

18. Durazzo, A.; Lucarini, M.; Camilli, E.; Marconi, S.; Gabrielli, P.; Lisciani, S.; Gambelli, L.; Aguzzi, A.; Novellino, E.; Santini, A.; et al. Dietary Lignans: Definition, Description and Research Trends in Databases Development. *Molecules* **2018**, *23*, 3251. [CrossRef] [PubMed]

19. Winstel, D.; Marchal, A. Lignans in Spirits: Chemical Diversity, Quantification, and Sensory Impact of (±)-Lyoniresinol. *Molecules* **2019**, *24*, 117. [CrossRef] [PubMed]

20. Sandberg, T.O.; Weinberger, C.; Smatt, J.-H. Molecular Dynamics on Wood-Derived Lignans Analyzed by Intermolecular Network Theory. *Molecules* **2018**, *23*, 1990. [CrossRef] [PubMed]

molecules

MDPI

Article

Lignans from Tujia Ethnomedicine Heilaohu: Chemical Characterization and Evaluation of Their Cytotoxicity and Antioxidant Activities

Yongbei Liu [1,†], Yupei Yang [1,†], Shumaila Tasneem [1], Nusrat Hussain [1,2],
Muhammad Daniyal [1] , Hanwen Yuan [1], Qingling Xie [1], Bin Liu [3], Jing Sun [4], Yuqing Jian [1],
Bin Li [1], Shenghuang Chen [1] and Wei Wang [1,2,4,*]

[1] TCM and Ethnomedicine Innovation & Development International Laboratory, Innovative Drug Research Institute, School of Pharmacy, Hunan University of Chinese Medicine, Changsha, 410208, China; ybliu2018@163.com (Y.L.); yangyupei24@163.com (Y.Y.); tasneemshum@gmail.com (S.T.); nusrat_hussain42@yahoo.com (N.H.) daniyaldani151@yahoo.com (M.D.); hanwyuan@hotmail.com (H.Y.); XieQL1992@163.com (Q.X.); cpujyq2010@163.com (Y.J.); libin_hucm@hotmail.com (B.L.); cshtyh@163.com (S.C.)

[2] H.E.J. Research Institute of Chemistry, International Center for Chemical and Biological Sciences, University of Karachi, Karachi-75270, Pakistan

[3] Hunan Province Key Laboratory of Plant Functional Genomics and Developmental Regulation, College of Biology, Hunan University, Changsha 410082, China; binliu2001@hotmail.com

[4] Shaanxi Key Laboratory of Basic and New herbal Medicament Research, Shaanxi Collaborative Innovation Center of Chinese Medicinal Resource Industrialization, Shaanxi University of Chinese Medicine, Xianyang 712046, China; ph.175@163.com

* Correspondence: wangwei402@hotmail.com; Tel.: +86-136-5743-8606

† These authors contributed equally to this work.

Received: 30 July 2018; Accepted: 24 August 2018; Published: 27 August 2018

Abstract: Heilaohu, the roots of *Kadsura coccinea,* has a long history of use in Tujia ethnomedicine for the treatment of rheumatoid arthritis and gastroenteric disorders, and a lot of work has been done in order to know the material basis of its pharmacological activities. The chemical investigation led to the isolation and characterization of three new (**1–3**) and twenty known (**4–23**) lignans. Three new heilaohulignans A-C (**1–3**) and seventeen known (**4–20**) lignans possessed dibenzocyclooctadiene skeletons. Similarly, one was a diarylbutane (**21**) and two were spirobenzofuranoid dibenzocyclooctadiene (**22–23**) lignans. Among the known compounds, **4–5, 7, 13–15** and **17–22** were isolated from this species for the first time. The structures were established, using IR, UV, MS and NMR data. The absolute configurations of the new compounds were determined by circular dichroism (CD) spectra. The isolated lignans were further evaluated for their cytotoxicity and antioxidant activities. Compound **3** demonstrated strong cytotoxic activity with an IC_{50} value of 9.92 μM, compounds **9** and **13** revealed weak cytotoxicity with IC_{50} values of 21.72 μM and 18.72 μM, respectively in the HepG-2 human liver cancer cell line. Compound **3** also showed weak cytotoxicity against the BGC-823 human gastric cancer cell line and the HCT-116 human colon cancer cell line with IC_{50} values of 16.75 μM and 16.59 μM, respectively. A chemiluminescence assay for antioxidant status of isolated compounds implied compounds **11** and **20,** which showed weak activity with IC_{50} values of 25.56 μM and 21.20 μM, respectively.

Keywords: lignans; heilaohu; tujia ethnomedicine; chemical characterization; cytotoxicity; antioxidant

1. Introduction

Kadsura coccinea (Lem.) A. C. Smith belongs to the medicinally important genus *Kadsura* from the Schisandraceae family. It is an evergreen climbing shrub, which is mainly distributed in south-western provinces of P. R. China [1]. Its leaves, fruits, stems and roots are used as medicine. The fruits have unique shapes and high nutritional and medicinal values [2]. The stems and roots are called Heilaohu in Tujia ethnomedicine for looking swarthy while dispelling wind effectively. The isolates of this plant mainly contain lignans, triterpenoids and essential oils. Bioactive lignans and triterpenoids from this plant are of special interest [3]. The compounds from genus *Kadsura* have been reported with different bioactivities including anti-tumor [4,5], anti-HIV [6–8], anti-inflammatory [9,10], inhibition of nitric oxide (NO) production [11,12] and other pharmacological effects.

The lignans from Heilaohu are very important due to their bioactivities and structural diversity. The lignans from this plant can be divided into four different categories on the basis of skeleton types: dibenzocyclooctadienes, spirobenzofuranoid dibenzocyclooctadienes, diarylbutanes, and aryltetralins lignans. Dibenzocyclooctadiene (two benzene rings sharing an eight membered ring neighborhood) is the most common basic skeleton in Heilaohu. Methoxy, hydroxyl and methylenedioxy are the most frequently found substituents at benzene rings, while other important substituents including acetyl-, angeloyl-, tigloyl-, propanoyl-, benzoyl-, cinnamoyl- and butyryl- groups are invariably presented at C-1, C-6 or C-9 [13–15]. Spirobenzofuranoid dibenzocyclooctadienes are rare in other genera and can be considered as the characteristic chemical constituents of genus *Kadsura*. This category features a furan-ring at C-14, C-15 and C-16 positions and a ketonic group at the C-1 or C-3 position [3], and the same connections on the eight membered ring located at the C-6 or C-9 position. Diarylbutanes and aryltetralins have previously been reported but are not very common in genus *Kadsura*, and most of them were found in the DCM ($CHCl_3$) layer and EtOAc layer.

This work was conducted to further explore lignans from Heilaohu. The chemical investigation led to the isolation and characterization of three new (**1**–**3**) and twenty known (**4**–**23**) lignans. The three new *Kadsura* lignans A–C (**1**–**3**) and seventeen known lignans, schizandrin (**4**) [16], binankadsurin A (**5**) [17], acetylbinankadsurin A (**6**) [18], isobutyroylbinankadsurin A (**7**) [19], isovaleroylbinankadsurin A (**8**) [19], kadsuralignan I (**9**) [20], kadsuralignan J (**10**) [20], kadsuralignan L (**11**) [21], kadsulignan N (**12**) [22], longipedunin B (**13**) [15], schisantherin F (**14**) [23], schizanrin D (**15**) [23], acetylgomisin R (**16**) [24], intermedin A (**17**) [25], kadsurarin (**18**) [14], kadsutherin A (**19**) [25] and kadsuphilol A (**20**) [26] possessed dibenzocyclooctadiene skeletons. Similarly, meso-dihydroguaiaretic acid dimethyl (**21**) [27] had a diarylbutane type. Schiarianrin E (**22**) [28] and schiarisanrin A (**23**) [29] contained spirobenzofuranoid dibenzocyclooctadiene lignan skeletons.

A literature survey revealed that kadsulignan I (**9**) exhibited inhibitory effects on LPS-induced NO production in BV-2 cells with IC_{50} value of 21.00 µM [30]. Kadsuralignan L (**11**) demonstrated moderate NO production inhibitory activity with an IC_{50} value of 52.50 µM [21]. Heilaohu has been used for the treatment of rheumatoid arthritis in traditional medicine for a long time, and a few of its isolated compounds have been used for their anti-inflammatory and cytotoxic activities [3]. With the aim of searching for natural compounds which are responsible for folk efficacy and medicinal application as anti-cancer agents and as anti-inflammatory agents, we employed a chemiluminescence assay for anti-oxidant activity to find out the anti-inflammatory properties of a compound. We also used a cytotoxicity assay against cancer cell lines, namely HepG-2 human liver cancer cells, BGC-823 human gastric cancer cells and HCT-116 human colon cancer cells, after the chemical characterization of compounds.

2. Results and Discussion

2.1. Structure Characterization of the Isolated Compounds from Heilaohu

Heilaohulignan A (**1**) (Figure 1) was obtained as an amorphous powder. Its molecular formula, $C_{26}H_{32}O_8$, was determined by [M + Na]$^+$ ion peak at *m/z* 495.1998 (calcd. 495.1995) in HR-ESI-MS, showing 11 degrees of unsaturation. The UV data, with absorption maxima at λ_{max} 242 nm, and its IR spectrum, with absorption bands at 3419 (OH), 1645 (C=C) and 1463 cm^{-1} (aromatic moiety), suggested that **1** is a dibenzocyclooctadiene lignan with a hydroxyl substitution.

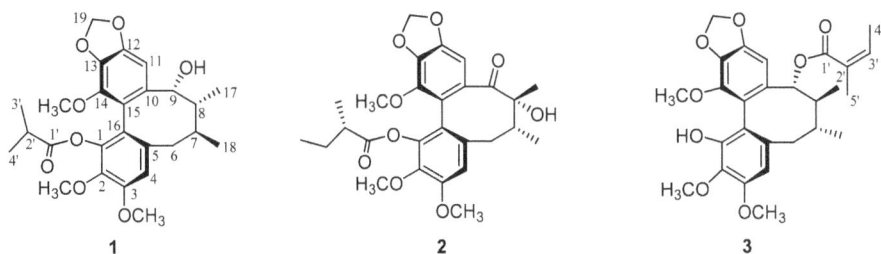

Figure 1. Structures of heilaohulignans A–C (**1–3**).

The ^{1}H- and ^{13}C-NMR spectra of **1** (Table 1) indicated the presence of 12 aromatic carbons (δ_C 141.7 (C-1), 138.8 (C-2), 151.5 (C-3), 113.0 (C-4), 134.8 (C-5), 102.4 (C-11), 148.8 (C-12), 135.0 (C-13), 140.3 (C-14), 118.2 (C-15) and 122.9 (C-16)) and two aromatic proton singlets at δ_H 6.71 (1H, s) and 6.32 (1H, s), which were assignable to H-4 and H-11, respectively. A butane chain was deduced on the cross-peaks of H-6 (δ_H 2.65, m), H-7 (δ_H 2.01, m), H-8 (δ_H 1.81, m) and H-9 (δ_H 4.73, s) in the ^{1}H-^{1}H COSY spectrum. In addition, in the HMBC spectrum, correlations were found between H-9 and C-10, C-8 and C-15, and between H-6 and C-4, C-7 and C-16. The functional moieties evident from the ^{1}H- and ^{13}C-NMR data included one methylenedioxy, three methoxy groups and four methyl groups; the presence of signals at δ_H 0.97 (d, *J* = 7.0 Hz, 3H), 1.09 (d, *J* = 7.0 Hz, 3H) and 2.61 (m, 1H), and δ_C 176.7 (C=O), 18.7 (CH$_3$), 18.7 (CH$_3$), 34.0 (CH) suggested the presence of an isobutyroyl group.

Table 1. ^{1}H- (600 MHz) and ^{13}C-NMR (150 MHz) data of compounds **1**, **2**, and **3** (CDCl$_3$).

Number	1		2		3	
	δ_H (ppm) *J* (Hz)	δ_C (ppm)	δ_H (ppm) *J* (Hz)	δ_C (ppm)	δ_H (ppm) *J* (Hz)	δ_C (ppm)
1	−	141.7	−	143.0	−	147.0
2	−	138.8	−	140.2	−	133.6
3	−	151.5	−	152.2	−	150.5
4	6.71 s	113.0	6.58 s	113.5	6.41 s	106.9
5	−	134.8	−	131.3	−	133.5
6	2.65 m	38.9	2.50 m, 3.03 m	34.6	2.66 m	38.6
7	2.01 m	35.1	2.04 m	43.0	2.12 m	34.8
8	1.81 m	43.0	−	80.9	2.10 m	41.7
9	4.73 s	82.8	−	207.3	5.62 s	82.9
10	−	134.8	−	135.4	−	136.0
11	6.32 s	102.4	6.52 s	100.7	6.54 s	103.0
12	−	148.8	−	148.7	−	148.9
13	−	135.0	−	136.9	−	136.1
14	−	140.3	−	141.7	−	141.1
15	−	118.2	−	117.7	−	119.0
16	−	122.9	−	121.6	−	117.1
17	1.01 d (7.3)	15.3	1.33 s	23.3	1.09 d (7.0)	19.8
18	1.17 d (7.3)	20.0	0.89 d (7.1)	14.8	1.61 dd (7.1, 1.1)	14.2
19	5.93 dd (8.9, 1.4)	101.0	5.96 s, 6.02 s	101.6	5.98 s, 5.93 s	101.2
1'	−	176.7	−	173.1	−	167.5
2'	2.61 dt (13.9, 6.9)	34.0	2.43 m	41.5	−	127.6
3'	0.97 d (7.0)	18.7	1.40 m, 1.62 m	26.8	6.02 d (1.5)	137.2
4'	1.09 d (7.0)	18.7	0.86 t (7.4)	11.7	1.47 s	11.8
5'	−	−	1.02 d (7.0)	16.9	0.97 d (7.1)	15.0
2-OCH$_3$	3.96 s	59.6	3.80 s	60.6	3.84 s	60.7
3-OCH$_3$	3.78 s	61.1	3.86 s	56.1	3.84 s	59.8
14-OCH$_3$	3.89 s	56.0	3.88 s	59.8	3.90 s	55.8

Further analysis of the HMBC spectrum (Figure 2) showed three methoxy groups (δ_H 3.96, 3.78, 3.89, 2-OCH$_3$, 3-OCH$_3$ and 14-OCH$_3$, respectively), with two secondary methyl groups (δ_C 15.3 and 20.0) assignable to CH$_3$-17 and CH$_3$-18, respectively, and one methylenedioxy group (δ_H 6.07, 6.02, each 1H, d, J = 1.5 Hz) located between C-12 and C-13. The NMR data was similar to a known compound, binankadsurin A (**5**) [17]. However, different carbon and proton chemical shifts for C-1′, C-2′, C-3′ and C-4′ indicated that the methyl group located at C-1 was substituted by an isobutyroyl group. Thus, the planar structure of compound **1** was established.

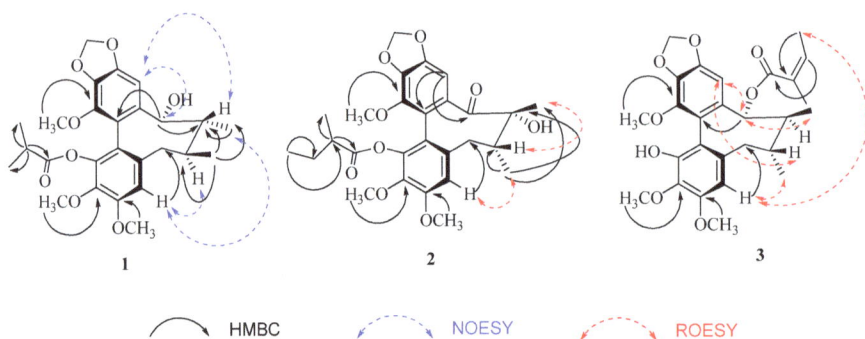

Figure 2. Key HMBC and NOESY correlations of heilaohulignan A (**1**) and ROESY correlations of heilaohulignans B–C (**2–3**).

The biphenyl group in **1** was found to have a twisted boat/chair configuration from its CD spectrum (Figure S63), which showed a negative Cotton effect around 250 nm and a positive value around 220 nm, favoring the *S*-biphenyl configuration as gomisin F [31,32] suggesting **1** possesses an *S*-biphenyl configuration [28]. The observed NOESY correlations (Figure 2) of δ_H 6.71 (H-4) and δ_H 2.01 (H-7), δ_H 1.01 (H$_3$-17), δ_H 6.32 (H-11) and δ_H 1.81 (H-8), δ_H 4.73 (H-9), indicated that CH$_3$-17 was α-oriented, and CH$_3$-18 and H-9 as β-oriented. Hence, 7*S*, 8*R*, and 9*R* configurations were confirmed at C-7, C-8, and C-9, respectively. Based on these data, the structure of **1** was unambiguously determined and was named as heilaohulignan A.

Heilaohulignan B (**2**) (Figure. 1) was obtained as an amorphous powder. Its molecular formula C$_{27}$H$_{32}$O$_9$ was determined by [M + COOH]$^-$ ion peak at m/z 545.2028 (calcd. 545.2026) in HR-ESI-MS. The UV absorption bands at 241 nm and IR absorption bands at 3446 (OH), 1704 (C=O) and 1457, 1579 cm^{-1} (aromatic ring) suggested **2** as a dibenzocyclooctadiene lignan possessing a hydroxy group and an ester.

The ^{1}H- and ^{13}C-NMR spectra of **2** (Table 1) supported a dibenzocyclooctadiene lignan basic skeleton with one methylenedioxy, three methoxy, and a 2-methylbutyryloxy (O-isovaleryl) substituents. The ^{1}H-NMR signals at δ_H 2.43 (m, H-2′), 1.40 (m, H-3′), 1.62 (m, H-3′), 0.86 (t, J = 7.4, H-4′), 1.02 (d, J = 7.0, H-5′) and ^{13}C-NMR signals at δ_C 173.1 (C-1′), 41.5 (C-2′), 26.8 (C-3′), 11.7 (C-4′), and 16.9 (C-5′) were assignable to a 2-methylbutyryloxy group. Comparison of the NMR data of **2** with a known lignan, kadoblongifolins A, showed great similarity [14]. The only difference was the presence of a 2-methylbutyryloxy (O-isovaleryl) group at C-1 in **2**.

The HMBC correlations (Figure 2) of methylenedioxy hydrogens (δ_H 5.96, 1H, s, OCH$_2$O-19a, 6.02, 1H, s, OCH$_2$O-19b) with carbons at δ_C 138.8 (C-12) and 151.5 (C-13) were used to locate its attachment to C-12 and C-13. The methoxy groups were located at C-2, C-3, and C-14, with one secondary methyl group (δ_C 14.8) assignable to CH$_3$-18 and one quaternary methyl group (δ_C 23.3) assignable to CH$_3$-17. The keto group position was confirmed at C-9 by HMBC correlations of H-11 (δ_H 6.52) and H$_3$-17 (δ_H 1.33) with C-9 (δ_C 207.3).

The CD curve of **2** (Figure S64) showed a positive Cotton effect around 205 nm and a negative Cotton effect around 254 nm, favoring the *S*-biphenyl configuration as gomisin F [31,32]. The ROESY (Rotating Frame Overhauser Effect) spectrum (Figure 2) of **2** showed cross-correlation peaks between δ_H 6.58 (H-4) and δ_H 0.89 (CH$_3$-18); δ_H 2.04 (H-7) and δ_H 1.33 (CH$_3$-17), which confirmed that CH$_3$-17 was β-oriented and CH$_3$-18 was α-oriented, thus supporting 7*R* and 8*R* configurations. The compound 2-methylbutyryl is derived from 2-methylbutyryl-CoA biosynthetically, in which the stereochemistry of 2-methyl group is *S*. As stereochemistry is retained, the configuration in the 2-methylbutyryl group was shown as *S*. Based on these spectral data, the structure of **2** was deduced and named as heilaohulignan B.

Heilaohulignan C (**3**) (Figure 1) was obtained as a yellow oil. Its molecular formula, C$_{27}$H$_{32}$O$_8$, was determined by [M + Na]$^+$ ion at *m/z* 507.1990 (calcd. 507.1995) in HR-ESI-MS, suggesting 12 degrees of unsaturation. The UV data, with absorption maxima at λ_{max} 242 nm, and its IR spectrum, with absorption bands at 3417 (-OH), 1700 (C=O) and 1613, 1503 cm^{-1} (aromatic moiety), suggested **3** as a dibenzocyclooctadiene lignan with a hydroxyl substitution.

The ^{1}H- and ^{13}C-NMR spectra of **3** (Table 1) indicated the presence of 12 aromatic carbons, two aromatic protons, one methylenedioxy and three methoxy groups, suggesting the presence of a biphenyl moiety. A butane chain was deduced on the cross-peaks of H-6 (δ_H 2.66, m), H-7 (δ_H 2.12, m), H-8 (δ_H 2.10, m) and H-9 (δ_H 5.62, s) in the ^{1}H-^{1}H COSY spectrum. In the HMBC spectrum (Figure 2.), two methyl groups (CH$_3$-17, CH$_3$-18) exhibited correlations with C-8 and C-9, and three methoxy groups at δ_H 3.84, 3.84 and 3.90 (2-OCH$_3$, 3-OCH$_3$ and 14-OCH$_3$) showed correlations with C-2, C-3, and C-14, respectively, confirming these substituted groups of positions undoubtedly. Thus, the planar structure of compound **3** was the same as angloybinankadsurin A [15]. However, the chemical shifts of C-4′ and C-5′ of **3** were around 4–5 ppm different from the known, which led to doubt about the stereochemistry of **3**.

The biphenyl group in **3** was determined to have an *S*-biphenyl configuration from its CD spectrum (Figure S65), identical to that of **1** and **2**. However, the ROESY experiment (Figure 2) revealed that cross-correlation peaks between δ_H 6.41 (H-4) and δ_H 0.97 (H$_3$-5′); δ_H 6.54 (H-11) and δ_H 2.12 (H-7), δ_H 5.62 (H-9); δ_H 5.62 (H-9) and δ_H 1.09 (H$_3$-17), δ_H 2.12 (H-7); δ_H 1.61 (H$_3$-18) and δ_H 1.47 (H$_3$-4′) confirmed that CH$_3$-17 was β-oriented and CH$_3$-18 was α-oriented, which were essentially different from the known angloybinankadsurin A [15], where CH$_3$-17 and CH$_3$-18 are both α-oriented. Thus *R*, *S*, and *R* configurations were confirmed at C-7, C-8, and C-9, respectively. The ROESY correlation peaks between δ_H 6.02 (H-3′) and δ_H 0.97 (H$_3$-5′), and comparison of data in the literature supported *Z*-configuration for the double bond in the angeloyloxy moiety. Based on these spectral data, the complete structure of **3** was established and it was named as heilaohulignan C.

The spectroscopic data of known compounds (Figures S20–S59) were in good agreement with those reported in the literature. Thus, the known compounds were identified as schizandrin (**4**), binankadsurin A (**5**), acetylbinankadsurin A (**6**), isobutyroylbinankadsurin A (**7**), isovaleroybinankadsurin A (**8**), kadsuralignan I (**9**), kadsuralignan J (**10**), kadsuralignan L (**11**), kadsulignan N (**12**), longipedunin B (**13**), schisantherin F (**14**), schizanrin D (**15**), acetylgomisin R (**16**), intermedin A (**17**), kadsurarin (**18**), kadsutherin A (**19**), kadsuphilol A (**20**), meso-dihydroguaiaretic acid dimethyl ether (**21**), schiarianrin E (**22**), and schiarisanrin A (**23**) (Figure S1).

For the chemical characterization of dibenzocyclooctadienes, there was little to distinguish among different compounds whether the substituents linked to C-1 or C-6/C-9. When the substituents such as acetyl-, angeloyl-, tigloyl-, propanoyl-, benzoyl-, cinnamoyl- and butyryl- groups connected to C-6/C-9, $\delta_{H-6/9}$ was displayed over 5.5 ppm and the relationship with C-1′ could be found in HMBC, while $\delta_{H-6/9}$ would be revealed around 4.7 ppm if substituents attached to C-1. For spirobenzofuranoid dibenzocyclooctadienes, $\delta_{C-1/3}$ with a ketonic group at 195 ppm nearby and δ_{C-16} around 65 ppm could be classified in ^{13}C-NMR. In addition, δ_{C-20} around 78 ppm (CH$_2$) is a typical signal in this compound.

2.2. Cytotoxic Activity of Isolated Compounds

Compounds **1–23** were assayed for their cytotoxic activity against the HepG-2 human liver cancer cell line, the BGC-823 human gastric cancer cell line and the HCT-116 human colon cancer cell line. The results are summarized in Table 2: heilaohulignan C (**3**) showed good cytotoxicity in HepG-2 human liver cancer cells with IC_{50} values of 9.92 μM, and weak cytotoxicity against BGC-823 human gastric cancer cells and HCT-116 human colon cancer cells with IC_{50} values of 16.75 μM and 16.59 μM, respectively. Meanwhile, in the HepG-2 human liver cancer cell line, kadsuralignan I (**9**) and longipedunin B (**13**) revealed weak cytotoxicity with IC_{50} values of 21.72 μM and 18.72 μM, respectively. The remaining compounds showed no cytotoxicity against the three cancer cell lines. Compounds **3**, **9** and **20** demonstrated good activity against all cells. Compounds **1–20**, bearing the same dibenzocyclooctadiene skeleton, indicate that spatial configuration and the relative configuration of structures may have an impact on bioactivities.

Table 2. Cytotoxicity data of compounds **3**, **9** and **13**.

Compound	Cell Lines		
	Hep G-2	HCT-116	BGC-823
3	9.92	16.59	16.75
9	21.72	NO	NO
13	18.72	NO	NO
Taxol	≤0.10	≤0.10	≤0.10

Results are expressed as IC_{50} in μM; Taxol used as a positive control; 'NO' = no activity.

2.3. Antioxidant Activity of Isolated Compounds

Compounds **1–23** were assayed for their antioxidant activity using a chemiluminescence assay. As shown in Table 3, kadsuralignan L (**11**) showed weak activity with an IC_{50} value of 25.56 μM, and kadsuphilol A (**20**) with an IC_{50} value of 21.20 μM. The remaining compounds exhibited no antioxidant activity.

Table 3. Antioxidant activity data of compounds **11** and **20**.

Compounds	Neutrophils IC_{50} (μM)
11	25.56
20	21.20
Vitamin E	77.29

Results are expressed as IC_{50} in μM; Vitamin E used as a positive control.

3. Materials and methods

3.1. Plant Material

Heilaohu were collected from Huaihua City of Hunan Province, China. The plant was identified by Wei Wang. It has been deposited at Sino-Pakistan TCM (Traditional Chinese Medicine) and the Ethnomedicine Research Center, School of Pharmacy, Hunan University of Chinese Medicine, Changsha, China.

3.2. General and Solvents

The HR-ESI-MS spectra were performed on Waters UHPLC-H-CLASS/XEVO G2-XS Qtof, Waters Corporation, Milford, MA, USA. NMR data were recorded on Bruker AV-600 spectrometers (Bruker Technology Co., Ltd., Karlsruhe, Germany) with TMS (Tetramethylsilane) as an internal standard.

Column chromatographic silica gel (80–100 mesh, 200–300 mesh and 300–400 mesh) was purchased from Qingdao Marine Chemical Inc., Qingdao, China. Semipreparative HPLC was performed on an Agilent 1100 liquid chromatograph (Agilent Technologies, Santa Clara, CA, USA) with an Agilent C_{18} (34 mm × 25 cm) column. Fractions were monitored by TLC, and spots were visualized by heating silica gel plates sprayed with 5% H_2SO_4 in vanillin solution. Petroleum ether (PE), hexane, ethyl acetate (EtOAc), ethanol, n-butanol (n-BuOH), methanol (MeOH) and dichloromethane (CH_2Cl_2) were purchased from Shanghai Titan Scientific Co., Ltd, Shanghai, China. Acetonitrile (HPLC grade) and methanol (HPLC grade) were from Merck KGaA, 64271 Darmstadt, Germany.

3.3. Experimental Procedures

Heilaohu (200 kg) was extracted twice with 80% ethanol for 2 h under reflux extraction. All extract solvents were evaporated under vacuum to obtain a crude extract (6 kg). Half of the extracts (3 kg) were suspended in water and partitioned with PE, CH_2Cl_2, EtOAc and n-BuOH, respectively.

The CH_2Cl_2 layer (945 g) was crudely separated on a silica gel column (6 kg, 25 cm × 75 cm) using gradient elution with cyclohexane/ethyl acetate/methanol (80:1:0, 20:1:0, 10:1:0, 5:1:0, 1:1:0, 0:1:0, 0:0:1, *v:v*) to afford twelve fractions. Fraction 5 (49.5 g) was subjected to a silica gel column (8 cm × 45 cm, 800 g), and eluted with cyclohexane/CH_2Cl_2 /EA (1:0:0, 80:1:0, 40:1:0, 20:1:0, 10:1:0, 5:1:0, 3:1:0, 2:1:0, 1:1:0, 0:1:0, 0:40:1, 0:20:1, 0:10:1, 0:5:1, *v:v:v*) to obtain twelve sub-fractions (E1–E12). Sub-fraction E6 (2.0 g) was repeated purified by a silica gel column (3 cm × 60cm, 40 g) eluted with hexane/CHCl₃/acetone (10:20:1, 20:10:1, 20:20:1, 10:10:1, *v:v:v*) to yield **2** (3 mg). Fraction 8 (10 g) was chromatographed by column chromatography on a silica gel (5 cm× 80 cm, 400 g) using the gradient system (CH_2Cl_2/methanol, 40:1, 20:1, 10:1, 5:1, 3:1, 1:1, 0:1, *v:v*) to afford ten fractions (H1–H10). Sub-fraction H7 (PE/CHCl₃/methanol, 80:1:0, 15.0 g) was repeat purified by a silica gel column (4 cm × 45cm, 100 g) eluted with PE/CHCl₃/methanol (40:1:0, 20:1:0, 10:1:0, 5:1:0, 3:1:0, 2:1:0, 1:1:0, 0:1:0, 0:40:1, 0:20:1, 0:10:1, 0:0:1, *v:v:v*) to afford **6** (800 mg). Sub-fraction H8 (15.0 g) was repeat purified by a silica gel column (4cm × 60 cm, 450 g) eluted with PE/acetone (40:1, 20:1, 10:1, 5:1, 3:1, 2:1:0, 1:1, *v:v*) to yield **5** (300 mg). Fraction 9 (53.9 g) was chromatographed by column chromatography on a silica gel (7cm× 60 cm, 500 g) using the gradient system (PE/EA,10:1, 5:1, 3:1, 2:1, 1:1, 0:1, *v:v*) to afford fourteen fractions (I1–I14). Sub-fraction I10 (0.5 g) was repeated purified by an RP-18 column eluted with methanol/water (40%, 50%, 60%, 70%, 80%, 90%, 100%) to yield **4** (5 mg). Sub-fraction I10-4 was purified by semi preparative HPLC with 73% MeOH-H_2O to obtain **17** (5 mg, t_R = 20.6 min). Sub-fraction I10-6 was purified by semi preparative HPLC with 71% MeOH-H_2O to obtain **18** (15 mg, t_R = 41.3 min). Sub-fraction I12 was purified by semi preparative HPLC with 72% MeOH-H_2O to yield **19** (15 mg, t_R = 78.1 min) and **20** (25 mg, t_R = 29.0 min).

The EtOAc layer (530 g) was separated into eight fractions (fraction 1–8) on a 80–100 mesh silica gel column (6.5 kg), using a step gradient elution with PE/EtOAc (10:0, 20:1, 9:1, 8:2, 7:3, 6:4, 1:1, 0:10). Fraction 3 (90 g) was applied to a silica gel column (200–300 mesh, 4.5 kg) with cyclohexane/EtOAc (10:0, 95:5, 90:1, 85:15, 8:2, 7:3, 6:4, 1:1), so as to afford 10 sub-fractions. Sub-fractions were subjected to repeated silica gel columns (isocratic elution and step gradient elution) and Sephadex LH-20 (MeOH/H_2O = 1:1) to give compounds **1** (11.9 mg), **7** (7.7 mg) and **8** (2.1 g), and the mini-fractions were conducted to semi preparative HPLC (MeOH-H_2O) to gain compound **3** (14.6 mg) (77% MeOH-H_2O), **9** (28.4 mg) (80% MeOH-H_2O), **10** (35.6 mg) (80%MeOH/H_2O), **13** (7.3 mg) (76%MeOH-H_2O) and **22** (9.0 mg) (80%MeOH-H_2O). Fraction 4 (60 g) was purified by a silica gel column (300–400 mesh, 4 kg) with PE/EtOAc (10:0, 10:1, 9:1, 8:2,7:3, 6:4, 1:1) to provide 12 sub-fractions. Fraction 5 (50 g) was chromatographed on a silica gel (300–400 mesh, 3.5 kg) to obtain 12 sub-fractions. Sub-fractions from fraction 4 and 5 were fractionated under the same chromatography conditions to obtain compounds **11** (130.1 mg), **12** (29.6 mg), **14** (7.3 mg), **15** (23.2 mg), **16** (2.7 mg), **21** (1.3 mg) and **23** (4.8 mg). The solvents of recrystallization of **7**, **8** and **9** were MeOH (HPLC grade), cyclohexane and hexane, respectively.

3.4. Spectroscopic Data of New Compounds

Heilaohulignan A (**1**): Amorphous powder, $[\alpha]_D^{25}$ − 160.0 (c = 0.0125, CHCl$_3$), UV (MeOH) λ_{max} (log ε): 242 (4.57) nm, IR (KBr) ν_{max} 3419, 1645, 1463, 1101, 721, 655 cm^{-1}; ^{1}H- and ^{13}C-NMR data, Table 1; $^+$HR-ESI-MS m/z 495.1998 ([M + Na]$^+$, calcd. 495.1995).

Heilaohulignan B (**2**):Amorphous powder, $[\alpha]_D^{25}$ + 40.0 (c = 0.10, CHCl$_3$), UV (MeOH) λ_{max} (log ε): 241(4.58) nm, IR (KBr) ν_{max} 3446, 2932, 2360, 1704, 1457, 1102; ^{1}H- and ^{13}C-NMR data, Table 1; $^-$HR-ESI-MS m/z 545.2028 ([M + COOH]$^-$, calcd. 545.2026).

Heilaohulignan C (**3**): Yellow oil, $[\alpha]_D^{25}$ + 48.0 (c = 0.09, CH$_3$OH), UV (MeOH) λ_{max} (log ε): 241 (4.59) nm; IR (KBr) ν_{max} 3417, 2945, 1700, 1457, 1368, 1248, 1108, 1025, ^{1}H- and ^{13}C-NMR data, Table 1; +HR-ESI-MS m/z 507.1990 ([M + Na]$^+$, calcd. 507.1995).

3.5. Cytotoxicity Assay

Cell viability was determined by MTT assay [33]. Taxol was used as a positive control. HepG-2 human liver cancer cells, BGC-823 human gastric cancer cells and HCT-116 human colon cancer cells were seeded at 6×10^3 cells/well in 96-well plates. Cells were allowed to adhere overnight, and then the media were replaced with fresh medium containing selected concentrations of the natural compounds dissolved in DMSO. After 48 h incubation, the growth of the cells was measured. The effect on cell viability was assessed as the percent cell viability compared with the untreated control group, which were arbitrarily assigned 100% viability. The compound concentration required to cause 50% cell growth inhibition (IC$_{50}$) was determined by interpolation from dose–response curves. All experiments were performed in triplicate.

3.6. Antioxidant Assay

Chemiluminescence (CL) [34] was applied to the antioxidant assay process. Chemiluminescence (CL) is a sensitive and accurate method for the measurement of the ability of samples to inhibit the generation of reactive oxygen species (ROS). The positive control was Vitamin E. In our study, we used phorbol 12-myristate 13-acetate (PMA) as stimulus for the production of different ROS by the phagocytic cells. PMA is activator of protein kinase C and an activator of nicotinamide adenine dinucleotide phosphate (NADPH) oxidase. Neutrophils stimulated with PMA give rise to robust chemiluminescence signals by a consequent increase in ROS production. The results were monitored by an Enspire Multimode Plate Reader, Perkin Elmer (EnSpire 2300, PerkinElmer, Singapore) as counts per second (CPS). Briefly, 40 μL diluted whole blood (1:25 dilution in sterile PBS, pH 7.4) or 40 μL poly morphonuclear neutrophils (PMN) (1×10^6/mL) suspended in hanks balanced salt solution (HBSS++), were incubated with different concentrations of compounds. The cells were stimulated with 40 μL of PMA followed by lucigenin as an enhancer (0.5 mM), and then HBSS++ was added to adjust the final volume to 200 μL. The final concentrations of the samples in the mixture were 2.5 μM, 5 μM, 10 μM, 20 μM and 40 μM. Tests were performed in white 96-well microplates which were incubated at 22 °C for 30 min. Control wells contained HBSS++ alone, lucigenin with PMA and cells but no test compounds, and cells with positive control. The inhibition percentage (%) for each concentration was calculated using the following formula:

$$\text{Inhibition percentage (\%)} = 100 - (\text{CPS test / CPS control}) \times 100$$

4. Conclusions

Phytochemical investigation on DCM and EtOAc fractions from Heilaohu were carried out. Twenty-three lignans were isolated and identified by spectroscopic techniques such as 1D-, 2D-NMR and HR-ESI-MS, including three new dibenzocyclooctadiene lignans, heilaohulignans A–C (**1–3**), together with 20 known compounds. Among the known compounds, 12 compounds (**4–5**, **7**, **13–15** and **17–22**) were isolated from this species for the first time.

All isolated compounds were evaluated for their cytotoxicities and antioxidant bioassays. The new dibenzocyclooctadiene heilaohulignans A and B (**1–2**) did not exhibit potential activity on evaluation of cytotoxicity and antioxidant activity. Heilaohulignan C (**3**) demonstrated good cytotoxicity with IC_{50} value of 9.92 µM against HepG-2 human liver cancer cell line, as well as weak cytotoxicity against BGC-823 human gastric cancer cells and HCT-116 human colon cancer cells with IC_{50} values of 16.75 µM and 16.59 µM, respectively. Compounds **9** and **13** revealed weak cytotoxicity with IC_{50} values of 21.72 µM and 18.72 µM, respectively in HepG-2 human liver cancer cells. The chemiluminescence assay implied that compounds **11** and **20** showed weak activity with IC_{50} values of 25.56 µM and 21.20 µM, respectively. Consequently, the underlying cytotoxicity and antioxidant mechanisms of dibenzocyclooctadiene lignans, as well as their main active constituents, need to be further investigated and clarified, providing the material basis on the relationship between traditional uses and modern pharmacological activities.

Supplementary Materials: The following are available online. Figure S1 is structures of compounds **1–23** isolated from Heilaohu. Figures S2–S19 are NMR data of new compounds (**1–3**); Figures S20–S59 are ^{1}H- and ^{13}C-NMR of known compounds (**4–23**); Figures S60–S62 HRESIMS spectrum of new compounds (**1–3**); Figures S63–S65 are CD spectrum of new compounds (**1–3**).

Author Contributions: W.W. (corresponding author) and S.C. conceived and designed the idea of the study; Y.L. and Y.Y. performed the paper writing; S.T. and M.D. participated in data analyses; N.H. contributed to writing and revision. H.Y. performed data processing; Q.X. contributed to charts and figures; B.L. (Bin Liu) designed the content of bioassays; J.S. participated in collection of literatures; Y.J. performed design of experiment process; B.L. (Bin Li) contributed to analysis of NMR data. All of the authors read and approved the final manuscript.

Funding: This study was supported by Hunan Province Universities 2011 Collaborative Innovation Center of Protection and Utilization of Huxiang Chinese Medicine Resources, Hunan Provincial Key Laboratory of Diagnostics in Chinese Medicine, and National Natural Science Foundation of China (81673579); Shaanxi innovative talents promotion plan, technological innovation team (2018TD-005).

Conflicts of Interest: The authors declare no conflict of interest.

References

1. Gao, X.M.; Pu, J.X.; Huang, S.X.; Lu, Y.; Lou, L.G.; Li, R.T.; Xiao, W.L.; Chang, Y.; Sun, H.D. Kadcoccilactones A–J, triterpenoids from *Kadsura coccinea*. *J. Nat. Product.* **2008**, *71*, 1182–1188. [CrossRef] [PubMed]
2. Jian, S.; Yao, J.; Huang, S.; Xing, L.; Wang, J.; García-García, E. Antioxidant activity of polyphenol and anthocyanin extracts from fruits of *Kadsura coccinea* (Lem.) A.C. Smith. *Food Chem.* **2009**, *117*, 276–281.
3. Liu, J.; Qi, Y.; Lai, H.; Zhang, J.; Jia, X.; Liu, H.; Zhang, B.; Xiao, P. Genus *Kadsura*, a good source with considerable characteristic chemical constituents and potential bioactivities. *Phytomedicine* **2014**, *21*, 1092–1097. [CrossRef] [PubMed]
4. Chen, D.F.; Zhang, S.X.; Kozuka, M.; Sun, Q.Z.; Feng, J.; Wang, Q.; Mukainaka, T.; Nobukuni, Y.; Tokuda, H.; Nishino, H.; et al. Interiotherins C and D, two new lignans from *Kadsura interior* and antitumor-promoting effects of related neolignans on epstein-barr virus activation. *J. Nat. Prod.* **2002**, *65*, 1242–1245. [CrossRef] [PubMed]
5. Xu, L.J.; Peng, Z.G.; Chen, H.S.; Wang, J.; Xiao, P.G. Bioactive triterpenoids from *Kadsura heteroclita*. *Chem. Biodivers.* **2010**, *7*, 2289–2295. [CrossRef] [PubMed]
6. Luo, X.; Shi, Y.M.; Luo, R.H.; Luo, S.H.; Li, X.N.; Wang, R.R.; Li, S.H.; Zheng, Y.T.; Du, X.; Xiao, W.L.; et al. Schilancitrilactones A–C: Three unique nortriterpenoids from *Schisandra lancifolia*. *Organ. Lett.* **2012**, *14*, 1286–1289. [CrossRef] [PubMed]
7. Pu, J.X.; Yang, L.M.; Xiao, W.L.; Li, R.T.; Lei, C.; Gao, X.M.; Huang, S.X.; Li, S.H.; Zheng, Y.T.; Huang, H.; et al. Compounds from *Kadsura heteroclita* and related anti-HIV activity. *Phytochemistry* **2008**, *69*, 1266–1272. [CrossRef] [PubMed]
8. Sun, R.; Song, H.C.; Wang, C.R.; Shen, K.Z.; Xu, Y.B.; Gao, Y.X.; Chen, Y.G.; Dong, J.Y. Cheminform abstract: Compounds from *Kadsura angustifolia* with anti-HIV activity. *Bioorg. Med. Chem. Lett.* **2011**, *21*, 961–965. [CrossRef] [PubMed]
9. Lin, L.C.; Shen, C.C.; Shen, Y.C.; Tsai, T.H. Anti-inflammatory neolignans from *Piper kadsura*. *J. Nat. Prod.* **2006**, *69*, 842–844. [CrossRef] [PubMed]

10. Kim, K.H.; Choi, J.W.; Ha, S.K.; Kim, S.Y.; Lee, K.R. Neolignans from *Piper kadsura* and their anti-neuroinflammatory activity. *Bioorg. Med. Chem. Lett.* **2010**, *20*, 409–412. [CrossRef] [PubMed]

11. Kim, K.H.; Choi, J.W.; Choi, S.U.; Ha, S.K.; Kim, S.Y.; Park, H.J.; Lee, K.R. The chemical constituents of *piper kadsura* and their cytotoxic and anti-neuroinflammtaory activities. *J. Enzyme Inhib. Med. Chem.* **2011**, *26*, 254–260. [CrossRef] [PubMed]

12. Mulyaningsih, S.; Youns, M.; Elreadi, M.Z.; Ashour, M.L.; Nibret, E.; Sporer, F.; Herrmann, F.; Reichling, J.; Wink, M. Biological activity of the essential oil of *Kadsura longipedunculata* (schisandraceae) and its major components. *J. Pharm. Pharmacol.* **2010**, *62*, 1037–1044. [CrossRef] [PubMed]

13. Gao, X.M.; Wang, R.R.; Niu, D.Y.; Meng, C.Y.; Yang, L.M.; Zheng, Y.T.; Yang, G.Y.; Hu, Q.F.; Sun, H.D.; Xiao, W.L.; et al. Bioactive dibenzocyclooctadiene lignans from the stems of *Schisandra neglecta*. *J. Nat. Prod.* **2013**, *76*, 1052–1057. [CrossRef] [PubMed]

14. Liu, H.T.; Xu, L.J.; Peng, Y.; Yang, J.S.; Yang, X.W.; Xiao, P.G. Complete assignments of ^{1}H- and ^{13}C-NMR data for new dibenzocyclooctadiene lignans from *Kadsura oblongifolia*. *Magn. Reson. Chem.* **2009**, *47*, 609–612. [CrossRef] [PubMed]

15. Ookawa, N.; Ikeya, Y.; Taguchi, H.; Yosioka, I. The constituents of *Kadsura japonica* Dunal. I. The structures of three new lignans, acetyl-, angeloyl- and caproyl-binankadsurin A. *Chem. Pharm. Bull.* **2008**, *29*, 123–127. [CrossRef]

16. Amujuri, D.; Siva, B.; Poornima, B.; Sirisha, K.; Sarma, A.V.S.; Nayak, V.L.; Tiwari, A.K.; Purushotham, U.; Babu, K.S. Synthesis and biological evaluation of schizandrin derivatives as potential anti-cancer agents. *Eur. J. Med. Chem.* **2018**, *149*, 182–192. [CrossRef] [PubMed]

17. Ookawa, N.; Ikeya, Y.; Sugama, K.; Taguchi, H.; Maruno, M. Dibenzocyclooctadiene lignans from *Kadsura japonica*. *Phytochemistry* **1995**, *39*, 1187–1191. [CrossRef]

18. Wang, N.; Li, Z.L.; Hua, H.M. Study on the chemical constituents of *Kadsura coccinea*. *Chin. J. Med. Chem.* **2012**, *41*, 195–197.

19. Li, L.N.; Xue, H.; Li, X. Three new dibenzocyclooctadiene lignans from *Kadsura longipedunculata*. *Planta Med.* **1991**, *57*, 169–171. [CrossRef] [PubMed]

20. Li, H.; Yang, W.Z. Kadsuralignans H–K from *Kadsura coccinea* and their nitric oxide production inhibitory effects. *J. Nat. Prod.* **2007**, *70*, 1999–2002. [CrossRef] [PubMed]

21. Hu, W.; Li, L.; Wang, Q.; Ye, Y.; Fan, J.; Li, H.X.; Kitanaka, S.; Li, H.R. Dibenzocyclooctadiene lignans from *Kadsura coccinea*. *J. Asian Nat. Prod. Res.* **2012**, *14*, 364–369. [CrossRef] [PubMed]

22. Liu, J.S.; Li, L. Kadsulignans L–N, three dibenzocyclooctadiene lignans from *Kadsura coccinea*. *Phytochemistry* **1995**, *38*, 241–245. [CrossRef]

23. Liu, J.S.; Ma, Y.T. Study on the composition of Schisandrae in Shennongjia. Separation and Identificaton of schisantherin F. *Acta Chim. Sin.* **1988**, *46*, 460–464.

24. Ban, N.K.; Thanh, B.V.; Kiem, P.V.; Minh, C.V.; Cuong, N.X.; Nhiem, N.X.; Huong, H.T.; Anh, H.T.; Eunjeon, P.; Donghwan, S.; et al. Dibenzocyclooctadiene lignans and lanostane derivatives from the roots of *Kadsura coccinea* and their protective effects on primary rat hepatocyte injury induced by t-butyl hydroperoxide. *Planta Med.* **2009**, *75*, 1253–1257. [CrossRef] [PubMed]

25. Li, H.M.; Lei, C.; Luo, Y.M.; Li, X.N.; Li, X.L.; Pu, J.X.; Zhou, S.Y.; Li, R.T.; Sun, H.D. Intermedins A and B: New metabolites from *Schisandra propinqua* var. Intermedia. *Arch. Pharm. Res.* **2008**, *31*, 684–687. [CrossRef] [PubMed]

26. Yaching, S.; Yuanbin, C.; Tingwei, L.; Chiaching, L.; Shorongshii, L.; Yaohaur, K.A.; Khalil, A.T. Kadsuphilols A–H, oxygenated lignans from *Kadsura philippinensis*. *J. Nat. Prod.* **2007**, *70*, 1139–1145.

27. Ward, R.S.; Satyanarayana, P.; Row, L.R.; Rao, B.V.G. The case for a revised structure for hypophyllanthin—An analysis of the ^{13}C-NMR spectra of aryltetralins. *Tetrahedron Lett.* **1979**, *20*, 3043–3046. [CrossRef]

28. Wu, M.D.; Huang, R.L.; Kuo, L.M.Y.; Hung, C.C.; Ong, C.W.; Kuo, Y.H. The Anti-HBsAg (Human Type B Hepatits, Surface Antigen) and Anti-HBeAg (Human Type B Hepatitis, e Antigen) C-(18) Dibenzocyclooctadiene Lignans from *Kadsura matsudai* and *Schizandra arisanensis*. *Cheminform* **2003**, *51*, 1233–1236.

29. Kuo, Y.H.; Kuo, L.M.Y.; Chen, C.F. Cheminform abstract: Four New C19 Homolignans, Schiarisanrins A, B, and D and Cytotoxic Schiarisanrin C, from *Schizandra arisanensis*. *J. Org. Chem.* **1997**, *28*, 3242–3245. [CrossRef]

30. Fang, L.; Xie, C.; Wang, H.; Jin, D.Q.; Xu, J.; Guo, Y.; Ma, Y. Lignans from the roots of *Kadsura coccinea* and their inhibitory activities on LPS-induced NO production. *Phytochem. Lett.* **2014**, *9*, 158–162. [CrossRef]
31. Taguchi, H.; Ikeya, Y. The constituents of schizandra chinensis baill. I. The structures of gomisin A, B and C. *Chem. Pharm. Bull.* **1975**, *23*, 3296–3298. [CrossRef]
32. Ikeya, Y.; Taguchi, H.; Yosioka, I.; Iitaka, Y.; Kobayashi, H. The constituents of schizandra chinensis baill. II. The structure of a new lignan, gomisin D. *Chem. Pharm. Bull.* **1979**, *27*, 1395–1401. [CrossRef] [PubMed]
33. Schauer, U.; Krolikowski, I.; Rieger, C.H. Detection of activated lymphocyte subsets by fluorescence and MTT staining. *J. Immunol. Methods* **1989**, *116*, 221–227. [CrossRef]
34. Tarpey, M.M.; Fridovich, I. Methods of detection of vascular reactive species: Nitric oxide, superoxide, hydrogen peroxide, and peroxynitrite. *Circ. Res.* **2001**, *89*, 224–236. [CrossRef] [PubMed]

Sample Availability: Samples of the compounds **1–23** are available from the authors.

molecules

MDPI

Article

Lignan Glycosides and Flavonoid Glycosides from the Aerial Portion of *Lespedeza cuneata* and Their Biological Evaluations

Jiwon Baek [1], Tae Kyoung Lee [1], Jae-Hyoung Song [2], Eunyong Choi [3], Hyun-Jeong Ko [2], Sanghyun Lee [4], Sang Un Choi [5], Seong Lee [6], Sang-Woo Yoo [7], Seon-Hee Kim [3] and Ki Hyun Kim [1],* ⓘ

[1] School of Pharmacy, Sungkyunkwan University, Suwon 16419, Korea; baekd5nie@gmail.com (J.B.); charmelon8@gmail.com (T.K.L.)
[2] College of Pharmacy, Kangwon National University, Chuncheon 24341, Korea; thdwohud@naver.com (J.-H.S.); hjko@kangwon.ac.kr (H.-J.K.)
[3] Sungkyun Biotech Co. Ltd., Suwon 16419, Korea; eychoi8812@sungkyunbiotech.co.kr (E.C.); seonhee31@gmail.com (S.-H.K.)
[4] Department of Integrative Plant Science, Chung-Ang University, Anseong 17546, Korea; slee@cau.ac.kr
[5] Korea Research Institute of Chemical Technology (KRICT), Daejeon 34114, Korea; suchoi@krict.re.kr
[6] Dankook University Hospital Research Institute of Clinical Medicine, Cheonan 31116, Korea; seonglee@empas.com
[7] Research & Development Center, Natural Way Co., Ltd., Pocheon 11160, Korea; nmrnmr@hanmail.net
* Correspondence: khkim83@skku.edu; Tel.: +82-31-290-7700

Received: 9 July 2018; Accepted: 30 July 2018; Published: 1 August 2018

Abstract: *Lespedeza cuneata* (Fabaceae), known as Chinese bushclover, has been used in traditional medicines for the treatment of diseases including diabetes, hematuria, and insomnia. As part of a continuing search for bioactive constituents from Korean medicinal plant sources, phytochemical analysis of the aerial portion of *L. cuneata* led to the isolation of two new lignan glycosides (**1,2**) along with three known lignan glycosides (**3–7**) and nine known flavonoid glycosides (**8–14**). Numerous analysis techniques, including 1D and 2D NMR spectroscopy, CD spectroscopy, HR-MS, and chemical reactions, were utilized for structural elucidation of the new compounds (**1,2**). The isolated compounds were evaluated for their applicability in medicinal use using cell-based assays. Compounds **1** and **4–6** exhibited weak cytotoxicity against four human breast cancer cell lines (Bt549, MCF7, MDA-MB-231, and HCC70) (IC$_{50}$ < 30.0 μM). However, none of the isolated compounds showed significant antiviral activity against PR8, HRV1B, or CVB3. In addition, compound **10** produced fewer lipid droplets in Oil Red O staining of mouse mesenchymal stem cells compared to the untreated negative control without altering the amount of alkaline phosphatase staining.

Keywords: *Lespedeza cuneata*; lignan glycoside; flavonoid glycoside; cytotoxicity; adipocyte and osteoblast differentiation

1. Introduction

Lespedeza cuneata (Dum. Cours.) G. Don. (Fabaceae), known as Chinese bushclover, is a warm-season, perennial legume that is widely distributed in Korea, China, and India [1]. This plant has been used in folk medicine for the treatment of diseases, including diabetes, hematuria, and insomnia, as well as for the protection of the kidneys, liver, and lungs [2,3]. Previous pharmacological studies of this medicinal plant have revealed that extracts of *L. cuneata* exhibit inhibition of inflammatory mediators in Lipopolysaccharide (LPS)-activated RAW264.7 cells and paw edema in carrageenan-stimulated rats [4], as well as hepatoprotective and antidiabetic effects [1,2,5,6]. A recent study of *L. cuneata* extract reported its in vitro cytotoxic effects against several cancer cell lines including HeLa, Hep3B, A549, and Sarcoma180 [7]. In terms of phytochemical components, it is a rich source of various compounds such as steroids, flavonoids, phenolics [3,6,8], phenylpropanoids [2,9], lignans [5,9], and phenyldilactones [10]. Among the constituents, lignans, and flavonoids are the main components of *L. cuneata*, and the lignans were found to have hepatoprotective [5] and anti-ulcerative colitis activities [9], and the flavonoids were reported to show hepatoprotective [6] and NO-inhibitory effects [11].

As part of a continuing search for bioactive constituents from Korean medicinal plant sources [12–14], the methanol (MeOH) extract of the aerial portion of *L. cuneata* was found to exhibit cytotoxic effects on human ovarian carcinoma cells [15]. In our recent study, bioassay-guided fractionation and repeated chromatography of the MeOH extract of *L. cuneata* resulted in isolation of (−)-9′-O-(α-L-rhamnopyranosyl)lyoniresinol, which suppresses the proliferation of A2780 human ovarian carcinoma cells through induction of apoptosis [15]. In the current study investigating bioactive compounds from the aerial portion of *L. cuneata*, further phytochemical analysis was carried out, which led to the isolation of two new lignan glycosides (**1,2**) along with three known lignan glycosides (**3–7**) and nine known flavonoid glycosides (**8–14**). Numerous analysis techniques, including 1D and 2D NMR spectroscopy, CD spectroscopy, HR-MS, and chemical reactions, were utilized for structural elucidation of the new compounds (**1,2**). Subsequently, we investigated the possible therapeutic effects of the isolated compounds using various cell-based assays. In this paper, we describe the isolation and structural characterization of compounds **1–14** (Figure 1), as well as the evaluation of their applicability to medicinal use including their cytotoxicity, antiviral activity, and their effects on the regulation of adipocyte and osteoblast differentiation.

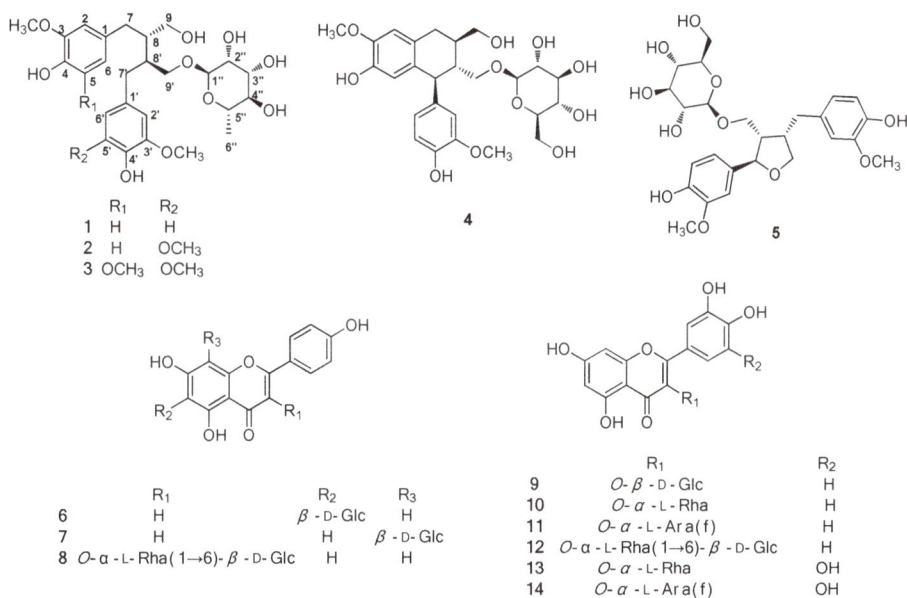

Figure 1. Chemical structures of compounds **1–14**. Glc, glucopyranosyl; Rha, rhamnopyranosyl; Ara(f), arabinofuranosyl.

2. Results and Discussion

2.1. Isolation of the Compounds

The dried aerial portion of *L. cuneata* was extracted with 80% MeOH to produce the methanolic extract, which was sequentially solvent-partitioned with hexane, CH_2Cl_2, EtOAc, and *n*-BuOH to obtain each solvent fraction. Phytochemical analysis of the EtOAc fraction using repeated column chromatography and high performance liquid chromatography (HPLC) purification led to the isolation of two new lignan glycosides (**1,2**) along with three known lignan glycosides (**3–7**) and nine known flavonoid glycosides (**8–14**) (Figure 1).

2.2. Structure Elucidation of the Compounds

Compound (**1**) was isolated as a colorless gum with an optical rotation of ($[\alpha]_D^{25}$ +24.0 (*c* 0.05, MeOH). The molecular formula was determined to be $C_{26}H_{36}O_{10}$ from the molecular ion peak $[M + H]^+$ at *m/z* 509.2384 (calculated for $C_{26}H_{37}O_{10}$ 509.2387) in positive mode High-resolution electrospray ionisation mass spectrometry (HRESIMS) and the NMR spectroscopic data (Table 1). The infrared (IR) spectrum exhibited absorptions of hydroxy groups (3351 cm^{-1}) and phenyl rings (1521 and 1455 cm^{-1}). The ^{1}H NMR spectrum (Table 1) showed signals from two sets of aromatic protons, one at δ_H 6.67 (1H, d, *J* = 8.0 Hz, H-5), 6.56 (1H, d, *J* = 2.0 Hz, H-2), and 6.53 (1H, dd, *J* = 8.0, 2.0 Hz, H-6) and another at δ_H 6.66 (1H, d, *J* = 8.0 Hz, H-5'), 6.54 (1H, d, *J* = 2.0 Hz, H-2'), and 6.53 (1H, dd, *J* = 8.0, 2.0 Hz, H-6'), as well as two methoxy groups at δ_H 3.74 (3H, s) and 3.73 (3H, s). The characteristic NMR data of **1**, combined with heteronuclear single quantum correlation (HSQC) data, also showed signals for four methylenes at δ_H 3.77 (1H, dd, *J* = 10.0, 6.0 Hz, H-9'a) and 3.33 (1H, m, H-9'b)/δ_C 69.7 (C-9'), δ_H 3.69 (1H, m, H-9a), and 3.48 (1H, dd, *J* = 11.0, 7.0 Hz, H-9b)/δ_C 62.6 (C-9), δ_H 2.67 (1H, dd, *J* = 14.0, 7.0 Hz, H-7a) and 2.56 (1H, dd, *J* = 14.0, 8.5 Hz, H-7b)/δ_C 35.6 (C-7), and δ_H 2.60 (2H, m, H-7')/δ_C 35.8 (C-7'), and two methines at δ_H 2.07 (1H, m, H-8')/δ_C 40.7 (C-8') and 1.94 (1H, m, H-8)/δ_C 44.1 (C-8), which are indicative of a secoisolariciresinol-type lignan [16,17]. In addition, characteristic rhamnose NMR signals were observed at δ_H 4.63 (1H, d, *J* = 1.5 Hz, H-1'') and 1.25 (3H, d, *J* = 6.0 Hz, H-6''), δ_C 102.0, 73.7, 72.4, 72.2, 69.9, and 17.8 [18]. These data suggest that compound **1** is a secoisolariciresinol-type lignan glycoside, and the ^{1}H and ^{13}C NMR spectra of **1** were highly similar to those of (−)-secoisolariciresinol-*O*-α-L-rhamnopyranoside [19]. The planar gross structure of **1** was established based on the ^{1}H-^{1}H correlation spectroscopy (COSY) and Heteronuclear multiple bond correlation (HMBC) spectral data (Figure 2). However, the absolute stereochemistry of **1** was not identical to (−)-secoisolariciresinol-*O*-α-L-rhamnopyranoside because compound **1** showed a positive optical rotation ($[\alpha]_D^{25}$ +24.0, *c* 0.05, MeOH) similar to chaenomiside F (compound **3**) ($[\alpha]_D^{25}$ +30.0, *c* 0.1, MeOH) [20] and (−)-secoisolariciresinol-*O*-α-L-rhamnopyranoside showed a negative rotation ($[\alpha]_D^{20}$ −49.5, *c* 0.30, acetone) [19]. Enzymatic hydrolysis was carried out to further confirm the absolute configuration of compound **1**, which yielded an aglycone and a rhamnose. The aglycone was determined to be (+)-secoisolariciresinol (**1a**) through LC/MS analysis with an *m/z* signal of 361.2 $[M − H]^-$ and a positive optical rotation ($[\alpha]_D^{25}$ +30.0, *c* 0.02, acetone) [16]. The CD spectrum of **1a** showed positive Cotton effects at 209, 223, and 288 nm, and negative effects at 216 and 230 nm, which is the first report of an experimental CD spectrum of (+)-secoisolariciresinol. The coupling constant (*J* = 1.5 Hz) of the anomeric proton of the rhamnose revealed the α-configuration of the anomeric proton [21]. The identity of L-rhamnose was established through LC/MS analysis of the rhamnose obtained from the enzymatic hydrolysis [22,23]. Thus, the structure of compound **1** was determined to be (+)-secoisolariciresinol-*O*-α-L-rhamnopyranoside.

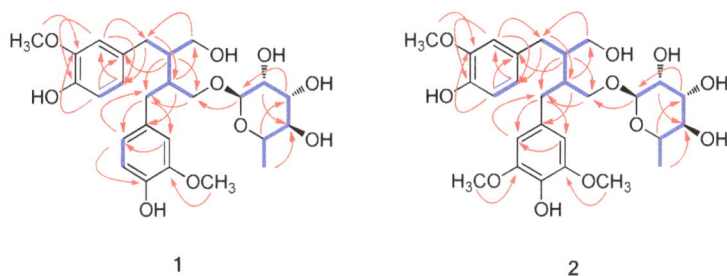

Figure 2. ^{1}H-^{1}H COSY (——) and key HMBC (⌒) correlations for **1** and **2**.

Table 1. ^{1}H and ^{13}C NMR data of **1** and **2** in CD$_3$OD (δ in ppm, 800 MHz for ^{1}H and 200 MHz for ^{13}C) [a].

Position	1		2	
	δ$_H$	δ$_C$	δ$_H$	δ$_C$
1		133.6 s		132.2 s
2	6.56 d (2.0)	113.0 d	6.54 d (2.0)	111.9 d
3	6.67 ^{α} d (8.0)	115.5 d	6.65 d (8.0)	114.2 d
4		145.4 s		144.5 s
5		148.9 s		147.5 s
6	6.53 dd (8.0, 2.0)	122.6 d	6.52 dd (8.0, 2.0)	121.3 d
7	2.67 dd (14.0, 7.0); 2.56 dd (14.0, 8.5)	35.6 t	2.69 dd (14.0, 6.5); 2.53 dd (14.0, 9.0)	34.5 t
8	1.94 m	44.1 d	1.92 m	42.5 d
9	3.69 m; 3.48 dd (11.0, 7.0)	62.6 t	3.71 m; 3.48 dd (11.0, 7.0)	61.2 t
1′		133.6 s		131.4 s
2′	6.54 d (2.0)	113.0 d	6.28 s	105.3 d
3′		148.8 s		147.6 s
4′		145.4 s		133.4 s
5′	6.66 ^{α} d (8.0)	115.5 d		147.6 s
6′	6.53 dd (8.0, 2.0)	122.6 d	6.28 s	105.3 d
7′	2.60 m	35.8 t	2.60 m	35.2 t
8′	2.07 m	40.7 d	2.08 m	39.3 d
9′	3.77 dd (10.0, 6.0); 3.33 m	69.7 t	3.79 dd (10.0, 6.0); 3.35 m	67.9 t
1″	4.63 d (1.5)	102.0 d	4.64 d (1.5)	100.7 d
2″	3.82 dd (3.5, 1.5)	72.2 d	3.81 dd (3.5, 1.5)	71.0 d
3″	3.68 dd (9.5, 3.5)	72.4 d	3.68 dd (9.5, 3.5)	71.1 d
4″	3.38 t (9.5)	73.7 d	3.38 t (9.5)	72.5 d
5″	3.62 dq (9.5, 6.0)	69.9 d	3.62 dq (9.5, 6.0)	68.7 d
6″	1.25 d (6.0)	17.8 q	1.25 d (6.0)	16.5 q
3-OCH$_3$	3.73 ^{β} s	55.8 q	3.72 s	54.7 q
3′-OCH$_3$	3.74 ^{β} s	55.8 q	3.74 s	55.1 q
5′-OCH$_3$			3.74 s	55.1 q

[a] *J* values are in parentheses and reported in Hz; ^{13}C NMR assignments based on ^{1}H-^{1}H COSY, HSQC, and HMBC experiments; $^{α, β}$ Exchangeable peaks.

Compound **2** was obtained as a colorless gum with a positive optical rotation value of $[α]_D^{25}$ +27.5 (*c* 0.04, MeOH). The molecular formula of **2** was determined to be C$_{27}$H$_{38}$O$_{11}$ from the molecular ion peak at *m/z* 537.2343 [M − H]$^−$ (calculated for C$_{27}$H$_{37}$O$_{11}$ 537.2336) in the negative mode HRESIMS and the NMR spectroscopic data (Table 1). The ultraviolet (UV) and IR spectra of **2** were almost identical to those of **1**. The ^{1}H and ^{13}C NMR spectra (Table 1) were also quite similar to those of **1**, with a noticeable difference being that the proton signals for a 1,3,4-trisubstituted aromatic ring in **1** were absent and the proton signals for a typical 1,3,4,5-tetrasubstituted aromatic ring (δ$_H$ 6.28 (2H, s)) and an overlapped signal for two methoxyl groups (δ_H 3.74 (6H, s)) was present in **2**. In light of these data, compound **2** was also deduced to be one of the secoisolariciresinol-type lignans like compound **1**, and the differences in the structure of **2** compared to compound **1** were confirmed through analysis of the ^{1}H-^{1}H COSY and HMBC data (Figure 2). Specifically, an HMBC correlation from the methoxyl group (δ_H 3.74) to C-3′/C-5′ (δ_C 147.6) was observed, which led to the assignment of the methoxyl group at C-3′/C-5′. The similarity between the characteristic CD curves of **1** (positive at 206, 229, and 285 nm and negative at 217 nm) and **2** (positive at 205, 233, and 283 nm and negative at 221 nm) revealed that the absolute configuration of **2** was identical to compound **1** as the 8S and 8′S form, which was also supported by the positive optical rotation value ($[\alpha]_D^{25}$ +27.5, *c* 0.04, MeOH) of **2** like that of **1**. Enzymatic hydrolysis was conducted to further confirm the absolute configuration of **2**, which yielded an aglycone (**2a**) and a rhamnose. As expected, the aglycone (**2a**) was determined to be (+)-seco-5′-methoxy-isolariciresinol using LC/MS analysis with an *m/z* signal of 393.2 [M + H]$^+$ and a positive optical rotation value of **2a** ($[\alpha]_D^{25}$ +25.5, *c* 0.02, acetone) [16]. The characteristic small coupling constant (*J* = 1.5 Hz) of the anomeric proton of the rhamnose at δ_H 4.64 indicated the α-configuration of the rhamnose [21], and L-rhamnose was confirmed using LC/MS analysis of the rhamnose obtained from the enzymatic hydrolysis of **2** [22,23]. Accordingly, the structure of compound **2** was determined to be (+)-seco-5′-methoxy-isolariciresinol-9′-*O*-α-L-rhamnopyranoside.

The known compounds were identified as chaenomiside F (**3**) [16,20], (+)-isolariciresinol 9-*O*-β-D-glucoside (**4**) [5], lariciresinol 9-*O*-β-D-glucopyranoside (**5**) [24], isovitexin (**6**) [25], vitexin (**7**) [26], nicotiflorin (**8**) [27], isoquercetin (**9**) [28], quercimelin (**10**) [29], avicularin (**11**) [30], rutin (**12**) [28], myricitrin (**13**) [31], and betmidin (**14**) [32,33], through comparison of their spectroscopic data, including ^{1}H and ^{13}C NMR, and physical data with previously reported values, as well as through LC/MS analysis.

2.3. Cytotoxic Activity of Isolated Compounds against Human Tumor Cell Lines

Based on the cytotoxic activity of the MeOH extract of *L. cuneata* in our recent study [15], the cytotoxic activities of the isolated compounds (**1**–**14**) were evaluated by determining their inhibitory effects on human tumor cell growth in human breast cancer cells (Bt549, MCF7, MDA-MB-231 and HCC70), using a sulforhodamine B (SRB) bioassay [12,34]. The results (Table S1) demonstrated that compound **1** showed cytotoxicity against Bt549, MDA-MB-231, and HCC70 cell lines with IC$_{50}$ values ranging from 24.38–26.16 μM. Compounds **4** and **5** exhibited cytotoxicity against MCF7 (IC$_{50}$: 28.08 μM) and HCC70 (IC$_{50}$: 24.81 μM) cell lines, respectively, and compound **6** showed cytotoxic activity against MCF7, MDA-MB-231, and HCC70 cell lines with IC$_{50}$ values ranging from 27.57–29.18 μM (Table S1). However, other compounds were inactive (IC$_{50}$ > 30.0 μM). Although recent studies of *L. cuneata* extract have reported that the extract showed cytotoxic effects against various cancer cell lines [7,15], the isolated compounds (**1**–**14**) did not appear to be responsible for the cytotoxicity.

2.4. Antiviral Activity of the Isolated Compounds against PR8, HRV1B, and CVB3 Infection

Recently, many studies exploring antiviral natural products and organic synthetic compounds have reported that a variety of flavonoids exhibit potent antiviral activity by inhibiting the early stages of viral infection, viral protein expression, and neuraminidase activity [35–37]. Therefore, we assessed the isolated compounds (**1**–**14**) for their antiviral activity against PR8, HRV1B, and CVB3 infection in A549, Vero, and HeLa cells, respectively. Less than 10% of the cells survived in the positive-control group (cells with virus only) after 48 hours of infection. In addition, cells treated with compounds **1**–**14** (10 μM) also had less than 10% survival. Because we could not identify any significant differences between the control and test groups, these results suggest that the compounds do not show significant antiviral activity against PR8, HRV1B, or CVB3.

*2.5. Regulatory Effects of Compound **10** on Differentiation into Adipocytes and Osteoblasts*

Mesenchymal stem cells (MSCs) in the bone marrow are pluripotent cells, which differentiate into osteocytes as well as adipocytes. Since microenvironmental changes such as hormones, immune responses, and metabolism cause alterations in the regulation of MSC differentiation, where alterations in the expression of the related genes might disturb the balance between osteoprogenitor and adipocyte progenitor cells in osteoporosis patients [38], natural products that are able to suppress MSC differentiation toward adipocytes and/or promote osteogenic differentiation of MSC would be promising in the management of postmenopausal osteoporosis. The biological activity of compound **10** was additionally tested regarding its effects on the differentiation of mouse MSCs into adipocytes or osteoblasts, since large amounts of compound **10** was isolated among the isolated compounds. Compound **10** was added to the MSC culture media during adipocyte differentiation. Compound **10** slightly reduced the formation of lipid droplets and resulted in somewhat fewer Oil Red O (ORO)-stained cells compared to the normally differentiated adipocytes (Figure 3A). However, ALP staining and ALP activity in the compound **10**-treated cells did not increase during the MSC differentiation into osteoblasts, in contrast to the positive control group treated with oryzativol A (Figure 3B). These results demonstrate that compound **10** marginally suppressed adipogenesis of MSCs but did not influence osteogenesis.

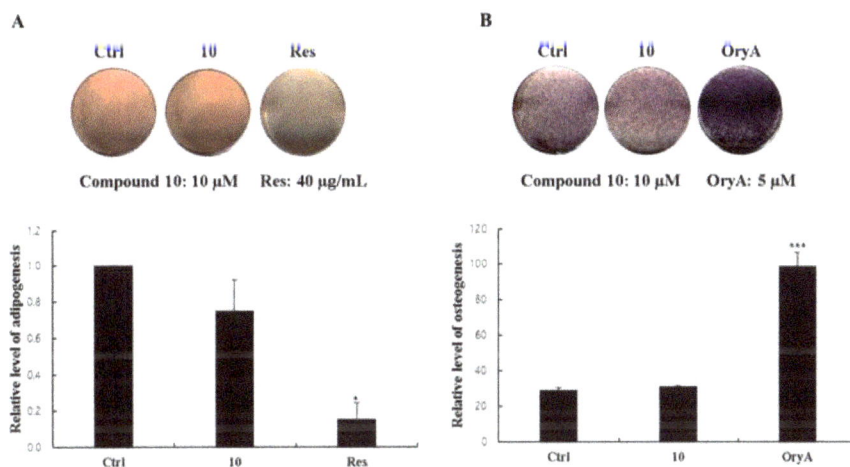

Figure 3. Reciprocal effects of compound **10** on the differentiation of MSCs into adipocytes or osteoblasts. Mouse mesenchymal stem cells (C3H10T1/2) were treated with 10 μM compound **10**. After adipogenic differentiation, the cells were stained with Oil Red O (ORO), and the number of stained lipid droplets was quantitatively evaluated (**A**). After osteoblast differentiation, the cells were stained for ALP levels, and the ALP activity was measured (**B**). Ctrl represents untreated negative control. For the positive controls, 40 micrograms of resveratrol (Res) was used for adipogenesis and 5 μM of oryzativol A (OryA) was added for osteogenesis. * denotes $0.01 \leq p \leq 0.05$ and *** denotes $p < 0.001$.

3. Materials and Methods

3.1. Plant Material

The aerial portions of *L. cuneata* were collected from Mt. Bangtae, Inje, Kangwon Province, Republic of Korea, in October 2016. The plant materials were identified by one of the authors, Prof. S. Lee. A voucher specimen (YKM-2016) was deposited at the herbarium of the School of Pharmacy, Sungkyunkwan University, Suwon, Republic of Korea.

3.2. Extraction and Isolation

The dried aerial portions of *L. cuneata* (4.2 kg) were extracted three times with 4.2 L of 80% MeOH for three days at room temperature and filtered. The resultant filtrate was evaporated under reduced pressure using a rotavap to obtain the MeOH extract (401.8 g), which was suspended in distilled H_2O (2 L) and successively solvent-partitioned with hexane, CH_2Cl_2, EtOAc, and n-BuOH (2.0 L $\times$ 3 for each) to yield the hexane- (20.6 g), CH_2Cl_2- (0.7 g), EtOAc- (12.7 g), and n-BuOH-soluble (69.3 g) fractions. The EtOAc-soluble fraction (12.7 g) was subjected to Diaion HP-20 column chromatography with a gradient solvent system of MeOH-H_2O (0–100% MeOH) to afford six fractions (A–F). Fraction D (5.4 g) was separated using RP-C18 column chromatography with a gradient solvent system of MeOH-H_2O (30–100% MeOH) to yield six sub-fractions (D_1–D_6). Sub-fraction D_3 (2.8 g) was fractionated using silica gel column chromatography with a gradient solvent system of CH_2Cl_2-MeOH-H_2O (15:1:0–9:3:0.5 $v/v/v$) to produce 10 sub-fractions (D_3-1–D_3-10). Sub-fraction D_3-7 (1.1 g) was separated using an RP-C18 column with 60% MeOH to produce four sub-fractions (D_3-71–D_3-74). Sub-fraction D_3-72 (506.7 mg) was subjected to silica gel column chromatography with a gradient solvent system of CH_2Cl_2-MeOH-H_2O (10:1:0–1:1:0.25, $v/v/v$) to give five sub-fractions (D_3-721–D_3-725). Sub-fraction D_3-722 (316.4 mg) was subjected to Sephadex LH-20 column chromatography with 100% MeOH to produce 10 sub-fractions (D_3-722A–D_3-722J). Sub-fraction D_3-722C (230.0 mg) was purified using semi-preparative HPLC with a Phenomenex Luna phenyl-hexyl column (18% MeCN, flow rate: 2 mL/min) to yield compound **5** (1.4 mg, t_R = 37.0 min). Sub-fraction D_3-73 (158.8 mg) was subjected to Sephadex LH-20 column chromatography with 100% MeOH to give 10 sub-fractions (D_3-73A–D_3-73J). Compounds **2** (0.7 mg, t_R = 49.5 min) and **3** (1.8 mg, t_R = 41.5 min) were obtained from sub-fraction D_3-73B (24.5 mg) using semi-preparative HPLC with a Phenomenex Luna phenyl-hexyl column (18% MeCN, flow rate: 2 mL/min). Compound **1** (7.6 mg, t_R = 61.0 min) was isolated from sub-fraction D_3-73C (44.7 mg) using semi-preparative HPLC with a Phenomenex Luna phenyl-hexyl column (18% MeCN, flow rate: 2 mL/min). Compound **14** (3.7 mg, t_R = 20.5 min) was obtained from sub-fraction D_3-73I (8.2 mg) using semi-preparative HPLC with a Phenomenex Luna phenyl-hexyl column (21% MeCN, flow rate: 2 mL/min). Sub-fraction D_3-74 (127.6 mg) was subjected to Sephadex LH-20 column chromatography with 100% MeOH to give eight sub-fractions (D_3-741–D_3-748). Compounds **9** (0.7 mg, t_R = 30.5 min) and **10** (32.8 mg, t_R = 48.0 min) were isolated from sub-fraction D_3-746 (42.3 mg) using semi-preparative HPLC with a Phenomenex Luna phenyl-hexyl column (18% MeCN, flow rate: 2 mL/min). Sub-fraction D_3-8 (515.0 mg) was subjected to RP-C18 column chromatography using a gradient solvent system of 40–60% MeOH to produce four sub-fractions (D_3-81–D_3-84). Sub-fraction D_3-82 (346.7 mg) was subjected to silica gel column chromatography with a gradient solvent system of CH_2Cl_2-MeOH (10:1–1:1, v/v) to give four sub-fractions (D_3-821–D_3-824). Sub-fraction D_3-822 (54.8 mg) was applied to Sephadex LH-20 column chromatography with 100% MeOH to produce six sub-fractions (D_3-822A–D_3-822F). Compound **4** (3.5 mg, t_R = 39.0 min) was purified from sub-fraction D_3-822A (16.3 mg) using semi-preparative HPLC with a Phenomenex Luna phenyl-hexyl column (15% MeCN, flow rate: 2 mL/min). Sub-fraction D_3-824 (78.1 mg) was separated using Sephadex LH-20 column chromatography with 100% MeOH to yield five sub-fractions (D_3-824A–D_3-824E). Sub-fraction D_3-824C (22.4 mg) was separated using semi-preparative HPLC with a Phenomenex Luna phenyl-hexyl column (16% MeCN, flow rate: 2 mL/min) to obtain compound **8** (2.3 mg, t_R = 72.5 min). Sub-fraction D_3-824D (37.3 mg) was separated using semi-preparative HPLC with a Phenomenex Luna phenyl-hexyl column (14% MeCN, flow rate: 2 mL/min) to obtain compound **13** (0.5 mg, t_R = 73.0 min), and compound **13**'s washing fraction D_3-824DW (20.5 mg) was collected. Compound **11** (1.0 mg, t_R = 49.5 min) was purified using semi-preparative HPLC with a Phenomenex Luna phenyl-hexyl column (18% MeCN, flow rate: 2 mL/min) from sub-fraction D_3-824DW (20.5 mg). Sub-fraction D_3-10 (132.7 mg) was applied to Sephadex LH-20 column chromatography with 80% MeOH to produce nine sub-fractions (D_3-101–D_3-109). Sub-fraction D_3-108 (50.3 mg) was further separated using semi-preparative HPLC with a Phenomenex Luna phenyl-hexyl column (38% MeOH, flow rate: 2 mL/min) to yield compound **12** (2.1 mg, t_R = 72.0 min). Finally, compounds **6** (0.6 mg, t_R = 37.0 min) and **7** (2.0 mg, t_R = 39.0 min) were isolated from sub-fraction D$_3$-109 (17.2 mg) using semi-preparative HPLC with a Phenomenex Luna phenyl-hexyl column (20% MeCN, flow rate: 2 mL/min).

3.2.1. (+)-Secoisolariciresinol-*O*-α-L-rhamnopyranoside (1)

Colorless gum; $[\alpha]_D^{25}$ +24.0 (*c* = 0.05, MeOH); ESIMS (negative mode) *m/z*: 507 [M − H]$^-$; HRESIMS (positive mode) *m/z*: 509.2384 [M + H]$^+$, calculated for C$_{26}$H$_{37}$O$_{10}$, 509.2387; UV (MeOH) λ_{max} nm (log ε): 205 (2.29), 233 (3.43), 283 (0.76); IR (KBr) ν_{max} cm^{-1}: 3703, 3351, 2947, 2833, 2513, 2302, 2047, 1521, 1455; CD (MeOH) λ_{max} nm (Δε): 206 (+19.2), 217 (−11.5), 229 (+10.3), 285 (+2.8); ^{1}H (CD$_3$OD, 800 MHz) and ^{13}C (CD$_3$OD, 200 MHz) NMR spectroscopic data, see Table 1.

3.2.2. (+)-Seco-5′-methoxy-isolariciresinol-9′-*O*-α-L-rhamnopyranoside (2)

Colorless gum; $[\alpha]_D^{25}$ +27.5 (*c* = 0.04, MeOH); ESIMS (negative mode) *m/z*: 537 [M − H]$^-$; HRESIMS (negative mode) *m/z*: 537.2343 [M − H]$^-$, calculated for C$_{27}$H$_{37}$O$_{11}$, 537.2341; UV (MeOH) λ_{max} nm (log ε): 205 (2.29), 233 (3.43), 283 (0.76); IR (KBr) ν_{max} cm^{-1}: 3705, 3340, 2945, 2831, 2512, 2302, 2045, 1516, 1453; CD (MeOH) λ_{max} nm (Δε): 205 (+11.5), 221 (−23.4), 233 (+13.8), 283 (+3.1); ^{1}H (CD$_3$OD, 800 MHz) and ^{13}C (CD$_3$OD, 200 MHz) NMR spectroscopic data, see Table 1.

3.3. Enzymatic Hydrolysis of Compounds 1,2

A solution of each compound (1.0 mg) in H$_2$O (1 mL) was individually hydrolyzed with naringinase (10 mg, from *Penicillium* sp.; ICN Biomedicals Inc., Irvine, CA, USA) at 40 °C for 36 h. Each reaction mixture was extracted with CH$_2$Cl$_2$ to yield the individual CH$_2$Cl$_2$ extract and a water phase. The CH$_2$Cl$_2$ extracts from compounds **1** and **2** were chromatographically separately with a Phenomenex Strata® C18-E column (2 g) using a gradient solvent system from 100% H$_2$O to 100% MeOH to give aglycones **1a** (0.3 mg) and **2a** (0.3 mg), respectively. The aglycone of **1a** was determined to be (+)-secoisolariciresinol using LC/MS analysis with an *m/z* signal of 361.2 [M − H]$^-$ and a positive optical rotation ($[\alpha]_D^{25}$ +30.0, *c* 0.02, acetone) [16]. The CD spectrum of **1a** showed positive Cotton effects at 209, 223, and 288 nm and negative effects at 216 and 230 nm. The aglycone of **2a** was determined to be (+)-seco-5′-methoxy-isolariciresinol using LC/MS analysis with an *m/z* signal of 393.2 [M + H]$^+$ and a positive optical rotation ($[\alpha]_D^{25}$ +25.5, *c* 0.02, acetone) [16]. After drying the water phase in vacuo, the residue was dissolved in anhydrous pyridine (200 μL) followed by the addition of L-cysteine methyl ester hydrochloride (0.6 mg). The reaction mixture was incubated at 60 °C for 1 h, then *O*-tolyl isothiocyanate (15 μL) was added and the mixture was incubated at 60 °C for 1 h. The reaction product was directly analyzed using LC/MS (0−35% MeCN for 30 min, flow rate: 0.3 mL/min) with an analytical Kinetex column (2.1 × 100 mm, 5 μm) (Agilent Technologies, Santa Clara, CA, USA). The L-rhamnose in compounds **1** and **2** was identified through comparison of the retention times with those of authentic sample (t_R = L-rhamnose 25.6 min).

3.4. Cytotoxicity Assay

A sulforhodamine B (SRB) bioassay was used to determine the cytotoxicity of each isolated compound against four cultured human tumor cell lines [12,34]. The assays were performed at the Korea Research Institute of Chemical Technology. All the cell lines used, Bt549, MCF7, MDA-MB-231, and HCC70, are human breast cancer cells. Etoposide (purity ≥ 98%, Sigma, St. Louis, MO, USA) was used as a positive control. The half maximal inhibitory concentrations (IC$_{50}$) of cancer cell growth are expressed as the mean from three distinct experiments.

3.5. Antiviral Activity Assay

Influenza A/PR/8 virus (PR8), human rhinovirus 1B (HRV1B), and coxsackievirus B3 (CVB3) were purchased from ATCC (American Type Culture Collection, Manassas, VA, USA). PR8, CVB3, and HRV1B were replicated in A549, Vero, and HeLa cells, respectively, at 37 °C. Antiviral activity was evaluated with the SRB method using cytopathic effect (CPE) reduction as previously reported [39].

3.6. Oil Red OStaining

At 6–8 days after differentiation, the adipocytes were fixed with 10% neutral buffered formalin (NBF) and stained with 0.5% Oil Red O (Sigma, St. Louis, MO, USA). To stop the reaction, cells were washed with distilled water three times. Stained cells were resolved with 1 mL of isopropanol and the colorimetric changes was measured at 520 nm to evaluate intra-cellular triglyceride content.

3.7. Alkaline Phosphatase (ALP) Staining and Activity

At 7–9 days after osteogenic differentiation, the medium was removed, and the cells were washed with 2 mM MgCl$_2$ solution. After incubation with AP buffer (100 mM Tris−HCl, pH 9.5, 100 mM NaCl, and 10 mM MgCl$_2$) for 15 min, the cells were treated in AP buffer containing 0.4 mg/mL of nitro-blue tetrazolium (NBT, Sigma) and 0.2 mg/mL of 5-bromo-4-chloro-3-indolyl phosphate (BCIP, Sigma) for 15 more minutes. To stop the reaction, the cells were exposed to 5 mM EDTA (pH 8.0) and fixed with 10% NBF for 1 h.

The differentiation into osteoblast was evaluated regarding ALP activity. The ALP activity was determined using an Alkaline Phosphatase Assay Kit (ab83369; Abcam, Cambridge, MA, USA). Briefly, the cell lysates were incubated with *p*-nitrophenyl phosphate (*p*-NPP) solution at RT for 1 h in the dark. After stopping the reaction, the optical density was measured at 405 nm using a SpectraMax M2/M2e Microplate Readers (Molecular Devices, San Jose, CA, USA).

4. Conclusions

In the present study, phytochemical analysis of the aerial portion of *L. cuneata* led to the isolation of two new lignan glycosides (**1,2**) along with three known lignan glycosides (**3–7**) and nine known flavonoid glycosides (**8–14**). All the isolated compounds were evaluated for their applicability for medicinal use using cell-based assays. Compounds **1** and **4–6** exhibited weak cytotoxicity against the breast cancer cell lines (Bt549, MCF7, MDA-MB-231 and HCC70) (IC$_{50}$ < 30.0 µM), while none of the isolated compound showed significant antiviral activity against PR8, HRV1B, or CVB3. In a mouse mesenchymal stem cell line, treatment with compound **10** resulted in fewer lipid droplets compared to the untreated negative without altering the amount of alkaline phosphatase staining.

Supplementary Materials: Supplementary materials are available online. General experimental procedures, 1D NMR, 2D NMR, HRESIMS, CD data of **1** and **2**, LC/MS analysis of **1** and **2**, and Table S1 are available free of charge on the Internet.

Author Contributions: H.J.K., S.L. (Sanghyun Lee), S.-H.K. and K.H.K. conceived and designed the experiments; J.B., T.K.L., J.-H.S., E.C. and S.U.C. performed the experiments; H.-J.K., S.L. (Sanghyun Lee), S.H.K., S.U.C. and K.H.K. analyzed the data; S.L. (Seong Lee), S.-W.Y. and S.U.C. contributed reagents/materials/analysis tools; J.B., S.-H.K. and K.H.K. wrote the paper.

Funding: This work was supported by the National Research Foundation of Korea (NRF) grant funded by the Korea government (MSIT) (2018R1A2B2006879) and by the Ministry of Education (NRF-2012R1A5A2A28671860). This work was also supported by the Korea Institute of Planning and Evaluation for Technology in Food, Agriculture, Forestry, and Fisheries (iPET) through the Technology Commercialization Support Program, funded by the Ministry of Agriculture, Food, and Rural Affairs (MAFRA) (816004-02-1-SB010), Korea.

Conflicts of Interest: The authors declare no conflict of interest.

References

1. Kim, M.S.; Sharma, B.R.; Rhyu, D.Y. Beneficial effect of *Lespedeza cuneata* (G. Don) water extract on streptozotocin-induced type 1 diabetes and cytokine-induced beta-cell damage. *Nat. Prod. Sci.* **2016**, 22, 175–179. [CrossRef]
2. Zhang, C.; Zhou, J.; Yang, J.; Li, C.; Ma, J.; Zhang, D.; Zhang, D. Two new phenylpropanoid glycosides from the aerial parts of *Lespedeza cuneata*. *Acta Pharm. Sin. B* **2016**, 6, 564–567. [CrossRef] [PubMed]

3. Min, J.Y.; Shim, S.H. Chemical constituents from *Lespedeza cuneata* G. Don (Leguminosae). *Biochem. Syst. Ecol.* **2016**, *66*, 293–296. [CrossRef]

4. Lee, H.; Jung, J.Y.; Hwangbo, M.; Ku, S.K.; Kim, Y.W.; Jee, S.Y. Anti-inflammatory effects of *Lespedeza cuneata* in vivo and in vitro. *Korea J. Herbol.* **2013**, *28*, 83–92. [CrossRef]

5. Zhang, C.F.; Zhou, J.; Yang, J.Z.; Li, C.J.; Ma, J.; Zhang, D.; Li, L.; Zhang, D.M. Three new lignanosides from the aerial parts of *Lespedeza cuneata*. *J. Asian Nat. Prod. Res.* **2016**, *18*, 913–920. [CrossRef] [PubMed]

6. Kim, S.M.; Kang, K.; Jho, E.H.; Jung, Y.J.; Nho, C.W.; Um, B.H.; Pan, C.H. Hepatoprotective effect of flavonoid glycosides from *Lespedeza cuneata* against oxidative stress induced by tert-butyl hyperoxide. *Phytother. Res.* **2011**, *25*, 1011–1017. [CrossRef] [PubMed]

7. Park, H.M.; Hong, J.H. Physiological activities of *Lespedeza cuneata* extracts. *Korea J. Food Preserv.* **2014**, *21*, 844–850. [CrossRef]

8. Deng, F.; Chang, J.; Zhang, J.S. New flavonoids and other constituents from *Lespedeza cuneata*. *J. Asian Nat. Prod. Res.* **2007**, *9*, 655–658. [CrossRef] [PubMed]

9. Zhou, J.; Li, C.J.; Yang, J.Z.; Ma, J.; Wu, L.Q.; Wang, W.J.; Zhang, D.M. Phenylpropanoid and lignan glycosides from the aerial parts of *Lespedeza cuneata*. *Phytochemistry* **2016**, *121*, 58–64. [CrossRef] [PubMed]

10. Jiang, W.; Ye, J.; Xie, Y.G.; Pan, Y.P.; Zheng, Y.; Qian, X.P.; Jin, H.Z. A new phenyldilactone from *Lespedeza cuneata*. *J. Asian Nat. Prod. Res.* **2016**, *18*, 200–205. [CrossRef] [PubMed]

11. Yoo, G.; Park, S.J.; Lee, T.H.; Yang, H.; Baek, Y.S.; Kim, N.; Kim, Y.J.; Kim, S.H. Flavonoids isolated from *Lespedeza cuneata* G. Don and their inhibitory effects on nitric oxide production in lipopolysaccharide-stimulated BV-2 microglia cells. *Pharmacogn. Mag.* **2015**, *11*, 651–656. [PubMed]

12. Yu, J.S.; Baek, J.; Park, H.B.; Moon, E.; Kim, S.Y.; Choi, S.U.; Kim, K.H. A new rearranged eudesmane sesquiterpene and bioactive sesquiterpenes from the twigs of *Lindera glauca* (Sieb. et Zucc.) Blume. *Arch. Pharm. Res.* **2016**, *39*, 1628–1634. [CrossRef] [PubMed]

13. Lee, D.; Kang, K.S.; Yu, J.S.; Woo, J.Y.; Hwang, G.S.; Eom, D.W.; Baek, S.H.; Lee, H.L.; Kim, K.H.; Yamabe, N. Protective effect of Korean Red Ginseng against FK506-induced damage in LLC-PK1 cells. *J. Ginseng Res.* **2017**, *41*, 284–289. [CrossRef] [PubMed]

14. Kim, S.; So, H.M.; Roh, H.S.; Kim, J.; Yu, J.S.; Lee, S.; Seok, S.; Pang, C.; Baek, K.H.; Kim, K.H. Vulpinic acid contributes to the cytotoxicity of *Pulveroboletus ravenelii* to human cancer cells by inducing apoptosis. *RSC Adv.* **2017**, *7*, 35297–35304. [CrossRef]

15. Baek, J.; Lee, D.; Lee, T.K.; Song, J.H.; Lee, J.S.; Lee, S.; Yoo, S.W.; Kang, K.S.; Moon, E.; Lee, S.; et al. (−)-9′-*O*-(α-L-Rhamnopyranosyl)lyoniresinol from *Lespedeza cuneata* suppresses ovarian cancer cell proliferation through induction of apoptosis. *Bioorg. Med. Chem. Lett.* **2018**, *28*, 122–128. [CrossRef] [PubMed]

16. Sugahara, T.; Yamauchi, S.; Kondo, A.; Ohno, F.; Tominaga, S.; Nakashima, Y.; Kishida, T.; Akiyama, K.; Maruyama, M. First stereoselective synthesis of meso-secoisolariciresinol and comparison of its biological activity with (+) and (−)-secoisolariciresinol. *Biosci. Biotechnol. Biochem.* **2007**, *71*, 2962–2968. [CrossRef] [PubMed]

17. Li, W.; Koike, K.; Liu, L.; Lin, L.; Fu, X.; Chen, Y.; Nikaido, T. New lignan glucosides from the stems of *Tinospora sinensis*. *Chem. Pharm. Bull.* **2004**, *52*, 638–640. [CrossRef] [PubMed]

18. Kang, H.R.; Lee, D.; Benndorf, R.; Jung, W.H.; Beemelmanns, C.; Kang, K.S.; Kim, K.H. Termisoflavones A-C., isoflavonoid glycosides from termite-associated *Streptomyces* sp. RB1. *J. Nat. Prod.* **2016**, *79*, 3072–3078. [CrossRef] [PubMed]

19. Chen, X.; Zhu, Q. Pregnane glycoside, lignan glycosides, triterpene glycosyl ester and flavonoid glycosides from *Rubus amabilis*. *Planta Med.* **2001**, *67*, 270–273. [CrossRef] [PubMed]

20. Kim, C.S.; Subedi, L.; Kim, S.Y.; Choi, S.U.; Kim, K.H.; Lee, K.R. Lignan glycosides from the twigs of *Chaenomeles sinensis* and their biological activities. *J. Nat. Prod.* **2015**, *78*, 1174–1178. [CrossRef] [PubMed]

21. He, W.J.; Fu, Z.H.; Zeng, G.Z.; Zhang, Y.M.; Han, H.J.; Yan, H.; Ji, C.J.; Chu, H.B.; Tan, N.H. Terpene and lignan glycosides from the twigs and leaves of an endangered conifer, *Cathaya argyrophylla*. *Phytochemistry* **2012**, *83*, 63–69. [CrossRef] [PubMed]

22. Kil, Y.S.; Kim, S.M.; Kang, U.; Chung, H.Y.; Seo, E.K. Peroxynitrite-scavenging glycosides from the stem bark of *Catalpa ovata*. *J. Nat. Prod.* **2017**, *80*, 2240–2251. [CrossRef] [PubMed]

23. Tanaka, T.; Nakashima, T.; Ueda, T.; Tomii, K.; Kouno, I. Facile discrimination of aldose enantiomers by reversed-phase HPLC. *Chem. Pharm. Bull.* **2007**, *55*, 899–901. [CrossRef] [PubMed]

24. Cho, H.K.; Suh, W.S.; Kim, K.H.; Kim, S.Y.; Lee, K.R. Phytochemical constituents of *Salsola komarovii* and their effects on NGF induction. *Nat. Prod. Sci.* **2014**, *20*, 95–101.

25. Rayyan, S.; Fossen, T.; Nateland, H.S.; Andersen, O.M. Isolation and identification of flavonoids, including flavone rotamers, from the herbal drug 'Crataegi folium cum flore' (hawthorn). *Phytochem. Anal.* **2005**, *16*, 334–341. [CrossRef] [PubMed]

26. Kim, D.K. Antioxidative constituents from the twigs of *Vitex rotundifolia*. *Biomol. Ther.* **2009**, *17*, 412–417. [CrossRef]

27. Park, S.Y.; Kim, J.S.; Lee, S.Y.; Bae, K.H.; Kang, S.S. Chemical constituents of *Lathyrus davidii*. *Nat. Prod. Sci.* **2008**, *14*, 281–288.

28. Han, J.T.; Bang, M.H.; Chun, O.K.; Kim, D.O.; Lee, C.Y.; Baek, N.I. Flavonol glycosides from the aerial parts of *Aceriphyllum rossii* and their antioxidant activities. *Arch. Pharm. Res.* **2004**, *27*, 390–395. [CrossRef] [PubMed]

29. Yang, N.Y.; Tao, W.W.; Duan, J.A. Antithrombotic flavonoids from the faeces of *Trogopterus xanthipes*. *Nat. Prod. Res.* **2010**, *24*, 1843–1849. [CrossRef] [PubMed]

30. Lee, M.H.; Son, Y.K.; Han, Y.N. Tissue factor inhibitory flavonoids from the fruits of *Chaenomeles sinensis*. *Arch. Pharm. Res.* **2002**, *25*, 842–850. [CrossRef] [PubMed]

31. Chung, S.K.; Kim, Y.C.; Takaya, Y.; Terashima, K.; Niwa, M. Novel flavonol glycoside, 7-O-methyl mearnsitrin, from *Sageretia theezans* and its antioxidant effect. *J. Agric. Food Chem.* **2004**, *52*, 4664–4668. [CrossRef] [PubMed]

32. Kim, H.J.; Woo, E.R.; Park, H. A novel lignan and flavonoids from *Polygonum aviculare*. *J. Nat. Prod.* **1994**, *57*, 581–586. [CrossRef]

33. Torres-Mendoza, D.; González, J.; Ortega-Barría, E.; Heller, M.V.; Capson, T.L.; McPhail, K.; Gerwick, W.H.; Cubilla-Rios, L. Weakly antimalarial flavonol arabinofuranosides from *Calycolpus warszewiczianus*. *J. Nat. Prod.* **2006**, *69*, 826–828. [CrossRef] [PubMed]

34. Yao, C.J.; Chow, J.M.; Chuang, S.E.; Chang, C.L.; Yan, M.D.; Lee, H.L.; Lai, I.C.; Lin, P.C.; Lai, G.M. Induction of Forkhead Class box O3a and apoptosis by a standardized ginsenoside formulation, KG-135, is potentiated by autophagy blockade in A549 human lung cancer cells. *J. Ginseng Res.* **2017**, *41*, 247–256. [CrossRef] [PubMed]

35. Chung, S.T.; Huang, Y.T.; Hsiung, H.Y.; Huang, W.H.; Yao, C.W.; Lee, A.R. Novel daidzein analogs and their in vitro anti-influenza activities. *Chem. Biodivers.* **2015**, *12*, 685–696. [CrossRef] [PubMed]

36. Argenta, D.F.; Silva, I.T.; Bassani, V.L.; Koester, L.S.; Teixeira, H.F.; Simões, C.M. Antiherpes evaluation of soybean isoflavonoids. *Arch. Virol.* **2015**, *160*, 2335–2342. [CrossRef] [PubMed]

37. Zhang, T.; Wu, Z.; Du, J.; Hu, Y.F.; Liu, L.; Yang, F.; Jin, Q. Anti-Japanese-encephalitis-viral effects of kaempferol and daidzin and their RNA-binding characteristics. *PLoS ONE* **2012**, *7*, e30259. [CrossRef] [PubMed]

38. Sui, B.D.; Hu, C.H.; Zheng, C.X.; Jin, Y. Microenvironmental views on mesenchymal stem cell differentiation in aging. *J. Den. Res.* **2016**, *95*, 1333–1340. [CrossRef] [PubMed]

39. Song, J.; Yeo, S.G.; Hong, E.H.; Lee, B.R.; Kim, J.W.; Kim, J.; Jeong, H.; Kwon, Y.; Kim, H.; Lee, S.; et al. Antiviral activity of hederasaponin B from hedera helix against enterovirus 71 subgenotypes C3 and C4a. *Biomol. Ther.* **2014**, *22*, 41–46. [CrossRef] [PubMed]

Sample Availability: Samples of the compounds are not available from the authors.

molecules

Review

Chemical Structures of Lignans and Neolignans Isolated from Lauraceae

Ya Li [1,*], Shuhan Xie [2], Jinchuan Ying [1], Wenjun Wei [1] and Kun Gao [1,*]

[1] State Key Laboratory of Applied Organic Chemistry, College of Chemistry and Chemical Engineering, Lanzhou University, Lanzhou 730000, China; yingjch16@lzu.edu.cn (J.Y.); weiwj14@lzu.edu.cn (W.W.)
[2] Lanzhou University High School, Lanzhou 730000, China; xieshzb@163.com
* Correspondences: liea@lzu.edu.cn (Y.L.); npchem@lzu.edu.cn (K.G.); Tel.: +86-931-8912500 (Y.L.)

Academic Editor: David Barker
Received: 9 November 2018; Accepted: 29 November 2018; Published: 30 November 2018

Abstract: Lauraceae is a good source of lignans and neolignans, which are the most chemotaxonomic characteristics of many species of the family. This review describes 270 naturally occurring lignans and neolignans isolated from Lauraceae.

Keywords: lignans; neolignans; Lauraceae; chemical components; chemical structures

1. Introduction

Lignans are widely distributed in the plant kingdom, and show diverse pharmacological properties and a great number of structural possibilities. The Lauraceae family, especially the genera of *Machilus*, *Ocotea*, and *Nectandra*, is a rich source of lignans and neolignans, and neolignans represent potential chemotaxonomic significance in the study of the Lauraceae. Lignans and neolignans are dimers of phenylpropane, and conventionally classified into three classes: lignans, neolignans, and oxyneolignans, based on the character of the C–C bond and oxygen bridge joining the two typical phenyl propane units that make up their general structures [1]. Usually, lignans show dimeric structures formed by a *β,β′*-linkage (8,8′-linkage) between two phenylpropanes units. Meanwhile, the two phenylpropanes units are connected through a carbon–carbon bond, except for the 8,8′-linkage, which gives rise to neolignans. Many dimers of phenylpropanes are joined together through two carbon–carbon bonds forming a ring, including an 8,8′-linkage and another carbon–carbon bond linkage; such dimers are classified as cyclolignans. When the two phenylpropanes units are linked through two carbon–carbon bonds, except for the 8,8′-linkage, this constitutes a cycloneolignan. Oxyneolignans also contain two phenylpropanes units which are joined together via an oxygen bridge. Herein, lignans and neolignans are classfied into five groups: lignans, cyclolignans, neolignans, cycloneolignans, and oxyneolignans on the basis of their carbon skeletons and cyclization patterns. The majority of lignans isolated from Lauraceae have shown only minor variations on well-known structures; for example, a different degree of oxidation in the side-chain and different substitutions in the aromatic moieties, including hydroxy, methoxy, and methylenedioxy groups. Since the nomenclature and numbering of the lignans and neolignans in the literature follow different rules, the trivial names or numbers of the compounds were used to represent them. Furthermore, the semi-systematic names of compounds and their corresponding names in the literature are summarized in the Supporting Information. Herein, we give a comprehensive overview of the chemical structures of lignans and neolignans isolated from Lauraceae.

2. Lignans

This section covers lignans formed by an 8,8′-linkage between two phenyl propane units, which are subclassified according to the pattern of the oxygen rings as depicted in Figure 1.

The semi-systematic names of those lignans without trivial names and their corresponding names in found in the literature are given in Table SI-1 (Supporting Information).

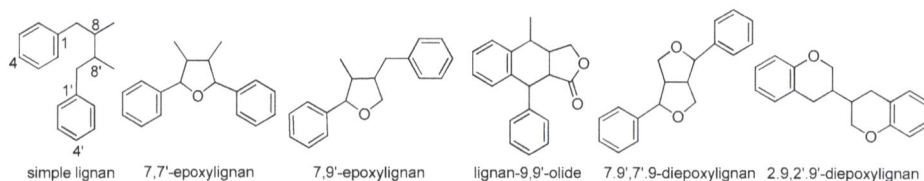

Figure 1. Subtypes of classical lignans.

2.1. Simple Lignans

Machilin A (**1**) was first obtained from the CHCl$_3$-soluble portion of the methanolic extract of the bark of *Machilus thunbergii* collected at Izu Peninsula, together with *meso*-dihydroguaiaretic acid (**2**). The absolute structure of machilin A (**1**) was determined to be 2*S* and 3*R* (meso-form) [2]. Yu and Ma et al. also reported that the bark of *M. thunbergii* contained machilin A (**1**) and *meso*-dihydroguaiaretic acid (**2**). Furthermore, *meso*-dihydroguaiaretic acid (**2**) was found to have significant neuroprotective activity against glutamate-induced neurotoxicity in primary cultures of rat cortical cells and exerted diverse hepatoprotective activity, perhaps by serving as a potent antioxidant [3,4]. Activity-guided fractionation of the dichloromethane extract of the bark of *M. thunbergii* not only led to the isolation of machilin A (**1**) and *meso*-dihydroguaiaretic acid (**2**), but also *meso*-austrobailignan-6 (**3**) and *meso*-monomethyl dihydroguaiaretic acid (**4**). It was reported that *meso*-dihydroguaiaretic acid (**4**) showed potent inhibitory activity against DNA topoisomerase I and II in vitro at a concentration of 100 μM with inhibition ratios of 93.6 and 82.1%, respectively. Furthermore, *meso*-austrobailignan-6 (**3**) was referred to as *threo*-austrobailignan-6 (**10**) in the article [5]. Two diastereomeric dibenzylbutane lignans ((**5**) and (**6**)) which exhibited selective inhibition against COX-2 (cyclooxygenase) were obtained from the leaves of *Ocotea macrophylla* Kunth, which were collected in Nocaima county, Colombia [6]. Besides machilin A (**1**) and *meso*-dihydroguaiaretic acid (**2**), oleiferin C (**7**) also were found in the stem bark of *M. thunbergii* collected at Ulleung-Do, Kyungbook, Korea. Moreover, *meso*-dihydroguaiaretic acid (**2**) and oleiferin C (**7**) induced an apoptotic effect in HL-60 cells via caspase-3 activation [7]. *meso*-Dihydroguaiaretic acid (**2**), *threo*-dihyidroguaiaretic acid (**8**), sauriol B (**9**), and *threo*-austrobailignan-6 (**10**) were isolated from the ethanolic extract of the bark of *Nectandra turbacensis* (Kunth) Nees [8]. The leaves and root bark of *N. turbacensis* (Kunth) Nees collected in the city of Santa Marta (Magdalena, Colombia) contained *meso*-monomethyl dihydroguaiaretic acid (**4**), *threo*-dihyidroguaiaretic acid (**8**), austrobailignan-5 (**11**), and schineolignin B (**17**) [9]. Lignan **12** was first obtained from the leaves of *Apollonias barbujana* collected in San Andrésy Sauces [10]. Compounds **13–16** were found to occur in the trunk wood of *N. puberula*. Proof of the absolute structure of compound **13** relied on its acid catalyzed cyclization into (-)-galbulin, a tetralin-type neolignan of known absolute stereochemistry [11] (Figure 2).

	R$_1$	R$_2$	R$_3$	R$_4$	R$_5$	R$_6$
1	-OCH$_2$O-	H	-OCH$_2$O-	H		
2	OCH$_3$	OH	H	OCH$_3$	OH	H
3	-OCH$_2$O-	H	OCH$_3$	OH	H	
4	OCH$_3$	OH	H	OCH$_3$	OCH$_3$	H
5	OCH$_3$	OCH$_3$	OCH$_3$	OCH$_3$	OCH$_3$	OCH$_3$

	R$_1$	R$_2$	R$_3$	R$_4$	R$_5$	R$_6$	R$_7$	R$_8$	R$_9$
6	OCH$_3$	OCH$_3$	OCH$_3$	OCH$_3$	OCH$_3$	OCH$_3$	H	H	H
7	-OCH$_2$O-	H	-OCH$_2$O-	H	H	H	βOH		
8	OCH$_3$	OH	H	OCH$_3$	OH	H	H	H	H
9	OCH$_3$	OH	OCH$_3$	OH	OH	OCH$_3$	H	H	H
10	OCH$_3$	OH	H	-OCH$_2$O-	H	H	H	H	
11	-OCH$_2$O-	H	-OCH$_2$O-	H	H	H	H		
12	OCH$_3$	OH	OCH$_3$	OCH$_3$	OH	OCH$_3$	OH	OH	H
13	OCH$_3$	OCH$_3$	H	OCH$_3$	OCH$_3$	H	H	H	αOH

	R$_1$	R$_2$
14	OCH$_3$	OH
15	-OCH$_2$O-	
16	OCH$_3$	OCH$_3$

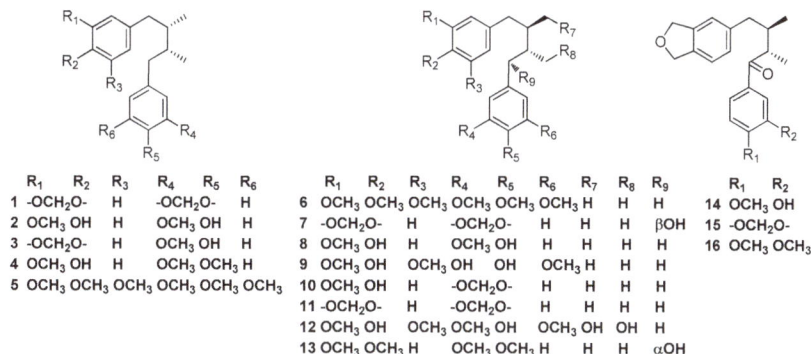

Figure 2. Chemical structures of simple lignans.

2.2. 7,7'-Epoxylignans

Nectandrin A (**19**) and nectandrin B (**20**) were first isolated from leaves and stems of *Nectandra rigida* Nees, along with galgravin (**18**) [12]. Nectandrin A (**19**) and nectandrin B (**20**), together with machilin F (**21**), machilin G (**22**), machilin H (**23**), and machilin I (**25**) were found to occur in the methanolic extract of the bark of *M. thunbergii* Sieb. et Zucc [13]. Galgravin (**18**), henriçine (**26**), and veraguensin (**29**) were found in the leaves and root bark of *N. turbacensis* (Kunth) Nees [9]. Zuonin B (**24**), machilin F (**30**), and nectandrin B (**20**) were obtained from the stem bark of *M. thunbergii* [7]. Galgravin (**18**) and veraguensin (**29**), together with the 2,5-phenyl ring disubstituted lignans **27** and **28** were described for the first time from an ethanolic extract of the leaves *Ocotea foetens* [14]. Veraguensin (**29**) was first reported to be isolated from *Ocotea veraguensis* [15,16], and this compound was also found in *N. puberula* [11]. Verrucosin (**30**) was first gained from the benzene extract of branch wood of *Urbanodendron verrucosum*, together with austrobailignan-7 (**31**) and calopiptin (**32**). The structure of verrucosin (**30**) was established by comparison with the synthetic racemate and by the preparation of a dimethyl ether followed by a comparison of spectral data with published data to determine the absolute structure [17]. (+)-Galbacin (**33**), (+)-galbelgin (**34**), nectandrin A (**19**), nectandrin B (**20**), and machilin-G (**22**) were found to occur in the dichloromethane extract of the bark of *M. thunbergii* Sieb. et Zucc. Furthermore, nectandrin B (**20**) showed potent inhibitory activity against DNA topoisomerase I and II in vitro at a concentration of 100 μM, with inhibition ratios of 79.1 and 34.3%, respectively [3–5]. Beilschminol B (**35**) was first obtained from the roots of *Beilschmiedia tsangii* [18]. Odoratisol C (**36**), odoratisol D (**37**), and machilin-I (**25**) were obtained from the air-dried bark of the Vietnamese medicinal plant *M. odoratissima* Nees [19] (Figure 3).

2.3. 7,9'-Epoxylignans

(-)-Parabenzoinol (**38**) was isolated from the fresh leaves of *Parabenzoin trilobum* Nakai, and its structure was elucidated by X-ray crystallographic analysis [20]. Actifolin (**39**) was identified in the stems of *Lindera obtusiloba*; moreover, its effect on tumor necrosis factor (TNF)-α and interleukin (IL)-6 as well as its inhibitory activity against histamine release were examined using human mast cells. Actifolin (**39**) suppressed the gene expressions of proinflammatory cytokines, TNF-α, and IL-6 in human mast cells [21,22] (Figure 4).

2.4. Lignan-9,9'-Olides

(-)-Parabenzlactone (**40**) and acetylparabenzylactone (**41**) were found in the fresh leaves of *P. trilobum* Nakai [23]. 5,6-Dihydroxymatairesinol (**42**) was found to occur in the methanolic extract of the stems of *L. obtusilob* [21] (Figure 4).

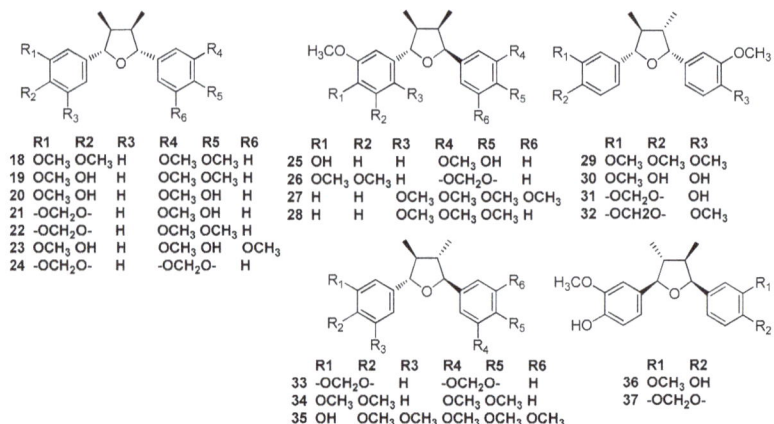

Figure 3. Chemical structures of 7,7′-epoxylignans.

2.5. 2.9,2′.9′-Diepoxylignans

The chromene dimer **43** was obtained from the ethanolic extract of the leaves of *Cinnamomu mparthenoxylon* (Jack) Meisn [24] (Figure 4).

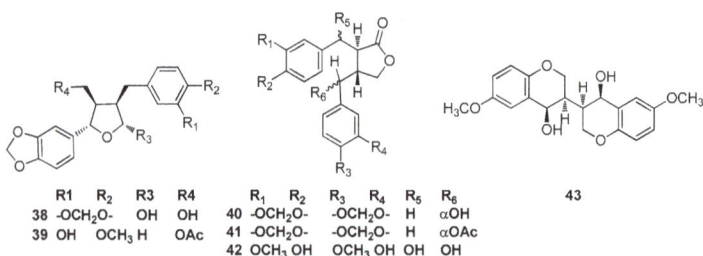

Figure 4. Chemical structures of 7,9′-epoxylignans, lignan-9,9′-olides, and 2.9,2′.9′-diepoxylignans.

2.6. 7.9′,7′.9-Diepoxylignans

The ethanol/H$_2$O (9:1) extract of the fruits of *L. armeniaca* contained magnolin (**44**) and eudesmin (**45**) [25]. Phytochemical studies revealed the presence of sesamin (**46**) and *O*-methylpiperitol (**47**) in the ethanolic extract of the fruit of calyces of *N. amazonurn* [26]. Sesamin (**46**) was also found to occur in the CH$_2$Cl$_2$ extract of the bark of *M. thunbergii* Sieb. et Zucc [3,4]. Magnolin (**44**), eudesmin (**45**), sesamin (**46**), and *O*-methylpiperitol (**47**) were all found to occur in *Persea pyrifolia* Nees and Mart. ex Nees [27]. Phytochemical investigations of the methanolic extract of the leaves of *A. barbujana* resulted in the isolation of demethylpiperitol (**48**) [10]. The ethanolic extract of *Pleurothyrium cinereum* also contained (+)-demethylpiperitol (**48**), as well as (+)-de-4″-*O*-methylmagnolin (**49**), which was found to be a potent COX-2/5-LOX dual inhibitor and platelet-activating factor (PAF)-antagonist (COX-2: IC$_{50}$ = 2.27 µM; 5-LOX: IC$_{50}$ = 5.05 µM; PAF: IC$_{50}$ = 2.51 µM) (**49**) [6,28]. (+)-Syringaresinol (**50**) was isolated from the stems of *C. reticulatum* Hay [29]. (+)-De-4″-*O*-methylmagnolin (**49**) and (+)-syringaresinol (**50**) both were found to occur in the methanolic extract of the stems of *Actinodaphne lancifolia* [30]. The leaves of *C. macrostemon* Hayata [31] and the stems of *C. burmanii* [32] both contained (+)-syringaresinol (**50**) and yangambin (**51**), which showed various pharmacological effects. Moreover, *C. burmanii* also contained (+)-sesamin (**46**) [32]. (+)-Demethoxyexcelsin (**52**), (+)-piperitol (**53**), and (+)-methoxypiperitol (**54**) were obtained from the bark and wood of *N. turbacensis*, together with (+)-sesamin (**46**) [33]. Epiyangambin (**55**), episesartemin (**56**), and yangambin (**51**) were isolated from the leaves of *O. duckei*, and yangambin (**51**) represented the major constituent [34]. Kwon et al. reported the isolation of (+)-syringaresinol (**50**) and pluviatilol (**57**) from the stems of *L. obtusilob*, and pluviatilol (**57**) showed cytotoxicity against a small panel of human tumor cell lines [21]. (+)-5-Demethoxyepiexcelsin (**58**) and (+)-epiexcelsin (**59**) were reported to be found in *Litsea verticillata* Hance, and (+)-5- demethoxyepiexcelsin (**58**) showed moderate anti-HIV activity with an IC$_{50}$ value of 16.4 μg/mL (42.7 μM) [35]. (+)-Xanthoxyol (**60**), (+)-syringaresinol (**50**), and pluviatilol (**57**) were obtained from the stems of *L. obtusiloba* Blume. The effect of these compounds on tumor necrosis factor (TNF)-α and interleukin (IL)-6 as well as their inhibitory activity against histamine release were examined using human mast cells. Pluviatilol (**57**) inhibited the release of histamine from mast cells [22]. 4-Keto-pinoresinol (**61**) was isolated from the ethanolic extract of the leaves and twigs of *Litsea chinpingensis* [36] (Figure 5).

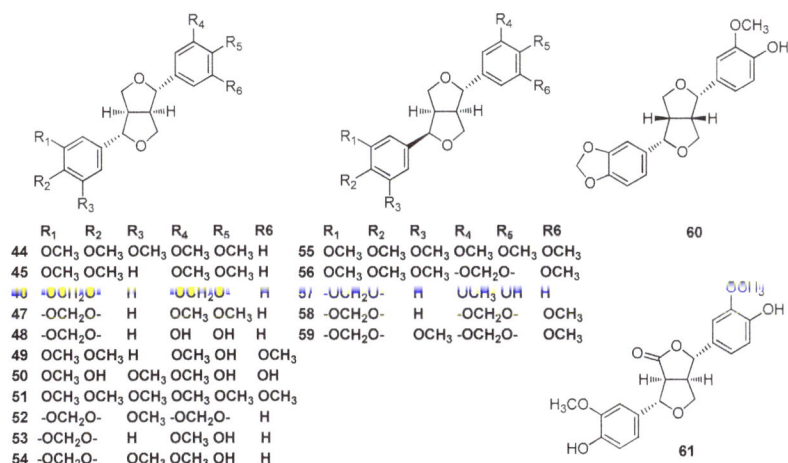

	R$_1$	R$_2$	R$_3$	R$_4$	R$_5$	R6		R$_1$	R$_2$	R$_3$	R$_4$	R$_6$	R6	
44	OCH$_3$	OCH$_3$	OCH$_3$	OCH$_3$	OCH$_3$	H	55	OCH$_3$	OCH$_3$	OCH$_3$	OCH$_3$	OCH$_3$	OCH$_3$	
45	OCH$_3$	OCH$_3$	H		OCH$_3$	OCH$_3$	H	56	OCH$_3$	OCH$_3$	OCH$_3$	-OCH$_2$O-		OCH$_3$	
46	-OCH$_2$O-		H	-OCH$_2$O-		H	57	-OCH$_2$O-		H	OCH$_3$	OH	H	
47	-OCH$_2$O-		H	OCH$_3$	OCH$_3$	H	58	-OCH$_2$O-		H	-OCH$_2$O-		OCH$_3$	
48	-OCH$_2$O-		H	OH	OH	H	59	-OCH$_2$O-		OCH$_3$	-OCH$_2$O-		OCH$_3$	
49	OCH$_3$	OCH$_3$	H		OCH$_3$	OH	OCH$_3$							
50	OCH$_3$	OH	OCH$_3$	OCH$_3$	OH	OH								
51	OCH$_3$	OCH$_3$	OCH$_3$	OCH$_3$	OCH$_3$	OCH$_3$								
52	-OCH$_2$O-		OCH$_3$	-OCH$_2$O-		H								
53	-OCH$_2$O-		H	OCH$_3$	OH	H								
54	-OCH$_2$O-		OCH$_3$	OCH$_3$	OH	H								

Figure 5. Chemical structures of 7.9′,7′.9-diepoxylignans.

3. Cyclolignans

There are three main types of cyclolignans isolated from nature, including 2,7′-cyclolignans, 2,2′-cyclolignans, and 7,7′-cyclolignans. Cyclolignans are not so common in Lauraceae. We have only retrieved less than 10 2,7′-cyclolignans isolated from Lauraceae. The semi-systematic names of those cyclolignans without trivial names and their corresponding names in the literature are given in Table SI-1 (Supporting Information).

2,7.'-Cyclolignans

(-)-Isoguaiacin (**62**) and (+)-guaiacin (**63**) were isolated from the extract of the bark of *M. thunbergii* Sieb. et Zucc. These two compounds showed significant neuroprotective activities against glutamate-induced neurotoxicity in primary cultures of rat cortical cells [3,4]. (+)-Otobaphenol (**64**) and cyclolignans **65** and **66** were isolated from the ethanolic extract of *P. cinereum* [28]. Cinnamophilin A (**67**) was first reported to be obtained from the methanolic extract of roots of *Cinnamomum philippinense* (Merr.) Chang [37]. (-)-Aristoligone (**68**), (-)-aristotetralone (**69**), and (-)-cagayanone A (**70**) were obtained from the ethanolic extract of the leaves and twigs of *L. chinpingensis* [36] (Figure 6).

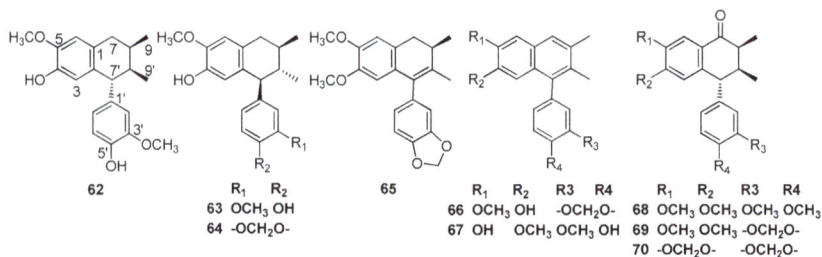

Figure 6. Chemical structures of 2,7'-cyclolignans.

4. Neolignans

Neolignans are widely distributed in the Lauraceae family, especially in the genera of *Aniba*, *Nectandra*, and *Ocotea*. The types of neolignans isolated from Lauraceae include 8,1'-neolignans, 8,3'-neolignans, 7,1'-neolignans, and 7,3'-neolignan (Figure 7). 3,3'-neolignans, which also exist in nature, have not been isolated from Lauraceae. The semi-systematic names of the abovementioned neolignans without trivial names and their corresponding names in the literature are given in Table SI-2 and SI-3 (Supporting Information).

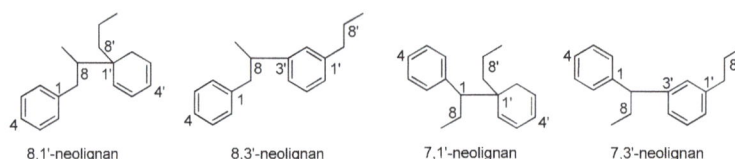

Figure 7. Subtypes of neolignans.

4.1. 8,1'-Neolignans

Burchellin (**71**) was first isolated from the trunk wood of *Aniba burchellii* Kosterm [38]. Burchellin (**71**) has also been found in the benzene extract of the trunk of an unclassified *Aniba* species collected in the vicinity of Manaus, Amazonas, along with compounds **72** and **73** [39,40]. Compounds **72** and **74** were obtained as a mixture from an unclassified Amazonian *Nectandra* species. As the analogous values for the mixture of compounds **72** and **74** were substantially identical to those of pure compound **72**, including the ORD (optical rotatory dispersion) curves, then the two compounds should have the same absolute configuration [41]. 3'-Methoxyburchellin (**75**) was first isolated from the stem bark of *O. veraguensis* [16]. Benzene extract of the trunk wood of *Aniba terminalis* [42] and ethanolic extract of the trunk wood of an *Aniba* species collected 130 km north of Manaus, Amazonas [43] both contained burchellin (**71**) and compound **76**. The trunk wood of *Ocotea catharinensis* yielded compound **77** [44,45]. Inspection of *Aniba simulans* revealed the occurrence of compounds **78–80** [46,47]. Armenin A (**81**) and armenin B (**82**) were first obtained from the benzene extract of the trunk wood of *Licaria armeniaca* [48]. The fruits of *L. armeniaca* yielded compounds **76** and **78** [25]. Compounds **74**, **78** and armenin C (**83**) were isolated from the fruits of *Aniba riparia* [49]. Compound **85** was isolated from the benzene extract of trunk wood of an unclassified *Aniba* species [50]. Canellin B (**86**) was first obtained from the benzene extract of the trunk wood of *Licaria canella* [51]. The trunk wood of an Amazonian *Aniba* species contained armenin A (**81**), armenin B (**82**), C (**83**), canellin B (**86**), canellin D (**87**), canellin E (**88**), porosin (**90**), and porosin B (**91**) [52]. Porosin (**90**) was first obtained from the wood of *Ocotea porosa* [53]. Porosin B (**91**) was first obtained from the branch wood of *U. verrucosum*, and porosin (**90**) also were found to exist in the same species [17]. The wood of *Ocotea catharinensis* yielded armenin B (**82**), canellin B (**86**), ferrearin C (**95**), ferrearin E (**96**), and compounds **92** and **93**. Moreover, the structures of compound **92** and ferrearin C (**95**) were certified by single-crystal X-ray analysis [45]. Ferrearin A (**99**) and ferrearin B (**100**) were first isolated from the trunk wood of the Amazonian *Aniba* *ferra* Kubitzki, together with compounds **85** and **92**. The relative structures of ferrearin A (**99**) and ferrearin B (**100**) were elucidated as structures of **97** and **98** [54], then revised as ferrearin A (**99**) and ferrearin B (**100**). Besides these two compounds, 3'-methoxyburchellin (**75**), compound (**77**), ferrearin C (**101**), and ferrearin D (**102**) were found to occur in the trunk wood of *Ocotea aciphylla* [55,56]. Burchelin (**71**), porosin (**90**), porosin B (**91**), and compounds **76, 89, 94, 103–106, 112**, and **113** all were identified in the trunk wood of *O. porosa*, collected from the Forest Reserve of the Botanical Institute, Sâo Paulo, Brazil [57,58]. Fifteen 8,1'-neolignans have been reported to be found in the bark and leaves of *O. porosa* harvested near Santa Maria, State of Rio Grande do Sul, Brazil, including burchellin (**71**), porosin (**90**), porosin B (**91**), and compounds **76, 89, 94**, and **105–113** [59] (Figure 8).

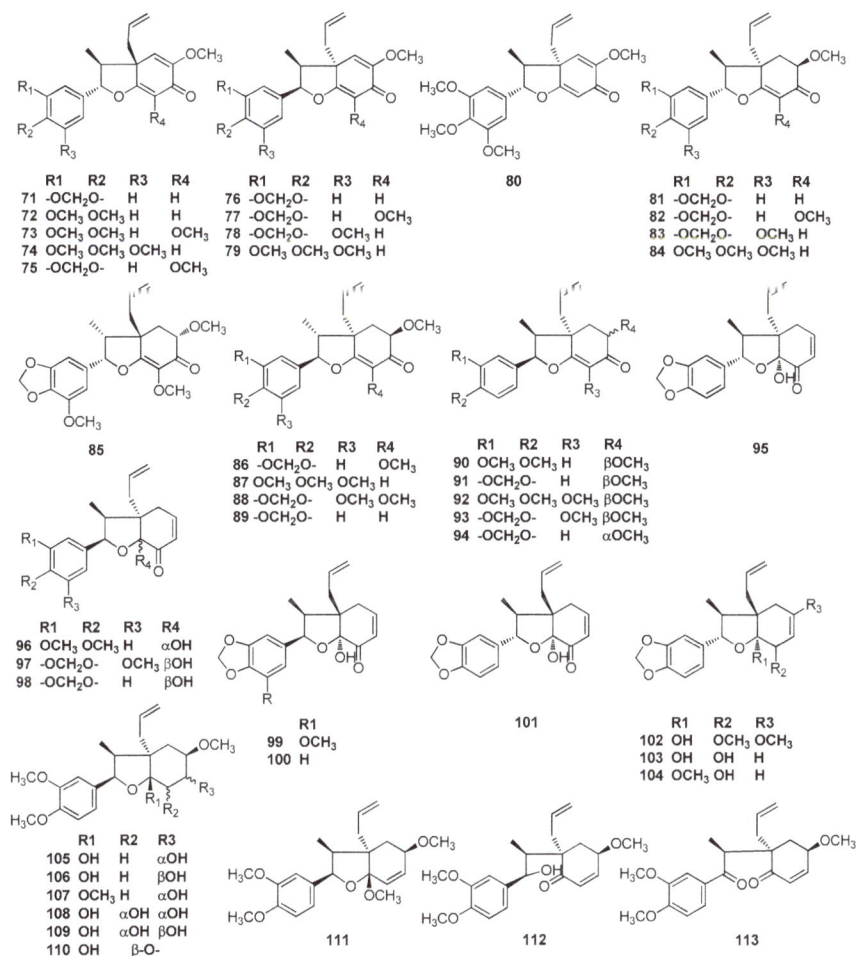

Figure 8. Chemical structures of 8,1'-neolignans.

4.2. 8,3'-Neolignans

The benzene extract of trunk wood of *Licaria aritu* Ducke [60] and the EtOH/H$_2$O (9:1) extract of the fruits of *N. glabrescens* contained licarin A (**114**) and licarin B (**115**) [26]. Licarin A (**114**) was also isolated from *N. rigida* Nees and is responsible for the major cytotoxic activity of crude extract of *N. rigida* Nees, displaying ED$_{50}$ vs. KB cancer cell line at 7.0 µg/mL [12]. Machilin B (**116**) was obtained from the methanolic extract of the bark of *M. thunbergii* [2], as well as licarin A (**114**) and licarin B (**115**). Licarin A (**114**) showed significant neuroprotective activities against glutamate-induced neurotoxicity in primary cultures of rat cortical cells and induced an apoptotic effect in HL-60 cells via caspase-3 activation [4,7]. Licarin A (**114**) and licarin D (**117**) were found in branch wood of the shrub *U. verrucosum* [17]. Obovatifol (**118**), odoratisol-A (**119**), and (-)-licarin A (**114**) were obtained from the air-dried bark of the Vietnamese medicinal plant *Machilius odoratissima* Nees [19]. Besides licarin A (**114**) and licarin B (**115**), machilusol A (**120**), machilusol B (**122**), machilusol C (**123**), machilusol D (**124**), machilusol E (**125**), machilusol F (**126**), and acuminatin (**127**) were isolated from the stem wood of *Machilus obovatifolia*. Machilusols A–F showed moderate cytotoxic activity [61]. The dichloromethane extract of the bark of *M. thunbergii* Sieb. et Zucc also contained (-)-acuminatin (**127**), together with licarin A (**114**). (-)-Acuminatin exerted diverse hepatoprotective activities, perhaps by serving as a potent antioxidant [3,5]. Dihydrodehydrodiconifery alcohol (**129**) was found in the ethanolic extract of the leaves and twigs of *L. chinpingensis* [36]. Compound **130** was first obtained from the benzene extract of the trunk of an *Aniba* species collected in the vicinity of Manaus, Amazonas, along with acuminatin (**127**) and licarin D (**116**) [40]. *A. burchellii* Kosterm contains compounds **130** and **131**. The determination of their absolute stereochemistry relied on spectra and a preparation by thermolysis as well as the acid isomerization of burchellin (**71**) [62]. Denudatin B (**132**), as well as (+)-licarin A (**114**), liliflol B (**121**), and (+)-acuminatin (**127**), have been found in leaves of *Nectandra amazonum* Nees [63]. Mirandin A (**133**) was proved to be the major neolignan of an unclassified *Nectandra* species, which grew at Rosa de Maio, a locality on the Manaus-Itacoatiara highway, Amazonas [41]. (+)-Mirandin A (**133**), (-)-licarin A (**114**), and (-)-licarin B (**115**) also were found to occur in the ethanolic extract of *P. cinereum* [28]. Licarin C (**128**), mirandin A (**133**), mirandin B (**134**), and compounds **135** and **146** were obtained from the benzene extract of *Nectundru mirunda* trunk wood [64]. Compounds **136** and **137** were found in the stem bark of *O. veraguensis* [16]. Furthermore, compound **136** also was found in the trunk wood of an *Aniba* species collected 130 km north of Manaus, Amazonas [43], and compound **137** was obtained from the wood of *O. catharinensis* [44] and the fruits of *O. veraguensis* [65]. Compounds **135**, **138**, **139**, **141**, and **143–145** were isolated from the benzene extract of *Anibu simulans* trunk wood [46]. The extract of EtOH/H$_2$O (9:1) of fruits of *L. armeniac* also provided compounds **136** and **138** [25]. Obovaten (**142**), perseal D (**159**), perseal C (**160**), and obovatinal (**161**) were first obtained from the leaves of *Persea obovatifolia*; together with obovatifol (**118**), these compounds showed significant cytotoxicity against P-388, KB16, A549, and HT-29 cancer cell lines in vitro [66,67]. Compound **148** was isolated from the benzene extract of the trunk wood of *A. terminalis* [42]. Lancifolins A–F (**149–154**) were obtained from branches of the shrub *Aniba lancifolia* Kubitzki et Rodrigues [68]. Neolignan ketone **156** was found to exist in the chloroform extract of the bark of *Ocotea bullata* [69]. Ocophyllals A (**157**) and ocophyllals B (**158**), which have a C-1' formyl side chain instead of a propenyl group, as well as (+)-licarin B (**115**) were observed to occur in the ethanolic extract from leaves of *O. macrophylla* [70]. Licarin A (**114**), licarin B (**115**), (7R,8S,1'R)-7,4'-epoxy-1'-methoxy-3,4-methylenedioxy-8,3'-neolign-8'-ene-6'(1'H)-one (**140**), and compounds **130**, **147**, **155**, **162**, and perseal F (**163**) were obtained from *O. porosa* [57–59]. Compound **162** also was found to occur in the dichloromethane extract of the bark of *M. thunbergii* SIEB. et ZUCC [5]. Meanwhile, perseal F (**163**) and perseal G (**164**) were present in the chloroform-soluble portion of the stem wood of *M. obovatifolia* [71] (Figure 9).

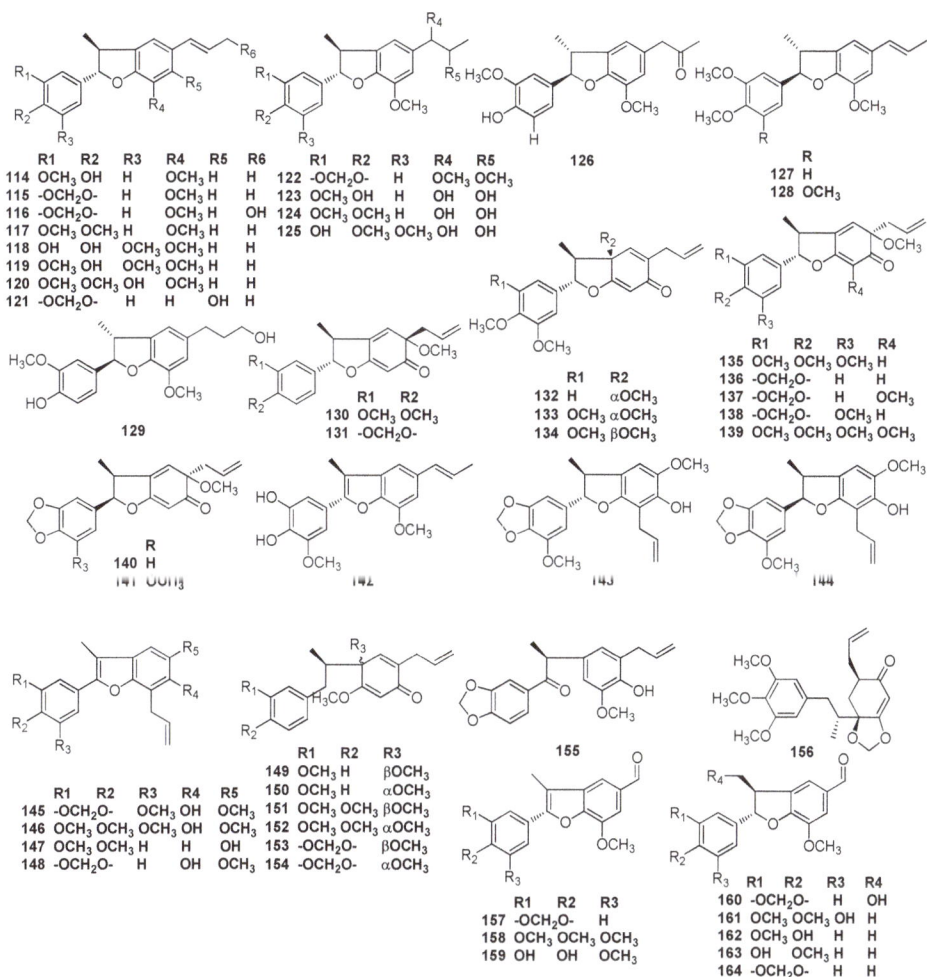

R1	R2	R3	R4	R5	R6
114	OCH₃	OH	H	OCH₃	H	H
115	-OCH₂O-		H	OCH₃	H	H
116	-OCH₂O-		H	OCH₃	H	OH
117	OCH₃	OCH₃	H		OCH₃	H
118	OH	OH	OCH₃	OCH₃	H	
119	OCH₃	OH	OCH₃	OCH₃	H	
120	OCH₃	OCH₃	OH	OCH₃	H	
121	-OCH₂O-		H	H	OH	H

R1	R2	R3	R4	R5
122	-OCH₂O-		H	OCH₃	OCH₃
123	OCH₃	OH	H	OH	OH
124	OCH₃	OCH₃	H	OH	OH
125	OH		OCH₃	OCH₃	OH	OH

126

R
127 H
128 OCH₃

R1	R2
130	OCH₃	OCH₃
131	-OCH₂O-	

129

R1	R2
132	H	αOCH₃
133	OCH₃	αOCH₃
134	OCH₃	βOCH₃

R1	R2	R3	R4
135	OCH₃	OCH₃	OCH₃	H
136	-OCH₂O-		H	H
137	-OCH₂O-		H	OCH₃
138	-OCH₂O-		OCH₃	H
139	OCH₃	OCH₃	OCH₃	OCH₃

R
140 H
141 OCH₃

142

143

144

R1	R2	R3	R4	R5
145	-OCH₂O-		OCH₃	OH	OCH₃
146	OCH₃	OCH₃	OCH₃	OH	OCH₃
147	OCH₃	OCH₃	H	H	OH
148	-OCH₂O-		H	OH	OCH₃

R1	R2	R3
149	OCH₃	H	βOCH₃
150	OCH₃	H	αOCH₃
151	OCH₃	OCH₃	βOCH₃
152	OCH₃	OCH₃	αOCH₃
153	-OCH₂O-		βOCH₃
154	-OCH₂O-		αOCH₃

155

156

R1	R2	R3
157	-OCH₂O-		H
158	OCH₃	OCH₃	OCH₃
159	OH	OH	OCH₃

R1	R2	R3	R4
160	-OCH₂O-		H	OH
161	OCH₃	OCH₃	OH	H
162	OCH₃	OH	H	H
163	OH	OCH₃	H	H
164	-OCH₂O-		H	H

Figure 9. Chemical structures of 8,3'-neolignans.

4.3. 7,1'-Neolignans

Licaria chrysophylla gave a considerable proportion of chrysophyllin A (**165**), which was the first type of 7,1'-neolignan to be obtained. Chrysophyllin B (**166**), chrysophyllon I-A (**167**), and chrysophyllon I-B (**168**) were also identified in *L. chrysophylla* [72,73]. The trunk wood of an Amazonian *Aniba* species collected in the vicinity of Manaus, Amazonas also contained chrysophyllin A (**165**) and chrysophyllin B (**166**) [52] (Figure 10).

4.4. 7,3'-Neolignans

Chrysophyllon II-A (**169**) and chrysophyllon II-B (**170**) belonging to the category of 7,3'-neolignans, both were found to occur in the bark, wood, and fruit calyces of *L. chrysophylla* [73] (Figure 10).

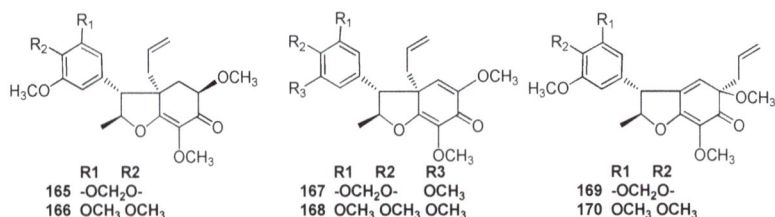

R1 R2
165 -OCH₂O-
166 OCH₃ OCH₃

R1 R2 R3
167 -OCH₂O- OCH₃
168 OCH₃ OCH₃ OCH₃

R1 R2
169 -OCH₂O-
170 OCH₃ OCH₃

Figure 10. Chemical structures of 7,1'-neolignans.

5. Cycloneolignans

Cycloneolignans are responsible for the chemotaxonomic characteristics of some genera in the Lauraceae family, such as *Aniba*, *Licaria*, and *Nectandra*. Most cycloneolignans isolated from Lauraceae belong to the categories of 7.3',8.1'-cycloneolignans and 7.3',8.5'-cycloneolignans (Figure 11). Only two 7.1',8.3'-cycloneolignans have been reported to be obtained from *O. bullata*. The semi-systematic names of those cycloneolignans without trivial names and their corresponding names in the literature are given in Table SI-4 (Supporting Information).

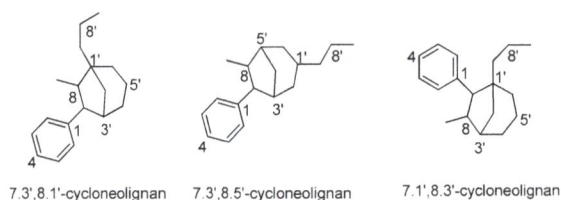

7.3',8.1'-cycloneolignan   7.3',8.5'-cycloneolignan   7.1',8.3'-cycloneolignan

Figure 11. Subtypes of cycloneolignans.

5.1. 7.3',8.1'-Cycloneolignans

Guianin (**171**) was first obtained from the wood of *Aniba guianensis* Aubl [74]. Meanwhile, 2'-epiguianin (**172**) was isolated from the leaves of *O. macrophylla* Kunth, which showed inhibition activity against the platelet-activating factor (PAF)-induced aggregation of rabbit platelets with an IC$_{50}$ value of 1.6 μM [70]. Fourteen 7.3',8.1'-cycloneolignans, including 3'-methoxyguianin (**173**) and compounds **174–176** and **178–187**, were isolated from the trunk wood of *O. porosa* collected from the mountainous Atlantic forest region of São Paulo State, where it is known as 'canela parda' [75,76]. Compound **192** was first obtained from the trunk wood of *O. porosa* collected from the Forest Reserve of Instituto BottInico (SHo Paulo, SP), along with guianin (**171**) and 2'-epiguianin (**172**) [58]. Compound **191** was first obtained from the benzene extract of *A. burchellii* Kosterm, together with guianin (**171**) [62]. Compounds **193–195** and **229** were found to be present in the benzene extract of the trunk of a unclassified *Aniba* species collected from the Ducke Forest Reserve, Manaus, Amazonas, as well as guianin (**171**) and 3'-methoxyguianin (**173**) [40]. Compounds **184**, **194–198**, and **202–207** were obtained from seed coat and dried fruit pulp of *O. veraguensis*. Since the author did not describe how to determine the absolute configuration of these compounds, their name should contain the addition of the prefix 'rel' [65]. Otherwise, compounds **188**, **191**, and **208–210** were isolated from the petrol and chloroform extract of the stem bark of *O. veraguensis* [16]. The trunk bark of *O. catharinensis* contained compounds **200**, **208**, **209**, as well as canellin-C (**212**) and 5-methoxycanellin-C (**213**), and the contents of all these compounds in the bark were over 0.01% [44,45]. Compounds **214–216**, **218**, and **228** were found to occur in the trunk wood of an *Aniba* species collected 130 km north of Manaus, Amazonas [43]. The benzene extract of trunk wood pertaining to an unclassified *Aniba* species collected from the Ducke Forest Reserve, Manaus yielded **217**, **219**, **221**, **222**, **224**, and methoxycanellin A (**226**) [50,77]. Compound **189** was first obtained from the ethanolic extract of wood of *Ocotea costulatum*, along with compound **221** [78]. The trunk wood of the Amazonian *A. ferra* Kubitzki contained *rel*-(7*S*,8*R*,1′*S*,2′*S*,3′*R*)-1′,2′-dihydro-2′-hydroxy-3,3′,5′-trimethoxy-4,5-methylenedioxy-7.3′,8.1′-cycloneolign-8′-ene- 4′(3′H)-one (**199**) and methoxycanellin A (**226**) [54]. Canellin A (**225**) and canellin C (**212**) were first obtained from the trunk wood of *L. canella* [51]. These two compounds have also been reported to be found in the trunk wood of *L. rigida* Kosterm. However, the relative structures of the compounds shown in the abovementioned article were different from those shown in other articles—the methyl occupied an exo-configuration and the aryl adopted an endo-configuration [79]. The trunk wood of the central Brazilian *O. aciphylla* also yielded canellin-A (**225**), as well as compound **208** and 3′-methoxyguianin (**173**) [55,56]. 2′-Epiguianin (**172**), compounds **177**, **190**–**192**, **211**, **220**, **223**, **227**, and *rel*-(7*R*,8*R*,1′*S*,3′*S*)-5′-methoxy-3,4-methylenedioxy-7.3′,8.1′-cycloneolign-8′-ene-2′,4′(1′H,3′H)-dione (**229**) were obtained from the bark and leaves of *O. porosa* [59]. Compounds **194**, **201**, **230**, and **231** were found in the extract of EtOH/H$_2$O (9:1) of the fruits of *L. armeniaca* [25]. The chloroform extract of the trunk wood of *L. armeniaca* yielded compound **232** [80] (Figure 12).

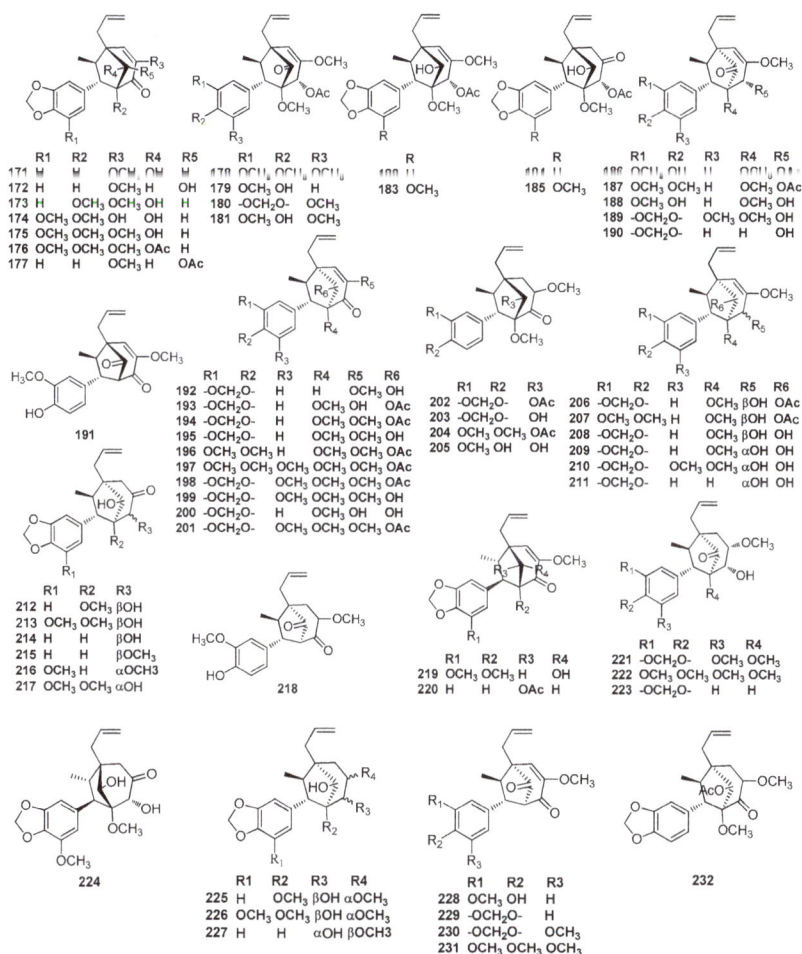

Figure 12. Chemical structures of 7.3′,8.1′-cycloneolignans.

5.2. 7.3',8.5'-Cycloneolignans

Macrophyllin B (**233**) was purified from an unclassified *Nectandra* species collected at Rosa de Maio, a locality on the Manaus-Itacoatiara highway (8 km), Amazonas [41]. Nectamazins A–C (**234–236**), macrophyllin B (**233**), denudanolide D (**237**), and kadsurenin C (**238**) isolated from leaves of *N. amazonum* Nees showed inhibition activity against the platelet-activating factor (PAF)-induced aggregation of rabbit platelets [63]. A phytochemical exploration of the leaves of *O. macrophylla* afforded ocophyllols A–C (**239–241**). Their absolute configurations were established by derivatizing them with (R)-and (S)-MTPA, and then analyzing the NMR data, as well as by a comparison of their circular dichroism (CD) data with that of a related compound whose absolute configuration was previously established by single-crystal X-ray analysis. Moreover, ocophyllols A–C (**239–241**) showed some inhibition activity against the platelet-activating factor (PAF)-induced aggregation of rabbit platelets [70]. Cinerin B (**242**), cinerin C (**243**), cinerin A (**244**), and cinerin D (**245**) were isolated from the leaves of *P. cinereum*. Again, their CD data was used to determine the absolute configuration of these compounds. Cinerin C (**243**) was the first known macrophyllin-type cycloneolignan, which was isolated from the trunk wood of *Licaria macrophylla* Kosterm and named as macrophyllin A [81]. Cinerin A–D also showed some inhibition activity against the platelet-activating factor (PAF)-induced aggregation of rabbit platelets [82]. Compound **246** and macrophyllin B (**233**) were identified in the ethanolic extract of leaves of *P. cinereum* [28] (Figure 13).

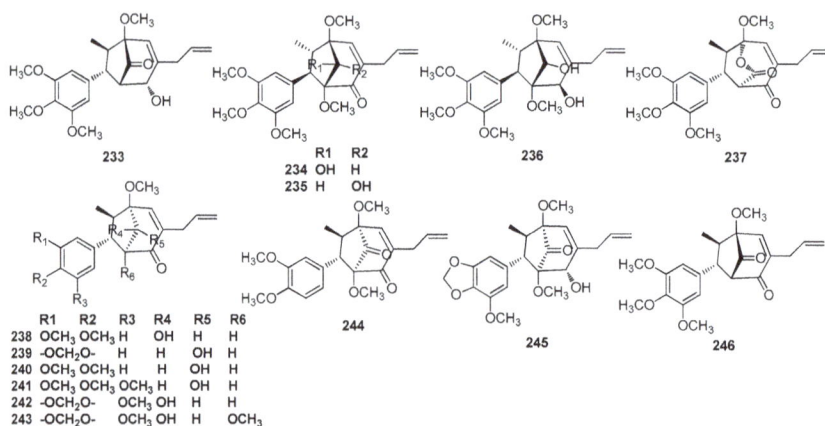

Figure 13. Chemical structures of 7.3',8.5'-cycloneolignans.

5.3. 7.1',8.3'-Cycloneolignans

Ocobullenone (**247**) was the first naturally occurring bicyclooctanoid found to exhibit the 7.1', 8.3' linkage, and it was isolated from the chloroform extract of the bark of *O. bullata* [83]. Iso-ocobullenone (**248**) was also isolated from the chloroform extract of the bark of *O. bullata*, and its structure was confirmed by single-crystal X-ray analysis [69] (Figure 14).

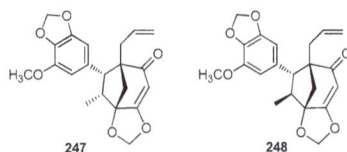

Figure 14. Chemical structures of 7.1',8.3'-cycloneolignans.

6. Oxyneolignans

An ether oxygen atom provides the linkage between the two phenylpropane units, giving rise to oxyneolignans. Oxyneolignans are rarely distributed in Lauraceae. Less than 10 oxyneolignans have been found to occur in the Lauraceae family, belonging to two categories: 7.3′,8.4′-dioxyneolignans and 8,4′-oxyneolignans (Figure 15).

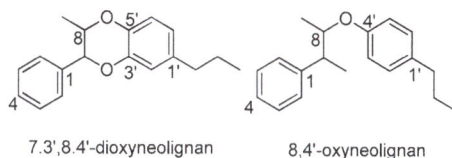

7.3′,8.4′-dioxyneolignan   8,4′-oxyneolignan

Figure 15. Subtypes of oxyneolignans.

6.1. 7.3′,8.4′-Dioxyneolignans

The trunk wood of *L. rigida* Kosterm contained eusiderin (**249**) and eusiderin B (**250**) [79]. The trunk wood of an unclassified *Aniba* species collected at the Ducke Forest Reserve, Manaus also yielded the benzodioxane-type neolignan eusiderin (**249**), eusiderin-F (**251**), and eusiderin-G (**252**) [50,77]. Eusiderin (**249**) was also found to be present in the ethanolic extract of wood of *O. cootulatum* [79] (Figure 16).

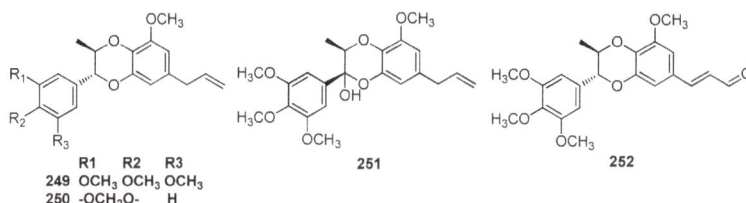

	R1	R2	R3
249	OCH₃	OCH₃	OCH₃
250	-OCH₂O-		H

Figure 16. Chemical structures of 7.3′,8.4′-dioxyneolignans.

6.2. 8,4′-Oxyneolignans

Machilin C (**253**), D (**254**), and E (**255**) were first obtained from the methanolic extract of the bark of *M. thunbergii* [2]. Odoratisol B was obtained from the air-dried bark of the Vietnamese medicinal plant *M. odoratissima* Nees. This compound showed the same relative structure as machilin C (**253**), but was termed odoratisol B in the article [19]. Perseal A (**256**) and perseal B (**257**), which have a C-1′ formyl side chain instead of a propenyl group, were isolated from the chloroform-soluble fraction of the leaves of *P. obovatifolia*. They showed significant cytotoxicity against P-388, KB 16, A549, and HT-29 cancer cell lines [67] (Figure 17).

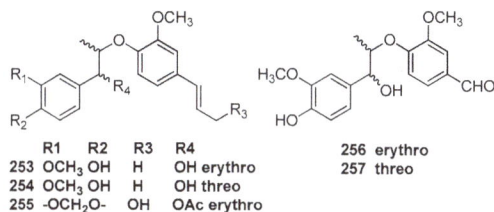

	R1	R2	R3	R4
253	OCH₃	OH	H	OH erythro
254	OCH₃	OH	H	OH threo
255	-OCH₂O-		OH	OAc erythro

256 erythro
257 threo

Figure 17. Chemical structures of 8,4′-oxyneolignans.

7. Uncommon Lignans

This section covers lignans and neolignans that contain uncommon skeletons. The molecular backbone of compounds **258–267** consists of a unique C6–C3 unit, and an ether oxygen atom provides the linkage between the phenyl and propyl groups. The semi-systematic names of those uncommon lignans without trivial names and their corresponding names in the literature are given in Table SI-5 (Supporting Information).

Compounds **258–261** were isolated from the benzene extract of *A. simulans* trunk wood [46]. Compounds **259** and **260** were also found in the fruits of *L. armeniaca* [25] and the trunk wood of *N. mirunda*, respectively [64]. Compound **262** was isolated from the benzene extract of the trunk wood of *A. terminalis* [42]. Compounds **263** and **264** have been found in the trunk bark of *O. catharinensis* [45]. The stem bark of *O. veraguensis* also yielded compound **263** [16]. Chrysophyllon III-A (**265**) and chrysophyllon III-B (**266**) have been found in trunk wood, bark, and fruit calyces of *L. chrysophylla* [73]. Compound **267** was obtained from the bark and leaves of *O. porosa* [59]. (+)-9'-*O*-trans-feruloyl-5,5'-dimethoxylariciresinol (**268**), which showed cytotoxicity against a small panel of human tumor cell lines with ED$_{50}$ values around 10 µg/mL, was isolated from the stems of *L. obtusiloba* Blume [21,22]. Turbacenlignan A (**269**), a 7,8-secolignan, was isolated from the leaves and root bark of *N. turbacensis* (Kunth) Nees [9]. Cinnaburmanin A (**270**) was isolated from the roots of *C. burmanii* [84] (Figure 18).

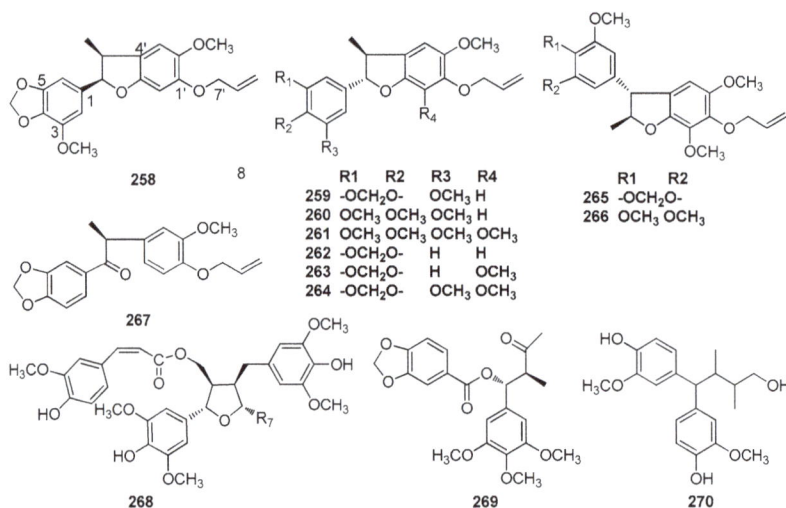

Figure 18. Chemical structures of uncommon neolignans.

8. Conclusions

A renewed interest in compounds isolated from natural resources has led to an enormous class of pharmacologically active compounds. Lignans and neolignans have been revealed to show significant pharmacological activities, including antitumor, anti-inflammatory, immunosuppression, cardiovascular, antioxidant, and antiviral activities [85,86]. The Lauraceae family, especially the genera of *Machilus*, *Ocotea*, and *Nectandra*, represents a rich source of lignans and neolignans. Moreover, neolignans are responsible for the potential chemotaxonomic significance found in the study of Lauraceae. Studies on lignans and neolignans in Lauraceae were mainly carried out in the 1980s. There have been more studies concerning the identification of lignans and neolignans in Lauraceae, but less on the biological activities of these compounds. Among the lignans and neolignans isolated from Lauraceae, the biological activities of sesamin and yangambin have been studied more, while there are relatively few articles published on other compounds. Sesamin, a 7.9',7'.9-diepoxylignan present in many species in the Lauraceae family such as *N. amazonurn*, *M. thunbergii*, *P. pyrifolia*, *C. burmanii*, and *N. turbacensis*, showed significant anticancer properties [87]. Yangambin (**51**) which was the major constituent of *O. duckei*, showed diverse biological activities [88]. Therefore, it is extremely urgent to expand the scope of research on the lignans and neolignans in Lauraceae, with the aim of discovering all biological activities of these compounds.

Supplementary Materials: The supplementary materials are available online at http://www.mdpi.com/1420-3049/23/12/3164/s1.

Author Contributions: Investigation, S.X. and J.Y.; Writing-Original Draft Preparation, W.W.; Writing-Review & Editing, Y.L.; Funding Acquisition, K.G.

Funding: This work was funded by the National Natural Science Foundation of China (21402074).

Conflicts of Interest: The authors declare no conflict of interest.

References

1. Teponno, R.B.; Kusari, S.; Spiteller, M. Recent advances in research on lignans and neolignans. *Nat. Prod. Rep.* **2016**, *33*, 1044–1094. [CrossRef] [PubMed]

2. Shimomura, H.; Sashida, Y.; Oohara, M. Lignans from *Machilus thunbergii*. *Phytochemistry* **1987**, *26*, 1513–1515. [CrossRef]

3. Yu, Y.U.; Kang, S.Y.; Park, H.Y.; Sung, S.H.; Lee, E.J.; Kim, S.Y.; Kim, Y.C. Antioxidant lignans from *Machilus thunbergii* protect CCl4–injured primary cultures of rat hepatocytes. *J. Pharm. Pharmacol.* **2000**, *52*, 1163–1169. [CrossRef] [PubMed]

4. Ma, C.J.; Sung, S.H.; Kim, Y.C. Neuroprotective lignans from the bark of *Machilus thunbergii*. *Planta Med.* **2004**, *70*, 79–80. [PubMed]

5. Li, G.; Lee, C.S.; Woo, M.H.; Lee, S.H.; Chang, H.W.; Son, J.K. Lignans from the bark of *Machilus thunbergii* and their DNA topoisomerases I and II inhibition and cytotoxicity. *Biol. Pharm. Bull.* **2004**, *27*, 1147–1150. [CrossRef]

6. Coybarrera, E.D.; Cucasuarez, L.E. In vitro anti-inflammatory effects of naturally–occurring compounds from two Lauraceae plants. *An. Acad. Bras. Cienc.* **2011**, *83*, 1397–1402. [CrossRef]

7. Park, B.Y.; Min, B.S.; Kwon, O.K.; Oh, S.R.; Ahn, K.S.; Kim, T.J.; Kim, D.Y.; Bae, K.; Lee, H.K. Increase of caspase-3 activity by lignans from *Machilus thunbergii* in HL-60 cells. *Biol. Pharm. Bull.* **2004**, *27*, 1305–1307. [CrossRef] [PubMed]

8. Macías-Villamizar, V.; Cuca-Suárez, L. Diaryldimethylbutane lignans and other constituents isolated from *Nectandra turbacensis* (Kunth) Nees (Lauraceae). *Rev. Colomb. Quim.* **2014**, *43*, 12–16.

9. Macías-Villamizar, V.; Cuca-Suárez, L.; González, F.V.; Rodríguez, S. Lignoids isolated from *Nectandra turbacensis* (Kunth) Nees (Lauraceae). *Rec. Nat. Prod.* **2016**, *10*, 654–658.

10. Pérez, C.; Almonacid, L.N.; Trujillo, J.M.; González, A.G.; Alonso, S.J.; Navarro, E. Lignans from *Apollonias barbujana*. *Phytochemistry* **1995**, *40*, 1511–1513. [CrossRef]

11. Moro, J.C.; Fernandes, J.B.; Vieira, P.C.; Yoshida, M.; Gottlieb, H.E. Neolignans from *Nectandra puberula*. *Phytochemistry* **1987**, *26*, 269–272. [CrossRef]

12. Quesne, P.W.L.; Larrahondo, J.E.; Raffauf, R.F. Antitumor plants. X. constituents of *Nectandra rigida*. *J. Nat. Prod.* **1980**, *43*, 353–359. [CrossRef] [PubMed]

13. Shimomura, H.; Sashida, Y.; Oohara, M. Lignans from *Machilus thunbergii*. *Phytochemistry* **1988**, *27*, 634–636. [CrossRef]

14. López, H.; Valera, A.; Trujillo, J. Lignans from *Ocotea foetens*. *J. Nat. Prod.* **1995**, *58*, 782–785. [CrossRef]

15. Crossley, N.S.; Djerassi, C. Naturally occurring oxygen heterocyclics. Part XI. Veraguensin. *J. Chem. Soc.* **1962**, 1459–1462. [CrossRef]

16. Khan, M.R.; Gray, A.I.; Waterman, P.G. Neolignans from stem bark of *Ocotea veraguensis*. *Phytochemistry* **1987**, *26*, 1155–1158. [CrossRef]

17. Dias, A.D.F.; Giesbrecht, A.M. Neolignans from *Urbanodendron verrucosum*. *Phytochemistry* **1982**, *21*, 1137–1139. [CrossRef]

18. Huang, Y.-T.; Chang, H.-S.; Wang, G.-J.; Lin, C.-H.; Chen, I.-S. Secondary metabolites from the roots of *Beilschmiedia tsangii* and their anti-inflammatory activities. *Int. J. Mol. Sci.* **2012**, *13*, 16430–16443. [CrossRef]

19. Phan, M.G.; Phan, T.S.; Matsunami, K.; Otsuka, H. New neolignans and lignans from Vietnamese medicinal plant *Machilus odoratissima* Nees. *Cheminform* **2006**, *54*, 380–383.

20. Takaoka, D.; Tani, H.; Nozaki, H. A new lignan, (-)-parabenzoinol, from *Parabenzoin trilobum* Nakai. *Chem. Lett.* **1995**, *24*, 915–916. [CrossRef]

21. Kwon, H.C.; Sang, U.C.; Lee, J.O.; Bae, K.H.; Zee, O.P.; Kang, R.L. Two new lignans from *Lindera obtusiloba* Blume. *Arch. Pharm. Res.* **1999**, *22*, 417–422. [CrossRef] [PubMed]

22. Choi, H.G.; Choi, Y.H.; Ji, H.K.; Kim, H.H.; Kim, S.H.; Kim, J.A.; Sang, M.L.; Na, M.K.; Lee, S.H. A new neolignan and lignans from the stems of *Lindera obtusiloba* Blume and their anti-allergic inflammatory effects. *Arch. Pharm. Res.* **2014**, *37*, 467–472. [CrossRef] [PubMed]

23. Wada, K.; Munakata, K. (-)-Parabenzlactone, a new piperolignanolide isolated from *Parabenzoin trilobum* Nakai. *Tetrahedron Lett.* **1970**, *11*, 2017–2019. [CrossRef]

24. Wei, X.; Li, G.H.; Wang, X.L.; He, J.X.; Wang, X.N.; Ren, D.M.; Lou, H.X.; Shen, T. Chemical constituents from the leaves of *Cinnamomum parthenoxylon* (Jack) Meisn. (Lauraceae). *Biochem. Syst. Ecol.* **2017**, *70*, 95–98. [CrossRef]

25. Barbosa-Filho, J.M.; Yoshida, M. Neolignans from the fruits of *Licaria armeniaca*. *Phytochemistry* **1987**, *26*, 319–321. [CrossRef]

26. Barbosa-Filho, J.M.; Yoshida, M. Lignoids from *Nectandra amazonum* and *N. glabrescens*. *Phytochemistry* **1989**, *28*, 1991. [CrossRef]

27. Batista, A.N.D.L.; Batista Junior, J.M.; López, S.N.; Furlan, M.; Cavalheiro, A.J.; Silva, D.H.S.; Bolzani, V.D.S.; Nunomura, S.M.; Yoshida, M. Aromatic compounds from three Brazilian Lauraceae species. *Quím. Nova* **2010**, *33*, 321–323. [CrossRef]

28. Barrera, E.D.C.; Suárez, L.E.C.; Suárez, L.E.C. Chemical constituents from *Pleurothyrium cinereum* (van der Werff) (Lauraceae) from Colombia. *Biochem. Syst. Ecol.* **2008**, *36*, 674–677. [CrossRef]

29. Lin, I.J.; Lo, W.L.; Chia, Y.C.; Huang, L.Y.; Cham, T.M.; Tseng, W.S.; Yeh, Y.T.; Yeh, H.C.; Wang, Y.D.; Chen, C.Y. Isolation of new esters from the stems of *Cinnamomum reticulatum* Hay. *Nat. Prod. Res.* **2010**, *24*, 775–780. [CrossRef] [PubMed]

30. Kim, M.R.; Jung, H.J.; Min, B.S.; Oh, S.R.; Kim, C.S.; Ahn, K.S.; Kang, W.S.; Lee, H.K. Constituents from the stems of *Actinodaphne lancifolia*. *Phytochemistry* **2002**, *59*, 861–865. [CrossRef]

31. Li, C.T.; Kao, C.L.; Li, H.T.; Huang, S.T.; Huang, S.C.; Chen, C.Y. Secondary metabolites from the leaves of *Cinnamomum macrostemon* Hayata. *Eur. J. Biomed. Pharm.Sci.* **2015**, *2*, 38–51.

32. Chen, C.; Hong, Z.; Yang, W.; Wu, M.; Huang, J.; Lee, J. A novel homosesquiterpenoid from the stems of *Cinnamomum burmanii*. *Nat. Prod. Res.* **2012**, *26*, 1218. [CrossRef] [PubMed]

33. Carvalho, M.G.D.; Yoshida, M.; Gottlieb, H.E. Lignans from *Nectandra turbacensis*. *Phytochemistry* **1987**, *26*, 265–267. [CrossRef]

34. Morais, L.C.S.L.; Almeida, R.N.; Dacunha, E.V.L.; Dasilva, M.S.; Barbosafilho, J.M.; Gray, A.I. Further lignans from *Ocotea duckei*. *Pharm. Biol.* **1999**, *37*, 144–147. [CrossRef]

35. Hoang, V.D.; Tan, G.T.; Zhang, H.J.; Tamez, P.A.; Hung, N.V.; Cuong, N.M.; Soejarto, D.D.; Fong, H.H.; Pezzuto, J.M. Natural anti-HIV agents– Part I: (+)-demethoxyepiexcelsin and verticillatol from *Litsea verticillata*. *Phytochemistry* **2002**, *59*, 325–329. [CrossRef]

36. Yang, L.J.; Wen, C.; Luo, Y.P.; Li, G.P.; Yang, X.D.; Liang, L. Lignans and ketonic compounds from *Litsea chinpingensis* (Lauraceae). *Biochem. Syst. Ecol.* **2009**, *37*, 696–698. [CrossRef]

37. Chen, C.Y.; Yeh, Y.T.; Hsui, Y.R. A new lignan from the roots of *Cinnamomum philippinense*. *Chem. Nat. Compd.* **2011**, *47*, 519–520. [CrossRef]

38. Lima, O.A.; Gottlieb, O.R.; Magalhães, M.T. Burchellin, a neolignan from *Aniba burchellii*. *Phytochemistry* **1972**, *11*, 2031–2037. [CrossRef]

39. Mourão, J.C.; Yoshida, M.; Mascarenhas, Y.P.; Rodrigues, M.; Rosenstein, R.D.; Tomita, K. Absolute configuration of the benzofuranoid neolignans. *Phytochemistry* **1977**, *16*, 1003–1006.

40. Fernandes, J.B.; Gottlieb, O.R.; Maia, J.G.S. Neolignans from an *Aniba* sp. *Phytochemistry* **1976**, *15*, 1033–1036. [CrossRef]

41. Filho, R.B.; Figliuolo, R.; Gottlieb, O.R. Neolignans from a *Nectandra* species. *Phytochemistry* **1980**, *19*, 659–662. [CrossRef]

42. Gottli, O.R.; Ferreira, S.Z.S. Neolignans from *Aniba terminalis*. *Phytochemistry* **1975**, *14*, 1825–1827. [CrossRef]

43. Juan, C.M.V.; Maia, J.S.; Yoshida, M. Neolignans from an *Aniba* species. *Phytochemistry* **1980**, *19*, 474–476.

44. Haraguchi, M.; Motidome, M.; Yoshida, M.; Gottlieb, O.R. Neolignans from *Ocotea catharinensis*. *Phytochemistry* **1983**, 22, 561–563. [CrossRef]

45. Ishige, M.; Motidome, M.; Yoshida, M.; Gottlieb, O.R. Neolignans from *Ocotea catharinensis*. *Phytochemistry* **1991**, 30, 4121–4128. [CrossRef]

46. Aiba, C.J.; Alvarenga, M.A.D.; Oscar, C.C.; Giesbrecht, A.M.; Pagliosa, F.M. Benzofuranoid neolignans from *Aniba simulans*. *Phytochemistry* **1977**, 16, 741–743. [CrossRef]

47. Aiba, C.J.; Fernandes, J.B.; Gottlieb, O.R.; Maia, J.G.S. Neolignans from an *Aniba species*. *Phytochemistry* **1975**, 14, 1597–1604. [CrossRef]

48. Aiba, C.J.; Maia, J.G.S.; Pagliosa, F.M.; Yoshida, M. Benzofuranoid neolignans from *Licaria armeniaca*. *Phytochemistry* **1978**, 17, 2038–2039. [CrossRef]

49. Barbosa-Filho, J.M.; Yoshida, M.; Barbosa, R. d. C.S.B.C.; Giesbrecht, A.M.; Young, M.C.M. Benzoyl esters and amides, styrylpyrones and neolignans from the fruits of *Aniba riparia*. *Phytochemistry* **1987**, 26, 2615–2617. [CrossRef]

50. Dias, S.M.C.; Fernandes, J.B.; Maia, J.G.S.; Gottlieb, H.E. Eusiderins and other neolignans from an *Aniba species*. *Phytochemistry* **1986**, 25, 213–217. [CrossRef]

51. Giesbrecht, A.M.; Franca, N.C.; Rocha, A.I.D. The neolignans of *Licaria canella*. *Phytochemistry* **1974**, 13, 2285–2293. [CrossRef]

52. Trevisan, L.M.V.; Yoshida, M. Hexahydrobenzofuranoid neolignans from an *Aniba species*. *Phytochemistry* **1984**, 23, 661–665. [CrossRef]

53. Aiba, C.J.; Filho, R.B.; Gottlieb, O.R. Porosin: A neolignan from *Ocotea porosa*. *Phytochemistry* **1973**, 12, 110–116. [CrossRef]

54. Andrade, C.H.S.; Filho, R.B. Neolignans from *Aniba ferrea*. *Phytochemistry* **1980**, 19, 1191–1194. [CrossRef]

55. Romoff, P.; Yoshida, M.; Gottlieb, O.R. Neolignans from *Ocotea aciphylla*. *Phytochemistry* **1984**, 23, 2101–2104. [CrossRef]

56. Felicio, J.D.A.; Motidome, M.; Yoshida, M. Further neolignans from *Ocotea aciphylla*. *Phytochemistry* **1986**, 25, 1707–1710. [CrossRef]

57. Dias, D.A.; Yoshida, M. Further neolignans from *Ocotea porosa*. *Phytochemistry* **1986**, 25, 2613–2616. [CrossRef]

58. Carvalho, M.G.D.; Yoshida, M.; Gottlieb, H.E. Bicyclooctanoid, carinatone and megaphone type neolignans from *Ocotea porosa*. *Phytochemistry* **1988**, 27, 2319–2323. [CrossRef]

59. David, J.M.; Yoshida, M.; Gottlieb, O.R. Neolignans from bark and leaves of *Ocotea porosa*. *Phytochemistry* **1994**, 36, 491–499. [CrossRef]

60. Aiba, C.J.; Corrêa, R.G.C.; Gottlieb, O.R. Natural occurrence of Erdtman's dehydrodiisoeugenol. *Phytochemistry* **1973**, 12, 1163–1164. [CrossRef]

61. Tsai, I.L.; Chen, J.H.; Duh, C.Y.; Chen, I.S. Cytotoxic neolignans from the stem wood of *Machilus obovatifolia*. *Planta Med.* **2000**, 66, 403–407. [CrossRef] [PubMed]

62. Alvarenga, M.A.D.; Brocksom, U.; Oscar, C.C.; Magalhães, M.T. Neolignans from *Aniba burchellii*. *Phytochemistry* **1977**, 16, 1797–1799. [CrossRef]

63. Coy Barrera, E.D.; Cuca Suárez, L.E. Errata: Three new 7.3′,8.5′-connected bicyclo [3.2.1]octanoids and other neolignans from leaves of *Nectandra amazonum* Nees. (Lauraceae). *Chem. Pharm. Bull.* **2009**, 57, 639–642. [CrossRef] [PubMed]

64. Aiba, C.J.; Gottlieb, O.R.; Pagliosa, F.M.; Yoshida, M.; Magalhães, M.T. Neolignans from *Nectandra miranda*. *Phytochemistry* **1977**, 16, 745–748. [CrossRef]

65. Dodson, C.D.; Stermitz, F.R.; Oscar, C.C.; Janzen, D.H. Neolignans from fruits of *Ocotea veraguensis*. *Phytochemistry* **1987**, 26, 2037–2040.

66. Tsai, I.L.; Hsieh, C.F.; Duh, C.Y. Additional cytotoxic neolignans from *Persea obovatifolia*. *Phytochemistry* **1998**, 48, 1371–1375. [CrossRef]

67. Tsai, I.L.; Hsieh, C.F.; Duh, C.-Y.; Ih-Sheng, C. Cytotoxic neolignans from *Persea obovatifolia*. *Phytochemistry* **1996**, 43, 1261–1263. [CrossRef]

68. Diaz, P.S.P.; Yoshida, M. Neolignans from *Aniba lancifolia*. *Phytochemistry* **1980**, 19, 285–288. [CrossRef]

69. Drewes, S.E.; Horn, M.M.; Sehlapelo, B.M.; Ramesar, N.; Field, J.S.; Shaw, S.; Sandor, P. Isoicobullenone and a neolignan ketone from *Ocotea bullatabark*. *Phytochemistry* **1995**, 38, 1505–1508. [CrossRef]

70. Coybarrera, E.D.; Cucasuárez, L.E.; Sefkow, M. PAF-antagonistic bicyclo[3.2.1]octanoid neolignans from leaves of *Ocotea macrophylla* Kunth. (Lauraceae). *Phytochemistry* **2009**, 70, 1309–1314. [CrossRef]

71. Tsai, I.L.; Chen, J.H.; Duh, C.Y.; Chen, I.S. Cytotoxic neolignans and butanolides from *Machilus obovatifolia*. *Planta Med.* **2001**, *67*, 559–561. [CrossRef] [PubMed]

72. Ferreira, Z.S.; Roque, N.C.; Gottlieb, H.E. An unusual porosin type neolignan from *Licaria chrysophylla*. *Phytochemistry* **1982**, *21*, 2756–2758. [CrossRef]

73. Lopes, M.N.; Silva, M.S.D.; José, M.B.F.; Ferreira, Z.S.; Yoshida, M. Unusual benzofuranoid neolignans from *Licaria chrysophylla*. *Phytochemistry* **1986**, *25*, 2609–2612. [CrossRef]

74. Bülow, M.V.V.; Franca, N.C.; Gottlieb, O.R.; Suarez, A.M.P. Guianin: A neolignan from *Aniba guianensis*. *Phytochemistry* **1973**, *12*, 1805–1808. [CrossRef]

75. Marques, M.O.M.; Gomes, M.C.C.P.; Yoshida, M. Bicyclo [3.2.1] octanoid neolignans from *Ocotea porosa*. *Phytochemistry* **1992**, *31*, 275–277. [CrossRef]

76. Gomes, M.C.C.P.; Yoshi, M.; Gottli, O.R.; Juan, C.; Martinez, C.; Gottlieba, H.E. Bicyclo(3.2.1)octane neolignans from an *Ocotea species*. *Phytochemistry* **1983**, *22*, 269–273. [CrossRef]

77. 77. Dias, S.M.C.; Fernandes, J.B.; Maia, J.G.S.; Gottlieb, O.R.; Gottlieb, H.E. Neolignans from an *Aniba* species. *Phytochemistry* **1982**, *21*, 1737–1740. [CrossRef]

78. Silva, W.D.D.; Braz-Filho, R. Bicyclooctanoid neolignans from *Ocotea costulatum*. *Phytochemistry* **1989**, *28*, 661–662. [CrossRef]

79. Fo, R.B.; Carvalho, M.G.D.; Maia, J.G.S.; Silva, M.L.D. Neolignans from *Licaria rigida*. *Phytochemistry* **1981**, *20*, 2049–2050.

80. Alegrio, L.V.; Fo, R.B.; Maia, J.G.S. Lignans and neolignans from *Licaria armeniaca*. *Phytochemistry* **1981**, *20*, 1963–1965. [CrossRef]

81. Franca, N.C.; Gottlieb, O.R.; Maia, J.G.S. Macrophyllin, a neolignan from *Licaria macrophylla*. *Phytochemistry* **1974**, *13*, 2839–2842. [CrossRef]

82. Coy, E.D.; Cuca, L.E.; Sefkow, M. Macrophyllin-type bicyclo[3.2.1]octanoid neolignans from the leaves of *Pleurothyrium cinereum*. *J. Nat. Prod.* **2009**, *72*, 1245–1248. [CrossRef] [PubMed]

83. Sehlapelo, B.M.; Drewes, S.E.; Sandor, P. Ocobullenone: A bicyclo[3.2.1]octanoid neolignan from *Ocotea bullata*. *Phytochemistry* **1993**, *32*, 1352–1353. [CrossRef]

84. Yuan, L.T.; Kao, C.L.; Chen, C.T.; Li, H.T.; Chen, C.Y. A new lignan from *Cinnamomum burmanii*. *Chem. Nat. Compd.* **2017**, *53*, 623–625. [CrossRef]

85. Pan, J.-Y.; Chen, S.-L.; Yang, M.-H.; Wu, J.; Sinkkonen, J.; Zou, K. An update on lignans: natural products and synthesis. *Nat. Prod. Rep.* **2009**, *26*, 1251–1292. [CrossRef] [PubMed]

86. Marcotullio, M.C.; Curini, M.; Becerra, J.X. An ethnopharmacological, phytochemical and pharmacological review on lignans from Mexican Bursera spp. *Molecules* **2018**, *23*, 1976. [CrossRef] [PubMed]

87. Majdalawieh, A.F.; Massri, M.; Nasrallah, G.K. A comprehensive review on the anti-cancer properties and mechanisms of action of sesamin, a lignan in sesame seeds (*Sesamum indicum*). *Eur. J. Pharmacol.* **2017**, *815*. [CrossRef] [PubMed]

88. Araújo, I.G.A.; Silva, D.F.; do Carmo de Alustau, M.; Dias, K.L.G.; Cavalcante, K.V.M.; Veras, R.C.; Barbosa-Filho, J.M.; Neto, M.A.; Bendhack, L.M.; de Azevedo Correia, N.; et al. Calcium influx inhibition is involved in the hypotensive and vasorelaxant effects induced by yangambin. *Molecules* **2014**, *19*, 6863–6876. [CrossRef] [PubMed]

molecules

MDPI

Review

An Ethnopharmacological, Phytochemical and Pharmacological Review on Lignans from Mexican *Bursera* spp.

Maria Carla Marcotullio * , **Massimo Curini and Judith X. Becerra**

Department of Pharmaceutical Sciences, University of Perugia, via del Liceo, 1-06123 Perugia, Italy;
massimo.curini@unipg.it (M.C.); jxb@email.arizona.edu (J.X.B.)
* Correspondence: mariacarla.marcotullio@unipg.it; Tel.: +39-075-585-5100

Received: 5 July 2018; Accepted: 5 August 2018; Published: 8 August 2018

Abstract: The genus *Bursera* belongs to the family Burseraceae and has been used in traditional Mexican medicine for treating various pathophysiological disorders. The most representative phytochemicals isolated from this genus are terpenoids and lignans. Lignans are phenolic metabolites known for their antioxidant, apoptotic, anti-cancer, anti-inflammatory, anti-bacterial, anti-viral, anti-fungal, and anti-protozoal properties. Though the genus includes more than 100 species, we have attempted to summarize the biological activities of the 34 lignans isolated from selected Mexican *Bursera* plants.

Keywords: *Bursera*; Burseraceae; lignans

1. Introduction

The genus *Bursera* Jacq. ex L. (family Burseraceae, order Sapindales), named after the Danish botanist Joachim Burser (1583–1639), is a monophyletic genus [1] that includes about 105 species of small trees and shrubs distributed from Southern U.S. to Peru and the Caribbean, particularly in Mexico (ca. 92 species) [2]. These plants are characterized by the production of resins that are exuded from the trunk and leaves and provide a chemical defense against specialized herbivores [3]. Once dried, the resin obtained by *Bursera* spp. is called "copal", a term which is also used to describe a large group of resins characterized by hardness and a relative high melting point, which also is found in other plants [4]. The loss of the essential oils and the oxidation and polymerization processes transform copal into amber.

The phytochemistry of this genus is characterized by the presence of volatile metabolites such as simple hydrocarbons and terpenoids as well as phenolics [5–8]. Among the compounds present in the volatile fraction, heptane, α- and β-pinene, β-phellandrene, and limonene are among the most frequent [5], whereas β-caryophyllene and germacrene D are the most common sesquiterpenes in the genus *Bursera* [8]. Cembrane and verticillane diterpenoids are often present [9–11]. Pentacyclic triterpenoids are largely present in the resin of several species, and the study of triterpenoidic composition of resins is important to define the botanical origin of archaeological samples of copal [12]. In *Bursera microphylla* resin, malabaricane triterpenoids were also found [11]. Leaves and branches of some *Bursera* also contain flavonoids [13–15] and luteolin 3′-*O*-rhamnoside is very common [13].

Lignans are naturally occurring plant phenolics, biosynthetically derived from phenylpropanoids, that are important components in foods and medicines; their chemical and biological properties have been reviewed [16]. The aim of this review is to summarize literature findings on the botanical characterization, distribution, ethnopharmacology, and biological activities of Mexican *Bursera* that produce lignans. Different *Bursera* species have been sorted according to, and synonyms are those reported in, the Plant List Database [17]. Unless otherwise specified, common names are those reported by Lemos and Rivera [18]. The phytochemistry was analyzed by data reported in the SciFinder database. Images of the species reported in this paper can be found on the Enciclovida web site [19].

2. Genus *Bursera*

Members of this genus are typically small- to medium-sized trees or shrubs, mostly dioecious, succulent, and with resin canals in vascularized tissues. Their leaves are deciduous, imparipinnate, or sometimes unifoliolate or trifoliolate (occasionally bipinnate). Their flowers are small, almost always unisexual, three- to six-merous. Their fruit is a dehiscent two- to three-valve drupe with a fleshy to coriaceous skin, and with a pyrene (the stone or pit that contains the seed), cartilaginous to bony, enveloped totally or partially by an arillate structure.

The taxonomy of the genus *Bursera* is based on morphological characteristics of fruit, bark, and leaves, as well as molecular data. Currently, there are two recognized subgenera: one subgenus named *Bursera* (previously called section Bursera) that includes species commonly known with the general vernacular name of "cuajiotes", and the other is called *Elaphrium* (previously called section Bullockia) that comprises species with the general common name of "copales" [20–22]. The most conspicuous difference between the subgenera is the bark: in subgenus *Bursera*, it tends to be colorful and exfoliating, whereas in *Bursera* subgenus *Elaphrium*, it is likely to be complete (not exfoliating), and grey or reddish grey. However, although bark helps in species identification because it is easy to see, whether the bark is complete or exfoliating is not an absolute difference between the two groups. Setting the bark aside, the most reliable distinction between the two subgenera is the number of locules in the ovary (three in subg. *Bursera* vs. two in *Elaphrium*), and the number of valves in the fruit (three in subg. *Bursera* vs. two in *Elaphrium*) [20,21]. Another distinguishing trait is the presence of well-developed cataphylls (small bract-like leaves that appear before "true leaves" and are short-lived) in subgenus *Elaphrium* and absent or very inconspicuous in subgenus *Bursera* [23]. Toledo further divided section *Bursera* into three groups that can be distinguished by the color of the exfoliating bark: mulatos, red cuajiotes, and yellow cuajiotes [24]. Furthermore, the section *Bullockia* was divided into two groups: pseudoaril-covered fruits group and partially covered fruits group [24]. In 1980, Gillet changed the name of section *Bullockia* into *Elaphrium* due to the fact that some characteristics of this section resemble those of *Elaphrium tomentosum* Jacq. [25]. Phylogeny studies by Becerra and Venable allowed the recognition of four different groups in section *Bursera*: the *simaruba* group (massive trees, trilobate cotyledons, red exfoliating bark, poor producing resin (*mulatos*)), the *microphylla* group (medium-sized trees or shrubs, multilobate cotyledons, yellow to red exfoliating bark, highly resinous), the *fagaroides* group (medium-sized trees or shrubs, multilobate cotyledons, highly resinous), and the *fragilis* group (medium-sized trees, multilobate cotyledons, red exfoliating bark, highly resinous (*cuajiotes*)) [22]; and two groups in *Elaphrium*: *copallifera* (seed completely or at least two-thirds covered by pseudoaril) and *glabrifolia* (seed partially covered or at least less than two-thirds by pseudoaril) [5]. *Bursera*'s flowers are small and inconspicuous, with few species-specific characteristics, occurring during bursts during the dry season. Thus, in their natural habitats, it is often easier to recognize them by their bark and leaf characteristics as well as the locations where they grow. The *Bursera* genus is closely related to the other two resin producing Burseraceae: *Boswellia* and *Commiphora*, and they differ mostly in their geographic distribution. *Boswellia* and *Commiphora* are present in desert parts of tropical Africa, Arabia, Pakistan, and India, whereas *Bursera* is distributed from the Southern U.S. to Peru and the Caribbean, and particularly in Mexico. The *Bursera* section shares some similarities with *Boswellia*, whereas *Elaphrium* is similar to *Commiphora* [26].

2.1. Traditional Uses, Phytochemistry, and Biological Activities

Most of the *Bursera* species that produce lignans are widely used by the Mexican native population. Although different *Bursera* species are used for different health issues, they are traditionally attributed medicinal properties including providing relief from pain, inflammation, rheumatism, and can help treat illnesses such as colds, skin tumors, polyps, and venereal diseases [27–30]. The following are reported lignan-producing *Bursera* plants from Mexico listed according to the subgenus they belong to and sorted alphabetically. Traditional uses, when found, and biological properties of the isolated compounds are described for each species.

2.2. Subgenus Elaphrium

2.2.1. Bursera citronella McVaugh and Rzed.

B. citronella (synonyms: none reported) is also known as *xochicopal* (Náhuatl name) or *lináloe*, and as *almárciga* in Spanish [31]. It is a 10 m tree with grey trunk bark and unifoliate or trifoliolate leaves, distributed in Western Mexico (Michoacan, Colima, Jalisco, and Guerrero). It belongs to the subsection *glabrifolia* [5]. The resin is mostly used as incense. It has been reported that *B. citronella* is used as antitussive in several regions in Mexico [32].

The phytochemistry of *B. citronella* has been studied by Koulman who isolated two lignans: hinokinin (**1**) and savinin (**2**) (Figure 1) [33]. Biological activities of hinokinin have been recently reviewed [34]. Cytotoxicity of hinokinin (**1**) has been tested by several authors [35–38]. Hinokinin (**1**) has been shown to have anti-inflammatory [39–42], immunosuppressive [43,44], antibacterial [45], and antiviral [46] properties. Hinokinin (**1**) was tested for several other biological activities, such as antispasmodic [47], neurite outgrowth-promoting in PC12 cells [48], antileukemic [49], antiproliferative [50], and neuroprotective activities [51]. This compound showed an interesting activity against *Trypanosoma cruzi*, [52–56], but with a low parasite selectivity [57]. Hinokinin has been chosen as a trypanosomicidal marker in *P. cubeba* [58]. In order to ascertain the safety of this compound toward mammalian cells, several studies have been performed [59–61]. The authors found that hinokinin did not increase DNA damage, demonstrating the absence of mutagenic and genotoxic activities. On the other hand, the results on the antimutagenic potential of this compound showed a strong inhibitory effect against some direct and indirect-acting mutagens.

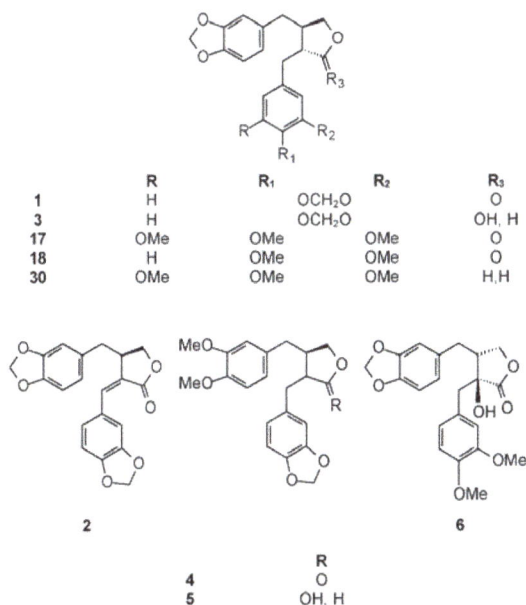

	R	R_1	R_2	R_3
1	H	OCH_2O		O
3	H	OCH_2O		OH, H
17	OMe	OMe	OMe	O
18	H	OMe	OMe	O
30	OMe	OMe	OMe	H,H

	R
4	O
5	OH, H

Figure 1. Structures of dibenzyl butyrolactone lignans isolated from *Bursera* spp.

Savinin (**2**), also called hibalactone [62], was isolated for the first time from *Juniperus* spp. [63]. It has been tested for different biological activities, such as cytotoxicity against several tumor cell lines [36,64–66]. Savinin (**2**) was shown to have an anti-inflammatory activity in several assays [40,44,67–69], but also interesting antinociceptive, anxiolytic, and antioxidant activities [70,71].

2.2.2. *Bursera cuneata* (Schltdl.) Engl.

B. cuneata (synonym: *Elaphrium cuneatum* Schltdl. L.) is a tree that grows up to 10 m in height with no-peeling grey-reddish bark. It has imparipinnate leaves of coriaceous texture with 3 to 13 leaflets, 6.5 cm long and 2.3 cm wide, and margin roughly serrated. Their flowers are clustered in inflorescences up to 8 cm long. Its flowers are white and its fruits are up to 1.2 cm long with a black pit, almost completely covered by a yellow or orange pseudoaril. It is native to Mexican oak-tropical deciduous forest transition zones from Jalisco to Oaxaca and is often known as copal or *copalillo* [72]. Although *B. cuneata* is characterized by seeds covered by pseudoaril, Becerra and Venable did not classify it into the *copallifera* group [22]. No medical uses have been reported for this species, but it is largely used as incense during sacred ceremonies and to prepare handcrafted objects.

Koulman isolated three lignans from this species: hinokinin (**1**), savinin (**2**), and cubebin (**3**) (Figure 1) [33]. Cubebin was first isolated by Chatterjee in 1968 from *Piper cubeba* [73] and then from several *Aristolochia* spp. Biological activities of cubebin have been recently reviewed by Cunha et al. [16]. In particular, trypanocidal activity of this compound against free amastigote forms of *Trypanosoma cruzi* has been studied by de Souza et al. [52], and Bastos et al. showed it is inactive against trypomastigote forms [74]. Notably, cubebin (**3**) is usually the starting material for the semi-synthetic preparation of hinokinin (**1**) and other lignans [52]. Recently, cubebin was proven to induce vasorelaxation via nitric oxide activation without prostacyclin involvement [75]. Because of its therapeutic potential, the effects of cubebin on mutagenicity and genotoxicity has been deeply studied by several research groups [76]. The authors found that cubebin (**3**) was cytotoxic at high doses (280 µM), but at lower concentrations, no cytotoxic, mutagenic, or proliferative effects were observed for this compound. The mutagenicity of cubebin (**3**), alone or in combination with doxorubicin (DXR), using standard (ST) and high bioactivation (HB) crosses of the wing Somatic Mutation And Recombination Test (SMART) in *Drosophila melanogaster* was also studied [77]. Even in this case, the effect of cubebin was dose-dependent. At lower doses (<1 mM), it reduces DXR toxicity, whereas at higher doses (>2.0 mM), it is cytotoxic. The biological activities of cubebin have been recently reviewed [78].

2.2.3. *Bursera excelsa* (Kunth) Engl.

B. excelsa (synonyms: *Bullockia sphaerocarpa* and *Elaphrium excelsum*) is commonly known as *tecomajaca*, or *copal santo*, *pom* (in the Maya language) [79] and *tecomahaca* in the Náhuatl language [29]. It belongs to the *copallifera* group. These are trees up to 8 m tall with grey non-peeling bark. Their leaves are 11 to 23 cm long and 6 to 10.5 cm wide, with winged raquis and 9 to 15 leaflets, hairy and margin conspicuously toothed. Their flowers are small, yellow, and densely hairy. Traditionally, it is used to treat tumors and muscle spasms [29]. It is widely distributed across Mexico from the state of Sinaloa to Chiapas. The phytochemistry of *B. excelsa* has been extensively studied [8,12,13]. Regarding lignan composition, Koulman isolated three compounds from this species: 3,4-dimethoxy-3′,4′-methylenedioxylignano-9,9′-lactone (**4**), 3,4-dimethoxy-3′,4′-methylenedioxylignano-9,9′-epoxylignan-9′-ol (DME) (**5**), and guayadequiol (**6**) (Figure 1) [33]. Compound **4** was named iso-bursehernin " . . . since the only difference between this compound and bursehernin is the placement of the aromatic groups in relation to the lactone ring" and compound **5** was named DME [33]. Iso-bursehernin (also called kusunokinin) was isolated from several plants, such as *Cinnamomun camphora* [80], *Virola* spp. [81,82], and from different species of *Haplophyllum* [83–85]. Compounds **4** and **5** were identified as active glioma inhibitors in a bioassay-guided isolation process from *Piper nigrum* fruits [86] and **4** selectively docked to *Leishmania mexicana* pyruvate kinase in a study to find potential antiprotozoal polyphenolic plant extracts [87]. Kusunokinin isolated from *P. nigrum* showed potent cytotoxic activity on breast cancer cells (MCF-7 and MDA-MB-468) with IC_{50} values of 1.18 and 1.62 µg/mL, respectively, but demonstrated lower cytotoxicity on normal breast cell lines (IC_{50} higher than 11 µg/mL). Cell cycle studies showed that this compound induced cell apoptosis and drove cells toward the G2/M phase. Moreover, it decreased topoisomerase II and Bcl-2. The authors observed an increasing in p53, p21, bax, cytochrome c, and caspase-8, -7, and -3 activities, except caspase-9, suggesting that kusunokinin has potent anticancer activity through the extrinsic pathway and G2/M phase arrest [88].

Lignans that contain a methylenedioxy group show high antifeedant or deterrent activity against insects. Polar substituents on the aromatic rings, such as hydroxyl or glycosyl groups, reduce this activity. Guayadequiol (**6**) was isolated for the first time from *Bupleurum salicifolium* [89]. No biological data for this compound have been reported. The hexane extract of *B. excelsa* was shown to possess in vivo anti-inflammatory activity [29].

2.2.4. *Bursera graveolens* (Kunth) Triana and Planch.

B. graveolens (synonyms: *Amyris caranifera*, *A. graveolens*, *B. andersonii*, *B. pilosa*, *B. tatamaco*, *Elaphrium graveolens*, *E. pilosum*, *Terebinthus graveolens*, and *T. pilosa*) is called copal and mizquixochicopalli in Náhuatl language [90]. Its Spanish common name is *palo santo* and it is native to the tropical dry forests from the Yucatan Peninsula of Mexico, south to Peru, and the Galapagos Islands of Ecuador. These are trees and sometimes shrubs up to 15 m tall, highly fragrant, with grey bark. Their leaves are imparipinnate, sometimes bipinnate, up to 30 cm long and 18 cm wide, with 7 to 11 leaflets. The leaflets are 3 to 9 cm long and 1 to 4 cm wide, of acuminate apex, and margin roughly serrated. Their small flowers are yellowish, white, or green and their fruits glabrous and are up to 1.0 cm long. Their seeds are black and about two-thirds covered by an orange-red pseudoaril. Traditionally, the alcoholic extract of the bark is used for rheumatism, and the bark infusion, as a digestive and for respiratory problems. In recent years, the resin and oils have been extracted from the wood by the perfume industry. From the active methanol extract of stems, Nakanishi et al. isolated a new aryltetralin lignan, burseranin (**7**), and picropolygamain (**8**) (Figure 2) along with known triterpenes, lupeol and epi-lupeol [91]. The two isolated lignans **7** and **8** showed important cytotoxic activity against the human HT1080 fibrosarcoma cell line. Both compounds exhibited potent inhibitory effects in comparison with adriamycin as a positive control (5.5 and 1.9 µg/mL vs. 0.1 µg/mL). Picropolygamain (**8**) was isolated for the first time in 1985 from *Commiphora incisa* resin [92] and later from *Bursera simaruba* [93]. This compound was shown to be active against LNCaP (androgen-sensitive human prostate adenocarcinoma) cell line (ED_{50} 1.1 µg/mL) during tests aimed at developing an in vivo Hollow Fiber Assay [94].

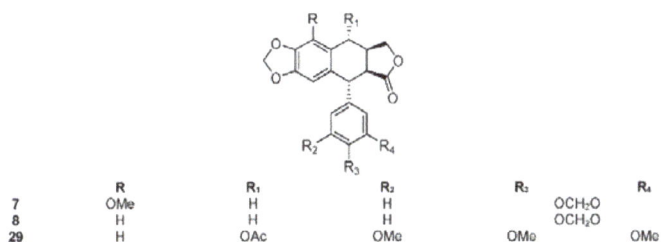

	R	R₁	R₂	R₃	R₄
7	OMe	H	H	OCH₂O	
8	H	H	H	OCH₂O	
29	H	OAc	OMe	OMe	OMe

Figure 2. Aryltetraline lignans (*picro* series) isolated from *Bursera* spp.

2.2.5. *Bursera penicillata* (Sessé and Moç. ex DC.) Engl.

B. penicillata (synonyms: *Amyris penicillata*, *Bursera mexicana*, *Elaphrium delpechianum*, *E. mexicanum*, *E. penicillatum*, *Terebinthus delpechiana*, and *T. mexicana*) belongs to the section *glabrifolia*. Its common names are *coyoluche*, *torote incienso*, and *torote copal* [95]. These are trees up to 12 m tall of grey or reddish grey bark, and are very fragrant, even sometimes from a distance. Their leaves are imparipinnate, 12 to 38 cm long, rachis-winged, and with 3 to 15 leaflets. The leaf blades are finely pubescent on both surfaces and the margins strongly toothed. Their flowers are small, white, and arranged in few to many inflorescences up to 14 cm long. Fruits are 1 to 1.3 cm long, 0.8 to 1.1 cm wide with a black pit, and partially covered by a red, orange, or pale pseudoaril. Endemic to Northwest Mexico, this species prospers in tropical deciduous forests and sporadically thornscrub and transition areas to oak woodland, from Southeastern Sonora and Southwest Chihuahua to Michoacan. According to Gentry, the leaves are used to treat the common cold and the gum is used for toothaches. It is also used as incense [96]. Koulman reported the presence of savinin (**2**) in this species [33].

2.2.6. *Bursera submoniliformis* Engl.

B. submoniliformis (synonyms: *Bursera subsessiliformis* Engl. and *Elaphrium submoniliforme* (Engl.) Marchand ex Engl.) is commonly known as *copal chino*. These trees are up to 12 m tall with grey to reddish gray bark. Their leaves are imparipinnate, up to 20 cm long and 7 cm wide, with 9 to 17 leaflets. The leaflets are velvety, 1.3 to 5 cm long, 0.5 to 2 cm wide, and have toothed margins. They have small white flowers that are arranged in inflorescences. Fruits are 7.5 to 12 mm long with a black pit almost or completely covered by a yellow or orange pseudoaril. Endemic to Mexico, this species inhabits tropical deciduous forests at altitudes of 500 to 1600 m of the Balsas and Papaloapan river basins in the states of Mexico, Michoacan, Guerrero, Puebla, Morelos, and Oaxaca. It belongs to the subsection *copallifera*. The gum resin is used to alleviate pain associated with flatulence and tooth-ache [97]. The only reference about the phytochemistry of *B. submoniliformis* is by Koulman, who reported the presence of savinin (**2**) in this species [33].

2.3. Section Bursera

2.3.1. *Bursera aptera* Ramirez

B. aptera (synonyms: *Elaphrium apterum* and *Terebinthus aptera*) belongs to the section *fagaroides*. Its common names are *cuajiote verde* [98,99] and *cuajiote blanco* (Náhuatl names). The species is distributed in Guerrero, Morelos, Oaxaca, and Puebla regions in Mexico [2]. These are shrubs or trees up to 10 m high with green trunks and bark that exfoliates in yellow or beige papyrus-like sheets. The leaves are glabrous, 2.5–7.5 cm long comprising 4 to 9 pairs of leaflets up to 15 mm long and 6 mm wide. The flowers are reddish, yellow, or white and the fruits are small, up to 7 mm long, greyish red when mature, and with a pit completely covered by a yellow or white papery pseudoaril.

Nieto-Yañez et al. evaluated the anti-leishmanial activity of a *B. aptera* methanolic extract. The extract showed strong activity against *Leishmania mexicana* both in the in vitro and in vivo tests. The gas chromatography-mass spectrometry (GC-MS) phytochemical analysis of the extract showed the presence of 11 compounds. Most of these compounds were fatty acids and fatty acid esters, but they revealed the presence of podophyllotoxin (**19**) (Figure 3) [100].

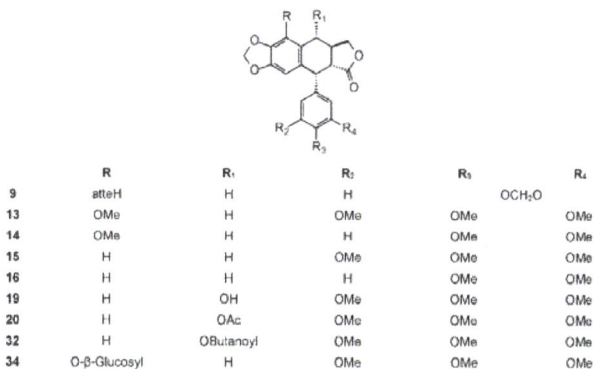

	R	R₁	R₂	R₃	R₄
9	atteH	H	H		OCH₂O
13	OMe	H	OMe	OMe	OMe
14	OMe	H	H	OMe	OMe
15	H	H	OMe	OMe	OMe
16	H	H	H	OMe	OMe
19	H	OH	OMe	OMe	OMe
20	H	OAc	OMe	OMe	OMe
32	H	OButanoyl	OMe	OMe	OMe
34	O-β-Glucosyl	H	OMe	OMe	OMe

Figure 3. Aryltetraline lignans isolated from *Bursera* spp.

2.3.2. *Bursera arida* (Rose) Standl.

B. arida (synonyms: *Elaphrium aridum* Rose and *Terebinthus arida* Rose) is endemic in the states of Oaxaca and Puebla [2]. It is commonly known as *zapotillo* [99]. These are small trees, often shrubs, of reddish-brown exfoliating bark with leaves up to 2.5 cm long and 1 cm wide, comprising 3 to 11 leaflets. Their flowers are very small, reddish, and solitary. Their fruits are solitary or in pairs, over short and pilose peduncles 1 to 2 mm long, with a seed completely covered by a pale yellow pseudoaril [101]. It belongs to the *microphylla* subsection. Traditionally, the plant latex is topically used for healing wounds and skin eruptions [102] in the Tehuacán-Cuicatlán valley. *Bursera arida* has different medicinal uses such as a disinfectant, cough suppressant, and antidepressant [103].

The phytochemistry of *B. arida* was studied by Ionescu, who prepared a chloroform extract of the stems, leaves, twigs, and bark, and found naringenin, β-sitosterol, betulonic acid, and four lignans: (+)-3-hydroxymethyl-5-methoxy-6,7-methylenedioxy-1-(3′,4′-methylenedioxybenzene)-1,2,3, 4-tetrahydronaphthalene-2-carboxylic acid lactone (**9**) (Figure 3), (+)-3-hydroxymethyl-6,7-methylenedioxy-1-(3′,4′-methylenedioxybenzene)-3,4-dihydronaphthalene-2-carboxylic acid lactone (**10**) (Figure 4), (+)-3-hydroxymethyl-6,7-methylenedioxy-1-(5′-methoxy-3′,4′-methylenedioxybenzene)-3,4-dihydronaphthalene-2-carboxylic acid lactone (**11**) (Figure 4), and 2,3-bis-(3,4-methylenedioxybenzyl)butane-1,4-diol diacetate (**12**) (Figure 5) [104]. Compound **12** has the structure of ariensin.

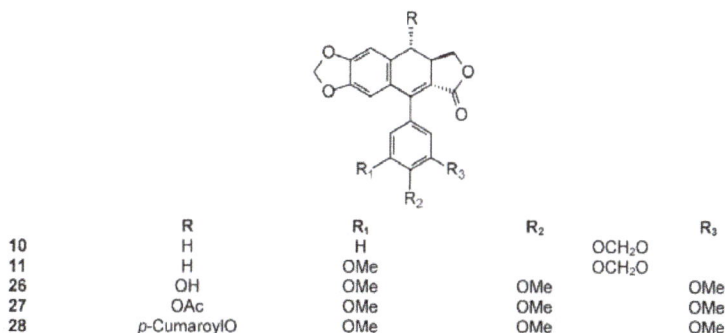

	R	R₁	R₂	R₃
10	H	H	OCH₂O	
11	H	OMe	OCH₂O	
26	OH	OMe	OMe	OMe
27	OAc	OMe	OMe	OMe
28	*p*-CumaroylO	OMe	OMe	OMe

Figure 4. 7′,8′-Dehydro-aryltetraline lignans isolated from *Bursera* spp.

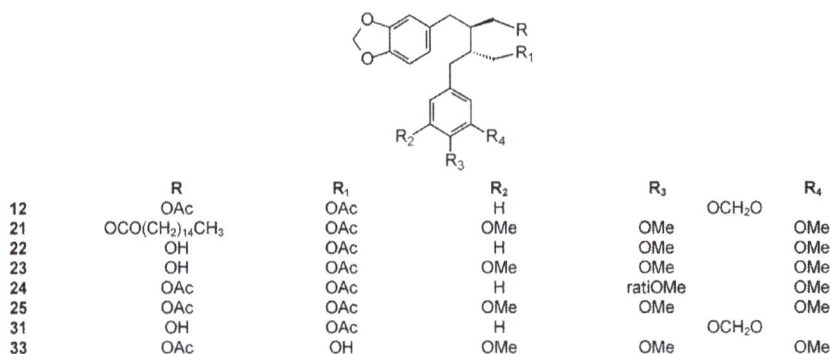

	R	R$_1$	R$_2$	R$_3$	R$_4$
12	OAc	OAc	H	OCH$_2$O	
21	OCO(CH$_2$)$_{14}$CH$_3$	OAc	OMe	OMe	OMe
22	OH	OAc	H	OMe	OMe
23	OH	OAc	OMe	OMe	OMe
24	OAc	OAc	H	ratiOMe	OMe
25	OAc	OAc	OMe	OMe	OMe
31	OH	OAc	H	OCH$_2$O	
33	OAc	OH	OMe	OMe	OMe

Figure 5. Dibenzylbutane diol lignans isolated from *Bursera* spp.

2.3.3. *Bursera ariensis* (Kunth) McVaugh and Rzed.

B. ariensis (synonyms *B. panosa* Engl., *B. sessiflora* Engl., *E. ariensis* Kunth, and *E. brachypodium* Rose). These are trees and sometimes shrubs between 2 and 8 m tall with greenish-gray trunks and bark that exfoliates in papery sheets that are yellowish or beige, sometimes with orange tones, with whitish resin that darkens upon contact with air. Their leaves are hairy especially when young, 5 to 22 cm long, and 2 to 7 cm wide, with a winged rachis and 5 to 9 leaflet pairs. Their flowers can be solitary but often develop in conglomerates at the end of branches of reddish-yellow color. Their fruits are 6 to 8 mm long, growing in thick conglomerates, with a pit completely covered by a yellow or orange pseudoaril. It is distributed in Mexico (Chiapas, Guerrero, Jalisco, and Oaxaca regions) where it is commonly known as *guande* and *cuajiote blanco* (Náhuatl name) [101] and used to treat colds and inflammation. It belongs to the *fagaroides* group and is poorly studied. From the acetonic extract of the bark, Hernandez isolated a new lignan named ariensin (**12**) [105]. Ariensin was shown to be active against the RAW246.7 murine cell line (IC$_{50}$ 9.8 µM) [11].

2.3.4. *Bursera fagaroides* (H.B.K.) Engl.

B. fagaroides, or "fragrant bursera" (*B. obovata* Turcz., *B. schaffneri* S. Wats) [2] is a dioecious shrub or tree, occasionally hermaphrodite, 0.5 to 10 m high, and highly resinous. The trunk is green with a bark that exfoliates in yellowish-gray papery sheets. The leaves are most often compound, with 5 to 13 leaflets, although occasionally they are unifoliolate or trifoliolate. They have whitish-green or yellow flowers, with a few of them arranged in small inflorescences or solitary. Male flowers are most often 5-merous (sometimes 3- and 4-). Female flowers are 3-merous. Fruits are typically 0.5 to 0.8 cm long with short peduncles no more than 2 mm long that terminate in a sharp point. When the pits mature, they are covered by a yellow or red pseudoaril. It is commonly known as *cuajiote amarillo* (in Morelos) or *pima bajo* [106]. This species belongs to the *fagaroides* group. Three subspecific variants of this species were recognized by McVaugh and Rzedowski (1965). *B. fagaroides* var. *fagaroides*, commonly known as *xixote* and *jiote* [72], are most often shrubs with leaves with serrated margins and are distributed in Northern, Central, and Western Mexico. *B. fagaroides* var. *purpusii* (common name *aceitillo*) [107] are most often trees with leaves of entire margin (not toothed) distributed throughout Southeastern Mexico. *B. fagaroides* var. *elongata,* common Mayo name *to'oro sahuali* [95], is from Northwestern Mexico, and has been classified by recent molecular studies as a separate taxon not belonging within *B. fagaoides* [108].

Bursera fagaroides is used to alleviate inflammation, skin tumors, and warts, and is perhaps the most studied *Bursera* species in terms of its chemistry and biological effects. The first report about *B. fagaroides* phytochemistry dates back to 1969, when Bianchi et al. isolated β-peltatin A-methyl ether (**13**) and the new 5′-demethoxy-β-peltatin A-methyl ether (**14**) (Figure 3) that showed activity against the Walker carcinoma 256 (WA16) tumor system from the chloroform extract of this plant. The ethanol extract from the dried exudate of *Bursera fagaroides* showed significant cytotoxic activity against the HT-29 (human colon adenocarcinoma) cell line. From this extract, Velazquez-Jimenez et al. isolated two aryltetraline lignans: (−)-deoxypodophyllotoxin (**15**) and (−)-morelensin (**16**) (Figure 3), and two dibenzylbutirolactone lignans: (−)-yatein (**17**) and (−)-5′-desmethoxy yatein (**18**) or bursehernin] (Figure 1) [109]. The authors determined the absolute configuration of these compounds by comparing the vibrational circular dichroism spectra of known podophyllotoxin and deoxypodophyllotoxin with those obtained by density functional theory calculations. Morelensin (**16**) was shown to be cytotoxic and antiproliferative against several cancer cell lines [35,110]. Yatein (**17**) was shown to be cytotoxic and antiproliferative to several cancer cell lines [111–114]. For example, it was tested against HL-60 (human promyelocytic leukemia cells), SMMC-7721 (human hepatoma), A-549 (human lung adenocarcinoma), MCF-7 (human breast cancer), and SW480 (human colon adenocarcinoma) cell lines using the MTT (3-(4,5-dimethylthiazol-2)2,5-difeniltetrazolium bromide) method as previously reported [115], with cisplatin as the positive control [116]. Yatein (**17**) showed significant cytotoxic activity against all tested cell lines being superior to cisplatin. Chen et al. tested yatein [117] against DLD-1 (human colorectal carcinoma), CCRF-CEM (human lymphoblastic leukemia), and IMR-32 (human neuroblastoma) cell lines. The study showed that yatein possesses similar cytotoxic activity to doxorubicin (positive control) against DLD-1 and CCRF-CEM cell lines [117]. Other studies showed that it is able to suppress herpes simplex virus type 1 (HSV-1) replication in HeLa cells in a plaque reduction assay. Doussot et al. studied the lignan profile and the antiproliferative activity of ethanol extracts from plants belonging to different species of *Linum*, *Callitris*, and *Juniperus*. They compared the activity of deoxypodophyllotoxin (**15**), podophyllotoxin (**19**), and yatein (**17**) against six human cancer cell lines: A549, U373 (glioblastoma), T98G (glioma), Hs683 (oligodendro-glioma), MCF7 (breast cancer), and SKMEL-28 (melanoma). The most active compound was deoxypodophyllotoxin ($IC_{50} < 0.01$ μM against all cell lines, not tested against U373), followed by podophyllotoxin ($IC_{50} = 0.03$ μM against all cell lines except U373 ($IC_{50} > 100$ μM). Yatein (**17**) showed an antiproliferative activity, but to a lesser extent ($IC_{50} = 30.9$, not tested, 26.5, 29.8, 31.9, and 39.6 μM, respectively) [118]. The inhibitory effect of yatein (**17**) on HSV-1 replication was concentration-dependent with an IC_{50} value of 30.6 ± 5.5 μM [119]. Furthermore, yatein (**17**) was demonstrated to be a potent CYP3A4 inhibitor and this study is of particular importance as **17** and other methylenedioxyphenol compounds were found to induce herb-drug interactions in clinical situations [120]. Yatein (**17**) showed other important biological activities, such as anti-platelet aggregation [121], and was shown to have moderate inhibitory activity against cytochrome P450 [122]. The bioactivity-guided separation of the hydroalcoholic extract of *B. fagaroides* var. *fagaroides* by Rojas-Sepúlveda et al. led to, besides the already isolated lignans, podophyllotoxin (**19**), burseranin (**7**), and acetyl podophyllotoxin (**20**) (Figure 3). All the isolated compounds were found to be active against tumor cell lines tested, especially 5′-demethoxy-β-peltatin A-methyl ether (**14**), which exhibited greater activity than camptothecin and podophyllotoxin against PC-3 ($ED_{50} = 1.0 \times 10^{-5}$ μg/mL) and KB ($ED_{50} = 1.0 \times 10^{-5}$ μg/mL) cell lines [123]. Furthermore, the cytotoxic and antitumor activity of the ethanol extract (70%) of *B. fagaroides* bark against L5178Y lymphoma cell line was tested. The antitumor activity was studied on BALB/c mice (2×10^4 cells L5178Y i.p). Treated animals (at 50 mg/kg/day over 15 days) showed a significant increase in survival compared with those treated with the placebo or without treatment [124].

Podophyllotoxin and deoxypodophyllotoxin are secondary metabolites of many plants [125]. The biological activities and the importance of podophyllotoxin (**19**), as the lead compound in the development of new anticancer agents, are well known [126–128]. The problem connected with its use is the scarce amount isolated from natural sources. For this reason, biotechnological production of this lignan has been studied [129]. Deoxypodophyllotoxin (**15**) is a promising anticancer agent [130,131].

Other diarylbutane lignans were isolated by Morales-Serna et al. from the chloroform extract of *B. fagaroides* resin, named by the author as 9′-acetyl-9-pentadecanoyl-dihydroclusin (**21**) (correctly, a hexadecanoyl derivative), 2,3-demethoxy-secoisolintetralin monoacetate (**22**), and dihydroclusin monoacetate (**23**), together with two known lignans: 2,3-demethoxy-secoisolintetralin diacetate (**24**) and dihydroclusin diacetate (**25**) [132] (Figure 5). Recently, Antúnez Mojica et al. isolated three new aryldihydronaphtalene-type lignans from the dichloromethane stem bark extract of *B. fagaroides* var. *fagaroides*: 7′,8′-dehydropodophyllotoxin (**26**), 7′,8′-dehydroacetylpodophyllotoxin (**27**), and 7′,8′-dehydro-*trans*-p-cumaroyl podophyllotoxin (**28**) (Figure 4), along with six known lignans: podophyllotoxin (**19**), acetylpodophyllotoxin (**20**), 5′-demethoxy-β-peltatin A methylether (**14**), acetylpicropodophyllotoxin (**29**) (Figure 2), burseranin (**7**), and hinokinin (**1**) [133]. The cytotoxic activity of the new isolated compounds **26–28** against the cancer cell lines KB, PC-3, MCF-7, and HF-6 was evaluated, which showed that all of them displayed good activity against KB, PC-3, and HF-6, but were not active against the MCF-7 cell line. When compared with podophyllotoxin (**19**) ($ED_{50} = 2.10 \times 10^{-4}$ μM), compounds **26** and **28** were most active against the PC-3 cell line displaying similar toxicity ($ED_{50} = 2.4 \times 10^{-5}$ and 2.42×10^{-5} μM, respectively), whereas compound **27** was less active ($ED_{50} = 0.06$ μM). Lignans **26–28** showed moderate activity against KB and HF-6 cell lines when compared to **19**. The cytotoxic activity of the other isolated compounds was already proven [123]. Acetylpicropodophyllotoxin (**29**) was previously isolated from *Hernandia ovigera* and it is a potent, selective inhibitor of type I insulin-like growth factor receptor (IGF-IR) [134].

Ornithine decarboxylases (ODC) are enzymes that catalyze the decarboxylation of ornithine to produce putrescine in the biosynthesis of polyamines. Polyamine metabolism is closely related with the progression of growth, proliferation, and cell regeneration. The in vitro effect of an ethanolic extract from the stem bark of *Bursera fagaroides* on ODC activity, and on the growth of *Entamoeba histolytica*, was studied by Rosas-Arreguín et al. using metronidazole and G418 as positive controls [135]. The authors found growth inhibition, with IC_{50} values in the order of 0.05 mg/mL. The ODC activity was inhibited by 12% at 4.0 mg/mL.

Gutiérrez-Gutiérrez et al., considering the use in Mexican traditional medicine of *B. fagaroides* as an antidiarrheic, investigated the in vitro anti-giardial activities of four podophyllotoxin-type lignans from *Bursera fagaroides* var. *fagaroides*: 5′-demethoxy-β-peltatin A methylether (**14**), acetylpodophyllotoxin (**20**), burseranin (**7**), and podophyllotoxin (**19**) [136]. They found that all lignans affected *Giardia* adhesion, but only compounds **14**, **19**, and **20** caused growth inhibition.

2.3.5. *Bursera microphylla* A. Gray

B. microphylla, or "elephant tree" (*Elaphrium microphyllum* (A. Gray) Rose, *Terebinthus microphylla* (A. Gray) Rose), is commonly known as *xoop* (Seri name) and *torote blanco* [95]. It is typically a small tree that grows up to 10 m tall, with a thickened trunk and thickened lower branches, with light gray to white peeling bark, with younger branches having a reddish color. The leaves are 3 to 8 cm long and have 7 to 35 small, linear, and glabrous leaflets. The leaflets are up to 1.5 cm long. The flowers are yellowish white or greenish and inconspicuous. The fruits are brownish red at maturity with a black pit completely covered by a yellow-orange pseudoaril [26]. It belongs to the *microphylla* group. The Seri Indians from Sonora, Mexico use the bark, leaves, flowers, and fruits to treat a variety of maladies such as inflammation, diarrhea, and venereal diseases. The first lignan isolated from *Bursera microphylla* was burseran (**30**) (Figure 1) [137]. Cole et al. showed that burseran has cytotoxic activity against human epidermoid carcinoma of the nasopharynx (9KB cell line) in a Cancer Chemotherapy National Service Center (CCNSC) test ($ED_{50} < 10$ μg/mL) acting as a spindle poison. Tomioka et al. synthetized *cis* (H_8′ α) and *trans* burseran and tested them in a cilia regeneration test in *Tetrahymena*, that is a useful model for studying the antitubulinic activity of spindle poisons [138]. Both *trans* and *cis* burseran have shown inhibitory activity, but the antitumor activity was higher for *trans* burseran [139]. An analysis of the chemical composition of the methanol extract of *B. microphylla* resin from the Sonora Desert (Mexico), revealed the presence of several known lignans: ariensin (**12**), burseranin (**7**), dihydroclusin diacetate (**25**), picropolygamain (**8**), desmethoxy-yatein (**18**), hemiariensin (**31**) (Figure 5), and dihydroclusin 9′-acetate (**23**); and two new ones: podophyllotoxin butanoate (**32**) and dihydroclusin 9-acetate (**33**) (Figures 3 and 5) [11,140]; in addition to burseran (**30**). Compound **32** was already known as a synthetic derivative of podophyllotoxin [141] but it was new as a natural product. Burseran (**30**) and dihydroclusin diacetate (**25**) were tested against human cancer cell lines: A549 (lung cancer), HeLa (cervix cancer), and PC-3 (prostate cancer), and on murine cell lines M12.C3.F6 (B cell lymphoma) and RAW264.7 (macrophages transformed by virus Abelson leukemia); which were found to be more active against murine cell lines in micromolar range (IC$_{50}$ 13.8, 36.3, and 2.5 μM, respectively). The anti-proliferative activities of dihydroclusin 9-acetate (**31**), dihydroclusin 9′-acetate (**23**), burseranin (**7**), picropolygamain (**8**), and hemiariensin (**31**) were evaluated on the human cancer cell lines A549, LS 180, and HeLa, and on the human non-cancer cell line ARPE-19. None of the evaluated compounds had statistically significant anti-proliferative effects with respect to dimethyl sulfoxide (DMSO) control on LS 180, A549, and ARPE cell lines. However, burseranin (**7**) and picropolygamain (**8**) had an interesting anti-proliferative activity on the gynecological cancer cell line, HeLa, with IC$_{50}$ values of 21.72 ± 1.03 and 9.31 ± 1.01 μM, respectively [140].

2.3.6. *Bursera morelensis* Ramírez

B. morelensis (synonym: *Elaphrium morelense* (Ramírez) Rose) is widely distributed in Mexico [101], where it is commonly known as *coabinillo* [99]. It is a tree up to 13 m tall with red bark that exfoliates in thin sheets. Its leaves are 5 to 11 cm long and 1.5 to 4.5 cm wide, with 15 to 51 linear leaflets. The flowers are yellow, pink, greenish, or white. The fruits are 0.5 to 1 cm long, with a pit completely covered by a pale yellow pseudoaril. It belongs to the *microphylla* group. The only two papers describing the phytochemistry of *B. morelensis* address the composition and the anti-inflammatory activity of the essential oil [142] and the isolation of deoxypodophyllotoxin (**15**) and morelensin (**16**) from the resin [110]. Morelensin (**16**), although highly active against the KB (epidermoid carcinoma) cancer cells, demonstrated only marginal activity against the porcine stable (PS) kidney cell line [110].

2.3.7. *Bursera roseana* Rzed., Calderón & Medina

B. roseana (synonyms: *Bursera acuminata* (Rose) Engl. and *Terebinthus acuminata* Rose) is a 12–20 m high tree with bark peeling in reddish-orange stripes. It is imparipinnate with 3 to 7 (sometimes 9) leaflets with a hairy underside but a bright and glabrous upper side. The leaflets are typically of oval shape, 4.5 to 15 cm long and 2 to 6 cm wide, ending in a long point. The flowers are white or greenish and the is fruit is glabrous, 0.9 to 1.2 cm long, with a pit completely covered by a pale pseudoaril [143].

This species grows in moist canyons in the transition zone between highland pine-oak forest and lowland tropical subdecidous forest. It is common in Nayarit, Zacatecas, Aguascalientes, Jalisco, Colima, Michoacán, Estado de México, and Guerrero [143]. It belongs to the *simaruba* group. The only phytochemical study on this species was reported by Koulman, who found 5′-desmethoxy yatein [bursehernin or (−)-*trans*-methylpluviatolide [144] (+)-*trans*-methylpluviatolide, which has been called dextrobursehernin [145] (**18**)], morelensin (**16**), deoxypodophyllotoxin (**15**), and β-peltatin-A methylether (**13**). Bursehernin (**18**) was studied by Ito et al. for its inhibitory effects on Epstein-Barr virus early antigen activation induced by 12-*O*-tetradecanoylphorbol 13-acetate in Raji cells [146]. The data reported demonstrated that bursehernin (**18**) was slightly weaker than β-carotene, which is commonly used in cancer prevention studies [147] so it might be a valuable antitumor promoter. Bursehernin (**18**) is able to inhibit the growth of *Neisseria gonorrhoeae* [148]. The trypanocidal activity of racemic mixtures of *cis*- and *trans*-bursehernin was evaluated in vitro against trypomastigote forms of two strains of *Trypanosoma cruzi*, and results showed that the racemic *cis*-stereoisomer was inactive, whereas the racemic *trans*-stereoisomer displayed trypanocidal activity, with an IC$_{50}$ ~89.3 μM. These results were different from those obtained for pure (−)-*trans*-methylpluviatolide by Bastos et al., but the difference could be ascribed to the use of the racemic mixture [74].

2.3.8. *Bursera schlechtendalii* Engl.

B. schlechtendalii (synonyms: *Bursera jonesii* Rose, *Elaphrium jonesii* (Rose) Rose, *Terebinthus jonesii* (Rose) Rose, and *Terebinthus schlechtendalii* (Engl.) Rose) is a small tree or shrub, 4–6 m high, with a strong turpentine smell, and known as *sak chakaj*. It has a glossy greyish pink bark that peels off in thin papery sheets; the branches are thick and stout. The leaves are simple (unifoliolate), often less than 6 cm long and 2.5 cm wide. The flowers are small, usually solitary, with yellow or reddish petals, and the fruits are 4 to 8 mm long with a pit completely covered by a yellow or red pseudoaril. It is used to treat the flu. It is found at altitudes of 200–400 m on dry rocky hillsides or in thickets in Southern Mexico and Guatemala. It belongs to the *fagaroides* group. In 1972, McDoniel et al. isolated from stems and leaves the chloroform extract of Mexican *B. schlechtendalii* Engl. yatein (**17**) and bursehernin (**18**).

2.3.9. *Bursera simaruba* (L.) Sarg.

B. simaruba or *"gumbo-limbo"* (*Bursera simaruba* var. *yucatanensis* Lundell) is commonly known as *yala-guito* [99]. It is a 6–15 m high tree with a peeling reddish bark that reveals a smooth grey underbark. The leaves are compound, bright, and mostly glabrous when mature. Leaves have 3 to 13 leaflets, 4 to 9 cm long and 1.8 to 3.5 cm wide. The small flowers have pink, pale yellow-green, or white petals and are arranged in inflorescences. The fruits are glabrous, red to brownish, 1 to 1.5 cm long, ending in a point, with a pit completely covered by a red pseudoaril. This is perhaps the most widespread species of *Bursera*, occurring from Southern Florida and the Caribbean, along both coasts of Mexico, to South America. Taken orally or as curative baths, the leaves and bark are attributed a variety of medicinal properties. It belongs to the *simaruba* group. In 1992, Peraza-Sánchez and Peña-Rodriguez isolated picropolygamain (**8**), which showed activity in the brine shrimp assay (LC_{50} = 52.2 ppm). Further in vitro evaluation against three human tumor cell lines (A-549, lung), MCF-7 (breast), and HT-29 (colon) showed that **8** has cytotoxic activity comparable to that of Adriamycin [93]. Noguera et al. isolated the anti-inflammatory β-peltatin A-methyl ether (**13**) from the leaf hexane extract of *B. simaruba*. It inhibited the carrageenan-induced rat paw edema, in a dose- and time-dependent manner (three hours = 9.55%, five hours = 34.37%, and seven hours = 35.6%) [149]. Maldini et al. studied the methanolic extract of *Bursera simaruba* bark and isolated 11 compounds, including lignans yatein (**17**), β-peltatin-*O*-β-D-glucopyranoside (**34**) (Figure 3), hinokinin (**1**), and bursehernin (**18**) [149].

3. Conclusions

Lignans are phenolic secondary metabolites characterized by a large variety of biological activities. Among these, the cytotoxic and anti-proliferative ones are perhaps the most common and most studied. Literature analysis of Mexican plants producing lignans *Burseara* spp., revealed that the most common lignan types are dibenzyl butyrolactones (nine compounds), picro aryltetraline derivatives (three compounds), aryltetraline derivatives (nine compounds), 7′,8′-dehydro-aryltetraline derivatives (five compounds), and dibenzylbutane diols (eight compounds). Notably, all the examined *Elaphrium* subgenus species produce only dibenzyl butyrolactones and picro aryltetraline derivatives, whereas *Bursera* subgenus produce all lignan types. The most common compound, up to now, appears to be bursehernin (**18**), which is present in five species belonging to the *Bursera* section. Hinokinin (**1**) and savinin (**2**) are also widespread (four species), both in the *Elaphrium* and *Bursera* sections. Compounds **3**, **8**, **12**, **13**, **15**, **16**, **17**, and **33** were isolated from three species.

Author Contributions: M.C.M. and J.X.B. contributed to the conception of the review; M.C. contributed to the constructive discussions; all authors have read and approved the manuscript.

Funding: This research received no external funding.

Conflicts of Interest: The authors declare no conflict of interest.

References

1. Becerra, J.X.; Noge, K.; Olivier, S.; Venable, D.L. The monophyly of *Bursera* and its impact for divergence times of Burseraceae. *Taxon* **2012**, *61*, 333–343.

2. Rzedowski, J.; Medina Lemos, R.; Calderón de Rzedowski, G. Inventario del conocimiento taxonómico, así como de la diversidad y del endemismo regionales de las especies mexicanas de *Bursera* (Burseraceae). *Acta Bot. Mex.* **2005**, *70*, 75–111. [CrossRef]

3. Becerra, J.X.; Venable, D.L.; Evans, P.H.; Bowers, W.S. Interactions between chemical and mechanical defenses in the plant genus *Bursera* and their implications for herbivores. *Am. Zool.* **2001**, *41*, 865–876.

4. Gigliarelli, G.; Becerra, J.X.; Curini, M.; Marcotullio, M.C. Chemical composition and biological activities of fragrant Mexican copal (*Bursera* spp.). *Molecules* **2015**, *20*, 22383–22394. [CrossRef] [PubMed]

5. Becerra, J.X. Evolution of Mexican Bursera (Burseraceae) inferred from ITS, ETS, and 5S nuclear ribosomal DNA sequences. *Mol. Phylogenet. Evol.* **2003**, *26*, 300–309. [CrossRef]

6. Becerra, J.X.; Noge, K.; Venable, D.L. Macroevolutionary chemical escalation in an ancient plant-herbivore arms race. *Proc. Natl. Acad. Sci. USA* **2009**, *106*, 18062–18066. [CrossRef] [PubMed]

7. Evans, P.H.; Becerra, J.X. Non-terpenoid essential oils from *Bursera chemapodicta*. *Flavour Fragr. J.* **2006**, *21*, 616–618. [CrossRef]

8. Noge, K.; Becerra, J.X. Germacrene D, a Common Sesquiterpene in the Genus Bursera (Burseraceae). *Molecules* **2009**, *14*, 5289–5297. [CrossRef] [PubMed]

9. Hernandez-Hernandez, J.D.; Garcia-Gutierrez, H.A.; Roman-Marin, L.U.; Torres-Blanco, Y.I.; Cerda-Garcia-Rojas, C.M.; Joseph-Nathan, P. Absolute configuration of cembrane diterpenoids from *Bursera multijuga*. *Nat. Prod. Commun.* **2014**, *9*, 1249–1252. [PubMed]

10. Hernández-Hernández, J.D.; Román-Marín, L.U.; Cerda-García-Rojas, C.M.; Joseph-Nathan, P. Verticillane Derivatives from *Bursera suntui* and *Bursera kerberi*. *J. Nat. Prod.* **2005**, *68*, 1598–1602. [CrossRef] [PubMed]

11. Messina, F.; Curini, M.; Di Sano, G.; Zadra, C.; Gigliarelli, G.; Rascon-Valenzuela, L.A.; Robles Zepeda, R.E.; Marcotullio, M.C. Diterpenoids and triterpenoids from the resin of *Bursera microphylla* and their cytotoxic activity. *J. Nat. Prod.* **2015**, *78*, 1184–1188. [CrossRef] [PubMed]

12. Lucero-Gómez, P.; Mathe, C.; Vieillescazes, C.; Bucio, L.; Belio, I.; Vega, R. Analysis of Mexican reference standards for *Bursera* spp. resins by Gas Chromatography–Mass Spectrometry and application to archaeological objects. *J. Archaeol. Sci.* **2014**, *41*, 679–690. [CrossRef]

13. Hernandez, J.D.; Garcia, L.; Hernandez, A.; Alvarez, R.; Roman, L.U. Luteolin and myricetin glycosides of Burseraceae. *Rev. Soc. Quim. Mex.* **2002**, *46*, 295–300.

14. Nakanishi, T.; Inatomi, Y.; Arai, S.; Yamada, T.; Fukatsu, H.; Murata, H.; Inada, A.; Matsuura, N.; Ubukata, M.; Murata, J.; et al. New luteolin 3′-O-acylated rhamnosides from leaves of *Bursera graveolens*. *Heterocycles* **2003**, *60*, 2077–2083.

15. Souza, M.P.; Machado, M.I.L.; Braz-Filho, R. Six flavonoids from *Bursera leptophloeos*. *Phytochemistry* **1989**, *28*, 2467–2470. [CrossRef]

16. Cunha, W.R.; Andrade e Silva, M.L.; Sola Veneziani, R.C.; Ambrósio, S.R.; Kenupp Bastos, J. Lignans: Chemical and biological properties. In *Phytochemicals-A Global Perspective of Their Role in Nutrition and Health*; Rao, V., Ed.; InTech: Vienna, Austria, 2012. [CrossRef]

17. The Plant List A Working List of All Plant Species. Available online: http://www.theplantlist.org/1.1/browse/A/Burseraceae/Bursera/ (accessed on 12 February 2015).

18. Medina Lemos, R.; Ramos Rivera, P. *El Género Bursera en México. Parte II*; Version 1.3; Comisión Nacional Para el Conocimiento y uso de la Biodiversidad: Mexico City, Mexico, 2017.

19. CONABIO Enciclovida CONABIO. Available online: http://www.enciclovida.mx/busquedas/resultados?utf8=%E2%9C%93&busqueda=basica&id=&nombre=bursera&button= (accessed on 29 July 2018).

20. Bullock, L.L. Notes on the Mexican species of the genus *Bursera*. *Bull. Misc. Inf. Kew* **1936**, 347–387.

21. McVaugh, R.; Rzedowski, J. Synopsis of the genus *Bursera* L. in western Mexico, with notes on the material of *Bursera* collected by Sessè & Mocino. *Kew Bull.* **1965**, *18*, 317–382.

22. Becerra, J.X.; Venable, D.L. Nuclear ribosomal DNA phylogeny and its implications for evolutionary trends in Mexican *Bursera* (Burseraceae). *Am. J. Bot.* **1999**, *86*, 1047–1057. [CrossRef] [PubMed]

23. Rzedowki, J. Notas sobre el genero *Bursera* (Burseraceae) en el Estado de Guerrero, (Mexico). *Am. Esc Nac Cienc. Biol. Mex.* **1968**, *7*, 17–36.

24. Toledo, C.A. *El Genero Bursera en el Estado Guerrero*; Universidad Nacional Autónoma de México: Mexico City, Mexico, 1982.

25. Gillet, J. *Commiphora* (Burseraceae) in South America and its relationship to *Bursera*. *Kew Bull.* **1980**, *34*, 569–587. [CrossRef]

26. Rzedowki, J.; Kruse, H. Algunas tendencias evolutivas en *Bursera* (Burseraceae). *Taxon* **1979**, *28*, 103–116. [CrossRef]

27. Hernández, F. *Historia de las Plantas de Nueva España*; Imprenta Universitaria: Mexico City, México, 1942.

28. Tripplett, K.J. *The Ethnobotany of Plant Resins in the Maya Cultural Region of Southern Mexico and Central America*; University of Texas: Austin, TX, USA, 1999.

29. Acevedo, M.; Nuñez, P.; Gónzalez-Maya, L.; CardosoTaketa, A.; Villarreal, M.L. Cytotoxic and anti-inflammatory activities of *Bursera* species from Mexico. *J. Clin. Toxicol.* **2015**, *5*, 1–8.

30. Alonso-Castro, A.J.; Villarreal, M.L.; Salazar-Olivo, L.A.; Gomez-Sanchez, M.; Dominguez, F.; Garcia-Carranca, A. Mexican medicinal plants used for cancer treatment: Pharmacological, phytochemical and ethnobotanical studies. *J. Ethnopharmacol.* **2011**, *133*, 945–972. [CrossRef] [PubMed]

31. CONABIO Biodiversidad Mexicana. Available online: http://www.biodiversidad.gob.mx/usos/copales/almarciga.html (accessed on 8 May 2018).

32. Waizel, J.; Waizel-Hayat, S. Antitussive plants used in Mexican traditional medicine. *Pharmacogn. Rev.* **2009**, *3*, 29–43.

33. Koulman, A. *Podophyllotoxin: A Study of the Biosynthesis, Evolution, Function and Use of Podophyllotoxin and Related Lignans*; University of Groningen: Groningen, The Netherlands, 2003.

34. Marcotullio, M.C.; Pelosi, A.; Curini, M. Hinokinin, an emerging bioactive lignan. *Molecules* **2014**, *19*, 14862–14878. [CrossRef] [PubMed]

35. Ikeda, R.; Nagao, T.; Okabe, H.; Nakano, Y.; Matsunaga, H.; Katano, M.; Mori, M. Antiproliferative constituents in Umbelliferae plants. IV. Constituents in the fruits of *Anthriscus sylvestris* Hoffm. *Chem. Pharm. Bull.* **1998**, *46*, 875–878. [CrossRef] [PubMed]

36. Chang, S.T.; Wang, D.S.Y.; Wu, C.L.; Shiah, S.G.; Kuo, Y.H.; Chang, C.J. Cytotoxicity of extractives from *Taiwania cryptomerioides* heartwood. *Phytochemistry* **2000**, *55*, 227–232. [CrossRef]

37. Chen, J.-J.; Huang, H.-Y.; Duh, C.-Y.; Chen, I.-S. Cytotoxic constituents from the stem bark of *Zanthoxylum pistaciiflorum*. *J. Chin. Chem. Soc.* **2004**, *51*, 659–663. [CrossRef]

38. Lin, R.-W.; Tsai, I.-L.; Duh, C.-Y.; Lee, K.-H.; Chen, I.-S. New lignans and cytotoxic constituents from *Wikstroemia lanceolata*. *Planta Med.* **2004**, *70*, 234–238. [PubMed]

39. Chen, J.-J.; Chung, C.-Y.; Hwang, T.-L.; Chen, J.-F. Amides and benzenoids from *Zanthoxylum ailanthoides* with inhibitory activity on superoxide generation and elastase release by neutrophils. *J. Nat. Prod.* **2009**, *72*, 107–111. [CrossRef] [PubMed]

40. Lee, D.-Y.; Seo, K.-H.; Jeong, R.-H.; Lee, S.-M.; Kim, G.-S.; Noh, H.-J.; Kim, S.-Y.; Kim, G.-W.; Kim, J.-Y.; Baek, N.-I. Anti-inflammatory lignans from the fruits of *Acanthopanax sessiliflorus*. *Molecules* **2013**, *18*, 41–49. [CrossRef] [PubMed]

41. Da Silva, R.; de Souza, G.H.B.; da Silva, A.A.; de Souza, V.A.; Pereira, A.C.; Royo, V.D.A.; e Silva, M.L.A.; Donate, P.M.; de Matos Araujo, A.L.S.; Carvalho, J.C.T.; et al. Synthesis and biological activity evaluation of lignan lactones derived from (−)-cubebin. *Bioorg. Med. Chem. Lett.* **2005**, *15*, 1033–1037. [CrossRef] [PubMed]

42. Lima, T.S.C.; Lucarini, R.; Volpe, A.C.; de Andrade, C.Q.J.; Souza, A.M.P.; Pauletti, P.M.; Januário, A.H.; Símaro, G.V.; Bastos, J.K.; Cunha, W.R.; et al. In vivo and in silico anti-inflammatory mechanism of action of the semisynthetic (−)-cubebin derivatives (−)-hinokinin and (−)-O-benzylcubebin. *Bioorg. Med. Chem. Lett.* **2017**, *27*, 176–179. [CrossRef] [PubMed]

43. Ramos, F.; Takaishi, Y.; Kawazoe, K.; Osorio, C.; Duque, C.; Acuna, R.; Fujimoto, Y.; Sato, M.; Okamoto, M.; Oshikawa, T.; et al. Immunosuppressive diacetylenes, ceramides and cerebrosides from *Hydrocotyle leucocephala*. *Phytochemistry* **2006**, *67*, 1143–1150. [CrossRef] [PubMed]

44. Zhang, W.; Yao, Z.; Zhang, Y.W.; Zhang, X.X.; Takaishi, Y.; Duan, H.Q. Immunosuppressive sesquiterpenes from *Buddleja daviddi*. *Planta Med.* **2010**, *76*, 1882–1887. [CrossRef] [PubMed]

45. Silva, M.L.A.; Coimbra, H.S.; Pereira, A.C.; Almeida, V.A.; Lima, T.C.; Costa, E.S.; Vinholis, A.H.C.; Royo, V.A.; Silva, R.; Filho, A.A.S.; et al. Evaluation of *Piper cubeba* extract, (−)-cubebin and its semi-synthetic derivatives against oral pathogens. *Phytoth. Res.* **2007**, *21*, 420–422. [CrossRef] [PubMed]

46. Huang, R.-L.; Huang, Y.-L.; Ou, J.-C.; Chen, C.-C.; Hsu, F.-L.; Chang, C. Screening of 25 compounds isolated from *Phyllanthus* species for anti-human Hepatitis B virus in vitro. *Phytother. Res.* **2003**, *17*, 449–453. [CrossRef] [PubMed]

47. Zhang, G.; Shimokawa, S.; Mochizuki, M.; Kumamoto, T.; Nakanishi, W.; Watanabe, T.; Ishikawa, T.; Matsumoto, K.; Tashima, K.; Horie, S.; et al. Chemical constituents of *Aristolochia constricta*: Antispasmodic effects of its constituents in guinea-pig ileum and isolation of a diterpeno-lignan hybrid. *J. Nat. Prod.* **2008**, *71*, 1167–1172. [CrossRef] [PubMed]

48. Kuroyanagi, M.; Ikeda, R.; Gao, H.Y.; Muto, N.; Otaki, K.; Sano, T.; Kawahara, N.; Nakane, T. Neurite outgrowth-promoting active constituents of the Japanese cypress (*Chamaecyparis obtusa*). *Chem. Pharm. Bull.* **2008**, *56*, 60–63. [CrossRef] [PubMed]

49. Adfa, M.; Rahmad, R.; Ninomiya, M.; Yudha, S.; Tanaka, K.; Koketsu, M. Antileukemic activity of lignans and phenylpropanoids of *Cinnamomum parthenoxylon*. *Bioorg. Med. Chem. Lett.* **2016**, *26*, 761–764. [CrossRef] [PubMed]

50. Cunha, N.L.; Teixeira, G.M.; Martins, T.S.D.; Souza, A.R.; Oliveira, P.F.; Símaro, G.V.N.; Rezende, K.C.S.; Gonçalves, N.L.D.S.; Souza, D.G.; Tavares, D.C.; et al. (−)-Hinokinin induces G2/M arrest and contributes to the antiproliferative effects of doxorubicin in breast cancer cells. *Planta Med.* **2016**, *82*, 530–538. [CrossRef] [PubMed]

51. Yoon, J.S.; Koo, K.A.; Ma, C.J.; Sung, S.H.; Kim, Y.C. Neuroprotective lignans from *Biota orientalis* leaves. *Nat. Prod. Sci.* **2008**, *14*, 167–170.

52. De Souza, V.A.; da Silva, R.; Pereira, A.C.; Royo Vde, A.; Saraiva, J.; Montanheiro, M.; de Souza, G.H.; da Silva Filho, A.A.; Grando, M.D.; Donate, P.M.; et al. Trypanocidal activity of (−)-cubebin derivatives against free amastigote forms of *Trypanosoma cruzi*. *Bioorg. Med. Chem. Lett.* **2005**, *15*, 303–307. [CrossRef] [PubMed]

53. Esperandim, V.R.; da Silva Ferreira, D.; Saraiva, J.; Silva, M.L.A.; Costa, E.S.; Pereira, A.C.; Bastos, J.K.; de Albuquerque, S. Reduction of parasitism tissue by treatment of mice chronically infected with *Trypanosoma cruzi* with lignano lactones. *Parasitol. Res.* **2010**, *107*, 525–530. [CrossRef] [PubMed]

54. Esperandim, V.R.; da Silva Ferreira, D.; Rezende, K.C.; Cunha, W.R.; Saraiva, J.; Bastos, J.K.; e Silva, M.L.; de Albuquerque, S. Evaluation of the in vivo therapeutic properties of (−)-cubebin and (−)-hinokinin against *Trypanosoma cruzi*. *Experim. Parasitol.* **2013**, *133*, 442–446. [CrossRef] [PubMed]

55. Saraiva, J.; Vega, C.; Rolon, M.; da Silva, R.; Andrade, E.S.M.L.; Donate, P.M.; Bastos, J.K.; Gomez-Barrio, A.; de, A.S. In vitro and in vivo activity of lignan lactones derivatives against *Trypanosoma cruzi*. *Parasitol. Res.* **2007**, *100*, 791–795. [CrossRef] [PubMed]

56. Saraiva, J.; Lira, A.A.M.; Esperandim, V.R.; da Silva Ferreira, D.; Ferraudo, A.S.; Bastos, J.K.; e Silva, M.L.A.; de Gaitani, C.M.; de Albuquerque, S.; Marchetti, J.M. (−)-Hinokinin-loaded poly(D,L-lactide-co-glycolide) microparticles for Chagas disease. *Parasitol. Res.* **2010**, *106*, 703–708. [CrossRef] [PubMed]

57. Sartorelli, P.; Carvalho, C.S.; Reimao, J.Q.; Lorenzi, H.; Tempone, A.G. Antitrypanosomal activity of a diterpene and lignans isolated from *Aristolochia cymbifera*. *Planta Med.* **2010**, *76*, 1454–1456. [CrossRef] [PubMed]

58. Haribabu, K.; Ajitha, M.; Mallavadhani, U.V. Quantitative estimation of (−)-hinokinin, a trypanosomicidal marker in *Piper cubeba*, and some of its commercial formulations using HPLC-PDA. *J. Pharm. Anal.* **2015**, *5*, 130–136. [CrossRef] [PubMed]

59. Andrade e Silva, M.L.; Cicarelli, R.M.; Pauletti, P.M.; Luz, P.P.; Rezende, K.C.; Januario, A.H.; da Silva, R.; Pereira, A.C.; Bastos, J.K.; de Albuquerque, S.; et al. *Trypanosoma cruzi*: Evaluation of (−)-cubebin derivatives activity in the messenger RNAs processing. *Parasitol. Res.* **2011**, *109*, 445–451. [CrossRef] [PubMed]

60. Resende, F.A.; Tomazella, I.M.; Barbosa, L.C.; Ponce, M.; Furtado, R.A.; Pereira, A.C.; Bastos, J.K.; Andrade, E.S.M.L.; Tavares, D.C. Effect of the dibenzylbutyrolactone lignan (−)-hinokinin on doxorubicin and methyl methanesulfonate clastogenicity in V79 Chinese hamster lung fibroblasts. *Mutat. Res.* **2010**, *700*, 62–66. [CrossRef] [PubMed]

61. Resende, F.A.; Barbosa, L.C.; Tavares, D.C.; de Camargo, M.S.; de Souza Rezende, K.C.; ML, E.S.; Varanda, E.A. Mutagenicity and antimutagenicity of (−)-hinokinin a trypanosomicidal compound measured by *Salmonella* microsome and comet assays. *BMC Complement. Altern. Med.* **2012**, *12*, 203. [CrossRef] [PubMed]

62. Masumura, M.; Okumura, F.S. Identity of hibalactone and savinin. *J. Am. Chem. Soc.* **1955**, *77*, 1906. [CrossRef]

63. Hartwell, J.L.; Johnson, J.M.; Fitzgerald, D.B.; Belkin, M. Podophyllotoxin from *Juniperus* species; savinin. *J. Am. Chem. Soc.* **1953**, *75*, 235–236. [CrossRef]

64. Woo, K.W.; Choi, S.U.; Park, J.C.; Lee, K.R. A new lignan glycoside from *Juniperus rigida*. *Arch. Pharm. Res.* **2011**, *34*, 2043–2049. [CrossRef] [PubMed]

65. Badawi, M.M.; Seida, A.A.; Kinghorn, A.D.; Cordell, G.A.; Farnsworth, N.R. Potential anticancer agents. XVIII. Constituents of *Amyris pinnata* (Rutaceae). *J. Nat. Prod.* **1981**, *44*, 331–334. [CrossRef]

66. Mansoor, T.A.; Borralho, P.M.; Luo, X.; Mulhovo, S.; Rodrigues, C.M.P.; Ferreira, M.-J.U. Apoptosis inducing activity of benzophenanthridine-type alkaloids and 2-arylbenzofuran neolignans in HCT116 colon carcinoma cells. *Phytomedicine* **2013**, *20*, 923–929. [CrossRef] [PubMed]

67. Cho, J.Y.; Park, J.; Kim, P.S.; Yoo, E.S.; Baik, K.U.; Park, M.H. Savinin, a lignan from *Pterocarpus santalinus* inhibits tumor necrosis factor-α production and T cell proliferation. *Biol. Pharm. Bull.* **2001**, *24*, 167–171. [CrossRef] [PubMed]

68. Lee, S.; Ban, H.S.; Kim, Y.P.; Kim, B.-K.; Cho, S.H.; Ohuchi, K.; Shin, K.H. Lignans from *Acanthopanax chiisanensis* having an inhibitory activity on prostaglandin E2 production. *Phytother. Res.* **2005**, *19*, 103–106. [CrossRef] [PubMed]

69. Mbaze, L.M.A.; Lado, J.A.; Wansi, J.D.; Shiao, T.C.; Chiozem, D.D.; Mesaik, M.A.; Choudhary, M.I.; Lacaille-Dubois, M.-A.; Wandji, J.; Roy, R.; et al. Oxidative burst inhibitory and cytotoxic amides and lignans from the stem bark of *Fagara heitzii* (Rutaceae). *Phytochemistry* **2009**, *70*, 1442–1447. [CrossRef] [PubMed]

70. Oliveira, T.L.S.; Leite, K.C.D.S.; de Macêdo, I.Y.L.; de Morais, S.R.; Costa, E.A.; de Paula, J.R.; de Souza Gil, E. Electrochemical behavior and antioxidant activity of hibalactone. *Int. J. Electrochem. Sci.* **2017**, *12*, 7956–7964. [CrossRef]

71. Oliveira, T.L.S.; Morais, S.R.; Sá, S.; Oliveira, M.G.; Florentino, I.F.; Silva, D.M.D.; Carvalho, V.V.; Silva, V.B.D.; Vaz, B.G.; Sabino, J.R.; et al. Antinociceptive, anti-inflammatory and anxiolytic-like effects of the ethanolic extract, fractions and Hibalactone isolated from *Hydrocotyle umbellata* L. (Acaricoba)—Araliaceae. *Biomed. Pharmacother.* **2017**, *95*, 837–846. [CrossRef] [PubMed]

72. Rzedowski, J.; Guevara-Féfer, F. *Flora del Bajío y de Regiones Adyacentes*; Instituto de Ecologia: Xalapa, Mexico, 1992; Volume 3.

73. Chatterjee, A.; Basa, S.C.; Ray, A.B. Spectral properties of the lignin cubebin. *J. Indian Chem. Soc.* **1968**, *45*, 723–725.

74. Bastos, J.K.; Albuquerque, S.; Silva, M.L.A. Evaluation of the trypanocidal activity of lignans isolated from the leaves of *Zanthoxylum naranjillo*. *Planta Med.* **1999**, *65*, 541–544. [CrossRef] [PubMed]

75. Carvalho, M.T.M.; Rezende, K.C.S.; Evora, P.R.B.; Bastos, J.K.; Cunha, W.R.; Andrade e Silva, M.L.; Celotto, A.C. The lignan (−)-Cubebin inhibits vascular contraction and induces relaxation via nitric oxide activation in isolated rat aorta. *Phytother. Res.* **2013**, *27*, 1784–1789. [CrossRef] [PubMed]

76. Niwa, A.M.; Marcarini, J.C.; Sartori, D.; Maistro, E.L.; Mantovani, M.S. Effects of (−)-cubebin (*Piper cubeba*) on cytotoxicity, mutagenicity and expression of p38 MAP kinase and GSTa2 in a hepatoma cell line. *J. Food Compos. Anal.* **2013**, *30*, 1–5. [CrossRef]

77. De Rezende, A.A.A.; Silva, M.L.A.; Tavares, D.C.; Cunha, W.R.; Rezende, K.C.S.; Bastos, J.K.; Lehmann, M.; de Andrade, H.H.R.; Guterres, Z.R.; Silva, L.P.; et al. The effect of the dibenzylbutyrolactolic lignan (−)-cubebin on doxorubicin mutagenicity and recombinogenicity in wing somatic cells of *Drosophila melanogaster*. *Food Chem. Toxicol.* **2011**, *49*, 1235–1241. [CrossRef] [PubMed]

78. Da Rocha Pissurno, A.P.; De Laurentiz, R.D.S. Cubebin: A small molecule with great potential. *Rev. Virtual Quim.* **2017**, *9*, 656–671. [CrossRef]

79. Cook, S. *The Forest of the Lacandon Maya: An Ethnobotanical Guide*; Springer: New York, NY, USA, 2016.

80. Takaoka, D.; Imooka, M.; Hiroi, M. Studies of lignoids in Lauraceae. III. A new lignan from the heart wood of *Cinnamomum camphora* Sieb. *Bull. Chem. Soc. Jpn.* **1977**, *50*, 2821–2822. [CrossRef]

81. Lopes, L.M.X.; Yoshida, M.; Gottlieb, O.R. The chemistry of Brazilian Myristicaceae. Part XIX. Dibenzylbutyrolactone lignans from *Virola sebifera*. *Phytochemistry* **1983**, *22*, 1516–1518. [CrossRef]

82. Kato, M.J.; Yoshida, M.; Gottlieb, O.R. Lignoids and arylalkanones from fruits of *Virola elongata*. *Phytochemistry* **1990**, *29*, 1799–1810. [CrossRef]

83. Gozler, T.; Gozler, B.; Patra, A.; Leet, J.E.; Freyer, A.J.; Shamma, M. Konyanin: A new lignan from *Haplophyllum vulcanicum*. *Tetrahedron* **1984**, *40*, 1145–1150. [CrossRef]

84. Evcim, U.; Gozler, B.; Freyer, A.J.; Shamma, M. Haplomyrtin and (−)-haplomyrfolin: Two lignans from *Haplophyllum myrtifolium*. *Phytochemistry* **1986**, *25*, 1949–1951. [CrossRef]

85. Sheriha, G.M.; Abouamer, K.; Elshtaiwi, B.Z.; Ashour, A.S.; Abed, F.A.; Alhallaq, H.H. Quinoline alkaloids and cytotoxic lignans from *Haplophyllum tuberculatum*. *Phytochemistry* **1987**, 26, 3339–3341. [CrossRef]

86. Rifai, Y. A new method for fast isolation of GLI inhibitory compounds. *Int. J. Pharma Res. Rev.* **2012**, 1, 28–30.

87. Ogungbe, I.V.; Erwin, W.R.; Setzer, W.N. Antileishmanial phytochemical phenolics: Molecular docking to potential protein targets. *J. Mol. Gr. Modell.* **2014**, 48, 105–117. [CrossRef] [PubMed]

88. Sriwiriyajan, S.; Sukpondma, Y.; Srisawat, T.; Madla, S.; Graidist, P. (−)-Kusunokinin and piperloguminine from *Piper nigrum*: An alternative option to treat breast cancer. *Biomed. Pharmacother.* **2017**, 92, 732–743. [CrossRef] [PubMed]

89. Gonzalez, A.G.; Estevez-Reyes, R.; Mato, C.; Estevez-Braun, A.M. Three lignans from *Bupleurum salicifolium*. *Phytochemistry* **1990**, 29, 1981–1983. [CrossRef]

90. Espinosa Organista, D.; Ramos Rivera, P.; Careaga Olvera, S.A. *El Género Bursera en México. Parte II*, 1.3 ed.; CONABIO: México City, México, 2017.

91. Nakanishi, T.; Inatomi, Y.; Murata, H.; Shigeta, K.; Iida, N.; Inada, A.; Murata, J.; Farrera, M.A.P.; Iinuma, M.; Tanaka, T.; et al. A new and known cytotoxic aryltetralin-type lignans from stems of *Bursera graveolens*. *Chem. Pharm. Bull.* **2005**, 53, 229–231. [CrossRef] [PubMed]

92. Provan, G.J.; Waterman, P.G. Picropolygamain: A new lignan from *Commiphora incisa* resin. *Planta Med.* **1985**, 51, 271–272. [CrossRef]

93. Peraza-Sanchez, S.R.; Pena-Rodriguez, L.M. Isolation of picropolygamain from the resin of *Bursera simaruba*. *J. Nat. Prod.* **1992**, 55, 1768–1771. [CrossRef] [PubMed]

94. Mi, Q.; Pezzuto, J.M.; Farnsworth, N.R.; Wani, M.C.; Kinghorn, A.D.; Swanson, S.M. Use of the in Vivo Hollow Fiber Assay in Natural Products Anticancer Drug Discovery. *J. Nat. Prod.* **2009**, 72, 573–580. [CrossRef] [PubMed]

95. Felger, R.S.; Johnson, M.B.; Wilson, M.F. *The Trees of Sonora, Mexico*; Oxford University Press: New York, NY, USA, 2001.

96. Gentry, H.S. *Rio Mayo Plants*; Carnegie Institution: Washington, DC, USA, 1942.

97. Garcia Martinez, L.E. *Aspectos Socio-Ecologicos Para el Manejo Sustentable del Coal en el Ejido de Acateyahualco, Gro*; UNAM: Morelia, Mexico, 2012.

98. Montúfar, A. *Los Copales Mexicanos y la Resina Sagrada del Templo Mayor de Tenochtitlan*; Instituto Nacional de Antropología e Historia: Mexico City, Mexico, 2007.

99. Dávila Aranda, P.; Ramos Rivera, P. *La Flora útil de dos Comunidades Indígenas del Valle de Tehuacán-Cuicatlán: Coxcatlán y Zapotitlán de las Salinas, Puebla*; Version 1.3; Comisión Nacional Para el Conocimiento y uso de la Biodiversidad: Mexico City, Mexico, 2017.

100. Nieto-Yañez, O.J.; Resendiz-Albor, A.A.; Ruiz-Hurtado, P.A.; Rivera-Yañez, N.; Rodriguez-Canales, M.; Rodriguez-Sosa, M.; Juarez-Avelar, I.; Rodriguez-Lopez, M.G.; Canales-Martinez, M.M.; Rodriguez-Monroy, M.A. In vivo and in vitro antileishmanial effects of methanolic extract from bark of *Bursera aptera*. *Afr. J. Tradit. Complement. Altern. Med.* **2017**, 14, 188–197. [CrossRef] [PubMed]

101. Rzedowski, J.; Medina Lemos, R.; Calderón de Rzedowski, G. Las especies de *Bursera* (Burseraceae) en la cuenca superior del rio Papaloapan (Mexico). *Acta Bot. Mex.* **2004**, 66, 23–151. [CrossRef]

102. Canales, M.; Hernández, T.; Caballero, J.; Vivar, A.R.D.; Avila, G.; Duran, A.; Lira, R. Informant consensus factor and antibacterial activity of the medicinal plants used by the people of San Rafael Coxcatlán, Puebla, México. *J. Ethnopharmacol.* **2005**, 97, 429–439. [CrossRef] [PubMed]

103. Gorgua Jiménez, G.; Nieto Yañez, O.; Ruiz Hurtado, P.A.; Rivera Yáñez, N.; Rodríguez Monroy, M.A.; Canales Martínez, M.M. *Bursera arida*. Evaluacion de su actividad antiparasitaria y caracterizacion fitoquimica del extracto metanolico. In *11 Reunion Internacional de Investigacion en Productos Naturales*; Navarrete, A., Ed.; Revista Latinoamericana de Quimica: San Carlos, Mexico, 2015.

104. Ionescu, F. *Phytochemical Investigation of Bursera Arida Family Burseraceae*; The University of Arizona: Tucson, Arizona, 1974.

105. Hernández, J.D.; Román, L.U.; Espiñeira, J.; Joseph-Nathan, P. Ariensin, a New Lignan from *Bursera ariensis*. *Planta Med.* **1983**, 47, 215–217. [CrossRef] [PubMed]

106. LLIFLE Encyclopedias of Living Form. Available online: http://www.llifle.com/Encyclopedia/ (accessed on 5 August 2018).

107. Avendaño Reyes, S.; Acosta Rosado, I. Plantas utilizadas como cercas vivas en el estado de Veracruz. *Madera Bosques* **2000**, 6, 55–71. [CrossRef]

108. Velazquez-Jimenez, R.; Torres-Valencia, J.M.; Cerda-Garcia-Rojas, C.M.; Hernandez-Hernandez, J.D.; Roman-Marin, L.U.; Manriquez-Torres, J.J.; Gomez-Hurtado, M.A.; Valdez-Calderon, A.; Motilva, V.; Garcia-Maurino, S.; et al. Absolute configuration of podophyllotoxin related lignans from *Bursera fagaroides* using vibrational circular dichroism. *Phytochemistry* **2011**, *72*, 2237–2243. [CrossRef] [PubMed]

109. Jolad, S.D.; Wiedhopf, R.M.; Cole, J.R. Cytotoxic agents from *Bursera morelensis* (Burseraceae): Deoxypodophyllotoxin and a new lignan, 5′-desmethoxydeoxypodophyllotoxin. *J. Pharm. Sci.* **1977**, *66*, 892–893. [CrossRef] [PubMed]

110. Chen, J.J.; Ishikawa, T.; Duh, C.Y.; Tsai, I.L.; Chen, I.S. New dimeric aporphine alkaloids and cytotoxic constituents of *Hernandia nymphaeifolia*. *Planta Med.* **1996**, *62*, 528–533. [CrossRef] [PubMed]

111. Novelo, M.; Cruz, J.G.; Hernández, L.; Pereda-Miranda, R.; Chai, H.; Mar, W.; Pezzuto, J.M. Cytotoxic Constituents from *Hyptis verticillata*. *J. Nat. Prod.* **1993**, *56*, 1728–1736. [CrossRef] [PubMed]

112. Donoso-Fierro, C.S.; Tiezzi, A.; Ovidi, E.; Ceccarelli, D.; Triggiani, D.; Mastrogiovanni, F.; Taddei, A.R.; Pérez, C.; Becerra, J.; Silva, M.; Passarella, D. Antiproliferative activity of yatein isolated from *Austrocedrus chilensis* against murine myeloma cells: Cytological studies and chemical investigations. *Pharm. Biol.* **2015**, *53*, 378–385. [CrossRef] [PubMed]

113. Liu, Y.; Young, K.; Rakotondraibe, L.H.; Brodie, P.J.; Wiley, J.D.; Cassera, M.B.; Callmander, M.W.; Rakotondrajaona, R.; Rakotobe, E.; Rasamison, V.E.; Tendyke, K.; et al. Antiproliferative compounds from *Cleistanthus boivinianus* from the Madagascar dry forest. *J. Nat. Prod.* **2015**, *78*, 1543–1547. [CrossRef] [PubMed]

114. Mosmann, T. Rapid colorimetric assay for cellular growth and survival: Application to proliferation and cytotoxicity assays. *J. Immunol. Methods* **1983**, *65*, 55–63. [CrossRef]

115. Wu, X.-D.; Wang, S.-Y.; Wang, L.; He, J.; Li, G.-T.; Ding, L.-F.; Gong, X.; Dong, L.-B.; Song, L.-D.; Li, Y.; et al. Labdane diterpenoids and lignans from *Calocedrus macrolepis*. *Fitoterapia* **2013**, *85*, 154–160. [CrossRef] [PubMed]

116. Chen, J.J.; Hung, H.C.; Sung, P.J.; Chen, I.S.; Kuo, W.L. Aporphine alkaloids and cytotoxic lignans from the roots of *Illigera luzonensis*. *Phytochemistry* **2011**, *72*, 523–532. [CrossRef] [PubMed]

117. Doussot, J.; Mathieu, V.; Colas, C.; Molinié, R.; Corbin, C.; Montguillon, J.; Moreno, Y.; Banuls, L.; Renouard, S.; Lamblin, F.; et al. Investigation of the Lignan Content in Extracts from *Linum, Callitris* and *Juniperus* Species in Relation toTheir In Vitro Antiproliferative Activities. *Planta Med.* **2017**, *83*, 574–581. [PubMed]

118. Kuo, Y.-C.; Kuo, Y.-H.; Lin, Y.-L.; Tsai, W.-J. Yatein from *Chamaecyparis obtusa* suppresses herpes simplex virus type 1 replication in HeLa cells by interruption the immediate-early gene expression. *Antivir. Res.* **2006**, *70*, 112–120. [CrossRef] [PubMed]

119. Usia, T.; Watabe, T.; Kadota, S.; Tezuka, Y. Metabolite-cytochrome P450 complex formation by methylenedioxyphenyl lignans of *Piper cubeba*: Mechanism-based inhibition. *Life Sci.* **2005**, *76*, 2381–2391. [CrossRef] [PubMed]

120. Chen, J.J.; Chang, Y.L.; Teng, C.M.; Chen, I.S. Anti-platelet aggregation alkaloids and lignans from *Hernandia nymphaeifolia*. *Planta Med.* **2000**, *66*, 251–256. [CrossRef] [PubMed]

121. Picking, D.; Chambers, B.; Barker, J.; Shah, I.; Porter, R.; Naughton, D.P.; Delgoda, R. Inhibition of cytochrome P450 activities by extracts of *Hyptis verticillata* Jacq.: Assessment for potential HERB-drug interactions. *Molecules* **2018**, *23*, 430. [CrossRef] [PubMed]

122. Rojas-Sepúlveda, A.M.; Mendieta-Serrano, M.; Mojica, M.Y.A.; Salas-Vidal, E.; Marquina, S.; Villarreal, M.L.; Puebla, A.M.; Delgado, J.I.; Alvarez, L. Cytotoxic podophyllotoxin type-lignans from the steam bark of *Bursera fagaroides* var. fagaroides. *Molecules* **2012**, *17*, 9506. [CrossRef] [PubMed]

123. Puebla-Pérez, A.M.; Huacuja-Ruiz, L.; Rodríguez-Orozco, G.; Villaseñor-García, M.M.; Miranda-Beltrán, M.D.L.L.; Celis, A.; Sandoval-Ramírez, L. Cytotoxic and antitumour activity from Bursera fagaroides ethanol extract in mice with L5178Y lymphoma. *Phytother. Res.* **1998**, *12*, 545–548. [CrossRef]

124. Guerram, M.; Jiang, Z.-Z.; Zhang, L.-Y. Podophyllotoxin, a medicinal agent of plant origin: Past, present and future. *Chin. J. Nat. Med.* **2012**, *10*, 161–169. [CrossRef]

125. Acton, Q.A. *Advances in Podophyllotoxin Research and Application: 2012 Edition: ScholarlyBrief*; Scholarly Editions: Atlanta, GA, USA, 2012.

126. Kamal, A.; Ali Hussaini, S.M.; Malik, M.S. Recent Developments Towards Podophyllotoxin Congeners as Potential Apoptosis Inducers. *Anti-Cancer Agents Med. Chem.* **2015**, *15*, 565–574. [CrossRef]

127. Lee, K.-H.; Xiao, Z. *Podophyllotoxins and Analogs*; CRC Press: Boca Raton, FL, USA, 2012; pp. 95–122.

128. Ionkova, I. *Podophyllotoxin and Related Lignans: Biotechnological Production by In Vitro Plant Cell Cultures*; CAB International: Wallingford, Oxfordshire, UK, 2010; pp. 138–155.

129. Khaled, M.; Jiang, Z.-Z.; Zhang, L.-Y. Deoxypodophyllotoxin: A promising therapeutic agent from herbal medicine. *J. Ethnopharmacol.* **2013**, *149*, 24–34. [CrossRef] [PubMed]

130. Guerram, M.; Jiang, Z.-Z.; Sun, L.; Zhu, X.; Zhang, L.-Y. Antineoplastic effects of deoxypodophyllotoxin, a potent cytotoxic agent of plant origin, on glioblastoma U-87 MG and SF126 cells. *Pharmacol. Rep.* **2015**, *67*, 245–252. [CrossRef] [PubMed]

131. Morales-Serna, J.A.; Cruz-Galicia, E.; Garcia-Rios, E.; Madrigal, D.; Gavino, R.; Cardenas, J.; Salmon, M. Three new diarylbutane lignans from the resin of *Bursera fagaroides*. *Nat. Prod. Res.* **2013**, *27*, 824–829. [CrossRef] [PubMed]

132. Mojica, M.A.; Leon, A.; Rojas-Sepulveda, A.M.; Marquina, S.; Mendieta-Serrano, M.A.; Salas-Vidal, E.; Villarreal, M.L.; Alvarez, L. Aryldihydronaphthalene-type lignans from *Bursera fagaroides* var. *fagaroides* and their antimitotic mechanism of action. *RSC Adv.* **2016**, *6*, 4950–4959. [CrossRef]

133. Gu, J.-Q.; Park, E.J.; Totura, S.; Riswan, S.; Fong, H.H.S.; Pezzuto, J.M.; Kinghorn, A.D. Constituents of the twigs of *Hernandia ovigera* that inhibit the transformation of JB6 murine epidermal cells. *J. Nat. Prod.* **2002**, *65*, 1065–1068. [CrossRef] [PubMed]

134. Rosas-Arreguin, P.; Arteaga-Nieto, P.; Reynoso-Orozco, R.; Villagomez-Castro, J.C.; Sabanero-Lopez, M.; Puebla-Perez, A.M.; Calvo-Mendez, C. *Bursera fagaroides*, effect of an ethanolic extract on ornithine decarboxylase (ODC) activity in vitro and on the growth of *Entamoeba histolytica*. *Exp. Parasitol.* **2008**, *119*, 398–402. [CrossRef] [PubMed]

135. Gutierrez-Gutierrez, F.; Puebla-Perez, A.M.; Gonzalez-Pozos, S.; Hernandez-Hernandez, J.M.; Perez-Rangel, A.; Alvarez, L.P.; Tapia-Pastrana, G.; Castillo-Romero, A. Antigiardial Activity of Podophyllotoxin-Type Lignans from *Bursera fagaroides* var. *fagaroides*. *Molecules* **2017**, *22*, 799. [CrossRef] [PubMed]

136. Cole, J.R.; Bianchi, E.; Trumbull, E.R. Antitumor agents from *Bursera microphylla* (Burseraceae) II: Isolation of a new lignan—Burseran. *J. Pharm. Sci.* **1969**, *58*, 175–176. [CrossRef] [PubMed]

137. Tomioka, K.; Ishiguro, T.; Mizuguchi, H.; Komeshima, N.; Koga, K.; Tsukagoshi, S.; Tsuruo, T.; Tashiro, T.; Tanida, S.; Kishi, T. Stereoselective reactions. XVII. Absolute structure-cytotoxic activity relationships of steganacin congeners and analogs. *J. Med. Chem.* **1991**, *34*, 54–57. [CrossRef] [PubMed]

138. Tomioka, K.; Koga, K. Stereoselective total synthesis of optically active trans- and *cis*-burseran. Determination of the stereochemistry of natural antitumor burseran. *Heterocycles* **1979**, *12*, 1523–1528. [CrossRef]

139. Gigliarelli, G.; Zadra, C.; Cossignani, L.; Robles Zepeda, R.E.; Rascon-Valenzuela, L.A.; Velazquez-Contreras, C.A.; Marcotullio, M.C. Two new lignans from the resin of *Bursera microphylla* A. gray and their cytotoxic activity. *Nat. Prod. Res.* **2017**, *18*, 1–6. [CrossRef] [PubMed]

140. Zhao, L.; Tian, X.; Fan, P.C.; Zhan, Y.J.; Shen, D.W.; Jin, Y. Separation, determination and identification of the diastereoisomers of podophyllotoxin and its esters by high-performance liquid chromatography/tandem mass spectrometry. *J. Chromatogr. A* **2008**, *1210*, 168–177. [CrossRef] [PubMed]

141. Alina, C.-M.C.; Rocio, R.-L.; Aurelio, R.-M.M.; Margarita, C.-M.M.; Angelica, R.-G.; Ruben, J.-A. Chemical composition and in vivo anti-inflammatory activity of *Bursera morelensis* Ramirez essential oil. *J. Essent. Oil-Bear. Plants* **2014**, *17*, 758–768. [CrossRef]

142. Rzedowski, J.; Medina Lemos, R.; Calderón de Rzedowski, G. Segunda restauración de *Bursera ovalifolia* y nombre nuevo para otro componente del complejo de *B. simaruba* (Burseraceae). *Acta Bot. Mex.* **2007**, *81*, 45–70. [CrossRef]

143. Buckingham, J. *Dictionary of Natural Products*; Taylor & Francis: Abingdon, UK, 1993.

144. Chang, C.-C.; Lien, Y.-C.; Liu, K.C.S.C.; Lee, S.-S. Lignans from *Phyllanthus urinaria*. *Phytochemistry* **2003**, *63*, 825–833. [CrossRef]

145. Ito, C.; Itoigawa, M.; Ogata, M.; Mou, X.Y.; Tokuda, H.; Nishino, H.; Furukawa, H. Lignans as anti-tumor-promoter from the seeds of *Hernandia ovigera*. *Planta Med.* **2001**, *67*, 166–168. [CrossRef] [PubMed]

146. Murakami, A.; Ohigashi, H.; Koshimizu, K. Anti-tumor promotion with food phytochemicals: A strategy for cancer chemoprevention. *Biosci. Biotechnol. Biochem.* **1996**, *60*, 1–8. [CrossRef] [PubMed]

147. Pettit, G.R.; Meng, Y.; Gearing, R.P.; Herald, D.L.; Pettit, R.K.; Doubek, D.L.; Chapuis, J.-C.; Tackett, L.P. Antineoplastic Agents. 522. *Hernandia peltata* (Malaysia) and *Hernandia nymphaeifolia* (Republic of Maldives). *J. Nat. Prod.* **2004**, *67*, 214–220. [CrossRef] [PubMed]

148. Noguera, B.; Diaz, E.; Garcia, M.V.; Feliciano, A.S.; Lopez-Perez, J.L.; Israel, A. Anti-inflammatory activity of leaf extract and fractions of *Bursera simaruba* (L.) Sarg (Burseraceae). *J. Ethnopharmacol.* **2004**, *92*, 129–133. [CrossRef] [PubMed]
149. Maldini, M.; Montoro, P.; Piacente, S.; Pizza, C. Phenolic compounds from *Bursera simaruba* Sarg. bark: Phytochemical investigation and quantitative analysis by tandem mass spectrometry. *Phytochemistry* **2009**, *70*, 641–649. [CrossRef] [PubMed]

Sample Availability: Samples of the compounds are not available from the authors.

molecules
MDPI

Article

9-Norlignans: Occurrence, Properties and Their Semisynthetic Preparation from Hydroxymatairesinol

Patrik Eklund * and Jan-Erik Raitanen

Johan Gadolin Process Chemistry Centre, Laboratory of Organic Chemistry, Åbo Akademi University,
Piispankatu 8, FIN-20500 Turku, Finland; jraitane@abo.fi
* Correspondence: paeklund@abo.fi; Tel.: +358-2-215-4720

Received: 8 November 2018; Accepted: 28 December 2018; Published: 9 January 2019

Abstract: Lignans, neolignans, norlignans and norneolignans constitute a large class of phenolic natural compounds. 9-Norlignans, here defined to contain a β–β' bond between the two phenylpropanoid units and to lack carbon number 9 from the parent lignan structure, are the most rarely occurring compounds within this class of natural compounds. We present here an overview of the structure, occurrence and biological activity of thirty-five 9-norlignans reported in the literature to date. In addition, we report the semisynthetic preparation of sixteen 9-norlignans using the natural lignan hydroxymatairesinol obtained from spruce knots, as starting material. 9-Norlignans are shown to exist in different species and to have various biological activities, and they may therefore serve as lead compounds for example for the development of anticancer agents. Hydroxymatairesinol is shown to be a readily available starting material for the preparation of norlignans of the imperanene, vitrofolal and noralashinol family.

Keywords: lignans; norlignans; 9-norlignans; semisynthesis; hydroxymatairesinol; bioactivity

1. Introduction

Many secondary metabolites, including lignans, flavonoids, and coumarins are formed from phenylpropanoids originating from the well-known shikimic acid pathway. The parent structures are usually further oxidized and arranged into various structures. Norlignans, a subclass of lignans lacking one or more carbon atoms, seem to be a rather unknown class of natural products. Compared to other phenylpropanoids their occurrence, biosynthetic pathway, and their properties are much less reported in the literature. Especially 9-norlignans (lacking carbon 9 from the parent lignan skeleton) with guaiacyl (3-methoxy-4-hydroxyphenyl) moieties have rarely been reported in the literature, although they would be expected to be common norlignans as guaiacyl lignans are the most abundant class of lignans. To the best of our knowledge, (+)-imperanene, vitrofolal E and F, noralashinol A and C are the only guaiacyl type 9-norlignans which have been reported as plant constituents [1–4]. The unnatural enantiomer, (−)-*R*-imperanene, dehydroxyimperanenes and dihydrodehydroxy-imperanenes have also been synthesized [5,6]. Normally, common substitution patterns are found in norlignans, but the more rare 2,4,5-trisubstituted phenyl moieties seem to be present in several 9-norlignans. In this paper we review the occurrence and properties of natural compounds belonging to the class of 9-norlignans, and in addition, we report the semisynthesis of 9-norlignans from the natural lignan hydroxymatairesinol.

1.1. Classification and Nomenclature of Norlignans

The term lignan is defined as two phenyl propane units coupled together by a β–β' bond, and if the same structural units are coupled in any other ways the product is called a neolignan. If the coupling between the units contains an ether function (fundamental parent structures) the compound is called an oxyneolignan. According to the present recommendations by the International Union of Pure and Applied Chemistry (IUPAC), the prefix nor (modification of the fundamental parent structure) is used when the lignan, neolignan or oxyneolignan lack one or more carbon atoms [7]. However, in the literature there are today over 60 compounds named as norlignans although the major part of these are lacking the β–β'coupling, which is the definition for the fundamental parent structure named lignan. In our opinion most of these structures should be named as norneolignans or noroxyneolignans (prefix nor + fundamental parent structure). The choice of the numbering of the lost carbon is also somewhat confusing. Numbering of the modified carbon skeleton (in this case nor) should be performed so that the modification is expressed by the lowest locant (number, unprimed). However, the choice of locants for the removal of a carbon atom (nor) is preferred in the order: unprimed, higher number. Consequently, there is a conflict when assigning structures to 9 or 7-norlignans or norneolignans. Although the exact definition of different norlignans seems a bit confusing and vague, we here define a norlignan as a plant-derived compound consisting of two phenylpropane units coupled in the propane moiety with a β–β' bond and missing one or more carbon atoms.

Moreover, we consider 9-norlignans to be derived from a parental lignan skeleton and thus to have a detectable β–β' coupling and to lack one (or more, in the case of di(bis)norlignans) of the terminal carbon atoms (carbon 9). In Figure 1, the fundamental structures of different norlignan and norneolignan structures coupled in the propane moiety are displayed (oxyneolignans and others are excluded, dotted lines in red indicate the missing carbon-9, the bold lines indicate the coupling in the propane moiety).

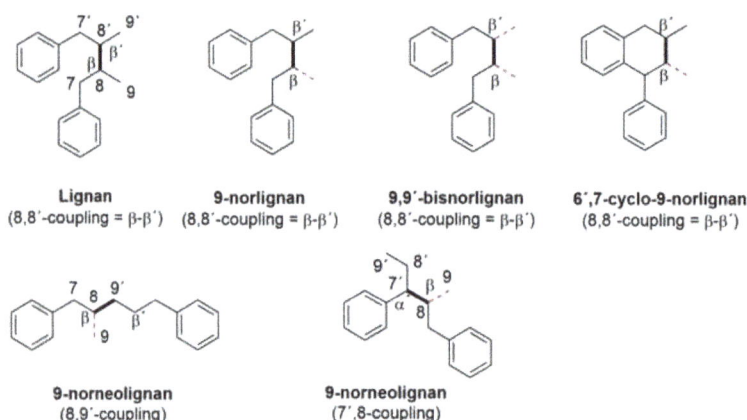

Figure 1. Fundamental structures for the lignan and different 9-norlignan and 9-norneolignan structures. Dotted (red lines) lines indicate the missing carbon-9 atom.

1.2. 9-Norlignans

According to the definition above, approximately 25 naturally occurring 9-norlignans can be found in the literature. These are yateresinol (**1**) [8], pachypostaudin A and B (**2,3**) [9], aglacin H (**4**) [10], vitrofolal A,B,E, and F (**5,6,7,8**) [2,11] descuraic acid (**9**) [12], cestrumoside (glycoside) (**10**) [13], (−)-justiflorinol (**11**) [14], (+)-virgatyne (**12**) [15], the arylnaphthalene derivative **13** [16] and the lactone **14** [17], compound **15** [18], *S*-(+)-imperanene (**16**) [1], 3-methoxy-imperanene (**18**) [19], noralashinol A and C (**19,20**) [3,4], tectonoelin A and B (**21,22**) [20], peperotetraphin (**23**) [21] the related cyclobutane carboxylates **24** and **25** [22] and hyperione A and B (**26,27**) [23] (Figure 2). In addition, some related 9-norlignans have been synthesized and some 9-norlignan like compounds have also been reported [6,24,25]. In Figure 3, the structures of compounds **28–35** are shown.

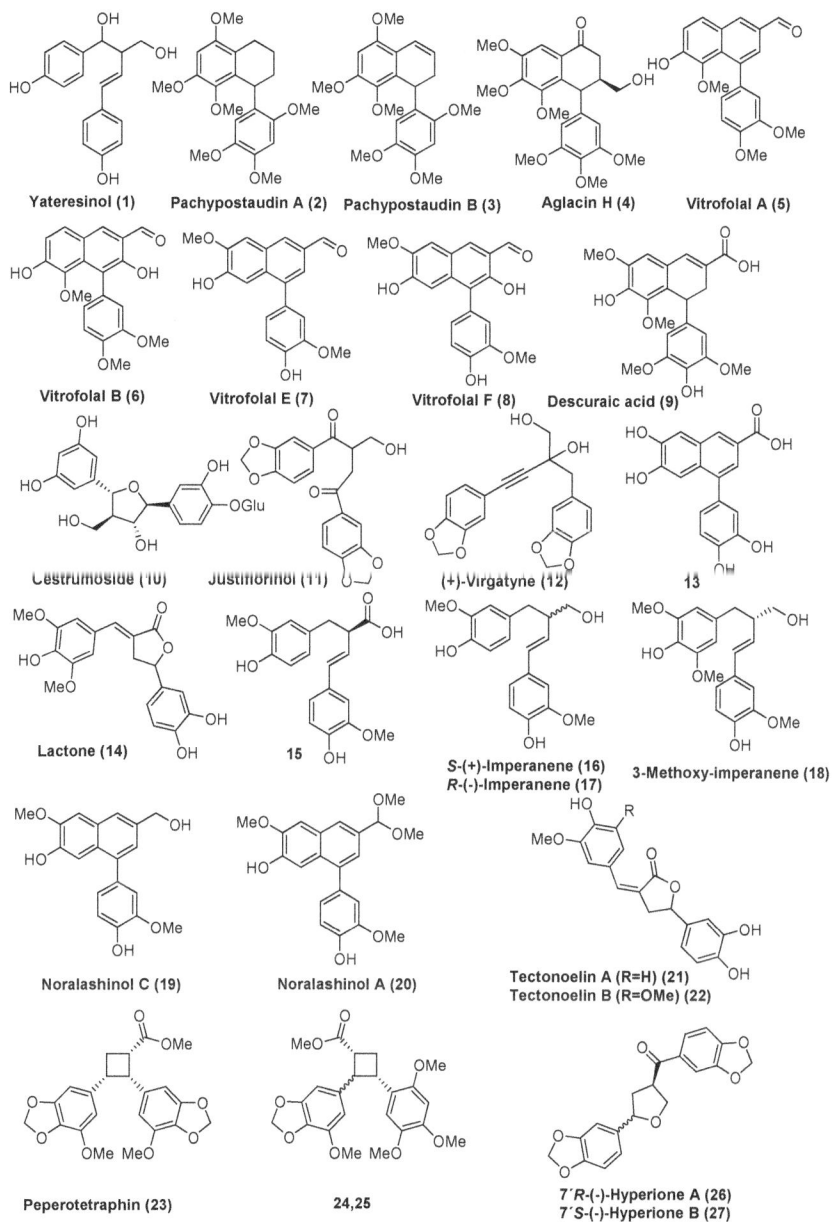

Figure 2. The structures of previously reported 9-norlignans with natural origin.

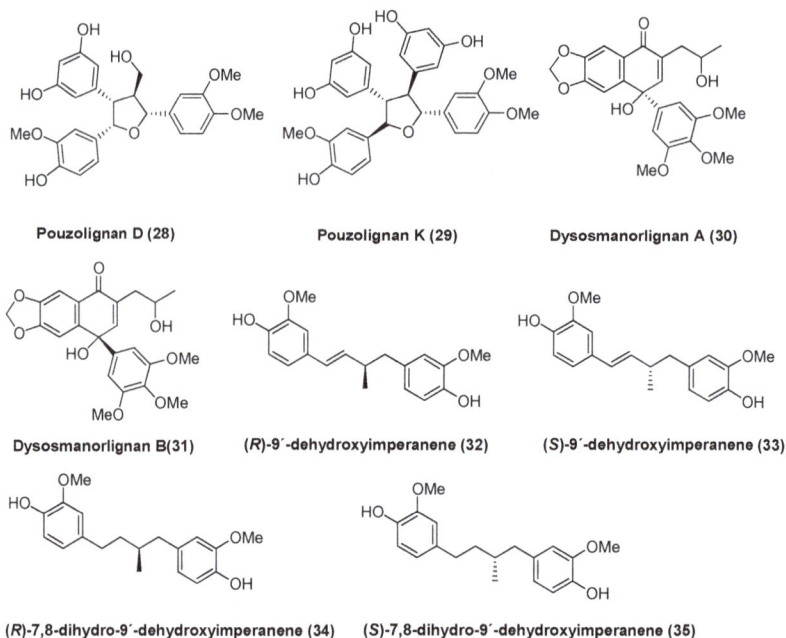

Pouzolignan D (28) Pouzolignan K (29) Dysosmanorlignan A (30)

Dysosmanorlignan B(31) (*R*)-9´-dehydroxyimperanene (32) (*S*)-9´-dehydroxyimperanene (33)

(*R*)-7,8-dihydro-9´-dehydroxyimperanene (34) (*S*)-7,8-dihydro-9´-dehydroxyimperanene (35)

Figure 3. Structures of synthetic 9-norlignans and 9-norlignan related compounds.

1.3. Occurrence and Biological Activity

Yateresinol (**1**) was first isolated from the heartwood of *Libocedrus yateensis* in 1979 by Ertman et al. [8]. It has also been found in *Cryptomerica japonica* D. Don and found to be part of the wood discoloring substances in sapwood [26]. Studies of butt-rot sugi wood suggested that the norlignans may be important for antifungal properties but no clear effect has been shown [27]. Yateresinol has been tested for cytotoxicity towards the human HL60 and Hepa G2 cancer cell lines [28] but was shown to be inactive with the half maximal inhibitory concentration (IC$_{50}$) values >20 µg/mL. The two 9,9´-bisnorlignans **2** and **3** (pachypostaudin A and B) have been isolated from stem bark of *Pachypodanthium staudtii* Engl. et. Diels [9]. The same bisnorlignans have also been isolated from bark of the related species *Pachypodanthium confine* Engl. et Diels [29]. Both species are trees growing in the tropical zone of west and central Africa, and have been used in traditional African medicine. However, no biological activity of the bisnorlignans has been reported. Aglacin H (**4**) has been isolated from the bark of *Aglaia cordata* collected in Indonesia [10]. No other sources, and no chemical and biological properties for this norlignan have so far been reported. Norlignans from *Vitex* species, namely vitrofolal A, B, E, and F (compounds **5–8**) are norlignans of the arylnaphthalene type. These compounds were first isolated from *Vitex rotundifolia* [2,11]. Compounds **5–8** have later also been isolated from the seeds [30] and **7** from the roots of *Vitex negundo* [31,32]. Compounds **7** and **8** have also been isolated from the fruits of *Vitex cannabifolia* [33]. Vitrofolal E (**7**) was also recently reported in the stem and bark of *Syringa pinnatifolia* [2,34]. Vitrofolals have been shown to have various biological activities. Compound **7** was shown to have antibacterial activity against methicillin-resistant *Staphylococcus aureus* at concentrations above 64 µg/mL [2]. Compounds **7** and **8** have also been reported to have antioxidant activity by inhibition of lipid peroxidation using the ferric thiocyanate method, and by scavenging of the free radical 2,2-diphenyl-1-picrylhydrazyl (DPPH). In both assays **7** and **8** showed comparable or slightly higher antioxidative activity than α-tocopherol, L-cysteine and butylated hydroxyanisole (BHA) [33,35]. Vitrofolal E (**7**) has also been shown to have moderate activity in the in vitro cholinesterase inhibition assay [31] and to have tyrosinase inhibitory activity in the same μM-range compared to known inhibitors [32]. Vitrofolal F (**8**) has been shown to have α-chymotrypsin (serine protease) inhibitory effect and has thus been suggested as a specific natural inhibitor of this enzyme [36]. The aryldihydronaphthoic acid derivative, descuraic acid (**9**), was isolated from seeds of *Descurainia Sophia* (L.) Webb ex Prantl, an annual/biennial herb [12]. Despite the structural similarities with vitrofolals, no biological properties of this compound have been published to date, however it is used in traditional medicine to alleviate common cold symptoms and to prevent asthma and oedema. The 9-norlignan glycoside cestrumoside (**10**), was recently identified from methanol extract of *Cestrum diurnum* L. leaves [13]. No biological effects for this compound were reported, although several species of this plant have been used in traditional Chinese medicine, especially for treatment of burns and swellings. However, recently **10** was claimed to be a protein kinase C inhibitor in an animal feed additive [37]. A study of the cytotoxic effects of the arylnaphthalene lignans from Vietnamese Acanthaceae, *Justicia patentiflora* revealed the structure of justiflorinol (**11**). However, the cytotoxic evaluation showed that **11** does not have the same cytotoxic effect as structurally related lignans found in this plant species [14]. Justiflorinol has also been isolated from leaves of *Piper sanguineispicum* and tested for its antileishmanial activity and for cytotoxicity on MCF7 and Vero cells. No significant biological activity was found [38]. (+)-Virgatyne (**12**) has been found in whole plant extracts of the Taiwanese annual plant *Phyllantus virgatus* Frost. F. (Euphorbiaceae) [15]. This plant has traditionally been used to protect the liver and **12** was tested for anti-hepatitis B virus in a MS-G2 cell line, but was shown to be completely inactive [39].

Different liverwort species have been shown to contain dihydronaphthalene and naphthalene lignans and norlignans. The arylnaphthalene derivative **13** was first isolated from *Pellia epiphylla* and the structure was confirmed by chemical synthesis [16]. The same compound (**13**) was later found in *Jamesoniella autumnalis* [40], *Bazzania trilobata* [41], *Lepidozia reptans* [42], *Lepidozia incurvata*, *Chiloscyphus polyanthos* and *Jungermannia exsertifolia* ssp. *cordifolia*. In the last three species conjugates with malic acid, shikimic acid and α-L-rhamnose were also isolated [43], but so far no biological activity has been reported.

The butyrolactone 9-norlignan (**14**) has been isolated from a methanol-water extract of leaves of *Cestrum parqui*, originally a South American shrub, but frequently occurring also in the Mediterranean region. Green cestrum (*Cestrum parqui*) has been shown to have phytotoxic effects and has been studied as a natural herbicide. Compound **14** showed phytotoxic effects on *Lactuca sativa* (lettuce) and *Lycopersicon esculentum* (tomato) by reducing the development of the plant [17].

Compound **15** was first identified in alkaline extraction (pulping) of Norway spruce (*Picea abies*) roots. Later it was also shown to be a degradation product of the abundant guaiacyl type dibenzylbutyrolactone lignan hydroxymatairesinol [18]. The degradation process has been studied in detail and **15** can be obtained by alkaline treatment of hydroxymatairesinol in nearly quantitative yields [5]. The carboxylic acid has not been reported as a constituent in other plants, but the corresponding alcohol ((−)-imperanene (**17**)), although the other enantiomer (+)-imperanene **16**, has been found in the rhizomes of *Imperata cylindrica*. (+)-Imperanene was shown to have platelet aggregation inhibitory effects and gave a complete inhibition, at 6×10^{-4} M concentration, of rabbit platelet aggregation induced by thrombin [1]. (+)-Imperanene, 3-methoxyimperanene (**18**) and their glucosides have also been detected in rum distillate wastewater. In the same report, these compounds were shown to be inhibitors of human tyrosinase activity (IC_{50} = 1.85 mM). (+)-Imperanene showed the highest activity, indicating that the guaiacyl (3-methoxy-4-hydroxyphenyl) substitution pattern is important for the activity [19]. Contrary to many other norlignans, imperanene has been a target for several synthetic works. Noralashinol C (**19**) and noralashinol A (**20**), structurally related to the arylnaphthalene norlignans vitrofolals, have been isolated from stem barks of *Syringa pinnatifolia*. Compound **20** seems to be the acetal of **7** (possibly formed by reaction with methanol). Compound **19** was inactive in a cytotoxicity assay using HepG2 hepatic cancer cells and **20** was inactive in the NO production inhibitory assay [3,4]. Tectonoelin A and B (**21,22**), and compound **14** have been isolated from the leaves of *Tectona grandis* and shown to have growth inhibition activity in the etiolated wheat coleoptiles bioassay test [20]. Interestingly, both green cestrum and teak leaves seem to contain these rare norlignans and display herbicidal effects. Cyclobutane-type norlignans peperoteraphin (**23**) and two related isomers (**24,25**) have been isolated from *Peperomia tetraphylla* (whole plant). Compounds **24** and **25** were tested for cytotoxicity in human liver cancer cell lines HepG2, human lung cancer cell lines A549, and human cervical cancer cell lines Hela. Moderate activity with IC_{50} values around 50 μM for all cell lines, were detected [21,22]. Hyperione A and B (**26,27**) have been isolated from *Hypericum chinense* (whole herb). *Hypericum* species are known for their antibacterial activity but compounds **26** and **27** showed no antibacterial activity [23]. The occurrence and biological activities of 9-norlignans are summarized in Table 1.

Table 1. 9-Norlignans, their occurrence and biological activity.

Compound	Number	Occurrence	Bioactivity	References
Yateresinol	1	*Libocedrus yateensis, Cryptomeria japonica*	Antifungal	[8,26]
Pachypostaudin A	2	*Pachypodanthium staudtii,*		[9,29]
Pachypostaudin B	3	*Pachypodanthium confine*		
Aglacin H	4	*Aglaia cordata*		[10]
Vitrofolal A	5		Antibacterial, antioxidant,	
Vitrofolal B	6	*Vitex rotundifolia, Vitex negundo,*	cholinesterase inhibition, tyrosinase	[2,11,30–34,
Vitrofolal E	7	*Vitex cannabifolia, Syringa pinnatifolia*	inhibitory effects, chymotrypsin	36]
Vitrofolal F	8		inhibitory effects	
Descuraic Acid	9	*Descurainia Sophia*		[12]
Cestrumoside	10	*Cestrum diurnum L.*		[13]
(−)-Justiflorinol	11	*Justicia patentiflora, Piper sanguineispicum*		[14,38]
(+)-Virgatyne	12	*Phyllantus virgatus*		[15]
Arylnaphthalene	13	*Pellia epiphylla, Jamesoniella autumnalis, Bazzania trilobata, Lepidozia reptans, Lepidozia incurvata, Chiloscyphus polyanthos, Jungermannia exsertifolia ssp. cordifolia*		[16,40–43]
Lactone	14	*Cestrum parqui*	Phytotoxic effects	[17]
Compound	15	*Picea abies*		[18]
(+)-Imperanene	16	*Imperata cylindrica, sugarcane rum distillate*	Platelet aggregation inhibitory effect, tyrosinase inhibitory effect	[1,19]
3-Methoxy-Imperanene	18	*Sugarcane rum distillate*	Tyrosinase inhibitory effect	[19]
Noralashinol C	19	*Syringa pinnatifolia*		[3,4]
Noralashinol A	20			
Tectonoelin A	21	*Tectona grandis*	Herbicidal activity	[20]
Tectonoelin B	22			
Peperoteraphin	23			
Cyclobutane carboxylate	24	*Peperomia tetraphylla*	Cytotoxic effect	[21,22]
Cyclobutane carboxylate	25			
Hyperione A	26	*Hypericum chinense*		[23]
Hyperione B	27			

In addition to these compounds, some 9-norlignan-like compounds (although containing additional carbon atoms) have been isolated (Figure 3). Pouzolignan D (**28**) and K (**29**) were isolated from the aerial parts of *Pouzolzia zeylanica* (L.) Benn. var. *microphylla* (wedd.) W.T. Wang [24] and dysosmanorlignans A (**30**) and B (**31**) from the roots of *Dysosma versipellis* [25]. No biological activity has been reported for these compounds.

Several of these naturally occurring lignans have been the topic for total synthesis, however, the synthetic methods are excluded from this literature review. Some semisynthetic methods as well as the synthesis of interesting derivatives related to the presented norlignans will be discussed briefly. For example, the synthesized dehydroxy- and dihydrodehydroxyimperanene isomers have been evaluated for their plant growth regulatory effects, for larvicidal, antifungal, antibacterial and cytotoxic activity, and compared to dihydroguaiaretic acid isomers. In most studies compounds **32–35** were less or equally active as dihydroguaiaretic acids. In the cytotoxicity study, using HL-60 and HeLa cells, compound **35** showed the highest activity with IC$_{50}$ values at approximately 6 μM, suggesting that norlignans may serve as lead compounds for anticancer agents [6].

As previously mentioned, imperanene derivatives can also be synthesized from the natural lignan hydroxymatairesinol and these compounds can be further transformed to vitrofolal and noralashinol type norlignans. To expand the substrate scope for investigation of 9-norlignans and to perform structure–activity relationship studies, we have undertaken the semisynthetic preparation of 15 different 9-norlignans with different functionalities using hydroxymatairesinol as starting material. In this strategy we chose to introduce flexibility or rigidity to the basic skeleton. We also varied the functionality at carbon-9 by the preparation of the carboxylic acid, the ester and the alcohol. In addition, the skeleton was also aromatized by oxidation and two different substitution patterns at the aromatic moieties were prepared. The semisynthesis of these compounds from hydroxymatairesinol is presented in detail below.

2. Results

Semisynthesis of Norlignan Derivatives from Hydroxymatairesinol

The knots of Norway spruce (*Picea abies* (L.) H. Karst) have been shown to contain about 10% (*w*/*w*) of the dibenzylbutyrolactone hydroxymatairesinol (**36**, Scheme 1). Methods for the separation of knots and the subsequent isolation of hydroxymatairesinol were developed during the last decade. Today hydroxymatairesinol is isolated on kg-scale, which certainly makes it one of the most readily available lignans in pure form for further studies. In our previous studies, we have shown that hydroxymatairesinol is degraded to norlignan derivative **15** by strong alkali. This retro-aldol-type reaction probably involves a formation of a quinone methide intermediate and the loss of formaldehyde [5].

During degradation, two of the stereocentra of hydroxymatairesinol are destroyed, but the third retains the R-configuration without isomerization, yielding an optically pure product in excellent yields, usually over 95%. Further esterification and reduction of the obtained product (**37**), by lithium aluminium hydride (LAH) affords the 9-norlignan (−)-imperanene (**17**), as previously reported. However, the structure of **15** offers many possibilities for modification of the structure, which enabled the preparation of a set of derivatives for structure-activity relationship (SAR) studies of their biological effects (Scheme 1). The double bond was shown to be susceptible for protonation and Friedel–Crafts ring closing to yield two diastereomers of the 6′-7-cyclo-9-norlignan (aryltetralin-type) **38** in a 3:4 ratio. The two diastereomers were not separable by column chromatography. Further modification of the carboxylic acid to the ester **39** and the alcohol **40** was performed equally as in the case of **15** [5]. Oxidation of the cyclonorlignan **39** by 2,3-dichloro-5,6-dicyano-1,4-benzoquinone (DDQ) was utilized to aromatize the aliphatic ring to obtain the corresponding arylnaphthalene vitrofolal-type norlignan **41**. This structure was also transformed into the acid **42** and the alcohol **43** by hydrolysis and reduction, respectively. Oxidation of **38** or **40** by DDQ was proven to be unsuccessful. In the case of **38**, the structure was easily decarboxylated during the oxidation. The double bond of **15** was also reduced by catalytic hydrogenation in a batch reactor using ethanol, Pd/C and hydrogen at a slightly elevated pressure (2 bar) to give **44**. Again, the ester **45** and the alcohol **46** were prepared equally as for the other structures (Scheme 1). The permethylated derivative **48** was obtained by methylation with MeI in dry acetone and K$_2$CO$_3$. The acid **47** and the alcohol **49** were prepared accordingly. With these simple chemical modifications, we obtained a set of compounds (derivatives of imperanenes, vitrofolals and noralashinols) with some structural alterations. In this strategy we chose to introduce both flexibility, by reducing the double bond, and rigidity by introducing the aliphatic ring, at the basic skeleton of imperanene. The nonflexible cyclic structures with a half-boat structure **38**–**40** adopted a certain non-planar conformation, and were therefore aromatized to yield a planar structure (compounds **41**–**43**) resembling bioactive 9-norlignans of the vitrofolal and noralashinol family. Finally, we chose to prepare the carboxylic acid with an ionizable group, the ester containing the carbonyl function and the alcohol, for comparison of the effects of the functionality at C-9′. Some of the derivatives were also submitted to methylation of the free phenolic groups for comparison of the 3-methoxy-4-hydroxyphenyl and 3,4-dimethoxyphenyl moieties. The liverwort norlignan **13** was obtained from **42** by treatment with AlCl₃ in pyridine.

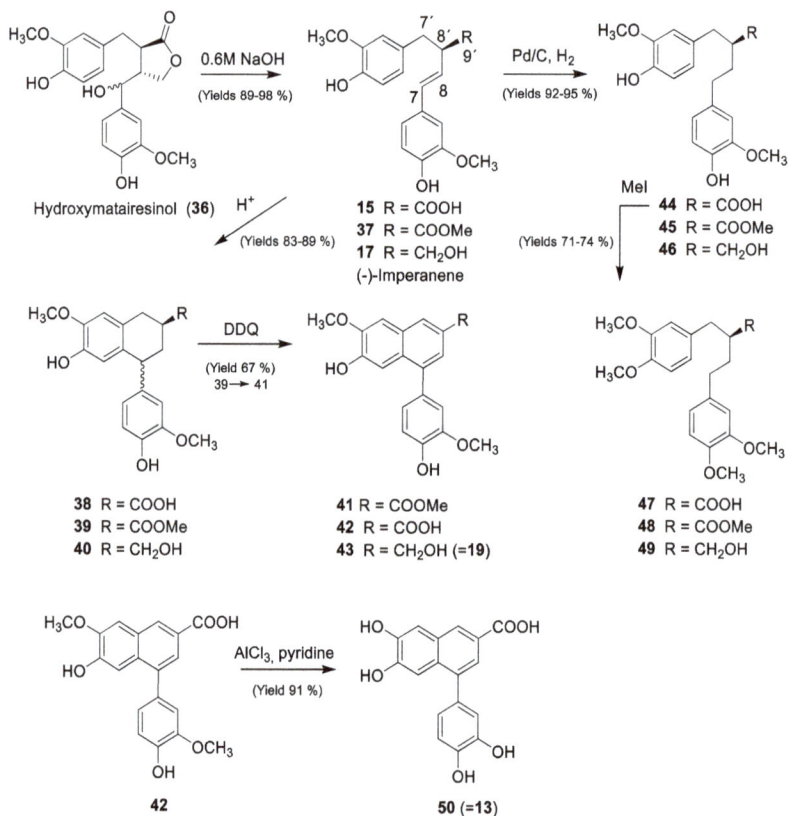

Scheme 1. Preparation of various 9-norlignans from hydroxymataresinol. This generalized scheme shows the fundamental chemical transformations. For more detailed information on the interconversions and transformations including reaction conditions, see the supplementary material.

Compounds **15, 44, 48, 37, 38, 40** and **41** have previously been evaluated as inhibitors of multidrug resistance protein 1 (MRP1)-mediated transport using human erythrocytes as model. This was part of a larger study on polyphenolic compounds comprising lignans, norlignans, stilbenes and flavonoids. Compound **15** was shown to be the most active one, with moderate activity at $IC_{50} = 50$ µM. The structure-activity relationship study showed that the carbonyl function at C-9′ is crucial for the activity. Furthermore, none of the norlignans showed detrimental effects up to 200 µM, indicating a large therapeutic width. These compounds could therefore be interesting as possible agents reversing multidrug resistance and to potentially sensitize cancer cells to anticancer drugs [44]. A similar set of compounds (**15, 38, 40, 41, 44** and **48**) were tested for estrogen and antiestrogen activity in the yeast estrogen assay. Compounds **40, 41** and **48** showed antiestrogenic effects, albeit at quite high concentrations (100–500 µM) [45]. Some of these norlignans were also studied for antimicrobial effects in comparison with the effective stilbene *trans*-pinosylvin (*trans*-3,5-dihydroxystilbene) and its monomethylated derivative (*trans*-3-hydroxy-5-methoxystilbene).

Although some effects were observed, it was concluded that the norlignans were ineffective against *Salmonella infantis*, *Listeria monocytogenes* and *Candida tropicalis* [46].

3. Discussion and Conclusions

Norlignans of the C-9 type are an uncommon group of polyphenols in nature. Many of these compounds can be found in plants used in traditional Asian medicine, however published scientific data of their bioactivities is in general rather scarce. In most cases, their biological effects have been tested in a few assays mainly for antioxidant, cytotoxic and antimicrobial activities. Clearly, a more broad and systematic screening of bioactivity should be conducted to evaluate these compounds as potential bioactive agents. 9-Norlignans belonging to the vitrofolal-, noralashinol-, and imperanene families have however been shown to be bioactive and some of them could find potential in treatment of cancer. Furthermore, these 9-norlignans have been shown to inhibit specific enzymes, which warrants further investigations of these compounds. In our semisynthetic approach using hydroxymatairesinol as starting material for the preparation of 9-norlignans, we have developed simple chemical transformations to unsaturated, saturated, cyclic and aromatic structures closely resembling those of the vitrofolal, noralashinol and imperanene family. The unsaturated compounds **15**, **17** and **37** could also be seen as polyphenolic compounds with both *trans*-stilbene-like, and lignan-like structures, which could make them interesting as antimicrobial compounds. The cyclic structures were obtained in high yields by treatment with formic acid or trifluoroacetic acid. However, this Friedel–Crafts-type reaction can also be performed in aqueous conditions, using mineral acids. The aromatization was introduced by reaction with DDQ with moderate yields. DDQ was superior for this reaction due to its benzylic hydride abstraction properties, yielding dehydrogenation of the product, rather than oxidative polymerization. Hydrogenation of the double bond was facile, proceeding in high yields. To broaden the substrate scope of the semisynthetic 9-norlignans, the initial carboxylic acid (of compound **15**) was esterified and reduced by conventional methods, in relatively high yields. By these interconversions the carboxylic acid, the methyl ester and the alcohol derivatives were obtained for all the basic structures. Some structures were also methylated at the phenolic positions, which gave us 16 norlignans with different structures available for screening of biological activity. This semisynthetic route is not restricted to these compounds and it offers numerous additional chemical modifications and functional group interconversions. The biological testing of some selected semisynthetic 9-norlignans showed that none of these compounds were highly active. However, compound **15** showed moderate activity as an inhibitor of MRP1-mediated transport. The activity was decreased for all other derivatives, which indicated that the double bond and the carboxylic acid function were important for the activity. Due to the limited scope of biotesting for these semisynthetic 9-norlignans, no generalized results of their bioactivity or results on structure-activity relationships can be proposed, without further screening of biological activity.

4. Materials and Methods

All commercially available chemicals were used as supplied by the manufacturers. Hydroxymatairesinol was isolated from Norway spruce (*Picea abies* (L.) Karst) knots by previously described methods [5]. Knots of Norway spruce were separated, ground and freeze-dried prior to extraction in a Soxhlet apparatus. The raw extract obtained with acetone-water (9:1 *v/v*), after the removal of lipophilic extractives with hexane was purified by flash chromatography (eluent CH_2Cl_2:EtOH 98:2 *v/v*) to yield hydroxymatairesinol.

GC analyses were performed on a standard gas chromatograph (Agilent Technologies, model 6850, Santa Clara, CA, USA) equipped with a HP-5 column and a flame ionization (FI) detector. The samples were silylated using hexamethyldisilazane-chlorotrimethylsilane in pyridine, prior to analyses. Gas chromatography-mass spectrometry (GCMS) analyses were performed essentially the same way (Agilent Technologies, model 7890A+5975C). High-resolution mass spectrometry (HRMS) were recorded using a Micro Q-TOF (Bruker, Billerica, MA, USA) with electrospray ionization (ESI) operated in positive mode or with a ZAB-Spec high-resolution MS-ESI instrument (Fisons Instruments, Ipswich, UK).

^{1}H- and ^{13}C-NMR spectra were recorded on an Avance instrument (Bruker) at 600.13 and 150.90 MHz, respectively. 2D NMR experiments (^{1}H-^{1}H COSY, HSQC, HMBC) were recorded using standard pulse sequences and chemical shifts are reported downfield from tetramethylsilane. Optical rotations were measured with a digital polarimeter model 241 (Perkin Elmer, Waltham, MA, USA) using a 1 dm, 1 mL cell.

The inhibition of multidrug resistance protein 1-mediated transport was studied by measurement of 2′,7′-bis-(carboxypropyl)-5(6)-carboxyfluorescein (BCPCF) efflux in human erythrocytes. Erythrocytes were loaded with 2′,7′-bis-(carboxypropyl)-5-(6)-carboxyfluorescein acetoxymethyl ester (BCPCF-AM) and incubated with or without the norlignans (10–200 mM). The extracellular BCPCF fluorescence intensity was then measured and the IC$_{50}$ values for BCPCF efflux were determined from dose-response curves. The detailed experimental procedure and results have previously been published [44]. The estrogen and antiestrogen activity was determined using the yeast estrogen assay consisting of a transformed yeast strain (*Saccharomyces cerevisiae*). Cells were pre-cultured and diluted in a medium containing chlorophenol red-β-D-galactopyranoside. The suspensions with compounds were incubated in 96-well plates for three days before the absorbance at 540 nm was measured. Estrogenic or anti-estrogenic activity was calculated by subtraction of absorbance at 620 nm from that of 540 nm. Dose curves were plotted and the IC$_{50}$ values were calculated. 17β-Estradiol was used as reference. A more detailed description has been published by Aehle et al. [45].

The antimicrobial activity tests were performed with *Listeria monocytogenes* L211, *Salmonella infantis* EELA72 and *Candida tropicalis* 4068 using an automated incubator and turbidity reader to monitor the microbial growth. The compounds were studied at equimolar concentrations (0.1–1 mM). The growth was calculated by subtraction of the turbidity of the test culture from that of the control and expressed as growth inhibition percentages. The detailed experimental procedure and results have previously been published [46].

Supplementary Materials: The following are available online. Full experimental procedures for compounds **36–50**, including analytical data.

Author Contributions: Conceptualization: P.E., methodology: P.E. and J.-E.R., formal analysis: P.E. and J.-E.R., investigation: P.E. and J.-E.R., resources: P.E., writing—original draft preparation: P.E., writing—review and editing: P.E. and J.-E.R., visualization: P.E., supervision: P.E., project administration: P.E., funding acquisition: P.E.

Funding: This research was part of the BIOSTIMUL (Bioactive and wood-associated stilbenes as multifunctional antimicrobial and health promoting agents) project. Financial support from the Finnish Funding Agency for Technology (Tekes), from the Magnus Ehrnrooth foundation and from the Liv och Hälsa foundation is gratefully acknowledged.

Acknowledgments: Annika Smeds is acknowledged for the HRMS analyses and Jan-Erik Lönnqvist for help with the syntheses.

Conflicts of Interest: The authors declare no conflicts of interest.

References

1. Matsunaga, K.; Shibya, M.; Ohizumi, Y. Imperanene, a novel phenolic compound with platelet aggregation inhibitory activity from *Imperata cylindrica*. *J. Nat. Prod.* **1995**, *58*, 138–139. [CrossRef]
2. Kawazoe, K.; Yutani, A.; Tamemoto, K.; Yuasa, S.; Shibata, H.; Higuti, T.; Takaishi, Y. Phenylnaphthalene Compounds from the Subterranean Part of *Vitex rotundifolia* and Their Antibacterial Activity Against Methicillin-Resistant *Staphylococcus aureus*. *J. Nat. Prod.* **2001**, *64*, 588–591. [CrossRef] [PubMed]
3. Su, G.; Bai, R.; Yu, X.; Cao, Y.; Yin, X.; Tu, P.; Chai, X. Noralashinol A, a new norlignan from stem barks of *Syringa pinnatifolia*. *Nat. Prod. Res.* **2016**, *19*, 2149–2153. [CrossRef] [PubMed]
4. Zhang, R.F.; Feng, X.; Su, G.Z.; Yin, X.; Yang, X.Y.; Zhao, Y.F.; Li, W.F.; Tu, P.F.; Chai, X.Y. Noralashinol B, a norlignan with cytotoxicity from stem barks of *Syringa pinnatifolia*. *J. Asian Nat. Prod. Res.* **2017**, *19*, 416–422. [CrossRef] [PubMed]

5. Eklund, P.; Riska, A.; Sjöholm, R. Synthesis of *R*-(−)-Imperanene from the Natural Lignan Hydroxymatairesinol. *J. Org. Chem.* **2002**, *67*, 7544–7546. [CrossRef] [PubMed]

6. Yamauchi, S.; Tanimura, R.; Nishiwaki, H.; Nishi, K.; Sugahara, T.; Maruyama, M.; Ano, Y.; Akiyama, K.; Kishida, T. Enantioselective syntheses of both enantiomers of 9′-dehydroxyimperanene and 7,8-dihydro-9′-dehydroxyimperanene and the comparison of biological activity between 9-norlignans and dihydroguaiaiaretic acids. *Bioorg. Med. Chem. Lett.* **2016**, *26*, 3019–3023. [CrossRef] [PubMed]

7. Moss, G.P. Nomenclature of Lignans and Neolignans (IUPAC Recommendations 2000). *Pure Appl. Chem.* **2000**, *72*, 1493–1523. [CrossRef]

8. Erdtman, H.; Harmata, J. Phenolic and Terpenoid Heartwood Constituents of *Libocedrus Yateensis*. *Phytochemistry* **1979**, *18*, 1495–1500. [CrossRef]

9. Ngadjui, B.T.; Lontsi, D.; Ayafor, J.F.; Sondengam, B.L. Pachypophyllin and pachypostaudins A and B: Three bisnorlignans from *Pachypodantum staudtii*. *Phytochemistry* **1989**, *28*, 231–234. [CrossRef]

10. Wang, B.-G.; Ebel, R.; Wang, C.-Y.; Wray, V.; Proksch, P. New methoxylated aryltetrahydronaphthalene lignans and a norlignan from *Aglaia cordata*. *Tetrahedron Lett.* **2002**, *43*, 5783–5787. [CrossRef]

11. Kawazoe, K.; Yutani, A.; Takaishi, Y. Aryl naphthalenes norlignans from *Vitex rotundifolia*. *Phytochemistry* **1999**, *52*, 1657–1659. [CrossRef]

12. Sun, K.; Li, X.; Li, W.; Liu, J.-M.; Wang, J.-H.; Yi, S. A new nor-lignan from the seeds of *Descurainia sophia*. *Nat. Prod. Res.* **2006**, *20*, 519–522. [CrossRef] [PubMed]

13. Mohamed, K.M.; Fouad, M.A.; Matsunami, K.; Kamel, M.S.; Otsuka, H. A new norlignan glycoside from *Cestrum diurnum* L. *Arkivoc* **2007**, *4*, 63–70. [CrossRef]

14. Gasplagas, G.; Van Hung, N.; Dignon, J.; Tholson, O.; Kruczynski, A.; Sévenet, T.; Guéritte, F. Cytotoxic Arylnaphthalene Lignans from a Vietnamese Acanthaceae, *Justicia patentiflora*. *J. Nat. Prod.* **2005**, *68*, 735–738. [CrossRef] [PubMed]

15. Huang, Y.-L.; Chen, C.-C.; Hsu, F.-L.; Chen, C.-F. Tannins, Flavonol Sulfonates, and a Norlignan from *Phyllantus virgatus*. *J. Nat. Prod.* **1998**, *61*, 1194–1197. [CrossRef] [PubMed]

16. Rishmann, M.; Mues, R.; Geiger, H.; Laas, H.J.; Eicher, T. Isolation and synthesis of 6,7-dihydroxy-4-(3,4-dihydroxyphenyl)naphthalene-2-carboxylic acid from *Pellia epiphylla*. *Phytochemistry* **1989**, *28*, 867–869. [CrossRef]

17. D'Abrosca, B.; Dellagreca, M.; Fiorentino, A.; Golino, A.; Minaco, P.; Zarrelli, A. Isolation and characterization of new lignans from the leaves of *Cestrum parqui*. *Nat. Prod. Res.* **2006**, *20*, 293–298. [CrossRef]

18. Ekman, R.; Sjöholm, R.T.; Sjöholm, R. A Degraded Lignan from Alkaline Hydrolysis of Norway Spruce Root Extractives. *Finn. Chem. Lett.* **1979**, *4*, 126–128.

19. Takara, K.; Iwasaki, H.; Ujihara, K.; Wada, K. Human Tyrosinase Inhibitor in Rum Distillate Wastewater. *J. Oleo Sci.* **2008**, *57*, 191–196. [CrossRef]

20. Lacret, R.; Varela, R.M.; Molinillo, J.M.G.; Nogueiras, C.; Macías, F.A. Tectonoelins, new norlignans from a bioactive extract of *Tectona grandis*. *Phytochem. Lett.* **2012**, *5*, 382–385. [CrossRef]

21. Li, Y.-Z.; Huang, J.; Gong, Z.; Tian, Z.-Q. A Novel Norlignan and a Novel Phenylpropanoid from *Peperomia tetraphylla*. *Helv. Chim. Acta* **2007**, *90*, 2222–2226. [CrossRef]

22. Li, Y.-Z.; Tong, A.P.; Huang, J. Two New Norlignans and a New Lignanamide from *Peperomia tetraphylla*. *Chem. Biodivers.* **2012**, *9*, 769–776. [CrossRef] [PubMed]

23. Wang, W.; Zeng, Y.H.; Osman, K.; Shinde, K.; Rahman, M.; Gibbons, S.; Mu, Q. Norlignans, Acylphloroglucinols, and a Dimeric Xanthone from *Hypericum chinense*. *J. Nat. Prod.* **2010**, *73*, 1815–1820. [CrossRef] [PubMed]

24. Zhong, C.-Q.; Tao, S.-H.; Yi, Z.-B.; Guo, L.-B.; Xie, Y.-F.; Chen, Y.-F. Four New Compounds from *Pouzolzia zeylanica* (L.) Benn. Var. *Microphylla*. *Heterocycles* **2015**, *91*, 1926–1936. [CrossRef]

25. Zheng, Y.; Xie, Y.-G.; Zhang, Y.; Li, T.; Li, H.-L.; Yan, S.-K.; Jin, H.-Z.; Zhang, W.-D. New norlignans and flavonoids of *Dysosma versipellis*. *Phytochem. Lett.* **2016**, *16*, 75–81. [CrossRef]

26. Takahashi, K.; Ogiyama, K. Phenols of discolored sugi (*Cryptomeria japonica* D. Don) sapwood II. Norlignans of discolored sugi sapwood collected in the Kyushu region. *Mokuzai Gakkaishi* **1985**, *31*, 28–38.

27. Noguchi, T.; Ohtani, Y.; Sameshima, K. Static defense components for sugi butt-rot disease. *Trans. Mater. Res. Soc. Jpn.* **2004**, *29*, 2479–2482.

28. Chen, T.-H.; Liau, B.-C.; Wang, S.-Y.; Jong, T.-T. Isolation and cytotoxicity of the lignanoids from *Chamaecyparis formosensis*. *Planta Med.* **2008**, *74*, 1806–1811. [CrossRef]

29. Mathouet, H.; Elomri, A.; Lameiras, P.; Daich, A.; Vérité, P. An alkaloid, two conjugate sesquiterpenes and a phenylpropanoid from *Pachypodanthium confine* Engl. and Diels. *Phytochemistry* **2007**, *68*, 1813–1818. [CrossRef]

30. Lou, Z.-H.; Li, H.-M.; Gao, L.-H.; Li, R.-T. Antioxidant lignans from the seeds of *Vitex negundo* var. *cannabifolia*. *J. Asian Nat. Prod. Res.* **2014**, *16*, 963–969. [CrossRef] [PubMed]

31. Haq, A.-U.; Malik, A.; Anis, I.; Khan, S.B.; Ahmed, E.; Ahmed, Z.; Nawaz, S.A.; Choudhary, M.I. Enzymes Inhibiting Lignans from *Vitex negundo*. *Chem. Pharm. Bull.* **2004**, *52*, 1269–1272.

32. Haq, A.-U.; Malik, A.; Khan, M.T.H.; Haq, A.-U.; Khan, S.B.; Ahmad, A.; Choudhary, M.I. Tyrosinase inhibitory lignans from the methanol extract of the roots of *Vitex negundo* Linn. And their structure-activity relationship. *Phytomedicine* **2006**, *13*, 255–260. [CrossRef]

33. Yamasaki, T.; Kawabata, T.; Masuoka, C.; Kinjo, J.; Ikeda, T.; Nohara, T.; Ono, M. Two new lignan glucosides from the fruit of *Vitex cannabifolia*. *J. Nat. Med.* **2008**, *62*, 47–51. [CrossRef] [PubMed]

34. Wang, Q.-H.; Huo, S.R.N.; Bao, Y.P.; Ao, W.L.J. Chemical constituents of *Syringa pinnatifolia* and its chemotaxonomic study. *Chem. Nat. Compd.* **2018**, *54*, 435–438. [CrossRef]

35. Ono, M.; Nishida, Y.; Masuoka, C.; Li, J.-C.; Okawa, M.; Ikeda, T.; Nohara, T. Lignan Derivatives and a Norditerpene from the seeds of *Vitex negundo*. *J. Nat. Prod.* **2004**, *67*, 2073–2075. [CrossRef] [PubMed]

36. Lodhi, M.A.; Haq, A.U.; Choudhary, M.I.; Malik, A.; Ahmad, S. Chymotrypsin inhibition studies on the lignans from *Vitex negundo* Linn. *J. Enzyme Inhib. Med. Chem.* **2008**, *23*, 400–405. [CrossRef] [PubMed]

37. Neufeld, K.; Neufeld, N. Animal Feed Additive Containing Diurnoside and/or Cestrumoside. PCT Int. Appl. WO 2017129732 A1 20170803, 3 August 2017.

38. Cabanillas, B.J.; Le Lamer, A.-C.; Castillo, D.; Arevalo, J.; Rojas, R.; Odonne, G.; Bourdy, G.; Moukarzel, B.; Sauvain, M.; Fabre, N. Caffeic Acid Esters and Lignans from *Piper sanguineispicum*. *J. Nat. Prod.* **2010**, *73*, 1884–1890. [CrossRef]

39. Huang, R.L.; Huang, Y.L.; Ou, J.C.; Chen, C.C.; Hsu, F.L.; Chang, C. Screening of 25 Compounds isolated from *Phyllantus* Species for Anti-Human Hepatitis B Virus In Vitro. *Phytother. Res.* **2003**, *17*, 449–453. [CrossRef]

40. Tazaki, H.; Adam, K.-P.; Becker, H. Five lignan derivatives from in vitro cultures of liverwort *Jamesoniella autumnalis*. *Phytochemistry* **1995**, *40*, 1671–1675. [CrossRef]

41. Martini, U.; Zapp, J.; Becker, H. Lignans from the liverwort *Bazzania Trilobata*. *Phytochemistry* **1998**, *49*, 1139–1146. [CrossRef]

42. Sanderson, S. Phytochemische Untersuchungen der Ledermoose *Jamesoniella rubricaulis* (Nees) Grolle und *Lepidozia reptans* (L.) Dum. Ph.D. Thesis, Universität des Saarlandes, Saarbrucken, Germany, 1995.

43. Cullman, F.; Schmidt, A.; Schuld, F.; Trennheuser, M.L.; Becker, H. Lignans from the liverworts *Lepidozia incurvata*, *Chiloscyphus polyanthos* and *Jungermannia exsertifolia* ssp. *cordifolia*. *Phytochemistry* **1999**, *52*, 1647–1650. [CrossRef]

44. Wróbel, A.; Eklund, P.; Bobrowska-Hägerstrand, M.; Hägerstrand, H. Lignans and norlignans Inhibit Multidrug Resistance Protein 1(MRP1/ABCC1)-mediated Transport. *Anticancer Res.* **2010**, *30*, 4423–4428. [PubMed]

45. Aehle, E.; Müller, U.; Eklund, P.C.; Willför, S.M.; Sippl, W.; Dräger, B. Lignans as food constituents with estrogen and antiestrogen activity. *Phytochemistry* **2011**, *72*, 2396–2405. [CrossRef] [PubMed]

46. Plumed-Ferrer, C.; Väkeväinen, K.; Komulainen, H.; Rautiainen, M.; Smeds, A.; Raitanen, J.-E.; Eklund, P.; Willför, S.; Alakomi, H.-L.; Saarela, M.; et al. The antimicrobial effects of wood-associated polyphenols on food pathogens and spoilage organisms. *Int. J. Food Microbiol.* **2013**, *164*, 99–107. [CrossRef] [PubMed]

molecules

MDPI

Review

Advances in the Synthesis of Lignan Natural Products

Xianhe Fang and Xiangdong Hu * [iD]

Key Laboratory of Synthetic and Natural Functional Molecule Chemistry of Ministry of Education, College of Chemistry & Materials Science, Northwest University, Xi'an 710127, China; xianhefang@stumail.nwu.edu.cn
* Correspondence: xiangdonghu@nwu.edu.cn; Tel.: +86-029-8153-5025; Fax: +86-029-8153-5026

Academic Editor: David Barker
Received: 7 November 2018; Accepted: 18 December 2018; Published: 19 December 2018

Abstract: Lignans comprise a family of secondary metabolites existing widely in plants and also in human food sources. As important components, these compounds play remarkable roles in plants' ecological functions as protection against herbivores and microorganisms. Meanwhile, foods rich in lignans have revealed potential to decrease of risk of cancers. To date, a number of promising bioactivities have been found for lignan natural products and their unnatural analogues, including antibacterial, antiviral, antitumor, antiplatelet, phosphodiesterase inhibition, 5-lipoxygenase inhibition, HIV reverse transcription inhibition, cytotoxic activities, antioxidant activities, immunosuppressive activities and antiasthmatic activities. Therefore, the synthesis of this family and also their analogues have attracted widespread interest from the synthetic organic chemistry community. Herein, we outline advances in the synthesis of lignan natural products in the last decade.

Keywords: natural products; total synthesis; lignan

1. Introduction

Lignans are a family of secondary metabolites widely distributed in plants and human food sources. The story of lignans can traced back to 1942, when Harworth introduced the term for the first time to describe this family [1]. It is known that lignans have remarkable ecological functions in plants, providing protection against herbivores and microorganisms [2–7]. The consumption of foods rich in lignans has potential to decrease of risk of cancers [8–11]. During its long research history, this family has exhibited attractive pharmacological activities [12–19], such as antibacterial [20], antiviral [21–24], antitumor [25–27], antiplatelet [28,29], phosphodiesterase inhibition [30,31], 5-lipoxygenase inhibition [32–34], HIV reverse transcription inhibition [35–37], cytotoxic [38], antioxidant [39], immunosuppressive [40] and antiasthmatic properties [31].

Lignan compounds have dimeric structures formed through a β,β′-linkage between two phenylpropane units with different degrees of oxidation on the side-chain and variable substitution patterns on the phenyl ring. Traditionally, lignans are divided into two classes: classical lignans and neolignans. It should be noted that the term lignan in the literature refers to classical lignans in most cases. Regarding the classification of classical lignans, four different types are reported. The first one arranged classical lignans into three subgroups: acyclic lignan derivatives, arylnaphthalene derivatives and dibenzocyclooctadiene derivatives [41]. The second type includes six subgroups: dibenzylbutanes, dibenzylbutyrolactones, arylnaphthalenes, dibenzocyclooctadienes, substituted tetrahydrofurans and 2,6-diarylfurofurans [9,14]. The third one is comprised of eight subgroups: furofurans, furans, dibenzylbutanes, dibenzylbutyrolactones, aryltetralins, arylnaphthalenes, dibenzocyclooctadienes and dibenzylbutyrolactols [8,42–45]. The fourth one includes seven subgroups of lignan scaffolds: cyclobutanes, tetrahydrofurans, furofurans, dibenzylbutanes, aryltetralins, cycloheptenes and dibenzocyclooctadienes [46].

The synthesis of lignans and their analogues is an active field in the synthetic organic chemistry community. Tremendous synthetic efforts on this family have been well documented by reviews [9–14,41,47–51]. In recent years, several nice reviews have outlined progress of particular topics related to the synthesis of furofuran lignans [48], arylnaphthalene lactone analogues [47] and aryltetralin glycosides [49]. The present review will focus on the papers on the synthesis of lignans published from 2008–2018. In order to avoid unnecessary duplication, we will not discuss works already presented in previous reviews.

For the convenience of introduction of advances in the synthesis of lignans, we discuss three subgroups in present review, namely, acyclic lignan derivatives, dibenzocyclooctadiene derivatives and arylnaphthalene derivatives.

2. Advances in the Synthesis of Acyclic Lignan Derivatives

In the last decade, synthetic progress in acyclic lignan derivatives is related to lignans featuring dibenzyl tetrahydrofuran, dibenzylbutyrolactone, and diphenyltetrahydrofuranfurofuran skeletons.

The synthesis of the acyclic lignan derivative (±)-paulownin (Scheme 1) was accomplished by Angle and coworkers in 2008 [52]. The key step is a formal [3 + 2]-cycloaddition between silyl ether **1** and aldehyde **2** in the presence of BF$_3$OEt$_2$ and 2,6-di-*tert*-butyl-4-methylphenol (DBMP), generating aryl tetrahydrofuran **3**. After oxidation and removal of the protecting group, the resulting product **4** was connected with imidate **5**, generating lactone **6**. The synthesis of (±)-paulownin was finished through photocyclization under a medium-pressure Hanovia lamp [53].

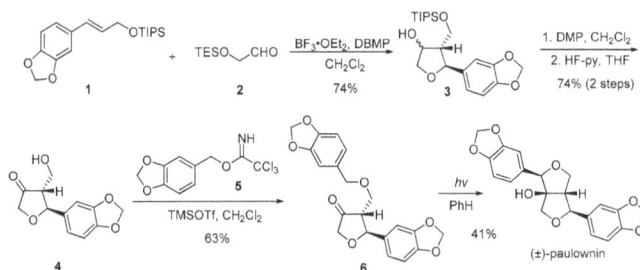

Scheme 1. Synthesis of (±)-paulownin. Adapted from Angle et al. [52].

In 2011, Barker and coworkers reported the total synthesis of (+)-galbelgin (Scheme 2) [54]. A stereoselective aza-Claisen rearrangement developed in their lab [55] afforded a reliable access to the original two stereocenters in chiral amide **8**. The subsequent nucleophilic addition from **11**, reduction, hydroxyl protection and double bond oxidative cleavage led to the formation of aldehyde **10**. The second nucleophilic addition from **11** afforded **12** with four adjacent stereocenters established. Methoxymethyl (MOM) group deprotection and cyclization completed the synthesis of (+)-galbelgin.

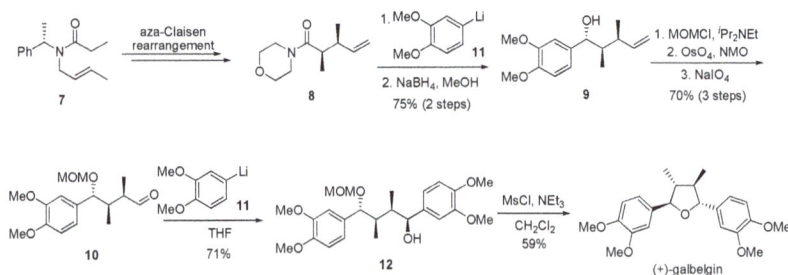

Scheme 2. Synthesis of (+)-galbelgin. Adapted from Barker et al. [54].

She and coworkers reported the total synthesis of beilschmin A and gymnothelignan N in 2014 (Scheme 3) [56]. Alcohol **14** was prepared by hydroxyl protection of chiral amide **13** and subsequent reduction. Aldehyde **16** was obtained after homologation and reduction. The nucleophilic addition of **17** and oxidation afforded ketone **18**. Dibenzyl tetrahydrofurans **20** was obtained after a highly stereoselective introduction of a methyl group, deprotection and reduction. The synthesis of beilschmin A was finished after the methylation. Inspired by a biosynthetic proposal from She's group, the challenging seven-membered ring skeleton in gymnothelignan N was constructed by an oxidative Friedel-Crafts reaction of compound **20** using phenyliodonium diacetate (PIDA) as the oxidant, finally affording gymnothelignan N.

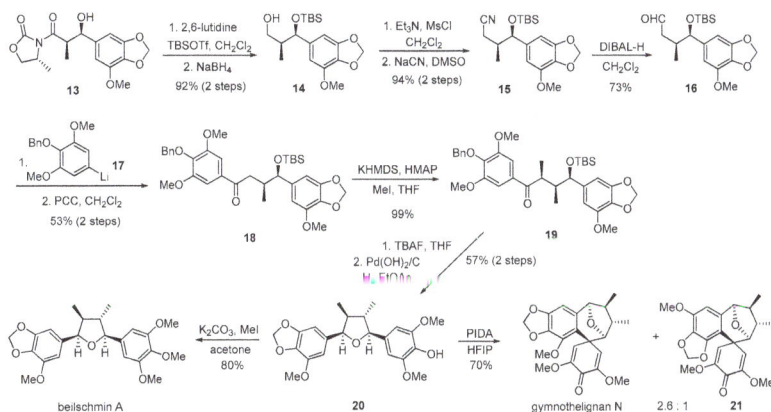

Scheme 3. Synthesis of beilschmin A and gymnothelignan N. Adapted from She et al. [56].

In 2015, Lump and coworkers reported a bioinspired total synthesis of (±)-tanegool and (±)-pinoresinol (Scheme 4) using [2 + 2] photodimerization and oxidative ring-opening as key steps [46]. The synthesis started with the esterification of ferulic acid. The resulting product **22** went through [2 + 2] photodimerization and reduction, generating diol **23** smoothly. The synthesis of (±)-tanegool through the expected oxidative ring-opening of **23** was achieved under different oxidative conditions. Moreover, synthesis of (±)-pinoresinol was also accomplished. Using the same strategy, *trans*-diester **26** was prepared from *cis*-diester **25**. Reduced product **27** was submitted to an oxidative ring-opening treatment using FeCl$_3$·6H$_2$O as the oxidant, finishing the synthesis of (±)-pinoresinol through an oxidative ring opening and two 5-*exo*-trig cyclization pathway.

As powerful synthetic tools, photoredox-catalyzed tranformations have received considerable attention in recent decades [57–59]. In 2015, MacMillian and coworkers developed an enantioselective α-alkylation of aldehydes using a combination of photoredox catalysis and enamine catalysis and achieved the asymmetric synthesis of (−)-bursehernin through this strategy (Scheme 5) [60]. Using Ru(bpy)$_3$Cl$_2$, chiral amine **33** and a compact fluorescent lamp (CFL) light source, the α-alkylation of aldehyde **28** with bromonitrile **29** generated chiral aldehyde **30** in excellent yields and excellent enantioselectivity. Subsequent reduction and cyclization afforded lactone **31**. The synthesis of (−)-bursehernin was achieved by a highly stereoselective alkylation between **31** and bromide **32**.

In 2017, Soorukram and coworkers reported the asymmetric synthesis of *ent*-fragransin C$_1$ (Scheme 6) [61]. Ketone **36** was produced by the nucleophilic addition of the aryllithium species generated from **35** to chiral Weinreb amide **34**. The following stereoselective reduction led to the formation of alcohol **37**. After hydroxyl protection, double bond oxidative cleavage and nucleophilic addition from aryllithium **39**, compound **40** was furnished in good diastereoselectivity. Followed by the deprotection and cyclization treatments, the tetrahydrofuran ring in **41** was established. Finally, the synthesis of *ent*-fragransin C$_1$ was accomplished through debenzylation under hydrogenation conditions.

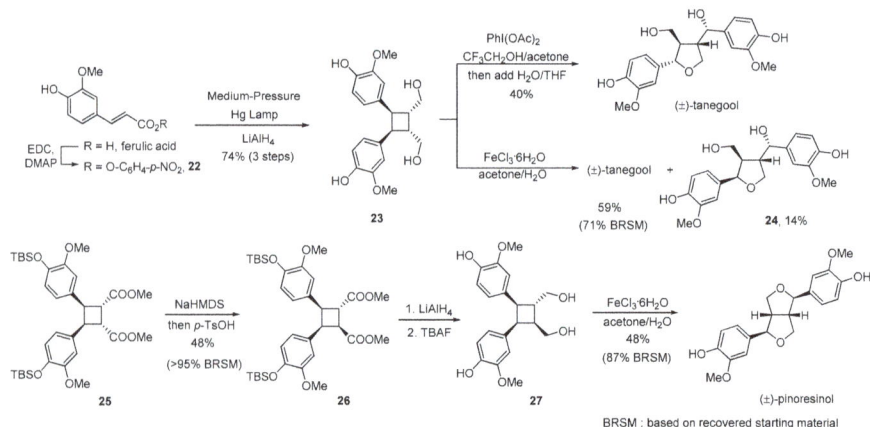

Scheme 4. Bioinspired synthesis of (±)-tanegool and (±)-pinoresinol. Adapted from Lump et al. [46].

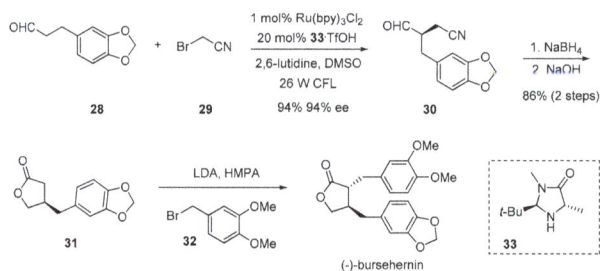

Scheme 5. Asymmetric synthesis of (−)-bursehernin. Adapted from MacMillian et al. [60].

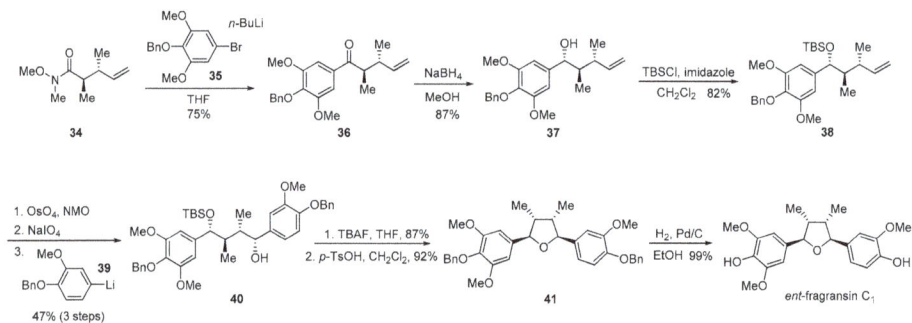

Scheme 6. Asymmetric synthesis of *ent*-fragransin C_1. Adapted from Soorukram et al. [61].

Based on a tandem nucleophilic addition/Ru-catalyzed isomerization/SET oxidation/radical dimerization strategy [62], Jahn and coworkers reported a bioinspired total synthesis of multiple lignans in 2018 (Scheme 7) [63]. Using bromide **42** as the substrate, a smooth unprecedented tandem 1,2-nucleophilic addition/Ru-catalyzed isomerization/SET oxidation/radical dimerization afforded 1,4-diketone **43** with acceptable diastereoselectivity. After the reduction, three different treatments of **43** led to the formation of **44** and **45** in varied ratios. With the removal of double *t*-butyldimethylsilyl (TBS) protecting groups, the synthesis of (±)-fragransin A_2 and (±)-odoratisol was achieved. Through the same strategy, Jahn and coworkers completed the synthesis of (±)-galbelgin, (±)-grandisin, (±)-galbacin, (±)-veraguensin, and (±)-beilschmin B.

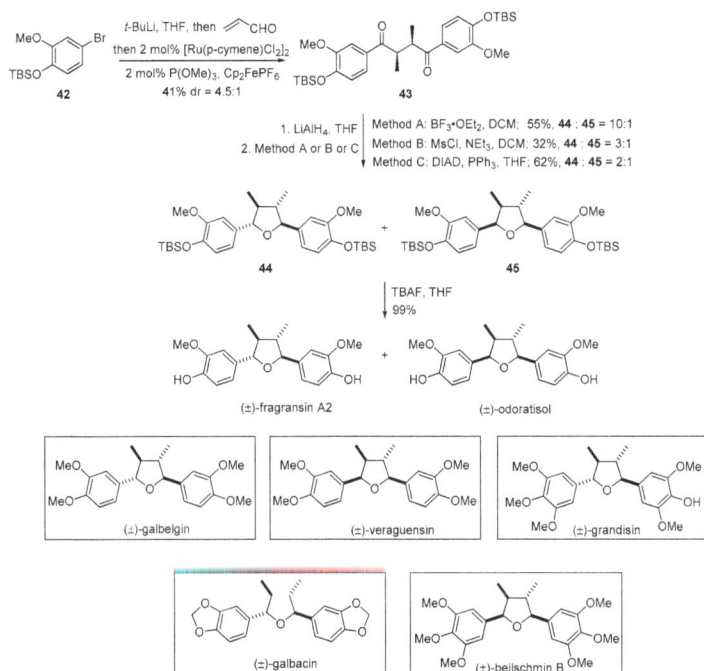

Scheme 7. Bioinspired synthesis of seven acyclic lignans. Adapted from Jahn et al. [63].

A Ni-catalyzed cyclization/cross-coupling strategy was developed and applied for the synthesis of (±)-kusunokinin, (±)-dimethylmetairesinol, (±)-bursehernin and (±)-yatein by Giri and coworkers in 2018 (Scheme 8) [64]. Ligand **48** was used for the Ni-catalyzed cyclization/cross-coupling between iodide **46** and aryl zinc reagent **47** followed by Jones oxidation, generating lactone **49** readily. Compound **49** was then connected with bromide **50** in a good diastereoselective manner, completing the synthesis of (±)-kusunokinin. The syntheses of (±)-dimethylmetairesinol, (±)-bursehernin and (±)-yatein were accomplished using the same protocol.

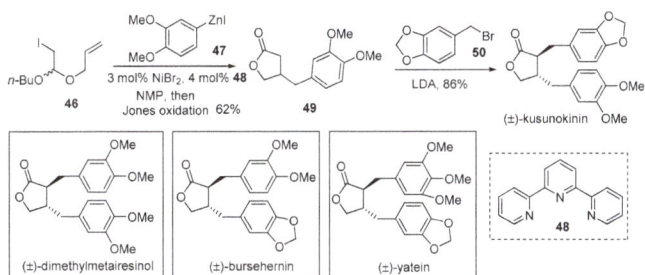

Scheme 8. Synthesis of four acyclic lignans. Adapted from Giri et al. [64].

3. Advances in Synthesis of Dibenzocyclooctadiene Serivatives

Dibenzocyclooctadiene derivative lignans feature a particular eight-membered ring containing a chiral biaryl axis. Members of this subgroup possess various substitution patterns with two aryl rings and different stereocenters on the aliphatic bridge.

In 2010, an interesting Ni-catalyzed enantioselective Ullmann coupling of bis-*ortho*-substituted arylhalides was developed and applied to the asymmetric synthesis of (+)-isoschizandrin by Lin and coworkers (Scheme 9) [65]. With the application of chiral ligand **56**, the Ni-catalyzed enantioselective Ullmann coupling of bromide **51** gave the axial chiral biaryl dial **52** with acceptable enantioselectivity. Aldehyde **55** was prepared after monoprotection, Wittig reaction and deprotection operations. The synthesis of (+)-isoschizandrin was accomplished according to Molander's cyclization protocol [66].

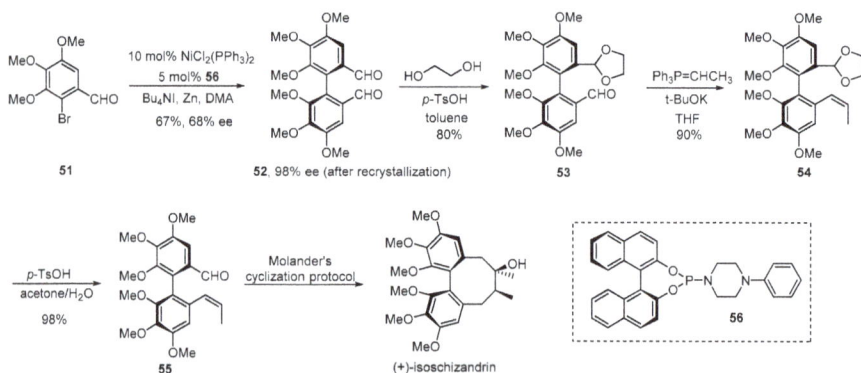

Scheme 9. Asymmetric synthesis of (+)-isoschizandrin. Adapted from Lin and Xu et al. [65].

Based on a double organocuprate oxidation strategy, Spring and coworkers reported the total synthesis of (±)-deoxyschizandrin in 2012 (Scheme 10) [67]. Symmetrical 1,3-diene **58** was prepared by the homo-coupling of alkenyl iodide **57** through a mild metalation, magnesio-cuprate transmetalation and subsequent oxidation using **61** as the oxidant [68]. Subsequent hydrogenation afforded **59** as a mixture of two diastereoisomers. After iodination, the expected iodide **60** was obtained. The synthesis of (±)-deoxyschizandrin was completed by an intramolecular organocuprate oxidation process, including metalation, magnesio-cuprate transmetalation and oxidation with **61**.

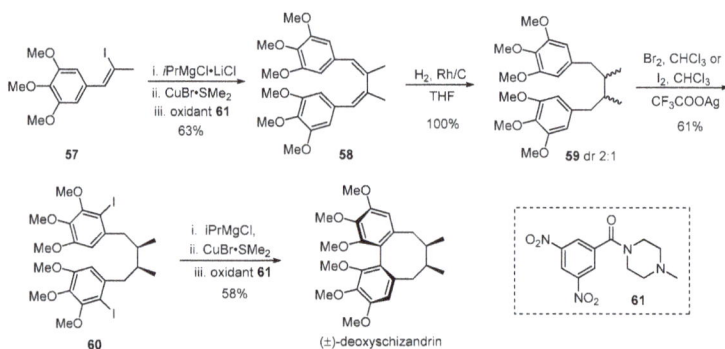

Scheme 10. Synthesis of (±)-deoxyschizandrin. Adapted from Spring et al. [67].

In 2013, the RajanBabu group reported a general synthetic approach to multiple dibenzocyclooctadienes lignans via an interesting borostannylative cyclization (Scheme 11) [69]. In the presence of PdCl$_2$(PPh$_3$)$_2$ and [B-Sn] reagent **70**, chiral diynyl precursor **62** was converted into dibenzocyclooctadiene **64** through a borostannylative cyclization and a subsequent acidification process [70].

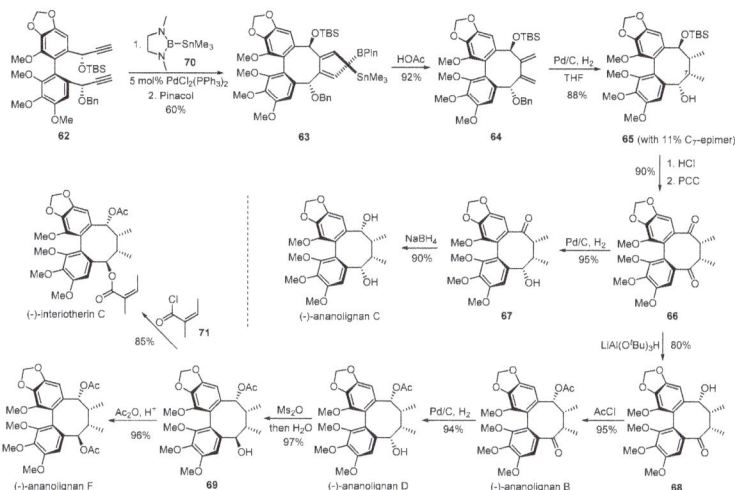

Scheme 11. Synthesis of five dibenzocyclooctadienes lignans. Adapted from RajanBabu et al. [69].

The subsequent hydrogenation afforded **65** as the major product. After the deprotection and oxidation, the general intermediate **66** was prepared. The synthesis of (−)-ananolignan C was achieved through two successive diastereoselective reductions of **66**. Meanwhile, the synthesis of (−)-ananolignan B was accomplished from the treatment of **66** with LiAl(O^{t}Bu)$_3$H and subsequent acetylation. The stereoselective hydrogenation of (−)-ananolignan B led to the formation of (−)-ananolignan D. The following configuration inversion of the hydroxyl group and actylation led to the synthesis of (−)-ananolignan F. In addition, (−)-interiotherin C can also be formed through the esterification of **69** and angeloyl chloride **71**.

Synthesis of other three lignans was reported by RajanBabu and coworkers in the same paper (Scheme 12) [69]. Oxidative cleavage of the right-bottom double bond of **64** was applied for the formation of ketone **72**. Diol **74** was obtained from the debenzylation and methyllithium 1,2-addition of **72**. After hydroxyl oxidation, diastereoselective reduction and benzoyl protection steps, compound **76** was obtained. Starting from **76**, synthesis of schizanrin F was achieved by TBS deprotection, oxidation, diastereoselective reduction and acetylation process. Starting from diol **74** again, compound **78** can be prepared by TBS deprotection, oxidation and double diastereoselective reduction. Finally, the synthesis of kadasuralignan B and tiegusanlin D was accomplished through acetylation and benzoylation of **78**, respectively.

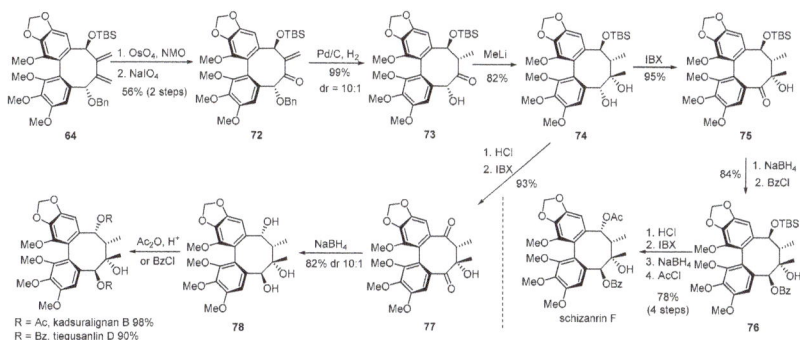

Scheme 12. Synthesis of another three dibenzocyclooctadienes lignans. Adapted from RajanBabu et al. [69].

In 2018, a mild and asymmetric synthetic route to (−)-gymnothelignan L was developed by She and coworkers through a Suzuki-Miyaura coupling and a bioinspired desymmetric transannular Friedel-Crafts cyclization strategy (Scheme 13) [71]. Iodide **80** was obtained from iodination of compound **79**. The Suzuki-Miyaura coupling of **80** and arylboronic acid **84** formed biphenyl compound **81**, which was transformed into **82** using DIBAL-H as the reducing agent. Under acidic conditions, a bioinspired desymmetric transannular Friedel-Crafts cyclization of **82** occurred readily, generating **83**. After removal of the benzyl protecting group, the synthesis of (−)-gymnothelignan L was completed. At almost the same time, a similar strategy was applied in total synthesis (−)-gymnothelignan V by Soorukram and coworkers [72].

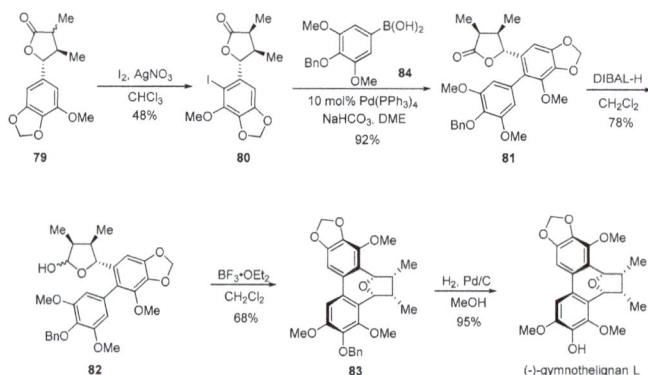

Scheme 13. Bioinspired asymmetric synthesis of (-)-gymnothelignan L. Adapted from She et al. [71].

4. Advances in the Synthesis of Arylnaphthalene Derivatives

In the literature, the arylnaphthalene derivative lignan subgroup includes arylnaphthalenes and aryltetralins. Structurally, these lignans have a substituted naphthalene core. It should be mentioned that, due to their excellent biological characters, several clinically used antitumor drugs are derived from the well-known member, podophyllotoxin, and its glycosides [49].

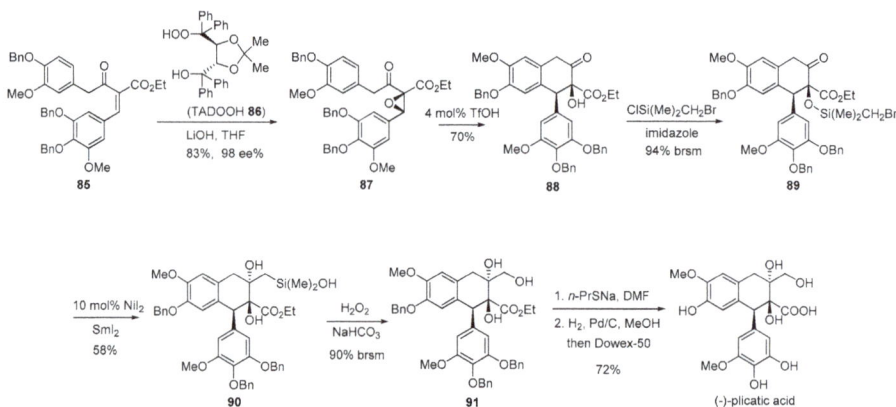

Scheme 14. Asymmetric synthesis of (−)-plicatic acid. Adapted from Deng et al. [73].

In 2009, Deng and coworkers reported an asymmetric total synthesis of (−)-plicatic acid (Scheme 14) [73]. The enantioselective epoxidation of trisubstitued olefin **85** was applied for the introduction of the original chiral stereocenters. Excellent enantioselectivity was obtained from the application of chiral (*S*,*S*)-TADOOH **86**. Intramolecular Friedel-Crafts reaction of **87** formed the six-membered ring of **88** effectively. Subsequent silylation and intramolecular Barbier reaction under SmI$_2$/NiI$_2$ conditions afforded a diastereoselective access to diol **90**. After Fleming-Tamao- Kumada oxidation [74,75], **91** was furnished. The synthesis of (−)-plicatic acid was completed following hydration and global debenzylation.

The stereoselective aza-Claisen rearrangement strategy developed by Barker and coworkers was not only effective for asymmetric synthesis of (+)-galbelgin (Scheme 2), but also for asymmetric synthesis of (−)-cyclogalgravin, and (−)-pycnanthulignenes A and B (Scheme 15) [54]. Through the aza-Claisen rearrangement strategy, alcohol **9** (Scheme 2) was prepared and submitted to hydroxyl protection and double bond oxidation, generating aldehyde **92**. The addition from aryllithium **11** gave compound **93**. The synthesis of (−)-cyclogalgravin was achieved through cyclization. Employing same protocol, Barker's group also finished an asymmetric synthesis of (−)-pycnanthulignenes A and B.

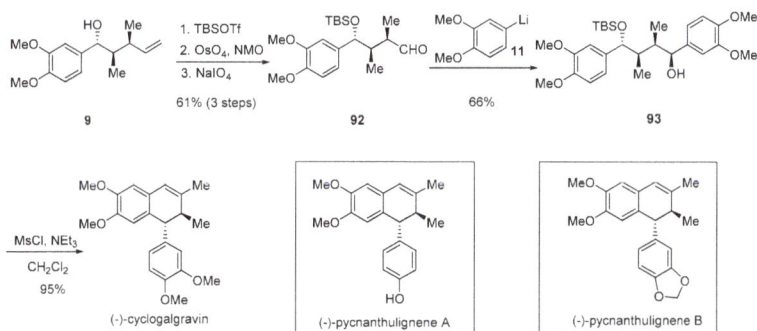

Scheme 15. Asymmetric synthesis of (−)-cyclogalgravin. Adapted from Barker et al. [54].

In 2012, Hong and coworkers reported the enantioselective total synthesis of (+)-galbulin using an organocatalytic asymmetric Michael-Michael-aldol cascade (Scheme 16) [76]. Under the promotion of Jørgensen-Hayashi catalyst **96**, ketoaldehyde **97** was readily prepared from the asymmetric Michael-Michael-aldol cascade of **94** and **95**. Compound **99** was produced by reduction, oxidation and epoxidation treatments of **97**. Following epoxide ring-opening and aromatization, compound **101** was obtained. The synthesis of (+)-galbulin was finally accomplished through selective methylation and dehydroxylation processes.

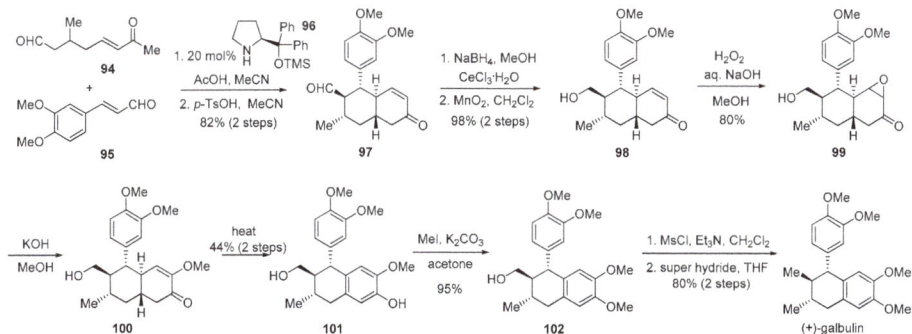

Scheme 16. Enantioselective synthesis of (+)-galbulin. Adapted from Hong et al. [76].

Peng and coworkers reported in 2013 the synthesis of sacidumlignan A employing Ueno-Stork radical cyclization and skeletal rearrangement strategy (Scheme 17) [77]. Alcohol **103** was connected with ethyl propenyl ether in the presence of bromine, readily generating **104**. The Ueno-Stork radical cyclization of **104** was enabled by Bu$_3$SnH and AIBN.

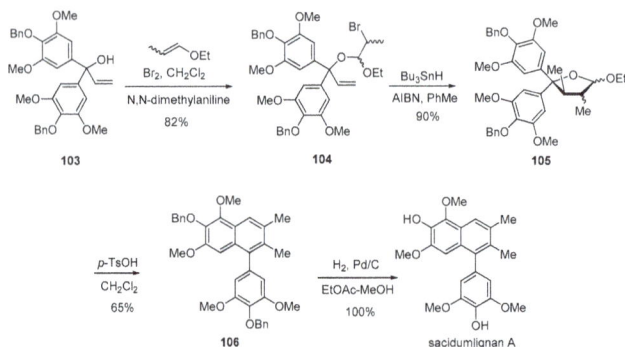

Scheme 17. Synthesis of sacidumlignan A. Adapted from Peng et al. [77].

The resulting **105** was submitted to a skeletal rearrangement, affording arylnaphthalene **106**. The synthesis of sacidumlignan A was achieved after benzyl deprotection.

The same year, Peng and coworkers also reported the total synthesis of (±)-cyclogalgravin and (±)-galbulin (Scheme 18) [78]. Cyclic acetal **108** was obtained from Ueno-Stork radical cyclization of **107**. Diol **110** was prepared by an oxidation, methylation and reduction process. Subsequent selective hydroxyl protection, dehydroxylation, deprotection and oxidation led to the generation of aldehyde **111**. Next an intramolecular Friedel-Crafts reaction was applied for the synthesis of (±)-cyclogalgravin. The synthesis of (±)-galbulin was readily achieved by the stereoselective hydrogenation of (±)-cyclogalgravin.

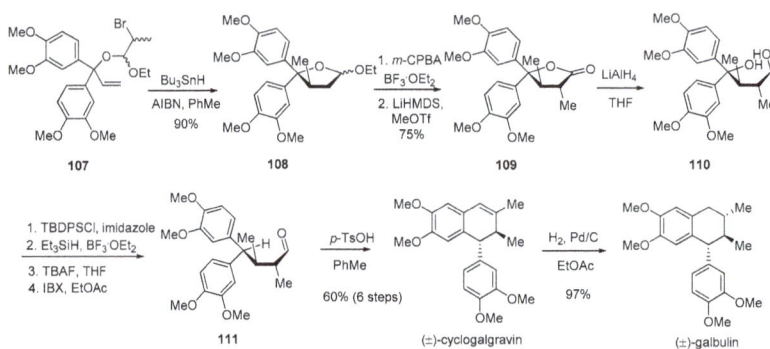

Scheme 18. Synthesis of (±)-cyclogalgravin and (±)-galbulin. Adapted from Peng et al. [78].

In 2013, Argade and coworkers reported a novel strategy to construct arylnaphthalene frameworks via Pd-promoted [2 + 2 + 2] cyclization (Scheme 19), which enabled the synthesis of justicidin B and retrojusticiding B [79]. Through a Pd-promoted [2 + 2 + 2] cyclization process, aryne precursor **112** was connected with diene **113**, generating arylnaphthalene **114**. After the regioselective hydrolysis of the ester group, the synthesis of justicidin B was achieved through a chemoselective reduction of the acid group using BH$_3$·SMe$_2$ and subsequent lactonization. Meanwhile, the synthesis of retrojusticidin B was achieved through the reduction of **115** and subsequent lactonization.

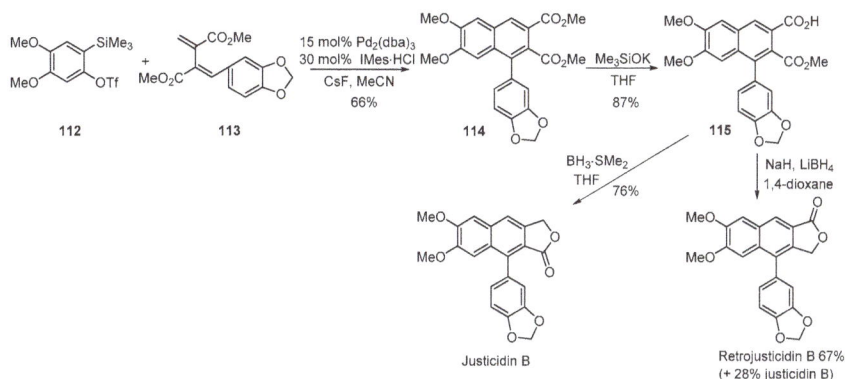

Scheme 19. Synthesis of justicidin B and retrojusticiding B. Adapted from Argade et al. [79].

Shia and coworkers reported in 2015 the synthesis of three arylnaphthalene derivative lignans using a Mn(III)-mediated free radical cyclization cascade (Scheme 20) [80]. Knoevenagel condensation of α-cyano ester **116** with aldehyde **117** and subsequent reduction was applied for the generation of α-cyano ester **118**. The following oxidative free radical cyclization cascade enabled by Mn(OAc)₃ afforded access to compound **119**. The synthesis of retrojusticidin B was accomplished after decyanation and aromatization operations. Using the same strategy, the synthesis of justicidin E and helioxanthin was completed.

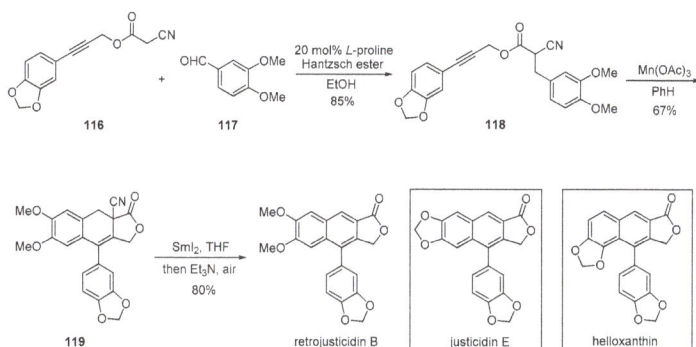

Scheme 20. Synthesis of arylnaphthalene derivative lignans. Adapted from Shia et al. 2015 [80].

In 2015, Narender and coworkers reported an interesting Ag-promoted radical addition/ cyclization process for the construction of highly substituted α-naphthol skeletons (Scheme 21) and the synthesis of three arylnaphthalene lignans [81]. Through the Ag-promoted radical addition/cyclization between ketoester **120** and aryl propiolate **121**, polysubstituted arynaphthol **122** was readily prepared. The synthesis of diphyllin was finished under known reductive-lactonization conditions [82]. The synthesis of justicidin A was then achieved through methylation of diphyllin. The synthesis of taiwanin E was also accomplished.

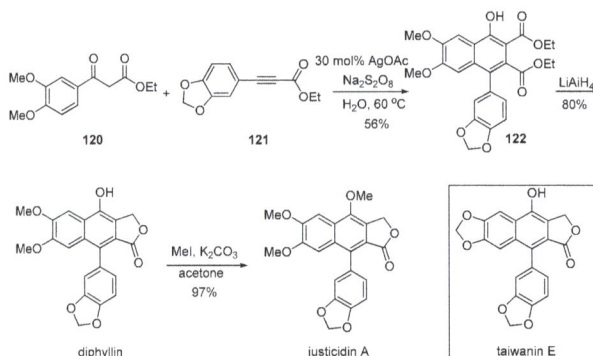

Scheme 21. Synthesis of arylnaphthalene lignans. Adapted from Narender et al. [81].

In 2017 Ham and coworkers reported the synthesis of seven arylnaphthalene derivative lignans based on a strategy involving Hauser-Kraus annulation and Suzuki-Miyaura cross-coupling (Scheme 22) [83]. In the presence of LiHMDS, the Hauser-Kraus annulation between cyanophthalide **123** and γ-crotonolactone **124** and subsequent protection treatment gave arylnaphthalene **125**. The synthesis of diphllin was finished by the subsequent Suzuki-Miyaura cross-coupling of **126** and potassium aryltrifluoroborate **127**. Justicidin A was produced by the methylation of diphllin. The syntheses of taiwannin E, chinensinaphthol justicidin C, justicidin D and cilinaphthalide B were also finished.

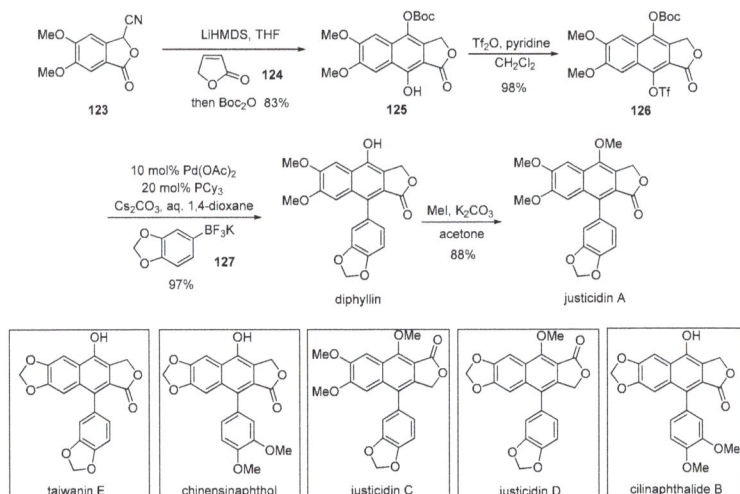

Scheme 22. Synthesis of seven arylnaphthalene lignans. Adapted from Ham et al. [83].

Hajra and coworkers reported the enantioselective total synthesis of (−)-podophyllotoxin and natural analogues in 2017 (Scheme 23) [84]. The L-proline-catalyzed asymmetric cross aldol reaction between 6-bromopiperonal **128** and aldehyde **129** introduced original stereocenters with excellent diastereoselectivity and excellent stereoselectivity at gram scale. Lactone **130** was obtained after reduction, lactonization and TBS protection operations. Z-Benzylidene lactone **132** was prepared through an aldol reaction between **130** and aldehyde **131**, and subsequent elimination. The intramolecular Heck reaction between trisubstituted Z-alkene motif and the bulky bromoarene motif in **132** happened smoothly, generating compound **133** in good yields. Notably, under different hydrogenation conditions, three stereoselective pathways of **133** led to the synthesis of (−)-podophyllotoxin, (−)-picropodophyllin, (+)-isopicropodophyllin, respectively. Meanwhile, the synthesis of (+)-isopicropodophyllone was achieved through the oxidation of (+)-isopicropodophyllin. The synthesis of (−)-isopodophyllotoxin can be accomplished through a TBS deprotection and reductive Heck reaction process from **132**.

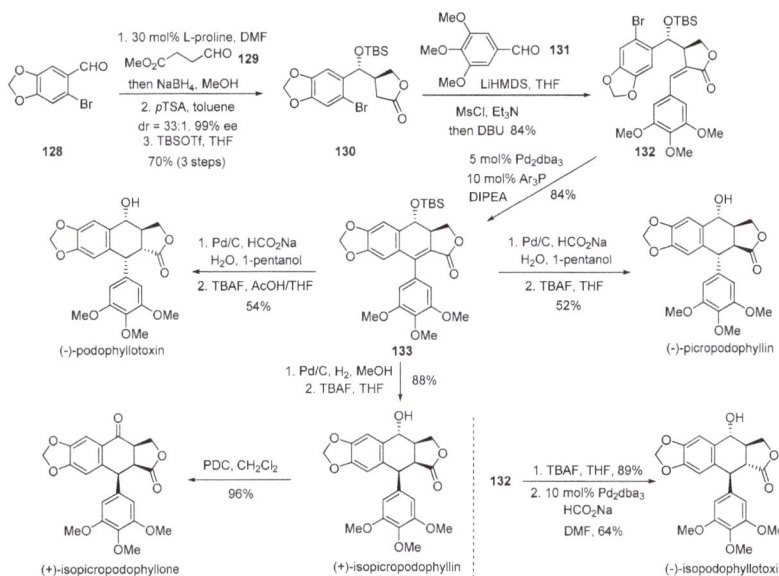

Scheme 23. Synthesis of (−)-podophyllotoxin and three natural analogues. Adapted from Hajra et al. [84].

Czarnocki and coworkers reported in 2018 the total synthesis of (+)-epigalcatin using a photocyclization process under continuous flow UV irradiation conditions (Scheme 24) [85]. Diester **134** was condensed with aldehyde **120** at basic conditions, affording *E,E*-bisbenzylidenesuccinic acid **135**. Through an amidation process, L-prolinol was introduced in amide **136** as a chiral auxiliary. Eight-membered ring compound **137** was prepared via following hydrolysis and macrolactonization. Under continuous flow irradiation with UV light, the photocyclization of **137** furnished **138** smoothly [86]. Remove of the chiral auxiliary, hydrogenation of the double bond and simultaneous reduction of formyl group led to the formation of ester **139**. The synthesis of (+)-epigalcatin was achieved through subsequent reductive transformations of the methyl ester of **139** into a methyl group via three steps.

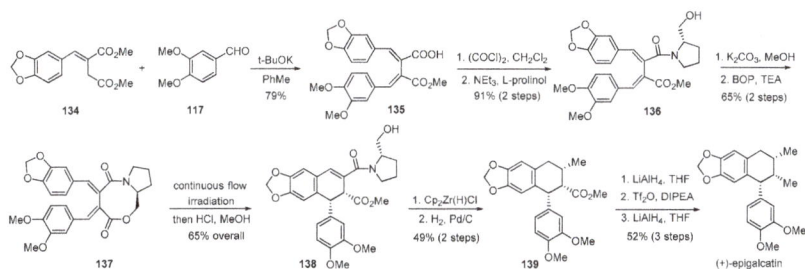

Scheme 24. Synthesis of (+)-epigalcatin. Adapted from Czarnocki et al. [85].

Peng and coworkers reported the total synthesis of (−)-podophyllotoxin and four natural analogues using a Ni-catalyzed reductive cascade in 2018 (Scheme 25) [87]. The asymmetric conjugated addition of **141** to chiral α,β-unsaturated amide **140** introduced the first stereocenter. Enol ether **144** was obtained after subsequent reduction, oxidation, acetal formation and elimination. β-Bromoacetal **145** was produced using 2,4,4,6-tetrabromo-2,5-cyclohexadienone (TBCD). With the application of the Ni-catalyzed reductive cascade [88], both **146** and **147** were produced in moderate yields. After the hydration and oxidation of **146**, the synthesis of (+)-deoxypicropodophyllin was accomplished. (+)-Isodeocypodophyllotoxin can be synthesized from the epimerization at C9a of (+)-deoxypicropodophyllin under basic conditions. With the radical bromination under visible-light irradiation [89], and further oxidation treatments, the synthesis of (−)-epipodophyllotoxin and (−)-podophyllotoxone was achieved in a stepwise fashion. The stereoselective reduction of (−)-podophyllotoxone using L-Selectride gave (−)-podophyllotoxin. Additionally, compound **147** can be transformed into Meyer's **150** intermediate for synthesis of (−)-picropodophyllin and (−)-picropodophyllone through three regular operations [90].

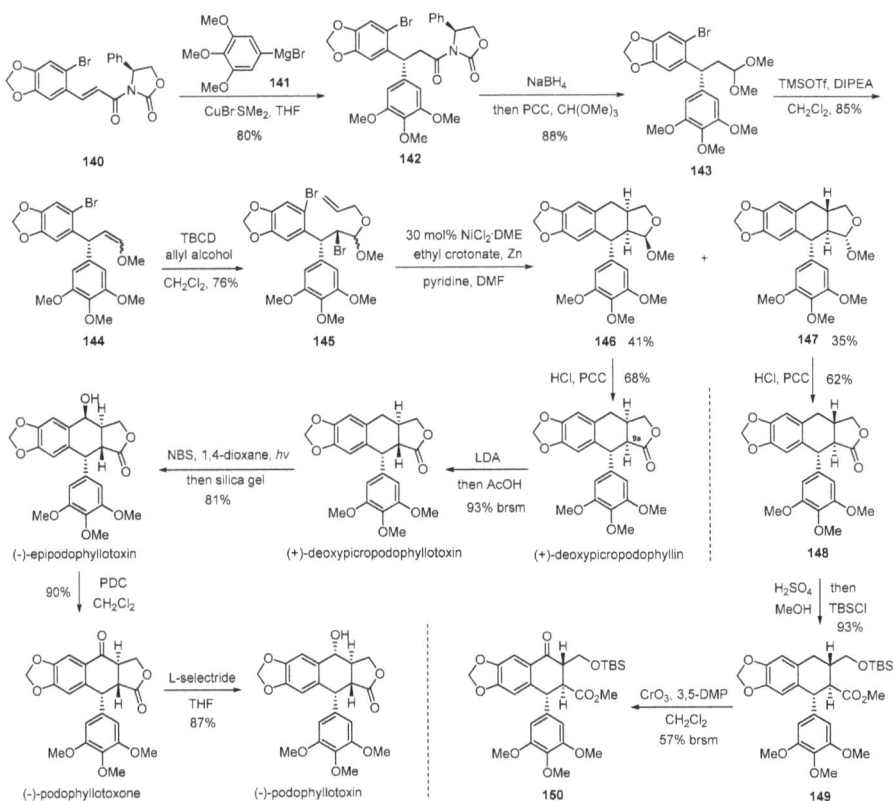

Scheme 25. Synthesis of (−)-podophyllotoxin and four natural analogues. Adapted from Peng et al. [87].

The Ni-catalyzed cyclization/cross-coupling has been verified as a suitable strategy for not only the synthesis of multiple acyclic lignan derivatives (Scheme 8) but also the synthesis of (±)-dimethylretrodendrin and (±)-collinusin (Scheme 26) by Giri and coworker [64]. Lactone **49** was obtanied from the Ni-catalyzed cyclization/cross-coupling process (Scheme 8). The stereoselective aldol reaction between **49** and aldehyde **117** and following intramolecular Friedel-Crafts reaction led to the formation of (±)-dimethylretrodendrin.

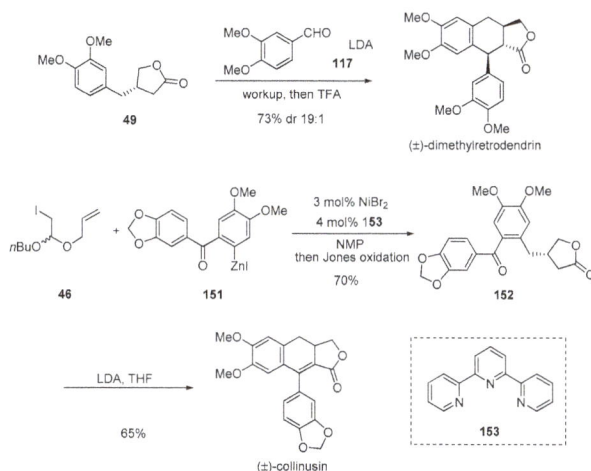

Scheme 26. Synthesis of (±)-dimethylretrodendrin and (±)-collinusin. Adapted from Giri et al. [64].

In the presence of **153**, the Ni-catalyzed cyclization/cross-coupling of **46** and **151** and following oxidation gave lactone **152**. The synthesis of (±)-collinusin was completed by subsequent intramolecular nucleophilic addition and dehydration.

In 2018, Belozerova and coworkers reported the synthesis of sevanol via an oxidative dimerization strategy (Scheme 27) [91]. Chiral ester **156** was prepared by esterification of acid **154** and chiral alcohol **155**. After the removal of two MOM protecting groups of **156**, compound **157** was submitted to FeCl$_3$-promoted oxidative dimerization, affording sevanol after the hydrolysis of all three ester groups.

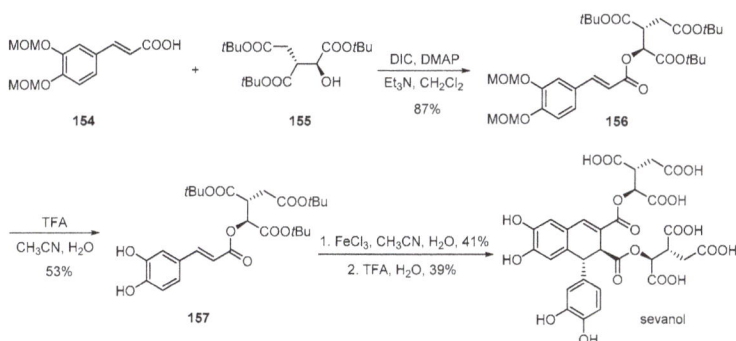

Scheme 27. Synthesis of sevanol. Adapted from Belozerova et al. [91].

Aria and coworkers reported total synthesis of (±)-isolariciresinol using a tandem Michael-aldol reaction in 2018 (Scheme 28) [92]. Alcohol **160** was obtained as a diastereomeric mixture from the tandem Michael-aldol reaction of dithiane **158**, lactone **124** and aldehyde **159** under basic conditions. After the cleavage of the dithiane substituent and TBS, the following cyclization furnished **162** as the major product.

Ester **163** was prepared from methanolysis of the lactone ring and TBS protection operations. The synthesis of (±)-isolariciresinol was achieved through the subsequent reduction and deprotection treatments.

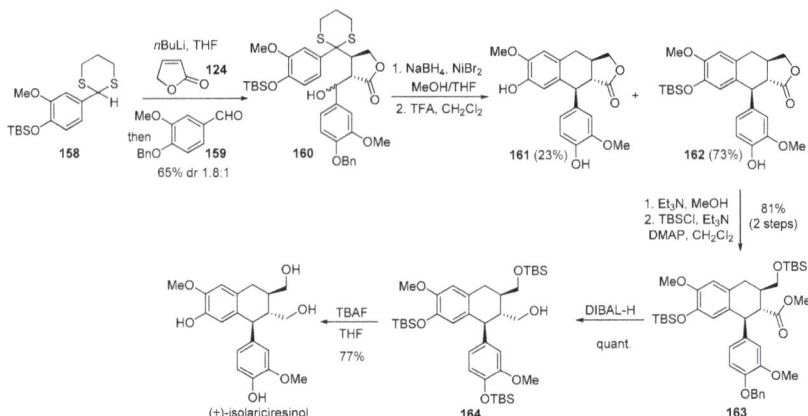

Scheme 28. Synthesis of (±)-isolariciresinol. Adapted from Aria et al. [92].

Barker and coworkers reported in 2017 the first total synthesis of (±)-ovafolinins A and B through the acyl-Claisen rearrangement developed in their lab and a cascade cyclization enabled by bulky protecting groups (Scheme 29) [93]. Notably, (±)-ovafolinins A and B have polycyclic skeletons rarely found in lignans. The acyl-Claisen rearrangement between acid chloride **165** and allylic morpholine **166** afforded amide **167** as a single diastereoisomer and in excellent yields. Alcohol **168** was prepared by hydration and reduction. Phenol **169** was introduced through a Mitsunobu reaction. Alcohol **170** was obtained after the oxidative cleavage of the double bond and following reduction. The subsequent *t*-butyldiphenylsilyl (TBDPS) protection, debenzylation and oxidation led to the formation of compounds **171** and **172** through a cascade cyclization enabled by the TBDPS group. The synthesis of (±)-ovafolinins A and B was achieved after subsequent hydrogenation and deprotection.

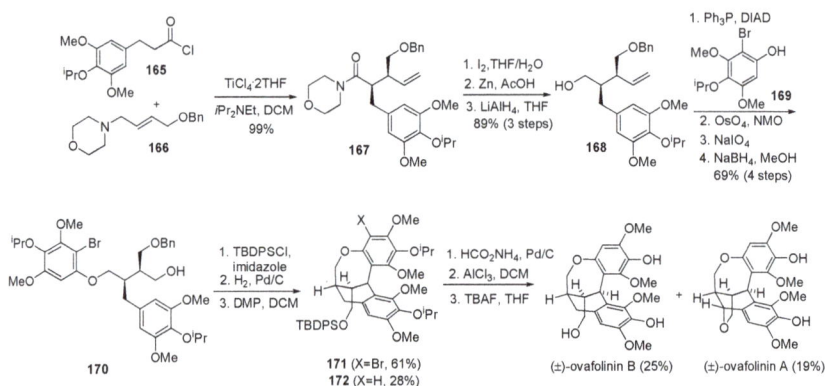

Scheme 29. Synthesis of (±)-ovafolinins A and B. Adapted from Barker et al. [93].

Taking advantage of the above achievement, Barker and coworkers reported the first asymmetric total synthesis of (+)-ovafolinins A and B (Scheme 30). Starting from acid chloride **165** again, chiral amide **174** was first prepared. Stereoselective allylation and dihydroxylation of the double bond led to the generation of lactone **176**. After the reduction and oxidative cleavage of the 1,2-diol motif, lactone **177** was formed by Fétizon oxidation. The introduction of the benzyloxymethyl group and the following reduction led to the formation of **179**. After TBDPS protection, Mitsunobu reaction with **181** and debenzylation, alcohol **180** was obtained. Chiral **172** was formed through a cascade cyclization under oxidative conditions. Finally, the first asymmetric synthesis of (+)-ovafolinins A and B was achieved after deprotection operations. Based on optical rotation comparisons between the synthetic samples and the natural compounds, Barker's group demonstrated that natural ovafolinins A and B were both isolated in scalemic mixtures. And the original stereochemical assignment of natural ovafolinin B was corrected.

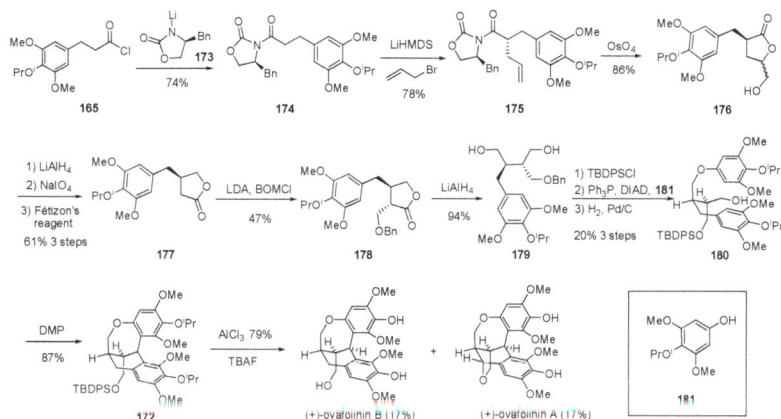

Scheme 30. Asymmetric synthesis of (+)-ovafolinins A and B. Adapted from Barker et al. [93].

Recently, we developed a new asymmetric synthetic route to (+)-ovafolinins A and B (Scheme 31) [94]. Starting from benzyl syringaldehyde **182**, bromide **183** was prepared after reduction and bromination. The diastereoselective alkylation of (S)-Taniguchi lactone **184** introduced two adjacent stereogenic centers in excellent stereoselectivity, affording lactone **185**. Subsequent double benzyl protection opened the lactone ring and generated ester **186**. Compound **189** was obtained from the reduction and connected with **188** through Mitsunobu reaction. After oxidative cleavage of double bond, aldehyde **190** was obtained. The polycyclic skeleton in **191** was constructed through a double Friedel-Crafts reaction of **190**. The synthesis of (+)-ovafolinin B was accomplished through the global debenzylation. And the synthesis of (+)-ovafolinin A was achieved through subsequent benzylic oxidation cyclization enabled by Cu(OAc)$_2$.

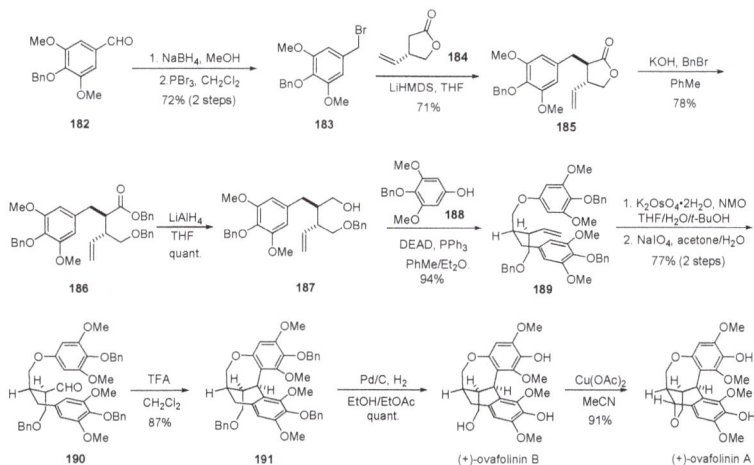

Scheme 31. Asymmetric synthesis of (+)-ovafolinins A and B. Adapted from Hu et al. 2018 [94].

5. Conclusions

In this review, we have summarized the advances in the synthesis of lignan natural products reported from 2008 to 2018. Synthetic progress in three areas was outlined: acyclic lignan derivatives, dibenzocycooctadiene derivative and arylnaphthalene derivatives. Novel synthetic methodologies had been applied for construction of challenging structures existing in lignan natural products. As the result, many elegant synthetic approaches to lignans had been developed. However, as a long term program, the promising biological features and development of concise synthetic approaches to lignan natural products and their analogues are continuing to attract more and more interest from the pharmaceutical industry and the organic synthesis community.

Funding: This research was funded by National Natural Science Foundation of China (21772153), and the Key Science and Technology Innovation Team of Shaanxi Province (2017KCT-37).

Conflicts of Interest: The authors declare no conflict of interest.

References

1. Harworth, R.D. Chemistry of the lignan group of natural products. *J. Chem. Soc.* **1942**, 448–456. [CrossRef]
2. Lampe, J.W.; Martini, M.C.; Kurzer, M.S.; Adlercreutz, H.; Slavin, J.L. Urinary lignan and isoflavonoid excretion in premenopausal women consuming flaxseed powder. *Am. J. Clin. Nutr.* **1994**, *60*, 122–128. [CrossRef]
3. Yamauchi, S.; Taniguchi, E. Synthesis and insecticidal activity of lignan analogs (1). *Agric. Biol. Chem.* **1991**, *55*, 3075–3084.
4. Yamauchi, S.; Taniguchi, E. Synthesis and insecticidal activity of lignan analogs II. *Biosci. Biotech. Biochem.* **1992**, *56*, 412–418. [CrossRef] [PubMed]
5. Yamauchi, S.; Nagata, S.; Taniguchi, E. Synthesis and insecticidal activity of haedoxan analogs. (Part IV). Effect on insecticidal activity of substituents at the 1,4-benzodioxanyl moiety of haedoxan. *Biosci. Biotech. Biochem.* **1992**, *56*, 1193–1197. [CrossRef]
6. Yamauchi, S.; Taniguchi, E. Synthesis and insecticidal activity of lignan analogs. Part 5. Influence on insecticidal activity of the 3-(3,4-methylenedioxyphenyl) group in the 1,4-benzodioxanyl moiety of haedoxan. *Biosci. Biotech. Biochem.* **1992**, *56*, 1744–1751. [CrossRef]
7. Nitao, J.K.; Johnson, K.S.; Scriber, J.M.; Nair, M.G. Magnolia virginiana neolignan compounds as chemical barriers to swallowtail butterfly host use. *J. Chem. Ecol.* **1992**, *18*, 1661–1671. [CrossRef]
8. Wang, C.T.; Makela, T.; Hase, H.; Adlercreutz, H.; Kurzer, M.S. Lignans and flavonoids inhibit aromatase enzyme in human preadipocytes. *J. Steroid Biochem. Molec. Biol.* **1994**, *50*, 205–212. [CrossRef]
9. Kirkman, L.M.; Lampe, J.W.; Campbell, D.R.; Martini, M.C.; Slavin, J.L. Urinary lignan and isoflavonoid excretion in men and women consuming vegetable and soy diets. *Nutr. Cancer* **1995**, *24*, 1–12. [CrossRef]
10. Makela, S.I.; Pylkkanen, L.H.; Santti, R.S.S.; Adlercreutz, H. Dietary soybean may be antiestrogenic in male mice. *J. Nutr.* **1995**, *125*, 437–445.
11. Adlercreutz, H.; Goldin, B.R.; Gorbach, S.L.; Hockerstedt, K.A.V.; Watanabe, S.; Hamalainen, E.K.; Markkanen, M.H.; Makela, T.H.; Wahala, K.T.; Hase, T.A.; et al. Soybean phytoestrogen intake and cancer risk. *J. Nutr.* **1995**, *125*, 757–770.
12. Xu, W.-H.; Zhao, P.; Wang, M.; Liang, Q. Naturally occurring furofuran lignans: Structural diversity and biological activities. *Nat. Prod. Res.* **2018**. [CrossRef] [PubMed]
13. Teponno, R.B.; Kusari, S.; Spiteller, M. Recent advances in research on lignans and neolignans. *Nat. Prod. Rep.* **2016**, *33*, 1044–1092. [CrossRef] [PubMed]
14. Pan, J.-Y.; Chen, S.-L.; Yang, M.-H.; Wu, J.; Sinkkonen, J.; Zou, K. An update on lignans: Natural products and synthesis. *Nat. Prod. Rep.* **2009**, *26*, 1251–1292. [CrossRef] [PubMed]
15. Ward, R.S. Lignans, neolignans and related compounds. *Nat. Prod. Rep.* **1999**, *16*, 75–96. [CrossRef]
16. Ward, R.S. Lignans, neolignans and related compounds. *Nat. Prod. Rep.* **1997**, *14*, 43–74. [CrossRef]
17. Ward, R.S. Lignans, neolignans, and related compounds. *Nat. Prod. Rep.* **1995**, *12*, 183–205. [CrossRef]
18. Ward, R.S. Lignans, neolignans, and related compounds. *Nat. Prod. Rep.* **1993**, *10*, 1–28. [CrossRef]
19. Whiting, D.A. Lignans and neolignans. *Nat. Prod. Rep.* **1985**, *2*, 191–211. [CrossRef]

20. Kawazoe, K.; Yutani, A.; Tamemoto, K.; Yuasa, S.; Shibata, H.; Higuti, T.; Takaishi, Y. Phenylnaphthalene compounds from the subterranean part of *Vitex rotundifolia* and their antibacterial activity against methicillin-resistant *Staphylococcus aureus*. *J. Nat. Prod.* **2001**, *64*, 588–591. [CrossRef]

21. Sagar, K.S.; Chang, C.C.; Wang, W.K.; Lin, J.Y.; Lee, S.S. Preparation and anti-HIV activities of retrojusticidin B analogs and azalignans. *Bioorg. Med. Chem.* **2004**, *12*, 4045–4054. [CrossRef] [PubMed]

22. Yeo, H.; Li, Y.; Fu, L.; Zhu, J.L.; Gullen, E.A.; Dutschman, G.E.; Lee, Y.; Chung, R.; Huang, E.S.; Austin, D.J.; et al. Synthesis and antiviral activity of helioxanthin analogues. *J. Med. Chem.* **2005**, *48*, 534–546. [CrossRef] [PubMed]

23. Li, Y.; Fu, L.; Yeo, H.; Zhu, J.L.; Chou, C.K.; Kou, Y.H.; Yeh, S.F.; Gullen, E.; Austin, D.; Cheng, Y.C. Inhibition of hepatitis B virus gene expression and replication by helioxanthin and its derivative. *Antiviral Chem. Chemother.* **2005**, *16*, 193–201. [CrossRef] [PubMed]

24. Janmanchi, D.; Tseng, Y.P.; Wang, K.C.; Huang, R.L.; Lin, C.H.; Yeh, S.F. Synthesis and the biological evaluation of arylnaphthalene lignans as anti-hepatitis B virus agents. *Bioorg. Med. Chem.* **2010**, *18*, 1213–1226. [CrossRef] [PubMed]

25. McDoniel, P.B.; Cole, J.R. Antitumor activity of Bursera schlechtendalii (Burseraceae). Isolation and structure determination of two new lignans. *J. Pharm. Sci.* **1972**, *61*, 1992–1994. [CrossRef] [PubMed]

26. Pelter, A.; Ward, R.S.; Satyanarayana, P.; Collins, P. Synthesis of lignan lactones by conjugate addition of thioacetal carbanions to butanolide. *J. Chem. Soc. Perkin Trans.* **1983**, *1983*, 643–647. [CrossRef]

27. Capilla, A.S.; Sánchez, I.; Caignard, D.H.; Renard, P.; Pujola, M.D. Antitumor agents. Synthesis and biological evaluation of new compounds related to podophyllotoxin, containing the 2,3-dihydro-1,4-benzodioxin system. *Eur. J. Med. Chem.* **2001**, *36*, 389–393. [CrossRef]

28. Chen, C.C.; Hsin, W.C.; Ko, F.N.; Huang, Y.L.; Ou, J.C.; Teng, C.M. Antiplatelet arylnaphthalide lignans from *Justicia procumbens*. *J. Nat. Prod.* **1996**, *59*, 1149–1150. [CrossRef]

29. Weng, J.R.; Ko, H.H.; Yeh, T.L.; Lin, H.C.; Lin, C.N. Two new arylnaphthalide lignans and antiplatelet constituents from Justicia procumbens. *Arch. Pharm.* **2004**, *337*, 207–212. [CrossRef] [PubMed]

30. Ukita, T.; Nakamura, Y.; Kubo, A.; Yamomoto, Y.; Takahashi, M.; Kotera, J.; Ikeo, T. Synthesis and in vitro pharmacology of a series of new chiral histamine H₃-receptor ligands: 2-(R and S)-amino-3-(1H-imidazol-4(5)-yl)propyl ether derivatives. *J. Med. Chem.* **1999**, *42*, 1293–1305. [CrossRef] [PubMed]

31. Iwasaki, T.; Kondo, K.; Kuroda, T.; Moritani, Y.; Yamagata, S.; Sugiura, M.; Kikkawa, H.; Kaminuma, O.; Ikezawa, K. Novel selective PDE IV inhibitors as antiasthmatic agents. synthesis and biological activities of a series of 1-aryl-2,3-bis(hydroxymethyl)naphthalene lignans. *J. Med. Chem.* **1996**, *39*, 2696–2704. [CrossRef] [PubMed]

32. Thérien, M.; Fitzsimmons, B.J.; Scheigetz, J.; Macdonald, D.; Choo, L.Y.; Guay, J.; Falgueyret, J.P.; Riendeau, D. Justicidin E: A new leukotriene biosynthesis inhibitor. *Bioorg. Med. Chem.* **1993**, *3*, 2063–2066. [CrossRef]

33. Delorme, D.; Ducharme, Y.; Brideau, C.; Chan, C.C.; Chauret, N.; Desmarais, S.; Dubé, D.; Falgueyret, J.P.; Fortin, R.; Guay, J.; et al. Dioxabicyclooctanyl naphthalenenitriles as nonredox 5-lipoxygenase inhibitors: structure−activity relationship study directed toward the improvement of metabolic stability. *J. Med. Chem.* **1996**, *39*, 3951–3970. [CrossRef] [PubMed]

34. Ducharme, Y.; Brideau, C.; Dubé, D.; Chan, C.C.; Falgueyret, J.P.; Gillard, J.W.; Guay, J.; Hutchinson, J.H.; McFarlane, C.S.; Riendeau, D.; et al. Naphthalenic lignan lactones as selective, nonredox 5-lipoxygenase inhibitors. Synthesis and biological activity of (methoxyalkyl)thiazole and methoxytetrahydropyran hybrids. *J. Med. Chem.* **1994**, *37*, 512–518. [CrossRef] [PubMed]

35. Chang, C.W.; Lin, M.T.; Lee, S.-S.; Liu, K.C.S.C.; Hsu, F.-L.; Lin, J.-Y. Differential inhibition of reverse transcriptase and cellular DNA polymerase-α activities by lignans isolated from Chinese herbs, Phyllanthus myrtifolius Moon, and tannins from Lonicera japonica Thunb and Castanopsis hystrix. *Antiviral Res.* **1995**, *27*, 367–374. [CrossRef]

36. Lee, S.S.; Lin, M.T.; Liu, C.L.; Lin, Y.Y.; Liu, K.C.S.C. Six lignans from *Phyllanthus myrtifolius*. *J. Nat. Prod.* **1996**, *59*, 1061–1065. [CrossRef] [PubMed]

37. Cow, C.; Leung, C.; Charlton, J.L. Antiviral activity of arylnaphthalene and aryldihydronaphthalene lignans. *Can. J. Chem.* **2000**, *78*, 553–561. [CrossRef]

38. Wu, S.J.; Wu, T.S. Cytotoxic arylnaphthalene lignans from Phyllanthus oligospermus. *Chem. Pharm. Bull.* **2006**, *54*, 1223–1225. [CrossRef]

39. Lu, H.; Liu, G.-T. Antioxidant activity of dibenzocyclooctene lignans isolated from Schisandraceae. *Planta Med.* **1992**, *58*, 311–313. [CrossRef]

40. Hirano, T.; Wakasugi, A.; Oohara, M.; Oka, K.; Sashida, Y. Suppression of mitogen-induced proliferation of human peripheral blood lymphocytes by plant lignans. *Planta Med.* **1991**, *57*, 331–334. [CrossRef]

41. Chang, J.; Reiner, J.; Xie, J. Progress on the chemistry of dibenzocyclooctadiene lignans. *Chem. Rev.* **2005**, *105*, 4581–4609. [CrossRef]

42. Satake, H.; Koyama, T.; Bahabadi, S.E.; Matsumoto, E.; Ono, E.; Murata, J. Essences in metabolic engineering of lignan biosynthesis. *Metabolites* **2015**, *5*, 270–290. [CrossRef] [PubMed]

43. Satake, H.; Ono, E.; Murata, J. Recent advances in the metabolic engineering of lignan biosynthesis pathways for the production of transgenic plant-based foods and supplements. *J. Agric. Food Chem.* **2013**, *61*, 11721–11729. [CrossRef] [PubMed]

44. Suzuki, S.; Umezawa, T. Biosynthesis of lignans and norlignans. *J. Wood Sci.* **2007**, *53*, 273–284. [CrossRef]

45. Umezawa, T. Diversity in lignan biosynthesis. *Phytochem. Rev.* **2003**, *2*, 371–390. [CrossRef]

46. Albertson, A.K.F.; Lumb, J.-P. A bio-inspired total synthesis of tetrahydrofuran lignans. *Angew. Chem. Int. Ed.* **2015**, *54*, 2204–2208. [CrossRef] [PubMed]

47. Zhao, C.; Rakesh, K.P.; Mumtaz, S.; Moku, B.; Asiri, A.M.; Marwani, H.M.; Manukumar, H.M.; Qin, H.-L. Arylnaphthalene lactone analogues: Synthesis and development as excellent biological candidates for future drug discovery. *RSC Adv.* **2018**, *8*, 9487–9502. [CrossRef]

48. Pohmakotr, M.; Kuhakarn, C.; Reutrakul, V.; Soorukram, D. Asymmetric synthesis of furofurans. *Tetrahedron Lett.* **2017**, *58*, 4740–4746. [CrossRef]

49. Sun, J.-S.; Liu, H.; Guo, X.-H.; Liao, J.-X. The chemical synthesis of aryltetralin glycosides. *Org. Biomol. Chem.* **2016**, *14*, 1188–1200. [CrossRef]

50. Sellars, J.D.; Steel, P.G. Advances in the synthesis of aryltetralin lignan lactones. *Eur. J. Org. Chem.* **2007**, *2007*, 3815–3828. [CrossRef]

51. Brown, R.C.D.; Swain, N.A. Synthesis of furofuran lignans. *Synthesis* **2004**, 811–827. [CrossRef]

52. Angle, S.R.; Choi, I.; Tham, F.S. Stereoselective synthesis of 3-alkyl-2-aryltetrahydrofuran-4-ols: Total synthesis of (±)-paulownin. *J. Org. Chem.* **2008**, *73*, 6268–6278. [CrossRef] [PubMed]

53. Kraus, G.A.; Chen, L. A total synthesis of racemic paulownin using a type II photocyclization reaction. *J. Am. Chem. Soc.* **1990**, *112*, 3464–3466. [CrossRef]

54. Rye, C.E.; Barker, D. Asymmetric synthesis of (+)-galbelgin, (−)-kadangustin J, (−)-cyclogalgravin and (−)-pycnanthulignenes A and B, three structurally distinct lignan classes, using a common chiral precursor. *J. Org. Chem.* **2011**, *76*, 6636–6648. [CrossRef] [PubMed]

55. Rye, C.E.; Barker, D. An acyl-Claisen approach to tetrasubstituted tetrahydrofuran lignans: Synthesis of fragransin A_2, talaumidin, and lignan analogues. *Synlett* **2009**, *2009*, 3315–3319.

56. Li, H.; Zhang, Y.; Xie, X.; Ma, H.; Zhao, C.; Zhao, G.; She, X. Bioinspired total synthesis of gymnothelignan N. *Org. Lett.* **2014**, *16*, 4440–4443. [CrossRef] [PubMed]

57. Skubi, K.L.; Blum, T.R.; Yoon, T.P. Dual catalysis strategies in photochemical synthesis. *Chem. Rev.* **2016**, *116*, 10035–10074. [CrossRef] [PubMed]

58. Prier, C.K.; Rankic, D.A.; MacMillan, D.W.C. Visible light photoredox catalysis with transition metal complexes: Applications in organic synthesis. *Chem. Rev.* **2013**, *113*, 5322–5363. [CrossRef] [PubMed]

59. Tucher, J.W.; Stephenson, C.R.J. Shining light on photoredox catalysis: Theory and synthetic applications. *J. Org. Chem.* **2012**, *77*, 1617–1622. [CrossRef]

60. Welin, E.R.; Warkentin, A.A.; Conrad, J.C.; MacMillian, D.W.C. Enantioselective α-alkylation of aldehydes by photoredox organocatalysis: Rapid access to pharmacophore fragments from β-cyanoaldehydes. *Angew. Chem. Int. Ed.* **2015**, *54*, 9668–9672. [CrossRef]

61. Chaimanee, S.; Pohmakotr, M.; Kuhakarn, C.; Reutrakul, V.; Soorukram, D. Asymmetric synthesis of ent-fragransin C_1. *Org. Biomol. Chem.* **2017**, *15*, 3985–3994. [CrossRef] [PubMed]

62. Jagtap, P.R.; Ford, L.; Deister, E.; Pohl, R.; Císařová, I.; Hodek, J.; Weber, J.; Mackman, R.; Bahador, G.; Jahn, U. Highly functionalized and potent antiviral cyclopentane derivatives formed by a tandem process consisting of organometallic, transition-metal-catalyzed, and radical reaction steps. *Chem. Eur. J.* **2014**, *20*, 10298–10304. [CrossRef] [PubMed]

63. Jagtap, P.R.; Císařová, I.; Jahn, U. Bioinspired total synthesis of tetrahydrofuran lignans by tandem nucleophilic addition/redox isomerization/oxidative coupling and cycloetherification reactions as key steps. *Org. Biomol. Chem.* **2018**, *16*, 750–755. [CrossRef]

64. KC, S.; Basnet, P.; Thapa, S.; Shrestha, B.; Giri, R. Ni-catalyzed regioselective dicarbofunctionalization of unactivated olefins by tandem cyclization/cross-coupling and application to the concise synthesis of lignan natural products. *J. Org. Chem.* **2018**, *83*, 2920–2936. [CrossRef] [PubMed]

65. Chen, W.-W.; Zhao, Q.; Xu, M.-H.; Lin, G.-Q. Nickel-catalyzed asymmetric Ullmann coupling for the synthesis of axially chiral tetra-ortho-substituted biaryl dials. *Org. Lett.* **2010**, *12*, 1072–1075. [CrossRef] [PubMed]

66. Molander, G.A.; George, K.M.; Monovich, L.G. Total synthesis of (+)-isoschizandrin utilizing a samarium(II) iodide-promoted 8-endo ketyl-olefin cyclization. *J. Org. Chem.* **2003**, *68*, 9533–9540. [CrossRef] [PubMed]

67. Zheng, S.; Aves, S.J.; Laraia, L.; Galloway, W.R.J.D.; Pike, K.G.; Wu, W.; Spring, D.R. A concise total synthesis of deoxyschizandrin and exploration of its antiproliferative effects and those of structurally related derivatives. *Chem. Eur. J.* **2012**, *18*, 3193–3198. [CrossRef] [PubMed]

68. Surry, D.S.; Su, X.B.; Fox, D.J.; Franckevicius, V.; Macdonald, S.J.F.; Spring, D.R. Synthesis of medium-ring and iodinated biaryl compounds by organocuprate oxidation. *Angew. Chem. Int. Ed.* **2005**, *44*, 1870–1873. [CrossRef]

69. Gong, W.; RajanBabu, T.V. Conformation and reactivity in dibenzocyclooctadienes (DBCOD). A general approach to the total synthesis of fully substituted DBCOD lignans via borostannylative cyclization of α,ω-diynes. *Chem. Sci.* **2013**, *4*, 3979–3985. [CrossRef]

70. Gong, W.; Singidi, R.R.; Gallucci, J.C.; RajanBabu, T.V. On the stereochemistry of acetylide additions to highly functionalized biphenylcarbaldehydes and multi component cyclization of 1,n diynes. Syntheses of dibenzocyclooctadiene lignans. *Chem. Sci.* **2012**, *3*, 1221–1230. [CrossRef]

71. Chen, P.; Huo, L.; Li, H.; Liu, L.; Yuan, Z.; Zhang, H.; Feng, S.; Xie, X.; Wang, X.; She, X. Bioinspired total synthesis of (−)-gymnothelignan L. *Org. Chem. Front.* **2018**, *5*, 1124–1128. [CrossRef]

72. Soorukram, D.; Pohmakotr, M.; Kuhakarn, C.; Reutrakul, V. Bioinspired asymmetric synthesis of (−)-gymnothelignan V. *J. Org. Chem.* **2018**, *83*, 4173–4179. [CrossRef] [PubMed]

73. Sun, B.-F.; Hong, R.; Kang, Y.-B.; Deng, L. Asymmetric total synthesis of (−)-plicatic acid via a highly enantioselective and diastereoselective nucleophilic epoxidation of acyclic trisubstitued olefins. *J. Am. Chem. Soc.* **2009**, *131*, 10384–10385. [CrossRef] [PubMed]

74. Tamao, K.; Ishida, N.; Tanak, T.; Kunada, M. Silafunctional compounds in organic synthesis. Part 20. Hydrogen peroxide oxidation of the silicon-carbon bond in organoalkoxysilanes. *Organometallics* **1983**, *2*, 1694–1696. [CrossRef]

75. Tamao, K.; Ishida, N.; Kunada, M. Chirality transfer in stereoselective synthesis. A highly stereoselective synthesis of optically active vitamin E side chains. *J. Org. Chem.* **1983**, *48*, 2122–2124.

76. Hong, B.-C.; Hsu, C.-S.; Lee, G.-H. Enantioselective total synthesis of (+)-galbulin *via* organocatalytic domino Michael–Michael–aldol condensation. *Chem. Commun.* **2012**, *48*, 2385–2387. [CrossRef]

77. Zhang, J.-J.; Yan, C.-S.; Peng, Y.; Luo, Z.-B.; Xu, X.-B.; Wang, Y.-W. Total synthesis of (±)-sacidumlignans D and A through Ueno–Stork radical cyclization reaction. *Org. Biomol. Chem.* **2013**, *11*, 2498–2513. [CrossRef] [PubMed]

78. Peng, Y.; Luo, Z.-B.; Zhang, J.-J.; Luo, L.; Wang, Y.-W. Collective synthesis of several 2,7′-cyclolignans and their correlation by chemical transformations. *Org. Biomol. Chem.* **2013**, *11*, 7574–7586. [CrossRef]

79. Patel, R.M.; Argade, N.P. Palladium-promoted [2 + 2 + 2] cocyclization of arynes and unsymmetrical conjugated dienes: Synthesis of justicidin B and retrojusticidin B. *Org. Lett.* **2013**, *15*, 14–17. [CrossRef]

80. Kao, T.-T.; Lin, C.-C.; Shia, K.-S. The total synthesis of retrojusticidin B, justicidin E, and helioxanthin. *J. Org. Chem.* **2015**, *80*, 6708–6714. [CrossRef]

81. Naresh, G.; Kant, R.; Narender, T. Silver(I)-Catalyzed regioselective construction of highly substituted α-naphthols and its application toward expeditious synthesis of lignan natural products. *Org. Lett.* **2015**, *17*, 3446–3449. [CrossRef] [PubMed]

82. Singh, O.V.; Tapadiya, S.M.; Deshmukh, R.G. Process for the Synthesis of Cleistanthin A. WO 2010009778 A2, 12 August 2010.

83. Kim, T.; Jeong, K.H.; Kang, K.S.; Nakata, M.; Ham, J. An optimized and general synthetic strategy to prepare arylnaphthalene lactone natural products from cyanophthalides. *Eur. J. Org. Chem.* **2017**, *2017*, 1704–1712. [CrossRef]

84. Hjra, S.; Garai, S.; Hazra, S. Catalytic Enantioselective Synthesis of (−)-Podophyllotoxin. *Org. Lett.* **2017**, *19*, 6530–6533. [CrossRef] [PubMed]

85. Lisiecki, K.; Czarnocki, Z. Flow photochemistry as a tool for the total synthesis of (+)-epigalcatin. *Org. Lett.* **2018**, *20*, 605–607. [CrossRef] [PubMed]

86. Lisiecki, K.; Krawczyk, K.K.; Roszkowski, P.; Maurin, J.K.; Czarnocki, Z. Formal synthesis of (−)-podophyllotoxin through the photocyclization of an axially chiral 3,4-bisbenzylidene succinate amide ester-a flow photochemistry approach. *Org. Biomol. Chem.* **2016**, *14*, 460–469. [CrossRef] [PubMed]

87. Xiao, J.; Cong, X.-W.; Yang, G.-Z.; Wang, Y.-W.; Peng, Y. Divergent asymmetric syntheses of podophyllotoxin and related family members via stereoselective reductive Ni-catalysis. *Org. Lett.* **2018**, *20*, 1651–1654. [CrossRef] [PubMed]

88. Xiao, J.; Cong, X.-W.; Yang, G.-Z.; Wang, Y.-W.; Peng, Y. Stereoselective synthesis of a Podophyllum lignan core by intramolecular reductive nickel-catalysis. *Chem. Commun.* **2018**, *54*, 2040–2043. [CrossRef] [PubMed]

89. Long, R.; Huang, J.; Shao, W.; Liu, S.; Lan, Y.; Gong, J.; Yang, Z. Asymmetric total synthesis of (−)-lingzhiol via a Rh-catalysed [3 + 2] cycloaddition. *Nat. Commun.* **2014**, *5*, 5707. [CrossRef] [PubMed]

90. Andrews, R.C.; Teague, S.J.; Meyers, A.I. Asymmetric total synthesis of (−)-podophyllotoxin. *J. Am. Chem. Soc.* **1988**, *110*, 7854–7858. [CrossRef]

91. Belozerova, O.A.; Deigin, V.I.; Khrushchev, A.Y.; Dubinnyi, M.A.; Kublitski, V.S. The total synthesis of sevanol, a novel lignan isolated from the thyme plant (Thymus armeniacus). *Tetrahedron* **2018**, *74*, 1449–1453. [CrossRef]

92. Sampei, M.; Arai, M.A.; Ishibashi, M. Total syntheses of schizandriside, saracoside and (±)-isolariciresinol with antioxidant activities. *J. Nat. Med.* **2018**, *72*, 651–654. [CrossRef] [PubMed]

93. Davidson, S.J.; Barker, D. Total synthesis of ovafolinins A and B: Unique poly-cyclic benzoxepin lignans through a cascade cyclization. *Angew. Chem. Int. Ed.* **2017**, *56*, 9483–9486. [CrossRef] [PubMed]

94. Fang, X.; Shen, L.; Hu, X. Asymmetric total synthesis of (+)-ovafolinins A and B. *Chem. Commun.* **2018**, *54*, 7539–7541. [CrossRef] [PubMed]

molecules

MDPI

Article

Modular Synthesis and Biological Investigation of 5-Hydroxymethyl Dibenzyl Butyrolactones and Related Lignans

Samuel J. Davidson [1], Lisa I. Pilkington [1], Nina C. Dempsey-Hibbert [2], Mohamed El-Mohtadi [2], Shiying Tang [2], Thomas Wainwright [2], Kathryn A. Whitehead [2] and David Barker [1,3,*]

[1] School of Chemical Sciences, University of Auckland, Auckamd 1010, New Zealand;
sdav134@aucklanduni.ac.nz (S.J.D.); lisa.pilkington@auckland.ac.nz (L.I.P.)

[2] School of Healthcare Science, Manchester Metropolitan University, Manchester M1 5GD, UK;
N.Dempsey-Hibbert@mmu.ac.uk (N.C.D.-H.); MOHAMED.EL-MOHTADI@stu.mmu.ac.uk (M.E.-M.);
SHIYING.TANG@stu.mmu.ac.uk (S.T.); thomas.wainwright@stu.mmu.ac.uk (T.W.);
K.A.Whitehead@mmu.ac.uk (K.A.W.)

[3] The MacDiarmid Institute for Advanced Materials and Nanotechnology, Wellington 6140, New Zealand

* Correspondence: d.barker@auckland.ac.nz; Tel.: +64-9-373-7599

Received: 12 November 2018; Accepted: 21 November 2018; Published: 22 November 2018

Abstract: Dibenzyl butyrolactone lignans are well known for their excellent biological properties, particularly for their notable anti-proliferative activities. Herein we report a novel, efficient, convergent synthesis of dibenzyl butyrolactone lignans utilizing the acyl-Claisen rearrangement to stereoselectively prepare a key intermediate. The reported synthetic route enables the modification of these lignans to give rise to 5-hydroxymethyl derivatives of these lignans. The biological activities of these analogues were assessed, with derivatives showing an excellent cytotoxic profile which resulted in programmed cell death of Jurkat T-leukemia cells with less than 2% of the incubated cells entering a necrotic cell death pathway.

Keywords: lignans; dibenzyl butyrolactones; anti-proliferative; acyl-Claisen; stereoselective synthesis

1. Introduction

Dibenzyl butyrolactone lignans **1** are a class of lignans which have been reported to exhibit a range of biological activities, including, but not limited to neuroprotective [1], anti-cancer [2,3], anti-inflammatory [2,4], and anti-aging effects (see Figure 1) [5]. Perhaps the most notable of these biological properties is their reported potent anti-proliferative activities; examples of this class include (−)-matairesinol **2** and (−)-arctigenin **3** which, along with their synthesized derivatives, have been shown to exhibit excellent activity against various cancer cell lines, including pancreatic, breast, endometrial, colorectal, lung, and bladder cancers [6–12].

Owing to their anti-cancer properties and their classification as drug-like compounds [13] extensive work has gone into the study of these compounds and their related analogues to explore and establish structure–activity relationships and the possible use of these lignans as lead compounds for therapeutics. Whilst previous work has explored the synthesis of these lignans and analogues thereof [14–16], mainly focusing on changing the substituents on the aryl rings [17], one area that has not been extensively investigated is the synthesis of C-5 substituted analogues of these butyrolactone lignans, represented by **4**.

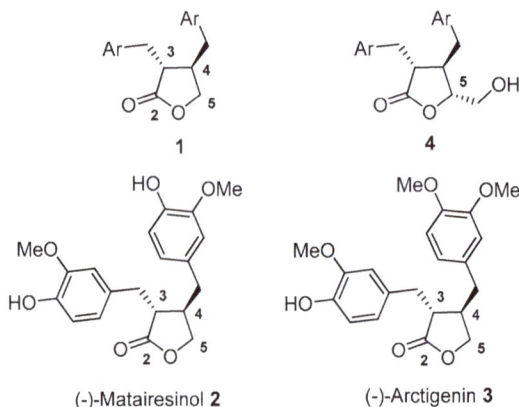

Figure 1. General structures of butyrolactone lignan **1**, natural dibenzylbutyrolactone lignans, (−)-matairesinol **2** and (−)-artigenin **3**, and 5-hydroxymethyl analogues **4**.

We have previously shown that the acyl-Claisen rearrangement can be used to prepare disubstituted morpholine pentenamides **5** with high diastereoselectivity at the C-3 and C-4 positions which correspond to the benzyl groups in the lactone scaffold (Figure 2) [18–22]. Furthermore, in our efforts to a prepare a number of different lignan scaffolds [18–36], we have used amides such as **5** to prepare compounds including tetrahydrofuran lignans (e.g., galbelgin **6**), aryltetralins (e.g., ovafolinin **7**) and aryl dihydronaphthalene lignans (e.g., (−)-pycananthuligene B **8**).

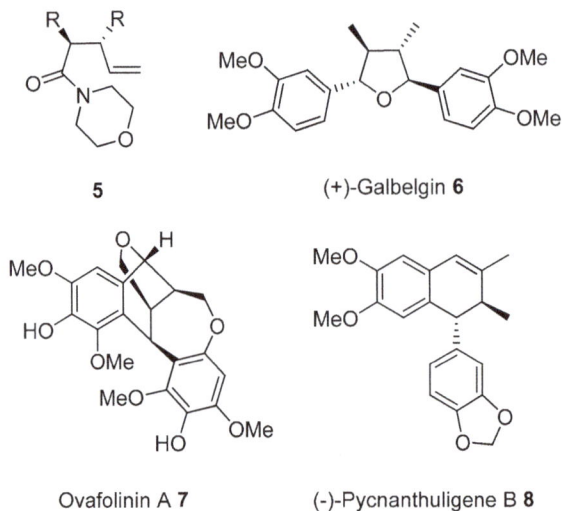

Figure 2. Use of amide **5**, the product of an acyl-Claisen rearrangement to access a number of lignan scaffolds and natural products **6–8**.

We wished to explore the usage of this methodology to synthesise butyrolactone lignans, as well as probe the effect of adding a substituent at the C-5 position on the biological activity. The route would be convergent and modular, allowing for simple modification of aromatic groups resulting in the synthesis of a number of analogues.

2. Results and Discussion

In order to utilise the acyl-Claisen rearrangement to prepare the desired lactones, the corresponding allylic morpholines and acid chlorides first needed to be synthesised. Allylic morpholines **9a** and **9b** were synthesised in five steps from 4-allyl-1,2-dimethoxybenzene **10** and safrole **11** (Scheme 1), respectively. Firstly, allylic benzenes **10** and **11** were dihydroxylated using catalytic osmium tetroxide giving **12** and **13**, followed by periodate cleavage to give aldehydes **14** and **15**. Aldehydes **14** and **15** were immediately used in a Wittig reaction with (carbethoxymethylene)-triphenylphosphorane to exclusively give the *E*-isomer of α,β-unsaturated esters **16** and **17**, in 55% and 56% yields, respectively, over three steps. The esters **16** and **17** were then reduced to allylic alcohols **18** and **19** using di-*iso*-butyl aluminium hydride (DIBAL-H) in excellent yields. Alcohols **18** and **19** were then converted to the corresponding allylic morpholines **9a** and **9b**, by first generating a mesylate in situ, which then underwent substitution to give allylic morpholines **9a** and **9b**.

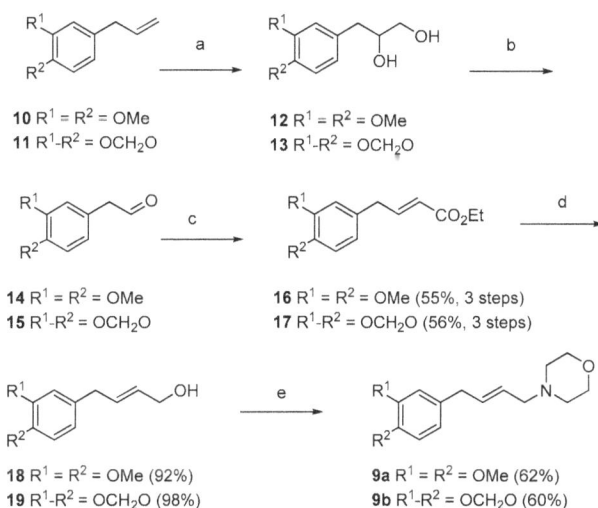

Scheme 1. (**a**) OsO$_4$ (0.1–0.3 mol%), *N*-methylmorpholine-*N*-oxide (3 eq.), ^{t}BuOH/H$_2$O (1:1), 4 days; (**b**) NaIO$_4$ (1.2 eq.), MeOH/H$_2$O (3:1), 0.5–2 h; (**c**) Ph$_3$PCHCO$_2$Et (1.1 eq.), CH$_2$Cl$_2$, 16 h; (**d**) **18**: DIBAL-H (3 eq.), CH$_2$Cl$_2$, −78 °C, 10 min, **19**: DIBAL-H (2.2 eq.), toluene, −10 °C, 10 min; (**e**) Et$_3$N (3 eq.), MsCl (1.2 eq.), morpholine (1.5 eq.), CH$_2$Cl$_2$, 0 °C, 2–18 h.

The required acid chlorides were then synthesised in four or five steps from commercially available benzaldehydes—piperonal **20**, 3,4,5-trimethoxybenzaldehyde **21** and vanillin **22** (Scheme 2). Benzaldehydes **20–22** first underwent a Wittig reaction with (carbethoxymethylene)triphenylphosphorane to give α,β-unsaturated esters **23–25** which were then hydrogenated using Pd on Carbon (10% *w/w*), giving saturated esters **26–28** in 88–94% yield over two steps. The phenol in **28** was protected as the benzyl ether, **29**, in 83% yield. Esters **26**, **27**, and **29** were hydrolysed using NaOH in methanol/water to the corresponding carboxylic acids **30**, **31**, and **32**, respectively, in 94–99% yields. Finally, chlorination of acids **30–32**, along with commercially available 3,4-dimethoxyphenyl propionic acid **33**, using oxalyl chloride gave acid chlorides **34a–d** in quantitative yields.

Acyl-Claisen rearrangements were undertaken using two allylic morpholines **9a** and **9b** which were reacted individually with the four acid chlorides **34a–d**, using TiCl$_4$·2THF as the Lewis acid, providing eight morpholine amides **35aa–bd** in 42–95% yields. All amides **35aa–bd** were obtained as single diastereomers with a *syn*-configuration between the C-2 and C-3 substituents (Scheme 3).

All amides **35aa–bd** then underwent dihydroxylation using osmium tetroxide and *N*-methylporpholine *N*-oxide (NMO) to give cyclized 5-hydroxymethyllactones **4aa–bd**.

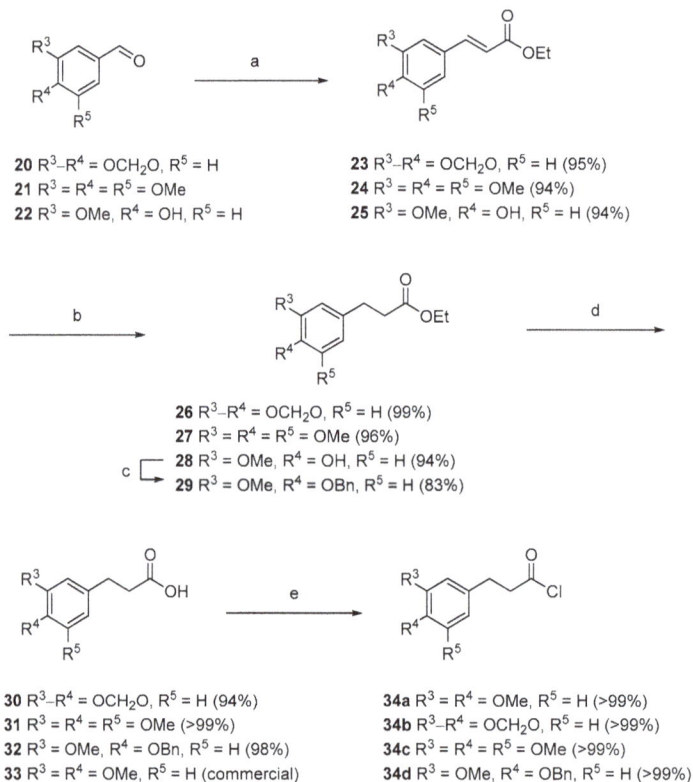

20 R^3–R^4 = OCH$_2$O, R^5 = H
21 R^3 = R^4 = R^5 = OMe
22 R^3 = OMe, R^4 = OH, R^5 = H

23 R^3–R^4 = OCH$_2$O, R^5 = H (95%)
24 R^3 = R^4 = R^5 = OMe (94%)
25 R^3 = OMe, R^4 = OH, R^5 = H (94%)

26 R^3–R^4 = OCH$_2$O, R^5 = H (99%)
27 R^3 = R^4 = R^5 = OMe (96%)
28 R^3 = OMe, R^4 = OH, R^5 = H (94%)
29 R^3 = OMe, R^4 = OBn, R^5 = H (83%)

30 R^3–R^4 = OCH$_2$O, R^5 = H (94%)
31 R^3 = R^4 = R^5 = OMe (>99%)
32 R^3 = OMe, R^4 = OBn, R^5 = H (98%)
33 R^3 = R^4 = OMe, R^5 = H (commercial)

34a R^3 = R^4 = OMe, R^5 = H (>99%)
34b R^3–R^4 = OCH$_2$O, R^5 = H (>99%)
34c R^3 = R^4 = R^5 = OMe (>99%)
34d R^3 = OMe, R^4 = OBn, R^5 = H (>99%)

Scheme 2. (**a**) Ph$_3$PCHCO$_2$Et (1.1 eq.), CH$_2$Cl$_2$, 3–20 h; (**b**) H$_2$, Pd/C (10% *w/w*), ethyl acetate, 1–2 h; (**c**) BnBr, K$_2$CO$_3$, CH$_3$CN, 65 h; (**d**) NaOH (4 eq.), MeOH/H$_2$O, 2.5 h; (**e**) (COCl)$_2$ (2 eq.), CH$_2$Cl$_2$, 1.5–4 h.

In all cases it was observed that only the 3,4-*trans*-4,5-*trans*-lactone was obtained. This configuration was confirmed through NOESY NMR analysis, depicted in Figure 3 with **4bb**. We propose that only this isomer was obtained due to the preferential cyclisation of the 3,4-*anti* diol **36**, leaving the polar uncyclised 3,4-*syn* diols **37** which were difficult to isolate. Upon dihydroxylation of amide **35bb** at a larger scale and following isolation of lactone **4bb** by column chromatography, a small sample of the corresponding uncyclised diol **37** was able to be isolated. This diol **37** was subsequently cyclised using 2 M H$_2$SO$_4$ in methanol to give the corresponding C-5 epimer, *epi*-**4bb**, confirming this hypothesis (Scheme 4).

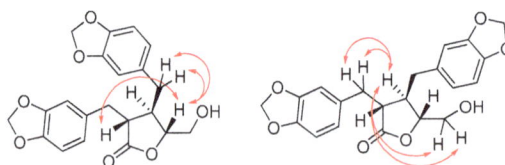

Figure 3. Selected NOESY correlations showing *trans,trans*-relationship of hydroxymethyl lactone lignan analogue **4bb**.

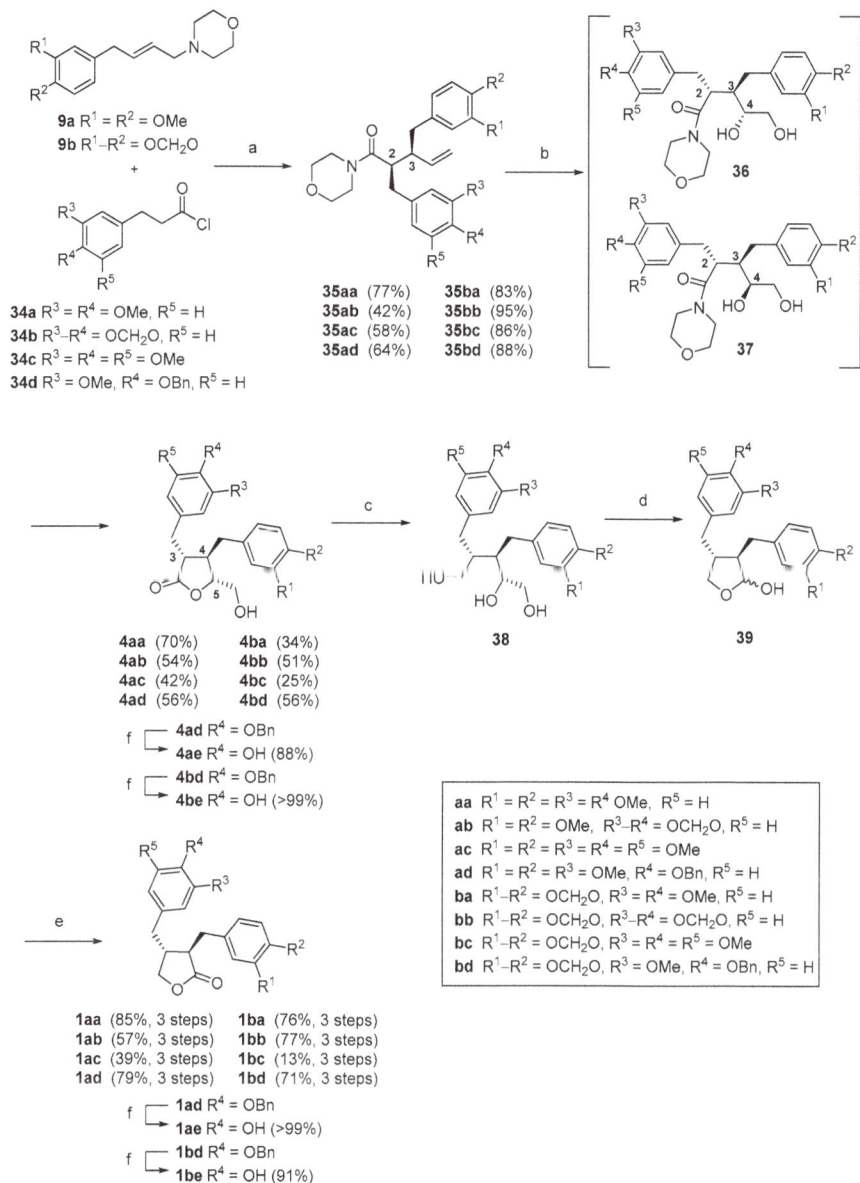

Scheme 3. (a) TiCl$_4$·2THF (100 mol%), ^{i}Pr$_2$NEt (1.5 eq.), acid chloride (1.2 eq.), CH$_2$Cl$_2$, 18–24 h; (b) OsO$_4$ (8 mol %), NMO (3 eq.), ^{t}BuOH/H$_2$O (1:1), 3–7 days; (c) LiAlH$_4$ (1.5 eq.), THF, 0.5–2 h; (d) NaIO$_4$ (1.2 eq.), MeOH/H$_2$O (3:1), 0.25–1 h; (e) Ag$_2$CO$_3$/Celite (2 eq.), toluene, reflux, 2–3 h; (f) H$_2$, Pd/C (10% w/w), MeOH, 10 min.

Finally, to deprotect the benzyl-protected lactones **4ad** and **4bd** to their respective alcohols, they were subjected to hydrogenolysis to give **4ae** and **4be** in excellent yields. Transformation of C-5 hydroxymethyl analogues **4** into dibenzylbutryolactone lignans **1** was achieved via reduction using LiAlH$_4$, to the corresponding triols **38aa–bd**, followed by periodate cleavage, forming lactols **39aa–bd**. These lactols **39aa–bd** were then oxidised using Fetizon's reagent [37,38] to give racemic samples of dibenzyl butyrolactone lignans **1aa–bd**, including known natural products arcitin **1aa**, bursehernin **1ab**, (3*R**,4*R**)-3-(3″,4″-dimethoxybenzyl)-4-(3′,4′,5′-trimethoxybenzyl)dihydrofuran-2(3*H*)-one **1ac**, kusunokinin **1ba**, hinokinin **1bb**, and isoyatein **1bc**. Additionally, phenolic lignans, buplerol **1ae**, and haplomyrfolin **1be** were produced by the debenzylation of **1ad** and **1bd**, respectively.

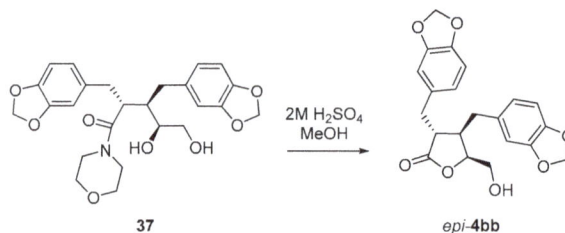

Scheme 4. Synthesis of *epi*-**4bb**.

Several of the synthesised compounds were then tested for their anti-microbial and cytotoxic activities. All tested compounds were found to be inactive against *Staphlycoccus aureus* and *Escherichia. coli*, showing no to little antimicrobial activity, while the compounds were shown to exhibit antiproliferative effects against Jurkat T-leukaemia cells, while also showing effects on cell cycle progression (Figure 4). While the synthesised naturally-occurring dibenzyl butyrolactones, arcitin **1aa**, bursehernin **1ab**, and (3*R**,4*R**)-3-(3″,4″-dimethoxybenzyl)-4-(3′,4′,5′-trimethoxybenzyl)dihydrofuran-2(3*H*)-one **1ac**, boasted the best activities, 5-hydroxymethyl analogue **4bb** had similar potency. Compound **4bb** was shown to have the best activity of all of the 5-hydroxymethyl analogues tested, inducing apoptosis, evidenced by the presence of cells in the early and predominantly in the late apoptotic cell cycle (Figure 4). Additionally the compounds demonstrated an effect on cell cycle progression. A significantly greater number of 4N cells were present following treatment with compound **4bb** in particular causing a significant increase in 4N cells (Figure 4D,E). During the cell cycle, DNA is replicated in the S-phase, going from 2N in G_1, to 4N by the end of this phase. The DNA content in cells then remains at 4N during G_2 and M phases, before cytokinesis at the M-phase. The observation that there was in increase in 4N cells indicates that it is likely these cells have arrested in G2/M and will not re-enter next G_1-phase after this mitotic slippage. This is in-line with published cell cycle data following treatment with other lignans [39,40]. Furthermore, our compounds showed minimal levels of necrosis, less than 2% (except **4ba** with 7%), suggesting that the cells are in fact entering programmed cell death cycles, which is considered the most effective and non-inflammatory mechanism of cancer-cell death.

In conclusion, the synthesis of dibenzyl butyrolactone lignans utilising the acyl-Claisen rearrangement has been accomplished and represent a new, modular, and convergent method towards the synthesis of this class of natural products. Furthermore, this route gives rise to the previously-unexplored 5-hydroxymethyl derivatives **4** of these natural products. The biological activities of this new set of derivatives were assessed, with one derivative in particular, **4bb**, showing a superior cytotoxic profile and resulting in cell cycle arrest and programmed cell death of Jurkat T-leukaemia cells with less than 2% of the incubated cells entering a necrotic cell death pathway.

A

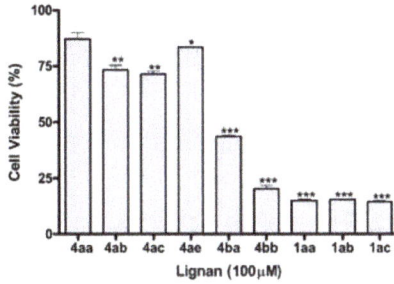

B

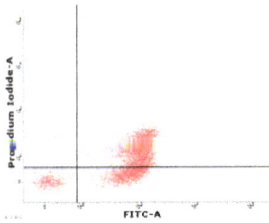

4bb:

Viable:	9.22%	Late Apoptosis:	59.71%
Early Apoptosis:	29.56%	Necrosis:	1.51%

C

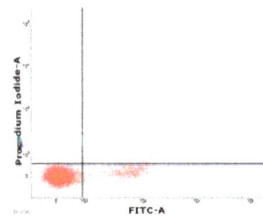

Control:

Viable:	85.22%	Late Apoptosis:	9.06%
Early Apoptosis:	4.97%	Necrosis:	0.76%

D

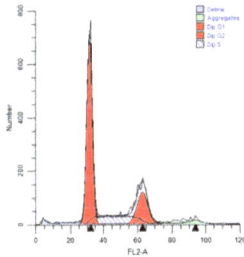

G0/G1: 63.02 ± 6.47   S: 20.33 ± 3.23   G2/M: 18.63 ± 2.06

E

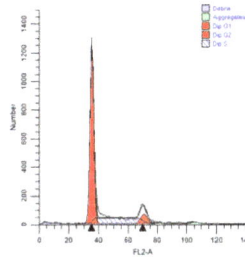

G0/G1: 68.05 ±0.007   S: 25.60 ± 1.66   G2/M: 8.3 ±1.46

Figure 4. (**A**) Cell survival (by a measure of metabolic activity) of Jurkat T-cell leukaemia cells incubated with 100 µM of lignans and lignan analogues for 48 h. The data represents means of triplicate experiments and is shown as means ± SEM ($n = 3$). The positive control (not shown) had a growth of 100%. Significance of the compound activity compared to the control is expressed: (*) p-value <0.05; (**) p-value <0.01; (***) p-value <0.001. (**B**) Dotplot showing the viability of Jurkat T-leukaemia cells after incubation with 100 µM **4bb** for 24 h followed by labelling with annexin V/propidium iodide and analysis using flow cytometry. Cells in the bottom-left quadrant represent viable cells, bottom-right quadrant are positive for annexin V and are in early apoptosis, top-right quadrant are double positive for annexin V and propidium iodide and are in late apoptosis, and top-left quadrant are only positive for propidium iodide and are undergoing necrosis. (**C**) Negative control showing the viability of vehicle-(DMSO) treated Jurkat T-leukaemia cells. (**D**) Cell cycle analysis of unsynchronized cells incubated in the presence of 100 µM **4bb** or **E:** vehicle for 24 h. DNA content of the cells was determined by flow cytometry. Percentage of cells in each stage of the cell cycle (average of three replicates ± SD is reported).

3. Experimental Section

3.1. General Methods

All reactions were carried out with oven-dried glassware and under a nitrogen atmosphere in dry, freshly distilled solvents unless otherwise noted. Diisopropylethylamine was distilled from CaH_2 and stored over activated 4Å molecular sieves. All melting points for solid compounds, given in degrees Celsius (°C), were measured using a Reicher–Kofler block and are uncorrected. Infrared (IR) spectra were recorded using a Perkin Elmer Spectrum1000 FT-IR spectrometer. The NMR spectra were recorded on a 400 MHz spectrometer. Chemical shifts are reported relative to the solvent peak of chloroform (δ 7.26 for ^{1}H and δ 77.16 ± 0.06 for ^{13}C). The ^{1}H-NMR data was reported as position (δ), relative integral, multiplicity (s, singlet; d, doublet; dd, doublet of doublets; ddd, doublet of doublet of doublets; dt, doublet of triplets; dq, doublet of quartets; t, triplet; td, triplet of doublets; q, quartet; m, multiplet), coupling constant (*J*, Hz), and the assignment of the atom. The ^{13}C-NMR data were reported as position (δ) and assignment of the atom. The NMR assignments were performed using COSY, HSQC and HMBC experiments. High-resolution mass spectroscopy (HRMS) was carried out by electrospray ionization (ESI) on a MicroTOF-Q mass spectrometer. Fetizon's reagent was prepared following a literature procedure [41]. Unless noted, chemical reagents were used as purchased.

3.2. Synthetic Methods

3.2.1. General Procedure A: Acyl-Claisen

To a stirred suspension of TiCl$_4$·2THF (1 mmol) in CH_2Cl_2 (5 mL), under an atmosphere of nitrogen, was added a solution of allylic morpholine (1 mmol) in CH_2Cl_2 (2.5 mL) followed by dropwise addition of ^{i}Pr$_2$NEt (1.5 mmol). After stirring for 10 min a solution of acid chloride (1.2 mmol) in CH_2Cl_2 (2.5 mL) was added dropwise and the resultant mixture stirred for the specified time. The reaction mixture was quenched with aqueous NaOH (12 mL, 1 M) and the aqueous phase extracted with CH_2Cl_2 (3 × 10 mL). The combined organic extracts were washed with brine (6 mL), dried (MgSO$_4$), the solvent removed in vacuo and the crude product purified by column chromatography.

3.2.2. General Procedure B: Dihydroxylation

To a stirred solution of morpholine pentenamide (1 mmol) in ^{t}BuOH/H$_2$O (1:1, 20 mL) or ^{t}BuOH/H$_2$O/THF (1:1:1, 30 mL) was added NMO (3 mmol). A solution of OsO$_4$ (0.08 mmol, 2.5% *w/v* in ^{t}BuOH) was then added dropwise and the resultant mixture stirred for the specified time. The mixture was quenched with saturated aqueous Na$_2$SO$_3$ (30 mL) and stirred for a further 1 h. The aqueous phase was extracted with ethyl acetate (3 × 20 mL), the combined organic extracts washed with aqueous KOH (5 mL, 1 M), dried (MgSO$_4$), the solvent removed in vacuo and the crude product purified by column chromatography.

3.2.3. General Procedure C: Lithium Aluminum Hydride Reduction

To a stirred suspension of LiAlH$_4$ (1.4 mmol) in THF (10 mL), under an atmosphere of nitrogen at 0 °C, was added a solution of lactone (1 mmol) in THF (10 mL) and the mixture stirred for the specified time. After warming to room temperature, the mixture was quenched with the addition of water (30 mL) and the aqueous phase extracted with ethyl acetate (3 × 40 mL). The combined organic extracts were washed with brine (25 mL), dried (MgSO$_4$), and the solvent removed in vacuo.

3.2.4. General Procedure D: Periodate Cleavage

To a stirred solution of triol (1 mmol) in MeOH/H$_2$O (3:1, 50 mL) was added NaIO$_4$ (1.2 mmol) and the resultant mixture stirred for the specified time. The reaction mixture was quenched with brine (40 mL) and extracted with ethyl acetate (3 × 80 mL). The organic layers were combined, washed with water (2 × 40 mL), dried (MgSO$_4$), and solvent removed in vacuo to give the crude product which was purified by column chromatography if necessary.

3.2.5. General Procedure E: Fétizon's Oxidation

To a stirred solution of lactol (1 mmol) in toluene (60 mL), under an atmosphere of nitrogen, was added Fétizon's reagent (2 mmol) and heated at reflux for the specified time. The reaction mixture was allowed to cool and filtered, the solvent removed in vacuo and the crude product purified by column chromatography.

3.2.6. General Procedure F: Benzyl Deprotection

To a stirred solution of benzyl ether (1 mmol) in MeOH (30 mL) was added 10% palladium on carbon (20% w/w) and the resultant mixture stirred under and atmosphere of hydrogen for the specified time. The reaction mixture was filtered through celite, washed with methanol (3 × 20 mL), the solvent removed in vacuo and the crude product purified by column chromatography if necessary (The ^{1}H and ^{13}C-NMR spectra of compounds in the Supplemental Materials).

(E)-Ethyl 4-(3′,4′-dimethoxyphenyl)but-2-enoate (**16**). To a stirred solution of NMO (7.9 g, 67.3 mmol) in H$_2$O/^{t}BuOH (1:1, 80 mL) was added 4-allyl-1,2-dimethoxybenzene **10** (3.86 mL, 22.4 mmol). A solution of OsO$_4$ (0.6 mL, 0.059 mmol, 2.5% w/v in ^{t}BuOH) was then added dropwise and the resulting mixture stirred at room temperature for 4 days. The mixture was then quenched with saturated aqueous Na$_2$SO$_3$ (100 mL) and stirred for 1 h. The mixture was extracted with ethyl acetate (3 × 50 mL), the organic layers combined, washed with aqueous KOH (1 M, 20 mL), and dried (MgSO$_4$). Solvent was removed in vacuo to give **12** (4.8 g, quant.) as a white solid which was used without further purification. To a stirred solution of diol **12** (4.8 g, 22.8 mmol) in methanol/H$_2$O (3:1, 100 mL) was added NaIO$_4$ (5.9 g, 27.4 mmol) and stirred for 30 min. The reaction mixture was then quenched with addition of brine (50 mL) and extracted with ethyl acetate (3 × 40 mL). The organic extracts were combined, washed with water (2 × 20 mL), and dried (MgSO$_4$). Solvent was removed in vacuo to give **14** (2.68 g, 65%) as a pale-yellow oil which was used without further purification. To a stirred solution of 2-(3,4-dimethoxyphenyl)acetaldehyde **14** (2.68 g, 14.8 mmol) in CH$_2$Cl$_2$ (100 mL), under an atmosphere of nitrogen, was added (carbethoxymethylene)triphenylphosphorane (5.7 g, 16.3 mmol) and the resulting mixture stirred for 16 h. Solvent was removed in vacuo and the crude product purified by column chromatography (3:1, hexanes, ethyl acetate) to give the title compound **16** (3.13 g, 84%) as a colourless oil. R$_f$ = 0.56 (2:1 hexanes, ethyl acetate). δ_H (400 MHz; CDCl$_3$) 1.27 (3H, t, J = 7.2 Hz, 1-OCH$_2$CH$_3$), 3.45 (2H, dd, J = 1.5, 6.7 Hz, 4-H), 3.86 (6H, s, 3′, 4′-H), 4.17 (2H, q, J = 7.2 Hz, 1-OCH$_2$CH$_3$), 5.80 (1H, td, J = 1.6, 15.5 Hz, 2-H), 6.67 (1H, d, J = 1.9 Hz, 2′-H), 6.71 (1H, dd, J = 1.9, 8.1 Hz, 6′-H), 6.81 (1H, d, J = 8.1 Hz, 5′-H), 7.07 (1H, td, J = 6.7, 15.5 Hz, 3-H). δ_C (100 MHz; CDCl$_3$) 14.3 (1-OCH$_2$CH$_3$), 38.1 (C-4), 55.9, 56.0 (3′, 4′-OCH$_3$), 60.3 (1-OCH$_2$CH$_3$), 111.5 (C-5′), 112.1 (C-2′), 120.8 (C-6′), 122.2 (C-2), 130.2 (C-1′), 147.6 (C-3), 147.9 (C-4′), 149.1 (C-3′), 166.6 (C-1). Values are in agreement with literature data [42].

(E)-4-(3′,4′-Dimethoxyphenyl)but-2-en-1-ol (**18**). To a stirred solution of ester **16** (1.0 g, 4.0 mmol) in CH$_2$Cl$_2$ (20 mL), under an atmosphere of nitrogen at −78 °C, was added DIBAL (12 mL, 1 M in cyclohexane) and the resulting mixture stirred for 10 min. The reaction mixture was quenched with addition of 2 M HCl until gas evolution ceased, the organic phase separated and the aqueous phase further extracted with CH$_2$Cl$_2$ (3 × 10 mL). The organic layers were combined then washed with water (10 mL) and dried (MgSO$_4$). Solvent was removed in vacuo and the crude product purified by column chromatography (1:1 hexanes, ethyl acetate) to give the title compound **18** (0.76 g, 92%) as a colourless oil. R$_f$ = 0.18 (2:1, hexanes, ethyl acetate). δ_H (400 MHz; CDCl$_3$) 3.30 (2H, d, J = 6.6 Hz, 4-H), 3.82 (3H, s, 4′-OCH$_3$), 3.83 (3H, s, 3′-OCH$_3$), 4.08 (2H, d, J = 5.6 Hz, 1-H), 5.64–5.69 (1H, m, 2-H), 5.78–5.83 (1H, m, 3-H), 6.68 (1H, s, 2′-H), 6.69 (1H, d, J = 8.0 Hz, 6′-H), 6.77 (1H, d, J = 8.0 Hz, 5′-H). δ_C (100 MHz; CDCl$_3$) 38.2 (C-4), 55.8 and 55.9 (3′ and 4′-OCH$_3$), 63.3 (C-1), 111.4 (C-5′), 112.0 (C-2′), 120.4 (C-6′), 130.2 (C-2),

131.6 (C-3), 132.7 (C-1′), 147.4 (C-4′), 148.9 (C-3′). IR: ν_{MAX} (film)/cm^{-1}; 3391 (broad), 2933, 2835, 1591, 1512, 1463, 1417, 1258, 1232, 1137, 1025, 971, 852, 806, 762. HRMS (ESI$^+$) Found [M + Na]$^+$ 231.0995; $C_{12}H_{16}NaO_3$ requires 231.0992.

(E)-4-(4-(3′,4′-Dimethoxyphenyl)but-2-en-1-yl)morpholine (**9a**). To a stirred solution of alcohol **18** (0.73 g, 3.5 mmol) in CH_2Cl_2 (20 mL), under an atmosphere of nitrogen at 0 °C, was added Et$_3$N (1.5 mL, 10.5 mmol) and stirred for 5 min. MsCl (0.48 mL, 4.2 mmol) was added and stirred for 10 min. Morpholine (0.50 mL, 5.3 mmol) was added and the mixture brought to room temperature and stirred for 2 h. Saturated aqueous NaHCO$_3$ (20 mL) and water (4 mL) was then added and the aqueous layer further extracted with CH_2Cl_2 (3 × 20 mL). The organic layers were then combined, dried (MgSO$_4$) and the solvent removed in vacuo. The crude product was purified by column chromatography (1:1 hexanes, ethyl acetate) to give the title compound 9a (0.60 g, 62%) as a colourless oil. R_f = 0.31 (1:2 hexanes, ethyl acetate). δ_H (400 MHz; CDCl$_3$) 2.41–2.44 (4H, m, O(CH$_2$CH$_2$)$_2$N), 2.96 (2H, d, J = 6.8 Hz, 1-H), 3.30 (2H, d, J = 6.7 Hz, 4-H), 3.68–3.71 (4H, m, O(CH$_2$CH$_2$)$_2$N), 3.83 (6H, s, 3′, 4′-OCH$_3$), 5.52–5.57 (1H, m, 3-H), 5.71–5.78 (1H, m, 2-H), 6.67–6.70 (2H, m, 2′ and 6′-H), 6.78 (1H, d, J = 7.9 Hz, 5′-H). δ_C (100 MHz; CDCl$_3$) 38.5 (C-4), 53.6 (O(CH$_2$CH$_2$)$_2$N), 55.8, 56.0 (3′, 4′-OCH$_3$), 61.1 (C-1), 67.0 (O(CH$_2$CH$_2$)$_2$N), 111.4 (C-5′), 111.9 (C-2′), 120.3 (C-6′), 127.1 (C-3), 132.8 (C-1′), 133.8 (C-2), 147.5 (C-4′), 149.0 (C-3′). IR: ν_{MAX} (film)/cm^{-1}; 2934, 2851, 1591, 1453, 1260, 1138, 1028, 976, 864, 805, 763. HRMS (ESI$^+$) Found [M + H]$^+$ 278.1762; $C_{16}H_{24}NO_3$ requires 278.1751.

(E)-Ethyl 4-(3′,4′-methylenedioxyphenyl)but-2-enoate (**17**). To a stirred solution of NMO (8.67 g, 74.0 mmol) in H_2O/^{t}BuOH (1:1, 80 mL) was added safrole **11** (4.0 mL, 27 mmol). A solution of OsO$_4$ (0.75 mL, 0.074 mmol, 2.5% w/v in ^{t}BuOH) was added dropwise and the resultant mixture stirred at room temperature for 17 h. The reaction mixture was quenched with saturated aqueous Na$_2$SO$_3$ (100 mL) and stirred for 1 h. The mixture was extracted with ethyl acetate (3 × 50 mL), the organic layers were combined, washed with aqueous KOH (1 M, 20 mL) and dried (MgSO$_4$). Solvent was removed in vacuo to give diol **13** (5.2 g, quant.) as a white solid which was used without further purification. To a stirred solution of diol **13** (5.2 g, 27 mmol) in methanol/H_2O (3:1, 100 mL) was added NaIO$_4$ (6.8 g, 32 mmol) and stirred for 2 h. The mixture was then quenched with addition of brine (50 mL) and extracted with ethyl acetate (3 × 50 mL). The organic extracts were combined, washed with water (2 × 20 mL), brine (10 mL), and dried (MgSO$_4$). Solvent was removed in vacuo to give aldehyde **15** (4.4 g, quant.) as a yellow oil which was used without further purification. To a stirred solution of 2-(3,4-methylenedioxyphenyl)acetaldehyde **15** (4.4 g, 27 mmol) in CH_2Cl_2 (50 mL), under an atmosphere of nitrogen, was added (carbethoxymethylene)triphenylphosphorane (10.4 g, 30 mmol) and the resulting mixture stirred for 16 h. Solvent was removed in vacuo and the crude product purified by column chromatography (19:1, hexanes, ethyl acetate) to give the title compound 17 (3.54 g, 56%) as a colourless oil. R_f = 0.73 (2:1 hexanes, ethyl acetate). δ_H (400 MHz; CDCl$_3$) 1.27 (3H, t, J = 7.2 Hz, 1-OCH$_2$CH$_3$), 3.42 (2H, dd, J = 6.6, 1.6 Hz, 4-H), 4.17 (2H, q, J = 7.2 Hz, 1-OCH$_2$CH$_3$), 5.78 (1H, dt, J = 15.5, 1.6 Hz, 2-H), 5.93 (2H, s, OCH$_2$O), 6.61 (1H, dd, J = 8.0, 2.0 Hz, 6′-H), 6.64 (1H, d, J = 2.0 Hz, 2′-H), 6.74 (1H, d, J = 8.0 Hz, 5′-H), 7.04 (1H, dt, J = 15.5, 6.6 Hz, 3-H). δ_C (100 MHz; CDCl$_3$) 14.4 (1-OCH$_2$CH$_3$), 38.2 (C-4), 60.4 (1-OCH$_2$CH$_3$), 101.1 (OCH$_2$O), 108.5 (C-5′), 109.4 (C-2′), 121.9 (C-6′), 122.4 (C-2), 131.5 (C-1′), 146.5 (C-4′), 147.5 (C-3), 148.0 (C-3′), 166.6 (C-1). Values are in agreement with literature data [43].

(E)-4-(3′,4′-Methylenedioxyphenyl)but-2-en-1-ol (**19**). To a stirred solution of ester **17** (3.2 g, 13.7 mmol) in toluene (100 mL), under an atmosphere of nitrogen at −10 °C, was added DIBAL (30 mL, 1 M in toluene) and the resultant mixture stirred for 10 min. The reaction mixture was quenched with addition of 2 M HCl until gas evolution ceased, the organic layer was separated and the aqueous phase further extracted with CH_2Cl_2 (3 × 50 mL). The organic layers were combined, washed with brine (30 mL) and dried (MgSO$_4$). Solvent was removed in vacuo and the crude product purified by column chromatography (3:1 hexanes, ethyl acetate) to give the title compound 19 (2.59 g, 98%) as a pale yellow oil. R_f = 0.42 (hexanes, ethyl acetate). δ_H (400 MHz; CDCl$_3$) 1.41 (1H, br s, 1-OH), 3.30 (2H, d, J =

6.6 Hz, 4-H), 4.12 (2H, br d, *J* = 4.5 Hz, 1-H), 5.64–5.72 (1H, m, 2-H), 5.77–5.85 (1H, m, 3-H), 5.92 (2H, s, OCH$_2$O), 6.63 (1H, dd, *J* = 7.9, 1.9 Hz, 6′-H), 6.67 (1H, d, *J* = 1.9 Hz, 2′-H), 6.73 (1H, d, 7.9 Hz, 5′-H). δ$_C$ (100 MHz; CDCl$_3$) 38.4 (C-4), 63.6 (C-1), 101.0 (OCH$_2$O), 108.3 (C-5′), 109.2 (C-2′), 121.4 (C-6′), 130.4, 131.8 (C-2, 3), 133.9 (C-1′), 146.0, 147.8 (C-3′, 4′). Values are in agreement with literature data [43].

(E)-4-(4-(3′,4′-Methylenedioxyphenyl)but-2-en-1-yl)morpholine (**9b**). To a stirred solution of alcohol **19** (1.66 g, 8.6 mmol) in CH$_2$Cl$_2$ (15 mL), under an atmosphere of nitrogen at 0 °C, was added Et$_3$N (3.6 mL, 25.9 mmol) and stirred for 5 min. MsCl (1.2 mL, 10.4 mmol) was added and stirred for 10 min. Morpholine (1.3 mL, 13.8 mmol) was added and the mixture brought to room temperature and stirred for 18 h. Saturated aqueous NaHCO$_3$ (25 mL) and water (5 mL) was added and the aqueous layer further extracted with CH$_2$Cl$_2$ (3 × 30 mL). The organic layers were combined, dried (MgSO$_4$) and the solvent removed in vacuo. The crude product was purified by column chromatography (2:1 hexanes, ethyl acetate) to give the title compound **9b** (1.4 g, 60%) as a pale yellow oil. R$_f$ = 0.39 (1:2 hexanes, ethyl acetate). δ$_H$ (400 MHz; CDCl$_3$) 2.43 (4H, br t, *J* = 4.7 Hz, NCH$_2$CH$_2$O), 2.96 (2H, d, *J* = 6.5 Hz, 1-H), 3.28 (2H, d, *J* = 7.0 Hz, 4-H), 3.71 (4H, t, *J* = 4.7 Hz, NCH$_2$CH$_2$O), 5.49–5.56 (1H, m, 2-H), 5.69–5.76 (1H, m, 3-H), 5.91 (2H, d, *J* = 2.0 Hz, OCH$_2$O), 6.61 (1H, dd, *J* = 7.5, 2.0 Hz, 6′-H), 6.65 (1H, d, *J* = 2.0 Hz, 2′-H), 6.72 (1H, d, *J* = 7.5 Hz, 5′-H). δ$_C$ (100 MHz; CDCl$_3$) 38.7 (C-4), 53.7 (NCH$_2$CH$_2$O), 61.2 (C-1), 67.1 (NCH$_2$CH$_2$O), 100.9 (OCH$_2$O), 108.3 (C-5′), 109.1 (C-2′), 121.4 (C-6′), 127.4 (C-2), 133.7 (C-3), 134.1 (C-1′), 146.0 (C-4′), 147.8 (C-3′). IR: ν$_{MAX}$ (film)/cm^{-1}; 2855, 1739, 1488, 1242, 1115, 1036, 926, 864, 736. HRMS (ESI$^+$) Found [M + H]$^+$ 262.1428; C$_{15}$H$_{20}$NO$_3$ requires 262.1438.

(E)-Ethyl-3-(3′,4′-methylenedioxyphenyl)prop-2-enoate (**23**).   To a stirred solution of piperonal **20** (5.0 g, 33 mmol) in CH$_2$Cl$_2$ (100 mL), under an atmosphere of nitrogen, was added (carbethoxymethylene)triphenylphosphorane (12.8 g, 37.0 mmol) and the resulting mixture stirred for 20 h. Solvent was then removed in vacuo and the crude product purified by column chromatography (3:1, hexanes, ethyl acetate) to give the title compound **23** (6.97 g, 95%) as a white solid. R$_f$ = 0.68 (2:1 hexanes, ethyl acetate). Melting point: 62–64 °C. δ$_H$ (400 MHz; CDCl$_3$) 1.32 (3H, t, *J* = 7.2 Hz, 1-OCH$_2$CH$_3$), 4.25 (2H, q, *J* = 7.2 Hz, 1-OCH$_2$CH$_3$), 6.00 (2H, s, -OCH$_2$O-), 6.25 (1H, d, *J* = 15.9 Hz, 2-H), 6.80 (1H, d, *J* = 8.0 Hz, 5′-H), 7.00 (1H, dd, *J* = 1.4, 8.0 Hz, 6′-H), 7.02 (1H, d, *J* = 1.4 Hz, 6′-H), 7.58 (1H, d, *J* = 15.9 Hz, 3-H). δ$_C$ (100 MHz; CDCl$_3$) 14.5 (1-OCH$_2$CH$_3$), 60.5 (1-OCH$_2$CH$_3$), 101.7 (-OCH$_2$O-), 106.6 (C-5′), 108.7 (C-2′), 116.4 (C-2), 124.5 (C-6′), 129.1 (C-1′), 144.4 (C-3), 148.5 (C-4′), 149.7 (C-3′), 167.3 (C-1). Values are in agreement with literature data [44].

(E)-Ethyl-3-(3′,4′,5′-trimethoxyphenyl)prop-2-enoate (**24**).   To a stirred solution of 3,4,5-trimethoxybenzaldehyde **21** (3.0 g, 15.3 mmol) in CH$_2$Cl$_2$ (100 mL), under an atmosphere of nitrogen, was added (carbethoxymethylene)triphenylphosphorane (5.9 g, 16.8 mmol) and the resulting mixture stirred for 3 h. Solvent was then removed in vacuo and the crude product purified by column chromatography (3:1, hexanes, ethyl acetate) to give the title compound **24** (4.0 g, 94%) as a white solid. R$_f$ = 0.52 (2:1 hexanes, ethyl acetate). Melting point: 64–66 °C. δ$_H$ (400 MHz; CDCl$_3$) 1.34 (3H, t, *J* = 7.2 Hz, 1-OCH$_2$CH$_3$), 3.87 (3H, s, 4′-OCH$_3$), 3.88 (6H, s, 3′-OCH$_3$), 4.26 (2H, q, *J* = 7.2 Hz, 1-OCH$_2$CH$_3$), 6.34 (1H, d, *J* = 15.9 Hz, 2-H), 6.75 (2H, s, 2′-H), 7.60 (1H, d, *J* = 15.9 Hz, 3-H). δ$_C$ (100 MHz; CDCl$_3$) 14.5 (1-OCH$_2$CH$_3$), 56.3 (3′-OCH$_3$), 60.6 (1-OCH$_2$CH$_3$), 61.1 (4′-OCH$_3$), 105.3 (C-2′), 117.7 (C-2), 130.1 (C-1′), 140.2 (C-4′), 144.7 (C-3), 153.6 (C-3′), 167.1 (C-1). Values are in agreement with literature data [45].

3-(3′,4′,5′-Trimethoxyphenyl)propionic acid (**31**). To a stirred solution of **24** (5.4 g, 19.4 mmol) in ethyl acetate (30 mL) was added 10% palladium on activated carbon (0.54 g, 10% *w/w*). The solution was flushed with an atmosphere of hydrogen and stirred for 2 h. The reaction mixture was then filtered through a plug of celite and washed with ethyl acetate, solvent was then removed in vacuo to give saturated ester **27** (5.23 g, 96%) which was then used without further purification.

To a stirred solution of ester **27** (5.1 g, 17.9 mmol) in methanol (30 mL) was added aqueous NaOH (72 mL, 1 M, 4 eq.) and stirred for 20 min. The mixture was then extracted with CH$_2$Cl$_2$ (10 mL) and the aqueous layer acidified with aqueous 2 M HCl. The aqueous phase was then extracted with ethyl acetate (3 × 50 mL), dried (MgSO$_4$) and solvent removed in vacuo to give the title compound **31** (4.6 g, quant.) as a white solid. R$_f$ = 0.15 (2:1 hexanes, ethyl acetate). Melting point: 104–105 °C. δ$_H$ (400 MHz; CDCl$_3$) 2.68 (2H, t, *J* = 7.8 Hz, 2-H), 2.90 (2H, t, *J* = 7.8 Hz, 3-H), 3.82 (3H, s, 4′-OCH$_3$), 3.84 (6H, s, 3′-OCH$_3$), 6.43 (2H, s, 2′-H). δ$_C$ (100 MHz; CDCl$_3$) 31.1 (C-2), 35.8 (C-3), 56.2 (3′-OCH$_3$), 61.0 (4′-OCH$_3$), 105.4 (C-2′), 136.0 (C-1′), 136.7 (C-4′), 153.4 (C-3′), 178.8 (C-1). Values are in agreement with literature data [46].

3-(3′,4′-Methylenedioxyphenyl)propionic acid (**30**). To a stirred solution of **23** (6.92 g, 31.4 mmol) in ethyl acetate (30 mL) was added 10% palladium on activated carbon (0.69 g, 10% *w/w*). The solution was flushed with an atmosphere of hydrogen and stirred for 1 h. The reaction mixture was then filtered through a plug of celite and washed with ethyl acetate, solvent was then removed in vacuo to give saturated ester **26** (6.9 g, 99%) which was then used without further purification.

To a stirred solution of ester **26** (6.74 g, 30.0 mmol) in methanol (30 mL) was added aqueous NaOH (121 mL, 1 M, 4 eq.) and stirred for 2.5 h. The mixture was then extracted with ethyl acetate (10 mL) and the aqueous layer acidified with aqueous 2 M HCl. The aqueous phase was then extracted with ethyl acetate (3 × 50 mL), dried (MgSO$_4$) and solvent removed in vacuo to give the title compound **30** (5.5 g, 94%) as a white solid. R$_f$ = 0.44 (2:1 hexanes, ethyl acetate). Melting point: 80–82°C. δ$_H$ (400 MHz; CDCl$_3$) 2.64 (2H, t, *J* = 7.7 Hz, 2-H), 2.88 (2H, t, *J* = 7.7 Hz, 3-H), 5.93 (2H, s, -OCH$_2$O-), 6.66 (1H, dd, *J* = 7.9, 1.4 Hz, 6′-H), 6.70 (1H, d, *J* = 1.4 Hz, 2′-H), 6.74 (1H, d, *J* = 7.9 Hz, 5′-H). δ$_C$ (100 MHz; CDCl$_3$) 30.5 (C-2), 36.1 (C-3), 101.0 (-OCH$_2$O-), 108.4 (C-2′), 108.9 (C-5′), 121.2 (C-6′), 134.1 (C-1′), 146.2 (C-3′), 147.8 (C-4′), 179.1 (C-1). Values are in agreement with literature data [47].

3-(3′-Methoxy-4′-benzyloxyphenyl)propionic acid (**32**).   To a stirred solution of vanillin **22** (3.0 g, 19.7 mmol) in CH$_2$Cl$_2$ (100 mL), under an atmosphere of nitrogen, was added (carbethoxymethylene)triphenylphosphorane (7.56 g, 21.7 mmol) and the resulting mixture stirred for 18 h. Solvent was then removed in vacuo and the crude product purified by column chromatography (2:1, hexanes, ethyl acetate) to give a 2:1 mixture of E and Z isomers of unsaturated ester **25** (4.13 g, 94%) as a yellow oil which was used immediately.

To a stirred solution of unsaturated ester **25** (4.13 g, 18.6 mmol) in ethyl acetate (30 mL) was added 10% palladium on activated carbon (0.4 g, 10% *w/w*). The solution was flushed with an atmosphere of hydrogen and stirred for 2 h. The reaction mixture was then filtered through a plug of celite and washed with ethyl acetate, solvent was then removed in vacuo to give saturated ester **28** (3.9 g, 94%) as a yellow oil which was then used without further purification. To a stirred solution of phenol **28** (3.75 g, 16.7 mmol) in acetonitrile (40 mL), under an atmosphere of nitrogen, was added K$_2$CO$_3$ (6.9 g, 50.0 mmol) and stirred for 10 min. Benzyl bromide (6.0 mL, 50.0 mmol) was then added and the resulting mixture allowed to stir for 65 h. The reaction mixture was then quenched with addition of water (50 mL) and extracted with CH$_2$Cl$_2$ (3 × 30 mL). The organic phases were combined, washed with water (2 × 10 mL) and dried (MgSO$_4$). Solvent was then removed in vacuo and the crude product purified by column chromatography (9:1 hexanes, ethyl acetate) to give benzyl ether **29** (4.38 g, 83%) as a colourless oil which was used immediately. To a stirred solution of ester **29** (4.3 g, 13.7 mmol) in methanol (30 mL) was added aqueous NaOH (55 mL, 1 M, 4 eq.) and stirred for 2.5 h. The mixture was then acidified with aqueous 2 M HCl, extracted with ethyl acetate (3 × 50 mL), dried (MgSO$_4$) and solvent removed in vacuo to give the title compound **32** (3.85 g, 98%) as a white solid. R$_f$ = 0.30 (2:1 hexanes, ethyl acetate). Melting point: 99–100°C. δ$_H$ (400 MHz; CDCl$_3$) 2.66 (2H, t, *J* = 7.7 Hz, 2-H), 2.90 (2H, t, *J* = 7.7 Hz, 3-H), 3.88 (3H, s, 3′-OCH$_3$), 5.13 (2H, s, 7′-H), 6.68 (1H, dd, *J* = 8.2, 2.0 Hz, 6′-H), 6.76 (1H, d, *J* = 2.0 Hz, 2′-H), 6.81 (1H, d, *J* = 8.2 Hz, 5′-H), 7.27–7.32 (1H, m, 11′-H), 7.34–7.39 (2H, m, 10′-H), 7.41–7.45 (2H, m, 9′-H). δ$_C$ (100 MHz; CDCl$_3$) 30.4 (C-2), 35.9 (C-3), 56.1 (3′-OCH$_3$), 71.3 (C-7′), 112.4 (C-2′), 114.5 (C-5′), 120.3 (C-6′), 127.4 (C-9′), 127.9 (C-11′), 128.7 (C-10′), 133.5 (C-1′), 137.4 (C-8′), 146.9 (C-4′), 149.8 (C-3′), 178.8 (C-1). Values are in agreement with literature data [48].

3-(3′,4′-Methylenedioxyphenyl)propanoyl chloride (**34b**). To a stirred solution of carboxylic acid **30** (0.22 g, 1.2 mmol) in CH_2Cl_2 (3 mL), under an atmosphere of nitrogen, was added oxalyl chloride (0.2 mL, 2.3 mmol) dropwise and the mixture stirred for 4 h. The solvent was removed in vacuo to give the title compound **34b** (0.24 g, quant.) as a green oil, which was placed under nitrogen and used without further purification.

3-(3′,4′-Dimethoxyphenyl)propanoyl chloride (**34a**). To a stirred solution of carboxylic acid **33** (0.24 g, 1.2 mmol) in CH_2Cl_2 (5 mL), under an atmosphere of nitrogen, was added oxalyl chloride (0.2 mL, 2.3 mmol) dropwise and the mixture stirred for 2.5 h. The solvent was removed in vacuo to give the title compound **34a** (0.26 g, quant.) as a yellow oil, which was placed under nitrogen and used without further purification.

3-(3′,4′,5′-Trimethoxyphenyl)propanoyl chloride (**34c**). To a stirred solution of carboxylic acid **31** (0.25 g, 1.2 mmol) in CH_2Cl_2 (3 mL), under an atmosphere of nitrogen, was added oxalyl chloride (0.2 mL, 2.3 mmol) dropwise and the mixture stirred for 1.5 h. The solvent removed in vacuo to give the title compound **34c** (0.27 g, quant.) as a green crystalline solid, which was placed under nitrogen and used without further purification.

3-(3′,4′-Methylenedioxyphenyl)propanoyl chloride (**34d**). To a stirred solution of carboxylic acid **32** (0.33 g, 1.2 mmol) in CH_2Cl_2 (3 mL), under an atmosphere of nitrogen, was added oxalyl chloride (0.2 mL, 2.3 mmol) dropwise and the mixture stirred for 4 h. The solvent was removed in vacuo to give the title compound **34d** (0.35 g, quant.) as a yellow oil, which was placed under nitrogen and used without further purification.

(2R*,3S*)-2-(3′,4′-Methylenedioxybenzyl)-3-(3″,4″-dimethoxybenzyl)-1-morpholinopent-4-en-1-one (**35ab**). Using general procedure A: Morpholine **9a** (0.57 g, 2.06 mmol), acid chloride **34b** (0.52 g, 2.47 mmol) and reaction time of 24 h. The crude product was purified by column chromatography (2:1 hexanes, ethyl acetate) to give the title compound **35ab** (0.39 g, 42%) as a pale-yellow amorphous solid. R_f = 0.58 (1:3, hexanes, ethyl acetate). Melting point: 114–116 °C. δ_H (400 MHz; CDCl$_3$) 2.57 (1H, dd, J = 13.6, 9.0 Hz, 7″-H$_A$), 2.66–2.73 (1H, m, 3-H), 2.77–2.85 (2H, m, 7′-H$_A$, OCH$_A$CH$_2$N), 2.85–2.94 (4H, m, 2-H, 7′-H$_B$, 7″-H$_B$, OCH$_2$CH$_A$N), 3.06 (1H, ddd, J = 13.3, 7.9, 3.3 Hz, OCH$_2$CH$_B$N), 3.27–3.41 (3H, m, OCH$_C$CH$_C$N, OCH$_B$CH$_2$N), 3.53–3.60 (1H, m, OH$_D$CH$_2$N), 3.67–3.75 (1H, m, OCH$_2$CH$_D$N), 3.85 (3H, s, 4″-OCH$_3$), 3.86 (3H, s, 3″-OCH$_3$), 4.88 (1H, dd, J = 16.9, 1.8 Hz, 5-H$_A$), 4.98 (1H, dd, J = 10.3, 1.8 Hz, 5-H$_B$), 5.85 (1H, ddd, J = 16.9, 10.3, 9.5 Hz, 4-H), 5.90 (1H, d, J = 1.3 Hz, OCH$_A$O), 5.91 (1H, d, J = 1.3 Hz, OCH$_B$O), 6.60 (1H, dd, J = 7.8, 1.6 Hz, 6′-H), 6.64 (1H, d, J = 1.6 Hz, 2′-H), 6.65–6.68 (2H, m, 2″, 6″-H), 6.70 (1H, d, J = 7.8 Hz, 5′-H), 6.77 (1H, d, J = 8.7 Hz, 5″-H). δ_C (100 MHz; CDCl$_3$) 37.4 (C-7′), 38.3 (C-7″), 42.0 (OCH$_2$CH$_{CD}$N), 46.4 (OCH$_2$CH$_{AB}$N), 46.6 (C-2), 48.5 (C-3), 56.0 (3′, 4′-OCH$_3$), 66.4 (OCH$_{AB}$CH$_2$N), 67.0 (OCH$_{CD}$CH$_2$N), 101.0 (OCH$_2$O), 108.4 (C-5′), 109.6 (C-2′), 111.1 (C-5″), 112.4 (C-2″), 116.8 (C-5), 121.3 (C-6″), 122.0 (C-6′), 132.3 (C-1″), 133.6 (C-1′), 139.3 (C-4), 146.2 (C-4′), 147.5 (C-4″), 147.7 (C-3′), 148.9 (C-3″), 172.6 (C-1). IR: ν_{MAX} (film)/cm^{-1}; 2963, 1631, 1515, 1488, 1442, 1236, 1031, 925, 807, 730. HRMS (ESI$^+$) Found [M + H]$^+$ 454.2241; $C_{26}H_{32}NO_6$ requires 454.2224.

(2R*,3S*)-2-(3′,4′,5′-Trimethoxybenzyl)-3-(3″,4″-dimethoxybenzyl)-1-morpholinopent-4-en-1-one (**35ac**). Using general procedure A: Morpholine **9a** (0.47 g, 1.7 mmol), acid chloride **34c** (0.53 g, 2.0 mmol) and a reaction time of 19 h. The crude product was purified by column chromatography (1:1 hexanes, ethyl acetate) to give the title compound **35ac** (0.50 g, 58%) as a yellow oil. R_f = 0.38 (1:3 hexanes, ethyl acetate). δ_H (400 MHz; CDCl$_3$) 2.59 (1H, dd, J = 13.6, 9.2 Hz, 7″-H$_A$), 2.67–2.74 (1H, m, 3-H), 2.78 (1H, ddd, J = 11.4, 7.8, 3.0 Hz, NCH$_2$CH$_A$O), 2.82–2.96 (5H, m, 2-H, 7′-H, 7″-H$_B$, NCH$_A$CH$_2$O), 3.06 (1H, ddd, J = 13.2, 7.8, 3.0 Hz, NCH$_B$CH$_2$O), 3.25–3.40 (3H, m, NCH$_B$CH$_2$O, NCH$_C$CH$_C$O), 3.54–3.61 (1H, m, NCH$_D$CH$_2$O), 3.67–3.73 (1H, m, NCH$_2$CH$_D$O), 3.80 (3H, s, 4′-OCH$_3$), 3.82 (6H, s, 3′-OCH$_3$), 3.85 (3H, s, 4″-OCH$_3$), 3.86 (3H, s, 3″-OCH$_3$), 4.90 (1H, dd, J = 17.0, 1.8 Hz, 5-H$_A$), 5.00 (1H, dd, J = 10.2, 1.8 Hz, 5-H$_B$), 5.87 (1H, ddd, J = 17.0, 10.2, 9.1 Hz, 4-H), 6.37 (2H, s, 2′-H), 6.66–6.70 (2H, m, 2″, 6″-H), 6.78 (1H, d, J = 8.7 Hz, 5″-H). δ_C (100 MHz; CDCl$_3$) 38.1 (C-7′), 38.3 (C-7″), 42.0 (NCH$_{CD}$CH$_2$O),

46.4 (NCH$_{AB}$CH$_2$O), 46.5 (C-2), 48.7 (C-3), 56.0 (3'', 4''-OCH$_3$), 56.3 (3'-OCH$_3$), 61.0 (4'-OCH$_3$), 66.4 (NCH$_2$CH$_{AB}$O), 66.9 (NCH$_2$CH$_{CD}$O), 106.2 (C-2'), 111.1 (C-5''), 112.5 (C-2''), 116.8 (C-5), 121.2 (C-6''), 132.3 (C-1''), 135.6 (C-1'), 136.8 (C-4'), 139.2 (C-4), 147.5 (C-4''), 148.8 (C-3''), 153.3 (C-3'), 172.6 (C-1). IR: ν_{MAX} (film)/cm^{-1}; 2940, 1632, 1589, 1459, 1236, 1123, 1028, 913, 735. HRMS (ESI$^+$) Found [M + Na]$^+$ 522.2474; C$_{28}$H$_{37}$NNaO$_7$ requires 522.2462.

(2R*,3S*)-2-(3',4'-Dimethoxybenzyl)-3-(3'',4''-dimethoxybenzyl)-1-morpholinopent-4-en-1-one (**35aa**). Using general procedure A: Morpholine **9a** (0.53 g, 1.91 mmol), acid chloride **34a** (0.52 g, 2.29 mmol) and a reaction time of 24 h. The crude product was purified by flash chromatography (1:3 hexanes, ethyl acetate) to give the title compound **35aa** (0.63 g, 77% yield) as a pale-yellow amorphous solid. R$_f$ = 0.42 (19:1 CH$_2$Cl$_2$, methanol). Melting point: 98–101 °C. δ$_H$ (400 MHz; CDCl$_3$) 2.55–2.63 (1H, m, 7''-H$_A$), 2.85–2.93 (1H, m, 7''-H$_B$), 2.67–2.85 (3H, m, 3-H, OCH$_2$CH$_{AB}$N), 3.29-3.37 (4H, m, OCH$_2$CH$_{CD}$N, OCH$_{AB}$CH$_2$N), 2.85–3.06 (3H, m, 2-H, 7'-H), 3.50–3.67 (2H, m, OCH$_{CD}$CH$_2$N), 3.83, 3.84, 3.85, 3.86 (12H, s, 3', 4', 3'', 4''-OCH$_3$), 4.89 (1H, dd, *J* = 17.1, 1.7 Hz, 5-H), 4.99 (1H, dd, *J* = 10.3, 1.9 Hz, 5-H), 5.82–5.91 (1H, m, 4-H), 6.67–6.69 (4H, m, 2', 6', 2'', 6''-H), 6.75–6.78 (2H, m, 5', 5''-H). δ$_C$ (100 MHz; CDCl$_3$) 37.2 (C-2), 38.2 (C-7''), 41.9, 46.5 (OCH$_2$CH$_2$N), 46.2 (C-7'), 48.5 (C-3), 55.8, 55.9 (3', 4', 3'', 4''-OCH$_3$), 66.3, 66.8 (OCH$_2$CH$_2$N), 111.0, 111.3 (C-5', 5''), 112.4, 112.6 (C-2', 2''), 116.6 (C-5), 120.9, 121.2 (C-6', 6''), 132.2, 132.3 (C-1', 1''), 139.2 (C-4), 147.3, 147.6 (4', 4''-OCH$_3$), 148.7, 148.8 (3', 3''-OCH$_3$), 172.6 (C-1). IR: ν_{MAX} (film)/cm^{-1}; 2935, 1628, 1591, 1462, 1260, 1155, 1027, 912, 857, 765. HRMS (ESI$^+$) Found [M + H]$^+$ 470.2537; C$_{27}$H$_{36}$NO$_6$ requires 470.2537

(2R*,3S*)-2-(3'-Methoxy-4'-benzyloxybenzyl)-3-(3'',4''-dimethoxybenzyl)-1-morpholino-pent-4-en-1-one (**35ad**). Using general procedure A: Morpholine **9a** (0.47 g, 1.7 mmol), acid chloride **34d** (0.62 g, 2.0 mmol) and a reaction time of 22 h. The crude product was purified by column chromatography (2:1 hexanes, ethyl acetate) to give the title compound **35ad** (0.59 g, 64%) as a yellow oil.

R$_f$ = 0.58 (1:3, hexanes, ethyl acetate). δ$_H$ (400 MHz; CDCl$_3$) 2.57 (1H, dd, *J* = 13.5, 9.0 Hz, 7''-H$_A$), 2.62–2.68 (1H, m, 3-H), 2.68–2.74 (1H, m, OCH$_A$CH$_2$N), 2.75–2.82 (1H, m, OCH$_2$CH$_A$N), 2.83–2.92 (4H, m, 2-H, 7'-H, 7''-H$_B$), 2.99 (1H, ddd, *J* = 13.3, 7.6, 3.2 Hz, OCH$_2$CH$_B$N), 3.20–3.32 (3H, m, OCH$_B$CH$_2$N, OCH$_C$CH$_C$N), 3.50–3.55 (1H, m, OCH$_D$CH$_2$N), 3.61–3.67 (1H, m, OCH$_2$CH$_D$N), 3.84 (3H, s, 3'-OCH$_3$), 3.85 (3H, s, 4''-OCH$_3$), 3.85 (3H, s, 3''-OCH$_3$), 4.88 (1H, dd, *J* = 17.1, 1.9 Hz, 5-H$_A$), 4.97 (1H, dd, *J* = 10.3, 1.9 Hz, 5-H$_B$), 5.13 (1H, s, 7'''-H), 5.85 (1H, ddd, *J* = 17.1, 10.3, 9.0 Hz, 4-H), 6.59 (1H, dd, *J* = 8.1, 1.9 Hz, 6'-H), 6.65–6.68 (2H, m, 2''-H, 6''-H), 6.69 (1H, d, *J* = 1.9 Hz, 2'-H), 6.74 (1H, d, *J* = 8.1 Hz, 5'-H), 6.77 (1H, d, *J* = 8.5 Hz, 5''-H), 7.25–7.30 (1H, m, 4'''-H), 7.32–7.37 (2H, m, 3'''-H), 7.38–7.42 (2H, m, 2'''-H). δ$_C$ (100 MHz; CDCl$_3$) 37.4 (C-7'), 38.3 (C-7''), 41.9 (OCH$_2$CH$_{CD}$N), 46.3 (OCH$_2$CH$_{AB}$N), 46.5 (C-2), 48.6 (C-3), 56.0, 56.2 (3', 3'', 4''-OCH$_3$), 66.4 (OCH$_{AB}$CH$_2$N), 66.9 (OCH$_{CD}$CH$_2$N), 71.2 (C-7'''), 111.1 (C-5''), 112.4 (C-2''), 113.3 (C-2'), 114.6 (C-5'), 116.7 (C-5), 120.9 (C-6'), 121.3 (C-6''), 127.3 (C-2'''), 127.9 (C-4'''), 128.7 (C-3'''), 132.4 (C-1''), 133.1 (C-1'), 137.3 (C-1'''), 139.3 (C-4), 146.7 (C-4'), 147.5 (C-4''), 148.8 (C-3''), 149.7 (C-3'), 172.7 (C-1). IR: ν_{MAX} (film)/cm^{-1}; 2936, 1736, 1633, 1513, 1454, 1261, 1140, 1028, 915, 733. HRMS (ESI$^+$) Found [M + Na]$^+$ 568.2671; C$_{33}$H$_{39}$NNaO$_6$ requires 568.2670.

(2R*,3S*)-2-(3',4'-Dimethoxybenzyl)-3-(3'',4''-methylenedioxybenzyl)-1-morpholinopent-4-en-1-one (**35ba**). Using general procedure A: Morpholine **9b** (0.25 g, 0.96 mmol), acid chloride **34a** (0.26 g, 1.2 mmol) and a reaction time of 21 h. The crude product was purified by column chromatography (1:1 hexanes, ethyl acetate) to give the title compound **35ba** (0.36 g, 83%) as a yellow oil.

R$_f$ = 0.50 (1:3 hexanes, ethyl acetate). δ$_H$ (400 MHz; CDCl$_3$) 2.56 (1H, dd, *J* = 13.4, 9.0 Hz, 7''-H$_A$), 2.62–2.70 (1H, m, 3-H), 2.75–2.94 (6H, m, 2-H, 7'-H, 7''-H$_B$, NCH$_A$CH$_A$O), 3.05 (1H, ddd, *J* = 13.6, 7.9, 3.1 Hz, NCH$_B$CH$_2$O), 3.28–3.41 (3H, m, NCH$_2$CH$_B$O, NCH$_C$CH$_C$O), 3.51–3.57 (1H, m, NCH$_2$CH$_D$O), 3.58–3.64 (1H, m, NCH$_D$CH$_2$O), 3.83 (3H, s, 3'-H), 3.84 (3H, s, 4'-H), 4.89 (1H, dd, *J* = 17.2, 1.9 Hz, 5-H$_A$), 4.99 (1H, dd, *J* = 10.2, 1.9 Hz, 5-H$_B$), 5.86 (1H, ddd, *J* = 17.2, 10.2, 9.1 Hz, 4-H), 5.92 (1H, d, *J* = 1.4 Hz, OCH$_A$O), 5.92 (1H, d, *J* = 1.4 Hz, OCH$_B$O), 6.58 (1H, dd, *J* = 7.9, 1.6 Hz, 6''-H), 6.64 (1H, d, *J* = 1.6 Hz, 2''-H), 6.66–6.70 (2H, m, 2', 6'-H), 6.71 (1H, d, *J* = 7.9 Hz, 5''-H), 6.76 (1H, d, *J* = 8.1 Hz, 5'-H).

δ_C (100 MHz; CDCl$_3$) 37.3 (C-7′), 38.5 (C-7″), 42.0 (NCH$_{AB}$CH$_2$O), 46.3 (NCH$_{CD}$CH$_2$O), 46.5 (C-2), 48.8 (C-3), 56.1 (3′, 4′-OCH$_3$), 66.4 (NCH$_2$CH$_A$BO), 66.9 (NCH$_2$CH$_{CD}$O), 101.0 (OCH$_2$O), 108.1 (C-5″), 109.6 (C-2″), 111.4 (C-5′), 112.7 (C-2′), 116.8 (C-5), 121.0 (C-6′), 122.1 (C-6″), 132.4 (C-1′), 133.7 (C-1″), 139.2 (C-4), 145.9 (C-4″), 147.6 (C-4′), 147.8 (C-3″), 149.0 (C-3′), 172.7 (C-1). IR: ν_{MAX} (film)/cm^{-1}; 2908, 1740, 1630, 1515, 1441, 1237, 1029, 923, 730. HRMS (ESI$^+$) Found [M + Na]$^+$ 476.2042; C$_{26}$H$_{31}$NNaO$_6$ requires 476.2044.

(2R*,3S*)-2-(3′,4′-Methylenedioxybenzyl)-3-(3″,4″-methylenedioxybenzyl)-1-morpholinopent-4-en-1-one (**35bb**). Using general procedure A: Morpholine **9b** (0.5 g, 1.91 mmol), acid chloride **34b** (0.49 g, 2.30 mmol) and a reaction time of 30 min. The crude product was purified by column chromatography (1:1 hexanes, ethyl acetate) to give the title compound **35bb** (0.798 g, 95%) as a pale-yellow solid. R$_f$ = 0.68 (1:3 hexanes, ethyl acetate). Melting point: 131–133 °C. δ_H (400 MHz; CDCl$_3$) 2.54 (1H, dd, J = 13.5, 8.9 Hz, 7″-H$_A$), 2.61–2.69 (1H, m, 3-H), 2.78–2.93 (6H, m, 2-H, 7′-H, 7″-H$_B$, NCH$_A$CH$_A$O), 3.06 (1H, ddd, J = 13.2, 7.8, 3.1 Hz, NCH$_B$CH$_2$O), 3.29–3.41 (3H, m, NCH$_2$CH$_B$O, NCH$_C$CH$_C$O), 3.53–3.61 (1H, m, NCH$_2$CH$_D$O), 3.66–3.74 (1H, m, NCH$_D$CH$_2$O), 4.89 (1H, dd, J = 17.0, 1.9 Hz, 5-H$_A$), 4.99 (1H, dd, J = 10.2, 1.9 Hz, 5-H$_B$), 5.85 (1H, ddd, J = 17.0, 10.2, 9.1 Hz, 4-H), 5.90 (1H, d, J = 1.4 Hz, 3′-OCH$_A$O), 5.91 (1H, d, J = 1.4 Hz, 3′-OCH$_B$O), 5.92 (1H, d, J = 1.5 Hz, 3″-OCH$_A$O), 5.93 (1H, d, J = 1.5 Hz, 3″-OCH$_B$O), 6.55–6.61 (2H, m, 6′, 6″-H), 6.62–6.64 (2H, m, 2′, 2″-H), 6.70, 6.71 (2 × 1H, 2 × d, J = 8.0 Hz, 5′, 5″-H). δ_C (100 MHz; CDCl$_3$) 37.4 (C-7′), 38.5 (C-7″), 42.0 (NCH$_{CD}$CH$_2$O), 46.4 (NCH$_{AB}$CH$_2$O), 46.5 (C-2), 48.8 (C-3), 66.4 (NCH$_2$CH$_{AB}$O), 67.0 (NCH$_2$CH$_{CD}$O), 101.0 (2 × OCH$_2$O), 108.2, 108.4 (C-5′, 5″), 109.6 (C-2′, 2″), 116.8 (C-5), 122.1 (C-6′, 6″), 133.6 (C-1′, 1″), 139.2 (C-4), 145.9, 146.2 (C-4′, 4″), 147.6, 147.7 (C-3′, 3″), 172.6 (C-1). IR: ν_{MAX} (film)/cm^{-1}; 2897, 1630, 1487, 1440, 1244, 1036, 925, 808, 730. HRMS (ESI$^+$) Found [M + Na]$^+$ 460.1722; C$_{25}$H$_{27}$NNaO$_6$ requires 460.1731.

(2R*,3S*)-2-(3′,4′,5′-Trimethoxybenzyl)-3-(3″,4″-methylenedioxybenzyl)-1-morpholinopent-4-en-1-one (**35bc**). Using general procedure A: Morpholine **9b** (0.25 g, 0.96 mmol), acid chloride **34c** (0.27 g, 1.2 mmol) and a reaction time of 18 h. The crude product was purified by column chromatography (1:1 hexanes, ethyl acetate) to give the title compound **35bc** (0.40 g, 86%) as a pale-yellow solid. R$_f$ = 0.55 (1:3 hexanes, ethyl acetate). Melting point: 104–106 °C. δ_H (400 MHz; CDCl$_3$) 2.56 (1H, dd, J = 13.4, 9.0 Hz, 7″-H$_A$), 2.62–2.70 (1H, m, 3-H), 2.75–2.95 (6H, m, 2-H, 7′-H, 7″-H$_B$, NCH$_A$CH$_A$O), 3.06 (1H, ddd, J = 13.2, 7.7, 3.0 Hz, NCH$_B$CH$_2$O), 3.25–3.40 (3H, m, NCH$_2$CH$_B$O, NCH$_C$CH$_C$O), 3.54–3.60 (1H, m, NCH$_2$CH$_D$O), 3.65–6.71 (1H, m, NCH$_D$CH$_2$O), 3.80 (3H, s, 4′-OCH$_3$), 3.82 (6H, s, 3′-OCH$_3$), 4.90 (1H, dd, J = 17.2, 1.9 Hz, 5-H$_A$), 5.00 (1H, dd, J = 10.2, 1.9 Hz, 5-H$_B$), 5.85 (1H, ddd, J = 17.2, 10.2, 9.0 Hz, 4-H), 5.92 (1H, d, J = 1.4 Hz, OCH$_A$O), 5.93 (1H, d, J = 1.4 Hz, OCH$_B$O), 6.36 (2H, s, 2′-H), 6.59 (1H, dd, J = 7.9, 1.6 Hz, 6″-H), 6.65 (1H, d, J = 1.6 Hz, 2″-H), 6.72 (1H, d, J = 7.9 Hz, 5″-H). δ_C (100 MHz; CDCl$_3$) 38.1 (C-7′), 38.5 (C-7″), 42.0 (NCH$_{CD}$CH$_2$O), 46.4 (C-2, NCH$_{AB}$CH$_2$O), 48.9 (C-3), 56.4 (3′-OCH$_3$), 61.1 (4′-OCH$_3$), 66.4 (NCH$_2$CH$_{AB}$O), 67.0 (NCH$_2$CH$_{CD}$O), 101.0 (OCH$_2$O), 106.2 (C-2′), 108.2 (C-5″), 109.6 (C-2″), 116.9 9 (C-5), 122.1 (C-6″), 133.6 (C-1″), 135.6 (C-1′), 136.9 (C-4′), 139.1 (C-4), 145.9 (C-4″), 147.7 (C-3″), 153.3 (C-3′), 172.6 (C-1). IR: ν_{MAX} (film)/cm^{-1}; 2922, 1632, 1589, 1490, 1240, 1120, 1036, 925, 730. HRMS (ESI$^+$) Found [M + Na]$^+$ 506.2145; C$_{27}$H$_{33}$NNaO$_7$ requires 506.2149.

(2R*,3S*)-2-(3′-Methoxy-4′-benzyloxybenzyl)-3-(3″,4″-methylenedioxybenzyl)-1-morpholinopent-4-en-1-one (**35bd**). Using general procedure A: Morpholine **9b** (0.25 g, 0.96 mmol), acid chloride **34d** (0.35 g, 1.2 mmol) and a reaction time of 18 h. The crude product was purified by column chromatography (1:1 hexanes, ethyl acetate) to give the title compound **35bd** (0.45 g, 88%) as a yellow oil.

R$_f$ = 0.67 (1:3 hexanes, ethyl acetate). δ_H (400 MHz; CDCl$_3$) 2.54 (1H, dd, J = 13.5, 8.9 Hz, 7″-H$_A$), 2.61–2.70 (2H, m, 3-H, NCH$_2$CH$_A$O), 2.73–2.91 (5H, m, 2-H, 7′-H, 7″-H$_B$, NCH$_A$CH$_2$O), 2.99 (1H, ddd, J = 13.2, 7.7, 3.0 Hz, NCH$_B$CH$_2$O), 3.20–3.35 (3H, m, NCH$_2$CH$_B$O, NCH$_C$CH$_C$O), 3.53 (1H, ddd, J = 11.0, 5.5, 2.5 Hz, NCH$_2$CH$_D$O), 3.62 (1H, ddd, J = 13.0, 5.5, 2.5 Hz, NCH$_D$CH$_2$O), 3.84 (3H, s, 3′-OCH$_3$), 4.88 (1H, dd, J = 17.0, 1.9 Hz, 5-H$_A$), 4.98 (1H, dd, J = 10.2, 1.9 Hz, 5-H$_B$), 5.13 (2H, s, 7‴-H), 5.84 (1H, ddd, J = 17.0, 10.2, 9.1 Hz, 4-H), 5.91 (1H, d, J = 1.4 Hz, OCH$_A$O), 5.92 (1H, d, J = 1.4 Hz,

OCH$_B$O), 6.57 (1H, dd, *J* = 8.0, 1.9 Hz, 6″-H), 6.59 (1H, dd, *J* = 8.2, 1.8 Hz, 6′-H), 6.64 (1H, d, *J* = 1.8 Hz, 2″-H), 6.69 (1H, d, *J* = 1.9 Hz, 2″-H), 6.71 (1H, d, *J* = 8.0 Hz, 5″-H), 6.75 (1H, d, *J* = 8.2 Hz, 5′-H), 7.25–7.30 (1H, m, 4‴-H), 7.32–7.37 (2H, m, 3‴-H), 7.38–7.43 (2H, m, 2‴-H). δ$_C$ (100 MHz; CDCl$_3$) 37.4 (C-7′), 38.5 (C-7″), 41.9 (NCH$_C$DCH$_2$O), 46.3 (NCH$_{AB}$CH$_2$O), 46.4 (C-2), 48.8 (C-3), 56.2 (3′-OCH$_3$), 66.3 (NCH$_2$CH$_{AB}$O), 66.9 (NCH$_2$CH$_C$DO), 71.2 (C-7‴), 100.9 (OCH$_2$O), 108.1 (C-5″), 109.6 (C-2″), 113.2 (C-2′), 114.5 (C-5′), 116.7 (C-5), 121.0 (C-6′), 122.1 (C-6″), 127.3 (C-2‴), 127.9 (C-4‴), 128.6 (C-3‴), 133.1 (C-1′) 133.6 (C-1″), 137.3 (C-1‴), 139.2 (C-4), 145.9 (C-4″), 146.7 (C-4′), 147.6 (C-3″), 149.7 (C-3′), 172.6 (C-1). IR: ν$_{MAX}$ (film)/cm^{-1}; 2920, 1630, 1489, 1231, 1114, 1034, 913, 729. HRMS (ESI$^+$) Found [M + Na]$^+$ 552.2354; C$_{32}$H$_{35}$NNaO$_6$ requires 552.2357.

(3R*,4R*)-3-(3′,4′-Methylenedioxybenzyl)-4-(3″,4″-dimethoxybenzyl)-5-(hydroxymethyl)dihydrofuran-2(3H)-one (**4ab**). Using general procedure B: Amide **35ab** (0.38 g, 0.84 mmol) in ^{t}BuOH/H$_2$O and a reaction time of 3 days. The crude product was purified by column chromatography (1:1 hexanes, ethyl acetate) to give the title compound **4ab** (180 mg, 54%) as a white foam. R$_f$ = 0.50 (19:1 CH$_2$Cl$_2$, methanol). δ$_H$ (400 MHz; CDCl$_3$) 1.79 (1H, t, *J* = 6.4 Hz, 6-OH), 2.36–2.44 (1H, m, 4-H), 2.51 (1H, dd, *J* = 13.7, 7.9 Hz, 7″-H$_A$), 2.58 (1H, dd, *J* = 13.7, 6.6 Hz, 7″-H$_B$), 2.68 (1H, ddd, *J* = 9.3, 7.0, 5.5 Hz, 3-H), 2.85 (1H, dd, *J* = 14.0, 7.0 Hz, 7′-H$_A$), 2.92 (1H, dd, *J* = 14.0, 5.5 Hz, 7′-H$_B$), 3.15 (1H, ddd, *J* = 12.5, 6.4, 5.1 Hz, 6-H$_A$), 3.54 (1H, ddd, *J* = 12.5, 6.4, 2.5 Hz, 6-H$_B$), 3.83 (3H, s, 3″-OCH$_3$), 3.85 (3H, s, 4″-OCH$_3$), 4.19 (1H, ddd, *J* = 8.0, 5.1, 2.5 Hz, 5-H), 5.92 (1H, d, *J* = 1.5 Hz, OCH$_A$H$_B$O), 5.93 (1H, d, *J* = 1.5 Hz, OCH$_A$H$_B$O), 6.47 (1H, d, *J* = 2.0 Hz, 2″-H), 6.57–6.60 (2H, m, 6′ and 6″-H), 6.61 (1H, d, *J* = 1.5 Hz, 2′-H), 6.71 (1H, d, *J* = 7.8 Hz, 5′-H), 6.77 (1H, d, *J* = 8.1 Hz, 5″-H). δ$_C$ (100 MHz; CDCl$_3$) 35.3 (C-7′), 38.7 (C-7″), 41.6 (C-4), 47.6 (C-3), 56.0 (3″-OCH$_3$, 4″-OCH$_3$), 63.2 (C-6), 84.1 (C-5), 101.2 (OCH$_2$O), 108.3 (C-5′), 109.7 (C-2′), 111.4 (C-5″), 112.0 (C-2″), 121.0 (C-6″), 122.5 (C-6′), 130.3 (C-1″), 131.6 (C-1′), 146.6 (C-4′), 148.0 (C-4″), 148.1 (C-3′), 149.3 (C-3″), 177.7 (C-2). IR: ν$_{MAX}$ (film)/cm^{-1}; 3496 (broad), 2936, 2254, 1760, 1515, 1489, 1442, 1239, 1025, 909, 809, 766. HRMS (ESI$^+$) Found [M + Na]$^+$ 423.1427; C$_{22}$H$_{24}$NaO$_7$ requires 423.1414.

(3R*,4R*)-3,4-bis(3′,4′-Dimethoxybenzyl)-5-(hydroxymethyl)dihydrofuran-2(3H)-one (**4aa**). Using general procedure B: Amide **35aa** (0.29 g, 0.61 mmol), in ^{t}BuOH/H$_2$O and a reaction time of 6 days. The crude product was purified by flash chromatography (1:1 hexanes, ethyl acetate) to give the title compound **4aa** (0.18 g, 70%) as a colourless oil. R$_f$ = 0.32 (19:1 CH$_2$Cl$_2$, methanol). δ$_H$ (400 MHz; CDCl$_3$) 2.39–2.44 (1H, m, 4-H), 2.53 (1H, dd, *J* = 13.7, 7.3 Hz, 7″-H$_A$), 2.58 (1H, dd, *J* = 13.7, 6.5 Hz, 7″-H$_B$), 2.64 (1H, br s, 6-OH), 2.71 (1H, ddd, *J* = 9.3, 6.7, 5.7 Hz, 3-H), 2.88 (1H, dd, *J* = 14.0, 6.7 Hz, 7′-H$_A$), 2.94 (1H, dd, *J* = 14.0, 5.5 Hz, 7′-H$_B$), 3.16 (1H, dd, *J* = 12.6, 4.9 Hz, 6-H$_A$), 3.53 (1H, dd, *J* = 12.6, 2.4 Hz, 6-H$_B$), 3.81, 3.83, 3.84 (12H, s, 3′, 4′, 3″, 4″-OCH$_3$), 4.15 (1H, ddd, *J* = 8.0, 4.9, 2.4 Hz, 5-H), 6.49 (1H, d, *J* = 1.9 Hz, 2″-H), 6.57 (1H, dd, *J* = 8.1, 1.9 Hz, 6″-H), 6.66–6.68 (2H, m, 2′, 6′-H), 6.73–6.80 (2H, m, 5′, 5″-H). δ$_C$ (100 MHz; CDCl$_3$) 35.0 (C-7′), 38.5 (C-7″), 41.6 (C-4), 47.5 (C-3), 55.8 (3′, 4′, 3″, 4″-OCH$_3$), 62.9 (C-6), 84.0 (C-5), 111.2, 111.4 (C-5′, 5″), 112.1 (C-2″), 112.6 (C-2′), 120.9 (C-6″), 121.4 (C-6′), 130.4 (C-1′, 1″), 147.9 (C-4′, 4″), 149.0 (C-3′, 3″), 178.0 (C-2). IR: ν$_{MAX}$ (film)/cm^{-1}; 3505 (br), 2938, 1761, 1591, 1514, 1465, 1259, 1156, 1025, 910, 808, 766, 647. HRMS (ESI$^+$) Found [M + H]$^+$ 417.1909; C$_{23}$H$_{29}$O$_7$ requires 417.1908.

(3R*,4R*)-3-(3′,4′,5′-Trimethoxybenzyl)-4-(3″,4″-dimethoxybenzyl)-5-(hydroxymethyl)dihydrofuran-2(3H)-one (**4ac**). Using general procedure B: Amide **35ac** (0.45 g, 0.90 mmol) in ^{t}BuOH/H$_2$O/THF and a reaction time of 3 days. The crude product was purified by column chromatography (1:1 hexanes, ethyl acetate) to give the title compound **4ac** (0.17 g, 42%) as a pale-yellow solid.

R$_f$ = 0.31 (19:1 CH$_2$Cl$_2$, methanol). Melting point: 141–142 °C. δ$_H$ (400 MHz; CDCl$_3$) 1.68 (1H, t, *J* = 6.5 Hz, 6-OH), 2.38–2.46 (1H, m, 4-H), 2.55 (1H, dd, *J* = 13.8, 8.2 Hz, 7″-H$_A$), 2.65 (1H, dd, *J* = 13.8, 5.9 Hz, 7″-H$_B$), 2.72 (1H, ddd, *J* = 9.7, 6.3, 5.7 Hz, 3-H), 2.90 (1H, dd, *J* = 14.0, 6.3 Hz, 7′-H$_A$), 2.95 (1H, dd, *J* = 14.0, 5.7 Hz, 7′-H$_B$), 3.15 (1H, ddd, *J* = 12.4, 5.1, 5.4 Hz, 6-H$_A$), 3.54 (1H, ddd, *J* = 12.4, 6.5, 2.5 Hz, 6-H$_B$), 3.82 (6H, s, 4′, 3″-OCH$_3$), 3.83 (6H, s, 3′-OCH$_3$), 3.85 (3H, s, 4″-OCH$_3$), 4.20 (1H, ddd, *J* = 8.2, 5.1, 2.5 Hz, 5-H), 6.38 (2H, s, 2′-H), 6.49 (1H, d, *J* = 2.0 Hz, 2″-H), 6.58 (1H, dd, *J* = 8.1, 2.0 Hz, 6″-H), 6.76 (1H, d, *J* = 8.1 Hz, 5″-H). δ$_C$ (100 MHz; CDCl$_3$) 35.7 (C-7′), 38.6 (C-7″), 41.8 (C-4), 47.7 (C-3), 56.0,

56.1 (3″, 4″-OCH$_3$), 56.3 (3′-OCH$_3$), 61.0 (4′-OCH$_3$), 63.2 (C-6), 83.9 (C-5), 106.5 (C-2′), 111.5 (C-5″), 112.2 (C-2″), 121.0 (C-6″), 130.3 (C-1″), 133.7 (C-1′), 137.2 (C-4′), 148.2 (C-4″), 149.3 (C-3″), 153.5 (C-3′), 177.7 (C-2). IR: ν_{MAX} (film)/cm^{-1}; 3527 (br), 2938, 1761, 1590, 1514, 1237, 1126, 1026, 735. HRMS (ESI$^+$) Found [M + Na]$^+$ 469.1839; C$_{24}$H$_{30}$NaO$_8$ requires 469.1833.

(3R*,4R*)-3-(3′-Methoxy-4′-benzyloxybenzyl)-4-(3″,4″-dimethoxybenzyl)-5-(hydroxymethyl)dihydrofuran-2(3H)-one (**4ad**). Using general procedure B: Amide **35ad** (0.59 g, 1.1 mmol) in ^{t}BuOH/H$_2$O and a reaction time of 7 days. The crude product was purified by column chromatography (1:1 hexanes, ethyl acetate) to give the title compound **4ad** (0.30 g, 56%) as a cloudy oil. R$_f$ = 0.27 (19:1 CH$_2$Cl$_2$, methanol). δ_H (400 MHz; CDCl$_3$) 1.57 (1H, t, *J* = 6.5 Hz, 6-OH), 2.34–2.42 (1H, m, 4-H), 2.50 (1H, dd, *J* = 13.5, 8.0 Hz, 7″-H$_A$), 2.59 (1H, dd, *J* = 13.5, 6.0 Hz, 7″-H$_B$), 2.70 (1H, ddd, *J* = 9.7, 6.2, 5.6 Hz, 3-H), 2.90 (1H, dd, *J* = 14.1, 6.2 Hz, 7′-H$_A$), 2.94 (1H, dd, *J* = 14.1, 5.6 Hz, 7′-H$_B$), 3.10 (1H, ddd, *J* = 12.5, 6.5, 5.2 Hz, 6-H$_A$), 3.48 (1H, ddd, *J* = 12.5, 6.5, 2.7 Hz, 6-H$_B$), 3.80 (3H, s, 3″-OCH$_3$), 3.86 (6H, s, 3′, 4″-OCH$_3$), 4.18 (1H, ddd, *J* = 8.3, 5.2, 2.7 Hz, 5-H), 5.12 (2H, s, 7‴-H), 6.46 (1H, d, *J* = 2.0 Hz, 2″-H), 6.56 (1H, dd, *J* = 8.0, 2.0 Hz, 6″-H), 6.61 (1H, dd, *J* = 8.1, 2.0 Hz, 6′-H), 6.72 (1H, d, *J* = 2.0 Hz, 2′-H), 6.75 (1H, d, *J* = 8.0 Hz, 5″-H), 6.79 (1H, d, *J* = 8.1 Hz, 5′-H), 7.25–7.30 (1H, m, 4‴-H), 7.31–7.36 (2H, m, 3‴-H), 7.39–7.42 (2H, m, 2‴-H). δ_C (100 MHz; CDCl$_3$) 35.1 (C-7′), 38.6 (C-7″), 41.7 (C-4), 47.6 (C-3), 56.0 (3″, 4″-OCH$_3$), 56.2 (3′-OCH$_3$), 63.3 (C-6), 71.3 (C-7‴), 84.0 (C-5), 111.5 (C-5″), 112.1 (C-2″), 113.3 (C-2′), 114.3 (C-5′), 121.0 (C-6″), 121.6 (C-6′), 127.4 (C-2‴), 128.0 (C-4‴), 128.7 (C-3‴), 130.3 (C-1″), 131.1 (C-1′), 137.2 (C-1‴), 147.2 (C-4′), 148.2 (C-4″), 149.3 (C-3″), 150.0 (C-3′), 177.8 (C-2). IR: ν_{MAX} (film)/cm^{-1}; 3523 (br), 2935, 1761, 1514, 1261, 1025, 911, 730. HRMS (ESI$^+$) Found [M + Na]$^+$ 515.2023; C$_{29}$H$_{32}$NaO$_7$ requires 515.2040.

(3R*,4R*,5S*)-4-(3″,4″-Dimethoxybenzyl)-3-(4′-hydroxy-3′-methoxybenzyl)-5-(hydroxymethyl)dihydrofuran-2(3H)-one (**4ae**). Using general procedure F: Benzyl ether **4ad** (0.27 g, 0.55 mmol) gave the title compound **4ae** (0.19 g, 88%) as a yellow solid. R$_f$ = 0.43 (19:1 CH$_2$Cl$_2$, methanol). Melting point: 183–185 °C. δ_H (400 MHz; CDCl$_3$) 1.63 (1H, t, *J* = 6.5 Hz, 6-OH), 2.34–2.43 (1H, m, 4-H), 2.53 (1H, dd, *J* = 13.8, 8.1 Hz, 7″-H$_A$), 2.62 (1H, dd, *J* = 13.8, 6.1 Hz, 7″-H$_B$), 2.69 (1H, dt, *J* = 9.5, 6.0 Hz, 3-H), 2.92 (2H, d, *J* = 6.0 Hz, 7′-H), 3.13 (1H, ddd, *J* = 12.5, 6.5, 5.3 Hz, 6-H$_A$), 3.51 (1H, ddd, *J* = 12.5, 6.5, 2.5 Hz, 6-H$_B$), 3.82 (3H, s, 3′-OCH$_3$), 3.84 (3H, s, 3″-OCH$_3$), 3.85 (3H, s, 4″-OCH$_3$), 4.19 (1H, ddd, *J* = 8.0, 5.3, 2.5 Hz, 5-H), 5.52 (1H, s, 4′-OH), 6.46 (1H, d, *J* = 2.0 Hz, 2″-H), 6.57 (1H, dd, *J* = 8.1, 2.0 Hz, 6″-H), 6.63 (1H, dd, *J* = 8.0, 1.9 Hz, 6′-H), 6.66 (1H, d, *J* = 1.9 Hz, 2′-H), 6.76 (1H, d, *J* = 8.1 Hz, 5″-H), 6.83 (1H, d, *J* = 8.0 Hz, 5′-H). δ_C (100 MHz; CDCl$_3$) 35.1 (C-7′), 38.6 (C-7″), 41.6 (C-4), 47.7 (C-3), 56.0, 56.1 (3′, 3″, 4″-OCH$_3$), 63.4 (C-6), 84.1 (C-5), 111.5 (C-5″), 111.9 (C-2′), 112.1 (C-2″), 114.4 (C-5′), 121.0 (C-6″), 122.3 (C-6′), 129.7 (C-1′), 130.4 (C-1″), 144.7 (C-4′), 146.8 (C-3′), 148.2 (C-4″), 149.3 (C-3″), 177.8 (C-2). IR: ν_{MAX} (film)/cm^{-1}; 3438 (br), 2937, 1755, 1514, 1236, 1155, 1025, 907, 723. HRMS (ESI$^+$) Found [M + Na]$^+$ 425.1564; C$_{22}$H$_{26}$NaO$_7$ requires 425.1571.

(3R*,4R*)-3,4-bis(3′,4′-Methylenedioxybenzyl)-5-(hydroxymethyl)dihydrofuran-2(3H)-one (**4bb**). Using general procedure B: Morpholine amide **35bb** (0.322 g, 0.74 mmol) in ^{t}BuOH/H$_2$O/THF and a reaction time of 5 days. The crude product was then purified by column chromatography (2:1 hexanes, ethyl acetate) to give the title compound **4bb** (0.145 g, 51%) as a pale-yellow oil.

R$_f$ = 0.59 (1:3 hexanes, ethyl acetate). δ_H (400 MHz; CDCl$_3$) 1.72 (1H, br, 6-OH), 2.32–2.41 (1H, m, 4-H), 2.47 (1H, dd, *J* = 13.7, 8.1 Hz, 7″-H$_A$), 2.56 (1H, dd, *J* = 13.7, 6.2 Hz, 7″-H$_B$), 2.65 (1H, ddd, *J* = 9.0, 7.5, 5.3 Hz, 3-H), 2.85 (1H, dd, *J* = 14.0, 7.5 Hz, 7′-H$_A$), 2.96 (1H, dd, *J* = 14.0, 5.3 Hz, 7′-H$_B$), 3.15 (1H, dd, *J* = 12.6, 4.9 Hz, 6-H$_A$), 3.54 (1H, dd, *J* = 12.6, 2.5 Hz, 6-H$_B$), 4.18 (1H, ddd, *J* = 7.7, 4.9, 2.5 Hz, 5-H), 5.93–5.95 (4H, m, 2 × OCH$_2$O), 6.45–6.49 (2H, m, 2″, 6″-H), 6.60 (1H, dd, *J* = 7.8, 1.7 Hz, 6′-H), 6.63 (1H, d, *J* = 1.7 Hz, 2′-H), 6.70 (1H, d, *J* = 7.8 Hz, 5″-H), 6.73 (1H, d, *J* = 7.8 Hz, 5′-H). δ_C (100 MHz; CDCl$_3$) 35.4 (C-7′), 38.9 (C-7″), 41.8 (C-4), 47.6 (C-3), 63.6 (C-6), 83.9 (C-5), 101.1, 101.2 (2 × OCH$_2$O), 108.4 (C-5′), 108.6 (C-5″), 109.2 (C-2″), 109.6 (C-2′), 121.9 (C-6″), 122.4 (C-6′), 131.5 (C-1′, 1″), 146.6 (C-4′, 4″),

148.0, 148.1 (C-3′, 3″), 177.6 (C-2). IR: ν_{MAX} (film)/cm^{-1}; 3432 (br), 2922, 1760, 1503, 1490, 1444, 1247, 1038, 927, 811. HRMS (ESI$^+$) Found [M + H]$^+$ 385.1279; $C_{21}H_{21}O_7$ requires 385.1282.

(3R*,4R*)-3-(3′,4′-Dimethoxybenzyl)-4-(3″,4″-methylenedioxybenzyl)-5-(hydroxymethyl)dihydrofuran-2(3H)-one (**4ba**). Using general procedure B: Morpholine amide **35ba** (0.336 g, 0.74 mmol) in ^{t}BuOH/H$_2$O/ THF and a reaction time of 4 days. The crude product was then purified by column chromatography (1:3 hexanes, ethyl acetate) to give the title compound **4ba** (0.103 g, 34%) as a pale yellow oil.

R_f = 0.48 (1:3 hexanes, ethyl acetate). δ_H (400 MHz; CDCl$_3$) 1.68 (1H, t, *J* = 6.6 Hz, 6-OH), 2.33–2.42 (1H, m, 4-H), 2.48 (1H, dd, *J* = 13.7, 7.9 Hz, 7″-H$_A$), 2.56 (1H, dd, *J* = 13.7, 6.3 Hz, 7″-H$_B$), 2.68 (1H, ddd, *J* = 9.3, 6.9, 5.4 Hz, 3-H), 2.89 (1H, dd, *J* = 14.0, 6.9 Hz, 7′-H$_A$), 2.96 (1H, dd, *J* = 14.0, 5.4 Hz, 7′-H$_B$), 3.15 (1H, ddd, *J* = 12.5, 6.6, 5.2 Hz, 6-H$_A$), 3.52 (1H, ddd, *J* = 12.5, 6.6, 2.6 Hz, 6-H$_B$), 3.85 (3H, s, 3′-OCH$_3$), 3.86 (3H, s, 4′-OCH$_3$), 4.18 (1H, ddd, *J* = 7.9, 5.2, 2.6 Hz, 5-H), 5.93 (1H, d, *J* = 1.4 Hz, OCH$_A$O), 5.94 (1H, d, *J* = 1.4 Hz, OCH$_B$O), 6.44 (1H, d, *J* = 1.6 Hz, 2″-H), 6.47 (1H, dd, *J* = 7.8, 1.6 Hz, 6″-H), 6.67 (1H, d, *J* = 2.2 Hz, 2′-H), 6.68–6.72 (2H, m, 6′, 5″-H), 6.79 (1H, d, *J* = 8.0 Hz, 5′-H). δ_C (100 MHz; CDCl$_3$) 35.2 (C-7′), 38.8 (C-7″), 41.7 (C-4), 47.6 (C-3), 56.0 (3′, 4′-OCH$_3$), 63.4 (C-6), 83.8 (C-5), 101.3 (OCH$_2$O), 108.5 (C-5′), 109.2 (C-2″), 111.3 (C-5″), 112.5 (C-2′), 121.6 (C-6′), 121.9 (C-6″), 130.3 (C-1′), 131.5 (C-1″), 146.7 (C-4″), 148.1 (C-4′, 3″), 149.2 (C-3′), 177.7 (C-2). IR: ν_{MAX} (film)/cm^{-1}; 3472 (br), 2933, 1760, 1516, 1490, 1242, 1157, 1028, 925, 810, 730. HRMS (ESI$^+$) Found [M + Na]$^+$ 423.1423; $C_{22}H_{24}NaO_7$ requires 423.1414.

(3R*,4R*)-3-(3′,4′,5′-Trimethoxybenzyl)-4-(3″,4″-methylenedioxybenzyl)-5-(hydroxymethyl)dihydrofuran-2(3H)-one (**4bc**). Using general procedure B: Morpholine amide **35bc** (0.372 g, 0.77 mmol) in ^{t}BuOH/H$_2$O/THF and a reaction time of 4 days. The crude product was then purified by column chromatography (1:3 hexanes, ethyl acetate) to give the title compound **4bc** (0.084 g, 25%) as a pale-yellow oil.

R_f = 0.38 (1:3 hexanes, ethyl acetate). δ_H (400 MHz; CDCl$_3$) 1.69 (1H, t, *J* = 6.6 Hz, 6-OH), 2.36–2.45 (1H, m, 4-H), 2.53 (1H, dd, *J* = 13.8, 7.6 Hz, 7″-H$_A$), 2.59 (1H, dd, *J* = 13.8, 6.8 Hz, 7″-H$_B$), 2.70 (1H, ddd, *J* = 9.5, 6.7, 5.4 Hz, 3-H), 2.87 (1H, dd, *J* = 14.0, 6.7 Hz, 7′-H$_A$), 2.93 (1H, dd, *J* = 14.0, 5.4 Hz, 7′-H$_B$), 3.22 (1H, ddd, *J* = 12.7, 6.6, 5.0 Hz, 6-H$_A$), 3.58 (1H, ddd, *J* = 12.7, 6.6, 2.5 Hz, 6-H$_B$), 3.82 (3H, s, 4′-OCH$_3$), 3.84 (6H, s, 3′-OCH$_3$), 3.85 (3H, s, 4″-OCH$_3$), 4.19 (1H, ddd, *J* = 7.9, 5.0, 2.5 Hz, 5-H), 5.94 (1H, d, *J* = 1.4 Hz, OCH$_A$O), 5.94 (1H, d, *J* = 1.4 Hz, OCH$_B$O), 6.37 (2H, s, 2′-H), 6.46 (1H, d, *J* = 1.8 Hz, 2″-H), 6.48 (1H, dd, *J* = 7.9, 1.8 Hz, 6″-H), 6.70 (1H, d, *J* = 7.9 Hz, 5″-H). δ_C (100 MHz; CDCl$_3$) 36.0 (C-7′), 38.8 (C-7″), 41.9 (C-4), 47.6 (C-3), 56.3 (3′-OCH$_3$), 61.1 (4′-OCH$_3$), 63.3 (C-6), 83.8 (C-5), 101.3 (OCH$_2$O), 106.5 (C-2′), 108.5 (C-5″), 109.2 (C-2″), 121.9 (C-6″), 131.4 (C-1″), 133.6 (C-1′), 137.1 (C-4′), 146.7 (C-4″), 148.2 (C-3″), 153.5 (C-3′), 177.7 (C-2). IR: ν_{MAX} (film)/cm^{-1}; 3475 (br), 2941, 1760, 1591, 1490, 1445, 1244, 1127, 1036, 926. HRMS (ESI$^+$) Found [M + Na]$^+$ 453.1519; $C_{23}H_{26}NaO_8$ requires 453.1520.

(3R*,4R*)-3-(3′-Methoxy-4′-benzyloxybenzyl)-4-(3″,4″-methylenedioxybenzyl)-5-(hydroxymethyl)dihydrofuran-2(3H)-one (**4bd**). Using general procedure B: Morpholine amide **35bd** (0.405 g, 0.77 mmol) in ^{t}BuOH/H$_2$O/ THF and a reaction time of 5 days. The crude product was then purified by column chromatography (1:3 hexanes, ethyl acetate) to give the title compound **4bd** (0.205 g, 56%) as a pale-yellow oil.

R_f = 0.58 (1:3 hexanes, ethyl acetate). δ_H (400 MHz; CDCl$_3$) 1.64 (1H, t, *J* = 6.6 Hz, 6-OH), 2.31–2.40 (1H, m, 4-H), 2.46 (1H, dd, *J* = 13.7, 7.9 Hz, 7″-H$_A$), 2.53 (1H, dd, *J* = 13.7, 6.3 Hz, 7″-H$_B$), 2.67 (1H, ddd, *J* = 9.2, 7.2, 5.3 Hz, 3-H), 2.87 (1H, dd, *J* = 14.0, 7.2 Hz, 7′-H$_A$), 2.95 (1H, dd, *J* = 14.0, 5.3 Hz, 7′-H$_B$), 3.13 (1H, ddd, *J* = 12.6, 6.6, 5.1 Hz, 6-H$_A$), 3.48 (1H, ddd, *J* = 12.6, 6.6, 2.6 Hz, 6-H$_B$), 3.86 (3H, s, 3′-OCH$_3$), 4.17 (1H, ddd, *J* = 7.8, 5.1, 2.6 Hz, 5-H), 5.13 (2H, s, 7‴-H), 5.93 (1H, d, *J* = 1.4 Hz, OCH$_A$O), 5.94 (1H, d, *J* = 1.4 Hz, OCH$_B$O), 6.43 (1H, d, *J* = 1.6 Hz, 2″-H), 6.45 (1H, dd, *J* = 7.9, 1.6 Hz, 6″-H), 6.64 (1H, dd, *J* = 8.2, 2.0 Hz, 6′-H), 6.68–6.70 (2H, m, 2′, 5″-H), 6.81 (1H, d, *J* = 8.2 Hz, 5′-H), 7.27–7.30 (1H, m, 4‴-H), 7.32–7.36 (2H, m, 3‴-H), 7.40–7.44 (2H, m, 2‴-H). δ_C (100 MHz; CDCl$_3$) 35.3 (C-7′), 38.8 (C-7″), 41.8 (C-4), 47.6 (C-3), 56.1 (3′-OCH$_3$), 63.4 (C-6), 71.3 (C-7‴), 83.8 (C-5), 101.3 (OCH$_2$O), 108.5 (C-5″), 109.2 (C-2″), 113.0 (C-2′), 114.4 (C-5′), 121.5 (C-6′), 121.9 (C-6″), 127.5 (C-2‴), 128.0 (C-4‴), 128.7 (C-3‴), 131.0 (C-1′), 131.5 (C-1″), 137.3 (C-1‴), 146.7 (C-4″), 147.2 (C-4′), 148.1 (C-3″), 150.0 (C-3′), 177.7 (C-2).

IR: ν_{MAX} (film)/cm^{-1}; 3471 (br), 2940, 1743, 1504, 1490, 1366, 1230, 1036, 926, 735. HRMS (ESI$^+$) Found [M + Na]$^+$ 499.1729; $C_{28}H_{28}NaO_7$ requires 499.1727.

(3R*,4R*,5S*)-4-(3'',4''-Methylenedioxybenzyl)-3-(4'-hydroxy-3'-methoxybenzyl)-5-(hydroxymethyl) dihydrofuran-2(3H)-one (**4be**). Using general procedure F: Benzyl ether **4bd** (0.02 g, 0.04 mmol) and a reaction time of 1 h. The crude product was then purified by column chromatography (1:3 hexanes, ethyl acetate) to give the title compound **4be** (0.017 g, quant.) as a colourless oil. R_f = 0.52 (1:3 hexanes, ethyl acetate). δ_H (400 MHz; CDCl$_3$) 1.74 (1H, br, 6-OH), 2.33–2.42 (1H, m, 4-H), 2.48 (1H, dd, J = 13.7, 8.0 Hz, 7''-H$_A$), 2.57 (1H, dd, J = 13.7, 6.2 Hz, 7''-H$_B$), 2.67 (1H, ddd, J = 9.4, 6.9, 5.5 Hz, 3-H), 2.88 (1H, dd, J = 14.0, 6.9 Hz, 7'-H$_A$), 2.94 (1H, dd, J = 14.0, 5.5 Hz, 7'-H$_B$), 3.15 (1H, br d, J = 12.6 Hz, 6-H$_A$), 3.52 (1H, br d, J = 12.6 Hz, 6-H$_B$), 3.86 (3H, s, 3'-OCH$_3$), 4.18 (1H, ddd, J = 8.0, 5.0, 2.5 Hz, 5-H), 5.54 (1H, s, 4'-OH), 5.93 (1H, d, J = 1.4 Hz, OCH$_A$O), 5.94 (1H, d, J = 1.4 Hz, OCH$_B$O), 6.45 (1H, d, J = 1.9 Hz, 2''-H), 6.47 (1H, dd, J = 7.7, 1.9 Hz, 6''-H), 6.63 (1H, dd, J = 8.0, 1.9 Hz, 6'-H), 6.67 (1H, d, J = 1.9 Hz, 2'-H), 6.70 (1H, d, J = 7.7 Hz, 5''-H), 6.84 (1H, d, J = 8.0 Hz, 5'-H). δ_C (100 MHz; CDCl$_3$) 35.3 (C-7'), 38.8 (C-7''), 41.7 (C-4), 47.7 (C-3), 56.1 (3'-OCH$_3$), 63.4 (C-6), 83.9 (C-5), 101.3 (OCH$_2$O), 108.6 (C-5''), 109.2 (C-2''), 111.8 (C-2'), 114.5 (C-5'), 121.9 (C-6''), 122.3 (C-6'), 129.6 (C-1'), 131.5 (C-1''), 144.7 (C-4'), 146.7 (C-3'), 146.8 (C-4''), 148.1 (C-3''), 177.8 (C-2). IR: ν_{MAX} (film)/cm^{-1}; 3449 (br), 2933, 1754, 1516, 1490, 1246, 1036, 926, 812. HRMS (ESI$^+$) Found [M + Na]$^+$ 409.1246; $C_{21}H_{22}NaO_7$ requires 409.1258.

(±)-Arcitin (**1aa**). Using general procedure C: Lactone **4aa** (0.16 g, 0.39 mmol) and a reaction time of 2 h to give triol **38aa** (0.17 g, quant.) as a colourless oil. Then using general procedure D: Triol **38aa** (0.16 g, 0.37 mmol) and a reaction time of 2.5 h to give lactol **39aa** (0.14 g, 97%) which was used without further purification. Then using general procedure E: Lactol **39aa** (0.054 g, 0.14 mmol) and a reaction time of 3 h. The crude product was purified by column chromatography (1:1, hexanes, ethyl acetate) to give the title compound **1aa** (0.05 g, 88%) as a pale yellow amorphous solid. R_f = 0.45 (19:1, CH$_2$Cl$_2$, methanol). Melting point: 114–116 °C [lit. [49] 113 °C]. δ_H (400 MHz; CDCl$_3$) 2.45–2.68 (4H, m, 8, 7', 8'-H), 2.92 (1H, dd, J = 14.3, 6.8 Hz, 7-H$_A$), 2.97 (1H, dd, J = 14.3, 5.5 Hz, 7-H$_B$), 3.82 (3H, s, 3'-OCH$_3$), 3.83 (3H, s, 3-OCH$_3$), 3.85–3.90 (7H, m, 4, 4'-OCH$_3$, 9'-H$_A$), 4.13 (1H, t, J = 7.0 Hz, 9'-H$_B$), 6.49 (1H, d, J = 1.9 Hz, 2'-H), 6.55 (1H, dd, J = 8.1, 1.9 Hz, 6'-H), 6.66 (1H, dd, J = 8.1, 1.9 Hz, 6-H), 6.69 (1H, d, J = 1.9 Hz, 2-H), 6.75 (1H, d, J = 8.1 Hz, 5-H), 6.77 (1H, d, J = 8.1 Hz, 5'-H). δ_C (100 MHz; CDCl$_3$) 34.5 (C-7), 38.2 (C-7'), 41.1 (C-8'), 46.6 (C-8), 55.8, 55.9 (3, 4, 3', 4'-OCH$_3$), 71.2 (C-9'), 111.1 (C-5), 111.4 (C-5'), 111.9 (C-2'), 112.4 (C-2), 120.6 (C-6'), 121.4 (C-6), 130.2 (C-1), 130.5 (C-1'), 147.9 (C-4'), 148.0 (C-4), 149.1 (C-3, 3'), 178.7 (C-9). IR: ν_{MAX} (film)/cm^{-1}; 2956, 1753, 1588, 1513, 1257, 1236, 1153, 1137, 1019, 825, 764. HRMS (ESI$^+$) Found [M + H]$^+$ 387.1806; $C_{22}H_{27}O_6$ requires 387.1802. Values are in agreement with literature data [50].

(±)-Bursehernin (**1a**). Using general procedure C: Lactone **4ab** (0.114 g, 0.28 mmol) and a reaction time of 30 min to give triol **38ab** (0.111 g, 97%) as a cloudy oil. Then using general procedure D: Triol **38ab** (0.111 g, 0.27 mmol) and a reaction time of 1 h to give lactol **39ab** (0.093 g, 91%) which was used without further purification. Then using general procedure E: Lactol **39ab** (0.093 g, 0.25 mmol) and a reaction time of 2 h. The crude product was purified by column chromatography (1:1, hexanes, ethyl acetate) to give the title compound **1ab** (0.06 g, 65%) as a pale-yellow oil. R_f = 0.66 (19:1, CH$_2$Cl$_2$, methanol). δ_H (400 MHz; CDCl$_3$) 2.41–2.62 (4H, m, 8, 7', 8'-H), 2.88 (1H, dd, J = 14.0, 6.9 Hz, 7-H$_A$), 2.96 (1H, dd, J = 14.0, 5.1 Hz, 7-H$_B$), 3.82 (3H, s, 3-OCH$_3$), 3.83–3.86 (4H, m, 4-OCH$_3$, 9'-H$_A$), 4.10 (1H, dd, J = 9.1, 6.9 Hz, 9'-H$_B$), 5.91 (1H, d, J = 1.4 Hz, OCH$_A$O), 5.92 (1H, d, J = 1.4 Hz, OCH$_B$O), 6.42 (1H, d, J = 1.5 Hz, 2'-H), 6.44 (1H, dd, J = 7.9, 1.5 Hz, 6'-H), 6.66 (1H, d, J = 1.9 Hz, 2-H), 6.67–6.70 (2H, m, 6, 5'-H), 6.78 (1H, d, J = 8.0 Hz, 5-H). δ_C (100 MHz; CDCl$_3$) 34.7 (C-7), 38.4 (C-7'), 41.2 (C-8'), 46.6 (C-8), 55.9 (3, 4-OCH$_3$), 71.2 (C-9'), 101.1 (OCH$_2$O), 108.4 (C-5'), 108.8 (C-2'), 111.2 (C-5), 112.3 (C-2), 121.4 (C-6), 121.6 (C-6'), 130.2 (C-1), 131.7 (C-1'), 146.4 (C-4'), 148.0 (C-3'), 148.1 (C-4), 149.2 (C-3), 178.7 (C-9). IR: ν_{MAX} (film)/cm^{-1}; 2907, 1764, 1514, 1489, 1442, 1240, 1025, 923, 808, 730. HRMS (ESI$^+$) Found [M + Na]$^+$ 393.1317; $C_{21}H_{22}NaO_6$ requires 393.1309. Values are in agreement with literature data [51].

(±)-*4-O-Methyl traxillagenin* (**1ac**). Using general procedure C: Lactone **4ac** (0.119 g, 0.27 mmol) and a reaction time of 45 min to give triol **38ac** (0.11 g, 90%) as a cloudy oil. The using general procedure D: Triol **38ac** (0.11 g, 0.24 mmol) and a reaction time of 15 min. The crude product was purified by column chromatography (1:2 hexanes, ethyl acetate) to give lactol **39ac** (0.06 g, 60%) as a colourless oil. Then using general procedure E: Lactol **39ac** (0.06 g, 0.15 mmol) and a reaction time of 3 h. The crude product purified by column chromatography (1:1, hexanes, ethyl acetate) to give the title compound **1ac** (0.044 g, 73%) as a white solid. R_f = 0.61 (19:1, CH_2Cl_2, methanol). Melting point: 126 °C. δ_H (400 MHz; CDCl$_3$) 2.44–2.66 (4H, m, 8, 7′, 8′-H), 2.91 (1H, dd, J = 14.1, 6.6 Hz, 7-H$_A$), 2.98 (1H, dd, J = 14.1, 5.4 Hz, 7-H$_B$), 3.79 (6H, s, 3′-OCH$_3$), 3.80 (6H, s, 4′-OCH$_3$), 3.83 (3H, s, 3-OCH$_3$), 3.84 (3H, s, 4-OCH$_3$), 3.87 (1H, dd, J = 9.2, 7.3 Hz, 9′-H$_A$), 4.14 (1H, dd, J = 9.2, 7.0 Hz, 9′-H$_B$), 6.19 (2H, s, 2′-H), 6.63 (1H, dd, J = 8.0, 2.0 Hz, 6-H), 6.70 (1H, d, J = 2.0 Hz, 2-H), 6.75 (1H, d, J = 8.0 Hz, 5-H). δ_C (100 MHz; CDCl$_3$) 34.6 (C-7), 39.0 (C-7′), 41.2 (C-8′), 46.7 (C-8), 56.0 (3, 4-OCH$_3$), 56.2 (3′-OCH$_3$), 60.9 (4′-OCH$_3$), 71.3 (C-9′), 105.7 (C-2′), 111.2 (C-5), 112.6 (C-2), 121.4 (C-6), 130.3 (C-1), 133.8 (C-1′), 137.0 (C-4′), 148.1 (C-4), 149.2 (C-3), 153.5 (C-3′), 178.7 (C-9). IR: ν_{MAX} (film)/cm^{-1}; 2938, 1764, 1590, 1509, 1460, 1237, 1123, 1014, 731. HRMS (ESI$^+$) Found [M + Na]$^+$ 439.1716; C$_{23}$H$_{28}$NaO$_7$ requires 439.1727. Values are in agreement with literature data [52].

(±)-*4′-O-Benzyl buplerol* (**1ad**). Using general procedure C: Lactone **4ad** (0.505 g, 1.02 mmol) and a reaction time of 3 h to give the triol **38ad** (0.472 g, 93%) as a cloudy oil. Then using general procedure D: Triol **38ad** (0.472 g, 0.95 mmol) and a reaction time of 30 min to give lactol **39ad** (0.416 g, 94%) as a white solid which was used without further purification. Then using general procedure E: Lactol **39ad** (0.416 g, 0.90 mmol) and a reaction time of 1.5 h. The crude product was purified by column chromatography (1:1, hexanes, ethyl acetate) to give the title compound **1ad** (0.374 g, 90%) as a pale-yellow oil. R_f = 0.52 (1:1, hexanes, ethyl acetate). δ_H (400 MHz; CDCl$_3$) 2.42–2.66 (4H, m, 8, 7′, 8′-H), 2.91 (1H, dd, J = 14.1, 6.2 Hz, 7-H$_A$), 2.95 (1H, dd, J = 14.1, 5.7 Hz, 7-H$_B$), 3.827, 3.829 (6H, 2 × s, 3, 3′-OCH$_3$), 3.85 (3H, s, 4-OCH$_3$), 3.83–3.88 (1H, m, 9′-H$_A$), 4.11 (1H, dd, J = 8.7, 7.0 Hz, 9′-H$_B$), 5.12 (2H, s, Ph-CH$_2$), 6.48 (1H, dd, J = 8.0, 2.0 Hz, 6′-H), 6.51 (1H, d, J = 2.0 Hz, 2′-H), 6.64 (1H, dd, J = 8.2, 2.0 Hz, 6-H), 6.68 (1H, d, J = 2.0 Hz, 2-H), 6.76 (1H, d, J = 8.2 Hz, 5-H), 6.77 (1H, d, J = 8.0 Hz, 5′-H), 7.27–7.32 (1H, m, Ph-*p*-H), 7.33–7.38 (2H, m, Ph-*m*-H), 7.40–7.44 (2H, m, Ph-*o*-H). δ_C (100 MHz; CDCl$_3$) 34.6 (C-7), 38.3 (C-7′), 41.2 (C-8′), 46.7 (C-8), 56.0 (3, 3′-OCH$_3$), 56.1 (4-OCH$_3$), 71.3, 71.4 (C-9′, Ph-CH$_2$), 111.3 (C-5), 112.5 (C-2), 112.6 (C-5′), 114.5 (C-5′), 120.7 (C-6′), 121.5 (C-6), 127.4 (Ph-*o*-C), 128.0 (Ph-*p*-C), 128.7 (Ph-*m*-C), 130.3 (C-1), 131.3 (C-1′), 137.3 (Ph-*i*-C), 147.2 (C-4′), 148.1 (C-4), 149.2 (C-3), 149.9 (C-3′), 178.8 (C-9). IR: ν_{MAX} (film)/cm^{-1}; 2935, 1763, 1512, 1260, 1233, 1140, 1014, 736, 697. HRMS (ESI$^+$) Found [M + Na]$^+$ 485.1934; C$_{28}$H$_{30}$NaO$_6$ requires 485.1935.

(±)-*Buplerol* (**1ae**). Using general procedure F: Lactone **1ad** (0.336 g, 0.73 mmol) and a reaction time of 3.5 h to give the title compound **1ae** (0.271 g, quant.) as a white solid. R_f = 0.33 (1:1, hexanes, ethyl acetate). Melting point: 101–103 °C. δ_H (400 MHz; CDCl$_3$) 2.42–2.66 (4H, m, 8, 7′, 8′-H), 2.90 (1H, dd, J = 14.1, 6.8 Hz, 7-H$_A$), 2.97 (1H, dd, J = 14.1, 5.3 Hz, 7-H$_B$), 3.81 (3H, s, 3-OCH$_3$), 3.83 (3H, s, 3′-OCH$_3$), 3.86 (4H, m, 4-OCH$_3$), 3.87 (1H, dd, J = 8.9, 7.1 Hz, 9′-H$_A$), 4.13 (1H, dd, J = 9.3, 7.1 Hz, 9′-H$_B$), 5.51 (1H, s, 4′-OH), 6.43 (1H, d, J = 1.9 Hz, 2′-H), 6.52 (1H, dd, J = 8.0, 1.9 Hz, 6′-H), 6.64–6.67 (2H, m, 2, 6-H), 6.77 (1H, d, J = 8.6 Hz, 5-H), 6.80 (1H, d, J = 8.0 Hz, 5′-H). δ_C (100 MHz; CDCl$_3$) 34.7 (C-7), 38.5 (C-7′), 41.3 (C-8′), 46.7 (C-8), 55.9, 56.0 (3, 3′, 4-OCH$_3$), 71.4 (C-9′), 111.1, 111.2 (C-5, 5′), 112.5 (C-2), 114.6 (C-2′), 121.5 (C-6, 6′), 129.9 (C-1′), 130.4 (C-1), 144.6 (C-4′), 146.7 (C-3′), 148.1 (C-4), 149.2 (C-3), 178.9 (C-9). IR: ν_{MAX} (film)/cm^{-1}; 3417, 2938, 1760, 1513, 1236, 1148, 1023, 812, 795. HRMS (ESI$^+$) Found [M + Na]$^+$ 395.1462; C$_{21}$H$_{24}$NaO$_6$ requires 395.1465. Values are in agreement with literature data [53].

(±)-*Kusunokinin* (**1ba**). Using general procedure C: Lactone **4ba** (0.082 g, 0.20 mmol) and a reaction time of 1 h to give the triol **38ba** (0.083 g, quant.) as a cloudy oil. Then using general procedure D: Triol **38ba** (0.083 g, 0.20 mmol) and a reaction time of 15 min to give lactol **39ba** (0.064 g, 84%) which was used without further purification. Then using general procedure E: Lactol **39ba** (0.056 g, 0.15 mmol) and a reaction time of 1 h. The crude product was purified by column chromatography (2:1, hexanes, ethyl acetate) to give the title compound **1ba** (0.051 g, 91%) as a colourless oil. R_f = 0.48 (1:1, hexanes, ethyl acetate). δ_H (400 MHz; CDCl$_3$) 2.44–2.65 (4H, m, 8, 7′, 8′-H), 2.84 (1H, dd, *J* = 14.1, 7.0 Hz, 7-H$_A$), 2.95 (1H, dd, *J* = 14.1, 5.1 Hz, 7-H$_B$), 3.82 (3H, s, 3′-OCH$_3$), 3.85 (3H, s, 4′-OCH$_3$), 3.87 (1H, dd, *J* = 9.2, 7.2 Hz, 9′-H$_A$), 4.14 (1H, dd, *J* = 9.2, 7.0 Hz, 9′-H$_B$), 5.92 (1H, d, *J* = 1.4 Hz, OCH$_A$O), 5.93 (1H, d, *J* = 1.4 Hz, OCH$_B$O), 6.48 (1H, d, *J* = 2.0 Hz, 2′-H), 6.55–6.60 (3H, m, 2, 6, 6′-H), 6.71 (1H, d, *J* = 7.7 Hz, 5-H), 6.76 (1H, d, *J* = 8.2 Hz, 5′-H). δ_C (100 MHz; CDCl$_3$) 34.9 (C-7), 38.4 (C-7′), 41.3 (C-8′), 46.6 (C-8), 55.9 (3′-OCH$_3$), 56.0 (4′-OCH$_3$), 71.4 (C-9′), 101.1 (OCH$_2$O), 108.3 (C-5), 109.6 (C-2), 111.4 (C-5′), 111.8 (C-2′), 120.8 (C-6′), 122.4 (C-6), 130.6 (C-1′), 131.5 (C-1), 146.6 (C-4), 148.0 (C-3, 4′), 149.2 (C-3′), 178.6 (C-9). IR: ν_{MAX} (film)/cm^{-1}; 2908, 1764, 1515, 1489, 1442, 1242, 1024, 912, 809, 729. HRMS (ESI$^+$) Found [M + Na]$^+$ 393.1301; C$_{21}$H$_{22}$NaO$_6$ requires 393.1309. Values are in agreement with literature data [50].

(±)-*Hinokinin* (**1bb**). Using general procedure C: Lactone **4bb** (0.12 g, 0.31 mmol) and a reaction time of 30 min to give the triol **38bb** (0.12 g, quant.) as a cloudy oil. Then using general procedure D: Triol **38bb** (0.121 g, 0.31 mmol) and a reaction time of 10 min to give lactol **39bb** (0.096 g, 86%) which was used without further purification. Then using general procedure E: Lactol **39bb** (0.089 g, 0.25 mmol) and a reaction time of 1 h. The crude product was purified by column chromatography (1:1, hexanes, ethyl acetate) to give the title compound **1bb** (0.08 g, 90%) as a pale-yellow oil. R_f = 0.73 (1:1, hexanes, ethyl acetate). δ_H (400 MHz; CDCl$_3$) 2.41–2.62 (4H, m, 8, 7′, 8′-H), 2.83 (1H, dd, *J* = 14.1, 7.2 Hz, 7-H$_A$), 2.98 (1H, dd, *J* = 14.1, 5.0 Hz, 7-H$_B$), 3.85 (1H, dd, *J* = 9.2, 7.1 Hz, 9′-H$_A$), 4.12 (1H, dd, *J* = 9.2, 6.9 Hz, 9′-H$_B$), 5.91–5.94 (4H, m, 2 × OCH$_2$O), 6.44–6.47 (2H, m, 2′, 6′-H), 6.59 (1H, dd, *J* = 7.9, 1.8 Hz, 6-H), 6.62 (1H, d, *J* = 1.8 Hz, 2-H), 6.69 (1H, d, *J* = 8.4 Hz, 5′-H), 6.72 (1H, d, *J* = 7.9 Hz, 5-H). δ_C (100 MHz; CDCl$_3$) 34.9 (C-7), 38.4 (C-7′), 41.4 (C-8′), 46.6 (C-8), 71.2 (C-9′), 101.1 (2 × OCH$_2$O), 108.4 (C-5, 5′), 108.9 (C-2′), 109.5 (C-2), 121.6 (C-6′), 122.3 (C-6), 131.5 (C-1), 131.7 (C-1′), 146.4 (C-4), 146.6 (C-4′), 148.0 (C-3, 3′), 178.5 (C-9). IR: ν_{MAX} (film)/cm^{-1}; 2901, 1764, 1488, 1441, 1242, 1015, 924, 808, 728. HRMS (ESI$^+$) Found [M + Na]$^+$ 377.0986; C$_{20}$H$_{18}$NaO$_6$ requires 377.0996. Values are in agreement with literature data [54].

(±)-*Isoyatein* (**1bc**). Using general procedure C: Lactone **4bc** (0.076 g, 0.18 mmol) and a reaction time of 1 h to give the triol **38bc** (0.077 g, >99%) as a cloudy oil. Then using general procedure D: Triol **38bc** (0.077 g, 0.18 mmol) and a reaction time of 1 h to give lactol **39bc** (0.057 g, 80%) which was used without further purification. Then using general procedure E: Lactol **39bc** (0.05 g, 0.12 mmol) and a reaction time of 3 h. The crude product was purified by column chromatography (1:1, hexanes, ethyl acetate) to give the title compound **1bc** (0.8 mg, 16%) as a pale-yellow oil. R_f = 0.55 (1:1, hexanes, ethyl acetate). δ_H (400 MHz; CDCl$_3$) 2.46–2.64 (4H, m, 8, 7′, 8′-H), 2.86 (1H, dd, *J* = 14.1, 7.0 Hz, 7-H$_A$), 2.98 (1H, dd, *J* = 14.1, 5.1 Hz, 7-H$_B$), 3.81 (6H, s, 3′-OCH$_3$), 3.82 (3H, s, 4′-OCH$_3$), 3.89 (1H, dd, *J* = 9.2, 7.0 Hz, 9′-H$_A$), 4.19 (1H, dd, *J* = 9.2, 6.8 Hz, 9′-H$_B$), 5.93 (1H, d, *J* = 1.5 Hz, OCH$_A$O), 5.94 (1H, d, *J* = 1.5 Hz, OCH$_B$O), 6.20 (2H, s, 2′-H), 6.58 (1H, dd, *J* = 7.9, 1.8 Hz, 6-H), 6.61 (1H, d, *J* = 1.8 Hz, 2-H), 6.71 (1H, d, *J* = 7.9 Hz, 5-H). δ_C (100 MHz; CDCl$_3$) 34.9 (C-7), 39.2 (C-7′), 41.4 (C-8′), 46.6 (C-8), 56.2 (3′-OCH$_3$), 61.0 (4′-OCH$_3$), 71.4 (C-9′), 101.2 (OCH$_2$O), 105.7 (C-2′), 108.3 (C-5), 109.6 (C-2), 122.4 (C-6), 131.5 (C-1), 133.8 (C-1′), 137.0 (C-4′), 146.7 (C-4), 148.1 (C-3), 153.5 (C-3′), 178.5 (C-9). IR: ν_{MAX} (film)/cm^{-1}; 2938, 1763, 1590, 1489, 1443, 1241, 1122, 1011, 927, 813, 732. HRMS (ESI$^+$) Found [M + Na]$^+$ 423.1400; C$_{22}$H$_{24}$NaO$_7$ requires 423.1414. Values are in agreement with literature data [55].

(±)-*4′-O-Benzyl haplomyrfolin* (**1bd**). Using general procedure C: Lactone **4bd** (0.18 g, 0.38 mmol) and a reaction time of 20 min to give the triol **38bd** (0.18 g, quant.) as a cloudy oil. Then using general procedure D: Triol **38bd** (0.18 g, 0.38 mmol) and a reaction time of 20 min to give lactol **39bd** (0.13 g, 76%) as a white solid which was used without further purification. Then using general procedure E: Lactol **39bd** (0.13 g, 0.28 mmol) and a reaction time of 2 h. The crude product was purified by column chromatography (3:1, hexanes, ethyl acetate) to give the title compound **1bd** (0.12 g, 94%) as a colourless oil. R_f = 0.65 (1:1, hexanes, ethyl acetate). δ_H (400 MHz; CDCl$_3$) 2.43–2.64 (4H, m, 8, 7′, 8′-H), 2.84 (1H, dd, *J* = 14.1, 7.0 Hz, 7-H$_A$), 2.94 (1H, dd, *J* = 14.1, 5.1 Hz, 7-H$_B$), 3.83 (3H, s, 3′-OCH$_3$), 3.87 (1H, dd, *J* = 9.1, 7.2 Hz, 9′-H$_A$), 4.14 (1H, dd, *J* = 9.1, 7.0 Hz, 9′-H$_B$), 5.12 (2H, s, 7″-H), 5.91 (1H, d,

J = 1.4 Hz, OCH$_A$O), 5.93 (1H, d, J = 1.4 Hz, OCH$_B$O), 6.49–6.52 (2H, m, 2′, 6′-H), 6.57 (1H, dd, J = 7.9, 1.8 Hz, 6-H), 6.59 (1H, d, J = 1.8 Hz, 2-H), 6.70 (1H, d, J = 7.9 Hz, 5-H), 6.78 (1H, d, J = 8.5 Hz, 5′-H), 7.27–7.32 (1H, m, 4″-H), 7.33–7.38 (2H, m, 3″-H), 7.41–7.45 (2H, m, 2″-H). δ_C (100 MHz; CDCl$_3$) 34.8 (C-7), 38.4 (C-7′), 41.3 (C-8′), 46.5 (C-8), 56.0 (3′-OCH$_3$), 71.2 (C-7″), 71.3 (C-9′), 101.1 (OCH$_2$O), 108.3 (C-5), 109.6 (C-2), 112.4 (C-2′), 114.4 (C-5′), 120.7 (C-6′), 122.4 (C-6), 127.4 (C-2″), 128.0 (C-4″), 128.6 (C-3″), 131.2 (C-1), 131.5 (C-1′), 137.2 (C-1″), 146.6 (C-4), 147.1 (C-4′), 148.0 (C-3), 149.9 (C-3′), 178.6 (C-9). IR: ν_{MAX} (film)/cm^{-1}; 2907, 1765, 1504, 1489, 1443, 1244, 1140, 1034, 911, 809, 730. HRMS (ESI$^+$) Found [M + Na]$^+$ 469.1612; C$_{27}$H$_{26}$NaO$_6$ requires 469.1622.

(±)-*Haplomyrfolin* (**1be**). Using general procedure F: Lactone **1bd** (0.119 g, 0.27 mmol) and a reaction time of 1.5 h. The crude product was purified by column chromatography (1:1 hexanes, ethyl acetate) to give the title compound **1be** (0.086 g, 91%) as a colourless oil. R$_f$ = 0.47 (1:1 hexanes, ethyl acetate). δ_H (400 MHz; CDCl$_3$) 2.43–2.63 (4H, m, 8, 7′, 8′-H), 2.84 (1H, dd, J = 14.1, 7.0 Hz, 7-H$_A$), 2.95 (1H, dd, J = 14.1, 5.2 Hz, 7-H$_B$), 3.83 (3H, s, 3′-OCH$_3$), 3.86 (1H, dd, J = 9.1, 7.2 Hz, 9′-H$_A$), 4.13 (1H, dd, J = 9.1, 7.0 Hz, 9′-H$_B$), 5.63 (1H, s, 4′-OH), 5.91 (1H, d, J = 1.4 Hz, OCH$_A$O), 5.92 (1H, d, J = 1.4 Hz, OCH$_B$O), 6.46 (1H, d, J = 1.9 Hz, 2′-H), 6.51 (1H, dd, J = 8.0, 1.9 Hz, 6′-H), 6.58 (1H, dd, J = 7.8, 1.7 Hz, 6-H), 6.60 (1H, d, J = 1.7 Hz, 2-H), 6.70 (1H, d, J = 7.8 Hz, 5-H), 6.80 (1H, d, J = 8.0 Hz, 5′-H). δ_C (100 MHz; CDCl$_3$) 34.8 (C-7), 38.3 (C-7′), 41.4 (C-8′), 46.5 (C-8), 55.9 (3′-OCH$_3$), 71.3 (C-9′), 101.1 (OCH$_2$O), 108.3 (C-5), 109.6 (C-2), 111.2 (C-2′), 114.6 (C-5′), 121.4 (C-6′), 122.4 (C-6), 129.9 (C-1′), 131.5 (C-1), 144.5 (C-4′), 146.5 (C-4), 146.7 (C-3′), 147.9 (C-3), 178.7 (C-9). IR: ν_{MAX} (film)/cm^{-1}; 3468, 2921, 1762, 1515, 1489, 1443, 1243, 1035, 907, 725. HRMS (ESI$^+$) Found [M + Na]$^+$ 379.1151; C$_{20}$H$_{20}$NaO$_6$ requires 379.1152. Values are in agreement with literature data [56].

4. Biological Assay Methods

4.1. Cell Culture

Jurkat E61 cells (ECACC) were maintained at 37 °C in RMPI media (Lonza) supplemented with 10% Foetal Bovine Serum (FBS) (Lonza) (10% RPMI) in a humidified environment of 5% CO$_2$ in air. Cells were routinely passaged to maintain a cell density of between 1×10^5 and 1×10^6/mL.

4.2. Drug Treatments

Lignans were diluted to stock concentrations of 30 mM in DMSO and further diluted to the working concentration in 10% RPMI. The DMSO diluted to the appropriate concentration was used as the vehicle-control. Cells were seeded at the relevant density per well depending upon the assay to be performed, in 100 µL volume of fresh 10% RPMI. Trypan blue exclusion method was used to assess viability prior to experiments and cell viability was always >95%. Lignans were added at 100 µL/well to the relevant wells. Cells were incubated at 37 °C in a humidified environment of 5% CO$_2$ in air for the indicated times. Dead cell controls were included in subsequent viability assays by treating cells with 50 µL/well EtOH (final concentration 50%) for 48 h. Apoptotic controls were included in subsequent apoptosis assays by exposing cells to a heat shock at 43 °C for 2 h. Positive controls for cell cycle analysis were included by treating cells with 0.5 µM camptothecin for 4 h to induce cell cycle arrest.

4.3. MTS Assay

Following treatments at a cell density of 1×10^5 cells/well, the samples were centrifuged at 500 g for 5 min and the supernatant was removed. A 100 µL/well volume of fresh 10 % RPMI was added. A 20 µL volume of MTS solution (Promega, G1112) was added to each well and the plate was incubated in the dark for 1 h at 37 °C. The absorbance was detected at 490 nm on a Synergy HT plate reader.

4.4. Annexin V/PI Assay

Following treatments at a cell density of 1×10^5 cells/well, the samples were centrifuged at 500 *g* for 5 min and the supernatant was removed. Cells were washed in 500 µL DPBS before addition of 100 µL of $1 \times$ Annexin V binding buffer (BD Biosciences). A 5 µL volume of FITC-conjugated Annexin V (BD Biosciences) and 10 µL Propidium Iodide (BD Biosciences) was added and the cells were incubated in the dark for 20 min. Samples were diluted by addition of 400 µL $1 \times$ Annexin V binding buffer before immediate analysis on an Accuri C6 Flow Cytometer (Becton Dickinson, Oxford, UK).

4.5. Cell Cycle Analysis

Following treatments at a cell density of 5×10^6/well, cells were centrifuged at 500 *g* for 5 min and the supernatant was removed. The remaining cell pellet was vortexed while simultaneously adding 500 µL of 70% ethanol dropwise, fixing the cells and minimising clumping. The samples were incubated at 4 °C for 30 min, and then centrifuged at 1000 *g* for 5 min. The supernatant was discarded, and the pellet was re-suspended in 500 µL DPBS. The samples were centrifuged again at 1000 *g* for 5 min, and the supernatant was removed a final time. The pellet was resuspended in 50 µL RNase A (100 µg/mL stock; Roche, UK) and 200 µL PI (50 µg/mL stock; Sigma, UK). The samples were analyzed on an Accuri C6 flow cytometer (Becton Dickinson) and data was modelled and interpreted using ModFit Analysis Software, version 5.0 (Verity Software House).

Supplementary Materials: The following are available online.

Author Contributions: Conceptualization, D.B. and N.C.D.-H.; Methodology, S.J.D., M.E.-M., S.T. and T.W.; Formal Analysis, L.I.P., D.B. and N.C.D.-H.; Investigation, S.J.D., M.E.-M., S.T. and T.W.; Writing-Original Draft Preparation, S.J.D. and L.I.P.; Writing-Review & Editing, D.B., L.I.P., S.J.D., K.A.W. and N.C.D.-H.; Supervision, D.B., K.A.W. and N.C.D.-H.; Project Administration, D.B.

Funding: This research received no external funding.

Conflicts of Interest: The authors declare no conflict of interest.

References

1. Jang, Y.P.; Kim, S.R.; Kim, Y.C. Neuroprotective dibenzylbutyrolactone lignans of Torreya nucifera. *Planta Med.* **2001**, *67*, 470–472. [CrossRef] [PubMed]

2. Marcotullio, M.C.; Pelosi, A.; Curini, M. Hinokinin, an emerging bioactive lignan. *Molecules* **2014**, *19*, 14862–14878. [CrossRef] [PubMed]

3. Su, S.; Cheng, X.; Wink, M. Natural lignans from Arctium lappa modulate P-glycoprotein efflux function in multidrug resistant cancer cells. *Phytomedicine* **2015**, *22*, 301–307. [CrossRef] [PubMed]

4. Yang, Y.-N.; Huang, X.-Y.; Feng, Z.-M.; Jiang, J.-S.; Zhang, P.-C. New butyrolactone type lignans from Arctii fructus and their anti-inflammatory activities. *J. Agric. Food Chem.* **2015**, *63*, 7958–7966. [CrossRef] [PubMed]

5. Su, S.; Wink, M. Natural lignans from Arctium lappa as antiaging agents in Caenorhabditis elegans. *Phytochemistry* **2015**, *117*, 340–350. [CrossRef] [PubMed]

6. Chang, H.; Wang, Y.; Gao, X.; Song, Z.; Awale, S.; Han, N.; Liu, Z.; Yina, J. Lignans from the root of Wikstroemia indica and their cytotoxic activity against PANC-1 human pancreatic cancer cells. *Fitoterpia* **2017**, *121*, 31–37. [CrossRef] [PubMed]

7. Maxwell, T.; Chun, S.Y.; Lee, K.S.; Kim, S.; Nam, K.S. The anti-metastatic effects of the phytoestrogen arctigenin on human breast cancer cell lines regardless of the status of ER expression. *Int. J. Oncol.* **2017**, *50*, 727–735. [CrossRef] [PubMed]

8. Maxwell, T.; Lee, K.S.; Kim, S.; Nam, K.S. Arctigenin inhibits the activation of the mTOR pathway, resulting in autophagic cell death and decreased ER expression in ER-positive human breast cancer cells. *Int. J. Oncol.* **2018**, *52*, 1339–1349. [CrossRef] [PubMed]

9. Huang, Q.; Qin, S.; Yuan, X.; Zhang, L.; Ji, J.; Liu, X.; Ma, W.; Zhang, Y.; Liu, P.; Sun, Z.; et al. Arctigenin inhibits triple-negative breast cancers by targeting CIP2A to reactivate protein phosphatase 2A. *Oncol. Rep.* **2017**, *38*, 598–606. [CrossRef] [PubMed]

10. Li, Q.C.; Liang, Y.; Tian, Y.; Hu, G.R. Arctigenin induces apoptosis in colon cancer cells through ROS/p38MAPK pathway. *J. Buon.* **2016**, *21*, 87–94. [PubMed]

11. Han, Y.H.; Kee, J.Y.; Kim, D.S.; Mun, J.G.; Jeong, M.Y.; Park, S.H.; Choi, B.M.; Park, S.J.; Kim, H.J.; Um, J.Y.; et al. Arctigenin inhibits lung metastasis of colorectal cancer by regulating cell viability and metastatic phenotypes. *Molecules* **2016**, *21*, 1135. [CrossRef] [PubMed]

12. Maimaitili, A.; Shu, Z.; Cheng, X.; Kaheerman, K.; Sikandeer, A.; Li, W. Arctigenin, a natural lignan compound, induces G0/G1 cell cycle arrest and apoptosis in human glioma cells. *Oncol. Lett.* **2017**, *13*, 1007–1013. [CrossRef] [PubMed]

13. Pilkington, L.I. Lignans: A chemometric analysis. *Molecules* **2018**, *23*, 1666. [CrossRef] [PubMed]

14. Amancha, P.K.; Liu, H.-J.; Ly, T.W.; Shia, K.-S. General approach to 2,3-dibenzyl-γ-butyrolactone lignans: Application to the total synthesis of (±)-5′-methoxyyatein, (±)-5′-methoxyclusin, and (±)-4′-hydroxycubebinone. *Eur. J. Org. Chem.* **2010**, *2010*, 3473–3480. [CrossRef]

15. Isemori, Y.; Kobayashi, Y. An approach to β-substituted γ-butyrolactones and its application to the synthesis of lignans. *Synlett* **2004**, 1941–1944. [CrossRef]

16. Ferrié, L.; Bouyssi, D.; Balme, G. Selective lewis acid catalyzed transformation (γ-butyrolactone versus cyclopropane) of 2-methoxy-4-benzyltetrahydrofuran derivatives. Efficient synthesis of lignan lactones. *Org. Lett.* **2005**, *7*, 3143–3146. [CrossRef] [PubMed]

17. Duan, S.; Huang, S.; Gong, J.; Shen, Y.; Zeng, L.; Feng, Y.; Ren, W.; Leng, Y.; Hu, Y. Design and synthesis of novel arctigenin analogues for the amelioration of metabolic disorders. *ACS Med. Chem. Lett.* **2015**, *6*, 386–391. [CrossRef] [PubMed]

18. Rye, C.; Barker, D. An acyl-Claisen approach to tetrasubstituted tetrahydrofuran lignans: Synthesis of fragransin A2, talaumidin, and lignan analogues. *Synlett* **2009**, 3315–3319. [CrossRef]

19. Barker, D.; Dickson, B.; Dittrich, N.; Rye, C.E. An acyl-Claisen approach to the synthesis of.lignans and substituted pyrroles. *Pure Appl. Chem.* **2012**, *84*, 1557–1565. [CrossRef]

20. Dickson, B.D.; Dittrich, N.; Barker, D. Synthesis of 2,3-syn-diarylpent-4-enamides via acyl-Claisen rearrangements of substituted cinnamyl morpholines: Application to the synthesis of magnosalicin. *Tetrahedron Lett.* **2012**, *53*, 4464–4468. [CrossRef]

21. Duhamel, N.; Rye, C.E.; Barker, D. Total Synthesis of ent-hyperione A and ent-hyperione B. *Asian J. Org. Chem.* **2013**, *2*, 491–493. [CrossRef]

22. Rye, C.E.; Barker, D. Asymmetric synthesis of (+)-galbelgin, (−)-kadangustin J, (−)-cyclogalgravin and (−)-pycnanthulignenes A and B, three structurally distinct lignan classes, using a common chiral precursor. *J. Org. Chem.* **2011**, *76*, 6636–6648. [CrossRef] [PubMed]

23. Pilkington, L.I.; Wagoner, J.; Polyak, S.J.; Barker, D. Enantioselective synthesis, stereochemical correction, and biological investigation of the rodgersinine family of 1,4-benzodioxane neolignans. *Org. Lett.* **2015**, *17*, 1046–1049. [CrossRef] [PubMed]

24. Pilkington, L.I.; Barker, D. Synthesis and biology of 1,4-benzodioxane lignan natural products. *Nat. Prod. Rep.* **2015**, *32*, 1369–1388. [CrossRef] [PubMed]

25. Pilkington, L.I.; Barker, D. Asymmetric synthesis and CD investigation of the 1,4-benzodioxane lignans eusiderins A, B, C, G, L, and M. *J. Org. Chem.* **2012**, *77*, 8156–8166. [CrossRef] [PubMed]

26. Pilkington, L.I.; Barker, D. Total synthesis of (−)-isoamericanin A and (+)-isoamericanol A. *Eur. J. Org. Chem.* **2014**, 1037–1046. [CrossRef]

27. Jung, E.; Pilkington, L.I.; Barker, D. Enantioselective synthesis of 2,3-disubstituted benzomorpholines: Analogues of lignan natural products. *J. Org. Chem.* **2016**, *81*, 12012–12022. [CrossRef] [PubMed]

28. Jung, E.; Dittrich, N.; Pilkington, L.I.; Rye, C.E.; Leung, E.; Barker, D. Synthesis of aza-derivatives of tetrahydrofuran lignan natural products. *Tetrahedron* **2015**, *71*, 9439–9456. [CrossRef]

29. Paterson, D.L.; Barker, D. Synthesis of the furo[2,3-b]chromene ring system of hyperaspindols A and B. *Beilstein J. Org. Chem.* **2015**, *11*, 265–270. [CrossRef] [PubMed]

30. Davidson, S.J.; Barker, D. Synthesis of various lignans via the rearrangements of 1,4-diarylbutane-1,4-diols. *Tetrahedron Lett.* **2015**, *56*, 4549–4553. [CrossRef]

31. Pilkington, L.I.; Barker, D. Synthesis of 3-methylobovatol. *Synlett* **2015**, *26*, 2425–2428.

32. Rye, C.E.; Barker, D. Asymmetric synthesis and anti-protozoal activity of the 8,4′-oxyneolignans virolin, surinamensin and analogues. *Eur. J. Med. Chem.* **2013**, *60*, 240–248. [CrossRef] [PubMed]

33. Tran, H.; Dickson, B.; Barker, D. Unexpected O-alkylation and ester migration in phenolic 2,3-diaryl-2,3-dihydrobenzo[b]furans. *Tetrahedron Lett.* **2013**, *54*, 2093–2096. [CrossRef]

34. Pilkington, L.I.; Song, S.M.; Fedrizzi, B.; Barker, D. Efficient total synthesis of (±)-isoguaiacin and (±)-isogalbulin. *Synlett* **2017**, *28*, 1449–1452.

35. Davidson, S.J.; Barker, D. Total synthesis of ovafolinins A and B: Unique polycyclic benzoxepin lignans through a cascade cyclization. *Angew. Chem. Int. Ed.* **2017**, *56*, 9483–9486. [CrossRef] [PubMed]

36. Davidson, S.J.; Pearce, A.N.; Copp, B.R.; Barker, D. Total synthesis of (−)-bicubebin A, B, (+)-bicubebin C and structural reassignment of (−)-cis-cubebin. *Org. Lett.* **2017**, *19*, 5368–5371. [CrossRef] [PubMed]

37. Kakis, F.J.; Fetizon, M.; Douchkine, N.; Golfier, M.; Mourgues, P.; Prange, T. Mechanistic studies regarding the oxidation of alcohols by silver carbonate on celite. *J. Org. Chem.* **1974**, *39*, 523–533. [CrossRef]

38. Fétizon, M.; Golfier, M.; Louis, J.-M. A new synthesis of lactones: Application to (±)-mevalonolactone. *J. Chem. Soc. D* **1969**, 1118–1119. [CrossRef]

39. Xin, H.; Kong, Y.; Wang, Y.; Zhou, Y.; Zhu, Y.; Li, D.; Tan, W. Lignans extracted from Vitex negundo possess cytotoxic activity by G2/M phase cell cycle arrest and apoptosis induction. *Phytomedicine* **2013**, *20*, 640–647. [CrossRef] [PubMed]

40. Bose, J.S.; Gangan, V.; Prakash, R.; Kumar Jain, S.; Kumar Manna, S. A dihydrobenzofuran lignan induces cell death by modulating mitochondrial pathway and G2/M cell cycle arrest. *J. Med. Chem.* **2009**, *52*, 3184–3190. [CrossRef] [PubMed]

41. Fetizon, M.; Balogh, V.; Golfier, M. Oxidations with silver carbonate/celite. V. Oxidations of phenols and related compounds. *J. Org. Chem.* **1971**, *36*, 1339–1341. [CrossRef]

42. Reddy, R.S.; Prasad, P.K.; Ahuja, B.B.; Sudalai, A. CuCN mediated cascade cyclization of 4-(2-bromophenyl)-2-butenoates: A high-yield synthesis of substituted naphthalene amino esters. *J. Org. Chem.* **2013**, *78*, 5045–5050. [CrossRef] [PubMed]

43. Sharma, P.; Ritson, D.J.; Burnley, J.; Moses, J.E. A synthetic approach to kingianin A based on biosynthetic speculation. *Chem. Commun.* **2011**, *47*, 10605–10607. [CrossRef] [PubMed]

44. Aldous, D.J.; Batsanov, A.S.; Yufit, D.S.; Dalençon, A.J.; Dutton, W.M.; Steel, P.G. The dihydrofuran template approach to furofuran synthesis. *Org. Biomol. Chem.* **2006**, *4*, 2912–2927. [CrossRef] [PubMed]

45. Tanoguchi, M.; Kashima, T.; Saika, H.; Inoue, T.; Arimoto, M.; Yamaguchi, H. Studies on the constituents of the seeds of hernandia ovigera L. VII.: Syntheses of (±)-hernolactone and (±)-hernandin. *Chem. Pharm. Bull.* **1989**, *37*, 68–72. [CrossRef]

46. Gomes, C.A.; Girão da Cruz, T.; Andrade, J.L.; Milhazes, N.; Borges, F.; Marques, M.P.M. Anticancer activity of phenolic acids of natural or synthetic origin: A structure-activity study. *J. Med. Chem.* **2003**, *46*, 5395–5401. [CrossRef] [PubMed]

47. Haga, Y.; Okazaki, M.; Shuto, Y. Systematic strategy for the synthesis of cyanobacterin and its stereoisomers. 1. Asymmetric total synthesis of dechloro-cyanobacterin and its enantiomer. *Biosci. Biotechnol. Biochem.* **2003**, *67*, 2183–2193. [CrossRef] [PubMed]

48. Li, D.; Zhao, B.; Sim, S.-P.; Li, T.-K.; Liu, A.; Liu, L.F.; LaVoie, E.J. 8,9-Methylenedioxybenzo[i]phenanthridines: Topoisomerase I-targeting activity and cytotoxicity. *Bioorg. Med. Chem.* **2003**, *11*, 3795–3805. [CrossRef]

49. Takei, Y.; Mori, K.; Matsui, M. Synthesis of *dl*-matairesinol dimethyl ether, dehydrodimethylconidendrin and dehydrodimethylretrodendrin from ferulic acid. *Agric. Biol. Chem.* **1973**, *37*, 637–641. [CrossRef]

50. Brown, E.; Daugan, A. Lignames: 10. Preparation des (R)-(+) et (S)-(−)-β-piperonyl et β-veratryl-γ-butyrolactones et leur utilisation dans la synthese totale de lignanes optiquement actifs. *Tetrahedron* **1989**, *45*, 141–154. [CrossRef]

51. Baran, P.S.; DeMartino, M.P. Intermolecular oxidative enolate heterocoupling. *Angew. Chem. Int. Ed.* **2006**, *45*, 7083–7086. [CrossRef] [PubMed]

52. Nishibe, S.; Okabe, K.; Hisada, S. Isolation of phenolic compounds and spectroscopic analysis of a new lignan from Trachelospermum asiaticum var. intermedium. *Chem. Pharm. Bull.* **1981**, *39*, 2078–2082. [CrossRef]

53. Gonzalez, A.G.; Estevez-Reyes, R.; Estevez-Braun, A.M. Buplerol and guayarol, new lignans from the seeds of bupleurum salicifolium. *J. Chem. Res.* **1990**, *21*, 220–221.

54. De Souza, V.A.; da Silva, R.; Pereira, A.C.; Royo, V.D.A.; Saraiva, J.; Montanheiro, M.; de Souza, G.H.B.; da Silva Filho, A.A.; Grando, M.D.; Donate, P.M.; et al. Trypanocidal activity of (−)-cubebin derivatives against free amastigote forms of Trypanosoma cruzi. *Bioorg. Med. Chem. Lett.* **2005**, *15*, 303–307. [CrossRef] [PubMed]

55. Badheka, L.P.; Prabhu, B.R.; Mulchandani, N.B. Dibenzylbutyrolactone lignans from Piper cubeba. *Phytochemistry* **1986**, *25*, 487–489. [CrossRef]
56. Evcim, U.; Gozler, B.; Freyer, A.J.; Shamma, M. Haplomyrtin and (−)-haplomyrfolin: Two lignans from haplophyllum myrtifolium. *Phytochemistry* **1986**, *25*, 1949–1951. [CrossRef]

Communication

Concise Synthesis of (+)-β- and γ-Apopicropodophyllins, and Dehydrodesoxypodophyllotoxin

Jian Xiao [1,2], Guangming Nan [1], Ya-Wen Wang [2] and Yu Peng [2,3,*]

[1] University and College Key Lab of Natural Product Chemistry and Application in Xinjiang,
    Yili Normal University, Yining 835000, China; xiaoj2012@lzu.edu.cn (J.X.);
    nanguangming02@sohu.com (G.N.)
[2] School of Life Science and Engineering, Southwest Jiaotong University, Chengdu 610031, China;
    ywwang@swjtu.edu.cn
[3] State Key Laboratory of Applied Organic Chemistry and College of Chemistry and Chemical Engineering,
    Lanzhou University, Lanzhou 730000, China
*   Correspondence: pengyu@lzu.edu.cn

Academic Editors: David Barker and Derek J. McPhee
Received: 30 October 2018; Accepted: 17 November 2018; Published: 21 November 2018

Abstract: Herein, we present an expeditous synthesis of bioactive aryldihydronaphthalene lignans (+)-β- and γ-apopicropodophyllins, and arylnaphthalene lignan dehydrodesoxypodophyllotoxin. The key reaction is regiocontrolled oxidations of stereodivergent aryltetralin lactones, which were easily accessed from a nickel-catalyzed reductive cascade approach developed in our group.

Keywords: aryldihydronaphthalene lignan; arylnaphthalene lignan; oxidation; synthesis

---

1. Introduction

Lignans are a class of secondary metabolites in various plants, and most of them have demonstrated interesting biological properties [1,2], thus attracting the attention of the synthetic chemists [3,4]. Some of 2,7'-cyclolignans such as 7,8,8',7'-tetrahydronaphthalene (THN), 7',8'-dihydronaphthalene (DHN) and 7'-arylnaphthalene types are exemplified in Scheme 1a. Hong and co-workers used organocatalytic domino Michael–Michael–aldol reactions to construct THN skeleton of galbulin and realized its first enantioselective synthesis [5]. Barker and co-workers completed the first asymmetric synthesis of (−)-cyclogalgravin based on a key construction of C2–C7' bond from in situ generated quinoid intermediate [6]. Notably, the other two structurally distinct class of lignans could also be obtained from a common precursor in their syntheses. Ramana et al. proposed a dehydrative cyclization of an aldehyde intermediate to build the DHN unit of sacidumlignan B, whose subsequent aromatization led to the synthesis of sacidumlignan A [7]. We were also involved in this fascinating field and achieved the synthesis of these three molecules through Ueno–Stork radical cyclization and Friedel–Crafts reaction [8,9]. However, almost all of the above syntheses applied stepwise strategies (i.e., a sequence of C2–C7', C8–C8', then C1–C7 bonds formation in our previous routes) for construction of the central core [10].

2. Results and Discussion

Recently, we completed a new synthesis of podophyllotoxin [11,12], an aryltetralin lignan used as building block for the chemotherapeutic drugs etoposide and teniposide. The key reaction is a Ni-catalyzed reductive tandem coupling [13–19] of dibromide *A* that led to the simultaneous construction of C8–C8' and C1–C7 bonds in THN framework of B (Scheme 1b). We envision that this aryltetralin lactone could serve as an advanced intermediate for the unified synthesis of the titled arylnaphthalene, DHN and THN lignans *C*, by means of the regioselective late-stage oxidation. Herein, we disclosed the preliminary results.

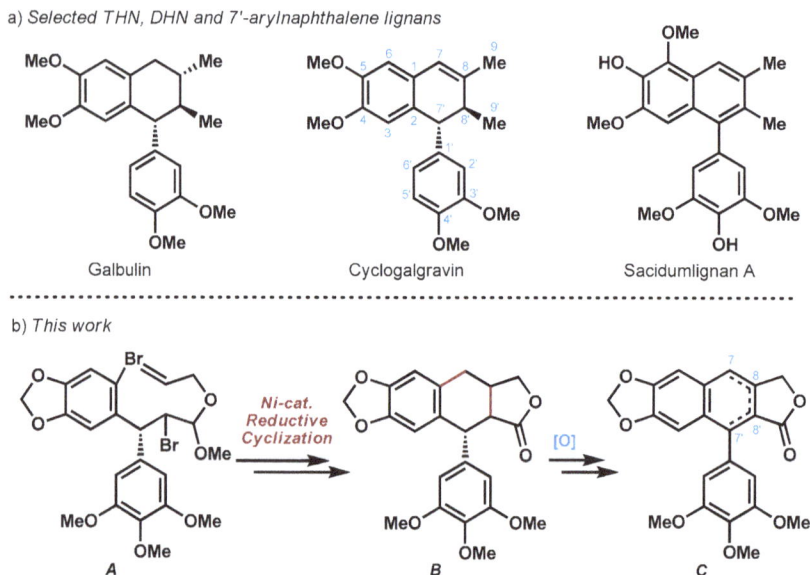

a) *Selected THN, DHN and 7'-arylnaphthalene lignans*

Galbulin

Cyclogalgravin

Sacidumlignan A

b) *This work*

Ni-cat. Reductive Cyclization

[O]

A

B

C

Scheme 1. (a) Several arylnaphthalene lignans and their DHN and THN derivatives; (b) Our synthetic logic.

Starting from the commercially available 6-bromopiperonal and 3,4,5-trimethoxyphenyl bromide, the chiral β-bromo acetal **1** was straightforwardly prepared as in gram-scale according to a known route [11]. Under a fully intramolecular reductive nickel-catalysis ligated by ethyl crotonate (Scheme 2), diastereodivergent (+)-deoxypicropodophyllin (**2**) and (+)-isodeoxypodophyllotoxin (**3**) were obtained in 50% overall yield after a conversion of acetal moiety to the corresponding lactone. With aryltetralin lactones **2** and **3** in hand, the designed regiocontrolled oxidation in central aliphatic ring could be executed (vide infra).

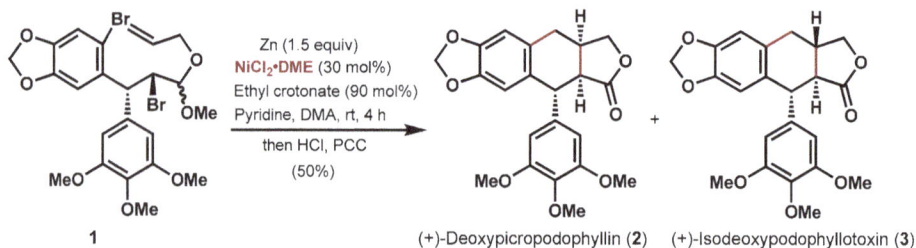

Zn (1.5 equiv)
NiCl$_2$·DME (30 mol%)
Ethyl crotonate (90 mol%)
Pyridine, DMA, rt, 4 h
then HCl, PCC
(50%)

1

(+)-Deoxypicropodophyllin (**2**)   (+)-Isodeoxypodophyllotoxin (**3**)

Scheme 2. Reductive tandem cyclization for tetralin lactones.

First of all, the increase of an unsaturation degree at either C8–C8' or C7'–C8' location was pursued in order to get (+)-β-apopicropodophyllin (**5**) and (+)-γ-apopicropodophyllin (**6**) quickly. As shown in Scheme 3, the introduction of a phenylselenyl group at C8' position of (+)-deoxypicropodophyllin (**2**) was done by an initial enolization and subsequent quench with phenylselenyl bromide (PhSeBr) at −78 °C. The generated products as two diastereoisomers (**4a** and **4b**) were separated by column chromatography on silica gel in 95% overall yield. The α-phenylselenide **4a** is supposed to adopt a pseudo-boat conformation, where the hydrogen atom at C8 is arranged *cis* to the -SePh. The requisite *syn*-elimination of phenylselenoxide in situ generated from oxidation of **4a** [20], eventually provided (+)-β-apopicropodophyllin (**5**) with in vivo insecticidal activity against the fifth-instar larvae of *Brontispa longissima* [21]. Its ¹H NMR spectral data (Table S2) and optical rotation were in agreement with the reported data by Toste and Meyers [22,23]. The structure was later unambiguously confirmed by its single-crystal analysis (Figure 1) [24]. In contrast, the hydrogen atom at C7' is oriented at *cis*-position of C8'-PhSe in the favored half-chair conformer of β-phenylselenide **4b**. Thus, a double bond within C7'–C8' was formed upon the subjection of **4b** to *m*-CPBA, therefore affording to (+)-γ-apopicropodophyllin (**6**) in 88% yield. As shown in Table S3, ¹H NMR spectra of the synthetic **6** was accord with the literature [25].

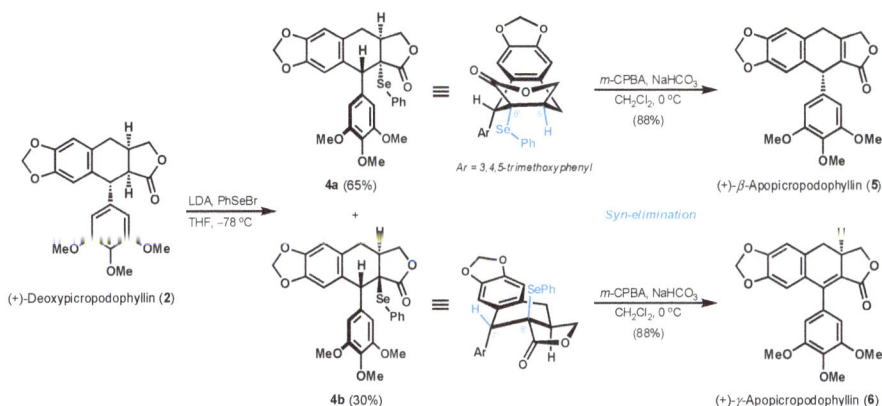

Scheme 3. Regiodivergent oxidation of (+)-deoxypicropodophyllin (**2**).

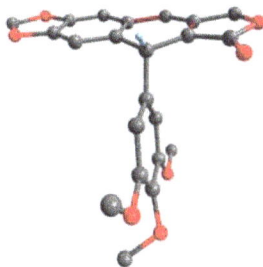

Figure 1. X-ray crystal structure of (+)-β-apopicropodophyllin (**5**), selected H atoms have been omitted for clarity.

Next, the potential aromatization within tetralin lactone was investigated. As shown in Scheme 4, one-step conversion of (+)-isodeoxypodophyllotoxin (**3**) to dehydrodesoxypodophyllotoxin (**7**) was realized in 56% yield promoted by a mixture of *N*-bromosuccinimide (NBS) and dibenzoyl peroxide (BPO) in refluxing CCl₄. The plausible mechanism of this tandem reaction would be radical bromination [26] catalyzed by BPO occurs firstly, and a fast elimination of the resulting labile benzylbromide followed by further oxidation, providing the central benzene ring in **7**. ¹H NMR spectra data (Table S4) of synthetic dehydrodesoxypodophyllotoxin was consistent with previous report [27].

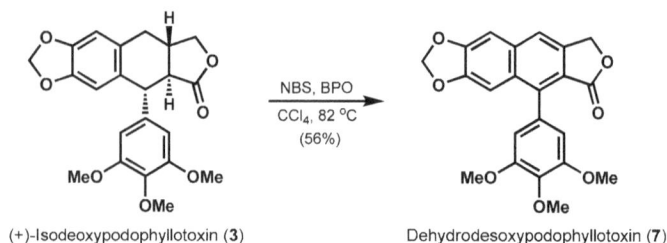

Scheme 4. One-step conversion of tetralin to arylnaphthalene skeleton.

3. Materials and Methods

3.1. General Procedure

For product purification by flash column chromatography, SiliaFlash P60 (particle size: 40–63 μm, pore size 60A) and petroleum ether (bp. 60–90 °C) were used. All solvents were purified and dried by standard techniques and distilled prior to use. All of experiments were conducted under an argon or nitrogen atmosphere in oven-dried or flame-dried glassware with magnetic stirring, unless otherwise specified. Organic extracts were dried over Na_2SO_4 or $MgSO_4$, unless otherwise noted. ^1H and ^{13}C-NMR spectra were taken on a Bruker AM-400, AM-600 and Varian mercury 300 MHz spectrometer with TMS as an internal standard and $CDCl_3$ as solvent unless otherwise noted. HRMS were determined on a Bruker Daltonics APEXII 47e FT-ICR spectrometer with ESI positive ion mode. The X-ray diffraction studies were carried out on a Bruker SMART Apex CCD area detector diffractometer equipped with graphite-monochromated Cu-Kα radiation source. Melting points were measured on Kofler hot stage and are uncorrected.

3.2. Synthesis of C9a-PhSe-Deoxypicropodophyllin (4a and 4b)

A solution of **2** [11] (100 mg, 0.25 mmol) in THF (8 mL) under argon was cooled to −78 °C, followed by the addition of freshly prepared LDA (0.5 mmol, 2.0 equiv). The stirred solution was maintained at this temperature for 20 min, and a solution of PhSeBr (118 mg, 0.5 mmol, 2.0 equiv) in THF (3 mL) was then added. The resulting mixture was stirred for 20 min at −78 °C, and then quenched by water (1 mL). The mixture was extracted with EtOAc (2 × 30 mL). The combined organic layers were washed with water (2 × 8 mL) and brine (8 mL) respectively, dried over Na_2SO_4, filtered and concentrated under reduced pressure. The crude product was purified by flash column chromatography (petroleum ether/EtOAc = 4:1 → petroleum ether/EtOAc =2:1) on silica gel to afford **4a** (90 mg, 65% yield) as a white solid and **4b** (42 mg, 30% yield) as a white solid. Characterization data for **4a**: R_f = 0.42 (petroleum ether/EtOAc = 1:1); ^1H-NMR (400 MHz, $CDCl_3$): δ = 7.48 (d, J = 8.0 Hz, 1H), 7.47 (d, J = 8.0 Hz, 1H), 7.40 (t, J = 7.2 Hz, 1H), 7.28 (t, J = 7.2 Hz, 2H), 6.68 (s, 1H), 6.61 (s, 2H), 6.56 (s, 1H), 5.88 (d, J = 1.2 Hz, 1H), 5.87 (d, J = 1.2 Hz, 1H), 4.49 (s, 1H), 4.10 (dd, J = 9.2, 7.6 Hz, 1H), 3.85 (s, 3H), 3.84 (s, 6H), 3.75 (dd, J = 5.2, 4.0 Hz, 1H), 3.48 (dd, J = 16.4, 8.4 Hz, 1H), 3.32–3.27 (m, 1H), 2.62 (d, J = 16.4 Hz, 1H) ppm; ^{13}C-NMR (100 MHz, $CDCl_3$): δ = 176.7, 152.9 (2C), 147.2, 146.9, 137.7 (2C), 137.3, 134.6, 131.8, 129.9, 129.1 (2C), 126.1, 126.0, 109.3, 108.8, 106.8 (2C), 101.0, 73.3, 60.9, 56.2 (2C), 53.9, 51.3, 41.5, 35.0 ppm; HRMS (ESI): calcd. for $C_{28}H_{30}NO_7Se^+$ [M + NH$_4$]$^+$: 572.1182, found: 572.1186.

3.3. Synthesis of (+)-β-Apopicropodophyllin (5)

To a stirred solution of **4a** (90 mg, 0.076 mmol) in CH_2Cl_2 (4 mL) was added *m*-CPBA (77%, 34.0 mg, 0.15 mmol, 2.0 equiv) at 0 °C followed by the addition of $NaHCO_3$ (12.6 mg, 0.15 mmol, 2.0 equiv). After stirring for 15 min, the reaction mixture was extracted with CH_2Cl_2 (3 × 20 mL). The combined organic layers were washed with saturated aqueous $NaHCO_3$ (4 × 5 mL), water (5 mL) and brine (5 mL) respectively, then dried over Na_2SO_4, filtered and concentrated under reduced pressure. The resulting residue was purified by flash column chromatography (petroleum ether/EtOAc = 3:1 → petroleum ether/EtOAc = 1:1) on silica gel to afford (+)-β-apopicropodophyllin (**5**) (56 mg, 88% yield) as a white solid. R_f = 0.37 (petroleum ether/EtOAc = 1:1); $[\alpha]_D^{20}$ = +92.04 (*c* = 1.00, CHCl$_3$), $[\alpha]_D^{23}$ = +65.1 (*c* = 2.72, CHCl$_3$)] [23]; m.p. 188–190 °C; ^{1}H-NMR (300 MHz, CDCl$_3$): δ = 6.72 (s, 1H), 6.63 (s, 1H), 6.37 (s, 2H), 5.954 (s, 1H), 5.947 (s, 1H), 4.90 (d, *J* = 17.4 Hz, 1H), 4.82 (d, *J* = 17.4 Hz, 1H), 4.81 (s, 1H), 3.86 (dd, *J* = 22.2, 3.9 Hz, 1H), 3.79 (s, 3H), 3.78 (s, 6H), 3.65 (dd, *J* = 22.2, 3.6 Hz, 1H) ppm; ^{13}C-NMR (100 MHz, CDCl$_3$): δ = 172.2, 157.2, 153.2 (2C), 147.3, 147.0, 138.3, 137.1, 129.7, 128.2, 123.8, 109.6, 107.7, 105.6 (2C), 101.3, 71.0, 60.8, 56.2 (2C), 42.8, 29.2 ppm.

This product (5 mg) was dissolved in EtOAc (1 mL) and hexane (2 mL). After three days, colorless single crystals were obtained by slow evaporation of solvents at room temperature.

3.4. Synthesis of (+)-γ-Apopicropodophyllin (**6**)

To a stirred solution of **4b** (42 mg, 0.16 mmol) in CH$_2$Cl$_2$ (3 mL) was added *m*-CPBA (77%, 72.0 mg, 0.32 mmol, 2.0 equiv) at 0 °C followed by the addition of NaHCO$_3$ (26.9 mg, 0.32 mmol, 2.0 equiv). After stirring for 15 min, the reaction mixture was extracted with CH$_2$Cl$_2$ (3 × 20 mL). The combined organic layers were washed with saturated aqueous NaHCO$_3$ (4 × 5 mL), water (5 mL) and brine (5 mL) respectively, then dried over Na$_2$SO$_4$, filtered and concentrated under reduced pressure. The resulting residue was purified by flash column chromatography (petroleum ether/EtOAc = 3:1 → petroleum ether/EtOAc = 1:1) on silica gel to afford (+)-γ-apopicropodophyllin (**6**) (26 mg, 88% yield) as a white solid. R_f = 0.23 (petroleum ether/EtOAc = 1:1); $[\alpha]_D^{20}$ = +27.03 (*c* = 1.00, CHCl$_3$), $[\alpha]_D^{19}$ = +25.0 (*c* = 1, CHCl$_3$)] [20], m.p. 200-200 °C, ^{1}H-NMR (500 MHz, CDCl$_3$): δ = 6.77 (s, 1H), 6.52 (brs, 3H), 5.97 (s, 2H), 4.70 (t, *J* = 8.7 Hz, 1H), 4.01 (t, *J* = 8.7 Hz, 1H), 3.92 (s, 3H), 3.83 (s, 6H), 3.39 (td, *J* = 15.9, 8.7 Hz, 1H), 2.94 (dd, *J* = 15.0, 6.9 Hz, 1H), 2.79 (dd, *J* = 15.6, 15.3 Hz, 1H) ppm; ^{13}C-NMR (150 MHz, CDCl$_3$): δ = 168.1, 152.7, 148.7 (2C), 147.3, 146.8, 138.1, 130.7 (2C), 129.9, 129.6, 119.9, 109.5, 108.6, 101.6 (2C), 70.9, 61.0, 56.2 (2C), 35.8, 33.3 ppm.

3.5. Synthesis of Dehydrodesoxypodophyllotoxin (**7**)

An oven-dried 10 mL round-bottom flask was charged with NBS (17.8 mg, 0.1 mmol, 1.0 equiv) and BPO (2.4 mg, 0.01 mmol, 0.1 equiv) at room temperature under argon, followed by the addition of a solution of **3** (40.0 mg, 0.1 mmol) in CCl$_4$ (3 mL). The reaction mixture was stirred for 2 h at 82 °C. The reaction solvent was then evaporated in vacuo. The resulting residue was purified by flash column chromatography (petroleum ether/EtOAc = 5:1 → petroleum ether/EtOAc = 2:1) on silica gel to afford dehydrodesoxypodophyllotoxin (**7**) (22.2 mg, 56% yield) as a white solid. R_f = 0.45 (petroleum ether/EtOAc = 1:1); m.p. 271–273 °C; ^{1}H-NMR (400 MHz, CDCl$_3$): δ = 7.70 (s, 1H), 7.21 (s, 1H), 7.12 (s, 1H), 6.55 (s, 2H), 6.09 (s, 2H), 5.38 (s, 2H), 3.97 (s, 3H), 3.84 (s, 6H) ppm; ^{13}C-NMR (150 MHz, CDCl$_3$): δ = 169.6, 153.0 (2C), 150.0, 148.7, 140.5, 139.8, 137.8, 134.6, 130.34, 130.30, 119.1, 118.7, 107.3 (2C), 103.8, 103.6, 101.8, 68.0, 61.0, 56.1 (2C) ppm.

4. Conclusions

In summary, a two-phase strategy was developed for the unified synthesis of (+)-β-apopicropodophyllin (**5**), (+)-γ-apopicropodophyllin (**6**), and dehydrodesoxypodophyllotoxin (**7**). In phase I, their tetrahydronaphthalene (THN) backbone was constructed by a Ni-catalyzed reductive cascade. In phase II, regioselective oxidation of stereodivergent tetralin lactone (**2** and **3**) gave arylnaphthalene lignan **7** and its dihydronaphthalene (DHN) congeners (**5** and **6**) efficiently.

Supplementary Materials: The following are available online. Copies of ^{1}H-, ^{13}C-NMR, and crystallographic information files (CIFs) for **5**.

Author Contributions: Y.P. conceived and designed the experiments; J.X. performed the experiments; J.X., G.N., Y.-W.W., and Y.P. analyzed the data; Y.-W.W. and Y.P. wrote the paper.

Funding: This work was supported by the Natural Science Foundation of China (nos. 21472075 and 21772078).

Conflicts of Interest: The authors declare no conflict of interest.

References

1. Ayres, D.C.; Loike, J.D. *Lignans, Chemical, Biological and Clinical Properties*; Cambridge University Press: Cambridge, UK, 1990.

2. Shi, J. *Lignans Chemistry*, 1st ed.; Chemical Industrial Press: Beijing, China, 2010; pp. 1–395, ISBN 978-7-122-06559-9.

3. Peng, Y. Lignans, lignins, and resveratrols. In *From Biosynthesis to Total Synthesis: Strategies and Tactics for Natural Products*; Zografos, A.L., Ed.; John Wiley & Sons, Inc.: Hoboken, NJ, USA, 2016; pp. 331–379.

4. Sellars, J.D.; Steel, P.G. Advances in the synthesis of aryltetralin lignan lactones. *Eur. J. Org. Chem.* **2007**, *2007*, 3815–3828. [CrossRef]

5. Hong, B.C.; Hsu, C.S.; Lee, G.H. Enantioselective total synthesis of (+)-galbulin via organocatalytic domino Michael-Michael-aldol condensation. *Chem. Commun.* **2012**, *48*, 2385–2387. [CrossRef] [PubMed]

6. Rye, C.E.; Barker, D. Asymmetric synthesis of (+)-galbelgin, (−)-kadangustin J, (−)-cyclogalgravin and (−)-pycnanthulignenes A and B, three structurally distinct lignan classes, using a common chiral precursor. *J. Org. Chem.* **2011**, *76*, 6636–6648. [CrossRef] [PubMed]

7. Route, J.K.; Ramana, C.V. Total synthesis of (−)-sacidumlignans B and D. *J. Org. Chem.* **2012**, *77*, 1566–1571. [CrossRef] [PubMed]

8. Zhang, J.J.; Yan, C.S.; Peng, Y.; Luo, Z.B.; Xu, X.B.; Wang, Y.W. Total synthesis of (±)-sacidumlignans D and A through Ueno-Stork radical cyclization reaction. *Org. Biomol. Chem.* **2013**, *11*, 2498–2513. [CrossRef] [PubMed]

9. Peng, Y.; Luo, Z.B.; Zhang, J.J.; Luo, L.; Wang, Y.W. Collective synthesis of several 2,7′-cyclolignans and their correlation by chemical transformations. *Org. Biomol. Chem.* **2013**, *11*, 7574–7586. [CrossRef] [PubMed]

10. Kocsis, L.S.; Brummond, K.M. Intramolecular Dehydro-Diels–Alder Reaction Affords Selective Entry to Arylnaphthalene or Aryldihydronaphthalene Lignans. *Org. Lett.* **2014**, *16*, 4158–4161. [CrossRef] [PubMed]

11. Xiao, J.; Cong, X.W.; Yang, G.Z.; Wang, Y.W.; Peng, Y. Divergent asymmetric syntheses of podophyllotoxin and related family members via stereoselective reductive Ni-catalysis. *Org. Lett.* **2018**, *20*, 1651–1654. [CrossRef] [PubMed]

12. Xiao, J.; Cong, X.W.; Yang, G.Z.; Wang, Y.W.; Peng, Y. Stereoselective synthesis of *podophyllum* lignans core by intramolecular reductive nickel-catalysis. *Chem. Commun.* **2018**, *54*, 2040–2043. [CrossRef] [PubMed]

13. Yan, C.S.; Peng, Y.; Xu, X.B.; Wang, Y.W. Nickel-mediated inter- and intramolecular reductive cross-coupling of unactivated alkyl bromides and aryl iodides at room temperature. *Chem. Eur. J.* **2012**, *18*, 6039–6048. [CrossRef] [PubMed]

14. Xu, X.B.; Liu, J.; Zhang, J.J.; Wang, Y.W.; Peng, Y. Nickel-mediated inter- and intramolecular C-S coupling of thiols and thioacetates with aryl iodides at room temperature. *Org. Lett.* **2013**, *15*, 550–553. [CrossRef] [PubMed]

15. Peng, Y.; Luo, L.; Yan, C.S.; Zhang, J.J.; Wang, Y.W. Ni-catalyzed reductive homocoupling of unactivated alkyl bromides at room temperature and its synthetic application. *J. Org. Chem.* **2013**, *78*, 10960–10967. [CrossRef] [PubMed]

16. Peng, Y.; Xu, X.B.; Xiao, J.; Wang, Y.W. Nickel-mediated stereocontrolled synthesis of spiroketals via tandem cyclization-coupling of β-bromo ketals and aryl iodides. *Chem. Commun.* **2014**, *50*, 472–474. [CrossRef] [PubMed]

17. Luo, L.; Zhang, J.J.; Ling, W.J.; Shao, Y.L.; Wang, Y.W.; Peng, Y. Unified synthesis of (−)-folicanthine and (−)-ditryptophenaline enabled by a nickel-mediated reductive dimerization at room temperature. *Synthesis* **2014**, *46*, 1908–1916. [CrossRef]

18. Peng, Y.; Xiao, J.; Xu, X.B.; Duan, S.M.; Ren, L.; Shao, Y.L.; Wang, Y.W. Stereospecific synthesis of tetrahydronaphtho[2,3-b]furans enabled by a nickel-promoted tandem reductive cyclization. *Org. Lett.* **2016**, *18*, 5170–5173. [CrossRef] [PubMed]

19. Xiao, J.; Wang, Y.W.; Peng, Y. Nickel-promoted reductive cyclization cascade: A short synthesis of a new aromatic strigolactone analogue. *Synthesis* **2017**, *49*, 3576–3581.

20. Uchiyama, M.; Kimura, Y.; Ohta, A. Stereoselective total syntheses of (±)-arthrinone and related natural compounds. *Tetrahedron Lett.* **2000**, *41*, 10013–10017. [CrossRef]

21. Zhang, J.; Liu, Y.Q.; Yang, L.; Feng, G. Podophyllotoxin derivatives show activity against *Brontispa longissima* larvae. *Nat. Prod. Commun.* **2010**, *5*, 1247–1250. [PubMed]

22. Kennedy-Smith, J.J.; Young, L.A.; Toste, F.D. Rhenium-catalyzed aromatic propargylation. *Org. Lett.* **2004**, *6*, 1325–1327. [CrossRef] [PubMed]

23. Andrews, R.C.; Teague, S.J.; Meyers, A.I. Asymmetric total synthesis of (−)-podophyllotoxin. *J. Am. Chem. Soc.* **1988**, *110*, 7854–7858. [CrossRef]

24. CCDC-1875746 (**5**) Contain the Supplementary Crystallographic Data for This Paper. These Data Can Be Obtained Free of Charge. Available online: http://www.ccdc.cam.ac.uk/conts/retrieving.html (accessed on 28 October 2018).

25. Kashima, T.; Tanoguchi, M.; Arimoto, M.; Yamaguchi, H. Studies on the constituents of the seeds of *Hernandia ovigera* L. VIII. Synthesis of (±)-desoxypodophyllotoxin and (±)-β-peltatin-A methyl ether. *Chem. Pharm. Bull.* **1991**, *39*, 192–194. [CrossRef]

26. Yamaguchi, H.; Arimoto, M.; Nakajima, S.; Tanoguchi, M.; Fukada, Y. Studies on the constituents of the seeds of *Hernandia ovigera* L.V. Syntheses of epipodophyllotoxin and podophyllotoxin from desoxypodophyllotoxin. *Chem. Pharm. Bull.* **1986**, *34*, 2056–2060. [CrossRef]

27. Nishii, Y.; Yoshida, T.; Asano, H.; Wakasugi, K.; Morita, J.-I.; Aso, Y.; Yoshida, E.; Motoyoshiya, J.; Aoyama, H.; Tanabe, Y. Regiocontrolled benzannulation of diaryl(*gem*-dichlorocyclopropyl)methanols for the synthesis of unsymmetrically substituted α-arylnaphthalenes: Application to total synthesis of natural lignan lactones. *J. Org. Chem.* **2005**, *70*, 2667–2678. [CrossRef] [PubMed]

28. Schrecker, A.W.; Hartwell, J.L. Components of podophyllin. IX. The structure of apopicropodophyllins. *J. Am. Chem. Soc.* **1952**, *74*, 5676–5683. [CrossRef]

Sample Availability: Samples of the compounds are not available from the authors.

molecules

MDPI

Review

Oxidative Transformations of Lignans

Patrik A. Runeberg[ID]**, Yury Brusentsev**[ID]**, Sabine M. K. Rendon**[ID] **and Patrik C. Eklund ***

Johan Gadolin Process Chemistry Center, Åbo Akademi University, Piispankatu 8, 20500 Turku, Finland;
patrik.runeberg@abo.fi (P.A.R.); ybrusent@abo.fi (Y.B.); srendon@abo.fi (S.M.K.R.)
* Correspondence: paeklund@abo.fi; Tel.: +358-2-215-4720

Academic Editor: David Barker
Received: 16 November 2018; Accepted: 29 December 2018; Published: 15 January 2019

Abstract: Numerous oxidative transformations of lignan structures have been reported in the literature. In this paper we present an overview on the current findings in the field. The focus is put on transformations targeting a specific structure, a specific reaction, or an interconversion of the lignan skeleton. Oxidative transformations related to biosynthesis, antioxidant measurements, and total syntheses are mostly excluded. Non-metal mediated as well as metal mediated oxidations are reported, and mechanisms based on hydrogen abstractions, epoxidations, hydroxylations, and radical reactions are discussed for the transformation and interconversion of lignan structures. Enzymatic oxidations, photooxidation, and electrochemical oxidations are also briefly reported.

Keywords: lignans; oxidation

---

1. Introduction

Lignans constitute a group of natural phenolics found in plant species. They have been identified in around 70 families in the plant kingdom, for example in trees, grasses, grains, and vegetables. Lignans are found in roots, rhizomes, stems, leaves, seeds, and fruits, from where they are usually isolated through extraction with an appropriate solvent. They have a diverse structure built up from two phenyl propane units with different degrees of oxidation in the propane moiety and different substitution patterns in the aromatic rings. Lignans have been shown to have a wide range of biological activities such as antibacterial and insecticidal effects in plants, and anti-cancerous, antiviral, anti-inflammatory, immunosuppressive, anti-diabetic, and antioxidant properties in mammals [1–6]. In addition, many lignans have been found in different foods and feeds, and have been associated with health benefits, such as antioxidant activity and anticancer properties [7–10]. The structure and occurrence of lignans have previously been extensively reviewed in the literature. In some of these reviews the chemical transformations and oxidative degradations have partially been described, mainly for determination of structures [6,11–16]. More recently, advances in the chemistry of lignans, including transformations and interconversions of different lignans, were reviewed by Ward [17]. Although synthetic modifications and interconversions have been extensively reported for lignans, a specific overview on oxidative transformations has not been published. Oxidative transformations of lignan structures can be found in various fields: in total synthesis, in biosynthesis, in antioxidant studies, in oxidative degradations, and as part of targeted chemical transformations. In this paper, we report an overview on oxidative transformations of lignan structures, where the focus is put on transformations targeting a specific structure, a specific reaction, or an interconversion of the lignan skeleton. Oxidative transformations related to studies of biosynthetic routes, antioxidant activities, and total syntheses for coupling phenylpropane units (creating the lignan skeleton) are mostly excluded. The purpose of this paper is to exemplify the transformations induced by different reagents or oxidation methods (Table 1). The presented text, figures, and schemes should be taken as representative examples rather than a comprehensive review of the literature. However, the most relevant literature in the field is cited.

Table 1. Various methods for oxidative transformations of lignans.

Non-Metal Ox.	Oxidative Transformation	Yields	Ref.
DDQ	Benzylic hydride abstraction and benzylic ring closures, Ar-Ar-coupling to dibenzocyclooctadiene, nucleophilic attack, alcohol oxidation, aromatization	~10–40%	[18–37]
*m*CPBA	Epoxidation, Baeyer-Villiger oxidation, oxidation of ethoxy-THF-lignans to lactones	~70–90%	[29,38–42]
peroxyl radical (ROO·)	Antioxidative radical scavenging to phenoxyl radicals and further radical couplings	No data	[43–46]
Azo compounds (AAPH, AIBN, ABTS)	Same as above	No data	[46–63]
DPPH	Same as above, radical 5-5 couplings to dimers and oxidation of benzylic alcohol to ketone	No data	[6,51,64–67]
TEMPO	Oxidation of benzylic alcohol to ketone	60%	[68]
BAIB and PIFA	phenolic hydroxyl oxidation to quinone type structures, Ar-Ar-coupling to dibenzocyclooctadienes, nucleophilic attack at *ipso*- or benzylic position	~10–80%	[69–74]
IBX	Aromatic demethylation, benzylic alcohol oxidation to carbonyl	~25–95%	[75–79]
Dess-Martin Periodinane (DMP)	Benzylic alcohol oxidation to carbonyl, 9,9′-diol oxidation to lactols	>90%	[80–89]
NaIO₄	Oxidation of guaiacyl or syringyl groups to demethylated *o*-quinones, Lemieux–Johnson oxidation	~50–95%	[90–100]
N-Bromosuccinimide (NBS)	Brominations (arylic or benzylic), benzylic CH₂ to ketone, benzylic ring closure and aromatization	75–90%	[35,101–107]
Dimethyldioxirane (DMDO)	Oxidative ring opening of furan rings	~80%	[108]
Nitrobenzene	Oxidative degradation to vanillin and vanillic acid	80–100%	[109–111]

Metal-Mediated Ox.	Oxidative Transformation	Yields	Ref.
Cr-(VI) (CrO₃, PCC, PDC)	Oxidation of benzylic alcohols to ketones, primary alcohols into lactones and carboxylic acid	~60–95%	[87,103,105, 112–120]
Pd, Au	Oxidation of benzylic alcohols in presence of free phenolic into (mainly) ketone	No data	[121–126]
MoOPH	α-Hydroxylation	~25–95%	[113,127–131]
MoCl₅	Ar-Ar-coupling to dibenzocyclooctadienes	~50–90%	[132,133]
VoF₃; V₂O₅; Tl₂O₃; RuO₅	Ar-Ar-coupling to dibenzocyclooctadienes, benzylic ring closure	~50–100%	[134–147]
MTO (catalyst)	Aromatic demethylation and quinone formation and simultaneous benzylic alcohol oxidation to ketones, benzylic cleavage, benzylic hydroxylation, oxidation of benzylic alcohol to ketone	~40–100%	[148–150]
Pb(OAc)₄	Benzylic acetoxylation	~30–70%	[151]
CeCl₃	α-Hydroxylation	71%	[152]

Other Oxidations		Oxidative Transformation	Yields	Ref.
	peroxidase	Benzylic ring closure	37–99%	[153–155]
	HRP	Radical 4-*O*-5 coupling	3.6%	[156]
Enzymatic ox.	SDH	Oxidation of diol to lactone	No data	[157]
	P450/CPR1	Arylic hydroxylation and rearrangement	No data	[158]
	CYP	Benzylic hydroxylation	90%	[159]
Electrochemical ox.		Demethylation, quinone formation, and benzylic ring closure	No data	[160]
		Ar-Ar-coupling to dibenzocyclooctadiene	>80%	[161]
Photooxidation		Benzylic: cleavage/alcohol oxidation to ketone/ring closure/nucleophilic attack	No data	[162]

2. Non-Metal-Mediated Oxidations

2.1. 2,3-Dichloro-5,6-dicyano-1,4-benzoquinone (or DDQ)

2,3-Dichloro-5,6-dicyano-1,4-benzoquinone (DDQ) has been used to promote benzylic functionalization of lignans by abstraction of a hydride from the benzylic position. The benzylic cation can then be a target for either intramolecular ring closure or nucleophilic attack. The nucleophilicity of the solvent can direct the reaction towards either benzylic ring closure to aryltetralins, or to aryl-aryl coupling to cyclooctadienes, or toward nucleophilic attack by the solvent. In the latter case, a nucleophilic solvent such as AcOH can function as an oxygen donor. As an example, (−)-dehydroxycubebin was oxidized by DDQ in AcOH, and reacted by either benzylic ring closure to an aryltetralin (**1**), or by *O*-acetylation through nucleophilic attack at the benzylic position (**2**) [18] (Scheme 1). When the solvent was changed to the more acidic and less nucleophilic trifluoroacetic acid (TFA), aryl-aryl coupling to a dibenzocyclooctadiene (**3**) was favored [19–27].

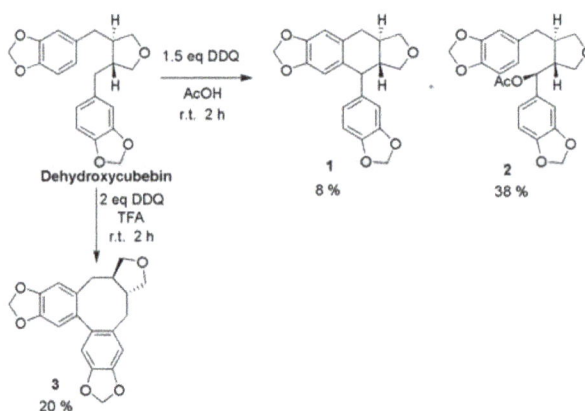

Scheme 1. Oxidation of dehydroxycubebin by DDQ. AcOH promotes benzylic functionalization while TFA promotes aryl-aryl coupling.

Another example of benzylic *O*-acetylation is shown in Scheme 2. Tomioka et al. reported that the dibenzocyclooctadiene lignan known as (+)-isostegane could be *O*-acetylated at the benzylic position by DDQ in acetic acid, forming (+)-steganacin (Scheme 2) [28].

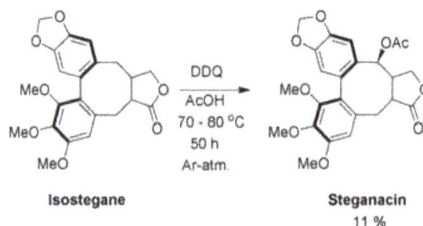

Scheme 2. Benzylic *O*-acetylation of (+)-isostegane by 2,3-Dichloro-5,6-dicyano-1,4-benzoquinone (DDQ) in AcOH.

A DDQ/TFA-mediated reaction of an epoxy-lignan (**5**), formed through mCPBA epoxidation of an olefinic lignan (**4**), did not form the cyclooctadiene as expected. Instead, TFA-induced epoxide opening and methyl or benzyl group rearrangement to two intermediates preceded DDQ mediated ring closure to **6** and **7** (Scheme 3) [29]. One intermediate underwent the previously described ring closure to a cyclooctadiene, while the other went through benzylic ring closure to a cyclohexanone, with subsequent ring closure to a benzofuran.

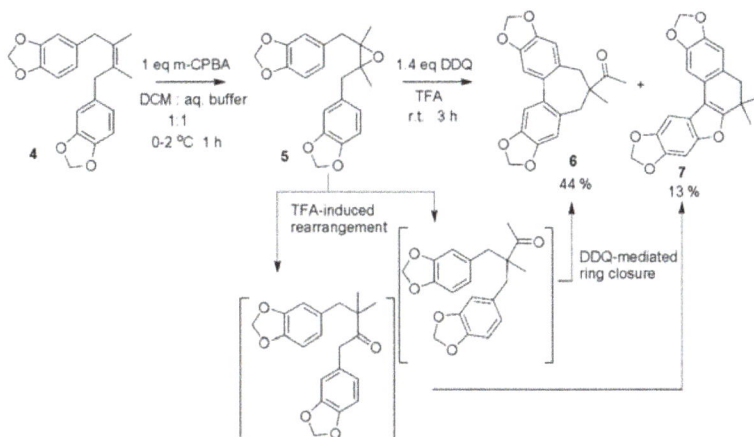

Scheme 3. TFA- and DDQ-mediated rearrangement and oxidative ring closure of an epoxide lignan

DDQ has also been used for benzylic dehydrogenation of lignan structures to olefins, and in some cases further aromatization to naphthalene type lignans [23,30–32]. One example of benzylic olefin formation is visualized in Scheme 4 by the DDQ oxidation of the natural lignan hydroxymatairesinol (HMR) [33,34]. Due to different stereoelectronic properties at the benzylic alcohol for the hydroxymatairesinol epimers, the oxidation reaction resulted in different major products, of which oxomatairesinol had the highest yield. A range of minor products, which are not shown here, were also formed in the reactions.

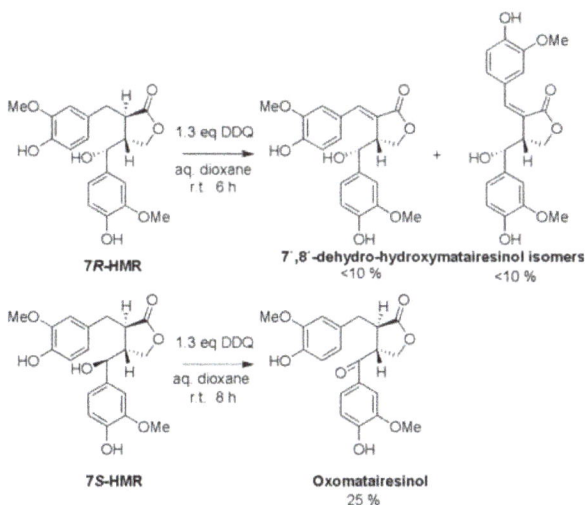

Scheme 4. Oxidation of hydroxymatairesinol (HMR) by DDQ is dependent on the stereochemistry of the benzylic alcohol position. The major products are shown here. A range of minor products were also formed in the reactions.

An example of DDQ-mediated aromatization of lignans is shown in Scheme 5, for the aromatization of isodeoxypodophyllotoxin [35].

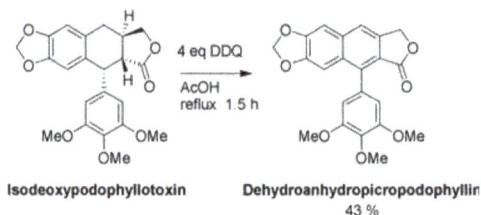

Scheme 5. DDQ-mediated aromatization of isodeoxypodophyllotoxin.

When gmelinol reacted with three equivalents of DDQ it underwent benzylic hydride abstraction and C-C-bond rearrangement, followed by an oxidative ring opening of an ether bridge to form an aldehyde, and finally, another benzylic hydride abstraction and formation of olefin (**8**) [36]. The reaction is visualized in Scheme 6.

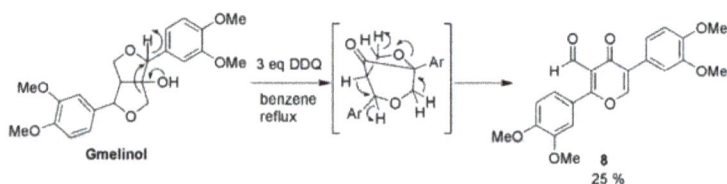

Scheme 6. Oxidation of gmelinol with three equivalents DDQ.

DDQ-mediated epimerization followed by nucleophilic ring closure at the benzylic position is visualized in Scheme 7 on a lignan (**9**) synthesized from gmelinol [37]. The epimerization step may go through benzylic hydride abstraction, leading to adjacent C-C bond cleavage and regeneration. One epimer (**10**) is favored as it can undergo nucleophilic attack at the positively charged benzylic position, forming product **11**.

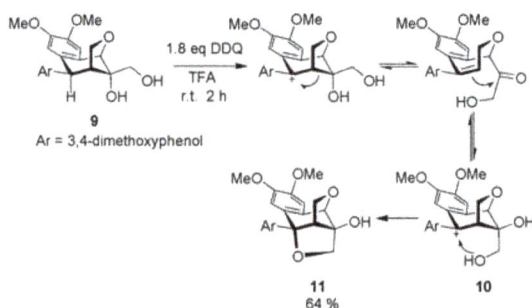

Scheme 7. DDQ-mediated epimerization and nucleophilic ring closure.

2.2. Meta-Chloroperoxybenzoic Acid (mCPBA)

Epoxidation of olefinic lignans with mCPBA has been reported, as previously shown in Scheme 3 [29,38].

The Bayer-Villiger oxidation of ketones with mCPBA has also been reported for lignans. The 3,7-dioxobicyclo[3,3,0]octane lignan **12** gave the corresponding dilactone product **13** (Scheme 8) [39,40]. In the total synthesis of taiwanin E, the last two steps consist of Baeyer-Villiger of an aldehyde (**14**) to the corresponding formate (**15**), followed by hydrolysis by MeOH and K_2CO_3 to the alcohol (Scheme 9) [41].

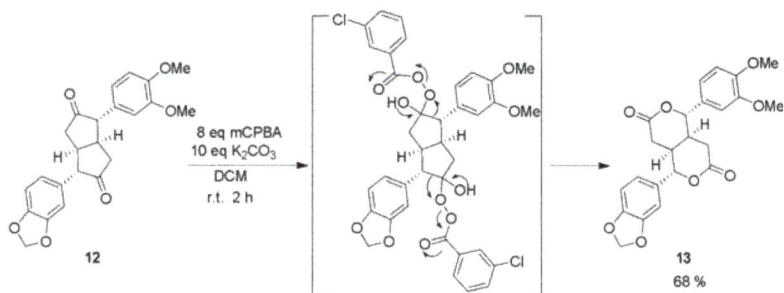

Scheme 8. Baeyer-Villiger oxidation of a diketone to a dilactone.

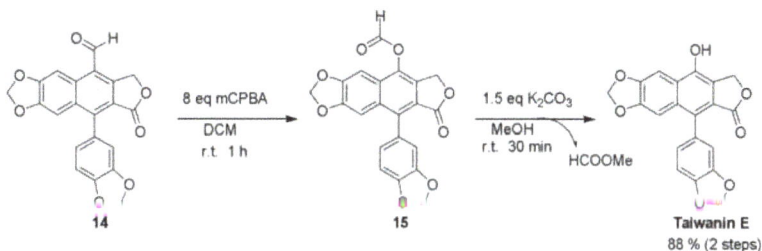

Scheme 9. Baeyer-Villiger oxidation of an aldehyde to the corresponding formate, which was further hydrolyzed to the alcohol (taiwanin E).

In addition, mCPBA and Lewis acid mediated oxidation of an ethoxy-THF-lignan (**16**) to a lactone (**17**) has been reported (Scheme 10) [42].

Scheme 10. Lewis acid and mCPBA mediated oxidation of an ethoxytetrahydrofuran to the corresponding lactone.

2.3. Oxidations by Peroxyl Radical

Radical scavenging of peroxyl radicals has been utilized for measuring antioxidant activity for natural phenolics, including flavonoids [43,44] and lignans [45,46]. The antioxidant activity of lignans is caused by their efficiency to scavenge radicals through hydrogen abstraction, forming phenoxyl radicals. However, in the process, the lignan structure is oxidized and undergoes various transformations. The highly antioxidant active lignan known as secoisolariciresinol was oxidized by two ethyl linoleate peroxide radicals induced by azobisisobutyronitrile (AIBN) to a quinone-like intermediate (**18**), which formed the lignan lariciresinol through trans-selective ring closure. As a side reaction, a peroxyl radical also undertook radical coupling at the *ipso*-position of the phenoxyl radical intermediate to **19** (Scheme 11) [45,46].

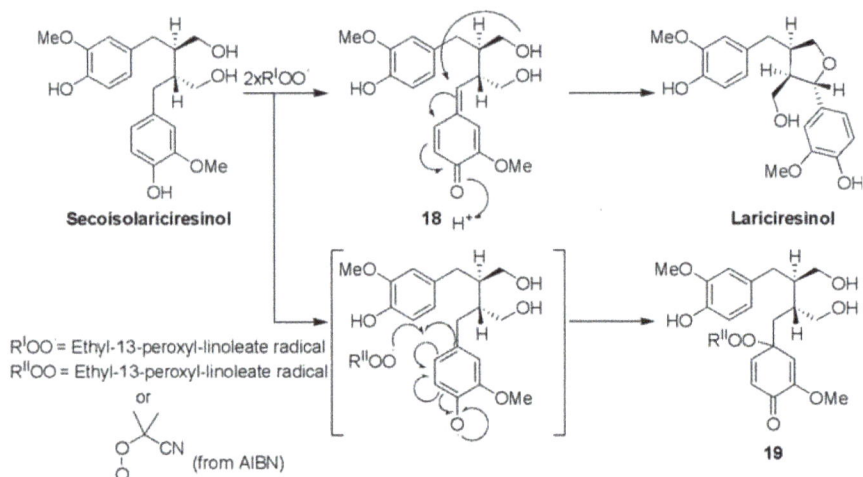

Scheme 11. Peroxyl radical mediated oxidation of the lignan secoisolariciresinol.

2.4. Azo Compounds (AAPH, AIBN and ABTS)

2,2′-Azobis(2-amidinopropan) dihydrochloride (AAPH or ABAP), azobisisobutyronitrile (AIBN), and 2,2′-azino-bis(3-ethylbenzothiazoline-6-sulfonic acid) (ABTS) are radical initiators often used in studies of oxidative stability of drugs and proteins, and of antioxidative activity in, among others, food supplements and natural products [47–49]. These azo compounds undergo thermal degradation under release of N_2, and formation of two amidino propane (from AAPH) or isobutyronitrile (from AIBN) radicals. These radicals can be transformed into other reactive radical species depending on the reaction environment, for example by oxygen addition into peroxyl radicals [50].

Studies on antioxidant activity on a range of lignans have been done using AAPH, but only in some cases the major oxidation products have been characterized [51–56]. Upon radical scavenging, the lignans form phenoxyl radicals with their radical delocalized over the phenolic ring, followed by radical coupling at either the *ipso-* or *meta*-position, or through the para phenoxyl radical. Alternatively, a second radical reaction yields a quinone methide intermediate that reacts further through an ionic mechanism, as shown in Scheme 11 for **18**. As an example, secoisolariciresinol primarily reacted through an aryl-aryl coupling, the so called 5-5 coupling, to dimer **20**, and through phenoxyl or *meta*-coupling with an amidino propane radical from AAPH to **21** and **22** respectively [57]. Further hydrolysis of the meta-coupled amidino propane unit formed a furanone product (**23**). Lariciresinol was also formed through a similar mechanism as described above (in Scheme 11) for peroxyl radicals. Uniquely for secoisolariciresinol, an alternative radical scavenging path is possible. In this path, an intermediate with alkoxyl radicals on one or both aliphatic hydroxyls, reacts through radical coupling with another hydroxyl radical, forming dimer **24**, or with water, forming hydroperoxides (**25**) (Scheme 12).

As described in the previous chapter, AIBN has been used as a radical initiator for peroxyl radicals in antioxidant studies on lignans [46]. As shown in Scheme 11, a possible antioxidant product is formed through *ipso*-coupling of the lignan with an AIBN derived peroxide. AIBN has also been used in total synthesis of a range of lignans, in aryl halide and $Ru_3SnH/AIBN$ initiated radical aryl-aryl couplings [58], and in $(Me_3Si)_3SiH/AIBN$ promoted radical conversion of a thionocarbonate to a lignan lactone [59]. The radical initiator ABTS has also been used for antioxidant studies on lignans [60–63], in the so called trolox equivalent antioxidant capacity (TEAC) assays [163]. However, no systematic study on the formed oxidation products by AIBN or ABTS has been done, excluding the product **19**.

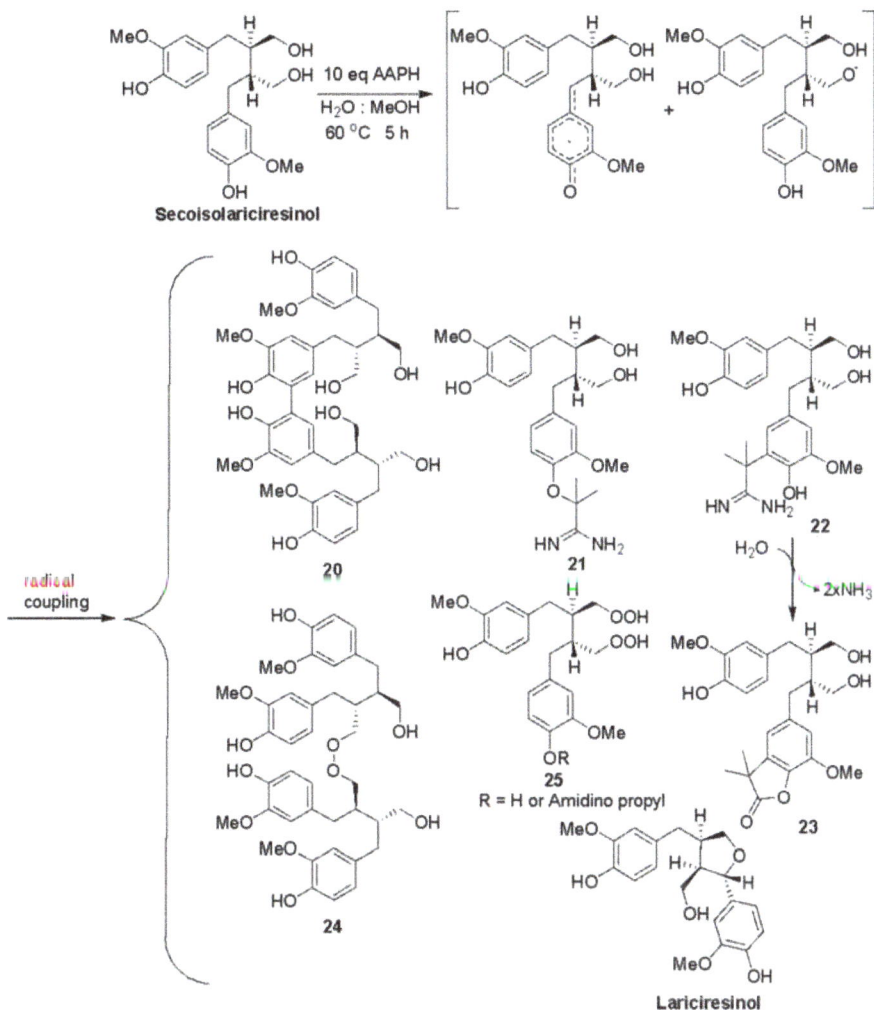

Scheme 12. Oxidation products formed from secoisolariciresinol after 2,2'-Azobis(2-amidinopropan) dihydrochloride (AAPH)-mediated radical scavenging.

2.5. 2,2-Diphenyl-1-picrylhydrazyl (DPPH)

DPPH is a stable nitrogen centered free radical that is commonly used for antioxidant assays [64,65]. It has been vastly used for antioxidant assays on lignans [6,66]. Eklund et al. [51] and Smeds et al. [67] investigated the antioxidant mechanism for the DPPH-initiated radical scavenging of different lignans. Upon radical scavenging, the lignan intermediates underwent further radical abstractions or aryl couplings to form dimers and oligomers by 5-5'- or 5-O-4'-couplings. An example is shown in Scheme 13 for the reaction between HMR and DPPH. The two major reaction paths, following the initial radical abstraction, were the intramolecular radical coupling forming an aryl-aryl dimer (**26**), or a second radical abstraction and rearrangement to the lignan oxomatairesinol. Further coupling of the dimeric structure also formed larger oligomers.

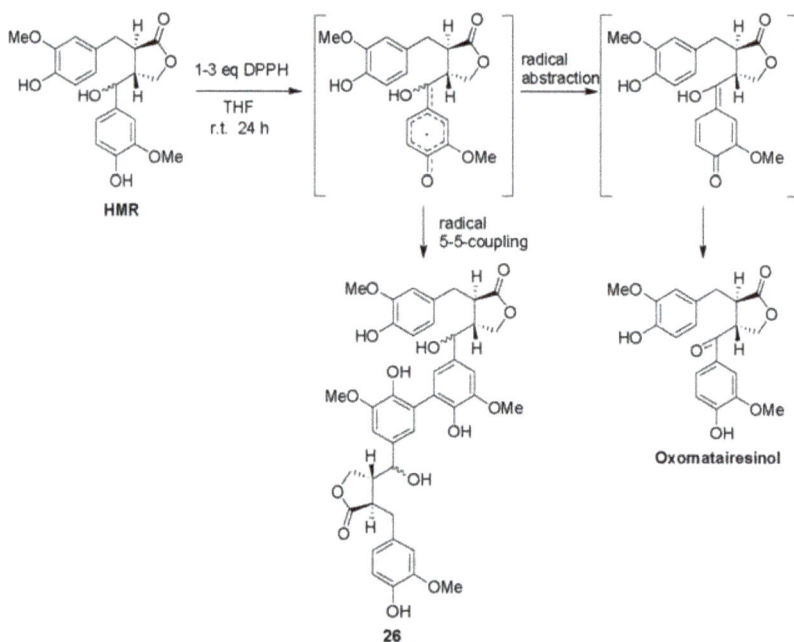

Scheme 13. Reaction products of HMR with the free radical DPPH.

2.6. 2,2,6,6-Tetramethyl-1-piperidinyloxy (TEMPO)

The free radical organocatalyst TEMPO has been used in combination with a number of different oxidants. For decades it has been used for oxidation of primary and secondary alcohols [164–166]. Surprisingly, TEMPO has not been widely used for the oxidation of lignans. However, a TEMPO-catalyzed oxidation of the benzylic hydroxyl on podophyllotoxin to the corresponding ketone (podophyllotoxone) has been reported (Scheme 14) [68]. Sodium periodate (NaIO$_4$) was used as the oxidant in the presence of NaBr as the co-catalyst.

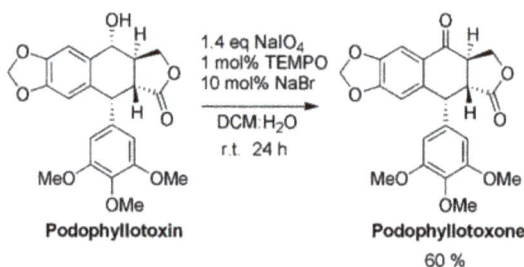

Scheme 14. 2,2,6,6-Tetramethyl-1-piperidinyloxy (TEMPO)-mediated oxidation of the benzylic alcohol of podophyllotoxin.

2.7. Hypervalent Iodine Reagents

The hypervalent iodine reagents BAIB, PIFA, IBX, DMP, and NaIO$_4$ have been used in numerous studies for mild oxidation of lignan structures. Efficient oxidation of primary and benzylic alcohols into carbonyls, and phenolics into quinones have been reported. In addition, PIFA has been used for arylic coupling to form benzocyclooctadienes.

2.7.1. [Bis(acetoxy)iodo]benzene (BAIB or PIDA) and [Bis(trifluoroacetoxy)iodo]benzene (PIFA)

Diphyllin was oxidized by BAIB or PIFA through phenolic hydroxyl oxidation and nucleophilic attack to a para-quinone type product (**27**) (Scheme 15) [69].

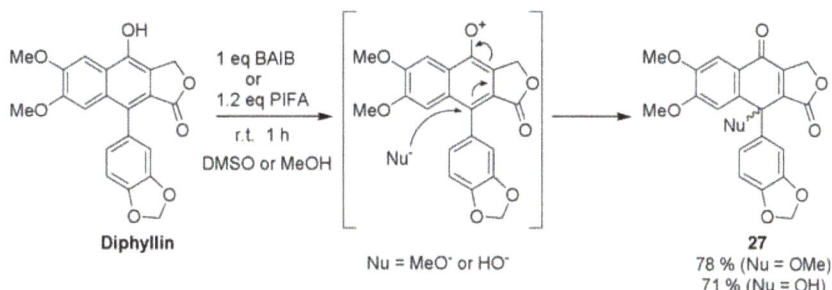

Scheme 15. Bis(trifluoroacetoxy)iodo benzene (PIFA) or Bis(acetoxy)iodo benzene (BAIB)-mediated oxidation of diphyllin.

Lignans have also been ring closed to cyclooctadienes by PIFA in TFE (Scheme 16, upper reaction path). The initial step is believed to be cyclization to a spirodienone intermediate (**28**) that undergoes rearrangement to the cyclooctadiene product (**29**). In the cases where a lignan has a para-hydroxyl group, in addition to the cyclooctadiene product, nucleophilic addition at ipso-position to **30** was reported when using a more nucleophilic solvent such as methanol (Scheme 16, lower reaction path) [70–73]. Other minor products were also formed.

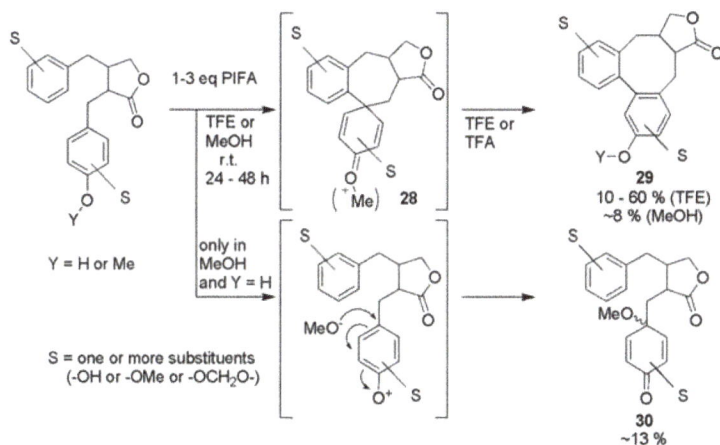

Scheme 16. Lignan oxidation by PIFA in TFE or MeOH. With TFE as the solvent, the major reaction was formation of the cyclooctadiene. In methanol, nucleophilic attack occurred as an additional reaction.

BAIB has also been utilized in the total synthesis of the natural lignan (±)-tanegool. The synthesis started from ferulic acid. Firstly, a diarylcyclobutanediol intermediate (**31**) was made through a three-step synthesis, involving para-nitro-esterification of the carboxylic acid, light induced [2 + 2] coupling, and finally reduction of the para-nitro esters with LAH. This intermediate was then oxidatively ring opened to a di-para-quinone methide (**32**) by BAIB in TFE and acetone, followed by 5-exo-trig cyclization, by one of the hydroxyls, to **33**. The cyclization was stereoselective due to steric hindrance of the bulky aromatic rings. Finally, hydration by addition of water gave (±)-tanegool (Scheme 17) [74].

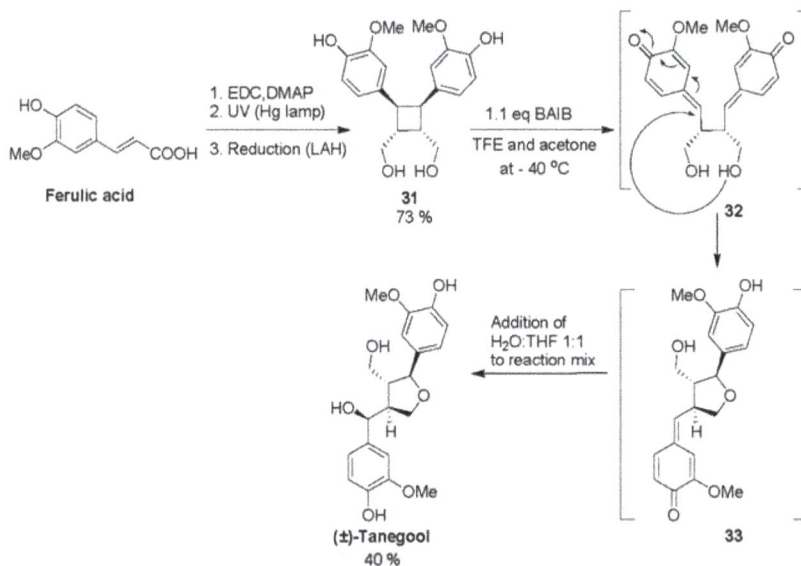

Scheme 17. Total synthesis of (±)-tanegool involving BAIB mediated oxidative ring opening.

2.7.2. 3-Iodobenzoic Acid (IBX)

A green and selective demethylation reaction, using IBX as the primary oxidant, has been reported for a number of lignans [75]. The substrates underwent oxidation to an *o*-quinone structure, followed by reduction by sodium hydrosulfite to the catechol products. The reaction was selective towards demethylation of phenolic methoxyl groups, leaving methyl esters intact. Scheme 18 shows the IBX oxidation of a methyl ester norlignan (**34**). The reaction mixture was acetylated prior to purification through silica gel column chromatography, giving **35**. The same reaction was also successful on the lignans hydroxymatairesinol (product yield = 55%), conidendrin (product yield = 70%), and lariciresinol (product yield = 40%), yielding demethylated products.

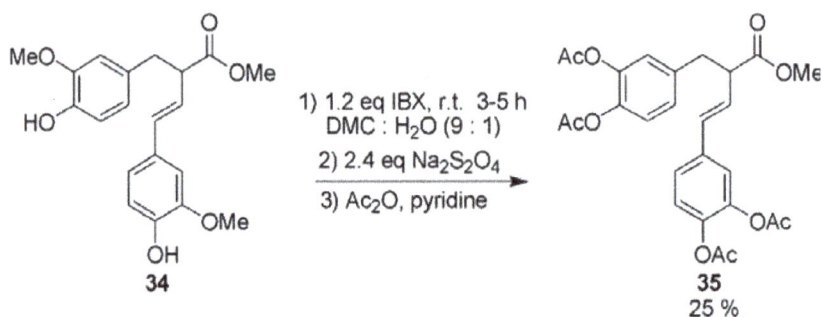

Scheme 18. Selective 3-Iodobenzoic Acid (IBX)-mediated demethylation of a norlignan.

When an aliphatic or benzylic alcohol was present, alcohol oxidation to a carbonyl outweighed the demethylation reaction [76]. Scheme 19 shows the selective oxidation of the benzylic alcohol in a synthetic lignan (**37**) to the corresponding ketone (**38**). Alternatively, after removal of the protective group, the product was oxidized to the diketone (**36**). Both of these reactions left the methoxyl groups intact [77]. Benzylic alcohol oxidations by IBX have also been reported on other lignans under similar reaction conditions. Both epi-aristoligone and magnolone have been prepared by this method [78,79] (Scheme 19).

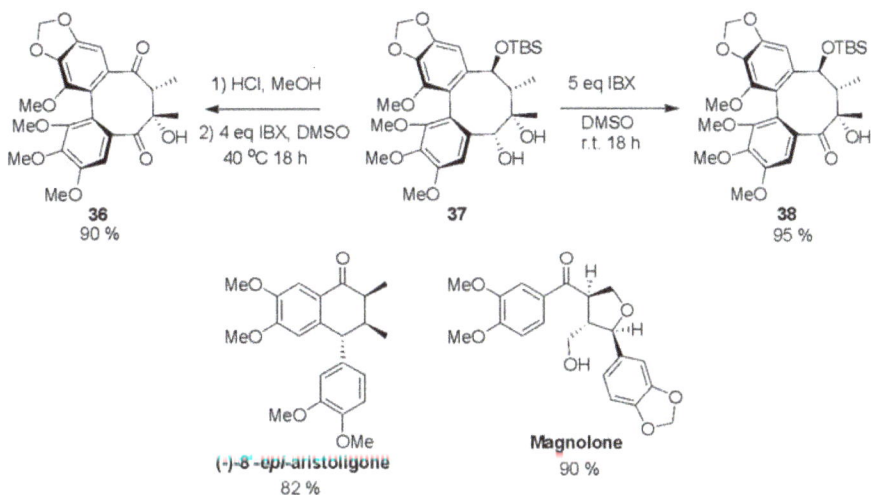

Scheme 19. Oxidation of benzylic alcohols by IBX.

2.7.3. Dess-Martin Periodinane (DMP)

DMP has been reported as an agent for mild and selective oxidation of alcohols to aldehydes and ketones. It has a higher solubility compared to its precursor IBX, and lacks the danger of explosion [80].

Oxidation of lignans baring benzylic or aliphatic hydroxyls to the corresponding ketones by DMP have been described for various lignans [81–86]. An example is shown in Scheme 20 for oxidation of isopicrosteganol to the atropisomers of picrosteganone [87]. Scheme 21a shows the oxidation of an aliphatic primary diol (**39**) with DMP [88]. The reaction quantitatively forms a lactol known as cis-cubebin through subsequent ring closure of the aldehyde. A similar ring closure has been reported by DMP oxidation of a lignan baring a hydroxyl group and a carboxylic acid (**40**), forming a hydroxy-butyrolactone (**41**, Scheme 21b) [89].

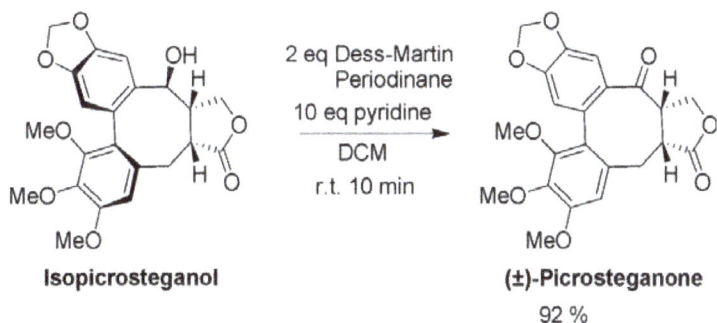

Scheme 20. Dess-Martin oxidation of the benzylic alcohol to the ketone.

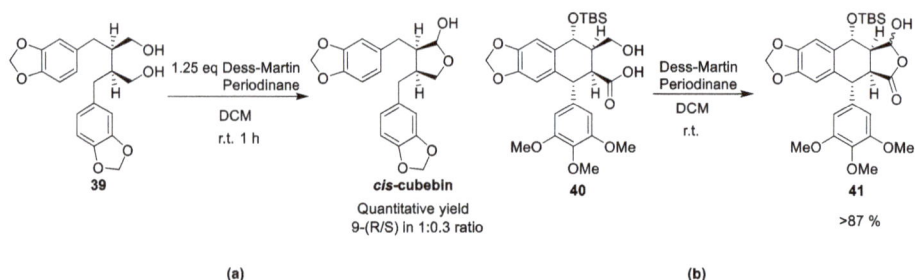

Scheme 21. Dess-Martin oxidation of (a) diol **39** to cis-cubebin; (b) hydroxyacid **40** to hydroxybutyrolactone **41**.

2.7.4. Sodium Periodate (NaIO$_4$)

In 1955, Adler et al. [90] reported NaIO$_4$ mediated oxidation of guaiacyl or syringyl groups, giving demethylation followed by formation of o-quinones. When a syringyl lignan (**42**) with aliphatic hydroxyls was oxidized with NaIO$_4$, it formed a 5-methoxyl o-quinone structure (**43**) with an ether bridge to the ortho-position (Scheme 22) [91]. For lignans without aliphatic hydroxyls, as for **44**, the o-quinone structures were formed without the ether bridge (**45**, Scheme 23) [92–94].

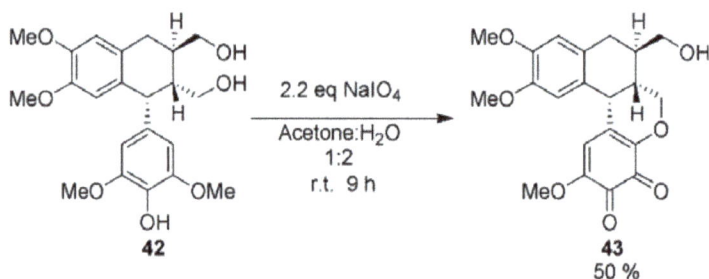

Scheme 22. NaIO$_4$-mediated oxidation to an ether bridged o-quinone structure.

Scheme 23. NaIO$_4$-mediated oxidation of syringyl-lignan forming an o-quinone structure.

In the total synthesis of sylvone, the final steps consisted of the so-called Lemieux–Johnson oxidation. A vicinal diol (**47**) was formed from an alkene (**46**) by OsO$_4$, and then a NaIO$_4$-mediated oxidative cleavage of the vicinal diol gave the ketone (sylvone) and formaldehyde (Scheme 24) [95]. Lemieux-Johnson oxidation has been used on other lignans as well, giving ketone products in high yields, as shown in Scheme 24 [96–100].

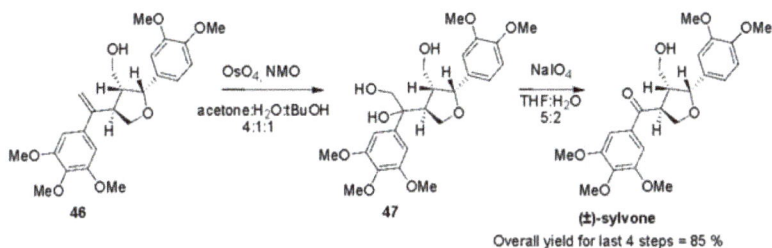

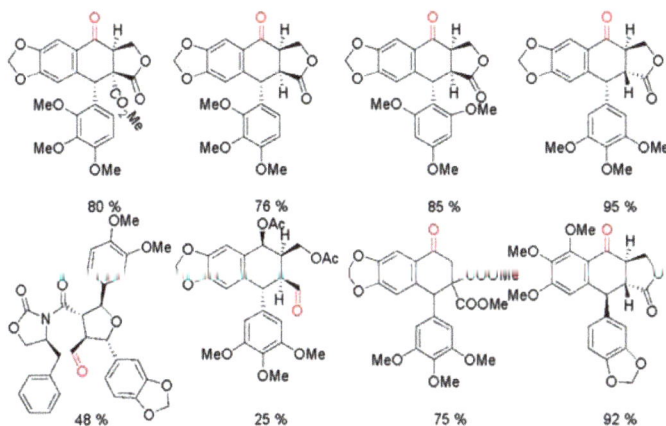

Scheme 24. Lemieux-Johnson oxidation in the synthesis of sylvone (**upper**). Other lignan structures where the same methodology has been applied (position for oxidation marked in red). The isolated overall yields for both steps are given (**lower**).

2.8. N-Bromosuccinimide (NBS)

The Wohl-Ziegler reaction using NBS and a radical initiator to promote arylic or benzylic bromination has been widely used, and also applied to lignans. Tomioka et al. reported a benzylic bromination of (−)-stegane by NBS in CCl₄, using benzoyl peroxide (BPO) as the radical initiator [101]. The formed 4-bromostegane yielded (−)-steganone after hydrolysis in aqueous THF, with an overall yield of 85% (Scheme 25). By further acetylation, the desired product (−)-steganacin was obtained in 72% yield. Similar results were also achieved starting from stegane, a stereoisomer of (+)-isostegane [35].

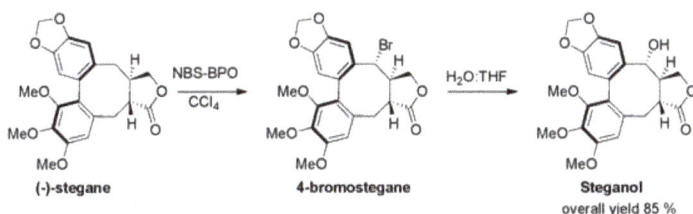

Scheme 25. Wohl-Ziegler bromination of (+)-isostegane followed by hydrolysis to (−)-steganol.

By using a solvent with increased polarity, NBS can also be used to oxidize alcohols to aldehydes and ketones [102]. This has been imposed on a photolytic NBS-mediated oxidation of lignan **48**, using dioxane as the solvent, forming a benzylic ketone (**49**) product in high yield (Scheme 26) [103].

The reaction was proposed to proceed through bromination of the benzylic position, followed by hydrolysis by water to a hydroxyl, and finally oxidation to the corresponding ketone.

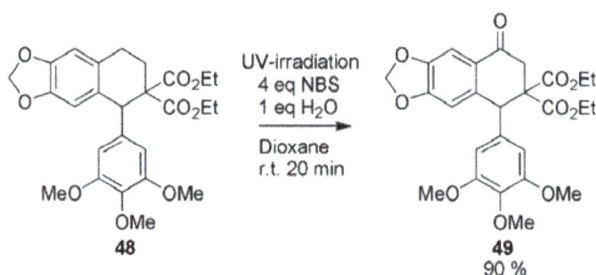

Scheme 26. UV and NBS-mediated formation of the benzylic ketone (**49**).

When a similar method was applied to deoxypodophyllotoxin by a reaction with NBS in DMF, only arylic bromination occurred (**50**, Scheme 27) [104]. To get the benzylic ketone, NBS-BPO oxidation in CCl$_4$ followed by hydrolysis was employed. The product epipodophyllotoxine was further oxidized by pyridinium chlorochromate (PCC) to podophyllotoxone. Interestingly, in addition to epipodophyllotoxine, the NBS-mediated reaction also yielded dehydroxypodophyllotoxin through oxidative aromatization [105].

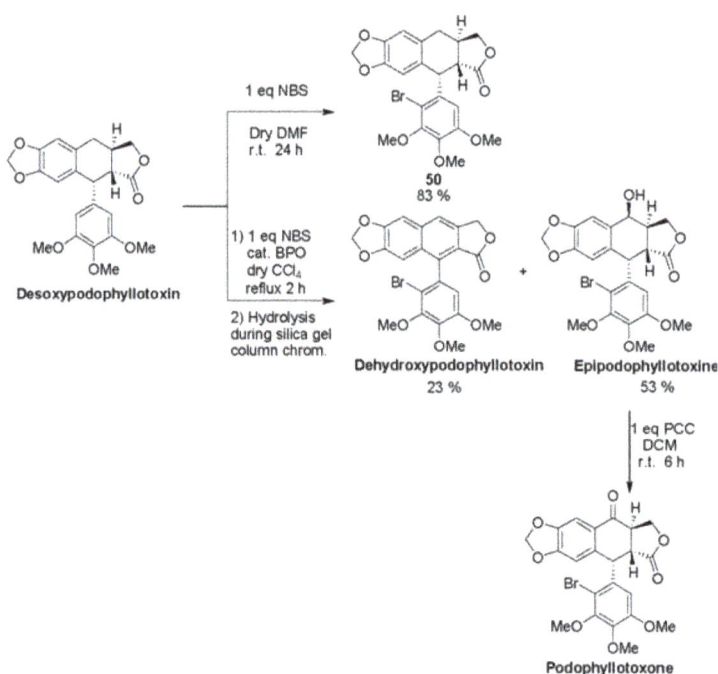

Scheme 27. Reaction of deoxypodophyllotoxin with NBS in DMF and CCl$_4$, and further oxidation of epipodophyllotoxine by PCC to the corresponding ketone (podophyllotoxone).

A similar naphthalene type lignan, known as justicidin B, was formed through NBS-treatment of jatrophan in CCl$_4$ (Scheme 28) [106,107]. The reaction includes trans-cis-isomerization, oxidative ring closure, and dehydrogenation.

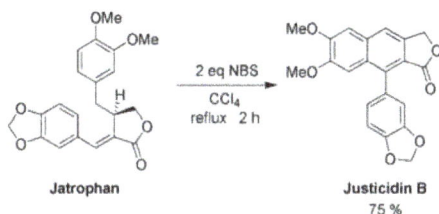

Scheme 28. Synthesis of justicidin B by NBS-oxidation of jetrophan.

2.9. Dimethyldioxirane (DMDO or DMD)

A highly selective ring opening of asarinin to a diastereomerically pure product (**52**) when using DMDO has been reported (Scheme 29) [108]. With one equivalent of DMDO, a selective ring opening was achieved at the C-7 with *R* configuration, while the other furan ring remained unaltered (**51**). DMDO was highly reactive towards the substrate, and the reaction needed to be performed at −20 °C. If more than one equivalent of DMDO was used, the reaction was believed to go through a radical mechanism yielding a mixture of products. However, after acetylation of the product, ring opening of the second furan ring was achieved under the same conditions as previously used, only with a slightly lower yield.

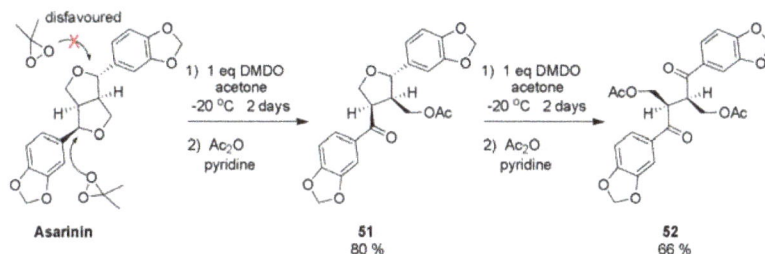

Scheme 29. Selective ring opening of asarinin by DMDO.

2.10. Nitrobenzene

Reports from the 1950s and 1960s show oxidative degradation of lignans to vanillin, vanillic acid, and syringaldehyde, using nitrobenzene as the oxidant. Nitrobenzene oxidation of thujaplicatin and dihydroxythujaplicatin, which are lignans found in western red cedar, gave vanillin as the sole product, but only in approximately 2% yield [109,110]. Thujaplicatin methyl ether, with the *meta*-hydroxyl protected as a methoxyl group, gave around 4% of both vanillin and syringaldehyde upon nitrobenzene oxidation. Another study systematically oxidized a range of lignans under the following conditions: 180 °C, 60 ml of 2 M NaOH and 8 mL nitrobenzene per gram of lignan. The reaction time was 2 h. The results are listed in Table 2 [111]. Similarly, nitrobenzene has also been used for the production of vanillin from industrial lignin [167].

Table 2. Results of nitrobenzene-mediated oxidative degradation of lignans.

Lignan	Vanillin %	Vanillic Acid %	Conversion %
Pinoresinol	31	9	81
Lariciresinol	63	5	100
Olivil	83	3	100
Matairesinol	15	2	100
Conidendrin	1	-	-
Isoolivil	3	-	-

3. Metal-Mediated Oxidations

Oxidation of lignans using metal reagents could be divided into methods which use equimolar amounts or catalytic amounts of the metal oxidant.

Catalytic reactions usually employ transition metals in heterogenous or homogenous conditions and may be sensitive for different functional groups or heteroatoms. Equimolar metal oxidants are widely used for all types of oxidation of lignans, and may involve one-electron oxidations or ionic mechanisms.

3.1. Chromium (VI) Oxidations

There are many examples of Chromium (VI) oxidation of benzylic alcohols to ketones in different lignan structures. Acyclic matairesinol derivatives, [112–114], cyclooctadiene structures (**53–56**) [87,115], and podophyllotoxin derivatives have all been oxidized by CrO_3 or PCC to the corresponding ketone in relatively high yields (60–95%) [103,105,116,117] (Scheme 30).

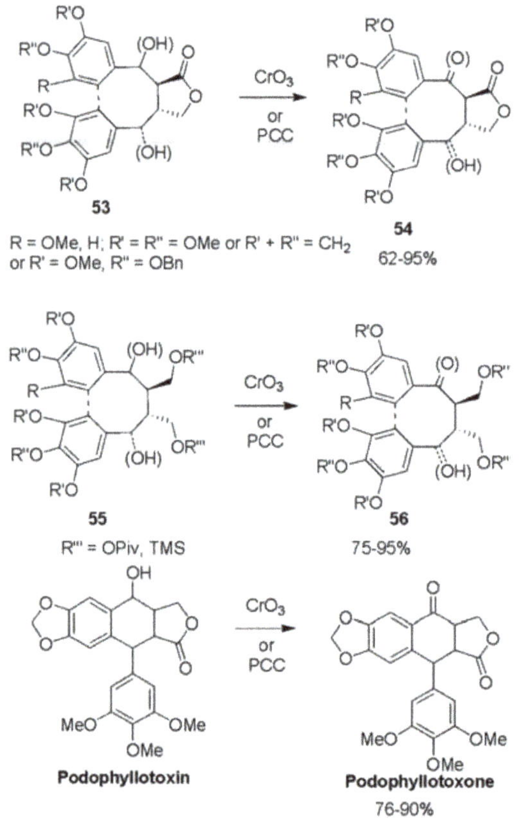

Scheme 30. Oxidation of the benzylic alcohol to the corresponding ketone by Cr(VI) oxidants. The dotted line corresponds to either the existence of the bond or the absence of the bond.

Cr(VI) mediated oxidation of primary alcohols in the lignan structures (**57**) has also been performed. Oxidation by PCC or CrO_3 led to lactones and carboxylic acids (**58** and **59**) [118–120] (Scheme 31).

There are no literature examples where chromium (VI) oxidants have successfully been used for oxidation of lignans in the presence of free phenolic groups. Therefore, it can be concluded that chromium mediated oxidations cannot be performed when phenolic groups are present in the structure.

Scheme 31. Oxidation of the primary alcohols by Cr(VI) oxidants. The dotted line corresponds to either the existence of the bond or the absence of the bond.

3.2. Palladium and Gold Mediated Oxidations

Oxidation of benzylic alcohols in the presence of free phenolic groups has been achieved by transition metal catalysis. Hydroxymatairesinol (HMR) has been shown to undergo catalytic dehydrogenation in mild conditions, using palladium on different supporting materials [121] (Scheme 32). The major oxidation product was oxomatairesinol, however, hydrogen formed during the dehydrogenation partially reacted with HMR by hydrogenolysis to give a significant amount of matairesinol as a side product. Later on, it was shown that the same transformation could also be performed using gold as a catalyst [121]. This method was much more selective towards oxomatairesinol [123] and showed faster conversion in the presence of oxygen [124–126]. It is noteworthy that one diastereomer of HMR gave oxomatairesinol much more selectively and with higher conversion rates, but some conidendrin was also formed as a side product depending on the conditions of the reaction.

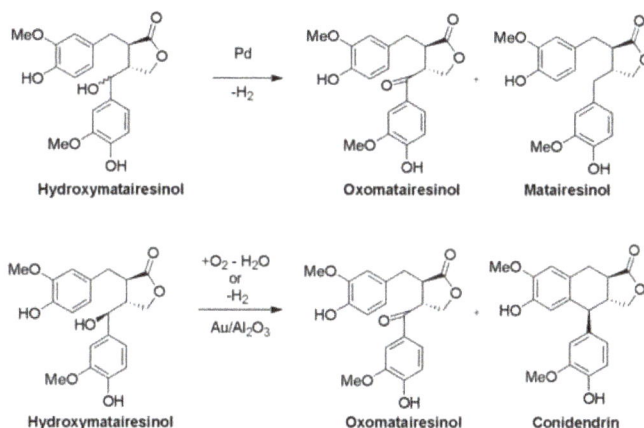

Scheme 32. Palladium and gold catalyzed oxidation (dehydrogenation) of hydroxymatairesinol.

3.3. Molybdenium Mediated Oxidations

Hydroxylation at the α-position of butyrolactone lignans (**60**) has been performed directly on the deprotonated ester by molecular oxygen [127,128]. However, this reaction proceeded much faster and with better selectivity when molybdenum reagents were used. For this purpose, oxodiperoxymolybdenum (pyridine) (hexamethylphosphoric triamide), $MoO_5 \cdot Py \cdot HMPA$ (MoOPH), also known as Vedejs' reagent was used [113,129–131]. The stereoselectivity of the oxidation was very much dependent on the base and the conditions used for the deprotonation. For example, if KHMDS was used, the diastereomeric excess of one isomer was just 11% [113,130]. At the same time, when KHMDS was used together with 18-crown-6, the selectivity raised to 64–99% [130,131] (Scheme 33).

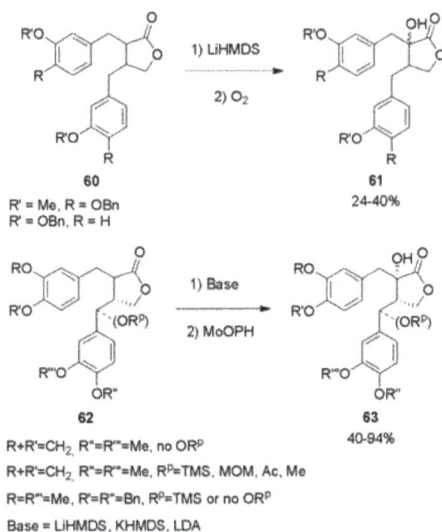

Scheme 33. Molybdenum catalyzed α-hydroxylation of butyrolactone lignans. The dotted line corresponds to either the existence of the bond or the absence of the bond.

Molybdenum reagents have also been used for the preparation of 2-2′-cyclolignan structures. Oxidative coupling of two aryls was performed with molybdenum pentachloride (MoCl5) [132,133]. The formation of the cyclooctadiene structure proceeded with excellent yields when the substrates were non-functionalized aliphatics, but the yields decreased dramatically with increasing degree of functionalization (Scheme 34).

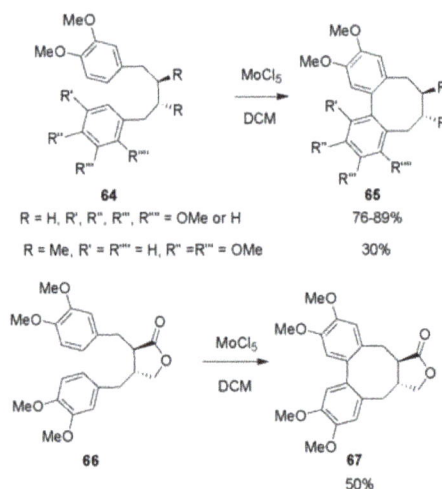

Scheme 34. Ar-Ar oxidative coupling by MoCl5.

3.4. Vanadium, Thallium and Ruthenium Oxidations

Oxidative coupling has also been reported using V, Tl, and Ru. Vanadinum oxofluoride in the presence of trifluoroacetic acid has shown good selectivity for the Ar-Ar oxidative coupling, forming the cyclooctadiene lignan structure Isostegnane in high yield [134] (Scheme 35).

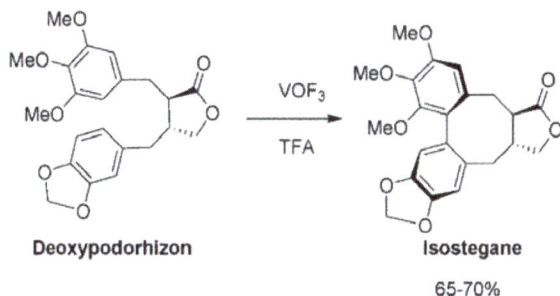

Scheme 35. Ar-Ar oxidative coupling by VOF$_3$.

Later it was also shown that thallium oxide (Tl$_2$O$_3$) or ruthenium oxide (RuO$_2$) also work well as oxidants in this reaction [135,136]. In fact, the reactions with ruthenium oxide showed higher yields than reactions with thallium oxide, and vanadium oxofluoride reactions showed slightly lower yields in the vast majority of the reported reactions [137–141]. High stereoselectivity was observed in most cases (Scheme 36).

In many of these reactions, 2-7'-cyclolignans (**74**) and (**77**) were formed as minor products by oxidation of the benzylic positions [140,142–144] (Scheme 37).

Interestingly, these reactions also proceeded in the presence of a free phenolic group. Yields up to 90% of **79** were obtained with matairesinol derivatives (**78**) [137,144–146] (Scheme 38).

It has also been reported by Planchenault et al. [145,147] that some metal oxidants other than Mo, V, Ru, and Tl, such as Mn(OAc)$_3$, Ce(OH)$_4$, Re$_2$O$_7$, Fe(OH)(OAc)$_2$, Co$_3$O$_4$, Ag(OCOCF$_3$)$_2$, CrO$_3$, IrO$_2$, Pr$_6$O$_{11}$, SeO$_2$, TeO$_2$ etc., are possible to use for high yield preparations of cyclooctadiene-lignans.

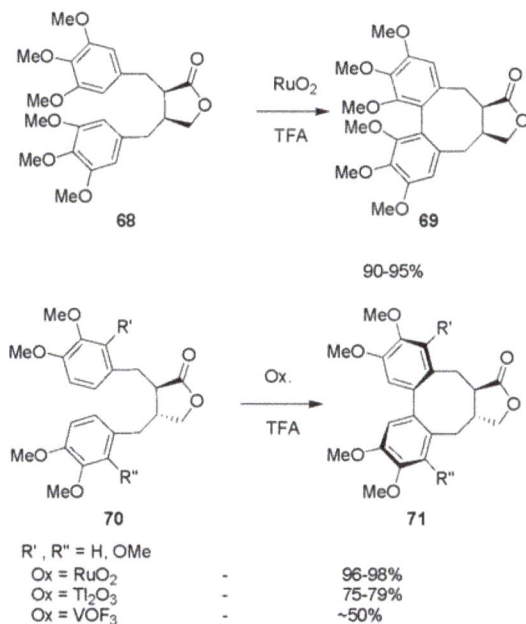

Scheme 36. Ar-Ar oxidative coupling by VOF$_3$, RuO$_2$, or Tl$_2$O$_3$.

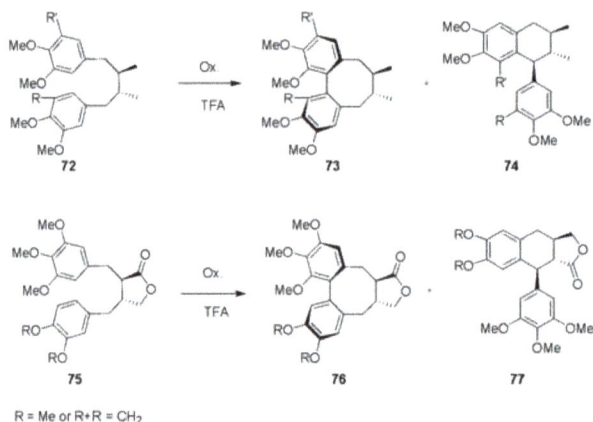

Scheme 37. Ru, Tl, and V mediated oxidative cyclizations, forming 2-2' and 2-7' cyclolignans.

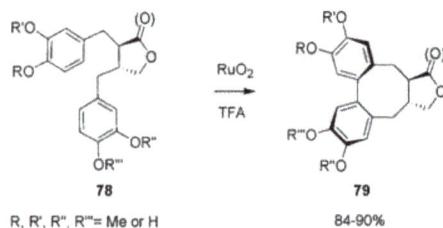

R, R', R'', R'''= Me or H

84-90%

Scheme 38. Oxidative coupling of matairesinol derivatives in the presence of free phenolic groups.

3.5. Methyl Trioxo-Rhenium (MTO) Catalyzed Oxidations

MTO catalyzed oxidations are interesting because of the diversity of possible transformations. For example, podophyllotoxin and related structures (**81**) were oxidized to quinone structures (**80 and 82**) by MTO [148] (Scheme 39). The reaction proceeded via demethylation, hydroxylations, and at the same time, the benzylic alcohol was oxidized.

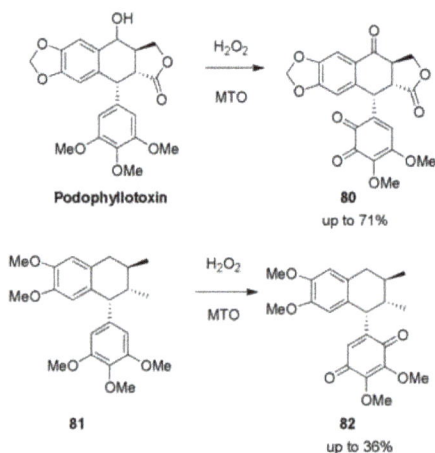

Scheme 39. Oxidation of podophyllotoxin and related structures by Methyl Trioxo-Rhenium (MTO).

When asaranin and sesaminin were treated with MTO in similar conditions, the reaction resulted in cleavage at the benzylic position to yield lactones (**83** and **84**) [149]. (Scheme 40).

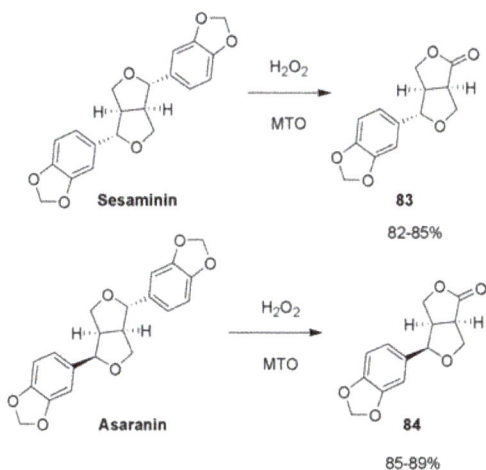

Scheme 40. Oxidation of asaranin and sesaminin with MTO.

MTO catalyzed reactions with lariciresinol, matairesinol and hydroxymatairesinol resulted in multiple benzylic hydroxylations, demethylations, oxidations, and cleavage of water [150] (Scheme 41, structures **85–90**).

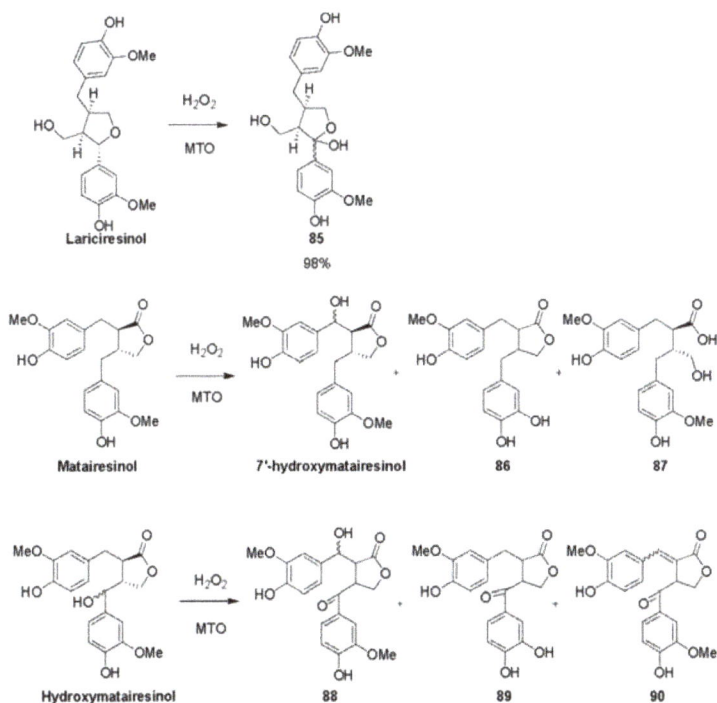

Scheme 41. Oxidation of lariciresinol, matairesinol, and hydroxymatairesinol by MTO (major products shown).

3.6. Other Metal Mediated Lignan Oxidations

Oxidative acetoxylation of matairesinol derivatives **91** and **93** has been accomplished using lead acetate [151] (Scheme 42). In the reactions of the *trans*-butyrolactone derivatives (**91**), one diastereomer of the 7-acetoxy product (**92**) was formed in high yield (60–73%). In the reaction of the cis lactone (**93**) in the same conditions, the reaction resulted in a mixture of the 7-acetoxy product (**94**) and the 2′-7-cyclolignan (**95**).

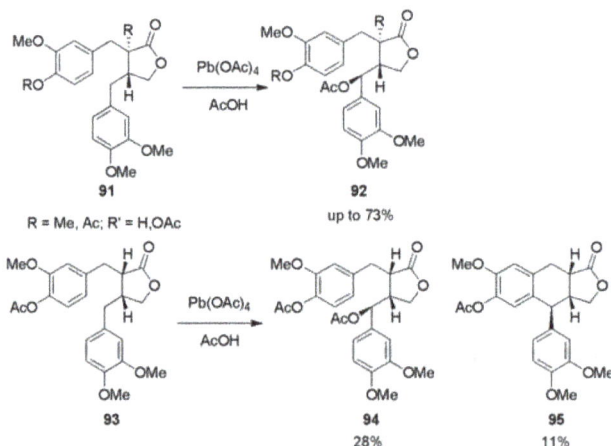

Scheme 42. Lead acetate acetoxylation of matairesinol derivatives.

In addition to molybdenum initiated α-hydroxylation of esters, cerium trichloride catalyzed α-hydroxylation has also been reported [152]. This method did not require deprotonation of the α-position. Coordination of cerium to two α-carbonyls directed the position of the oxidation. (Scheme 43).

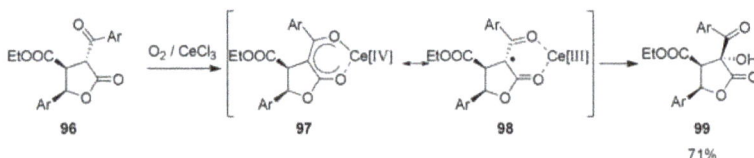

Scheme 43. Cerium trichloride mediated α-hydroxylation.

4. Other Oxidation Methods

4.1. Enzymatic Oxidations

Although enzymatic oxidations have mostly been studied for elucidation of biosynthetic routes for many lignans, some attempts for targeted oxidations in vitro have been reported.

Laccases are copper-containing oxidases which have been used in, among others, wood, pulp, and paper industries [168]. They play a key role in the dimerization of 4-hydroxycinnamic acids in the biosyntheses of natural lignans [22,169]. Lignans have also been used as substrates in laccase activity studies [170,171]. Laccase oxidations of lignans baring guaiacyl groups have led to efficient polymerization [172–174]. The mechanism is believed to involve hydrogen atom abstraction to form phenoxyl radicals, which then undergo intermolecular radical couplings [175].

Peroxidases are another group of oxidases which are also involved in the biosynthesis of lignans through β-β coupling of two phenylpropanoid units [22,169,176,177]. They typically use hydrogen peroxide as cofactor in the oxidation reactions. A range of different peroxidases has been used for selective ring closure of butyrolactone lignans (ex. **100**) with up to quantitative yields of the aryltetralin product (ex. **101**) (Scheme 44). The reactions were performed with both immobilized cell cultures and freely suspended plant cell cultures, and in absence of foreign hydrogen peroxide. Although the reaction worked also in hexane, quantitative ring closure only occurred when the reaction was done in B5 medium [153,154]. A similar reaction, but with H_2O_2 as a cofactor and ethanol as solvent, has also been reported on the same lignan [155].

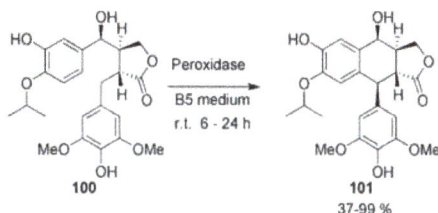

Scheme 44. Enzymatic, peroxidase-mediated ring closure.

A study of 4-*O*-5 units in softwood lignin used pinoresinol, among others, as a lignin model compound. A pinoresinol dimer (**102**) coupled through a 4-*O*-5 bond, was synthesized using horseradish peroxidase (HRP) and a hydrogen peroxide-urea complex (Scheme 45) [156]. Although sufficient for the study, the isolated yield was very low.

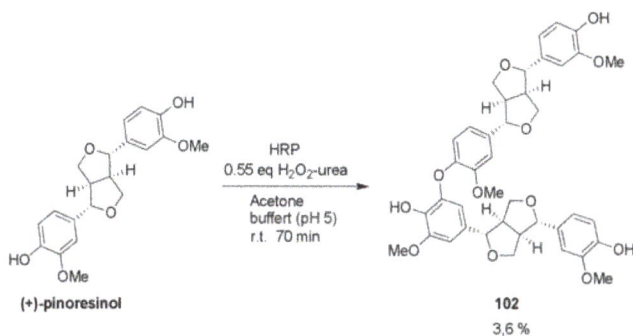

Scheme 45. Peroxidase and H_2O_2- mediated 4-*O*-5-coupling.

In the biosynthesis of matairesinol, secoisolariciresinol is oxidized by an enzyme known as secoisolariciresinol dehydrogenase (SDH) [178,179]. The reaction proceeds through an oxidative ring closure of the diol to a lactol, and further oxidation to the lactone. In a study on the mode of catalysis of SDH, the reaction was done in vitro, in a 5 mM scale, using isolated SDH and NAD^+ in a Tris-HCl buffer (Scheme 46) [157].

Enzymatic oxidation of sesamin has been studied in vitro, using yeast cells expressing some of the enzymes found in sesame seeds [158]. Yeast strains containing a combination of a P450 enzyme (CYP92B14) and *Sesamum indicum* cytochrome P450 oxidoreductase 1 (CPR1) in a synthetic defined (SD) medium gave, via oxidation and rearrangement, sesaminol and sesamolin (Scheme 47), two lignans found in sesame seeds.

Escherichia coli expressing human hepatic enzymes (CYP) have been used for the bioconversion of deoxypodophyllotoxin into epipodophyllotoxin (Scheme 48) [159]. The objective of the study was to investigate a possible route for industrial production of epipodophyllotoxin, which is a valuable precursor for the pharmaceutical industry. High conversions were achieved, but the reactions were done in scales up to only 0.5 µmols. The high cost of up scaling was presented as the bottleneck that needed to be tackled before a possible industrial production. A large scale isolation of deoxypodophyllotoxin, from wild chervil (*Anthriscus sylvestris*), had already been accomplished [180].

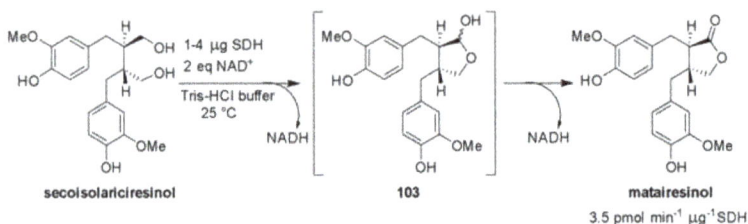

Scheme 46. Enzymatic transformation of secolariciresinol into matairesinol.

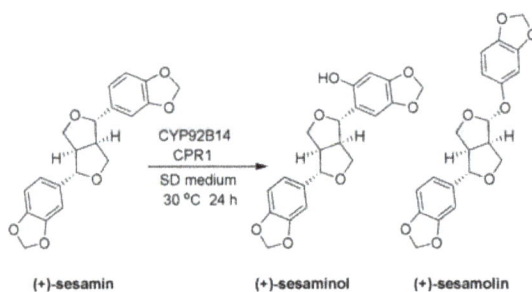

Scheme 47. Enzymatic transformation of sesamin into sesaminol and sesamolin.

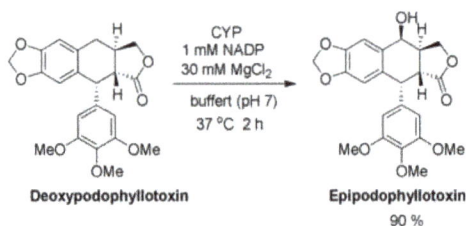

Scheme 48. Enzymatic bioconversion of deoxypodophyllotoxin into epipodophyllotoxin.

4.2. Electrochemical Oxidations

An acetoxy lignan (**104**) (Scheme 49) has been electrochemically oxidized to an *o*-quinone (**105**) at a constant anode potential of 0.725 V using a platinum gauze working electrode, a platinum wire counter electrode and an Ag/Ag⁺ reference electrode. The electrolyte was a solution of 0.01 M tetrabutylammonium tetraboroflourate in acetonitrile, with solid potassium carbonate as a buffer. The product was not isolated, but further reduced to a catechol, which was isolated and characterized [160].

Dibenzocyclooctadienes have been synthesized by electrochemical oxidation of butyrolactone lignans (**106**) (Scheme 50). A platinum working electrode, a platinum counter electrode, a Ag/AgBr, Et₄NBr reference electrode, and a 0.1 M Et₄NClO₄-CH₃CN supporting electrolyte was used. The products (**107**) were obtained in high yields, more than 80% in both cases [161].

The lignan hibalactone has been electrochemically oxidized using voltammetry with a three-electrode system, consisting of a glassy carbon working electrode, a platinum counter electrode, and a Ag/AgCl counter electrode. The lignan showed a single quasi-reversible electro-oxidation, with the alkene bonded to the lactone ring being the most probable site of oxidation [181]. The lignans *epi*-guaiacin, guaiacin, verrucosin, and nectandrin B have been isolated from *Iryanthera juruensis* fruits and electrochemically oxidized during an investigation of their antioxidative properties (Figure 1). A glassy carbon working electrode, a carbon counter electrode, and a Ag/AgCl reference electrode were used [182]. Honokiol and magnolol were electrochemically oxidized using an acetylene black nanoparticle-modified glassy carbon electrode as a working electrode, with a platinum wire counter electrode and a saturated calomel reference electrode. The results indicated that the oxidation of honokiol was reversible and involved two electrons. The oxidation of magnolol was irreversible [183].

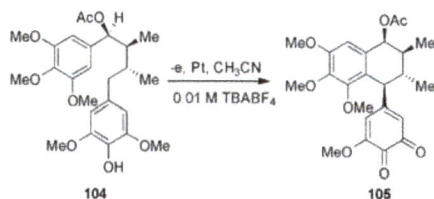

Scheme 49. Electrochemical oxidation of an acetoxy lignan to an *o*-quinone.

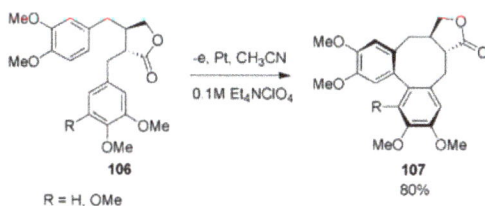

Scheme 50. Synthesis of dibenzocyclooctadienes by electrochemical oxidation of butyrolactone lignans.

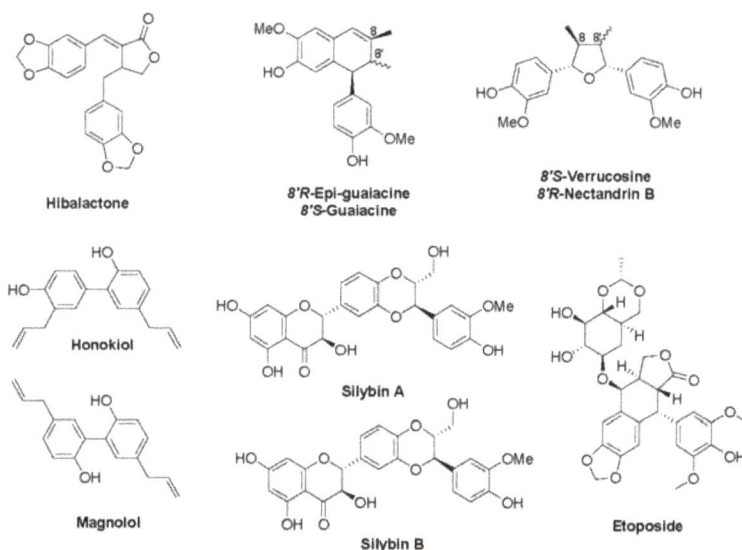

Figure 1. The structures of the lignans hibalactone, *epi*-guaiacin, guaiacin, verrucosin, nectandrin B, honokiol and magnolol, the glycoside etoposide, and the flavonolignan silybin.

Foodstuff and plant extracts containing phenolic compounds including lignans have been electrochemically oxidized during the determination of the phenolic content [184–187] and the antioxidant capacity [188,189]. The drug etoposide, which is a semisynthetic epipodophyllotoxin glycoside, has been oxidized electrochemically using a carbon paste working electrode, a Ag/AgCl/3 M KCl reference electrode, and a platinum wire counter electrode, with a supporting electrolyte of varying pH. The oxidation involves the transfer of two electrons and proceeds in one voltammetric oxidation step at a pH < 4.0 and two voltammetric oxidation steps at a pH > 4.0. In the first and reversible oxidation step, one electron is transferred, resulting in a stable radical. The product in the second oxidation step is an unstable cation, which rapidly converts into the *o*-quinone [190]. The flavonolignan silybin and its derivatives has been electrochemically oxidized using various methods [191–195].

4.3. Photooxidations

The lignans hydroxymatairesinol, allohydroxymatairesinol, α-conidendrin, and oxo-matariresinol have been used as substrates for light-irradiation experiments in different solvents. The products of light-irradiation of hydroxymatairesinol that were either isolated or detected were allohydroxymatairesinol, oxomatairesinol, α-conidendrin, allo-7′-methoxymatairesinol, 7′-methoxymatairesinol, and vanillin (Scheme 51). The irradiation of allo-hydroxymatairesinol formed the reaction products hydroxymatairesinol, oxomatairesinol, α-conidendrin, allo-7′-methoxymatairesinol, 7′-methoxymatairesinol, and vanillin. Oxomatairesinol was formed from the irradiation of hydroxymatairesinol, and vanillin from the irradiation of oxomatairesinol [162].

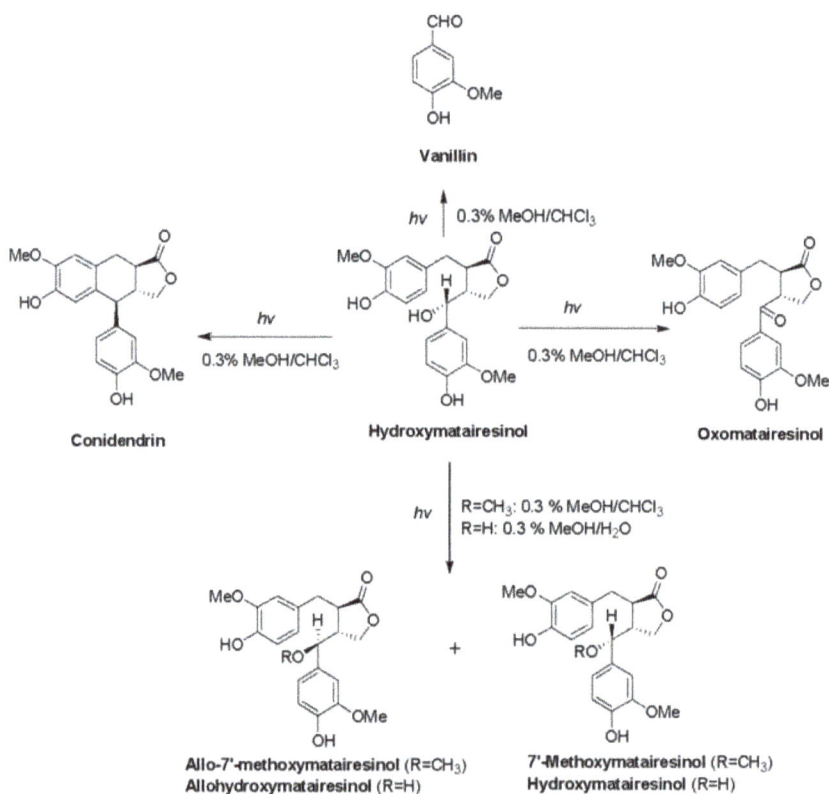

Scheme 51. The photooxidation reactions of hydroxymatairesinol.

5. Conclusions

Lignans are oxidatively transformed in a number of processes and conditions. In nature, oxidations are mostly related to biosynthetic pathways, or processes where lignans act as primary antioxidant scavenging free radicals. When it comes to targeted oxidative transformations or studies between lignans and oxidants, the reports in the literature can be divided into non-metal mediated and metal mediated oxidations. The oxidative transformations include the oxidation of alcohols, both primary and benzylic, or benzylic functionalization by halogenation, hydroxylation, dehydrogenation, or by ring closing reactions. Oxidative Ar-Ar couplings forming cyclooctadienes have been extensively reported and can be quite selectively accomplished with several reagents in high yields. α-Hydroxylations and epoxidations have also been reported, although to a much lesser extent and these reactions may warrant further investigations. Radical mediated oxidations often result in polymerization or radical addition reactions giving a wide range of products with poor selectivity, especially in the presence of free phenolic groups. Although there are over 200 papers reporting the oxidative transformations of lignans, we can conclude that there is still room for further investigations, especially concerning the selectivity. In addition, electrochemical oxidations and photooxidations are rather unexplored and could be a future area of research.

Funding: The authors would like to thank The Swedish Cultural Foundation in Finland (Svenska Kulturfonden) and Viktoriastiftelsen for financial support.

Conflicts of Interest: The authors declare no conflict of interest.

References

1. Zhang, J.; Chen, J.; Liang, Z.; Zhao, C. New Lignans and Their Biological Activities. *Chem. Biodivers.* **2014**, *11*, 1–54. [CrossRef] [PubMed]
2. Saleem, M.; Kim, H.J.; Ali, M.S.; Lee, Y.S. An update on bioactive plant lignans. *Nat. Prod. Rep.* **2005**, *22*, 696–716. [CrossRef] [PubMed]
3. Harmatha, J.; Dinan, L. Biological activities of lignans and stilbenoids associated with plant-insect chemical interactions. *Phytochem. Rev.* **2003**, *2*, 321–330. [CrossRef]
4. Talaei, M.; Pan, A. Role of phytoestrogens in prevention and management of type 2 diabetes. *World J. Diabetes* **2015**, *6*, 271–283. [CrossRef]
5. Deo, S.; Utane, R.; Khubalkar, R.; Thombre, S. Extraction and isolation, synthesis, physiological activity of 1-phenyl naphthalene and its derivatives: A review. *Pharma Innov. J.* **2017**, *6*, 21–30.
6. Teponno, R.B.; Kusari, S.; Spiteller, M. Recent advances in research on lignans and neolignans. *Nat. Prod. Rep.* **2016**, *33*, 1044–1092. [CrossRef] [PubMed]
7. Adlercreutz, H. Lignans and Human Health. *Crit. Rev. Clin. Lab. Sci.* **2007**, *44*, 483–525. [CrossRef]
8. Witkowska, A.M.; Waśkiewicz, A.; Zujko, M.E.; Szcześniewska, D.; Stepaniak, U.; Pająk, A.; Drygas, W. Are Total and Individual Dietary Lignans Related to Cardiovascular Disease and Its Risk Factors in Postmenopausal Women? A Nationwide Study. *Nutrients* **2018**, *10*, 865. [CrossRef]
9. Kiyama, R. Biological effects induced by estrogenic activity of lignans. *Trends Food Sci. Technol.* **2016**, *54*, 186–196. [CrossRef]
10. Bobe, G.; Murphy, G.; Albert, P.S.; Sansbury, L.B.; Lanza, E.; Schatzkin, A.; Cross, A.J. Dietary lignan and proanthocyanidin consumption and colorectal adenoma recurrence in the Polyp Prevention Trial. *Int. J. Cancer* **2012**, *130*, 1649–1659. [CrossRef]
11. Hearon, W.M.; MacGregor, W.S. The Naturally Occurring Lignans. *Chem. Rev.* **1955**, *55*, 957–1068. [CrossRef]
12. Rao, C. *Chemistry of Lignans*; Andhra University Press: Visakhapatnam, India, 1978.
13. Ayres, D.C.; Loike, J.D. *Lignans: Chemical, Biological, and Clinical Properties*; Phillipson, J.D., Ayres, D.C., Baxter, H., Eds.; Cambridge University Press: Cambridge, UK, 1990; ISBN 0521304210.
14. Whiting, D.A. Lignans, neolignans, and related compounds. *Nat. Prod. Rep.* **1987**, *4*, 499–525. [CrossRef] [PubMed]
15. Ward, R.S. Lignans, neolignans and related compounds. *Nat. Prod. Rep.* **1999**, *16*, 75–96. [CrossRef]

16. Kariyappa, A.K.; Ramachandrappa, R.K.; Seena, S. Lignans: Insight to Chemistry and Pharmacological Applications-An Overview. *Chem Sci Rev Lett.* **2015**, *4*, 1157–1165.

17. Ward, R.S. Recent Advances in the Chemistry of Lignans. *Stud. Nat. Prod. Chem.* **2000**, *24*, 739–798. [CrossRef]

18. Pelter, A.; Ward, R.S.; Venkateswarlu, R.; Kamakshi, C. Oxidative transformations of lignans - reactions of dihydrocubebin and a derivative with DDQ. *Tetrahedron* **1991**, *47*, 1275–1284. [CrossRef]

19. Venkateswarlu, R.; Kamakshi, C.; Moinuddin, S.G.; Subhash, P.V.; Ward, R.S.; Pelter, A.; Hursthouse, M.B.; Light, M.E. Transformations of lignans, Part V. Reactions of DDQ with a gmelinol hydrogenolysis product and its derivatives. *Tetrahedron* **1999**, *55*, 13087–13108. [CrossRef]

20. Ward, R.S.; Hughes, D.D. Oxidative cyclisation of cis- and trans-2,3-dibenzylbutyrolactones using phenyl iodonium bis(trifluoroacetate) and 2,3-dichloro-5,6-dicyano-1,4-benzoquinone. *Tetrahedron* **2001**, *57*, 5633–5639. [CrossRef]

21. Charlton, J.L.; Chee, G.-L. Asymmetric synthesis of lignans using oxazolidinones as chiral auxiliaries. *Can. J. Chem.* **1997**, *75*, 1076–1083. [CrossRef]

22. Magoulas, G.; Papaioannou, D.; Magoulas, G.E.; Papaioannou, D. Bioinspired Syntheses of Dimeric Hydroxycinnamic Acids (Lignans) and Hybrids, Using Phenol Oxidative Coupling as Key Reaction, and Medicinal Significance Thereof. *Molecules* **2014**, *19*, 19769–19835. [CrossRef]

23. Satyanarayana, P.; Venkateswarlu, S.; Viswanatham, K.N. Oxidative aryl-aryl, aryl-benzyl coupling of lignans-reactions of phyllanthin and haloderivatives with TTFA, DDQ, Li/THF: synthesis of dibenzocyclooctadiene system and phyltetralin. *Tetrahedron* **1991**, *47*, 8277–8284. [CrossRef]

24. Pelter, A.; Ward, R.S.; Jones, D.M.; Maddocks, P. Asymmetric synthesis of lignans of the dibenzylbutanediol and tetrahydrodibenzocyclooctene series. *J. Chem. Soc. Perkin Trans. 1* **1993**, 2631–2637. [CrossRef]

25. Pelter, A.; Ward, R.S.; Jones, D.M.; Maddocks, P. Asymmetric syntheses of lignans of the dibenzylbutyrolactone, dibenzylbutanediol, aryltetraun and dibenzocyclooctadiene series. *Tetrahedron: Asymmetry* **1992**, *3*, 239–242. [CrossRef]

26. Chattopadhyay, S.K.; Rao, K.V. Chemistry of saururus cernuus IV: cyclooctadiene systems derived from austrobailignan-5. *Tetrahedron* **1987**, *43*, 669–678. [CrossRef]

27. Venkateswarlu, R.; Kamakshi, C.; Moinuddin, S.G.; Subhash, P.V.; Ward, R.S.; Pelter, A.; Coles, S.J.; Hursthouse, M.B.; Light, M.E. Transformations of lignans. Part 4: Oxidative and reductive rearrangements of dibenzocyclooctadiene and spirodienone lignans. *Tetrahedron* **2001**, *57*, 5625–5632. [CrossRef]

28. Tomioka, K.; Ishiguro, T.; Koga, K. First asymmetric total synthesis of (+)-steganacin determination of absolute stereochemistry. *Tetrahedron Lett.* **1980**, *21*, 2973–2976. [CrossRef]

29. Chang, J.; Xie, J. Synthesis of some new lignans and the mechanism of intramolecular nonphenolic oxidative coupling of aromatic compounds. *Sci. China Ser. B Chem.* **2000**, *43*, 323–330. [CrossRef]

30. Tilve, S.G.; Torney, P.S.; Patre, R.E.; Kamat, D.P.; Srinivasan, B.R.; Zubkov, F.I. Domino Wittig-Diels Alder reaction: Synthesis of carbazole lignans. *Tetrahedron Lett.* **2016**, *57*, 2266–2268. [CrossRef]

31. Ward, R.S.; Satyanarayana, P.; Gopala Rao, B.V. Reactions of Aryltetralin Lignans with DDQ—An Example of DDQ Oxidation of Allylic Ether Groups. *Tetrahedron Lett.* **1981**, *22*, 3021–3024. [CrossRef]

32. Yamaguchi, H.; Arimoto, M.; Tanoguchi, M.; Ishida, T.; Inoue, M. Studies on the constituents of the seeds of *Hernandia ovigera* L. III. Structures of two new lignans. *Chem. Pharm. Bull. (Tokyo)* **1982**, *30*, 3212–3218. [CrossRef]

33. Eklund, P.C.; Sjöholm, R.E. Oxidative transformation of the natural lignan hydroxymatairesinol with 2,3-dichloro-5,6-dicyano-1,4-benzoquinone. *Tetrahedron* **2003**, *59*, 4515–4523. [CrossRef]

34. Taskinen, A.; Eklund, P.; Sjöholm, R.; Hotokka, M. The molecular structure and some properties of hydroxymatairesinol. An ab initio study. *J. Mol. Struct. THEOCHEM* **2004**, *677*, 113–124. [CrossRef]

35. Ishiguro, T.; Mizuguchi, H.; Tomioka, K.; Koga, K. Stereoselective reactions. VIII. Stereochemical requirement for the benzylic oxidation of lignan lactone. A highly selective synthesis of the antitumor lignan lactone steganacin by the oxidation of stegane. *Chem. Pharm. Bull.* **1985**, *33*, 609–617. [CrossRef]

36. Ward, R.S.; Pelter, A.; Jack, I.R.; Satyanarayana, P.; Rao, B.V.G.; Subrahmanyam, P. Reactions of paulownin, gmelinol and gummadiol with 2,5-dichloro-5,6-dicyanobenzoquinone. *Tetrahedron Lett.* **1981**, *22*, 4111–4114. [CrossRef]

37. Ward, R.S.; Pelter, A.; Venkateswarlu, R.; Kamakshi, C.; Subhash, P.V.; Moinuddin, S.G.; Hursthouse, M.B.; Coles, S.J.; Hibbs, D.E. Transformations of lignans, part IV. Acid-catalysed rearrangements of gmelinol with BF₃-etherate and study of a product with a unique lignan skeleton formed by further oxidation with DDQ. *Tetrahedron* **1999**, *55*, 13071–13086. [CrossRef]

38. Chang, J.B.; Xie, J.X. Synthesis of New Schizandrin Analogues. *Chinese Chem. Lett.* **2001**, *12*, 667–670.

39. Orito, K.; Sasaki, T.; Suginome, H. Photoinduced molecular transformations. 158. A total synthesis of (.+-.)-methyl piperitol: An unsymmetrically substituted 2,6-diaryl-3,7-dioxabicyclo [3.3.0] octane lignan. *J. Org. Chem.* **1995**, *60*, 6208–6210. [CrossRef]

40. Suginome, H.; Orito, K.; Yorita, K.; Ishikawa, M.; Shimoyama, N.; Sasaki, T. Photoinduced Molecular Transformations. 157. A New Stereo- and Regioselective Synthesis of 2,6-Diaryl-3,7 dioxabicyclo[3.3.0]octane Lignans Involving a β-Scission of Alkoxyl Radicals as the Key Step. New Total Syntheses of (±)-Sesamin, (±)-Eudesmin, and. (+)-Yangambin. *J. Org. Chem.* **1995**, *60*, 3052–3064. [CrossRef]

41. Sato, Y.; Tamura, T.; Mori, M. Arylnaphthalene Lignans through Pd-Catalyzed[2+2+2] Cocyclization of Arynes and Diynes: Total Synthesis of Taiwanins C and E. *Angew. Chem. Int. Ed.* **2004**, *43*, 2436–2440. [CrossRef]

42. Peng, Y.; Xiao, J.; Xu, X.-B.; Duan, S.-M.; Ren, L.; Shao, Y.-L.; Wang, Y.-W. Stereospecific Synthesis of Tetrahydronaphtho[2,3-*b*]furans Enabled by a Nickel-Promoted Tandem Reductive Cyclization. *Org. Lett.* **2016**, *18*, 5170–5173. [CrossRef]

43. Sang, S.; Tian, S.; Wang, H.; Stark, R.E.; Rosen, R.T.; Yang, C.S.; Ho, C.-T. Chemical studies of the antioxidant mechanism of tea catechins: radical reaction products of epicatechin with peroxyl radicals. *Bioorg. Med. Chem.* 2003, 11, 3371–3378. [CrossRef]

44. Fujisawa, S.; Kadoma, Y. Comparative study of the alkyl and peroxy radical scavenging activities of polyphenols. *Chemosphere* **2006**, *62*, 71–79. [CrossRef] [PubMed]

45. Spatafora, C.; Daquino, C.; Tringali, C.; Amorati, R. Reaction of benzoxanthene lignans with peroxyl radicals in polar and non-polar media: Cooperative behaviour of OH groups. *Org. Biomol. Chem.* **2013**, *11*, 4291–4294. [CrossRef] [PubMed]

46. Masuda, T.; Akiyama, J.; Fujimoto, A.; Yamauchi, S.; Maekawa, T.; Sone, Y. Antioxidation reaction mechanism studies of phenolic lignans, identification of antioxidation products of secoisolariciresinol from lipid oxidation. *Food Chem.* **2010**, *123*, 442–450. [CrossRef]

47. Betigeri, S.; Thakur, A.; Raghavan, K. Use of 2,2'-Azobis(2-Amidinopropane) Dihydrochloride as a Reagent Tool for Evaluation of Oxidative Stability of Drugs. *Pharm. Res.* **2005**, *22*, 310–317. [CrossRef] [PubMed]

48. Pisoschi, A.M.; Negulescu, G.P. Methods for Total Antioxidant Activity Determination: A Review. *Biochem. Anal. Biochem.* **2011**, *1*. [CrossRef]

49. Amorati, R.; Zotova, J.; Baschieri, A.; Valgimigli, L. Antioxidant Activity of Magnolol and Honokiol: Kinetic and Mechanistic Investigations of Their Reaction with Peroxyl Radicals. *J. Org. Chem.* **2015**, *80*, 10651–10659. [CrossRef]

50. Werber, J.; Wang, Y.J.; Milligan, M.; Li, X.; Ji, J.A. Analysis of 2,2'-Azobis (2-Amidinopropane) Dihydrochloride Degradation and Hydrolysis in Aqueous Solutions. *J. Pharm. Sci.* **2011**, *100*, 3307–3315. [CrossRef]

51. Eklund, P.C.; Långvik, O.K.; Wärnå, J.P.; Salmi, T.O.; Willför, S.M.; Sjöholm, R.E. Chemical studies on antioxidant mechanisms and free radical scavenging properties of lignans. *Org. Biomol. Chem.* **2005**, *3*, 3336. [CrossRef]

52. Hosseinian, F.S.; Muir, A.D.; Westcott, N.D.; Krol, E.S. Antioxidant capacity of flaxseed lignans in two model systems. *J. Am. Oil Chem. Soc.* **2006**, *83*, 835. [CrossRef]

53. Kang, M.-H.; Naito, M.; Sakai, K.; Uchida, K.; Osawa, T. Mode of action of sesame lignans in protecting lowdensity lipoprotein against oxidative damage in vitro. *Life Sci.* **1999**, *66*, 161–171. [CrossRef]

54. Hodaj, E.; Tsiftsoglou, O.; Abazi, S.; Hadjipavlou-Litina, D.; Lazari, D. Lignans and indole alkaloids from the seeds of Centaurea vlachorum Hartvig (Asteraceae), growing wild in Albania and their biological activity. *Nat. Prod. Res.* **2017**, *31*, 1195–1200. [CrossRef] [PubMed]

55. Alphonse, P.A.S.; Aluko, R.E. Anti-carcinogenic and anti-metastatic effects of flax seed lignan secolariciresinol diglucoside (SDG). *Discov. Phytomedicine* **2015**, *2*, 12–17. [CrossRef]

56. Hosseinian, F.F.H. Antioxidant Properties of Flaxseed Lignans Using In Vitro Model Systems. Ph.D. Thesis, University of Saskatchewan, Saskatoon, SK, Canada, 2006.

57. Hosseinian, F.S.; Muir, A.D.; Westcott, N.D.; Krol, E.S. AAPH-mediated antioxidant reactions of secoisolariciresinol and SDG. *Org. Biomol. Chem.* **2007**, *5*, 644–654. [CrossRef] [PubMed]

58. Narasimhan, N.S.; Aidhen, I.S. Radical mediated intramolecular arylation using tributyltinhydride/aibn: A formal synthesis of steganone. *Tetrahedron Lett.* **1988**, *29*, 2987–2988. [CrossRef]

59. Fischer, J.; Reynolds, A.J.; Sharp, L.A.; Sherburn, M.S. Radical Carboxyarylation Approach to Lignans. Total Synthesis of (−)-Arctigenin, (−)-Matairesinol, and Related Natural Products. *Org. Lett.* **2004**, *6*, 1345–1348. [CrossRef] [PubMed]

60. Huang, X.-X.; Bai, M.; Zhou, L.; Lou, L.-L.; Liu, Q.-B.; Zhang, Y.; Li, L.-Z.; Song, S.-J. Food Byproducts as a New and Cheap Source of Bioactive Compounds: Lignans with Antioxidant and Anti-inflammatory Properties from Crataegus pinnatifida Seeds. *J. Agric. Food Chem.* **2015**, *63*, 7252–7260. [CrossRef]

61. Chen, J.-J.; Wei, H.-B.; Xu, Y.-Z.; Zeng, J.; Gao, K. Antioxidant Lignans from the Roots of Vladimiria muliensis. *Planta Med.* **2013**, *79*, 1470–1473. [CrossRef]

62. Lee, E.-J.; Chen, H.-Y.; Hung, Y.-C.; Chen, T.-Y.; Lee, M.-Y.; Yu, S.-C.; Chen, Y.-H.; Chuang, I.-C.; Wu, T.-S. Therapeutic window for cinnamophilin following oxygen–glucose deprivation and transient focal cerebral ischemia. *Exp. Neurol.* **2009**, *217*, 74–83. [CrossRef] [PubMed]

63. Mocan, A.; Schafberg, M.; Crişan, G.; Rohn, S. Determination of lignans and phenolic components of Schisandra chinensis (Turcz.) Baill. using HPLC-ESI-ToF-MS and HPLC-online TEAC: Contribution of individual components to overall antioxidant activity and comparison with traditional antioxidant assays. *J. Funct. Foods* **2016**, *24*, 579–594. [CrossRef]

64. Sharma, O.P.; Bhat, T.K. DPPH antioxidant assay revisited. *Food Chem.* **2009**, *113*, 1202–1205. [CrossRef]

65. Szabo, M.; Idiţoiu, C.; Chambre, D.; Lupea, A. Improved DPPH determination for antioxidant activity spectrophotometric assay. *Chem. Pap.* **2007**, *61*, 214–216. [CrossRef]

66. Yamauchi, S.; Masuda, T.; Sugahara, T.; Kawaguchi, Y.; Ohuchi, M.; Someya, J.; Akiyama, J.; Tominaga, S.; Yamawaki, M.; Kishida, T.; Akiyama, K.; Maruyama, M. Antioxidant Activity of Butane Type Lignans, Secoisolariciresinol, Dihydroguaiaretic Acid, and 7,7′-Oxodihydroguaiaretic Acid. *Biosci. Biotechnol. Biochem.* **2008**, *72*, 2981–2986. [CrossRef] [PubMed]

67. Smeds, A.I.; Eklund, P.C.; Monogioudi, E.; Willför, S.M. Chemical characterization of polymerized products formed in the reactions of matairesinol and pinoresinol with the stable radical 2,2-diphenyl-1-picrylhydrazyl. *Holzforschung* **2012**, *66*, 283–294. [CrossRef]

68. Lei, M.; Hu, R.-J.; Wang, Y.-G. Mild and selective oxidation of alcohols to aldehydes and ketones using NaIO$_4$/TEMPO/NaBr system under acidic conditions. *Tetrahedron* **2006**, *62*, 8928–8932. [CrossRef]

69. Venkateswarlu, R.; Kamakshi, C.; Subhash, P.V.; Moinuddin, S.G.A.; Rama Sekhara Reddy, D.; Ward, R.S.; Pelter, A.; Gelbrich, T.; Hursthouse, M.B.; Coles, S.J.; et al. Transformations of lignans. Part 11: Oxidation of diphyllin with hypervalent iodine reagents and reductive reactions of a resulting 1-methoxy-1-aryl-4-oxonaphthalene lactone. *Tetrahedron* **2006**, *62*, 4463–4473. [CrossRef]

70. Pelter, A.; Ward, R.S.; Abd-El-Ghani, A. Preparation of dibenzocyclooctadiene lignans and spirodienones by hypervalent iodine oxidation of phenolic dibenzylbutyrolactones. *J. Chem. Soc. Perkin Trans. 1* **1992**, 2249–2251. [CrossRef]

71. Ward, R.S.; Pelter, A.; Abd-El-Ghani, A. Preparation of tetrahydrodibenzocyclooctene lignans and spirodienones by hypervalent iodine oxidation of phenolic dibenzylbutyrolactones. *Tetrahedron* **1996**, *52*, 1303–1336. [CrossRef]

72. Pelter, A.; Satchwell, P.; Ward, R.S.; Blake, K. Effective, direct biomimetic synthesis of dibenzocyclooctene lignans by hypervalent iodine oxidation of phenolic dibenzylbutyrolactones. *J. Chem. Soc. Perkin Trans. 1* **1995**, 2201–2202. [CrossRef]

73. Ward, R.S. Different strategies for the chemical synthesis of lignans. *Phytochem. Rev.* **2003**, *2*, 391–400. [CrossRef]

74. Albertson, A.K.F.; Lumb, J.P. A bio-inspired total synthesis of tetrahydrofuran lignans. *Angew. Chem. Int. Ed.* **2015**, *54*, 2204–2208. [CrossRef] [PubMed]

75. Bernini, R.; Barontini, M.; Mosesso, P.; Pepe, G.; Willför, S.M.; Sjöholm, R.E.; Eklund, P.C.; Saladino, R. A selective de-O-methylation of guaiacyl lignans to corresponding catechol derivatives by 2-iodoxybenzoic

acid (IBX). The role of the catechol moiety on the toxicity of lignans. *Org. Biomol. Chem.* **2009**, *7*, 2367–2377. [CrossRef] [PubMed]

76. Ozanne, A.; Pouységu, L.; Depernet, D.; François, B.; Quideau, S. A Stabilized Formulation of IBX (SIBX) for Safe Oxidation Reactions Including a New Oxidative Demethylation of Phenolic Methyl Aryl Ethers. *Org. Lett.* **2003**, *5*, 2903–2906. [CrossRef] [PubMed]

77. Gong, W.; RajanBabu, T.V. Conformation and reactivity in dibenzocyclooctadienes (DBCOD). A general approach to the total synthesis of fully substituted DBCOD lignans via borostannylative cyclization of α,ω-diynes. *Chem. Sci.* **2013**, *4*, 3979. [CrossRef]

78. Reddel, J.C.T.; Lutz, K.E.; Diagne, A.B.; Thomson, R.J. Stereocontrolled Syntheses of Tetralone- and Naphthyl-Type Lignans by a One-Pot Oxidative [3,3] Rearrangement/Friedel-Crafts Arylation. *Angew. Chem. Int. Ed.* **2014**, *53*, 1395–1398. [CrossRef] [PubMed]

79. Pandey, G.; Luckorse, S.; Budakoti, A.; Puranik, V.G. Synthesis of optically pure 2,3,4-trisubstituted tetrahydrofurans via a two-step sequential Michael-Evans aldol cyclization strategy: total synthesis of (+)-magnolone. *Tetrahedron Lett.* **2010**, *51*, 2975–2978. [CrossRef]

80. Tohma, H.; Kita, Y. Hypervalent Iodine Reagents for the Oxidation of Alcohols and Their Application to Complex Molecule Synthesis. *Adv. Synth. Catal.* **2004**, *346*, 111–124. [CrossRef]

81. Koprowski, M.; Bałczewski, P.; Owsianik, K.; Różycka-Sokołowska, E.; Marciniak, B. Total synthesis of (±)-epithuriferic acid methyl ester via Diels–Alder reaction. *Org. Biomol. Chem.* **2016**, *14*, 1822–1830. [CrossRef]

82. Jung, E.-K.; Dittrich, N.; Pilkington, L.I.; Rye, C.E.; Leung, E.; Barker, D. Synthesis of aza-derivatives of tetrahydrofuran lignan natural products. *Tetrahedron* **2015**, *71*, 9439–9456. [CrossRef]

83. Meresse, P.; Magiatis, P.; Bertounesque, E.; Monneret, C. Synthesis and antiproliferative activity of retroetoposide. *Bioorg. Med. Chem. Lett.* **2003**, *13*, 4107–4109. [CrossRef]

84. Meresse, P.; Monneret, C.; Bertounesque, E. Synthesis of podophyllotoxin analogues: δ-lactone-containing picropodophyllin, podophyllotoxin and 4'-demethyl-epipodophyllotoxin derivatives. *Tetrahedron* **2004**, *60*, 2657–2671. [CrossRef]

85. Roulland, E.; Magiatis, P.; Arimondo, P.; Bertounesque, E.; Monneret, C. Hemi-synthesis and Biological Activity of New Analogues of Podophyllotoxin. *Bioorg. Med. Chem.* **2002**, *10*, 3463–3471. [CrossRef]

86. Reynolds, A.J.; Scott, A.J.; Turner, C.I.; Sherburn, M.S. The Intramolecular Carboxyarylation Approach to Podophyllotoxin. *J. Am. Chem. Soc.* **2003**, *125*, 12108–12109. [CrossRef] [PubMed]

87. Monovich, L.G.; Le Huérou, Y.; Rönn, M.; Molander, G.A. Total synthesis of (-)-steganone utilizing a samarium(II) iodide promoted 8-endo ketyl-olefin cyclization. *J. Am. Chem. Soc.* **2000**, *122*, 52–57. [CrossRef]

88. Davidson, S.J.; Pearce, A.N.; Copp, B.R.; Barker, D. Total Synthesis of (−)-Bicubebin A, B, (+)-Bicubebin C and Structural Reassignment of (−)-*cis*-Cubebin. *Org. Lett.* **2017**, *19*, 5368–5371. [CrossRef]

89. Roulland, E.; Bertounesque, E.; Huel, C.; Monneret, C. Synthesis of picropodophyllin homolactone. *Tetrahedron Lett.* **2000**, *41*, 6769–6773. [CrossRef]

90. Adler, E.; Hernestam, S.; Boss, E.; Caglieris, A. Estimation of Phenolic Hydroxyl Groups in Lignin. I. Periodate Oxidation of Guaiacol Compounds. *Acta Chem. Scand.* **1955**, *9*, 319–334. [CrossRef]

91. LaLonde, R.T.; Ramdayal, F.D.; Sarko, A.; Yanai, K.; Zhang, M. Modes of Methyleneoxy Bridging and Their Effect on Tetrahydronaphthalene Lignan Cytotoxicity. *J. Med. Chem.* **2003**, *46*, 1180–1190. [CrossRef]

92. Ayres, D.C.; Ritchie, T.J. Lignans and related phenols. Part 18. The synthesis of quinones from podophyllotoxin and its analogues. *J. Chem. Soc. Perkin Trans. 1* **1988**, 2573–2578. [CrossRef]

93. Zhi, X.; Yu, X.; Yang, C.; Ding, G.; Chen, H.; Xu, H. Synthesis of 4β-acyloxypodophyllotoxin analogs modified in the C and E rings as insecticidal agents against Mythimna separata Walker. *Bioorg. Med. Chem. Lett.* **2014**, *24*, 765–772. [CrossRef]

94. Swigor, J.E.; Haynes, U.J. Synthesis of 4-o-(4,6-ethylidene-α-D-glucopyranosyl)-4'-demethyl-3'-O-[14C]methyl-4-epipodophyllotoxin. *J. Label. Compd. Radiopharm.* **1990**, *28*, 137–141. [CrossRef]

95. Nasveschuk, C.; Rovis, T. A Rapid Total Synthesis of (±)-Sylvone. *Synlett* **2008**, 126–128. [CrossRef]

96. Vitale, M.; Prestat, G.; Lopes, D.; Madec, D.; Kammerer, C.; Poli, G.; Girnita, L. New Picropodophyllin Analogs via Palladium-Catalyzed Allylic Alkylation–Hiyama Cross-Coupling Sequences. *J. Org. Chem.* **2008**, *73*, 5795–5805. [CrossRef] [PubMed]

97. Stadler, D.; Bach, T. Concise Stereoselective Synthesis of (−)-Podophyllotoxin by an Intermolecular Iron(III)-Catalyzed Friedel-Crafts Alkylation. *Angew. Chem. Int. Ed.* **2008**, *47*, 7557–7559. [CrossRef] [PubMed]

98. Mingoia, F.; Vitale, M.; Madec, D.; Prestat, G.; Poli, G. Pseudo-domino palladium-catalyzed allylic alkylation/Mizoroki–Heck coupling reaction: a key sequence toward (±)-podophyllotoxin. *Tetrahedron Lett.* **2008**, *49*, 760–763. [CrossRef]

99. Pullin, R.D.C.; Sellars, J.D.; Steel, P.G. Silenes in organic synthesis: a concise synthesis of (±)-epi-picropodophyllin. *Org. Biomol. Chem.* **2007**, *5*, 3201–3206. [CrossRef] [PubMed]

100. Akindele, T.; Marsden, S.P.; Cumming, J.G. Stereocontrolled Assembly of Tetrasubstituted Tetrahydrofurans: A Concise Synthesis of Virgatusin. *Org. Lett.* **2005**, *7*, 3685–3688. [CrossRef] [PubMed]

101. Tomioka, K.; Ishiguro, T.; Iitaka, Y.; Koga, K. Asymmetric total synthesis of natural (−)-and unnatural (+)-steganacin: Determination of the absolute configuration of natural antitumor steganacin. *Tetrahedron* **1984**, *40*, 1303–1312. [CrossRef]

102. Fan, J.-C.; Shang, Z.-C.; Liang, J.; Liu, X.-H.; Liu, Y. The oxidation of alcohols to aldehydes and ketones with *N* -bromosuccinimide in polyethylene glycol: an experimental and theoretical study. *J. Phys. Org. Chem.* **2008**, *21*, 945–953. [CrossRef]

103. Kende, A.S.; Liebeskind, L.S.; Mills, J.E.; Rutledge, P.S.; Curran, D.P. Oxidative aryl-benzyl coupling. A biomimetic entry to podophyllin lignan lactones. *J. Am. Chem. Soc.* **1977**, *99*, 7082–7083. [CrossRef]

104. Yamaguchi, H.; Nakajima, S.; Arimoto, M.; Tanoguchi, M.; Ishida, T.; Inoue, M. Studies on the constituents of the seeds of *Hernandia ovigera* L. IV. Syntheses of β-peltatin-A and -B methyl ethers from desoxypodophyllotoxin. *Chem. Pharm. Bull. (Tokyo)* **1984**, *32*, 1754–1760. [CrossRef]

105. Yamaguchi, H.; Arimoto, M.; Nakajima, S.; Tanoguchi, M.; Fukada, Y. Studies on the constituents of the seeds of *Hernandia ovigera* L. V Syntheses of epipodophyllotoxin and podophyllotoxin from desoxypodophyllotoxin. *Chem. Pharm. Bull. (Tokyo)* **1986**, *34*, 2056–2060. [CrossRef]

106. Ghosal, S.; Banerjee, S. Synthesis of retrochinensin; a new naturally occurring 4-aryl-2,3-naphthalide lignan. *J. Chem. Soc. Chem. Commun.* **1979**, *0*, 165–166. [CrossRef]

107. Banerji, J.; Das, B.; Chatterjee, A.; Shoolery, J.N. Gadain, a lignan from *jatropha gossypifolia*. *Phytochemistry* **1984**, *23*, 2323–2327. [CrossRef]

108. Mincione, E.; Sanetti, A.; Bernini, R.; Felici, M.; Bovicelli, P. Selective oxidation of lignan compounds by dimethyldioxirane. Diastereoselective opening of asarinin furo-furan skeleton. *Tetrahedron Lett.* **1998**, *39*, 8699–8702. [CrossRef]

109. MacLean, H.; Murakami, K. Lignans of western red cedar (*Thuja plicata* Donn) IV. Thujaplicutin and Thujaplicutin Methyl Ether. *Can. J. Chem.* **1966**, *44*, 1541–1545. [CrossRef]

110. MacLean, H.; Macdonald, B.F. Lignans of western red cedar (*Thuja plicata* Donn). VII. Dihydroxythujaplicatin. *Can. J. Chem.* **1967**, *45*, 739–740. [CrossRef]

111. Leopold, B.; Malmström, I.-L.; Holtermann, H.; Sörensen, J.S.; Sörensen, N.A. Nitrobenzene Oxidation of Compounds of the Lignan Type. *Acta Chem. Scand.* **1951**, *5*, 936–940. [CrossRef]

112. Wada, K.; Munakata, K. (−) Parabenzlactone, a new piperolignanolide isolated from nakai. *Tetrahedron Lett.* **1970**, *11*, 2017–2019. [CrossRef]

113. Yamauchi, S.; Sugahara, T.; Nakashima, Y.; Okada, A.; Akiyama, K.; Kishida, T.; Maruyama, M.; Masuda, T. Radical and Superoxide Scavenging Activities of Matairesinol and Oxidized Matairesinol. *Biosci. Biotechnol. Biochem.* **2006**, *70*, 1934–1940. [CrossRef]

114. Yamauchi, S.; Hayashi, Y.; Nakashima, Y.; Kirikihira, T.; Yamada, K.; Masuda, T. Effect of benzylic oxygen on the antioxidant activity of phenolic lignans. *J. Nat. Prod.* **2005**, *68*, 1459–1470. [CrossRef] [PubMed]

115. Brown, E.; Robin, J.P. A new route to the bis-benzocyclooctadiene lignan skeleton: Total syntheses of (±) picrostegane, (±) isopicrostegane and (±) isostegane. *Tetrahedron Lett.* **1978**, *19*, 3613–3616. [CrossRef]

116. Hajra, S.; Garai, S.; Hazra, S. Catalytic Enantioselective Synthesis of (−)-Podophyllotoxin. *Org. Lett.* **2017**, *19*, 6530–6533. [CrossRef] [PubMed]

117. Xiao, J.; Cong, X.-W.; Yang, G.-Z.; Wang, Y.-W.; Peng, Y. Divergent Asymmetric Syntheses of Podophyllotoxin and Related Family Members via Stereoselective Reductive Ni-Catalysis. *Org. Lett.* **2018**, *20*, 1651–1654. [CrossRef] [PubMed]

118. Ogawa, M.; Ogihara, Y. Studies on the constituents of *enkianthus nudipes*. V. A new lignan xyloside from the stems. *Chem. Pharm. Bull. (Tokyo)* **1976**, *24*, 2102–2105. [CrossRef]

119. Brusentsev, Y.; Sandberg, T.; Hotokka, M.; Sjöholm, R.; Eklund, P. Synthesis and structural analysis of sterically hindered chiral 1,4-diol ligands derived from the lignan hydroxymatairesinol. *Tetrahedron Lett.* **2013**, *54*, 1112–1115. [CrossRef]

120. Brusentsev, Y.; Hänninen, M.; Eklund, P. Synthesis of Sterically Hindered Chiral 1,4-Diols from Different Lignan-Based Backbones. *Synlett* **2013**, *24*, 2423–2426. [CrossRef]

121. Markus, H.; Plomp, A.J.; Sandberg, T.; Nieminen, V.; Bitter, J.H.; Murzin, D.Y. Dehydrogenation of hydroxymatairesinol to oxomatairesinol over carbon nanofibre-supported palladium catalysts. *J. Mol. Catal. A Chem.* **2007**, *274*, 42–49. [CrossRef]

122. Simakova, O.A.; Murzina, E.V.; Mäki-Arvela, P.; Leino, A.-R.; Campo, B.C.; Kordás, K.; Willför, S.M.; Salmi, T.; Murzin, D.Y. Oxidative dehydrogenation of a biomass derived lignan – Hydroxymatairesinol over heterogeneous gold catalysts. *J. Catal.* **2011**, *282*, 54–64. [CrossRef]

123. Simakova, O.A.; Smolentseva, E.; Estrada, M.; Murzina, E.V.; Beloshapkin, S.; Willför, S.M.; Simakov, A.V.; Murzin, D.Y. From woody biomass extractives to health-promoting substances: Selective oxidation of the lignan hydroxymatairesinol to oxomatairesinol over Au, Pd, and Au–Pd heterogeneous catalysts. *J. Catal.* **2012**, *291*, 95–103. [CrossRef]

124. Simakova, O.A.; Murzina, E.V.; Leino, A.-R.R.; Mäki-Arvela, P.; Willför, S.; Murzin, D.Y. Gold Catalysts for Selective Aerobic Oxidation of the Lignan Hydroxymatairesinol to Oxomatairesinol: Catalyst Deactivation and Regeneration. *Catal. Letters* **2012**, *142*, 1011–1019. [CrossRef]

125. Prestianni, A.; Ferrante, F.; Simakova, O.A.; Duca, D.; Murzin, D.Y. Oxygen-Assisted Hydroxymatairesinol Dehydrogenation: A Selective Secondary-Alcohol Oxidation over a Gold Catalyst. *Chem. Eur. J.* **2013**, *19*, 4577–4585. [CrossRef] [PubMed]

126. López, M.; Simakova, O.A.; Murzina, E.V.; Willför, S.M.; Prosvirin, I.; Simakov, A.; Murzin, D.Y. Gold particle size effect in biomass-derived lignan hydroxymatairesinol oxidation over Au/Al$_2$O$_3$ catalysts. *Appl. Catal. A Gen.* **2015**, *504*, 248–255. [CrossRef]

127. Khamlach, K.; Dhal, R.; Brown, E. Total syntheses of (−)-trachelogenin, (−)-nortrachelogenin and (+)-wikstromol. *Tetrahedron Lett.* **1989**, *30*, 2221–2224. [CrossRef]

128. Mäkelä, T.H.; Kaltia, S.A.; Wähälä, K.T.; Hase, T.A. α,β-Dibenzyl-γ-butyrolactone lignan alcohols: Total synthesis of (±)-7′-hydroxyenterolactone, (±)-7′-hydroxymatairesinol and (±)-8-hydroxyenterolactone. *Steroids* **2001**, *66*, 777–784. [CrossRef]

129. Chen, B.-C.; Zhou, P.; Davis, F.A.; Ciganek, E. α-Hydroxylation of Enolates and Silyl Enol Ethers. In *Organic Reactions*; John Wiley & Sons, Inc.: Hoboken, NJ, USA, 2003; pp. 1–356. ISBN 9780471264187.

130. Moritani, Y.; Fukushima, C.; Ukita, T.; Miyagishima, T.; Ohmizu, H.; Iwasaki, T. Stereoselective syntheses of cis- and trans-isomers of α-hydroxy-α,β-dibenzyl-γ-butyrolactone lignans: New syntheses of (±)-trachelogenin and (±)-guayadequiol. *J. Org. Chem.* **1996**, *61*, 6922–6930. [CrossRef] [PubMed]

131. Moritani, Y.; Ukita, T.; Hiramatsu, H.; Okamura, K.; Ohmizu, H.; Iwasaki, T. A highly stereoselective synthesis of 3-hydroxy-1-aryltetralin lignans based on the stereoselective hydroxylation of α,β-dibenzyl-γ-butyro-lactones: the first synthesis of (±)-cycloolivil. *J. Chem. Soc., Perkin Trans. 1* **1996**, *0*, 2747–2753. [CrossRef]

132. Kramer, B.; Fröhlich, R.; Waldvogel, S.R. Oxidative Coupling Reactions Mediated by MoCl$_5$ Leading to 2,2′-Cyclolignans: The Specific Role of HCl. *Eur. J. Org. Chem.* **2003**, *18*, 3549–3554. [CrossRef]

133. Kramer, B.; Averhoff, A.; Waldvogel, S.R. Highly selective formation of eight-membered-ring systems by oxidative cyclization with molybdenum pentachloride—An environmentally friendly and inexpensive access to 2,2′-cyclolignans. *Angew. Chem. Int. Ed.* **2002**, *41*, 2981–2982. [CrossRef]

134. Damon, R.E.; Schlessinger, R.H.; Blount, J.F. A short synthesis of (.+-.)-isostegane. *J. Org. Chem.* **1976**, *41*, 3772–3773. [CrossRef]

135. Landais, Y.; Lebrun, A.; Robin, J.P. Ruthenium(IV) tetrakis(trifluoroacetate), a new oxidizing agent. II. A new access to schizandrins skeleton using biaryl oxidative coupling of cis-substituted butanolides. *Tetrahedron Lett.* **1986**, *27*, 5377–5380. [CrossRef]

136. Landais, Y.; Robin, J.-P. Le tétrakis (trifluoroacétate) de ruthénium(IV), nouveau catalyseur a température ambiante du couplage biarylique oxydant non phénolique -première synthèse totale biomimétique du néoisostégane-. *Tetrahedron Lett.* **1986**, *27*, 1785–1788. [CrossRef]

137. Robin, J.P.; Landais, Y. Ruthenium(IV) dioxide in fluoro acid medium. An efficient biaryl phenol coupling process, exemplified with a biomimetic access to the skeleton of steganacin from presteganes. *J. Org. Chem.* **1988**, *53*, 224–226. [CrossRef]

138. Landais, Y.; Robin, J.-P.; Lebrun, A. Ruthenium dioxide in fluoro acid medium: I. A new agent in the biaryl oxidative coupling. Application to the synthesis of non phenolic bisbenzocyclooctadiene lignan lactones. *Tetrahedron* **1991**, *47*, 3787–3804. [CrossRef]

139. Robin, J.-P.; Landais, Y. Ruthenium dioxide in fluoro acid medium: II. Application to the formation of steganes skeleton by oxidative phenolic coupling. *Tetrahedron* **1992**, *48*, 819–830. [CrossRef]

140. Dhal, R.; Landais, Y.; Lebrun, A.; Lenain, V.; Robin, J.-P. Ruthenium dioxide in fluoro acid medium V. Application to the non phenolic oxidative coupling of diarylbutanes. Conformational studies of cis and trans deoxyschizandrins. *Tetrahedron* **1994**, *50*, 1153–1164. [CrossRef]

141. Pettit, G.R.; Meng, Y.; Gearing, R.P.; Herald, D.L.; Pettit, R.K.; Doubek, D.L.; Chapuis, J.-C.; Tackett, L.P. Antineoplastic Agents. 522. *Hernandia peltata* (Malaysia) and *Hernandia nymphaeifolia* (Republic of Maldives). *J. Nat. Prod.* **2004**, *67*, 214–220. [CrossRef]

142. Landais, Y.; Lebrun, A.; Lenain, V.; Robin, J.-P. Synthesis of diarylbutanes from cordigerines and reinvestigation of their oxidative couplings in deoxyschizandrins.—An unusual formation of phenyltetralin lignans -. *Tetrahedron Lett.* **1987**, *28*, 5161–5164. [CrossRef]

143. Cambie, R.; Craw, P.; Rutledge, P.; Woodgate, P. Oxidative Coupling of Lignans. III. Non-Phenolic Oxidative Coupling of Deoxypodorhizon and Related Compounds. *Aust. J. Chem.* **1988**, *41*, 897–918. [CrossRef]

144. Burden, J.; Cambie, R.; Craw, P.; Rutledge, P.; Woodgate, P. Oxidative Coupling of Lignans. IV. Monophenolic Oxidative Coupling. *Aust. J. Chem.* **1988**, *41*, 919–933. [CrossRef]

145. Planchenault, D.; Dhal, R.; Robin, J.-P. New agents of biaryl oxidative coupling in fluoro acid medium. VI. Application to the synthesis of phenolic bisbenzocyclooctadiene lignans. *Tetrahedron* **1995**, *51*, 1395–1404. [CrossRef]

146. Ward, R.S.; Hughes, D.D. Oxidative cyclisation of cis- and trans-2,3-dibenzylbutyrolactones using ruthenium tetra(trifluoroacetate). *Tetrahedron* **2001**, *57*, 4015–4022. [CrossRef]

147. Planchenault, D.; Dhal, R.; Robin, J.-P. Synthesis of non-phenolic bisbenzocyclooctadiene lignan lactones and aporphinic alkaloids, by oxidative coupling with new agents in fluoro acid medium. IV. *Tetrahedron* **1993**, *49*, 5823–5830. [CrossRef]

148. Saladino, R.; Fiani, C.; Belfiore, M.C.; Gualandi, G.; Penna, S.; Mosesso, P. Methyltrioxorhenium catalysed synthesis of highly oxidised aryltetralin lignans with anti-topoisomerase II and apoptogenic activities. *Bioorganic Med. Chem.* **2005**, *13*, 5949–5960. [CrossRef] [PubMed]

149. Saladino, R.; Fiani, C.; Crestini, C.; Argyropoulos, D.S.; Marini, S.; Coletta, M. An Efficient and Stereoselective Dearylation of Asarinin and Sesamin Tetrahydrofurofuran Lignans to Acuminatolide by Methyltrioxorhenium/H₂ O₂ and UHP Systems. *J. Nat. Prod.* **2007**, *70*, 39–42. [CrossRef]

150. Bernini, R.; Gualandi, G.; Crestini, C.; Barontini, M.; Belfiore, M.C.; Willför, S.; Eklund, P.; Saladino, R. A novel and efficient synthesis of highly oxidized lignans by a methyltrioxorhenium/hydrogen peroxide catalytic system. Studies on their apoptogenic and antioxidant activity. *Bioorg. Med. Chem.* **2009**, *17*, 5676–5682. [CrossRef] [PubMed]

151. Nishibe, S.; Chiba, M.; Sakushima, A.; Hisada, S.; Yamanouchi, S.; Takido, M.; Sankawa, U.; Sakakibara, A. Introduction of an alcoholic hydroxyl group into 2,3-dibenzylbutyrolactone lignans with oxidizing agents and carbon-13 nuclear magnetic resonance spectra of the oxidation products. *Chem. Pharm. Bull. (Tokyo)* **1980**, *28*, 850–860. [CrossRef]

152. Pohmakotr, M.; Pinsa, A.; Mophuang, T.; Tuchinda, P.; Prabpai, S.; Kongsaeree, P.; Reutrakul, V. General strategy for stereoselective synthesis of 1-substituted Exo,Endo-2,6-diaryl-3,7-dioxabicyclo[3.3.0]octanes: Total synthesis of (±)-gmelinol. *J. Org. Chem.* **2006**, *71*, 386–389. [CrossRef]

153. Takemoto, M.; Aoshima, Y.; Stoynov, N.; Kutney, J.P. Establishment of *Camellia sinensis* cell culture with high peroxidase activity and oxidative coupling reaction of dibenzylbutanolides. *Tetrahedron Lett.* **2002**, *43*, 6915–6917. [CrossRef]

154. Takemoto, M.; Fukuyo, A.; Aoshima, Y.; Tanaka, K. Synthesis of Lyoniresinol with Combined Utilization of Synthetic Chemistry and Biotechnological Methods. *Chem. Pharm. Bull. (Tokyo)* **2006**, *54*, 226–229. [CrossRef]

155. Kutney, J.P.; Xinyao, D.; Naidu, R.; Stoynov, N.M.; Takemoto, M. Biotransformation of Dibenzylbutanolides by Peroxidase Enzymes. Routes to the Podophyllotoxin Family. *Heterocycles* **1996**, *42*, 479–484. [CrossRef]

156. Yue, F.; Lu, F.; Ralph, S.; Ralph, J. Identification of 4–O–5-Units in Softwood Lignins via Definitive Lignin Models and NMR. *Biomacromolecules* **2016**, *17*, 1909–1920. [CrossRef] [PubMed]

157. Moinuddin, S.G.A.; Youn, B.; Bedgar, D.L.; Costa, M.A.; Helms, G.L.; Kang, C.; Davin, L.B.; Lewis, N.G. Secoisolariciresinol dehydrogenase: mode of catalysis and stereospecificity of hydride transfer in Podophyllum peltatum. *Org. Biomol. Chem.* **2006**, *4*, 808–816. [CrossRef] [PubMed]

158. Murata, J.; Ono, E.; Yoroizuka, S.; Toyonaga, H.; Shiraishi, A.; Mori, S.; Tera, M.; Azuma, T.; Nagano, A.J.; Nakayasu, M.; Mizutani, M.; Wakasugi, T.; Yamamoto, M.P.; Horikawa, M. Oxidative rearrangement of (+)-sesamin by CYP92B14 co-generates twin dietary lignans in sesame. *Nat. Commun.* **2017**, *8*, 2155. [CrossRef] [PubMed]

159. Vasilev, N.P.; Julsing, M.K.; Koulman, A.; Clarkson, C.; Woerdenbag, H.J.; Ionkova, I.; Bos, R.; Jaroszewski, J.W.; Kayser, O.; Quax, W.J. Bioconversion of deoxypodophyllotoxin into epipodophyllotoxin in E. coli using human cytochrome P450 3A4. *J. Biotechnol.* **2006**, *126*, 383–393. [CrossRef]

160. Krauss, A.S.; Taylor, W.C. Intramolecular Oxidative Coupling of Aromatic Compounds. III. Monophenolic Substrates. *Aust. J. Chem.* **1992**, *45*, 925–933. [CrossRef]

161. Fernandes, J.B.; Fraga, R.L.; Capelato, M.D.; Vierira, P.C.; Yoshida, M.; Kato, M.J. Synthesis of Dibenzocyclooctadienes by Anodic Oxidation. *Synth. Commun.* **1991**, *21*, 1331–1336. [CrossRef]

162. Kawamura, F.; Miyachi, M.; Kawai, S.; Ohashi, H. Photodiscoloration of western hemlock (Tsuga heterophylla) sapwood III Early stage of photodiscoloration reaction with lignans. *J. Wood Sci.* **1998**, *44*, 47–55. [CrossRef]

163. Litescu, S.C.; Eremia, S.A.V.; Tache, A.; Vasilescu, I.; Radu, G.-L. The Use of Oxygen Radical Absorbance Capacity (ORAC) and Trolox Equivalent Antioxidant Capacity (TEAC) Assays in the Assessment of Beverages' Antioxidant Properties. In *Processing and Impact on Antioxidants in Beverages*; Preedy, V., Ed.; Elsevier: Amsterdam, The Netherlands, 2014; pp. 245–251. ISBN 978-0-12-404738-9.

164. Golubev, V.A.; Rozantsev, E.G.; Neiman, M.B. Some reactions of free iminoxyl radicals with unpaired electron participation. *Izv. Akad. Nauk SSSR Ser. Khim.* **1965**, *11*, 1927–1936. [CrossRef]

165. Merbouh, N.; Bobbitt, J.M.; Brückner, C. Preparation of Tetramethylpiperidine-1-oxoammonium Salts And Their Use As Oxidants In Organic Chemistry. A Review. *Org. Prep. Proced. Int.* **2004**, *36*, 1–31. [CrossRef]

166. Tojo, G.; Fernández, M.I. TEMPO-Mediated Oxidations. In *Oxidation of Primary Alcohols to Carboxylic Acids*; Springer: New York, NY, USA, 2007; pp. 79–103.

167. Nandanwar, R.A.; Chaudhari, A.R.; Ekhe, J.D. Nitrobenzene Oxidation for Isolation of Value Added Products from Industrial Waste Lignin. *J. Chem. Biol. Phys. Sci.* **2016**, *6*, 501–513.

168. Pezzella, C.; Guarino, L.; Piscitelli, A. How to enjoy laccases. *Cell. Mol. Life Sci.* **2015**, *72*, 923–940. [CrossRef] [PubMed]

169. Huis, R.; Morreel, K.; Fliniaux, O.; Lucau-Danila, A.; Fénart, S.; Grec, S.; Neutelings, G.; Chabbert, B.; Mesnard, F.; Boerjan, W.; Hawkins, S. Natural hypolignification is associated with extensive oligolignol accumulation in flax stems. *Plant Physiol.* **2012**, *158*, 1893–1915. [CrossRef] [PubMed]

170. Glazunova, O.A.; Trushkin, N.A.; Moiseenko, K.V.; Filimonov, I.S.; Fedorova, T.V. Catalytic Efficiency of Basidiomycete Laccases: Redox Potential versus Substrate-Binding Pocket Structure. *Catalysts* **2018**, *8*, 152. [CrossRef]

171. Moya, R.; Saastamoinen, P.; Hernández, M.; Suurnäkki, A.; Arias, E.; Mattinen, M.-L. Reactivity of bacterial and fungal laccases with lignin under alkaline conditions. *Bioresour. Technol.* **2011**, *102*, 10006–10012. [CrossRef] [PubMed]

172. Buchert, J.; Mustranta, A.; Tamminen, T.; Spetz, P.; Holmbom, B. Modification of Spruce Lignans with Trametes hirsuta Laccase. *Holzforschung* **2002**, *56*, 579–584. [CrossRef]

173. Maijala, P.; Mattinen, M.-L.; Nousiainen, P.; Kontro, J.; Asikkala, J.; Sipilä, J.; Viikari, L. Action of fungal laccases on lignin model compounds in organic solvents. *J. Mol. Catal. B Enzym.* **2012**, *76*, 59–67. [CrossRef]

174. Mattinen, M.-L.; Maijala, P.; Nousiainen, P.; Smeds, A.; Kontro, J.; Sipilä, J.; Tamminen, T.; Willför, S.; Viikari, L. Oxidation of lignans and lignin model compounds by laccase in aqueous solvent systems. *J. Mol. Catal. B Enzym.* **2011**, *72*, 122–129. [CrossRef]

175. Mattinen, M.-L.; Struijs, K.; Suortti, T.; Mattila, I.; Kruus, K.; Willför, S.; Tamminen, T.; Vincken, J.-P. Modification of Lignans by Trameter Hirsuta Laccase. *BioResources* **2009**, *4*, 482–496. [CrossRef]

176. Frív, I.; Siverio, J.M.; González, C.; Trujillo, J.M.; Pérez, J.A. Purification of a new peroxidase catalysing the formation of lignan-type compounds. *Biochem. J.* **1991**, *273 (Pt 1)*, 109–113. [CrossRef]

177. Hazra, S.; Chattopadhyay, S. An overview of lignans with special reference to podophyllotoxin, a cytotoxic lignan. *Chem. Biol. Lett.* **2016**, *3*, 1–8.

178. Nakatsubo, T.; Mizutani, M.; Suzuki, S.; Hattori, T.; Umezawa, T. Characterization of *Arabidopsis thaliana* Pinoresinol Reductase, a New Type of Enzyme Involved in Lignan Biosynthesis. *J. Biol. Chem.* **2008**, *283*, 15550–15557. [CrossRef]

179. Xia, Z.Q.; Costa, M.A.; Pelissier, H.C.; Davin, L.B.; Lewis, N.G. Secoisolariciresinol dehydrogenase purification, cloning, and functional expression. Implications for human health protection. *J. Biol. Chem.* **2001**, *276*, 12614–12623. [CrossRef] [PubMed]

180. Van Uden, W.; Bos, J.A.; Boeke, G.M.; Woerdenbag, H.J.; Pras, N. The Large-Scale Isolation of Deoxypodophyllotoxin from Rhizomes of *Anthriscus sylvestris* Followed by Its Bioconversion into 5-Methoxypodophyllotoxin β- d -Glucoside by Cell Cultures of Linum flavum. *J. Nat. Prod.* **1997**, *60*, 401–403. [CrossRef]

181. Oliveira, T.L.S.; de Siqueira Leite, K.C.; de Macêdo, I.Y.L.; de Morais, S.R.; Costa, E.A.; de Paula, J.R.; de Souza Gil, E. Electrochemical Behavior and Antioxidant Activity of Hibalactone. *Int. J. Electrochem. Sci.* **2017**, *12*, 7596–7964. [CrossRef]

182. Silva, D.H.S.; Pereira, F.C.; Zanoni, M.V.B.; Yoshida, M. Lipophyllic antioxidants from *Iryanthera juruensis* fruits. *Phytochemistry* **2001**, *57*, 437–442. [CrossRef]

183. Yang, X.; Gao, M.; Hu, H.; Zhang, H. Electrochemical detection of honokiol and magnolol in traditional Chinese medicines using acetylene black nanoparticle-modified Electrode. *Phytochem. Anal.* **2011**, *22*, 291–295. [CrossRef] [PubMed]

184. Enache, T.A.; Amine, A.; Brett, C.M.A.; Oliveira-Brett, A.M. Virgin olive oil ortho-phenols—electroanalytical quantification. *Talanta* **2013**, *105*, 179–186. [CrossRef]

185. Natale, A.; Nardiello, D.; Palermo, C.; Muscarella, M.; Quinto, M.; Centonze, D. Development of an analytical method for the determination of polyphenolic compounds in vegetable origin samples by liquid chromatography and pulsed amperometric detection at a glassy carbon electrode. *J. Chromatogr. A* **2015**, *1420*, 66–73. [CrossRef]

186. Fernández, E.; Vidal, L.; Canals, A. Rapid determination of hydrophilic phenols in olive oil by vortex-assisted reversed-phase dispersive liquid-liquid microextraction and screen-printed carbon electrodes. *Talanta* **2018**, *181*, 44–51. [CrossRef]

187. Apetrei, I.M.; Apetrei, C. Voltammetric e-tongue for the quantification of total polyphenol content in olive oils. *Food Res. Int.* **2013**, *54*, 2075–2082. [CrossRef]

188. Przygodzka, M.; Zielińska, D.; Ciesarová, Z.; Kukurová, K.; Zieliński, H. Comparison of methods for evaluation of the antioxidant capacity and phenolic compounds in common spices. *LWT - Food Sci. Technol.* **2014**, *58*, 321–326. [CrossRef]

189. Rivas Romero, M.P.; Estévez Brito, R.; Rodríguez Mellado, J.M.; González-Rodríguez, J.; Ruiz Montoya, M.; Rodríguez-Amaro, R. Exploring the relation between composition of extracts of healthy foods and their antioxidant capacities determined by electrochemical and spectrophotometrical methods. *LWT* **2018**, *95*, 157–166. [CrossRef]

190. Radi, A.-E.; Abd-Elghany, N.; Wahdan, T. Electrochemical study of the antineoplastic agent etoposide at carbon paste electrode and its determination in spiked human serum by differential pulse voltammetry. *Chem. Pharm. Bull. (Tokyo)* **2007**, *55*, 1379–1382. [CrossRef] [PubMed]

191. Hassan, E.M.; Khamis, E.F.; El-Kimary, E.I.; Barary, M.A. Development of a differential pulse voltammetric method for the determination of Silymarin/Vitamin E acetate mixture in pharmaceuticals. *Talanta* **2008**, *74*, 773–778. [CrossRef] [PubMed]

192. El-Desoky, H.S.; Ghoneim, M.M. Stripping voltammetric determination of silymarin in formulations and human blood utilizing bare and modified carbon paste electrodes. *Talanta* **2011**, *84*, 223–234. [CrossRef] [PubMed]

193. Mpanza, T.; Sabela, M.I.; Mathenjwa, S.S.; Kanchi, S.; Bisetty, K. Electrochemical Determination of Capsaicin and Silymarin Using a Glassy Carbon Electrode Modified by Gold Nanoparticle Decorated Multiwalled Carbon Nanotubes. *Anal. Lett.* **2014**, *47*, 2813–2828. [CrossRef]

194. Gažák, R.; Svobodová, A.; Psotová, J.; Sedmera, P.; Přikrylová, V.; Walterová, D.; Křen, V. Oxidised derivatives of silybin and their antiradical and antioxidant activity. *Bioorganic Med. Chem.* **2004**, *12*, 5677–5687. [CrossRef]

195. Kosina, P.; Kren, V.; Gebhardt, R.; Grambal, F.; Ulrichova, J.; Walterova, D. Antioxidant properties of silybin glycosides. *Phytother Res* **2002**, *16*, 33–39. [CrossRef]

molecules

Article

Lignans from *Bursera fagaroides* Affect In Vivo Cell Behavior by Disturbing the Tubulin Cytoskeleton in Zebrafish Embryos

Mayra Antúnez-Mojica [1], Andrés M. Rojas-Sepúlveda [2], Mario A. Mendieta-Serrano [3],
Leticia Gonzalez-Maya [4], Silvia Marquina [5], Enrique Salas-Vidal [3,*] and Laura Alvarez [5,*]

[1] CONACYT-Centro de Investigaciones Químicas-IICBA, Universidad Autónoma del Estado de Morelos, Cuernavaca 62209, Morelos, Mexico; myam@uaem.mx

[2] Facultad de Ciencias, Universidad Antonio Nariño, Armenia Quindío 63003, Colombia; andres.rojas@uan.edu.co

[3] Departamento de Genética del Desarrollo y Fisiología Molecular, Instituto de Biotecnología, Universidad Nacional Autónoma de México, Cuernavaca 62210, Morelos, Mexico; mbimams@nus.edu.sg

[4] Facultad de Farmacia, Universidad Autónoma del Estado de Morelos, Cuernavaca 62209, Morelos, Mexico; letymaya@uaem.mx

[5] Centro de Investigaciones Químicas-IICBA, Universidad Autónoma del Estado de Morelos, Cuernavaca 62209, Morelos, Mexico; smarquina@uaem.mx

* Correspondence: esalas@ibt.unam.mx (E.S.-V.); lalvarez@uaem.mx (L.A.); Tel./Fax: +52-77-7329-7997 (L.A.)

Academic Editor: David Barker

Received: 27 November 2018; Accepted: 17 December 2018; Published: 20 December 2018

Abstract: By using a zebrafish embryo model to guide the chromatographic fractionation of antimitotic secondary metabolites, seven podophyllotoxin-type lignans were isolated from a hydroalcoholic extract obtained from the steam bark of *Bursera fagaroides*. The compounds were identified as podophyllotoxin (**1**), β-peltatin-A-methylether (**2**), 5′-desmethoxy-β-peltatin-A-methylether (**3**), desmethoxy-yatein (**4**), desoxypodophyllotoxin (**5**), burseranin (**6**), and acetyl podophyllotoxin (**7**). The biological effects on mitosis, cell migration, and microtubule cytoskeleton remodeling of lignans **1–7** were further evaluated in zebrafish embryos by whole-mount immunolocalization of the mitotic marker phospho-histone H3 and by a tubulin antibody. We found that lignans **1**, **2**, **4**, and **7** induced mitotic arrest, delayed cell migration, and disrupted the microtubule cytoskeleton in zebrafish embryos. Furthermore, microtubule cytoskeleton destabilization was observed also in PC3 cells, except for **7**. Therefore, these results demonstrate that the cytotoxic activity of **1**, **2**, and **4** is mediated by their microtubule-destabilizing activity. In general, the in vivo and in vitro models here used displayed equivalent mitotic effects, which allows us to conclude that the zebrafish model can be a fast and cheap in vivo model that can be used to identify antimitotic natural products through bioassay-guided fractionation.

Keywords: *Bursera fagaroides*; podophyllotoxin-type lignans; cell cycle; cell migration; epiboly; microtubules; F-actin; cancer; lignans

1. Introduction

Lignans are a big family of secondary metabolites biosynthesized in plants through the shikimic acid pathway that represent an important class of compounds in cancer research studies [1,2]. One of the most representative groups of lignans are the aryltetralin lignans [3]. Into this group of compounds, podophyllotoxin (**1**) is the most known because of its use as an effective antiviral for genital warts treatment. Podophyllotoxin inhibits tubulin polymerization inducing the mitotic arrest of cancer cells [4]. In vitro, it binds to the colchicine binding site [5,6]. Because of the unacceptable gastrointestinal toxicity, many semisynthetic podophyllotoxin derivatives were developed and tested for anticancer activity [7]. Etoposide and teniposide are glycosylated derivatives of podophyllotoxin developed in the 1970s. These derivatives are used in many clinical chemotherapeutic regimens, because of their easy administration and good toleration. However, the action mechanism through which they exert their effect is different from that of podophyllotoxin (**1**), since these semisynthetic drugs break DNA by the interaction with the topoisomerase II enzyme [1].

Our research team has characterized an interesting traditional Mexican medicinal plant named *Bursera fagaroides*, which is popularly used as an antitumoral. Our previous studies described the isolation of seven aryltetralin-type lignans including podophyllotoxin (**1**), β-peltatin-A-methylether (**2**), 5′-desmethoxy-β-peltatin-A-methylether (**3**), desmethoxy-yatein (**4**), desoxypodophyllotoxin (**5**), burseranin (**6**), and acetyl podophyllotoxin (**7**) [8], as well as three aryldihydronaphtalene-type lignans, i.e., 7′,8′-dehydropodophyllotoxin (**8**), 7′,8′-dehydroacethyl podophyllotoxin (**9**), and 7′,8′-dehydro *trans-p*-cumaroylpodophyllotoxin (**10**) (Figure 1) [9]. The cytotoxic evaluation of these compounds indicated that all lignans showed potent cytotoxic activity against a panel of four human cancer cell lines (KB, HF-6, MCF-7, and PC-3), being selective against prostate cancer cells (PC-3).

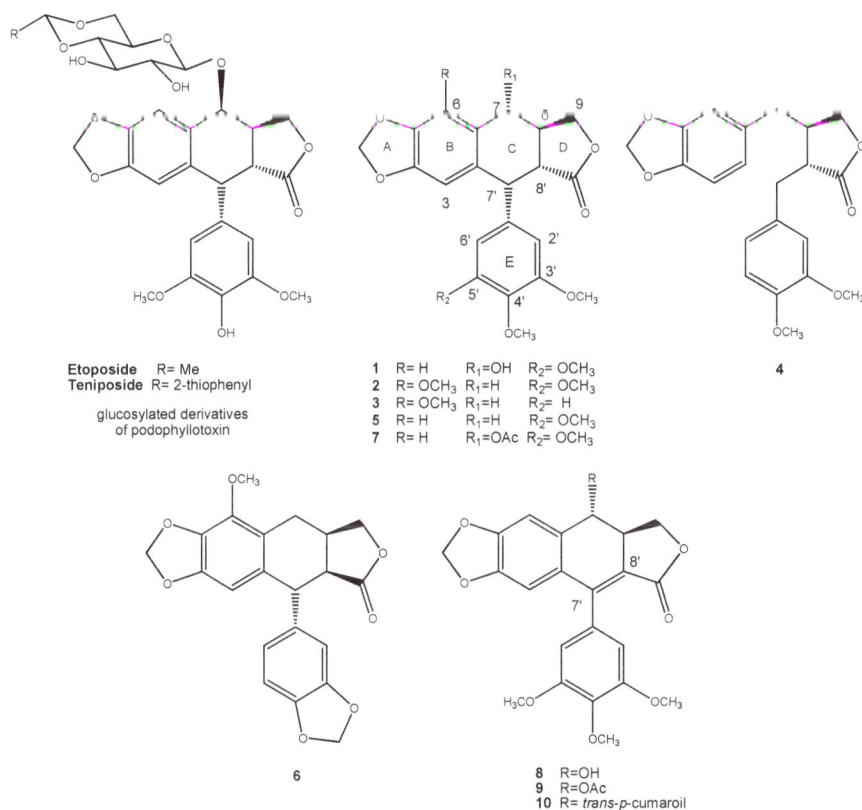

Figure 1. Glycosylated derivatives of podophyllotoxin (etoposide and teniposide) and podophyllotoxin-type lignans **1–10**: podophyllotoxin (**1**), β-peltatin-A-methylether (**2**), 5′-desmethoxy-β-peltatin-A-methylether (**3**), desmethoxy-yatein (**4**), desoxypodophyllotoxin (**5**), burseranin (**6**), acetyl podophyllotoxin (**7**), 7′,8′-dehydropodophyllotoxin (**8**), 7′,8′-dehydroacethyl podophyllotoxin (**9**), and 7′,8′-dehydro *trans-p*-cumaroylpodophyllotoxin (**10**), isolated from the steam bark of *Bursera fagaroides*.

Other studies proved that lignans **3** and **7** are more potent than the semisynthetic podophyllotoxin drug etoposide (14.1 and 7.6 µg/mL, respectively) in the human breast cancer cell line BT-549 [10]. Additionally, compounds **3**, **6**, and **7** showed anti-giardial activity in vitro, provoking morphological alterations in *Giardia* parasite [11]. In a deeper molecular recognition study of lignans **3**, **7**, and **9**, the action mechanism consisting in disrupting microtubule networks and cell cycle arrest in G2/M phase in the human carcinoma lung cell line A549 was demonstrated. Also, we established that these compounds interact with the tubulin colchicine binding site with a high binding constant (K_b) [6].

Recently, in vivo studies performed with the aryldihydronaphtalene-type lignans (**8–10**) in the developing zebrafish embryo model, demonstrated that these compounds promote mitotic arrest, delay cell migration and disrupt microtubules [9].

Continuing the study of this important medicinal plant, in this work, we analyzed the hydroalcoholic extract of *B. fagaroides* steam bark, employing developing zebrafish embryos as a model with the aim to identify compounds able to affect the tubulin cytoskeleton and to find a faster and easier biological model to discover new destabilizing anticancer drugs.

2. Results and Discussion

2.1. In Vivo Analysis of Mitotic Arrest in Zebrafish Embryos

Histone H3 phosphorylated at serine 10 (H3S10ph) has long been used to identify mitotic cell nuclei in cultured cell lines [12] as well as in whole embryos such as those of *Xenopus* and zebrafish [13–15]. In particular, 24 h-old zebrafish embryos have been used to screen the effects of small chemical molecules on the cell cycle and the induction of mitotic arrest using whole-mount immunohistochemistry of H3S10ph [16].

First, we evaluated the effect of a *B. fagaroides* hydroalcoholic extract (HA) on mitotic arrest in whole 24 h post fertilization (hpf) zebrafish embryos. Nocodazole and aphidicolin were used as controls of inhibition of DNA polymerase and tubulin depolymerization, respectively [17,18]. The aphidicolin-treated embryos exhibited considerably fewer H3S10ph-positive nuclei compared with the control embryos incubated with the vehicle (DMSO). In contrast, nocodazole-treated embryos appeared to have substantially more H3S10ph-positive nuclei compared with the control embryos (Figure S1).

The HA extract of *B. fagaroides* did not affect the number of mitotic cells, as measured by the number of H3S10ph-positive nuclei with respect to that of the control (Figure 2, Figure S2, and Table S1). Nevertheless, in our previous study, this extract showed potent in vitro cytotoxic activity and strong in vivo antitumor activity against L5178Y lymphoma in mice [8]. Therefore, we considered that the lack of effect of the HA extract in zebrafish embryos could be due to poor bioavailability because of the low concentration of the active compounds present in the extract. For this reason, we decided to test the effect of its chromatographic fractions. Therefore, successive chromatographic fractionations of the HA extract were performed, which afforded four fractions, two of which (F-1 and F-2) were analyzed (Figure S2). These two fractions were the most active when the fractionation was guided by the cytotoxic activity. In contrast, F-3 and F-4 were not evaluated in this work because of their lack of cytotoxic activity reported by Rojas et al. [8]. Fractions F-1 and F-2 increased the number of H3S10ph-positive nuclei by 4.08- and 4.48-fold, respectively (Figure 2 and Table S1). The treated embryos that exhibited increased H3S10ph levels also displayed abnormal morphology; these embryos were curved ventrally (Figure S2), like the embryos treated with nocodazole.

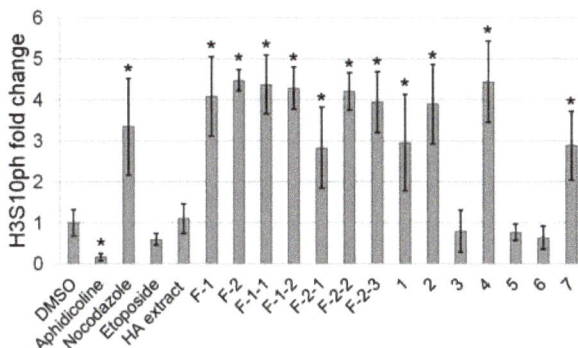

Figure 2. Staining of mitotic nuclei by phospho-histone H3 (H3S10ph) in zebrafish embryos treated with the hydrocoholic extract fractions and the pure compounds: Podophyllotoxin (**1**), β-peltatin-A-methylether (**2**), 5′-desmethoxy-β-peltatin-A-methylether (**3**), desmethoxy-yatein (**4**), desoxypodophyllotoxin (**5**), burseranin (**6**), and acetyl podophyllotoxin (**7**). The fold changes were determined in comparison with the control (DMSO-treated). Values represent means ± s.d. * Significant differences ($p < 0.001$). These results are also shown in Table S1.

Chromatographic purification of fractions F-1 and F-2 afforded seven lignans identified as podophyllotoxin (**1**), β-peltatin-A-methylether (**2**), 5′-desmethoxy-β-peltatin-A-methylether (**3**), desmethoxy-yatein (**4**), desoxypodophyllotoxin (**5**), burseranin (**6**), and acetyl podophyllotoxin (**7**) by direct comparison with authenticated samples obtained in previous work (Figure 1) [8].

Next, we analyzed the effect of all the isolated podophyllotoxin-type lignans (**1**–**7**) on the H3S10ph marker and morphology in zebrafish embryos (Figure 3). The results showed that compounds **1**, **2**, **4**, and **7** significantly increased the H3S10ph levels, by 2.95-, 3.89-, 4.44- and 2.88-fold, respectively (Figure 2 and Table S1).

In contrast, compounds **3**, **5**, and **6** did not affect the number of mitotic cells or the circularity of the embryos. Zebrafish embryos treated with podophyllotoxin (**1**) exhibited H3S10ph levels 2.95-fold higher than those of the control embryos (Figure 2). Many semisynthetic derivatives of podophyllotoxin with anticancer activity have been described, such as etoposide. We asked whether etoposide had any effect in this model. The results revealed that etoposide appeared to slightly decrease the number of H3S10ph-positive nuclei compared with control embryos, but this difference was not statistically significant (Figure 2 and Table S1).

This difference could be attributed to the mechanism of action by which these compounds act. It is well known that etoposide interacts with DNA and inhibits DNA-topoisomerase II, while podophyllotoxin (**1**) and its congeners (**3**, **6**, and **7**) inhibit tubulin assembly [6].

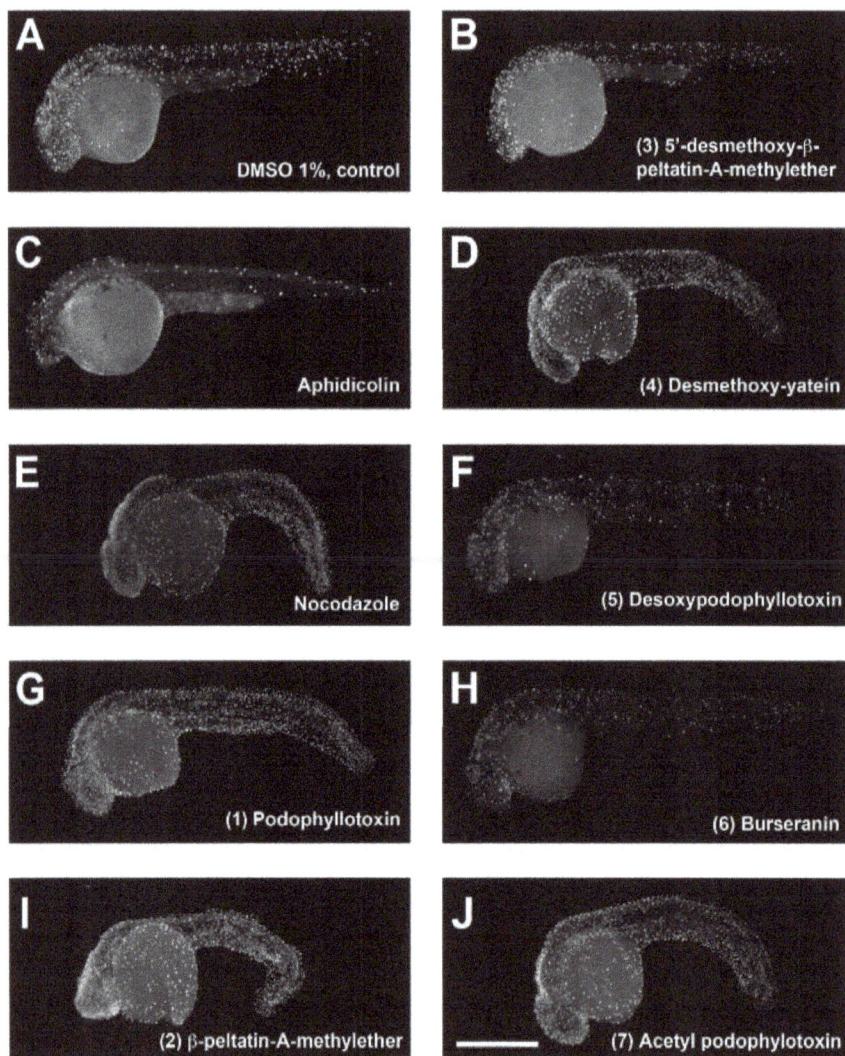

Figure 3. Whole-mount immunolocalization of H3S10ph in zebrafish embryos treated with *B. fagaroides* lignans. Wild-type 24 h post-fertilization zebrafish embryos were immunostained for H3S10ph after a 6 h treatment with different compounds. (**A**) Dimethyl sulfoxide (control), (**B**) Aphidicolin, (**C**) Nocodazole, (**D**) Podophyllotoxin (**1**), (**E**) β-peltatin-A-methylether (**2**), (**F**) 5′-desmethoxy-β-peltatin-A-methylether (**3**), (**G**) Desmethoxy-yatein (**4**), (**H**) Desoxypodophyllotoxin (**5**), (**I**) Burseranin (**6**), (**J**) Acetyl podophylotoxin (**7**). The images were visualized by confocal microscopy. Scale bar, 500 μm.

2.2. Characterization of Compounds that Affect Cell Migration in the Zebrafish Model

Previous studies have shown that when the microtubule cytoskeleton is compromised in early zebrafish embryos by ultraviolet light radiation or by nocodazole treatment, epiboly migration is severely affected [19,20]. The effect on epiboly migration has been used to screen libraries of compounds that destabilize microtubules [21]. By using this approach, we analyzed the effect of pure lignans (**1**–**7**) on epiboly migration in zebrafish embryos. Embryo treatment started at the sphere stage, and all treatment groups were fixed at the same time point when the control DMSO-treated embryos reached 90% of epiboly. Compounds **1**, **2**, **4**, and **7** induced epiboly delay and larger blastoderm cells and nuclei, as evidenced by phalloidin alexa and SYTOX green staining of the embryos (Figure 4). These effects show similarities to those of nocodazol treatment, although this destabilizing microtubule compound showed a more penetrant phenotype. Compounds **3** and **5** delayed epiboly too but did not show effects on the size of blastoderm cells or in their nuclei (Figure 4).

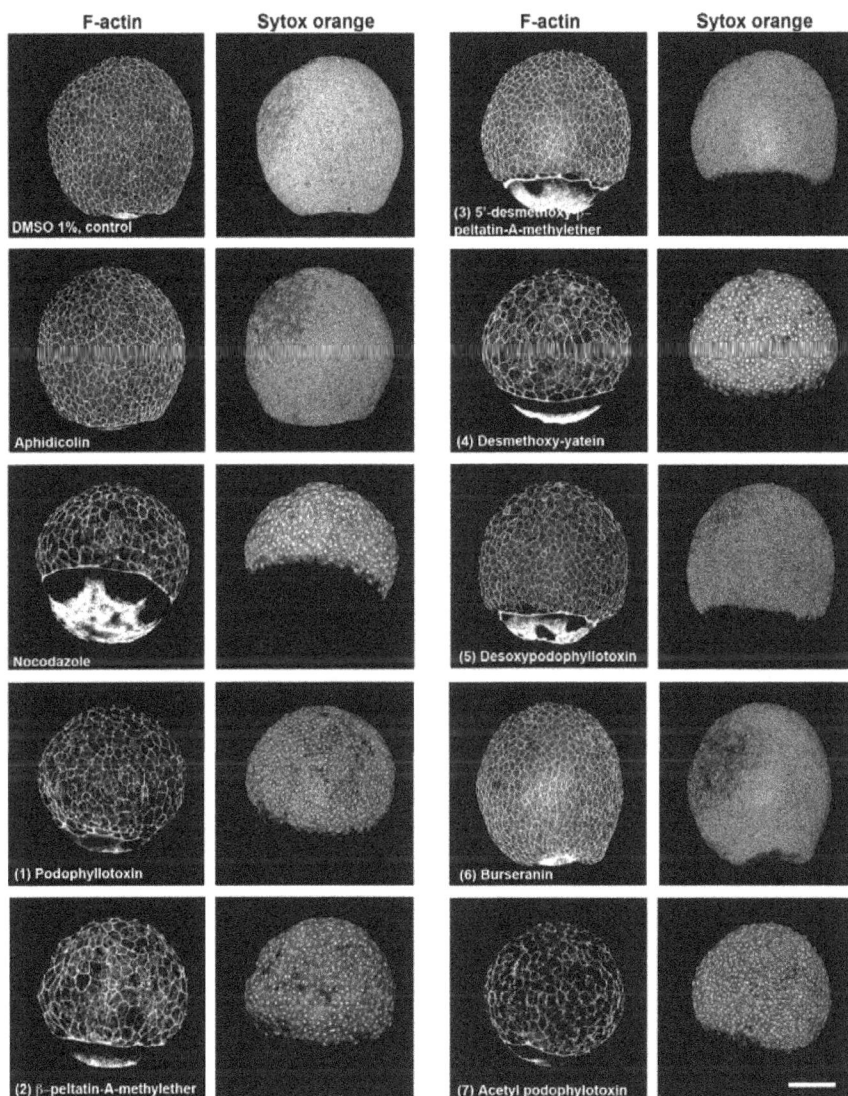

Figure 4. Immunolocalization of actin filaments and nuclei in zebrafish embryos after treatment with lignans (**1**)–(**7**), dimethyl sulfoxide (negative control), Aphidicolin, and Nocodazole (positive controls). Sphere-stage zebrafish embryos were treated with different compounds until control embryos reached 90% epiboly, fixed, processed for actin and nuclei staining, and visualized by confocal microscopy. Scale bar, 250 μm.

2.3. In Vivo and In Vitro Analysis of Compounds that Affect Microtubules

To corroborate that these lignans possess microtubule destabilizing activity, we decided to perform a tubulin immunolocalization assay in embryos treated with lignans **1**–**7**. Starting from the sphere stage, embryos in all treatment groups were fixed when they reached 50% of epiboly. This stage was chosen to perform tubulin immunolocalization since the yolk cells present large and conspicuous microtubules that are readily visualized by confocal microscopy [22].

The results showed that, again, compounds **1**, **2**, **4**, and **7** induced microtubules destabilization similar to nocodazol (Figure 5). On the other hand, zebrafish embryos treated with compound **3** showed some degree of disorganized microtubules, and compound **5**-treated embryos presented less abundant arrays of microtubules, although neither treatment showed the degree of microtubule destabilization found in embryos treated with compounds **1**, **2**, **4**, and **7** (Figure 5).

Finally, we asked if the identified lignans also destabilize the microtubule cytoskeleton in cancer cells when exposed in vitro to these compounds. Specifically, we tested the effect of the pure lignans **1**–**7** in the PC3 human prostate cancer cell line, as this was the most susceptible cancer cell line when evaluated by Rojas et, al. [8]. We found that **1**, **2**, and **4** destabilized microtubules; in contrast **3**, **5**, **6**, and **7** did not affect the microtubule cytoskeleton (Figure 6).

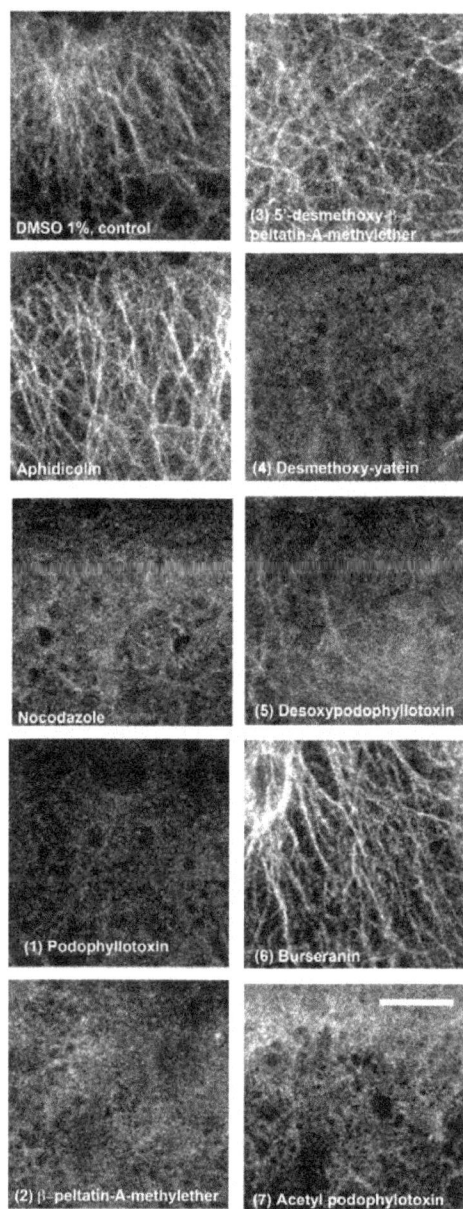

Figure 5. Whole-mount immunolocalization of yolk cell microtubules in zebrafish embryos after treatment with lignans (**1**)–(**7**). Sphere-stage zebrafish embryos were treated with different compounds until all embryos reached 50% epiboly; they were then fixed, processed for microtubule fluorescent immunolocalization, and visualized by confocal microscopy. Scale bar, 25 μm.

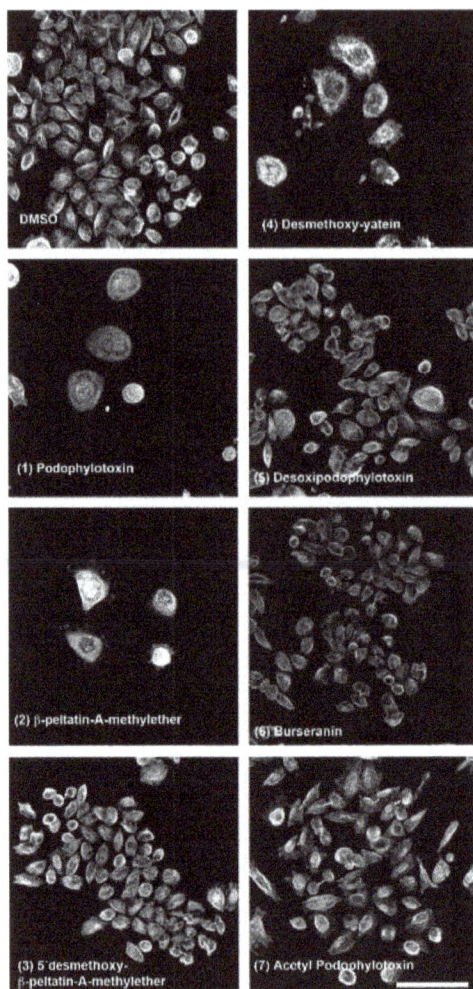

Figure 6. Whole-mount immunolocalization of cancer PC3 cells microtubules as viewed by confocal microscopy after treatment with DMSO (control) and lignans (**1**)–(**7**). Scale bar, 100 μm.

These results indicated that six of the seven *B. fagaroides* lignans showed equivalent effects to those observed in whole zebrafish embryos. Only acetyl podophyllotoxin (**7**) presented a dissimilar effect in both systems, severely affecting microtubules in zebrafish embryos (Figure 4) without showing any effect on the microtubule cytoskeleton of PC3 cells in the assay (Figure 6).

In general, lignans with major conformational mobility, such as **1**, **2**, **4**, and **7**, isolated in this work, showed a better in vivo effect than the planar lignans **8**–**10** isolated previously [9].

3. Materials and Methods

3.1. Plant Material, Extraction, and Isolation

The bark of *B. fagaroides* (Kunth) Engl., was collected in the village of Capula between Zacapu and Quiroga, Michoacán, Mexico. Its identification was made in the herbarium of the Instituto Mexicano del Seguro Social (registration number-12 051 IMSSM) and the Institute of Botany, University of Guadalajara (IBUG-140 748). The voucher specimens were deposited in both herbariums.

The bark of *B. fagaroides* (1700 g) was dried at room temperature, and the dry material was extracted by triplicate with hexane, followed by MeOH/H$_2$O 70:30 to yield 40.5 g hydroalcoholic extract (HA). Then, the extract was fractionated by percolation, eluting with mixtures of n-hexane/acetone/MeOH with increasing polarity, to yield: F-1, 1.09 g (8:2:0), F-2, 1.45 g (6:4:0), F-3, 16 g (0:100:0), and F-4, 18.5 g (0:1:1).

The fractionation of F-1 led to the purification of podophyllotoxin (**1**) (4.92 mg), β-peltatin-A-methylether (**2**) (3.5 mg), and 5′-desmethoxy-β-peltatin-A-methylether (**3**) (7.9 mg), while desmethoxy-yatein (**4**) (8.7 mg), desoxypodophyllotoxin (**5**) (4.3 mg), burseranin (**6**) (15 mg), and acetyl podophyllotoxin (**7**) (6.2 mg) were obtained from F-2. This procedure was performed by RP–HPLC as described by Rojas-Sepulveda et al. [8]. The pure lignans were characterized by direct comparison (¹H NMR, co-TLC, Table S2) with authenticated samples available in our laboratory [8].

3.2. Fish Maintenance and Strains

Zebrafish (*Danio rerio*) embryos were obtained from natural crosses from wild-type and AB–TU–WIK hybrid lines. Embryo stages were determined by morphological criteria [23]. Zebrafish were handled in compliance with the local animal welfare regulations, EU Directive 2010/63/EU indications [24], and zebrafish use was approved by the Ethics Committee of Instituto de Biotecnología, UNAM.

3.3. Chemical Treatment of Zebrafish Embryos

Treatments were performed as previously described [9,16]. In brief, the compounds for the chemical treatments were diluted in anhydrous DMSO (276855, Sigma-Aldrich, Saint Louis, MO, USA). Plant extracts and pure compounds for screening were tested at a standard final concentration of 200 μg/mL in water. The control compounds were aphidicolin (10 μg/mL, A0781, Sigma-Aldrich, Saint Louis, MO, USA) and nocodazole (10 μg/mL, M1404, Sigma-Aldrich, Saint Louis, MO, USA). Etoposide was obtained from Sigma-Aldrich (E1383). To test the effect on mitotic arrest, 10 zebrafish embryos of 24 h of postfertilization age, for each treatment, were incubated at 28 °C in egg water (60 μg/mL of "Instant Ocean" sea salts in distilled water) for 6 h in 48-well plates [25]. Three microliters of each compound stock solution were added to a total volume of 300 μL at the beginning of the incubation. Control embryos were treated with DMSO alone at 1% final concentration (v/v) and processed for immunohistochemistry as described below. To test the effect on cell migration and on F-actin, zebrafish embryos at sphere stage were incubated until the control embryos reached 90% epiboly to bud stage. Finally, to test the effect on the tubulin cytoskeleton, zebrafish embryos at sphere stage were incubated until the control embryos reached 60% epiboly. Embryos in all treatments were stage-matched, fixed, and processed as described below.

3.4. Immunofluorescense and Immunohistochemistry

Whole-mount fluorescent immunostaining against the mitotic marker serine 10 phospho-histone H3 (H3S10ph) in zebrafish embryos was used as previously described [13]. For immunohistochemistry, embryos were processed as for fluorescent immunostaining except that after overnight incubation with the primary anti-phospho-histone H3 antibody, the embryos were processed with R.T.U. Vectastain Universal Quick Kit (PK-7800, Vector, Burlingame, CA, USA) and developed in diaminobenzidine with nickel contrast substrate (SK-4100, Vector, Burlingame, CA, USA). For the analysis of the effects on microtubules, the embryos were incubated with the different compounds, fixed at 60% epiboly, and processed as described.

The primary antibody used was a mouse anti-α-tubulin monoclonal antibody (T9026, Sigma-Aldrich, Saint Louis, MO, USA), and a goat anti-mouse coupled to Alexa Fluor 647 (A21235, Molecular Probes, Eugene, OR, USA) was used as a secondary antibody. For the analysis of cell migration, all embryos were fixed at the same time when the controls reached the tail bud stage. Afterwards, they were washed three times, stained for 1 h at room temperature with SYTOX orange (S11368, Molecular Probes, Eugene, OR, USA) diluted 1:2000 in blocking buffer to visualize DNA and nuclei, and counter-stained with phalloidin Alexa Fluor 488 (A12379, Molecular Probes, Eugene, OR, USA). The embryos were then washed three times, mounted, and imaged by confocal microscopy.

3.5. Fluorescence Microscopy

Th fluorescent signals corresponding to H3S10ph-positive cells in whole zebrafish embryos where imaged by fluorescence microscopy with a 5× objective, 0.15 N.A. Plan-Neofluar under a Zeiss Axioscop Microscope. Image stack (10 to 14 images per embryo) corresponding to different focal planes were acquired with a CoolSNAP cfd CCD camera (Roper Scientific, Tucson Arizona, AZ, USA), controlled by MicroManager 1.5 software (NIH, Bethesda, MD USA). Image resolution was 1392 × 1040, and stack images were saved in 8-bit multi-image TIFF file.

3.6. Confocal Laser Scanning Microscopy

Zebrafish embryos stained with the specified fluorescent dyes were visualized on a Zeiss LSM 510 META confocal inverted microscope (Carl Zeiss, Jena, Germany) with a Plan-Neofluar 10× (0.3 N.A.) objective, a Plan-Neofluar 20× (0.5 N.A.) objective, or a Plan-Neofluar 40× objective (0.75 N.A.). Double-stained embryos (SYTOX orange and Phalloidin Alexa Fluor 488) were simultaneously excited at 488 nm and 543 nm and visualized on a FluoView FV1000 confocal microscope coupled to an up-right BX61WI Olympus microscope with a Plan FLN 10× (0.3 N.A.) objective or a Plan FLN 20× (0.75 N.A.) objective. The pinhole aperture was maintained at 105. Serial optical sections were obtained with a z-step of 8 μm. The images were processed with the public domain software ImageJ [26] the LSM Image Browser from Zeiss and Adobe Photoshop. To quantify the H3S10ph-positive nuclei in each embryo, focused images were made binary by thresholding to highlight in black the H3S10ph-positive nuclei and in white the background, and automatically quantified by the "analyze particles" command in ImageJ software. Embryo contour was delineated on each image, and the circularity was measured in Image J. The Student´s t-test was performed in Microsoft Excel for statistical analysis.

3.7. Immunofluorescence of α-Tubulin in PC3 Cells

PC3 cells were grown in RPMI medium supplemented with 10% FCS. In total, 3×10^4 cells were cultured in 24-well culture plates containing slides and allowed to attach overnight at 37 °C in 5% CO_2. The cells were then treated with compounds **1–7** at their IC_{50} values (0.95, 0.085, 1.0×10^{-5}, 1.7×10^{-3}, 2.0×10^{-3}, 2.0×10^{-3}, and 5.0×10^{-3} μg/mL, respectively) at 37 °C for 72 h. The cells were fixed with PFA (paraformaldehyde) 4% in PEM buffer (PIPES 100 mM pH 6.9, EGTA 5 mM, $MgCl_2$ 2 mM). After 15 min, $PFA/NaHCO_3$ was added, and the cells incubated for 45 min at room temperature. The slides were rinsed with PBS and treated with 0.1% Triton X-100 (Sigma Aldrich, Saint Louis, MO, USA), then incubated with a primary anti-α-tubulin antibody (1:300, Sigma Aldrich, Saint Louis, MO, USA) overnight at 4 °C. A secondary anti-mouse Alexa Fluor 647 antibody (1:1000, Molecular Probes, Eugene, OR, USA) was added, and the slides were incubated for 1 h at 37 °C. The cells were stained with 0.4 μg/mL of 4, 6-diamidino-2-phenylindole (DAPI, Molecular Probes, Eugene, OR, USA) in PBS for 10 min, mounted, and imaged by confocal microscopy as recently described [27].

4. Conclusions

In this work, following a bioassay-guided fractionation by using the zebrafish model, we isolated seven antimitotic active lignans (**1–7**) from an antitumoral hydroalcoholic extract of *B. fagaroides*. We found by using immunoassays in zebrafish embryos that **1**, **2**, **4**, and **7** lignans induced mitotic arrest, delayed cell migration, and disrupted the microtubule cytoskeleton. These results enabled us to demonstrate that microtubule destabilization is the mechanism of action through which these compounds exert their cytotoxic activity.

Furthermore, microtubule array disruption was demonstrated also in PC3 cells. In general, equivalent antimitotic effects were observed in both in vivo and in vitro models.

Finally, we can conclude that the in vivo zebrafish model developed in this work is a suitable, faster, and cheap model to identify antimitotic natural products through bioassay-guided fractionation.

Supplementary Materials: Whole-mount immunolocalization of phospho-histone-H3 (H3S10ph) in zebrafish embryos. Table quantification of H3S10ph by the extract and fractions. ^{1}H NMR table of pure compounds (**1–7**).

Author Contributions: M.A.-M., A.M.R.-S., S.M., and L.A. isolated and identified lignans. M.Y.A., A.M.R.-S., M.A.M.-S., and E.S.-V. performed the experiments and image analysis; all authors participated in designing the experiments; L.A., E.S.-V., L.G., and S.M.-B designed the project; E.S.-V., L.A., and M.A-M. wrote the paper.

Funding: Financial support obtained from UNAM (Grants IX201110, IN205612, IN210316) and partial financial support from CONACYT, Mexico (Grants No. 82851 and LN 251613). M.Y.A.-M., A.M.R.-S., and M.A.M.-S. acknowledge CONACYT fellowship support (253315, 219701, and 323762, respectively).

Acknowledgments: Confocal microscopy service provided by Laboratorio Nacional de Microscopía Avanzada, Instituto de Biotecnología, Universidad Nacional Autónoma de México. The authors thank Laboratorio Nacional de Estructura de Macromoléculas (Conacyt 294406) for the spectroscopic and mass analyses.

Conflicts of Interest: The authors declare no conflict of interest.

References

1. Ayres, D.C.; Loike, J.D. *Lignans Chemical, Biological and Clinical Properties*; Cambridge University Press: Cambridge, UK, 1990.
2. Teponno, R.B.; Kusari, S.M. Recent advances in research on lignans and neolignans. *Nat. Prod. Rep.* **2016**, *33*, 1044–1092. [CrossRef] [PubMed]
3. Pilkington, L. Lignans: A Chemometric Analysis. *Molecules* **2018**, *23*, 1666. [CrossRef] [PubMed]
4. Tseng, C.J.; Wang, Y.J.; Liang, Y.C.; Jeng, J.H.; Lee, W.S.; Lin, J.K.; Chen, C.H.; Liu, I.C.; Ho, Y.S. Microtubule damaging agents induce apoptosis in HL 60 cells and G2/M cell cycle arrest in HT 29 cells. *Toxicology* **2002**, *175*, 123–142. [CrossRef]
5. Gordaliza, M.; Garcia, P.A.; del Corral, J.M.; Castro, M.A.; Gomez-Zurita, M.A. Podophyllotoxin: Distribution, sources, applications and new cytotoxic derivatives. *Toxicon* **2004**, *44*, 441–459. [CrossRef] [PubMed]
6. Antunez-Mojica, M.; Rodríguez-Salarichs, J.; Redondo-Horcajo, M.; León, A.; Barasoain, I.; Canales, A.; Cañada, F.J.; Jiménez-Barbero, J.; Alvarez, L.; Díaz, J.F. Structural and Biochemical Characterization of the Interaction of Tubulin with Potent Natural Analogues of Podophyllotoxin. *J. Nat. Prod.* **2016**, *79*, 2113–2121. [CrossRef] [PubMed]
7. Hande, K.R. Etoposide: Four decades of development of a topoisomerase II inhibitor. *Eur. J. Cancer* **1998**, *34*, 1514–1521. [CrossRef]
8. Rojas-Sepulveda, A.M.; Mendieta-Serrano, M.; Mojica, M.Y.; Salas-Vidal, E.; Marquina, S.; Villarreal, M.L.; Puebla, A.M.; Delgado, J.I.; Alvarez, L. Cytotoxic Podophyllotoxin Type-Lignans from the Steam Bark of Bursera fagaroides var. fagaroides. *Molecules* **2012**, *17*, 9506–9519. [CrossRef]
9. Antunez-Mojica, M.; León, A.; Rojas-Sepúlveda, A.M.; Marquina, S.; Mendieta-Serrano, M.; Salas-Vidal, E.; Villareal, M.L.; Alvarez, L. Aryldihydronaphthalene-type lignans from Bursera fagaroides var. fagaroides and their antimitotic mechanism of action. *RSC Adv.* **2016**, *6*, 4950–4959. [CrossRef]
10. Peña-Morán, O.; Villareal, M.L.; Álvarez, L.; Meneses-Acosta, A.; Rodríguez-López, V. Cytotoxicity, Post-Treatment Recovery, and Selectivity Analysis of Naturally Occurring Podophyllotoxins from Bursera fagaroides var. fagaroides on Breast Cancer Cell Lines. *Molecules* **2016**, *21*, 1013. [CrossRef]
11. Gutiérrez-Gutiérrez, F.; Puebla-Pérez, A.M.; González-Pozos, S.; Hernández-Hernández, J.M.; Pérez-Rangel, A.; Alvarez, L.; Tapia-Pastrana, G.; Castillo-Romero, A. Antigiardial Activity of Podophyllotoxin-Type Lignans from Bursera fagaroides var. *fagaroides*. *Molecules* **2017**, *22*, 799. [CrossRef]
12. Hendzel, M.J.; Wei, Y.; Mancini, M.A.; Van Hooser, A.; Brinkley, B.R.; Bazett-Jones, D.P.; Allis, C.D. Mitosis-specific phosphorylation of histone H3 initiates primarily within pericentromeric heterochromatin during G2 and spreads in an ordered fashion coincident with mitotic chromosome condensation. *Chromosoma* **1997**, *106*, 348–360. [CrossRef] [PubMed]
13. Mendieta-Serrano, M.A.; Schnabel, D.; Lomeli, H.; Salas-Vidal, E. Cell proliferation patterns in early zebrafish development. *Anat. Rec.* **2013**, *296*, 759–773. [CrossRef] [PubMed]

14. Saka, Y.; Smith, J.C. Spatial and temporal patterns of cell division during early Xenopus embryogenesis. *Dev. Biol.* **2001**, *229*, 307–318. [CrossRef] [PubMed]
15. Zhang, L.; Kendrick, C.; Julich, D.; Holley, S.A. Cell cycle progression is required for zebrafish somite morphogenesis but not segmentation clock function. *Development* **2008**, *135*, 2065–2070. [CrossRef] [PubMed]
16. Murphey, R.D.; Stern, H.M.; Straub, C.T.; Zon, L.I. A chemical genetic screen for cell cycle inhibitors in zebrafish embryos. *Chem. Biol. Drug Des.* **2006**, *68*, 213–219. [CrossRef] [PubMed]
17. Urbani, L.; Sherwood, S.W.; Schimke, R.T. Dissociation of nuclear and cytoplasmic cell cycle progression by drugs employed in cell synchronization. *Exp. Cell Res.* **1995**, *219*, 159–168. [CrossRef] [PubMed]
18. Ikegami, R.; Zhang, J.; Rivera-Bennetts, A.K.; Yager, T.D. Activation of the metaphase checkpoint and an apoptosis programme in the early zebrafish embryo, by treatment with the spindle-destabilising agent nocodazole. *Zygote (Cambridge, England)* **1997**, *5*, 329–350. [CrossRef]
19. Solnica-Krezel, L.; Driever, W. Microtubule arrays of the zebrafish yolk cell: Organization and function during epiboly. *Development (Cambridge, England)* **1994**, *120*, 2443–2455.
20. Strahle, U.; Jesuthasan, S. Ultraviolet irradiation impairs epiboly in zebrafish embryos: Evidence for a microtubule-dependent mechanism of epiboly. *Development (Cambridge, England)* **1993**, *119*, 909–919.
21. Moon, H.S.; Jacobson, E.M.; Khersonsky, S.M.; Luzung, M.R.; Walsh, D.P.; Xiong, W.; Lee, J.W.; Parikh, P.B.; Lam, J.C.; Kang, T.W.; et al. A novel microtubule destabilizing entity from orthogonal synthesis of triazine library and zebrafish embryo screening. *J. Am. Chem. Soc.* **2002**, *124*, 11608–11609. [CrossRef]
22. Jesuthasan, S.; Stahle, U. Dynamic microtubules and specification of the zebrafish embryonic axis. *Curr. Biol.* **1997**, *7*, 31–42. [CrossRef]
23. Kimmel, C.B.; Ballard, W.W.; Kimmel, S.R.; Ullmann, B.; Schilling, T.F. Stages of embryonic development of the zebrafish. *Dev. Dyn.* **1995**, *203*, 253–310. [CrossRef] [PubMed]
24. Strahle, U.; Scholz, S.; Geisler, R.; Greiner, P.; Hollert, H.; Rastegar, S.; Schumacher, A.; Selderslaghs, I.; Weiss, C.; Witters, H.; et al. Zebrafish embryos as an alternative to animal experiments—A commentary on the definition of the onset of protected life stages in animal welfare regulations. *Reprod. Toxicol.* **2012**, *33*, 128–132. [CrossRef] [PubMed]
25. Westerfield, M. *The Zebrafish Book: A Guide for the Laboratory Use of Zebrafish (Danio Rerio)*; Institute of Neuroscience, University of Oregon: Eugene, OR, USA, 2000.
26. Abramoff, M.D.; Magelhaes, P.J.; Ram, S.J. Image Processing with ImageJ. *Biophotonics Int.* **2004**, *11*, 36–42.
27. Alejandre-García, I.; Álvarez, L.; Cardoso-Taketa, A.; González-Maya, L.; Antúnez, M.; Salas-Vidal, E.; Díaz, J.F.; Marquina-Bahena, S.; Villarreal, M.A. Cytotoxic Activity and Chemical Composition of the Root Extract from the Mexican Species Linum scabrellum: Mechanism of Action of the Active Compound 6-Methoxypodophyllotoxin. *Evid. Based Complement. Alternat. Med.* **2015**, *2015*, 298463. [CrossRef] [PubMed]

Sample Availability: Samples of the compounds **1–7** are available from the authors.

molecules

Article

Combined In Vitro Studies and in Silico Target Fishing for the Evaluation of the Biological Activities of *Diphylleia cymosa* and *Podophyllum hexandrum*

Marina Pereira Rocha [1,2], Priscilla Rodrigues Valadares Campana [2,3],
Denise de Oliveira Scoaris [3], Vera Lucia de Almeida [3], Julio Cesar Dias Lopes [4],
Julian Mark Hugh Shaw [5] and Claudia Gontijo Silva [1,3,*]

[1] Servico de Biotecnologia Vegetal, Fundacao Ezequiel Dias (FUNED), Belo Horizonte 30510-010, MG, Brazil; procha.marina@gmail.com
[2] Departamento de Produtos Farmaceuticos FAFAR-UFMG, Belo Horizonte 31270-901, MG, Brazil; prvcampana@gmail.com
[3] Servico de Fitoquimica e Prospeccao Farmaceutica, Fundacao Ezequiel Dias, Belo Horizonte 30510-010, MG, Brazil; deniscoaris@gmail.com (D.d.O.S.); veluca2002@gmail.com (V.L.d.A.)
[4] Chemoinformatics Group (NEQUIM), Departamento de Quimica, Universidade Federal de Minas Gerais, Belo Horizonte 31270-901, MG, Brazil; jlopes.ufmg@gmail.com
[5] Science and Collections, Royal Horticultural Society, Wisley, Working, Surrey GU23 6QB, UK; julianshaw@rhs.org.uk
* Correspondence: claudia.gontijo@funed.mg.gov.br; Tel.: +55-31-3314-4791

Academic Editor: David Barker
Received: 22 November 2018; Accepted: 10 December 2018; Published: 13 December 2018

Abstract: This paper reports the in silico prediction of biological activities of lignans from *Diphylleia cymosa* and *Podophyllum hexandrum* combined with an in vitro bioassays. The extracts from the leaves, roots and rhizomes of both species were evaluated for their antibacterial, anticholinesterasic, antioxidant and cytotoxic activities. A group of 27 lignans was selected for biological activities prediction using the Active-IT system with 1987 ligand-based bioactivity models. The in silico approach was properly validated and several ethnopharmacological uses and known biological activities were confirmed, whilst others should be investigated for new drugs with potential clinical use. The extracts from roots of *D. cymosa* and from rhizomes and roots of *P. hexandrum* were very effective against *Bacillus cereus* and *Staphylococcus aureus*, while podophyllotoxin inhibited the growth of *Staphylococcus aureus* and *Escherichia coli*. *D. cymosa* leaves and roots showed anticholinesterasic and antioxidant activities, respectively. The evaluated extracts showed to be moderately toxic to THP-1 cells. The chromatographic characterization indicated that podophyllotoxin was the major constituent of *P. hexandrum* extract while kaempferol and its hexoside were the main constituents of *D. cymosa* leaves and roots, respectively. These results suggest that the podophyllotoxin could be the major antibacterial lignan, while flavonoids could be responsible for the antioxidant activity.

Keywords: lignans; in silico studies; podophyllotoxin; antibacterial activity; acetylcholinesterase inhibitors; antioxidant activity; cytotoxicity

1. Introduction

Lignans are a large group of phenylpropanoid dimers with a different degree of oxidation in the side-chain and a different substitution in the aromatic group [1,2]. They are classified in groups according to their oxygenation and cyclization patterns. The most prominent member of this group of natural products is podophyllotoxin (PTOX, **1**). Its antitumour activity prompted several studies, and resulted in the introduction of successful clinical drugs. This aryltetralin lignan is a lead compound for the semi-synthetic derivatives etoposide (**27**), teniposide (**28**), and etopophos (**29**) (Figure 1), which have an important role in cancer therapy [3,4]. In addition, analogues of podophyllotoxin (**1**) were evaluated for the treatment of rheumatoid arthritis, psoriasis, and malaria with good results [3,5]. Furthermore, there are reviews published referring to the semisynthesis of PTOX derivatives, applications, mode of action and structure-activity relationships [6–8].

Arylnaphatalene lignans such as diphyllin (**24**) and its glycosides **25** and **26** (Figure 1) were isolated from some traditional medicinal plants and have been reported to possess a wide range of pharmacological activities, including antitumour, anti-leshmania, antifungal, antiviral and antibacterial [9,10].

Currently, there are few plant sources of podophyllotoxin (**1**) and its related lignans occur in a particular taxonomic group, but in low amounts. Podophyllotoxin (**1**) is still obtained from wild *Podophyllum* populations, and this is a major constraint in supplying the lignan to the pharmaceutical industry that is under pressure to meet demand. To overcome this situation, several studies focussing on its production by biotechnological strategies and synthetic approaches have been reported [11,12].

Podophyllotoxin (**1**) is found in the rhizomes, roots and leaves of both *Podophyllum hexandrum* Royle and *Diphylleia cymosa* Michaux (Berberidaceae), while the occurrence of diphyllin (**24**) is reported in the latter species but not in *Podophyllums* [13]. *P. hexandrum* and *D. cymosa* are herbaceous perennials found growing in moist shady conditions [14], and are known for their medicinal use in American and Asian cultures. Both genera are taxonomically closely related, and some common features are their habitat, morphology, karyotype and chemical profile, whilst the differences are related to floral biology [13,15,16]. A study based on four molecular markers and morphology confirms the close relationship between *Diphylleia* and *Podophyllum* [17]. *P. hexandrum* is sometimes treated as a monotypic genus *Sinopodophyllum* [18].

P. hexandrum is commonly named as the Himalayan Mayapple or Indian Mayapple and Indian *Podophyllum*. There are ethnobotanical records based on its healing properties in Asian culture for the treatment of skin cancers as well as due to its purgative, emetic, cytotoxicity, antitumour and antileukaemic properties [11,19]. Overall, the parts used for medicinal purposes are mainly the rhizomes, roots and fruits. *D. cymosa* has been called the Southern Mayapple and Umbrella Leaf [20]. There is an account that describes the American Cherokee Indians using an infusion of the plant as a diuretic, antiseptic, diaphoretic and for the treatment of smallpox [21]. According to an earlier clinical study, the resin demonstrated none of the biological properties associated with *Podophyllum* [22]. Considerable interest has been centered on *P. hexandrum* due to the PTOX content and its related lignans. With regard to *D. cymosa*, the species has been investigated less than the *Podophyllums* and only a few studies have been reported [13,23,24].

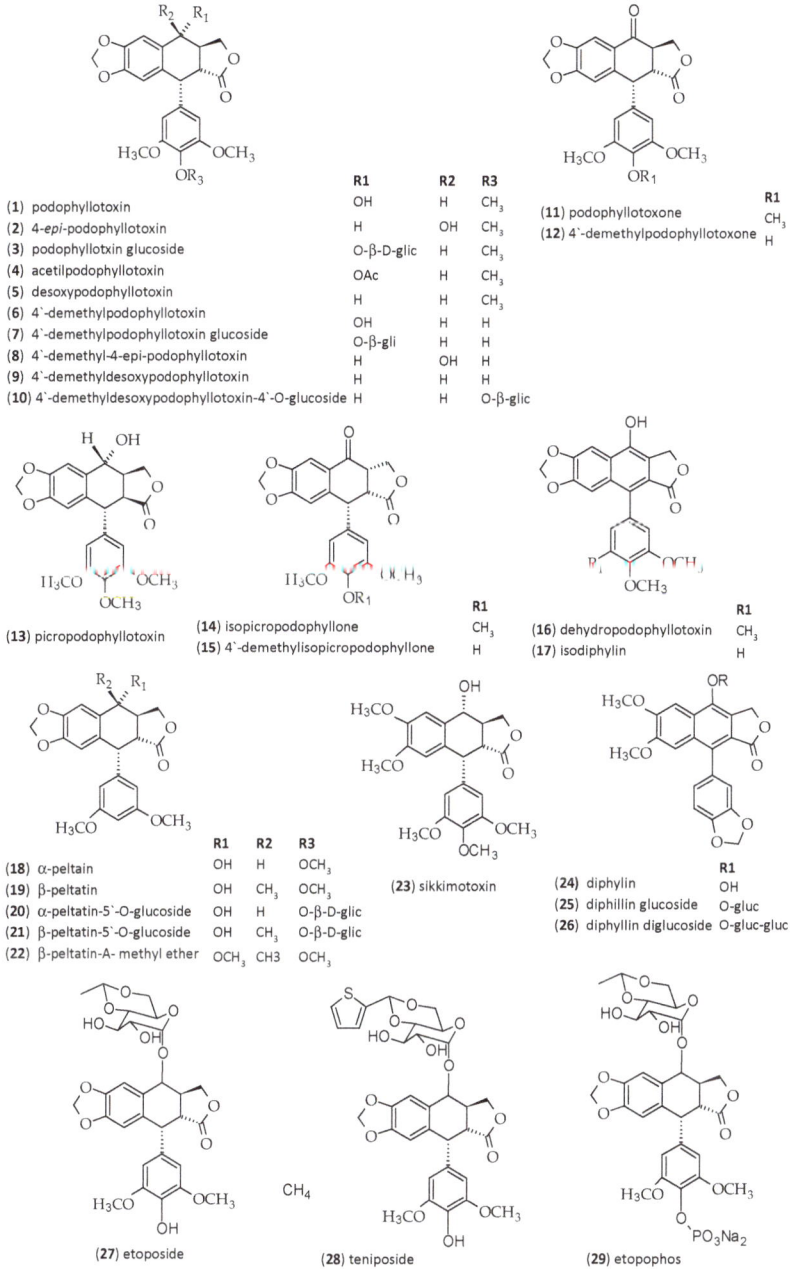

Figure 1. Chemical structures of lignans and podophyllotoxin derivatives [13,25].

This paper reports on the evaluation of ethanolic extracts from the leaves and roots of *D. cymosa*, and from the rhizomes and roots of *P. hexandrum* for the antibacterial activity, inhibition of AChE, as well as for the antioxidant activity and cytotoxicity combined with an in silico target fishing approach. The latter was used to predict new activities for known lignans from both species. The lignans profiling was based on the chromatographic analyses (UPLC-DAD-ESI-MS/MS) which were included with the aim of identifying the major phenolic components in the extracts (Figure 2).

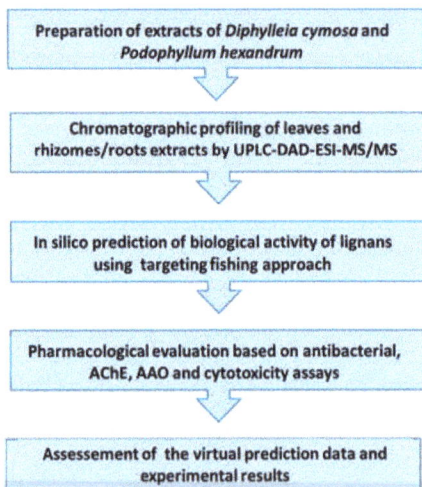

Figure 2. Workflow of the methodology performed in this study.

2. Results

A growing amount of work has been applied to investigating *P. hexandrum* due to its content of PTOX (**1**) and related lignans. However, it is surprising to note that studies into *D. cymosa* are extremely limited when compared with the *Podophyllum* species, even though this species is endemic in the Southern Appalachian Mountains of the Eastern North America [22,26].

There has been a decrease in the wild populations of *P. hexandrum* in India due to the over collection of rhizomes and roots of this species [27]. In addition, the species shows a short season of availability, and thus plants are limited in the field. A number of studies have been undertaken to achieve mass propagation [28] as well as to establish plant derived-cultures for the production of podophyllotoxin [29,30], whereas their low yields were far from meeting commercial needs. Enhancement of the lignan was attempted by other systems, including transgenic cultures, addition of a precursor feeding to the culture medium, the use of an elicitor such as methyl jasmonate, and the production by endophytes [11]. With regard to the latter, there are reports on the production of PTOX (**1**) by the endophytic fungi *Fusarium solani* [31] and *Trametes hirsuta* [32] isolated from *P. hexandrum*. These species of endophytes could be a promising source for large-scale production of PTOX, whereas the yields must be improved.

2.1. Chromatographic Profiling by UPLC-DAD-ESI-MS/MS

Several lignans have been already reported for *D. cyomsa* and *P. hexandrum*. From the leaves and roots of *D. cymosa*, Broomhead and Dewick [13] isolated the lignans PTOX (**1**), 4′-demethyl-podophyllotoxin (**6**), 4′-demethyldesoxypodophyllotoxin (**9**), diphyllin (**24**), diphyllin glucoside (**25**), diphyllin diglucoside (**26**) and 4′-demethyldesoxypodophyllotoxin 4-O-glucoside (**10**) (Figure 1), this being the only study reporting the isolation and characterization of compounds from *D. cymosa*.

A phytochemical study of *P. hexandrum* led to the isolation of the lignans PTOX (**1**), 4′-demethylpodophyllotoxin (**6**), 4′-demethyldesoxypodophyllotoxin (**9**), PTOX glucoside (**3**), desoxy-podophyllotoxin (deoxypodophyllotoxin, **5**), 4′-demethylpodophyllotoxin glucoside (**7**), 4′-demethylisopicropodophyllone (**15**), podophyllotoxone (**11**), 4′-demethylpodophyllotoxone

(**12**), picropodophyllotoxin (**13**), isopicropodophyllone (**14**), 4′-demethyldeoxypodophyllotoxin (4′-demethyldesoxypodophyllotoxin, **9**), α-peltatin (**18**) and β-peltatin (**19**) (Figure 1) [13].

In this study, the chemical characterization of *D. cymosa* and *P. hexandrum* was performed using UPLC-DAD-ESI-MS. The obtained chromatographic profiles indicated the presence of compounds of different polarities in the EtOH extracts of *D. cymosa* and *P. hexandrum* (Figure 3). The crude extracts as well as lignans previously isolated from *P. hexandrum* such as PTOX (**1**), deoxypodophyllotoxin (**5**), 4′-demethylpodophyllotoxin (**6**), podophyllotoxone (**11**), α-peltatin (**18**) and β-peltatin (**19**) were evaluated in the same chromatographic conditions. The UV and ESI⁺-MS spectra for the reference compounds podophyllotoxin (**1**) and α-peltatin (**18**) are presented in Figure 4 (Figure 4A,B). The UV and MS spectra for all reference compounds and identified chromatographic peaks are available in the Supplementary Material.

The chromatographic profile obtained by UPLC-DAD for the EtOH extracts of *D. cymosa* leaves (Figure 3A), *D. cymosa* roots (Figure 3B) and *P. hexandrum* rhizomes and roots (Figure 3C) showed peaks with UV absorption spectra with λ_{max} around 260–290 nm which is compatible with the chemical structure of lignans due to conjugation of the aromatic rings. It was possible to identify peaks with UV absorption pattern characteristic of aryltetralin lignans related to podophyllotoxin (peaks 4, 8, 11 and 13, λ_{max} at 290 nm) and peltatins (peaks 6 and 10, λ_{max} at 275 nm), and arylnaphtalene lignans, such as diphyllin (peaks 5, 7 and 12; λ_{max} at ca. 260 nm). Peaks with UV spectra characteristic of other phenolic compounds were also identified, such as phenolic acids (peak 1, λ_{max} at 246, 295 and 326 nm) and flavonoids (peaks 2, 3 and 9, λ_{max} at 265 and ca. 350 nm), which are compounds with more conjugated chromophores.

UPLC-ESI-MS/MS analyses were carried out in order to identify the major constituents of the extracts (Table 1). Chromatographic peak 1 (RT = 1.96 min) showed *m/z* at 353 and 355 in the negative and positive ionization modes, respectively. The UV absorption profile of compound **1** was indicative of phenolic acids (λ_{max} ~ 295 and 326 nm). The fragmentation of the parent ion at *m/z* 353 in a MS2 experiment in the negative mode afforded daughter ions at *m/z* 191 [M − H − caffeoyl]⁻, 179 [M − H − quinic]⁻, and 173 [quinic acid − H − H₂O]⁻ suggesting the identity of compound **1** as one of the regioisomers of caffeoylquinic acid. The fragmentation pattern of caffeoylquinic acids have been extensively described [33].

The chromatographic peaks 2, 3 and 9 had a UV absorption profile indicative of flavonols, with λ_{max} ~ 255–265 and 350 nm. MS spectra associated to peak 2 (RT = 2.91 min) showed a peak of [M − H]⁻ at *m/z* 463 and [M + H]⁺ at *m/z* 465. The MS2 fragmentation of the parent ion at at *m/z* 465 in the positive mode afforded a daughter ion at *m/z* 303 [M + H − hexose]⁺, indicating that peak 2 could correspond to a hexoside of the flavonol quercetin. The observed fragmentation pattern of the ion at *m/z* 303 generated a peak at *m/z* 165, which is indicative of the presence of a hydroxyl group at C_3 of flavonols. A similar UV absorption profile was observed for peak 3 (RT = 3.17 min), which showed a MS spectra with a peak of [M − H]⁻ at *m/z* 447 and [M + H]⁺ at *m/z* 449. The MS2 fragmentation of the parent ions at *m/z* 447 and *m/z* 449 afforded daughter ions at *m/z* 285 [M − H − hexose] and 287 [M + H − hexose] in the negative and positive ionization modes, respectively. The 16 a.m.u. difference observed between peaks 2 and 3, along with the similarity in the UV spectra and in the fragmentation pattern obtained in MS2 experiments, suggest that peak 2 is a hexoside of the flavonol kaempferol. This is the major peak observed in the chromatogram of *D. cymosa* leaves (Figure 3A). The compound corresponding to peak 9 (RT = 4.53 min) showed a UV and MS profiles similar to those observed for peak 3. The MS spectra registered for this compound presented a peak at *m/z* 285 corresponding to the deprotonated molecule [M − H]⁻. The fragmentation pattern of the parent ion at *m/z* 285 was similar to that observed for peak 3, suggesting that peak 9 corresponds to the aglicone kaempferol. This flavonol is the major constituent of the EtOH extract of *D. cymosa* roots.

Analysis of the UV and MS spectra associated with peaks 4 (RT = 3.73 min) and 8 (RT = 4.46 min) showed λ_{max} at 290 nm, indicative of aryltetralin lignans. MS spectra associated with peaks 4 and 8 showed signals of [M + H]⁺ at *m/z* 577 and 415, respectively. The same pattern was observed in the negative ionization mode, with signals of $[M - H]^-$ at m/z 575 and 459 $[M - H + \text{formiate}]^-$ for peaks 4 and 8, respectively. The difference of 162 Da between the two compounds indicates the presence of a hexose residue. The fragmentation of the parent ion at m/z 415 in the positive ionization mode generated the daughter ion at m/z 247 which is compatible with the neutral loss of a trimethoxybenzyl group ($C_9H_{12}O_3$). These results, along with the analysis of the isolated lignan podophyllotoxin in the same conditions, allowed the identification of compound **8** as podophyllotoxin (**1**) and compound **4** as an *O*-hexosyl derivative of podophyllotoxin. PTOX (**1**) is the major component of the EtOH extract of *P. hexandrum* rhizomes and roots. On the other hand, this lignan was not found in the extract of *D. cymosa* roots.

Chromatographic peaks 5 (RT = 3.84 min), 7 (RT = 4.11 min) and 12 (RT = 5.52 min) showed UV spectra similar to those observed for arylnaphtalene lignans (λ_{max} at 275 nm). The MS spectra registered for these compounds showed peaks at m/z 543, for compounds **5** and **7**, and m/z 379 for compound **12**, corresponding to the deprotonated molecules $[M - H]^-$ in the negative mode. The difference of 162 Da between the compound **12** and compounds **5** and **7** indicates the presence of a hexose residue in the latter. The fragmentation of ion at m/z 379 in the negative mode originated the daughter ions at m/z 319 and 391. These fragments were reported for the lignan diphyllin [34], which suggest that compound **12** is diphyllin (**24**) while compounds **5** and **7** are *O*-hexosyl derivatives of diphyllin.

In the UPLC profile of the EtOH extract of rhizomes and roots of *P. hexandrum*, the peak eluted at 3.84 min (peak 5′) showed a different UV absorption profile, with λ_{max} at 287 nm, characteristic of aryltetralin lignans. The MS spectra associated with this compound showed a peak at m/z 399 $[M - H]^-$ and 401 $[M + H]^+$ in the negative and positive ionization modes, respectively. Analysis of the lignan 4′-demethylpodophyllotoxin (**6**) in the same conditions allowed us to assign this lignan as the compound responsible for peak 5′.

The compound corresponding to peaks 6 (RT = 3.92 min) and 10 (RT = 4.62 min) showed UV absorption pattern similar to what is described for peltatins. The MS spectra associated with chromatographic peak 6 peak showed peaks at m/z 399 and 401 for the deprotonated and protonated molecules, respectively. The MS2 fragmentation of the parent ion at m/z 401, in the positive ionization mode, generated the daughter ion at m/z 247 which is compatible with the neutral loss of a trimethoxybenzyl group ($C_9H_{12}O_3$). These results, along with the analysis of the lignan α-peltatin (**18**) in the same conditions, allowed the identification of compound **6** as α-peltatin (**18**). The MS spectra associated with peak 10 showed peaks at m/z 413 and 415 for the deprotonated and protonated molecules, respectively. The analysis of the isolated lignan β-peltatin (**19**) in the same conditions, allowed the identification of compound **10** as β-peltatin (**14**).

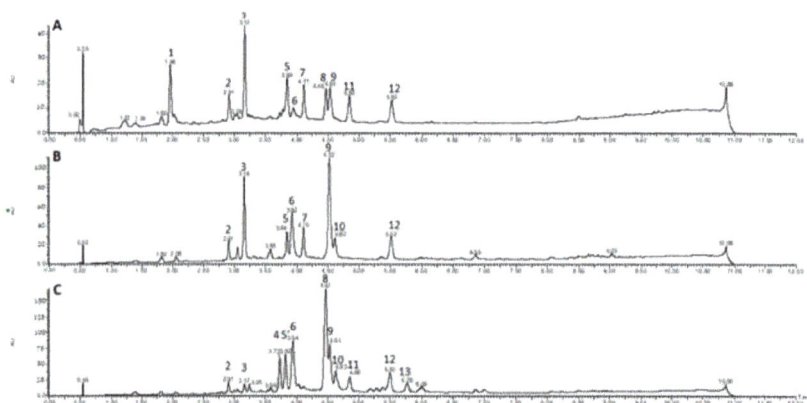

Figure 3. Chromatographic profiles obtained by UPLC-DAD for the EtOH extracts of *D. cymosa* leaves (**A**) and roots (**B**) and *P. hexandrum* rhizomes and roots (**C**).

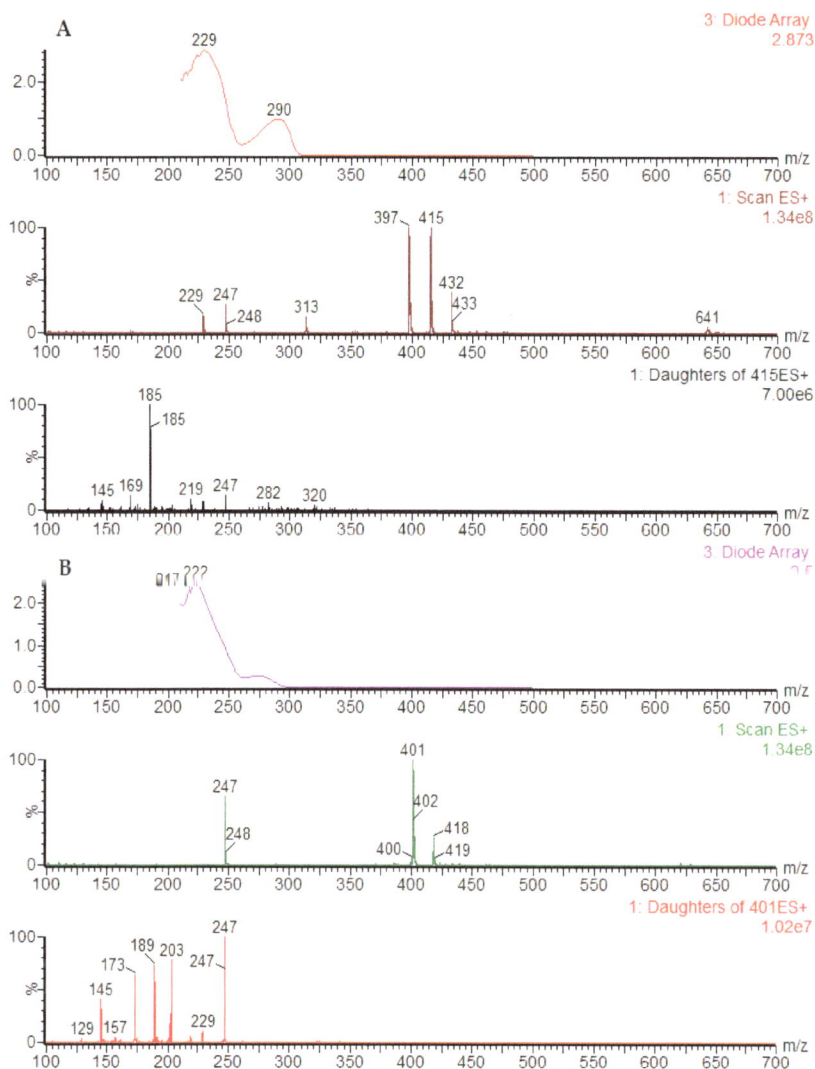

Figure 4. UV, ESI/MS and ESI-MS/MS in the positive ionization mode spectra obtained for podophyllotoxin (**A**) and α-peltatin (**B**).

The UV spectra observed for the compound of peak 11 was very similar to that observed for peaks 4 and 8, suggesting that compound **11** (RT = 4.86) is an aryltetralin lignan. The MS spectrum registered for this compound presented peaks at *m/z* 411 and 413 in the negative and positive modes, respectively. The MS2 fragmentation of the parent ion at *m/z* 413 was very similar to that observed for the lignan podophyllotoxone (**11**) (RT = 5.28 min), suggesting that the compound corresponding to peak 11 could be its isomer isopicropodophyllone (**14**).

The peak 13 (RT = 5.78), which is only present in the EtOH extract of *P. hexandrum* rhizomes and roots, showed UV spectra compatible with aryltetralin lignans and the MS spectra presented peaks at *m/z* 444 [M − H + formiate]$^-$ and 399 [M + H]$^+$. These results, along with the results obtained for the purified compound allowed the identification of compound **13** as desoxypodophyllotoxin (**5**).

Table 1. Identification of phenolic compounds in the EtOH extrats of *D. cymosa* leaves, *D. cymosa* roots and *P. hexandrum* rhizomes and roots by UPLC-DAD-ESI-MS/MS in negative and positive ionization modes.

Peak No.	Sample	RT (min)	Identity	[M − H]⁻ Parent Ion	[M + H]⁺ Parent Ion	MS2 Fragments Negative Mode (Daughter Ion)	MS2 Fragments Positive Mode (Daughter Ion)	UV (nm)
1	*D. cymosa* leaves	1.96	caffeoylquinic acid	353	355	191 [M − H − caffeoyl] 179 [M − H − quinic]	-	217, 246, 295, 326
2	*D. cymosa* leaves *D. cymosa* roots *P. hexandrum* rhizomes and roots	2.91	quercetin hexoside	463	465		303 [M + H − hexose]	255, 352
3	*D. cymosa* leaves *D. cymosa* roots *P. hexandrum* rhizomes and roots	3.17	kaempferol hexoside	447	449	285 [M − H − hexose]	287 [M + H − hexose]	265, 348
4	*P. hexandrum* rhizomes and roots	3.73	podophyllotoxin hexoside	575621 [M + formiate]⁻	577	413 [M − H − hexose]	-	290
5	*D. cymosa* leaves *D. cymosa* roots	3.84	diphyllin hexoside	-	543		381 [M + H − hexose] 363, 333, 319	261
5′	*P. hexandrum* rhizomes and roots	3.84	4′-demethylPTOX	399	401			287
6	*D. cymosa* leaves *D. cymosa* roots *P. hexandrum* rhizomes and roots	3.92	α-peltatin	399	401		247 [M + H − $C_9H_{12}O_3$]	275
7	*D. cymosa* leaves *D. cymosa* roots	4.11	diphyllin hexoside	-	543		381 [M + H − hexose] 363, 333, 319	261
8	*D. cymosa* leaves *P. hexandrum* rhizomes and roots	4.46	podophyllotoxin	459 [M + formiate]⁻	415		247 [M + H − $C_9H_{12}O_3$] 185 (247-H_2O-CO_2)	290
9	*D. cymosa* leaves *D. cymosa* roots *P. hexandrum* rhizomes and roots	4.53	kaempferol	285	-			265, 366
10	*D. cymosa* roots *P. hexandrum* rhizomes and roots	4.62	β-peltatin	413	415		247 [M + H − $C_9H_{12}O_3$] 185 (247-H_2O-CO_2)	275
11	*P. hexandrum* rhizomes and roots	4.86	isopicropodophyllone	411	413		-	290
12	*D. cymosa* leaves *D. cymosa* roots	5.52	diphyllin	379	381	319, 291, 275	363, 333, 319	261
13	*P. hexandrum* rhizomes and roots	5.78	deoxypodophyllotoxin	444 [M + formiate]⁻	-		-	291

190

Molecules **2018**, *23*, 3303

2.2. In Silico Prediction of Biological Activity of Lignans

The Active-IT system was composed, at the time the calculations were run, of 1987 biological activity datasets modeled with SVM and Naïve Bayes machine learning methods. About 1815 datasets were obtained directly from the PubChem Bioassay database and the remaining 172 datasets were obtained from different sources, including the combination of several PubChem datasets. Some of these modeled datasets were reported before, such as AMES [35], AChE [36] and antifungal/antibacterial activities [37], however most remain unpublished. The complete description of all modeled datasets is far beyond the scope of this paper and will not be discussed in detail.

The target fishing approach was performed using the Active-IT programme [36]. Before being able to perform predictions of biological activities of the lignans with the Active-IT system, we first made an ultimate validation using some known activities of these compounds. About 12 of the 27 lignans (compounds from **1** to **27**) used in this study appear in one or more datasets from the PubChem Bioassay. Among the 1815 PubChem PubChem Bioassay datasets within the Active-IT system about 243 have one or more lignans of the series, with a total of 309 activity points, with 128 classified as active and 181 classified as inactive. For example, the podophyllotoxin (**1**) appears in 195 different datasets. As expected, the activity predictions of lignans using these models produced excellent results with an AUC of 0.96 for SVM and 0.82 for Naïve Bayes (Figure 5A and Table 2).

Therefore, we decided the best approach for validation was to re-build all of these 243 models, excluding the lignans that appear in each dataset and repeat the prediction. The details of all these models, as well as their internal validation are included in the Supplementary Material (Table S3). The calculated values Pa-Pi of lignans that were excluded from the models are shown in Table S4 of the Supplementary Material. As expected the prediction was a little worse, but the results are still very good, considering most datasets have a cell-based format, with an AUC of 0.71 for SVM and 0.73 for Naïve Bayes (Figure 5A and Table 2). It is worth pointing out that while the SVM method experienced a large decrease in AUC (0.25 AUC units) the Naïve Bayes had a far smaller decrease (0.09 AUC units). This is an evidence that the Naïve Bayes method has a smaller dependence from the input and is less unresponsive to small variations of the dataset composition. The SVM has a lower resilience as it is much more dependent on the input dataset.

However, the validation with AUC only tells us whether the global prediction was accurate or not. In the chemoinformatics it is more important to define whether the most probable active compounds appear in the highest positions in the ranking.

Table 2. Results of the area under the ROC curve (AUC) for biological activity prediction of lignans. Only models with AUC > 0.5 were considered. For activity classes the experimental data points were merged into 125 activity classes. In this case, all 27 lignans have had their activity predicted and were used to build the activity class score.

Method	Dataset	Full Experimental		Activity Classes	
		AUC	Number of Data Points	AUC	Number of Data Points
Naïve Bayes	NEW [a]	0.730 ± 0.037	292	0.767 ± 0.054	125
Naïve Bayes	OLD [b]	0.816 ± 0.031	292	0.803 ± 0.050	125
SVM	NEW [a]	0.710 ± 0.037	305	0.673 ± 0.062	125
SVM	OLD [b]	0.962 ± 0.014	305	0.868 ± 0.041	125

[a] Datasets where all lignans were excluded; [b] Datasets that include some lignans.

Several metrics can be used to decide the better cutoff to be applied in a classification schema. We used the three metrics based on a contingency matrix to decide the best cutoff to be used:

(1) F-score is a measure of the accuracy of the test, calculated by the harmonic mean of recall or sensitivity [TP/(TP + FN)] and precision [TP/(TP + FP)];

(2) Matthews Correlation Coefficient (MCC) is a balanced measure of the quality of binary classification and is the most informative single score to establish the quality of a binary classifier prediction in a confusion matrix context [38];

(3) Enrichment Factor (EF) is a measure of how many more active compounds we find relative to a random distribution, it is calculated from the proportion of true active compounds selected in relation to the proportion of true active compounds in the entire dataset [39].

The complete results of Pa-Pi values calculated for all lignans, as well the predictions made, are included in the Supplementary material (Table S4). The best cutoff for the classification of lignan compounds as active or inactive was Pa-Pi ≥ 0.15 for both SVM and Naïve Bayes methods, using a positive lower limit of Pa-Pi as an additional filter. In Table 3 are presented a summary of the results when this threshold value is applied. In this table it is possible to note that SVM outperforms Naïve Bayes method for both datasets in the prediction of active compounds, and both methods present a higher specificity as few inactive compounds are misclassified as active (low false positive rate). This draws attention to the fact that the misclassified inactive compounds are almost the same for datasets with (OLD) or without the lignans (NEW) and, despite the fact that there are several active missed, both methods present a good precision as most of the compounds predicted as active are really active. Both methods lost performance when the lignans are deleted from the modeling dataset, but the effect is more pronounced in SVM, as discussed before for AUC results.

The main approach used in this work to make predictions about the biological activities of lignans was the grouping of several models in activity classes and this procedure must be validated as well. The 243 datasets used in the validation were classified in 137 activity classes, with the more populated being the models associated with leukemia cancer with 16 different datasets. From these 137 activity classes only 125 showed a predominance of active (47 cases) or inactive compounds (78 cases), and 12 showed an equal number of active and inactive. Thus, the predominance of active or inactive lead us to classify 47 classes as "probable active" as the lignans are prone to be active and 78 classes as "probable inactive" as the lignans are prone to be inactive.

To make the prediction, all 27 lignans were submitted to 243 models and the activity class score was calculated as described in the Material and Methods section. To build this score, the individual Pa-Pi values were not taken in account, only the number of cases where the value of Pa-Pi is above the threshold and with a positive minimum value.

Table 3. Validation of biological activity prediction of lignans. For all methods the threshold was defined at Pa-Pi = 0.15. Compounds with Pa-Pi better or equal to the threshold are classified as active and compounds with Pa-Pi below the threshold are classified as inactive. Compounds with a Pa-Pi minimum below zero are classified as inactive, no matter the mean Pa-Pi. In the rows named as "Both" the compound was predicted as active if any method predicted it as active. The metrics F-score, MCC and EF were used to define the best threshold (full data not shown).

Method	Dataset	TP	FP	TN	FN	F-Score	MCC	EF
Naïve Bayes	NEW [a]	30	17	151	94	0.35	0.19	1.54
SVM	NEW [a]	46	21	159	79	0.48	0.30	1.66
Both	NEW [a]	60	33	148	68	0.54	0.31	1.56
Naïve Bayes	OLD [b]	51	17	151	73	0.53	0.36	1.81
SVM	OLD [b]	106	18	162	19	0.85	0.75	2.06
Both	OLD [b]	112	33	148	16	0.82	0.68	1.87

[a] Datasets where all lignans were excluded; [b] Datasets that include some lignans.

The best threshold was determined by analyzing the value of MCC for each threshold (Table 3), being Pa-Pi ≥ 0.3 for Naïve Bayes and Pa-Pi ≥ 0.25 for SVM. The score of each activity class, in these cases, was calculated by the number of lignans classified as active among all the models belonging to the same class divided by the number of lignans used in the calculations (27) and the square root of the number of models of this class. The division by the square root of the number of models works like a normalization process because as the number of models in the same class increases more compounds are prone to be predicted as active for this class. From our experience, if we divided by the number of models, those classes with a large number of models are penalized, and if we take only the number of compounds the classes with a large number of models are privileged. The division by square root brings some balance to the prediction.

Using this approach we reconstructed the ROC curves using the scores of activity classes and the results are show in the Figure 5B and Table 3. As we can see the values of AUC metric although a little smaller than those obtained when we use the Pa-Pi scores of each individual compound are high enough to say that this approach sounds viable and allow us to use it to classify the whole set of lignans instead of analyzing a much larger number of results when we consider each compound independently. It is worth noting that all lignans were used in this activity class validation and not only those for which experimental results are available. Another interesting result was that for Naïve Bayes the AUC increases when the data points are grouped into activity classes in relation to the analysis of all experimental data.

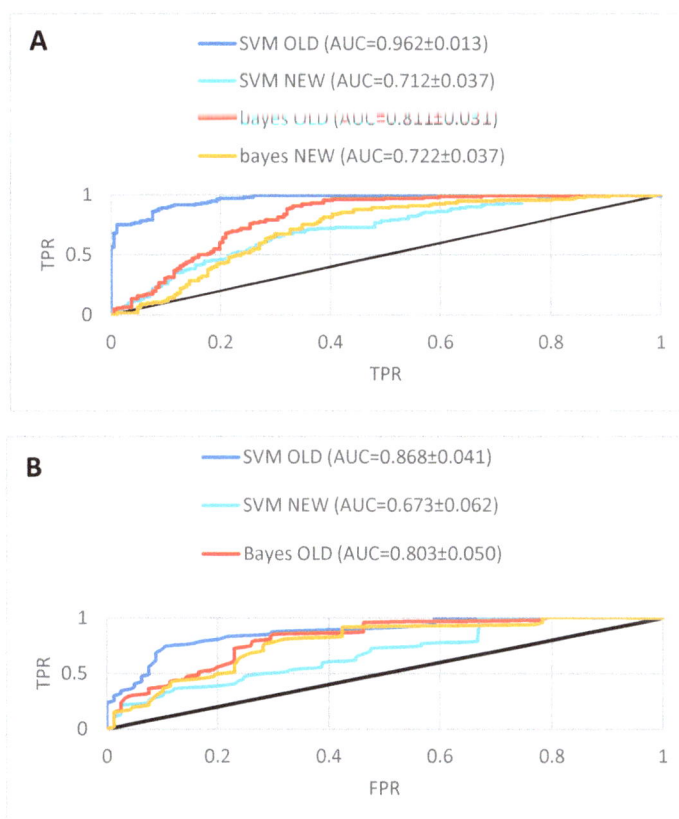

Figure 5. Validation of biological activities prediction of lignans. (**A**) ROC curves obtained in the prediction using all 309 experimental data available for lignans. The values of Pa-Pi scores were used to rank each data point. (**B**) ROC curves obtained for prediction after grouping the experimental data in 125 activity classes (12 classes with an equal number of active and inactive were not considered). The rank was built with an activity class score calculated with the number of lignans with Pa-Pi scores above the threshold, divided by the square root of the number of models belonging to the same class.

A partial list of predicted biological activities as well as the targets associates with the diseases are described in the supplementary material (Table S1 for SVM, Table S2 for Naïve Bayes and Table S3 for results when both methods are taken together). In Table 4 is presented a summary about the activities predicted until the 50 position in the ranking of 753 activity classes using both machine-learning methods (Table S3 of the Supplementary Material).

2.3. Antibacterial Activity

Infectious diseases, especially those caused by bacteria, are a major concern in several countries mostly due to antimicrobial resistance, which is a global public health issue that could hold back the control of many bacterial diseases [40]. Therefore, the search for new antimicrobial compounds from natural sources is immensely valuable. The in silico target fishing showed a probable antibacterial activity for aryltetralin lignans, as presented previously.

In our study, the ethanolic extracts from roots of *D. cymosa* and from rhizomes and roots of *P. hexandrum* were the most effective extracts against Gram-positive bacteria, reaching total inhibition of the microbial growth against *B. cereus* and proeminent inhibition rates (58.43 ± 2.7% and 67.56 ± 0.8%, respectively) against *S. aureus* (Table 5). PTOX also presented antibacterial activity against *S. aureus* (53.34 ± 8.8%) and *E. coli* (51.57 ± 9.08%). No antibacterial activity of the extracts and PTOX was observed against the Gram-negative pathogens EHEC, *P. aeruginosa* and *Salmonella* Typhi.

v

Table 4. In vitro biological activities of extracts and lignans isolated from *Diphylleia cymosa* and *Podophyllum hexandrum*, ethnopharmacological uses and predicted activities using both machine learning methods.

Studies	Etnopharmacological Uses, Predicted and Observed Biological Activities	Literature
Ethnopharmacological uses	*P. hexandrum*: Treatment of diarrhoea and liver problems, to promote conception, eye treatment, chronic constipation, hepatic stimulant, antitumour, purgative, cholagogue and purgative	[11,24]
	D. cymosa: Diuretic, antiseptic, diaphoretic and for the treatment of smallpox	[21,24]
Biological activities reported in the literature	Antitumour, insecticidal, antimalarial, fungicidal, antiviral, anti-inflammatory, neurotoxic, immunosuppressive, antirheumatic, antispasmogenic and hypolipidemic properties	[41,42]
Predicted biological activities using both machine learning methods	**ADMET** Cytochrome P450 3A4 (30/753), human intestinal absorption (48/753)	[43]
	Antibacterial *S. typhimurium* (3/753), *M. tuberculosis* (4/753), *P. aeruginosa* (18/753), *S. aureus* (23/753)	[42]
	Antifungal *C. albicans* (12/753)	[44]
	Antiparasitic *Plasmodium falciparum* (5/753), *Trypanosoma* (15/753), *Caenorhabditis elegans* (16/753), *G. lamblia* (44/753)	[45]
	Antitumour Apel Endonuclease (6/753), Agonist of p53 (7/753), GLI family zinc finger 1 (11/753), TOR pathway (13/753), RecQ-Like Dna Helicase 1 (RECQ1) (14/753), Microphthalmia-associated transcription factor (17/753), Hsf1 protein (22/753), serine/threonine-protein kinase 33 (25/753), miR-21 (27/753), sentrin-specific protease 8 (31/753), Acute myelogenous leukemia (35/753), Steroid receptor coactivator 3 (40/753), dual specificity protein phosphatase 3 (43/753), leukemia (45/753)	[41,42]
	Antivirus HIV-1 (9/753), herpes (42/753)	[46–49]
	Cytotoxicity and genotoxicity Lymphoblastoid (2/753), ATAD5 (10/753), isogenic chicken DT40 (19/753), HEK293 (32/753), MAGI-CCR5 (50/753)	[41,42]
	Endocrine disorders Muscleblind-like protein 1 (26/753), estrogen receptor alpha agonist (34/753), estrogen receptor alpha antagonist (38/753), androgen receptor antagonist (46/753),	[50]
	Lipid disorders Regulator of G-protein signaling 8 (28/753), 1-acylglycerol- 3-phosphate O-acyltransferase ABHD5 (41/753)	[51,52]
	Neuropathies Sphingosine 1-phosphate receptor 1 (20/753), regulator of G-protein signaling 4 (33/753), peripheral myelin protein 22 (36/753), Mitochondria permeability (47/753), DNA damage-inducible transcript 3 protein (49/753)	[53]
	Others Angiogenesis (21/753), osteoporosis (25/753) Anti-Inflammatory (1/753)	[5,40–42,54–56]

Numbers in parenthesis indicate the rank position among 753 activity classes.

Table 5. Antibacterial activity of EtOH extracts (250 µg/mL) from *D. cymosa* and *P. hexandrum* and podophyllotoxin.

Samples	*S. aureus* Mean ± SD	*B. cereus* Mean ± SD	*E. coli* Mean ± SD	EHEC *E. coli* Mean ± SD	*P. aeruginosa* Mean ± SD	*Salmonella* Typhi Mean ± SD
D. cymosa (leaves)	24.76 ± 5.8	0	34.30 ± 2.33	0	23.18 ± 9.94	24.88 ± 3.93
D. cymosa (roots)	67.56 ± 0.8	100.68 ± 0.22	44.68 ± 8.80	16.61 ± 2.13	20.41 ± 8.82	41.33 ± 8.32
P. hexandrum (rhizomes and roots)	58.43 ± 2.7	100.66 ± 0.28	49.28 ± 6.97	18.53 ± 5.21	22.51 ± 15.91	42.52 ± 6.40
Podophyllotoxin	53.34 ± 8.8	0	51.57 ± 9.08	33.00 ± 7.76	30.19 ± 9.39	31.62 ± 7.54

2.4. Anticholinesterasic Activity

We evaluated the AChE inhibitory activity of *D. cymosa* and *P. hexandrum* extracts as well as of PTOX. All extracts showed promising inhibition of AChE activity in the quantitative assay. The EtOH extract of leaves from *D. cymosa* showed the higher percentage of inhibition at 400 µg/mL (Table 6). In the bioautographic assay, the extract of *P. hexandrum* showed a higher intensity of white spots, and this result did not correspond to the quantitative assay. This may have happened due to the possibility of false-positive results when the evaluated extract presents some classes of secondary metabolites, such as tannins and phenolics, which do not directly inhibit the enzyme but can induce enzymatic denaturation [57,58]. On the other hand, the in silico studies did not confirm the AChE inhibition activity (position 577 in 753 models) by the group of 27 lignans investigated in the present study.

Table 6. In vitro evaluation of antioxidant and anticholinesterase activity of ethanolic extracts from *D. cymosa*, *P. hexandrum* and podophyllotoxin.

Samples	Antioxidant Activity (IC$_{50}$-µg/mL)			Inhibition of AChE		Cytotoxicity (CC$_{50}$)
	β-Carotene/linoleic Acid	DPPH Radical Sequestration	TBARS Assay	TLC	Microplate (%I)	
D. cymosa (leaves)	19.48 ± 5.90	133.94 ± 25.60	>50	+	64.22 ± 4.87	368.0 ± 13.8
D. cymosa (roots)	20.76 ± 1.76	43.77 ± 6.69	10.20 ± 1.46	-	40.86 ± 3.70	100.0 ± 5.3
P. hexandrum (rhizomes and roots)	30.70 ± 2.12	24.66 ± 4.45	13.66 ± 1.35	++++	47.04 ± 3.17	338.9 ± 15.1
Podophyllotoxin	>200	>200	>50	-	32.73 ± 5.38	400 ± 10.3
Quercetin	0.3 ± 0.1	NA	NA	NA	NA	NA
Pyrogallol	NA	1.14 ± 0.15	NA	NA	NA	NA
Propylgalate	NA	NA	<20	NA	NA	NA
Physostigmine	NA	NA	NA	++++	89.81 ± 1.16	NA

- not active; + slight inhibition, ++++ strong inhibition.

2.5. Antioxidant Activity

Although, inhibition of AChE is still considered as the main therapeutic strategy to treat Alzheimer's disease, other events are implicated in the physiopathology of this disease. The role of oxygen reactive species (ROS) have been extensively investigated. Therefore, we evaluated the antioxidant potential of the EtOH extracts of *D. cymosa* leaves, roots and *P. hexandrum* rhizomes and roots.

The extract of leaves from *D. cymosa* showed antioxidant activity only in the β-carotene/linoleic acid co-oxidation assay, while the extract of roots showed promising antioxidant activity in all evaluated models. The extract of *P. hexandrum* roots and rhizomes presented significant antioxidant potential in all evaluated models. However, the isolated lignan PTOX (**1**) showed no antioxidant activity (Table 6).

2.6. Cytotoxicity

The results of the cytotoxicity assay, performed in THP-1 cells using the SRB method, indicated a moderate toxicity for the extracts of leaves from *D. cymosa* and of roots and rhizomes from *P. hexandrum*, which showed cell viability lower than 80% at 200 and 400 µg/mL, with CC_{50} of 368.0 ± 13.8 and 338.9 ± 15.1 µg/mL (Table 6). On the other hand, the extract of roots from *D. cymosa* was cytotoxic at all tested concentrations, with a CC_{50} value of 100.0 ± 5.3 µg/mL. The lignan PTOX (**1**) showed a CC50 value of 400.0 ± 10.3 µg/mL (Table 6). In silico prediction indicated a small probability of lignans to present THP-1 cytotoxicity (position 316 in 753 activity classes).

3. Discussion

A broad range of biological activities have been associated with lignans, thus making them an interesting class of secondary metabolites. Even though lignans are known for their toxicity, other biological activities of lignan-rich plant extracts are worth investigating.

The chromatographic characterization of the phenolic content of the EtOH extracts of *D. cymosa* leaves, *D. cymosa* roots and *P. hexandrum* rhizomes and roots indicated a similar chemical composition for the three extracts. The caffeoylquinic acid was only identified in *D. cymosa* leaves, while PTOX hexoside, deoxypodohyllotoxin (**5**) and 4′-demethylpodohyllotoxin (**6**) were only found in *P. hexandrum*. PTOX was found in the EtOH extracts of *D. cymosa* leaves and *P. hexandrum* rhizomes and roots, being the major constituent of the latter. The EtOH extract of roots from *D. cymosa* showed the best antioxidant potential among the evaluated extracts. These results can be explained by the presence of kaempferol as the major constituent of this extract. Numerous studies have shown that flavonoids, such as quercetin and kaempferol, and their heterosides have a wide range of biological activities, including antioxidant, anti-inflammatory and antimicrobial activities [59,60]. Recently, Wang and co-workers [61] observed a DPPH and ABTS radical scavenging activity for kaempferol as well as an inhibition of concanavalin A (Con A)-induced NO or ROS production in LPS-induced RAW 264.7 macrophage cells [61]. In another study, kaempferol was able to scavenge the superoxide anion, hypochlorous acid, chloramine and nitric oxide [62] as well as showed scavenging ability on superoxide anion produced by electrochemical reduction of oxygen [63].

The lignan content of *Podophyllum* and *Diphylleia* species varies both qualitatively and quantitatively, according to the data previously published. UPLC-DAD-MS methods are largely employed for the identification and quantification of lignans in the aforementioned species [64–67]. Sharma and Arora identified four aryltetralin lignans in the MeOH extract of rhizomes from *P. hexandrum* [64]. In another study, Sharma and Kumar evaluated the extracts of leaves and roots of *P. hexandrum* obtained from different locations by HPLC-ESI-MS, and found that the podophyllotoxin content was twice as high in the roots in comparison with the content found in the leaves of *P. hexandrum* found in high altitudes [65]. Avula e coworkers evaluated the content of podophyllotoxin (**1**), 4′-demethylpodophyllotoxin (**6**), α-peltatin (**18**) and β-peltatin (**19**) in samples from *P. peltatum*. 4′-demethylpodophyllotoxin (**6**) and α-peltatin (**18**) were the main lignans observed for this species, while the content of PTOX varied from 0.004–0.77% when plants colletcted from various colonies within the same site were evaluated [66]. UPLC-ESI-MS methods can also be employed for the pharmacokinetic studies. The lignans podophyllotoxin (**1**), 4-*epi*-podophyllotoxin (**2**), and 4′-demethylpodophyllotoxin (**6**) were simultaneously evaluated in rat plasma using a UPLC-ESI-MS method after oral administration of the EtOH extract of *Diphylleia sinensis*, 367 mg/kg, to Wistar rats [67]. It is noteworthy to mention that this is the first report of the characterization of flavonoids and caffeoylquinic acid in *D. cymosa*.

Some predictions of biological activities observed in this study were in accordance with the ethnopharmacological uses for both plant species, as well as for the other isolated *podophyllum* lignans. However, some new predict activities such as angiogenesis, osteoporosis, myotonic dystrophy and autoimmune diseases were also observed (Table 4). A reasonable agreement could be noted between predictions made with SVM and Naïve Bayes modeling methods, although the use of both methods can produce results that are more reliable, as indicated in the validation step. Thus, all predictions discussed below will be based in a unique rank of 753 activity classes, where the best rank between both methods and the averaged rank were used to produce the final ordered list.

The in silico approach showed a high probability of lignans have an anti-inflammatory activity (position 1 in 753 activity classes). This activity has been described before for lignans [11]. The most probable target related with this activity is NF-κB activation. The nuclear factor NF-κB pathway has been considered a classical proinflammatory signaling pathway [68].

The cytotoxic activity of lignans has been much explored, as well as their mechanism of action. The results from in silico prediction showed a low cytotoxicity against THP-1 cells (316/753), which is in accordance with our experimental results. Podophyllotoxin have been evaluated in several models using THP-1 cells, with moderate to low toxicity reported. The effect of podophyllotoxin on IL-1β and TNF expression was evaluated using THP-1 cells at 10 μM and no cytotoxicity was observed [69]. However, according with our calculations, lignans could present cytotoxicity against human lymphoblastic (position 2 in 753 activity classes), isogenic chicken DT 40 (position 19 in 753 activity classes), HEK293 (position 32 in 753 activity classes) and MAGI-CCR5 (position 50 in 753 activity classes) cells. Derivatives have been synthetized [11] to obtain new antitumour compounds. Many targets involved in cancer therapy have also been predicted as potential targets for lignans, such as AP1 endonuclease (position 6 in 753 activity classes), tumor antigen p53 (position 7 in 753 activity classes), GLI family zinc (position 11 in 753 activity classes), RecQ-like DNA helicase 1 (position 14 in 753 activity classes). Other targets can be found in Table 4 and Table S3 of Supplementary Material.

The lignans also showed high probability to be active against *Salmonella typhimurum* (position 3 in 753 activity classes), *Mycobacterium tuberculosis* (position 4 in 753 activity classes), *Staphylococcus aureus* (position 23 in 753 activity classes), *Pseudomonas aeruginosa* (position 18 in 753 activity classes) and *Escherichia coli* (position 52 in 753 activity classes). In this study, all these microorganisms were evaluated, with the exception of *M. tuberculosis*. These predictions are in agreement with ours in vitro results from the antibacterial assay. The activity against *M. tuberculosis* is consistent with the ethnopharmacological use in the Eastern world folk medicine [42].

Regarding the in vitro antibacterial activity, the ethanolic extracts from rhizomes and roots of *P. hexandrum* and from roots of *D. cymosa* were the most effective samples against Gram-positive bacteria. PTOX (1) showed significative activity against *S. aureus* and *E. coli*, while no antibacterial activity was observed against the other Gram-negative pathogens. The antibacterial activity of lignans have already been reported. Nanjundaswamy and coleagues [70] reported a relevant antibacterial activity of two synthetic precursors of PTOX against *E. coli*, *P. aeruginosa* and *Salmonella* Typhi. Other authors have also indicated antibacterial activity of extracts from the rhizomes of *P. hexandrum* [71] and analogues of PTOX against *P. aeruginosa* [72], contrasting the results of the present study. This finding supports the ethnopharmacologial uses of both plant species. No mention of antimicrobial activity from *D. cymosa* was reported so far.

Interestingly, PTOX did not elicit a higher inhibition of the growth of most of the evaluated microorganisms, in comparison with the crude extracts. This is probably due to the synergistic effect of secondary metabolites in the crude extracts.

Considering *S. typhimurum*, one target related to the predicted activity is the *PhoP* regulation. This target is composed by two genes *PhoP* and *PhoQ*, associated with virulence, survival inside the macrophages and defensing resistance of *S. typhimurum* [73]. Against *M. tuberculosis*, the possible target is a transaminase BioA, an enzyme involved in biotin biosynthesis, representing a potential target to develop new antitubercular agents [74]. The putative target involved in the activity against *S. aureus* is the Quorum sensing (QS), defined by Reuter and co-workers [75] as the exchange of chemical signals in bacterial populations, that depends on the bacterial density. QS is responsible for virulence in the clinically relevant bacteria. It has been suggested as a promising target for developing new anti-infective compounds. It was not found a specific target related to *E. coli*.

The extracts of roots from *D. cymosa* and rhizomes and roots of *P. hexandrum* showed the best antioxidant potential, while the extract of leaves from *D. cymosa* showed an anticholinesterasic activity. All the extracts showed moderate toxicity to the THP-1 cells, and the cytotoxic activity observed for the extracts of leaves of *D. cymosa* and rhizomes and roots of *P. hexandrum* were similar to that observed for PTOX (**1**), which did not exhibit antioxidant and anticholinesterase activity. These results indicate that PTOX (**1**) could be the major cytotoxic lignan in these extracts, while other phenolic constituents could be responsible for the antioxidant and anticholinesterase activities observed. However, in the extract of roots of *D. cymosa*, which showed the lowest CC_{50} value, the lignan PTOX (**1**) was not found, indicating that other unidentified minor compound is the responsible for the observed cytotoxicity.

According to the in silico prediction (position 49 in 753 models), the lignans can act DNA damage-inducible transcript 3 protein (C/EBP homologous protein, CHOP) which has been proposed as a target of treatments for some neurodegenerative diseases as Alzheimer's diseases [52]. Furthermore, according to Naïve Bayes prediction (Table S2 in the Supplementary Material) the lignans can possibly act over the protein Tau indicating a potential application to treat the Alzheimer's disease and other diseases [76].

Another important effect of lignans is the immunosuppressive activity what can be associated to their uses as therapeutic agents against psoriasis and rheumatoid arthritis, as well to prevent the acute rejection of transplanted organs [77,78]. One putative target that could explain this activity is the sphingosine 1-phosphate receptor 1, predict as potential target for lignans (position 20 in 753 activity class), due to its involvement in immune system modulation [79].

The anti-viral activity of lignans is well known since it was first cited in 1942 as a treatment for veneral wart (*Condyloma acuminatum*), an ailment caused by a papilloma virus [80]. There have been reported effects against HIV, herpes simplex, influenza, vaccinia viruses, and measles [42]. These results also confirm our in silico predictions as the activities against HIV-1 and herpes appear in position 9 and 42, respectively, among 753 activity classes.

The hypolipidemic properties of lignans, as reported by Iwasaki and co-workers [81] are in consonance with our prediction as potential target for these compounds the 1-acylglycerol-3-phosphate *O*-acyltransferase, a protein activator of the lipase Atgl [82], predict in position 41 among 753 activity classes.

There are no reports in literature of the evaluation of the anticholinesterase activity of Berberidaceae plants or of the lignan PTOX (**1**), but other lignans have already demonstrated in vitro AChE inhibitory activity [83]. In the study by Hung and co-workers [84], sixteen lignans were isolated from *Schizandra chinensis* and were evaluated for the inhibition of AChE in vitro. Among the evaluated compounds, only five were active with IC_{50} lower than 15 µM. Schisandrin was evaluated in vivo by Itoh et al. [85] and was active at 3 mg/kg. El-Hassan et al. [86] demonstrated that Syringaresinol inhibited AChE in vitro with an IC_{50} value of 200 µg/mL. Still exploring the investigation of AChE inhibition activity by lignans, Salleh et al. [87], isolated five lignans from the stem extract of *Beilschmiedia pulverulenta*, which were evaluated in vitro in the microplate inhibition assay for the AChE, showing have IC_{50} values in the range of 179.8 to 504 µM.

Regarding the antioxidant capacity of these compounds, studies by Wang et al. [88] evaluated the ability of the extracts of *S. chinensis* and *S. sphenanthera* to scavenge the DPPH radical. The authors suggested that variations in lignan content between the extracts lead to different antioxidant activities. The *S. chinensis* species showed higher activity due to the higher content of the lignans Schisandrol A and B and Schinsandrin B.

In a study by Dar et al. [63], the antioxidant capacity of PTOX (**1**) was evaluated in the DPPH sequestration and the TBARS lipid peroxidation assays. In both experiments, PTOX presented IC_{50} value higher than 250 µg/mL.

These results, along with the data found in literature, indicates that PTOX (**1**) is probably not involved in the antioxidant activity observed for *P. hexandrum* and *D. cymosa* extracts, but other lignans or other phenolic constituents may account for the observed activity.

4. Materials and Methods

4.1. Plant Material

The leaves and roots of *D. cymosa* plants were collected at the University of Nottingham (Nottingham, UK) in June, July and August 1995. The authentication of plants was confirmed by Julian MH Shaw (Senior Registrar, Horticultural Taxonomy, Royal Horticultural Society, Wisley, Working, UK). Dried rhizomes and roots of *P. hexandrum* were purchased from United Chemical and Allied Products, Calcutta, India. Both plants were kindly provided by Dr Paul M. Dewick (University of Nottingham, Nottingham, UK). The purified lignans used in this study were isolated and identified in a previous work [24].

4.2. Preparation of Extracts

Powdered material (4.8 g) of *D. cymosa* (leaves and roots) and *P. hexandrum* (rhizomes and roots) was extracted with 150 mL ethanol 92.8° by sonication for 10 min at room temperature. The solutions were filtered and the solvent was removed under reduced pressure in a rotary evaporator. The process was repeated three times, and the combined extracts were dried in a water bath (40 °C) to yield the ethanolic crude extracts. The latter were 15.83% yield for leaves and 14.03% for roots of *D. cymosa* respectively, as well as 8.33% for rhizomes and roots of *P. hexandrum*. All samples were stored at 4 °C in amber flasks until required.

4.3. Chromatographic Characterization of D. cymosa and P. hexandrum Extracts by UPLC-DAD-ESI-MS/MS

The UPLC-ESI-MS profiles were obtained in a UPLC system coupled with DAD and ESI-TQ-MS detectors. The samples were prepared at 1.0 mg/mL using MeOH, centrifuged (10,000 rpm, 10 min) and then filtered through 0.22 μm PTFE filters. A portion of 3 μL of each sample was injected into the chromatographic system.

4.3.1. Chromatographic Conditions

The elution was carried out using a gradient elution of 0.1% formic acid in deionized water with (A) and 0.1% formic acid in acetonitrile (B), in a gradient elution from 5 to 36.5% of B in 3.5 min, 36.5–54.5% from 3.5 to 7.5 min, 54.5 to 95% from 7.5 to 9.0 min, with a final isocratic period at 95% B from 9.0 to 10.0 min. The analyses were performed in an Acquity-HSS-ODS (150 × 4.0 mm, 1.8 μm) C-18 column at 40 °C.

4.3.2. Mass Spectrometric Conditions

The mass spectrometer (Waters TQ-XS, Milford, DE, USA) was operated in negative and positive electrospray ionization modes and spectra were recorded by scanning the mass range from m/z 100 to 1000 in both MS and MS/MS modes. Nitrogen was used as drying, nebulising and collision gas. Drying gas flow rate was 12 L/min. The heated capillary temperature was set at 350 °C and nebulizer pressure at 45 psi. The source parameters such as capillary voltage (VCap), fragmentor, skimmer and octapole voltages were set at 3500 V, 175 V, 65 V and 750 V, respectively. For the MS/MS analysis, a ramp of collision energies, from 15 to 70 eV, was used. The obtained data were processed using the MassLynx (version B 04.00) software (Waters, Milford, DE, USA).

4.4. In Silico Studies

4.4.1. In Silico Prediction of Biological Activity of Lignans

The molecular descriptors used to build the models were the multi-conformational 3-point pharmacophore fingerprints produced by in-house software 3D-Pharma [89]. Each conformation of each compound was treated separately, and its heavy atoms were converted to potential pharmacophore points (PPP) which could be one or more of the following six types: hydrogen bond donor, hydrogen bond acceptor, positively charged, negatively charged, aromatic and lipophilic. For each conformation all combinations of three pharmacophore points in the 3D space (triplets) were calculated to compose a pharmacophore fingerprint. The union of uni-conformational fingerprints produce a unique modal fingerprint for each compound [90] which was used for all subsequent calculations. In the cases where the datasets were obtained from PubChem Bioassay the conformations were downloaded from PubChem Compound [91], in all other cases the conformations were produced with OMEGA software from OpenEye with standard options limited to a maximum of 10 conformations [92,93].

The multi-conformation (modal) pharmacophore fingerprint of active and inactive compounds of each dataset were submitted to the in-house software ExCVBA [94] to build and validate machine learning models using support vector machine (SVM) and Naïve Bayes approaches. Each dataset was used to produce SVM and Naïve Bayes ensemble of models through recurrent stratified random partition of the original dataset to produce a training set composed of 70% of the original dataset and a validation set composed of 30% of the original dataset. This process was repeated at least 30 times and the average scores of each compound over the models in which it appears in the validation set was used to assess the modeling performance with the area under the Receiver Operating Characteristic curve (AUC-ROC), as well for activity prediction of new compounds, as described below.

The calculation of AUC-ROC was performed as defined in Equation (1) with the rank sum of active compounds, which is also called Mann-Whitney U test:

$$AUC = 1 - \frac{1}{N_a} \sum_{j=1}^{N_a} \frac{(r_j - j)}{N_i} \tag{1}$$

where r_j is the rank of the *j*th active, N_a, and N_i are the number of active and inactive compounds, respectively. When ties occur between active and inactive the rank of the active were scaled by interpolation to avoid any bias. The expected standard error of AUC in this paper follows the proposition of Nicholls [95] (Equation (2)), based on Hanley [96] approximation for an 'typical' ROC curve:

$$AUC = w \pm t_{95\%} \sqrt{\frac{w^2(1-w)/(1+w)}{N_a} + \frac{w(1-w)^2/(2-w)^2}{N_i}} \tag{2}$$

where w is the observed AUC. In the estimation of *t*-statistic at 95% ($t_{95\%}$) the number of degrees of freedom, ν, follows the proposition of Nicholls [97] (Equation (3)), from the variances of actives and inactives and using the Welch–Satterthwaite formula:

$$\nu_{eff}^{AUC} = \frac{\left(\frac{AUC}{1+AUC}N_i + \frac{1-AUC}{2-AUC}N_a\right)^2}{\frac{\left(\frac{AUC}{1+AUC}N_i\right)^2}{N_a-1} + \frac{\left(\frac{1-AUC}{2-AUC}N_a\right)^2}{N_i-1}} \tag{3}$$

The SVM models were built with LibSVM [98] software with linear kernel option. The cost C, which is a penalty parameter applied to misclassified compounds on the training data, was selected with exponentially growing sequences from 2^{-12} to 2^{+6}, by means of a 5-fold cross-validation (CV) using the Power Metric [39,99] at $\chi = TPR + FPR = 0.5$ as an optimization objective metric to assure early recovery of active compounds. The Naïve Bayes model was produced using Perl module from CPAN repository [100] which was incorporated into the ExCVBA software (NEQUIM, Belo Horizonte, Brazil).

In the prediction phase the modal multi-conformational pharmacophore fingerprints of the new compounds were submitted to SVM or Naïve Bayes model ensemble and the average raw scores obtained were converted into probabilities through comparison with the score distribution of active and inactive compounds used to build the models (validation sets only), producing a measure of belonging to these two subsets [101].

Considering the SVM or Naïve Bayes score of the new compound as the threshold, the probability of it being active (Pa) is equal to the fraction of active compounds with a worse score (FNR) than the compound under prediction (Equation (4)) and the probability of being inactive (Pi) is equal to the fraction of inactive compounds with a better score (FPR) than the lignan under prediction (Equation (5)), as described elsewhere [36]:

$$Pa = \frac{FN}{N_a} = FNR \tag{4}$$

$$Pi = \frac{FP}{N_i} = FPR \tag{5}$$

where N_a and N_i are the number of active compounds and the number of inactive compounds; FN is the number of active compounds with worse scores than the threshold; and FP is the number of inactive compounds with better scores than the threshold. For each model ensemble, the difference between the Pa and Pi (Pa-Pi) was used to evaluate the potential activity of the modeled compounds. Although the variance of Pa-Pi, as well its limits, can be analytically estimated from the variances of Pa and Pi, as described before [36], in this work we used a new approach, as described below.

The variance of the SVM or Naïve Bayes scores of each compound when it appears in the validation sets and the standard error of the mean (SEM) (Equation (6)) were used to build a better estimation of the limits of the prediction:

$$SEM = \frac{SD}{\sqrt{N}} \tag{6}$$

where SD is the standard deviation, defined as the root mean squared of the variance. Accordingly, the limits of scores, computed at 95% of confidence interval, were estimated by Equation (7):

$$score_{limits} = score_{mean} \pm t_{95\%} * SEM \tag{7}$$

The values of mean score were used to calculate the mean value of Pa and Pi, while the maximum and minimum values were used to estimate their upper and lower limits, as exemplified in the Figure 6.

In the prediction phase, the scores of each unseen compound over all 30 models of the ensemble are averaged and the limits are calculated with 95% of confidence interval. The average score is used to calculated the Pa-Pi mean from the average value of Pa minus the average value of Pi. The upper limit of the score is used to calculate the upper limit of Pa-Pi, using the higher value of Pa minus the lowest value of Pi, whereas the lower limit is used to calculate the lowest value of Pa-Pi, using the lower value of Pa minus the higher value of Pi. The limits of Pa-Pi calculated in this way provide a better confidence interval of the prediction, although much larger than the analytical estimate it is more useful as it can be used as an applicability domain approximation, likewise the proposition of Norinder and co-workers [102]. If the new compound looks like an outlier the score variance over all 30 models can be very large and this will reflect in the range of values of Pa-Pi. If the lower value of Pa-Pi falls below zero the compound cannot be predicted as active.

Molecules **2018**, *23*, 3303

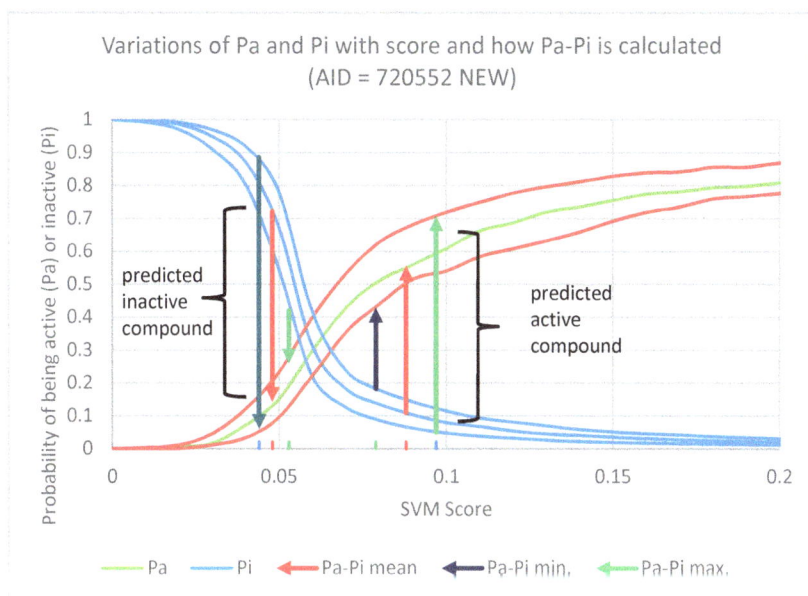

Figure 6. Variation of Pa and Pi probabilities with the SVM score in the model for agonists of the P53 signaling pathway (AID = 720552) together with its upper and lower limits. Calculation of the limits of Pa-Pi for each new compound involve the upper and lower limits of the score over all 30 models of the ensemble. Maximum Pa-Pi will be calculated from the higher value of Pa minus lower value of Pi and minimum Pa-Pi will be calculated from the lower value of Pa minus higher value of Pi. If this value falls below zero the compound is considered inactive.

4.4.2. In Silico Prediction of Putative Activity Classes of Lignans

To predict potential biological activities of lignans used in this study, all 1987 datasets of Active-IT system were grouped into 924 activity classes. This grouping approach will make it easier to predict the larger datasets of compounds with a common substructure. To make the prediction of most probable activity classes all data points are filtered with a pre-defined threshold and, additionally, a positive minimum limit of Pa-Pi. In the next step, the instances that pass the filters are counted. The final activity class score is calculated by the sum of instances that pass the filter among all models belonging to the same class, divided by the number of compounds and by the square root of the number of models of the same class.

4.5. *Evaluation of Antibacterial Activity*

Antibacterial susceptibility was performed using the modified microdilution method for bacteria [36,103], against Gram-positive (*Staphylococcus aureus* ATCC 25923 and *Bacillus cereus* ATCC 11778) and Gram-negative recognized pathogens (*Escherichia coli* ATCC 11775, Enterohemorrhagic *E. coli* ATCC 43895-EHEC, *Pseudomonas aeruginosa* ATCC 10145 and *Salmonella choleraesuis* subs. *choleraesuis* sorotype Typhi BM/Panama-TY2).

The experiments were performed in 96 well microplates. The extracts and isolated substance solubilized in DMSO (Vetec^{TM}) at 50 mg/mL were diluted in Mueller Hinton Broth (MHB, Difco) at a concentration of 500 μg/mL. The bacterial isolates, criopreserved at −80 °C were grown in Tryptic Soy Agar (TSA, Acumedia$^®$) plates, at 37 °C for 18–24 h. The inoculum was adjusted in saline solution in a spectrophotometer at 625 nm, to a concentration of 1–5 × 108 CFU/mL and then diluted in MHB to 1–5 × 105 CFU/mL.

Volumes of 100 µL of the inoculum were added to wells containing 100 µL of the extracts at 500 mg/mL, in triplicate (one well intended for extract control), resulting in a final concentration of 250 µg/mL. Chloramphenicol at 50 µg/mL and 0.5% DMSO were used as positive and negative controls, respectively. Evaluation of microbial growth was carried out by adding the inoculum to a well containing only MHB. Sterility of the culture medium was also confirmed by incubation in the assay plate. Assays were performed in triplicate. The microplates were incubated at 37 °C for 24 h. As an indicator of microbial growth, 20 µL of 2,3,5-triphenyltetrazolium chloride (TTC, Sigma-Aldrich, St. Louis, MO, USA) at 5 mg/mLwere added to each well. The plates were incubated at 37 °C for 3 h and then the TTC was solubilized with 100 µL of sodium lauryl sulfate solution in isopropanol 7 µg/mL and measured in a microplate reader at 485 nm. The result was expressed as the percentage of inhibition compared with the microbial growth control [104], and was considered positive for antibacterial activity for those extracts with an inhibition higher than 50%.

4.6. Inhibition of Acetylcholinesterase

4.6.1. Bioautographic Assay

The bioautographic assay was performed by thin layer chromatography (TLC) according to the method proposed by Marston, Kissling and Hostettman [105]. The extracts were solubilized in methanol (Merck) at 20 mg/mL, and a portion of 10 µL of this solution was applied on a TLC plate (silica gel 60 F254, Merck, Darmstadt, Germany), which was eluted with CHCl3:MeOH (9:1). The acetylcholinesterase enzyme (*Electropardus electricous* type VI-S, Sigma-Aldrich, St. Louis, MO, USA) solution was prepared by diluting 1 KU of the enzyme in 30 mL of Tris/HCl buffer, pH 7.8, with the addition of 30 mg of bovine serum albumin (BSA). The enzyme solution was sprayed on the TLC plate and incubated in a humid chamber at 37 °C for 20 min. The substrate solution was prepared using 1-naphthyl acetate (2.5 mg/mL) and Fast Blue B salt (2.5 mg/mL), and sprayed on the TLC plate after the incubation period. The purple staining shows the enzymatic activity and the appearance of white bands after 5 min indicates inhibition of AChE activity. The result was expressed by the intensity of the white bands observed.

4.6.2. Microplate Assay

AChE inhibition assay was performed on a 96-well microplate according to Ellman's method [106] with adaptations. Initially, the extracts were solubilized on MeOH, at 20 mg/mL, and diluted to 4 mg/mL using Tris/HCl buffer solution (50 mM, pH 8.0). A portion of 25 µL of the samples (4 mg/mL) were added to the wells of the microplate, as well as 50 µL of 0.1% BSA solution in Tris/HCl Buffe (pH 8.0), 125 µL of DTNB in Tris/HCl (3 mM) containing NaCl (10 mM) and MgCl2 (20 mM). Subsequently, 25 µL of the ATCI aqueous solution (15 mM) was added to all wells. Then, a background reading was performed at 405 nm using microplate reader (Multiskan Go, Thermo Scientific, Whaltham, MA, USA). The enzymatic reaction was initiated after the addition of 25 µL of AChE solution (AChE from *Electropardus electricous* type VI-S, 0.2 U/mL, Sigma-Aldrich, St. Louis, MO, USA) in Tris/HCl buffer (50 mM, pH 8.0) containing 0.1% BSA. The kinetic cycle was performed in a period of 25 min, with readings at 5 min intervals, at 405 nm. Physostigmine (Eserine, Sigma, St. Louis, MO, USA) was employed as the positive control. A blank was performed in the same assay conditions using MeOH. The assay was performed in triplicate and the % of inhibition (%I) was calculated according to the following equation: %I = [(a − b)/a] × 100, where a = ΔA/min of control; b = ΔA/min of test sample; ΔA = change in absorbance between time *x* and time zero. Extracts with enzyme %I higher than 40% at 400 µg/mL were considered promising.

4.7. Evaluation of Antioxidant Activity

4.7.1. β-Carotene/Linoleic Acid Co-Oxidation Assay

The evaluation of the antioxidant activity using the β-carotene/linoleic acid co-oxidation system was performed according to Duarte-Almeida et al. [107], with adaptations. The extracts were solubilized in MeOH (2.2 mg/mL) and further diluted to concentrations ranging from 3.125 to 200 μg/mL. An aliquot of 25 μL of sample solutions were added in the 96-well microplate. 25 mg of linoleic acid and 100 mg Tween 20 (Sigma, St. Louis, MO, USA) were added to a round-bottom flask containing 1 mL of a β-carotene solution (1 mg/mL, Sigma, St. Louis, MO, USA) in chloroform. The latter was removed in a rotatory evaporator and, then, 50 mL of aerated water was added to the flask, affording the β-carotene emulsion. Portions of 250 μL of the freshly prepared emulsion were added to the wells corresponding to the odd columns of a microplate. A blank emulsion was prepared as described above, without the addition of the β-carotene solution, and portions of 250 μL were added to the even columns of the microplate. MeOH and quercetin (20 μg/mL, Sigma, St. Louis, MO, USA) were employed as negative and positive controls, respectively. Readings were performed immediately at 470 nm. After the first reading, the microplate was incubated at 45 °C and the kinetic cycle was performed with readings at 15 min intervals for a total of 120 min. The antioxidant activity was expressed as % of inhibition of lipid peroxidation (%I), using the formula (I% = Ac (initial abs − Final abs) − Aam (final abs − Initial abs)/Ac × 100). Assays were performed in triplicate and IC_{50} values were determined by non-linear regression using GraphPad Prism, version 6.0 (GraphPad Software, San Diego, CA, USA). It was considered promising when the extracts exhibited IC_{50} values lower than 50 μg/mL.

4.7.2. DPPH Radical Scavenger Activity

The evaluation of the antioxidant activity using the 2,2-diphenyl-1-picrylhydrazyl radical (DPPH) was performed according to the method proposed by Mensor et al. [108], with adaptations. The samples were dissolved in MeOH at concentrations ranging from 1.0 to 200 μg/mL. A portion of 250 μL of each sample was added to a 96-well microplate. MeOH and pyrogallol (50 μg/mL) were employed as negative and positive controls, respectively. Then, 100 μL of either MeOH (blank wells) or DPPH solution (120 μg/mL, reaction wells) was added to the microplate. The reading was performed on a microplate reader (EL808IU-Biotek model, Biotek, Winooski, VT, USA), at 5 min intervals for a total of 45 min, at 515 nm. The percentage of radical scavenging activity (%RSA) was calculated using the following equation: % RSA = ((AC − AS)/AC) × 100, where AC is the absorbance of control and AS is the absorbance of samples taken at 35 min. The results were expressed in EC_{50} (effective concentration for 50% capture of the radicals) determined by non-linear regression using GraphPad Prism, version 6.0. In this assay, extracts showing EC_{50} values lower than 15 μg/mL were considered promising.

4.7.3. Thiobarbituric Acid Reactive Substances (TBARS) Assay

The evaluation of antioxidant activity using the TBARS assay [109], with adaptations. Briefly, the samples were solubilized in methanol (50.0 to 31.25 μg/mL) and a portion of 25 μL of each extract solution was added to 10 mL test tubes in triplicate (two reaction tubes and one control). The tubes corresponding to the blank, negative and positive controls were prepared by adding 25 μL of deionized water, MeOH or propylgalate (Sigma, St. Louis, MO, USA, 0.1 mM), respectively. The phospholipid liposomes were prepared using bovine brain extract type VII (Sigma Aldrich, St. Louis, MO, USA, 10 mg/mL) in phosphate buffered saline (PBS) solution. The suspension was subjected to an ultrasonic bath until the formation of the liposomes. Then, 50 μL of the liposome suspension was added to the tubes corresponding to the blank, negative and positive controls, and sample reaction tubes. In sequence, 125 μL of PBS was added to all tubes. The lipid peroxidation was initiated by the addition of 25 μL of $FeCl_3$ solution (Vetec, Rio de Janeiro, Brazil, 1 mM) and 25 μL of ascorbic acid solution (Vetec, Rio de Janeiro, Brazil, 1 mM) in the tubes corresponding to the negative control, positive control and sample reaction. The blank tubes were prepared by adding 75 µL of deionized water. All tubes were incubated for 20 min at 37 °C, and after that, 25 µL of BHT (di-tert-butyl methyl phenol, 2% in ethanol), 125 µL of thiobarbituric acid (Merck, Darmstadt, Germany, 1% in 50 mM NaOH), and 125 µL of hydrochloric acid (1%) were added in all tubes. All tubes were shaken vigorously and incubated for 30 min at 80–90 °C. Then, 500 µL of *n*-butanol (Vetec) was added to all tubes which were shaken and placed in the centrifuge for 10 min at 3600 rpm. The butanolic phase was removed and transferred to a 96-well microplate, and the absorbance was recorded on a Multiskan Go microplate reader (Thermo Scientific, Whaltham, MA, USA) at 532 nm. The percentage of inhibition (%I) of lipid peroxidation was calculated as: %I = ((AC − AS)/AC) × 100 where AC is the absorbance of control, AS is the absorbance of samples. Assays were performed in triplicate and IC_{50} values were determined by non-linear regression using GraphPad Prism, version 6.0 (GraphPad Software, San Diego, CA, USA). Extracts showing IC_{50} values lower than 50 µg/mL were considered promising.

4.8. Evaluation of Cytotoxicity in THP-1 Cells

The assay was performed following the sulforhodamine B (SRB) method [110]. THP-1 cells (acute human monocytic leukemia cells, ATCC TIB-202) were cultured in a RPMI medium supplemented with 10% fetal bovine serum. A cell suspension was prepared at the density of 1×10^6 cells/mL with 2 µL of PMA (phorbol myristyl acetate, Sigma, St. Louis, MO, USA, 30 µg/mL). A portion of 100 µL of this suspension was transferred to the wells of a microplate. The plate was incubated at 37 °C, in a 5% CO_2 atmosphere, for about 12 to 16 h to allow cells to adhere. After that, 100 µL of each sample (400, 200, 100, 50 and 25 µg/mL) were added to the wells. The plate was then incubated at 37 °C for 18 h. After incubation, 100 µL of cold trichloroacetic acid (TCA, 10%) solution was added in each well. The plate was incubated at 4 °C for 1 h. The supernatant was discarded and the wells were washed three times with 200 µL of distilled water. Then, the wells were allowed to dry for 24 h. After this period, 100 µL of sulforhodamine B solution (0.057%) was added to each well and the plate was kept at room temperature for 30 min. After that, the wells were washed with 200 µL of 1% acetic acid solution, four times. The plate was allowed to dry for 24 h. In sequence, 100 µL of 10 mM Tris Base solution (pH 10.5) was added to all wells. The plate was then stirred for 5 min inside the microplate reader. The optical density was measured at 510 nm in a microplate reader. A toxicity control of the sample diluent was performed. Cell viability was calculated according to the formula (A − B/C − B) × 100, where A, B and C corresponds to the absorbance measured for the samples, blank and negative control, respectively. The extracts were considered non-toxic when cell viability was higher than 80%.

4.9. Statistical Analysis

The results were expressed as mean ± SD of three independent experiments and IC_{50} values were determined by non-linear regression using GraphPad Prism, version 6.0 (GraphPad Software, San Diego, CA, USA). The statistical significance of differences was evaluated using one-way ANOVA in comparison with control groups. Results were considered different when $p < 0.05$.

5. Conclusions

In this study the extracts from the leaves and roots of *D. cymosa* and from the rhizomes and roots of *P. hexandrum* showed antibacterial activity against *B. cereus* and *S. aureus*. On the other hand, podophyllotoxin inhibited the growth of *S. aureus* and *E. coli*. It is important to highlight that the antimicrobial activity of *D. cymosa* was not reported before. *D. cymosa* leaves showed anticholinesterase and antioxidant activities, while the extracts of roots showed antioxidant activities in all evaluated models. The extracts from the rhizomes and roots of *P. hexandrum* presented antioxidant activities in two models used except in the DDPH assay. Additionally, the evaluated extracts from both species were shown to be moderately toxic to THP-1 cells.

According to the chromatographic profiles, the presence of PTOX (**1**), deoxypodophyllotoxin (**5**), 4′-demethylpodophyllotoxin (**6**), podophyllotoxone (**11**), α-peltatin (**18**) and β-peltatin (**19**) were characterized in the EtOH extracts of *D. cymosa* and *P. hexandrum*. These lignans were previously isolated from both Berberidaceae species. Podophyllotoxin (**1**) was the major constituent of *P. hexandrum* extract while kaempferol and its hexoside were the main constituents of *D. cymosa* leaves and roots, respectively. To the best of our knowledge, this is the first report of the characterization of flavonoids and caffeoylquinic acid in *D. cymosa*.

Furthermore, it might be useful to investigate whether *D. cymosa* could be a possible source of podophyllotoxin, and how this could be achieved in the future.

Our Active-IT system proved to be very useful in predicting a broad spectra of biological activities. It was well validated in relation to both the direct score of probabilities to be active or inactive and indirect activity class score. Both approaches produced AUC-ROC values higher than 0.7, even for datasets where no information about lignans was provided. In particular, our in silico studies using machine learning methods was very effective in confirming both ethnopharmacological uses and biological activities of *D. cymosa* and *P. hexandrum* extracts. Moreover, the prediction results suggest that extracts of *D. cymosa* and *P. hexandrum* could provide insights in the research against Alzheimer's, antimicrobial and anti-inflammatory diseases. In addition, new predicted activities against diseases related to the endocrine system, lipidic disorders, neuropathies, osteoporosis, as well as antiangiogenic should be investigated in the search for new drugs with a clinical use. The aforementioned activities and their associated targets should be more fully explored with the aim of obtaining new uses for known lignans as well as to contributing to the understanding of the mechanism of the actions of these natural compounds from *D. cymosa* and *P. hexandrum*. It would be desirable if the new predicted activities could attract more attention from researchers and students, and hopefully the results would be helpful to the worldwide community.

Supplementary Materials: The following are available online at http://www.mdpi.com/1420-3049/23/12/3303/s1, Figures S1 and S2: Chromatographic profiles obtained by UPLC for the EtOH extract of leaves (S1) and roots (S2) from *D. cymosa* with detection by DAD and ESI-MS, Figure S3: Chromatographic profiles obtained by UPLC for the EtOH extract of rhizomes and roots from *P. hexandrum*, with detection by DAD and ESI-MS, Figures S4–S22: DAD, ESI-MS and ESI-MS/MS spectra obtained online by UPLC-DAD-ESI-MS/MS for peaks 1-2-3-4-5-5′, 4′-demethylpodophyllotoxin, peak 6, alpha-peltatin, peaks 7–8, podophyllotoxin, peaks 9–10, beta-peltatin, peaks 11–12–13, deoxypodophyllotoxin, respectively; Table S1: Biological activities better classified by the SVM method for the 27 lignans analyzed, Table S2: Biological activities better classified by the Naïve Bayes method for the 27 lignans analyzed, Table S3: Best predict activity classes for lignans using both methods SVM and Naïve Bayes, Table S4: Bioassays used to validate the bioactivity predictions of lignans, Table S5: Validation of lignans bioactivity prediction with new models built without lignans, Table S6: Main targets predicted associated to anti-inflammatory, antibacterial and anti-protozoa activities of lignans, Table S7: Targets related to anti-inflammatory effects, cytotoxicity THP-1, *Salmonella typhimurium*, *Mycobacterium tuberculosis*, *Plasmodium falciparum*, *Pseudomonas aeruginosa* and *Escherichia coli*.

Author Contributions: C.G.S. selected the plant species, designed the study, as well as prepared the extracts and isolated the lignans; M.P.R. carried out the assays for the inhibition of AChE, antioxidant activity, and cytotoxicity; P.R.V.C. characterized the extracts by UPLC-DAD-ESI-MS/MS analysis; D.O.S. performed the antibacterial assays; V.L.A. and J.C.D.L. were responsible for the in silico studies; J.M.H.S. was responsible for propagating the rhizomes of *D. cymosa* in the Pharmacy Garden at the University of Nottingham, and for the taxonomic authentication of plants. The authors wrote and approved the final manuscript.

Funding: M.P. Rocha received financial support by Fundação Ezequiel Dias (261/2015), and V.L. Almeida (249299/2013-5) and J.C.D. Lopes (202407/2014-4) received their fellowships from Conselho Nacional de Desenvolvimento Científico e Tecnológico (CNPq) in the Programme Science Without Borders.

Acknowledgments: The authors are grateful to P.M. Dewick (University of Nottingham, UK) for providing plant material. The authors are also grateful to F.C. Braga for allowing the use of the Phytochemistry laboratory for the conduction of the cytotoxicity assay. The authors acknowledge A.W.A. Linghorn for revising the English language in the manuscript.

Conflicts of Interest: The authors declare that there is no conflict of interest.

References

1. Teponno, R.B.; Kusari, S.; Spiteller, M. Recent advances in research on lignans and neolignans. *Nat. Prod. Rep.* **2016**, *33*, 1044–1092. [CrossRef] [PubMed]
2. Dewick, P.M. The shikimate pathway: Aromatic amino acids and phenylpropanoids. In *Medicinal Natural Products: A Biosynthetic Approach*, 3rd ed.; Wiley: Chichester, UK, 2009; pp. 151–156. ISBN 978-0-470-74167-2.
3. Gordaliza, M.; García, P.A.; del Corral, J.M.M.; Castro, M.A.; Gómez-Zurita, M.A. Podophyllotoxin: Distribution, sources, applications and new cytotoxic derivatives. *Toxicon* **2004**, *44*, 441–459. [CrossRef] [PubMed]
4. De Luca, V.; Salim, V.; Atsumi, S.M.; Yu, F. Mining the biodiversity of plants: A revolution in the making. *Science* **2012**, *336*, 1658–1661. [CrossRef] [PubMed]
5. Bohlin, L.; Rosén, B. Podophyllotoxin derivatives: Drug discovery and development. *Drug Discov. Today* **1996**, *1*, 343–351. [CrossRef]
6. Apers, S.; Vlietinck, A.; Pieters, L. Lignans and neolignans as lead compounds. *Phytochem. Rev.* **2003**, *2*, 201–217. [CrossRef]
7. Lv, M.; Xu, H. Recent advances in semisynthesis, biosynthesis, biological activities, mode of action, and structure-activity relationship of podophyllotoxins: An update (2008–2010). *Mini-Rev. Med. Chem.* **2011**, *11*, 901–909. [CrossRef]
8. Liu, Y.-Q.; Tian, J.; Qian, K.; Zhao, X.-B.; Morris-Natschke, S.L.; Yang, L.; Nan, X.; Tian, X.; Lee, K.-H. Recent progress on C-4-modified podophyllotoxin analogs as potent antitumor agents. *Med. Res. Rev.* **2015**, *35*, 1–62. [CrossRef]
9. Hui, J.; Zhao, Y.; Zhu, L. Synthesis and in vitro anticancer activities of novel aryl-naphthalene lignans. *Med. Chem. Res.* **2012**, *21*, 3994–4001. [CrossRef]
10. Hemmati, S.; Hassan, S. Justicidin B: A promising bioactive lignan. *Molecules* **2016**, *21*, 820. [CrossRef]
11. Silva, C.G.; De Almeida, V.L.; Campana, P.R.V.; Rocha, M.P. Plant cell cultures as producers of secondary metabolites: *Podophyllum* lignans as a model. In *Transgenesis and Secondary Metabolism, Reference Series in Phytochemistry*, 1st ed.; Jha, S., Ed.; Springer International Publishing: Basel, Switzerland, 2017; pp. 67–102. ISBN 978-3-319-28668-6.
12. Li, M.; Ge, L.; Kang, T.; Suna, P.; Xing, H.; Yang, D.; Zhang, J.; Paré, P.W. High-elevation cultivation increases anti-cancer podophyllotoxin accumulation in *Podophyllum hexandrum*. *Ind. Crops Prod.* **2018**, *121*, 338–344. [CrossRef]
13. Broomhead, A.J.; Dewick, P.M. Tumour-inhibitory aryltetralin lignans in *Podophyllum versipelle*, *Diphylleia cymosa* and *Diphylleia grayi*. *Phytochemistry* **1990**, *29*, 3831–3837. [CrossRef]
14. Shaw, J.M.H. Review of other herbaceous Berberidaceae. Part II. The genus Podophyllum. Part III. In *The Genus Epimedium and Other Herbaceous Berberidaceae Including the Genus Podophyllum*; Green, P.S., Mathew, B., Eds.; Kew Publishing: London, UK, 2002; pp. 211–314. ISBN 1-84246-039-0.
15. Meacham, C.A. Phylogeny of the Berberidaceae with an evaluation of classifications. *Syst. Bot.* **1980**, *5*, 149–172. [CrossRef]
16. Shaw, J.M.H. A Taxonomic Revision of Podophyllum L. (Berberidaceae). Master's Thesis, University of Nottingham, Nottingham, UK, 1996.
17. Wang, W.; Lu, A.-M.; Ren, Y.; Endress, M.E.; Chena, Z.-D. Phylogeny and classification of Ranunculales: Evidence from four molecular loci and morphological data. *Perspect. Plant Ecol. Syst.* **2009**, *11*, 81–110. [CrossRef]
18. Ying, T.S. On *Dysosma* and *Sinopodophyllum* Ying gen. nov. of the Berberidaceae. *Acta Phytotaxon. Sin.* **1979**, *17*, 15–22.
19. Chaurasia, O.P.; Ballah, B.; Tayade, A.; Kumar, R.; Kumar, G.P.; Singh, S.B. *Podophyllum* L.: An endangered and anticancerous medicinal—An overview. *Indian J. Tradit. Knowl.* **2012**, *11*, 234–241.
20. Lloyd, J.U.; Lloyd, C.G. Diphylleia cymosa. *Drugs Med. North Am.* **1887**, *2*, 120–121.
21. Moerman, D.E. Medicinal Plants of Native America. In *Technical Reports, No. 19*; University of Michigan Museum of Antropology: Ann Arbor, MI, USA, 1986; Volume 1, p. 354.

22. Foster, E. Phytogeographic and botanical considerations of medicinal plants in Eastern Asia and Eastern North America. In *Herbs, Spices, and Medicinal Plants: Recents, Advances in Botany, Horticulture, and Pharmacology*; Cracker, L.E., Sim, J.E., Eds.; Food Product Press: New York, NY, USA, 1989; Volume 4, pp. 115–144. ISBN 1-56022-857-1.

23. Broomhead, A.J.; Rahman, M.M.A.; Dewick, P.M.; Jackson, D.E.; Lucas, J.A. Biosynthesis of *Podophyllum* lignans 5: Matairesinol as precursor of *Podophyllum* lignans. *Phytochemistry* **1991**, *30*, 1489–1492. [CrossRef]

24. Silva, C.G. Tissue Culture and Phytochemical Studies of *Podophyllum*, *Diphylleia* and *Passiflora* Species. Ph.D. Thesis, University of Nottingham, Nottingham, UK, 2000.

25. Botta, B.; Delle Monache, G.; Misiti, D.; Vitali, A.; Zappia, G. Aryltetralin lignans: Chemistry, pharmacology and biotransformations. *Curr. Med. Chem.* **2001**, *8*, 1363–1381. [CrossRef]

26. Barbour, E.R. Exploring the Implications of Climate Change for the Range of an Endemic Plant Species: Threats and Conservation Options. Honors Project. Smith College: Northampton, MA, USA, 2014. Available online: https://scholarworks.smith.edu/theses/28/ (accessed on 1 December 2018).

27. Choudhary, D.K.; Kaul, B.L.; Khan, S. Cultivation and conservation of *Podophyllum hexandrum*—An overview. *J. Med. Aromat. Plant Sci.* **1998**, *20*, 1071–1073.

28. Nadeem, M.; Palni, L.M.S.; Purohit, A.N.; Pandey, H.; Nandi, S.K. Propagation and conservation of *Podophyllum hexandrum* Royle: An important medicinal herb. *Biol. Conserv.* **2000**, *92*, 121–129. [CrossRef]

29. Majumder, A.; Jha, S. Characterization of podophyllotoxin yielding cell lines of *Podophyllum hexandrum*. *Caryologia* **2009**, *62*, 220–235.

30. Majumder, A.; Jha, S. Biotechnological approaches for the production of potential anticancer leads podophyllotoxin and paclitaxel. An overview. *J. Biol. Sci.* **2009**, *1*, 46–69.

31. Nadeem, M.; Ram, M.; Alam, P.; Ahmad, M.M.; Mohammad, A.; Al-Qurainy, F.; Khan, S.; Abdin, M.Z. *Fusarium solani*, P1, a new endophytic podophyllotoxin-producing fungus from roots of *Podophyllum hexandrum*. *Afr. J. Microbiol. Res.* **2012**, *6*, 2493–2499. [CrossRef]

32. Puri, S.C.; Nazir, A.; Chawla, R.; Arora, R.; Riyaz-Ul-Hasan, S.; Amna, T.; Ahmed, B.; Verma, V.; Singh, S.; Sagar, R.; et al. The endophytic fungus *Trametes hirsuta* as a novel alternative source of podophyllotoxin and related aryltetralin lignans. *J. Biotechnol.* **2006**, *122*, 494–510. [CrossRef]

33. Gobbo-Neto, L.; Lopes, N.P. Online identification of chlorogenic acids, sesquiterpene lactones, and flavonoids in the Brazilian arnica *Lychnophora ericoides* Mart. (Asteraceae) leaves by HPLC-DAD-MS and HPLC-DAD-MS/MS and a validated HPLC-DAD method for their simultaneous analysis. *J. Agric. Food Chem.* **2008**, *56*, 1193–1204. [CrossRef]

34. Suman, T.; Elangomathavan, R.; Kasipandi, M.; chakkaravarthi, K.; Tamilven, D.; Parimelazhagan, T. Diphyllin: An effective anticandidal agent isolated from *Cleistanthus collinus* leaf extract. *J. Basic Appl. Sci.* **2018**, *5*, 130–137. [CrossRef]

35. Santos, F.M.; Lopes, J.C.D. Prediction of AMES mutagenicity of small molecules through pharmacophore fingerprints and SVM Modeling. In Proceedings of the 50th International Conference on Medicinal Chemistry, Rouen, France, 2–4 July 2014. [CrossRef]

36. Rocha, M.P.; Campana, P.R.V.; Scoaris, D.O.; Almeida, V.L.; Lopes, J.C.D.; Silva, A.F.; Pieters, L.; Silva, C.G. Biological activities of extracts from *Aspidosperma subincanum* Mart. and in silico prediction for inhibition of acetylcholinesterase. *Phytother. Res.* **2018**, *32*, 2021–2033. [CrossRef]

37. Briñez-Ortega, E.; Almeida, V.L.; Lopes, J.C.D.; Castellanos, A.E.B. Partial inclusion of bis (1,10-phenanthroline) silver (I) salicylate in β-cyclodextrin: Spectroscopic characterization, in vitro and in silico antimicrobial evaluation. *An. Acad. Bras. Ciênc.* **2018**. submitted for publication.

38. Chicco, D. Ten quick tips for machine learning in computational biology. *BioData Min.* **2017**, *10*, 1–17. [CrossRef]

39. Lopes, J.C.D.; Dos Santos, F.M.; Martins-José, A.; Augustyns, K.; De Winter, H. The power metric: A new statistically robust enrichment-type metric for virtual screening applications with early recovery capability. *J. Cheminform.* **2017**, *9*, 7. [CrossRef]

40. Ferri, M.; Ranucci, E.; Romagnoli, P.; Giaccone, V. Antimicrobial resistance: A global emerging threat to public health systems. *Crit. Rev. Food Sci. Nutr.* **2017**, *57*, 2857–2876. [CrossRef]

41. Yu, X.; Che, Z.; Xu, H. Recent advances in the chemistry and biology of podophyllotoxins. *Chem. Eur. J.* **2017**, *23*, 4467–4526. [CrossRef] [PubMed]

42. Liu, Y.Q.; Yang, L.; Tian, X. Podophyllotoxin: Current perspectives. *Curr. Bioact. Compd.* **2007**, *3*, 37–66.

43. Song, J.H.; Sun, D.X.; Chen, B.; Ji, D.H.; Pu, J.; Xu, J.; Tian, F.D.; Guo, L. Inhibition of CYP3A4 and CYP2C9 by podophyllotoxin: Implication for clinical drug–drug interactions. *J. Biosci.* **2011**, *36*, 879–885. [CrossRef] [PubMed]

44. Kumar, K.A.; Singh, S.K.; Kumar, B.S.; Doble, M. Synthesis, antifungal activity evaluation and QSAR studies on podophyllotoxin derivatives. *Cent. Eur. J. Chem.* **2007**, *5*, 880–897.

45. Medicinal Uses for Podophyllotoxins. Available online: https://patents.google.com/patent/US4788216 (accessed on 1 December 2018).

46. Lee, C.T.L.; Lin, V.C.K.; Zhang, S.X.; Zhu, X.K.; VanVliet, D.; Hu, H.; Beers, S.A.; Wang, Z.Q.; Cosentino, L.M.; Morris-Natschke, S.L.; et al. Anti-AIDS agents. 29 Anti-HIV activity of modified podophyllotoxin derivatives. *Bioorg. Med. Chem. Lett.* **1997**, *7*, 2897–2902. [CrossRef]

47. Zhu, X.K.; Guan, J.; Xiao, Z.Y.; Cosentino, L.M.; Lee, K.H. Anti-AIDS agents. Part 61: Anti-HIV activity of new podophyllotoxin derivatives. *Bioorg. Med. Chem.* **2004**, *12*, 4267–4273. [CrossRef]

48. Hammonds, T.R.; Denyer, S.P.; Jackson, D.E.; Irving, W.L. Studies to show that with podophyllotoxin the early replicative stages of herpes simplex virus type 1 depend upon functional cytoplasmic microtubules. *J. Med. Microbiol.* **1996**, *45*, 167–172. [CrossRef]

49. Castro, M.A.; Miguel del Corral, J.M.; Gordaliza, M.; Gomez-Zurita, M.A.; Luz de La Puente, M.; Betancur-Galvis, L.A.; Sierra, J.; Feliciano, A.S. Synthesis, cytotoxicity and antiviral activity of podophyllotoxin analogues modified in the E-ring. *Eur. J. Med. Chem.* **2003**, *38*, 899–911. [CrossRef]

50. Takeuchi, S.; Takahashi, T.; Sawada, Y.; Iida, M.; Matsuda, T.; Kojima, H. Comparative study on the nuclear hormone receptor activity of various phytochemicals and their metabolites by reporter gene assays using Chinese hamster ovary cells. *Biol. Pharm. Bull.* **2009**, *32*, 195–202. [CrossRef]

51. Iwasaki, T.; Kondo, K.; Nishitani, T.; Kuroda, T.; Hirakoso, K.; Ohtani, A.; Takashima, K. Synthesis and hypolipidemic activity of diesters of arylnaphthalene lignan and their heteroaromatic analogs. *Chem. Pharm. Bull.* **1995**, *43*, 1701–1705. [CrossRef]

52. Yang, S.H.; Li, W.; Sumien, N.; Forster, M.; Simpkins, J.W.; Liu, R. Alternative mitochondrial electron transfer for the treatment of neurodegenerative diseases and cancers: Methylene blue connects the dots. *Prog. Neurobiol.* **2017**, *157*, 273–291. [CrossRef] [PubMed]

53. Xu, P.; Sun, Q.; Wang, X.J.; Zhang, S.G.; An, S.S.; Cheng, J.; Gao, R.; Xiao, H. Pharmacological effect of deoxypodophyllotoxin: A medicinal agent of plant origin, on mammalian neurons. *NeuroToxicology* **2010**, *31*, 680–686. [CrossRef] [PubMed]

54. Carlstrom, K.; Hedin, P.J.; Jonsson, L.; Lerndal, T.; Lien, J.; Weitoft, T.; Axelson, M. Endocrine effects of the podophyllotoxin derivative drug CPH 82 (Reumacon w) in patients with rheumatoid arthritis. *Scand. J. Rheumatol.* **2000**, *29*, 89–94. [PubMed]

55. Pugh, N.; Khan, I.A.; Moraes, R.M.; Pasco, D.S. Podophyllotoxin lignans enhance IL-1β but suppress TNF-α mRNA expression in LPS-treated monocytes. *Immunopharmacol. Immunotoxicol.* **2001**, *23*, 83–95. [CrossRef] [PubMed]

56. Kim, Y.; Kim, S.B.; You, Y.J.; Ahn, B.Z. Deoxypodophyllotoxin; the cytotoxic and antiangiogenic component from *Pulsatilla koreana*. *Planta Med.* **2002**, *68*, 271–274. [CrossRef] [PubMed]

57. Seidl, C.; Correia, B.L.; Stinghen, A.E.M.; Santos, A.M.C. Acetylcholinesterase inhibitory activity of uleine from *Himantanthus lancifolius*. *Z. Naturforsch.* **2010**, *65c*, 440–444. [CrossRef]

58. Trevisan, M.T.S.; Macedo, F.V.V.; Meent, M.V.; Rhee, I.K.; Verpoorte, R. Screening for acetylcholinesterase inhibitors from plants to treat Alzheimer's disease. *Quim. Nova* **2003**, *26*, 301–304. [CrossRef]

59. Kumar, S.; Pandey, A. Chemistry and biological activities of flavonoids: An overview. *Sci. World J.* **2013**, *2013*, 162750. [CrossRef]

60. Chen, A.Y.; Chen, Y.C. A review of the dietary flavonoid, kaempferol on human health and cancer chemoprevention. *Food Chem.* **2013**, *138*, 2099–2107. [CrossRef]

61. Wang, J.; Fang, X.; Ge, L.; Cao, F.; Zhao, L.; Wang, Z. Antitumor, antioxidant and anti-inflammatory activities of kaempferol and its corresponding glycosides and the enzymatic preparation of kaempferol. *PLoS ONE* **2018**, *13*, E0197563. [CrossRef]

62. Vellosa, J.C.R.; Regasini, L.O.; Khalil, N.M.; Bolzani, V.S.; Khalil, O.A.K.; Manente, F.A.; Pasquini Neto, H.; Oliveira, O.M.M.F. Antioxidant and cytotoxic studies for kaempferol, quercetin and isoquercitrin. *Eclet. Quim.* **2011**, *36*, 7–20. [CrossRef]

63. Dar, R.A.; Brahman, P.K.; Khurana, N.; Wagay, J.A.; Lone, Z.A.; Ganaie, M.A.; Pitre, K.S. Evaluation of antioxidant activity of crocin, podophyllotoxin and kaempferol by chemical, biochemical and electrochemical assays. *Arab. J. Chem.* **2017**, *10*, S1119–S1128. [CrossRef]

64. Sharma, E.; Arora, B.S. Identification of aryltetralin lignans from Podophyllum hexandrum using hyphenated techniques. *Int. J. Pharm. Sci. Drug Res.* **2015**, *7*, 83–88.

65. Sharma, E.; Kumar, A. Identification and quantification of podophyllotoxin from *Podophyllum hexandrum* by ESI-LC/MS/MS. *Int. J. Appl. Phis. Biochem. Res.* **2015**, *5*, 1–8.

66. Avula, B.; Wang, Y.H.; Moraes, R.M.; Khan, I.A. Rapid analysis of lignans from leaves of *Podophyllum peltatum* L. samples using UPLC-UV-MS. *Biomed. Chromatogr.* **2011**, *25*, 1230–1236. [CrossRef] [PubMed]

67. Zhao, C.; Zhang, N.; He, W.; Li, R.; Shi, D.; Pang, L.; Dong, N.; Xu, H.; Ji, H. Simultaneous determination of three major lignans in rat plasma by LC-MS/MS and its application to a pharmacokinetic study after oral administration of *Diphylleia sinensis* extract. *Biomed. Chromatogr.* **2014**, *28*, 463–467. [CrossRef] [PubMed]

68. Lawrence, T. The nuclear factor NF-*k*B pathway in inflammation. *Cold Spring Harb. Perspect. Biol.* **2009**, *1*, a001651. [CrossRef] [PubMed]

69. Sultan, P.; Shawl, A.S.; Jan, A.; Hamid, B.; Irshad, H. Germplasm conservation and quantitative estimation of podophyllotoxin and related glycosides of Podophyllum hexandrum. *Crop Sci.* **2014**, *12*, 267–274.

70. Nanjundaswamy, N.; Satishi, S.; Rai, K.M.L.; Shashikanth, S.; Raveesha, K.A. Antibacterial activity of synthetic precursors of podophyllotoxin. *Int. J. Biomed. Sci.* **2007**, *3*, 112–115. [PubMed]

71. Ahmad, T.; Salam, M.D. Antimicrobial activity of methanolic and aqueous extracts of *Rheum emodi* and *Podophyllum hexandrum*. *Int. J. Pharm. Sci. Rev. Res.* **2015**, *30*, 182–185.

72. Umesha, B.; Basavaraju, Y.B.; Mahendra, C.; Okiramakonda, D.D.; Rao, K.P.; Krishna, M.H. Synthesis and biological activity of novel nitrogen containing analogues of podophyllotoxin. *Indo Am. J. Pharm. Res.* **2014**, *4*, 905–914.

73. Miller, S.I.; Mekalanos, J.J. Constitutive expression of the phoP regulon attenuates Salmonella virulence and survival within macrophages. *J. Bacteriol.* **1990**, *172*, 2485–2490. [CrossRef]

74. Shi, C.; Geders, T.W.; Park, S.W.; Wilson, D.J.; Boshoff, H.I.; Abayomi, O.; Barry, C.E., III; Schnappinger, D.; Finzel, B.C.; Aldrich, C.C. Mechanism-based inactivation by aromatization of the transaminase BioA involved in biotin biosynthesis in *Mycobaterium tuberculosis*. *J. Am. Chem. Soc.* **2011**, *133*, 18194–18201. [CrossRef]

75. Reuter, K.; Steinbach, A.; Helms, V. Interfering with Bacterial Quorum Sensing. *Perspect. Med. Chem.* **2016**, *8*, 1–15. [CrossRef]

76. Bolós, M.; Llorens-Martín, M.; Jurado-Arjona, J.; Hernández, F.; Rábano, A.; Avila, J. Direct evidence of internalization of Tau by microglia in vitro and in vivo. *J. Alzheimers Dis.* **2016**, *50*, 77–87. [CrossRef]

77. Gordaliza, M.; Faircloth, G.T.; Castro, M.A.; Miguel del Corral, J.M.; Ljpez-Vazquez, M.L.; San Feliciano, A. Immunosuppressive cyclolignans. *J. Med. Chem.* **1996**, *39*, 2865–2868. [CrossRef]

78. Gordaliza, M.; Castro, M.A.; Miguel del Corral, J.M.; López-Vazquez, M.L.; San Feliciano, A.; Faircloth, G.T. In vico immunosuppressive activity of some cyclolignans. *Bioorg. Med. Chem. Lett.* **1997**, *72*, 2781–2786. [CrossRef]

79. Sharma, N.; Akhade, A.S.; Qadri, A. Sphingosine-1-phosphate suppresses TLR-induced CXCL8 secretion from human T cells. *J. Leukoc. Biol.* **2013**, *93*, 521–528. [CrossRef]

80. Kaplan, I.W. Condylomata acuminata. *New Orleans Med. Surg. J.* **1942**, *94*, 388–390.

81. Iwasaki, T.; Kondo, K.; Nishitani, T.; Kuroda, T.; Hirakoso, K.; Ohtani, A.; Takashima, K. Arylnaphtalene lignans as novel series of hypolipidemic agents raising high-density lipoprotein level. *Chem. Pharm. Bull.* **1995**, *43*, 1701–1705. [CrossRef]

82. Granneman, J.G.; Moore, H.P.H.; Mottillo, E.P.; Zhu, Z. Functional interactions between Mldp (LSDP5) and Abhd5 in the control of intracellular lipid accumulation. *J. Biol. Chem.* **2009**, *284*, 3049–3057. [CrossRef]

83. Barbosa-Filho, B.M.J.; Medeiros, K.C.P.; Diniz, M.F.F.M.; Batista, L.M.; Athayde-Filho, P.F.; Silva, M.S.; Cunha, E.V.L.; Almeida, J.R.G.S.; Quintans-Junior, L.J. Natural products inhibitors of the enzyme acetylcholinesterase. *Braz. J. Pharmacogn.* **2006**, *16*, 258–285. [CrossRef]

84. Hung, T.T.; Na, M.; Min, B.S.; Ngoc, T.M.; Lee, I.; Zhang, X.; Bae, K. Acetylcholinesterase inhibitory effect of lignans isolated from *Schizandra chinensis*. *Arch. Pharm. Res.* **2007**, *30*, 685–690. [CrossRef]

85. Itoh, K.; Ishige, A.; Hosoya, E. Cerebral Function Improving Drug. Patent-PCT Int Appl-89 08. 21 September 1989. Available online: https://worldwide.espacenet.com/publicationDetails/originalDocument?CC=EP&NR=0400148A1&KC=A1&FT=D&ND=3&date=19901205&DB=EPODOC&locale=en_EP (accessed on 1 December 2018).

86. El-Hassan, A.; El-Sayed, M.; Hamed, A.I.; Rhee, I.K.; Ahmed, A.A.; Zeller, K.P.; Verpoorte, R. Bioactive constituents of *Leptadenia arborea*. *Fitoterapia* **2003**, *74*, 184–187. [CrossRef]

87. Salleh, W.M.N.H.W.; Ahmad, F.; Yen, K.H.; Zulkifli, R.M. Anticholinesterase and anti-inflammatory constituents from *Beilschmiedia pulverulenta*. *Nat. Prod. Sci.* **2016**, *22*, 225–230. [CrossRef]

88. Wang, X.; Yu, J.; Li, W.; Wang, C.; Li, H.; Ju, W.; Chen, J.; Sun, J. Characteristics and antioxidant activity of lignans in *Schisandra chinensis* and *Schisandra sphenanthera* from different locations. *Chem. Biodivers.* **2018**, *15*, e1800030. [CrossRef]

89. Horizonte, B. 3D-Pharma: Uma Ferramenta Para Triagem Virtual Baseada em Fingerprints de Farmacóforos. Available online: http://www.bibliotecadigital.ufmg.br/dspace/handle/1843/BUBD-9DKHDA (accessed on 1 December 2018).

90. Shemetulskis, N.E.; Weininger, D.; Blankley, C.J.; Yang, J.J.; Humblet, C. Stigmata: An algorithm to determine structural commonalities in diverse datasets. *J. Chem. Inf. Comput. Sci.* **1996**, *36*, 862–871. [CrossRef]

91. Bolton, E.E.; Kim, S.; Bryant, S.H. PubChem3D: Conformer generation. *J. Cheminform.* **2011**, *3*, 4. [CrossRef]

92. Hawkins, P.C.D.; Nicholls, A. Conformer Generation with OMEGA: Learning from the data set and the analysis of failures. *J. Chem. Inf. Model.* **2012**, *52*, 2919–2936. [CrossRef]

93. Hawkins, P.C.D.; Skillman, A.G.; Warren, G.L.; Ellingson, B.A.; Stahl, M.T. Conformer generation with OMEGA: Algorithm and validation using high quality structures from the protein databank and Cambridge structural database. *J. Chem. Inf. Model.* **2010**, *50*, 572–584. [CrossRef]

94. Santos, F.M.; De Winter, H.; Augustyns, K.; Lopes, J.C.D. Use of Extensive Cross-Validation and Bootstrap Application (ExCVBA) for Molecular Modeling of Some Pharmacokinetics Properties. Available online: https://www.researchgate.net/profile/Julio_Lopes2/publication/282644866_2015_-_Poster_OpenTox-_Use_of_Extensive_Cross-Validation_and_Bootstrap_Application_ExCVBA_for_Molecular_Modeling_of_Some_Pharmacokinetics_Properties/data/561518bd08aec622441191cc/2015-Poster-OpenTox-Use-of-Extensive-Cross-Validation-and-Bootstrap-Application-ExCVBA-for-Molecular-Modeling-of-Some-Pharmacokinetics-Properties.pdf (accessed on 1 December 2018).

95. Nicholls, A. What do we know?: Simple statistical techniques that help. *Methods Mol. Biol.* **2011**, *672*, 531–581. [CrossRef]

96. Hanley, J.A.; McNeil, B.J. The meaning and use of the area under a receiver operating characteristic (ROC) curve. *Radiology* **1982**, *143*, 29–36. [CrossRef]

97. Nicholls, A. Confidence limits, error bars and method comparison in molecular modeling. Part 1: The calculation of confidence intervals. *J. Comput. Aided Mol. Des.* **2014**, *28*, 887–918. [CrossRef]

98. Chang, C.-C.; Lin, C.-J. LIBSVM: A Library for Support Vector Machines. ACM Transactions on Intelligent Systems and Technology, 2:27:1-27:27. 2011. Available online: http://www.csie.ntu.edu.tw/~||cjlin/libsvm (accessed on 1 December 2018).

99. De Winter, H.; Lopes, J.C.D. Reply to the comment made by Šicho, Voršilák and Svozil on 'The Power metric: A new statistically robust enrichment-type metric for virtual screening applications with early recovery capability'. *J. Cheminform.* **2018**, *10*, 14. [CrossRef]

100. CPAN 2017. The Comprehensive Perl Archive Network. Available online: http://search.cpan.org/perldoc?Algorithm%3A%3ANaiveBayes (accessed on 19 November 2018).

101. Filimonov, D.A.; Lagunin, A.A.; Gloriozova, T.A.; Rudik, A.V.; Druzhilovskii, D.S.; Pogodin, P.V.; Poroikov, V.V. Prediction of the biological activity spectra of organic compounds using the pass online web resource. *Chem. Heterocycl. Compd.* **2014**, *50*, 444–457. [CrossRef]

102. Norinder, U.; Carlsson, L.; Boyer, S.; Eklund, M. Introducing conformal prediction in predictive modeling. A transparent and flexible alternative to applicability domain determination. *J. Chem. Inf. Model.* **2014**, *54*, 1596–1603. [CrossRef]

103. CLSI. *Methods for Dilution Antimicrobial Susceptibility Tests for Bacteria That Grow Aerobically*, Approved Standard, 9th ed.; Clinical and Laboratory Standards Institute: Wayne, PA, USA, 2012; ISBN 1-56238-783-9.

104. Fukuda, M.; Ohkoshi, E.; Makino, M.; Fujimoto, Y. Studies on the constituents of the leaves of *Baccharis dracunculifolia* (Asteraceae) and their cytotoxic activity. *Chem. Pharm. Bull.* **2006**, *54*, 1465–1468. [CrossRef]

105. Marston, A.; Kissling, J.; Hostettmann, K. A rapid TLC bioautography method for the detection of acetylcholinesterase and butyrylcholinesterase inhibitors in plants. *Phytochem. Anal.* **2002**, *13*, 51–54. [CrossRef]

106. Ellman, G.L.; Courtney, K.D.; Junior, V.A.; Featherstone, R.M. A new and rapid colorimetric determination of acetylcholinesterase activity. *Biochem. Pharmacol.* **1961**, *7*, 88–95. [CrossRef]

107. Duarte-Almeida, J.M.; Santos, R.J.; Genovese, M.I.; Lajolo, F.M. Avaliação da atividade antioxidante utilizando o sistema β-caroteno/ácido linoleico e método de sequestro de radicais DPPH. *Cienc. Tecnol. Aliment.* **2006**, *26*, 446–452. [CrossRef]

108. Mensor, L.L.; Menezes, F.S.; Leitão, G.G.; Reis, A.S.; Santos, T.C.; Coube, C.S.; Leitão, S.G. Screnning of Brazilian plant extracts for antioxidant activity by the use of DPPH free radical method. *Phytother. Res.* **2001**, *15*, 127–130. [CrossRef]

109. Ohkawa, H.; Ohishi, N.; Yagi, K. Assay for lipid peroxides in animal tissues by thiobarbituric acid reaction. *Anal. Biochem.* **1979**, *95*, 351–358. [CrossRef]

110. Vichai, V.; Kirtikara, K. Sulforhodamine B colorimetric assay for cytotoxicity screening. *Nat. Protoc.* **2006**, *1*, 1112–1116. [CrossRef]

Sample Availability: Samples of the compounds are not available from the authors.

Article

Targeted Lignan Profiling and Anti-Inflammatory Properties of *Schisandra rubriflora* and *Schisandra chinensis* Extracts

Agnieszka Szopa [1,*] , **Michał Dziurka** [2] , **Angelika Warzecha** [1] , **Paweł Kubica** [1] ,
Marta Klimek-Szczykutowicz [1] and **Halina Ekiert** [1]

[1]   Chair and Department of Pharmaceutical Botany, Medical College, ul. Medyczna 9, 30-688 Kraków, Poland;
    a.warzecha@student.uj.edu.pl (A.W.); p.kubica@uj.edu.pl (P.K.);
    marta.klimek-szczykutowicz@doctoral.uj.edu.pl (M.K.-S.); mfekiert@cyf-kr.edu.pl (H.E.)

[2]   Polish Academy of Sciences, The Franciszek Górski Institute of Plant Physiology, ul. Niezapominajek 21,
    30-239 Kraków, Poland; m.dziurka@ifr-pan.edu.pl

*   Correspondence: a.szopa@uj.edu.pl; Tel.: +48-12-620-54-30; Fax: +48-620-54-40

Received: 7 November 2018; Accepted: 22 November 2018; Published: 27 November 2018

Abstract: *Schisandra rubriflora* is a dioecious plant of increasing importance due to its lignan composition, and therefore, possible therapeutic properties. The aim of the work was lignan profiling of fruits, leaves and shoots of female (F) and male (M) plants using UHPLC-MS/MS. Additionally, the anti-inflammatory activity of plant extracts and individual lignans was tested in vitro for the inhibition of 15-lipooxygenase (15-LOX), phospholipases A2 (sPLA$_2$), cyclooxygenase 1 and 2 (COX-1; COX-2) enzyme activities. The extracts of fruits, leaves and shoots of the pharmacopoeial species, *S. chinensis*, were tested for comparison. Twenty-four lignans were monitored. Lignan contents in *S. rubriflora* fruit extracts amounted to 1055.65 mg/100 g DW and the dominant compounds included schisanhenol, aneloylgomisin H, schisantherin B, schisandrin A, gomisin O, angeloylgomisin O and gomisin G. The content of lignan in leaf extracts was 853.33 (F) and 1106.80 (M) mg/100 g DW. Shoot extracts were poorer in lignans—559.97 (F) and 384.80 (M) mg/100 g DW. Schisantherin B, schisantherin A, 6-*O*-benzoylgomisin O and angeloylgomisin H were the dominant compounds in leaf and shoot extracts. The total content of detected lignans in *S. chinensis* fruit, leaf and shoot extracts was: 1686.95, 433.59 and 313.83 mg/100 g DW, respectively. Gomisin N, schisandrin A, schisandrin, gomisin D, schisantherin B, gomisin A, angeloylgomisin H and gomisin J were the dominant lignans in *S. chinensis* fruit extracts were. The results of anti-inflammatory assays revealed higher activity of *S. rubriflora* extracts. Individual lignans showed significant inhibitory activity against 15-LOX, COX-1 and COX-2 enzymes.

Keywords: *Schisandra rubriflora*; *Schisandra chinensis*; red-flowered Chinese magnolia vine; Chinese magnolia vine; lignans; phytochemical analysis; UHPLC-MS/MS; anti-inflammatory activity; LOX; COX; sPLA$_2$

---

1. Introduction

Schisandra rubriflora (Franch.) Rehd. et Wils, is a rare and little-known plant species of the genus *Schisandra* beyond East Asian phytotherapy. *S. rubriflora* occurs at natural sites in the western Sichuan province of China. It is an endemic species that occurs only in this region [1,2]. *S. rubriflora* cultivations outside the East Asian region are rare, but attempts have recently been made to grow this species in Europe, including Poland [2,3].

S. rubriflora is a dioecious vine reaching about 3–4 meters in height [1]. *S. rubriflora* leaves are characterized by elliptical to obovate-elliptical shape, 7–11.5 cm long and 2.5–5.5 cm wide. The leaves are sharp-edged, rarely blunt-edged, and leaf blade edges are finely serrated. Mature berry-shaped fruits of *S. rubriflora*, collected in the hanging ears, are dark red in color, the size of peas, sitting on about 5–8 cm long stalks [1].

Schisandra chinensis (Turcz.) Baill. is a related species, much better known in terms of medicinal properties, for which cultivation methods have been developed (with cultivations in Europe and America) [1,4,5]. The description of the raw material, i.e., the fruit of the Chinese magnolia vine—*Schisandrae chinensis fructus*—appeared for the first time in 2008 in the European Pharmacopoeia 6 [6]. The raw material has been used for many years in the official health care of Asian countries [7–10]. It is a pharmacopoeial species also known in the USA [11]. A World Health Organization (WHO) monograph is also devoted to this plant [12]. *Schisandrae chinensis* fruit extracts show valuable, proven, therapeutic properties. These include: anti-inflammatory, anti-tumor, and anti-ulcer properties, anti-bacterial and anti-fungal activity; additionally, they can act hepatoprotectively, adaptogenically and ergogenically; these extract also have antioxidant and detoxification properties [4,13,14].

Scientific information about therapeutic properties of *S. rubriflora* fruits is less available, and its monograph is not listed in any of the world pharmacopoeias [2]. This species is known in the traditional Chinese medicine as a sedative and toning agent, and its fruits are still consumed locally. There are indications regarding the use of this species in the treatment of hepatitis, chronic gastroenteritis and neurasthenia [2,15]. The biological activity of compounds contained in the fruit of this species, described only by Chinese research groups, is limited mainly to the anti-HIV-1 properties, resulting from the inhibition of HIV-1 replication in H9 lymphocytes [16,17]. According to available sources, compounds belonging to the group of dibenzocyclooctadiene lignans as well as nortriterpenoids and bisnortriterpenoids are both responsible for anti-HIV-1 activity [16,18,19]. Furthermore, extracts from *S. rubriflora* shoots have been shown to effectively reduce the level of GPT (glutamin-pyruvate transaminase) in the blood, which may be useful in the treatment of liver and bile duct diseases [2,20].

Valuable biological properties and therapeutic applications resulting from them are conditioned by the unique chemical composition of *S. chinensis* [4,21]. Lignans are the main group of secondary metabolites specific to this genus, among which the main role is played by dibenzocyclooctadiene lignans [21,22]. The majority of scientific research has focused on this group of metabolites. They are referred to as "schisandra lignans" due to the characteristic, complicated chemical structure of these compounds as well as the occurrence limited only to this genus. Schisandrin, gomisin A, deoxyschisandrin and schisantherin A and B are listed as the most important from the group of dibenzocyclooctadiene lignans (Figure 1). Recent studies have reported the identification of ever new structures from the group of lignans and their derivatives [22–24]. The available data show that dibenzocyclooctadiene lignans, and their derivatives, specific only for *S. rubriflora*, such as schirubrin A-D, rubrilignans A and B or rubrisandrin A and B are also present in *S. rubriflora* [16,18,25].

There are several studies on the anti-inflammatory activity of *S. chinensis* fruit extracts [26,27] and some individual lignans [28–31], but there are no studies on *S. chinensis* leaf and shoot extracts. importantly, these investigations have not yet been performed in *S. rubriflora* species. Moreover, there are no studies comparing the results obtained for complex plant material to the results obtained for pure lignans. In this work, we attempt to assess the anti-inflammatory potential of plant extracts and compare it with anti-inflammatory properties of pure lignan samples.

The present study introduces phytochemical characteristics of lignan contents using the UHPLC-MS/MS method in *S. rubriflora*, including the division of the material into female (F) and male (M) specimens of soil-grown plants. The results were compared to the analyses of pharmacopoeial species–*S. chinensis*—performed for comparison purposes.

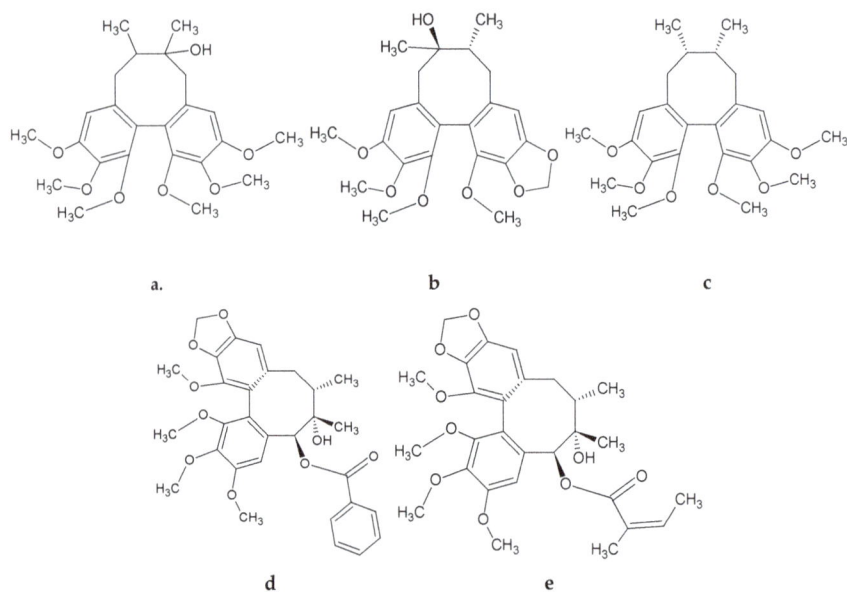

Figure 1. Example structural formulas of abundant *S. rubriflora* dibenzocyclooctadiene lignans: (**a**) schisandrin; (**b**) gomisin A; (**c**) deoxyschisandrin; (**d**) schisantherin A; (**e**) schisantherin B. Structural formulas drown in: ACD/ChemSkech (Freeware), version 12.00, Advanced Chemistry Development, Inc., Toronto, ON, Canada, www.acdlabs.com, 2010.

Moreover, the anti-inflammatory potential of *S. rubriflora* fruits, leaves and shoots of F and M specimens was studied for the first time using estimations based on the inhibition of eicosanoid-generating enzymes; these included cyclooxygenases (COX-1 and COX-2), lipoxygenase (LOX) and secretory phospholipase A_2 (sPLA$_2$), reducing the concentrations of prostanoids and leukotrienes. Additionally, the analyses involved individual lignans as well as an artificially created "average sample of *S. rubriflora* lignan composition". Comparative studies with *S. chinensis* shoot, leaf and fruit extracts were also conducted in this study.

2. Results

2.1. Schisandra Rubriflora Lignan Profile

The UHPLC-MS/MS method was used for both qualitative and quantitative analyses of the extracts tested (Tables S1 and S2, Figure S1). Twenty-four lignans were quantified in all analyzed samples, representing four lignan groups: dibenzocyclooctadiene lignans (schisantherin A and B, schisandrin, schisandrin C, gomisin A, D, G, J, N, O, 6-O-benzoylgomisin O, schisandrin A, rubrisandrin A, epigomisin O, schisanhenol, rubriflorin A, angeloylgomisin H and O), aryltetralin lignan (wulignan A_1), dibenzylbutane lignans (pregomisin, mesodihydroguaiaretic acid), tetrahydrofuran lignan (fragransin A_2) and dihydrobenzofuran neolignans (licarin A and B) [32–34].

The total lignan content in the analyzed fruit extracts of *S. rubriflora* was 1055.65 mg/100 g DW. Quantitatively dominant compounds were: Schisanhenol (268.02 mg/100 g DW), aneloylgomisin H (185.10 mg/100 g DW), schisantherin B (118.07 mg/100 g DW), schisandrin A (104.32 mg/100 g DW), gomisin O (103.64 mg/100 g DW), angeloylgomisin O (76.88 mg/100 g DW) and gomisin G (66.39 mg/100 g DW) (Table 1).

Table 1. The lignan contents [mg/100g DW] ± SD (*n* =5) in fruit and leaf and shoot female (F) and male (M) extracts of *Schisandra rubriflora*.

Lignans	Fruits	Leaves		Shoots	
		F	M	F	M
Wulignan A$_1$	19.39 ± 1.35	0.04 ± 0.001	0.04 ± 0.001	0.03 ± 0.002	0.07 ± 0.002
Rubrisandrin A	0.07 ± 0.001	0.06 ± 0.001	0.06 ± 0.001	0.06 ± 0.002	0.10 ± 0.001
Rubriflorin A	traces	traces	traces	traces	traces
Schisandrin	6.57 ± 0.35	8.15 ± 0.25	5.69 ± 0.13	4.01 ± 0.005	2.25 ± 0.01
Gomisin D	3.52 ± 0.16	16.45 ± 0.76	116.51 ± 24.28	8.25 ± 0.42	20.26 ± 0.44
Gomisin J	5.40 ± 0.27	0.97 ± 0.01	0.76 ± 0.03	0.46 ± 0.01	0.36 ± 0.004
Gomisin A	0.75 ± 0.01	6.40 ± 0.18	4.20 ± 0.07	2.99 ± 0.02	1.65 ± 0.01
Gomisin G	66.35 ± 12.37	11.13 ± 1.01	8.23 ± 0.12	5.25 ± 0.04	3.67 ± 0.02
Licarin B	1.98 ± 0.08	0.41 ± 0.06	0.24 ± 0.02	0.19 ± 0.04	0.12 ± 0.01
Epigomisin O	7.46 ± 0.32	10.62 ± 0.41	7.83 ± 0.12	4.91 ± 0.06	3.15 ± 0.01
Gomisin O	103.64 ± 26.25	22.90 ± 1.29	2.81 ± 0.01	12.07 ± 0.20	1.82 ± 0.01
Mesodihydroguaiaretic acid	1.03 ± 0.02	0.34 ± 0.01	0.32 ± 0.004	0.17 ± 0.001	0.16 ± 0.004
Schisantherin A	27.19 ± 3.00	226.80 ± 16.70	107.17 ± 1.66	84.35 ± 4.64	24.27 ± 1.51
Schisantherin B	118.07 ± 18.42	291.47 ± 51.98	104.26 ± 18.16	239.11 ± 38.00	169.04 ± 49.85
Dehydroisoeugenol	0.41 ± 0.004	0.73 ± 0.01	0.44 ± 0.01	0.33 ± 0.004	0.23 ± 0.01
Schisanhenol	268.02 ± 43.12	2.05 ± 0.06	2.73 ± 0.01	1.13 ± 0.003	2.53 ± 0.03
Schisandrin A	104.32 ± 10.11	0.22 ± 0.002	0.38 ± 0.004	0.12 ± 0.001	0.50 ± 0.002
Fragransin A$_2$	0.01 ± 0.002	0.01 ± 0.002	Nd *	nd	0.004 ± 0.002
Pregomisin	0.003 ± 0.001	traces	traces	0.01 ± 0.001	traces
Gomisin N	19.20 ± 0.66	1.58 ± 0.02	2.21 ± 0.03	0.81 ± 0.01	1.08 ± 0.01
6-O-Benzoylgomisin O	35.28 ± 3.55	134.51 ± 5.91	564.62 ± 33.66	72.38 ± 4.77	52.18 ± 2.63
Schisandrin C	4.96 ± 0.12	0.44 ± 0.03	0.19 ± 0.02	0.28 ± 0.002	0.09 ± 0.01
Angeloylgomisin H	185.10 ± 27.55	100.83 ± 7.89	129.28 ± 13.66	105.80 ± 9.04	74.73 ± 1.54
Angeloylgomisin O	76.88 ± 4.55	17.21 ± 0.26	48.80 ± 3.66	17.25 ± 0.23	26.53 ± 0.46
Total content	1055.65 ± 152.26	853.33 ± 86.85	1106.80 ± 78.33	559.97 ± 57.50	384.80 ± 56.56

Shaded parts indicates the highest quantities of given compounds. *—nd—not detected.

The presence of twenty-four and twenty-three lignans was found in the analyzed extracts of female (F) and male (M) *S. rubriflora* specimens, respectively (Table 1). No fragransin A_2 was found in leaf extracts of male specimens. The total contents of the tested group of compounds in leaf extracts were: F—853.33 mg/100 g DW and M—1106.80 mg/100 g DW.

Qualitatively dominant compounds in F leaf extracts were: Schisantherin B (291.47 mg/100 g DW), schisantherin A (226.80 mg/100 g DW), 6-*O*-benzoylgomisin O (134.51 mg/100 g DW) and angeloylgomisin H (100.83 mg/100 g DW). Qualitatively dominant in M leaf extracts were: 6-*O*-benzoylgomisin O (564.62 mg/100 g DW), angeloylgomisin H (129.28 mg/100 g DW), gomisin D (116.51 mg/100 g DW), schisantherin A (107.17 mg/100 g DW), schisantherin B (104.28 mg/100 g DW) and angeloylgomisin O (48.80 mg/100 g DW) (Table 1).

The presence of twenty-four and twenty-three lignans were found in the analyzed shoot extracts of female (F) and male (M) *S. rubriflora* specimens, respectively. These were the same compounds that were identified in fruit and leaf extracts (Table 1). The total contents of the tested group of compounds in shoot extracts were: F—559.97 mg/100 g DW and M—384.80 mg/100 g DW.

Qualitatively the dominant compounds in F shoot extracts were: Schisantherin B (239.11 mg/100 g DW), angeloylgomisin H (105.80 mg/100 g DW), schisantherin A (84.35 mg/100 g DW) and 6-O-benzoylgomisin O (72.38 mg/100 g DW). Fragransin A_2 was not found in these extracts. Qualitatively dominant compounds in M shoot extracts were: schisantherin B (169.04 mg/100 g DW), angeloylgomisin H (74.73 mg/100 g DW) and 6-O-benzoylgomisin O (52.18 mg/100 g DW) (Table 1).

2.2. *Schisandra Chinensis Lignan Profile*

The UHPLC-MS/MS analysis of lignans in fruit, leaf and shoot extracts of *Schisandra chinensis* was performed for comparative purposes. When comparing the results, qualitative similarities and quantitative differences were found between the extracts tested (Table 2). In all analyzed samples, twenty-four lignans were quantified, representing four lignan groups: dibenzocyclooctadiene lignans (schisantherin A and B, schisandrin, schisandrin C, gomisin A, D, G, J, N, O, 6-*O*-benzoylgomisin O, schisandrin A, rubrisandrin A, epigomisin O, schisanhenol, rubriflorin A, angeloylgomisin H and O), aryltetralin lignan (wulignan A_1), dibenzylbutane lignans (pregomisin, mesodihydroguaiaretic acid) and tetrahydrofuran lignan (fragransin A_2). In addition, dihydrobenzofuran neolignans (licarin A and B) were also found in the analyzed extracts (Table 2).

The total contents of detected lignans in fruit, leaf and shoot extracts were equal to: 1686.95, 433.59 and 313.83 mg/100 g DW, respectively. Qualitatively the dominant compounds in *S. chinensis* fruits were: gomisin N (259.05 mg/100 g DW), schisandrin A (212.50 mg/100 g DW), schisandrin (206.08 mg/100 g DW), gomisin D (195.22 mg/100 g DW), schisantherin B (195.82 mg/100 g DW), gomisin A (177.94 mg/100 g DW), angeloylgomisin H (161.90 mg/100 g DW) and gomisin J (142.35 mg/100 g DW). Fragrasin A_2 was not detected in the fruit extract (Table 2).

The amounts of individual compounds were lower in leaf and shoot extracts than in fruit extracts. Rubriflorin A was detected only in trace amounts in the leaf extract. The dominant lignans in both leaf and shoot extracts were: Schisantherin B, gomisin A, gomisin N and angeloylgomisin H, and their quantities were equal to 102.47 and 35.27; 73.82 and 36.29; 55.06 and 62.69; and 47.34 and 44.84 mg/100 g DW, respectively (Table 2).

Table 2. The lignan contents (mg/100g DW) $\pm$ SD (n =5) in fruit, leaf and shoot extracts of *Schisandra chinensis*.

Lignans	Fruits	Leaves	Shoots
Wulignan A$_1$	0.15 $\pm$ 0.03	0.03 $\pm$ 0.001	0.04 $\pm$ 0.001
Rubrisandrin A	0.03 $\pm$ 0.002	0.04 $\pm$ 0.001	0.03 $\pm$ 0.001
Rubriflorin A	0.01 $\pm$ 0.001	traces	0.01 $\pm$ 0.001
Schisandrin	206.08 $\pm$ 22.32	32.51 $\pm$ 3.14	32.87 $\pm$ 4.14
Gomisin D	195.22 $\pm$ 15.63	9.62 $\pm$ 1.96	11.33 $\pm$ 1.12
Gomisin J	142.35 $\pm$ 19.12	18.06 $\pm$ 3.11	13.22 $\pm$ 0.54
Gomisin A	177.94 $\pm$ 20.14	73.82 $\pm$ 8.41	36.29 $\pm$ 2.41
Gomisin G	44.56 $\pm$ 5.44	12.18 $\pm$ 2.14	11.69 $\pm$ 1.44
Licarin B	0.37 $\pm$ 0.02	0.03 $\pm$ 0.001	0.03 $\pm$ 0.001
Epigomisin O	3.16 $\pm$ 0.09	1.01 $\pm$ 0.07	0.91 $\pm$ 0.80
Gomisin O	4.08 $\pm$ 1.21	5.35 $\pm$ 0.55	4.45 $\pm$ 0.12
Mesodihydroguaiaretic acid	0.46 $\pm$ 0.09	0.38 $\pm$ 0.06	0.42 $\pm$ 0.07
Schisantherin A	31.32 $\pm$ 3.25	3.86 $\pm$ 0.98	2.22 $\pm$ 0.14
Schisantherin B	185.82 $\pm$ 20.39	102.47 $\pm$ 4.87	35.27 $\pm$ 3.12
Dehydroisoeugenol	0.16 $\pm$ 0.05	0.36 $\pm$ 0.09	0.41 $\pm$ 0.04
Schisanhenol	9.60 $\pm$ 1.88	1.00 $\pm$ 0.07	0.91 $\pm$ 0.02
Schisandrin A	212.50 $\pm$ 18.45	17.74 $\pm$ 1.02	13.89 $\pm$ 1.21
Fragransin A$_2$	nd	0.02 $\pm$ 0.001	0.01 $\pm$ 0.001
Pregomisin	traces	traces	traces
Gomisin N	259.05 $\pm$ 30.88	55.06 $\pm$ 4.52	62.69 $\pm$ 4.98
6-*O*-Benzoylgomisin O	33.64 $\pm$ 2.89	10.83 $\pm$ 2.01	7.48 $\pm$ 1.21
Schisandrin C	18.54 $\pm$ 2.15	37.20 $\pm$ 2.77	29.94 $\pm$ 4.23
Angeloylgomisin H	161.90 $\pm$ 15.65	47.34 $\pm$ 3.45	44.84 $\pm$ 2.27
Angeloylgomisin O	65.56 $\pm$ 5.99	4.67 $\pm$ 0.87	4.89 $\pm$ 0.84
Total content	1686.95 $\pm$ 185.67	433.59 $\pm$ 40.09	313.83 $\pm$ 28.70

Shaded parts indicates the highest quantities of given compounds. nd—not detected.

2.3. Anti-Inflammatory Activity

The following plant material extracts were tested for anti-inflammatory activity: Fruits and leaves of *Schisandra rubriflora* and *Schisandra chinensis* as well as selected most abundant lignans present in plant samples: 6-*O*-benzoylgomisin O, schisandrin, gomisin D, gomisin N and schisantherin A. Extracts from the shoots were not assayed for their anti-inflammatory activity, due to the relatively low lignan contents, determined in phytochemical studies, compared to leaf and fruit extracts (Tables 1 and 2).

The tests were based on the in vitro inhibition of 15-lipooxygenase (15-LOX), phospholipase A$_2$ (sPLA$_2$), cyclooxygenase-1 (COX-1) and cyclooxygenase-2 (COX-2) enzymes.

Plant material extracts showed moderate inhibition of 15-LOX and relatively high inhibitory activity against COX-1, COX-2 and sPLA$_2$ (Table 3). Evaluation of 15-LOX inhibition showed that *S. rubriflora* fruit and leaf extracts moderately inhibited this enzyme: 22%—fruits, 38%—F leaves, 42%—M leaves at 17.5 µg/mL. For *S. chinensis*, the activity was lower: 25%—fruits (17.5 µg/mL) and 31%—leaves (175.0 µg/mL) (Table 3).

The sPLA$_2$ enzyme inhibition assay showed that fruit and leaf extracts of *S. rubriflora* inhibited its activity to about 62–65% at 175.0 µg/mL. Inhibition percentage for fruit and leaf extracts of *S. chinensis* was lower: 25% and 49%, respectively (at 175.0 µg/mL) (Table 3).

The most promising results were obtained for in vitro inhibitory COX-1 and COX-2 enzyme activities. *S. rubriflora* fruit extracts (at 17.4 µg/mL) inhibited COX-1 and COX-2 activities in 71% and 48%, respectively. Leaf extracts showed higher activity at 175.0 µg/mL, and the inhibition was 86% and 82% (F), and 96% and 90% (M), respectively (Table 3). *S. chinensis* extracts exhibited lower activity. The percentage of COX-1 and COX-2 inhibition was 59% and 66% for fruits, and 69% and 77% for leaves, respectively (Table 3).

Evaluation of anti-inflammatory properties of individual lignan solutions and the average sample of lignan composition (MIX) (Table S3) showed that they were not active against the sPLA$_2$ enzyme (Table 4). All studied lignans, i.e., 6-*O*-benzoylgomisin O, schisandrin, gomisin D, gomisin N and schisantherin A, as well as their MIX sample, showed from 49% to 57% 15-LOX inhibitory activity at 0.175 μg/mL (Table 4). The highest inhibition for COX-1 was estimated for schisandrin—62% at 1.75 μg/mL, schisantherin A—74% at 0.175 μg/mL, and for the average sample of lignan composition—61% at 1.75 μg/mL (Table 4). The highest inhibition for COX-2 was detected for schisandrin—54% at 1.75 μg/mL, gomisin D—62% at 1.75 μg/mL, gomisin N—70% at 0.175 μg/mL and for the MIX sample—56% at 0.175 μg/mL (Table 4).

3. Discussion

Twenty-four lignans were identified from four chemical lignan groups in all analyzed samples of both plant species: Dibenzocyclooctadiene lignans (schisantherin A and B, schisandrin, schisandrin C, gomisin A, D, G, J, N, O, 6-*O*-benzoylgomisin O, schisandrin A, rubrisandrin A, epigomisin O, schisanhenol, rubriflorin A, angeloylgomisin H and O), aryltetralin lignan (wulignan A$_1$), dibenzylbutane lignans (pregomisin, mesodihydroguaiaretic acid), and tetrahydrofuran lignan (fragransin A$_2$). In addition, the presence of dihydrobenzofuran neolignans (licarin A and B) was also found in the analyzed extracts. Until now, there have been no reports on the detection of so many lignan compounds in *S. rubriflora* fruit, leaf and shoot extracts, including the differentiation on male and female specimens (Table 1, Tables S1 and S2, Figure S1).

Schisanhenol was quantitatively predominant in the analyzed *S. rubriflora* fruit extracts (268.02 mg/100 g DW), and its content was: 1.45-, 2.27-, 2.57-, 2.59-, 3.48- and 4.04-fold higher, respectively, than the content of the remaining dominant compounds: Angeloylgomisin H, schisantherin B, schisandrin A, gomisin O, angeloylgomisin O and gomisin G (Table 1).

Twenty-four and twenty-three lignans were found in both leaf and shoot extracts of F and M *S. rubriflora* specimens, respectively. Fragransin A$_2$ was found in the extracts from leaves and shoots of F specimens, while it was not detected in analogous extracts from M specimens. The total content of lignans in leaf extracts of F specimens was 1.30-fold lower compared to the content in leaf extracts of M specimens (Table 1). The following compounds were predominant in F specimen leaf extracts: schisantherin A and B, 6-*O*-benzoylgomisin O, and angeloylgomisin H. The contents of schisantherin A and B were: 2.12- and 2.80-fold higher, respectively, compared to their contents in leaf extracts of M. specimens. The contents of 6-*O*-benzoylgomisin O and angeloylgomisin H in leaf extracts of M specimens were: 4.20-, 1.28-fold higher, respectively, than in leaf extracts of F specimens (Table 1). The most dominant compounds in leaf extracts of M specimens were: 6-*O*-benzoylgomisin O, angeloylgomisin H, gomisin D, angeloylgomisin O, schisantherin A and B. Gomisin D and angeloylgomisin O contents in leaf extracts of M. specimens were: 7.08-, 2.84-fold higher, respectively, than in leaf extracts of F specimens (Table 1).

The total lignan content in shoot extracts of F specimens was 1.46-fold higher than in the extracts from M specimens (Table 1). The quantitatively dominant compounds in shoot extracts of F specimens were: Schisantherin B, angeloylgomisin H, schisantherin A, and 6-*O*-benzoylgomisin O: 1.41-, 1.42-, 3.48- and 1,39-fold higher, respectively, compared to shoot extracts of M specimens. Schisantherin B, angeloylgomisin H, and 6-*O*-benzoylgomisin O were predominant in the extracts of M specimens (Table 1).

Table 3. In vitro inhibition activity of studied *S. rubriflora* and *S. chinensis* extracts solutions against 15-LOX, COX-1, COX-2, and sPLA$_2$.

Plant Extracts		Concentration (µg/mL)	15-LOX		COX-1		COX-2		sPLA$_2$	
			% Inh	SD	% Inh	SD	% Inh	SD	% Inh	SD
Schisandra rubriflora	Fruits	175.0	1	0.1	32	3.5	52	5.7	62	2.5
		17.5	22	1.5	71	7.8	48	5.2	25	1.0
	Leaves F	175.0	38	2.7	96	10.6	90	9.9	64	2.6
		17.5	42	2.9	51	5.6	40	4.4	54	2.2
	Leaves M	175.0	37	2.6	86	9.5	82	9.0	65	2.7
		17.5	38	2.7	58	6.4	49	5.4	55	2.3
Schisandra chinensis	Fruits	175.0	no inhibition	0.0	59	6.5	66	7.3	25	1.0
		17.5	25	1.7	34	3.8	49	5.4	8	0.3
	Leaves	175.0	31	2.2	69	7.6	77	8.4	49	2.0
		17.5	28	2.0	51	5.6	53	5.8	25	1.0
Control inhibitor	NDGA	30.2 (100 µM)	23	2.0	-	-	-	-	-	-
	Ibuprofen	2.1 (10 µM)	-	-	23	2.5	21	2.0	-	-
	Thioetheramide-PC	73.6 (100 µM)	-	-	-	-	-	-	91	4.0

% Inh—percent of enzyme activity inhibition; SD—standard deviation (*n* = 3).

Table 4. In vitro inhibition activity of selected pure lignan solutions against 15-LOX, COX-1, COX-2, and sPLA2.

Lignans		Concentration (µg/mL)	15-LOX		COX-1		COX-2		sPLA$_2$	
			% Inh	SD	% Inh	SD	% Inh	SD	% Inh	SD
6-O-Benzoylgomisin O		1.75	18	1.2	35	4	47	5	no inhibition	-
		0.175	49	3.4	19	2	47	5	no inhibition	-
Schisandrin		1.75	31	2.0	62	5.6	54	4.9	no inhibition	-
		0.175	57	4.0	40	4.4	50	5.5	no inhibition	-
Gomisin D		1.75	31	1.8	42	4.2	62	6.4	3	0.1
		0.175	53	3.7	34	3.7	58	6.4	9	0.4
Gomisin N		1.75	16	1.1	42	4.3	58	7.0	6	0.3
		0.175	54	3.8	40	4.4	70	7.6	no inhibition	-
Schisantherin A		1.75	31	1.6	38	5.8	48	5.3	8	0.3
		0.175	55	3.8	74	8.1	31	3.4	1	0.0
Average sample lignan composition (mix)		1.75	20	1.4	61	6.7	43	4.7	no inhibition	-
		0.175	53	3.7	32	3.6	56	6.1	2	0.1
Control inhibitor	NDGA	30.2 (100 µM)	23	2.0	-	-	-	-	-	-
	Ibuprofen	2.1 (10 µM)	-	-	23	2.5	21	2.0	-	-
	Thioetheramide-PC	73.6 (100µM)	-	-	-	-	-	-	91	4.0

% Inh—percent of enzyme activity inhibition; SD—standard deviation (n = 3).

This work presents for the first time such a wide determination of lignan contents in the extracts of *S. rubriflora*, taking into account the division into the material originating from female (F) and male (M) specimens (leaves and shoots and fruits). Studies on lignan composition of *S. rubriflora* were performed before only by Chinese teams [16–18,25,35]. Importantly, these studies were only qualitative, no quantitative data were found, and the authors did not distinguish extracts in terms of gender. In 2006, Chen et al. [18] isolated fruit extracts of *S. rubriflora* and detected following compounds by the ^{1}H NMR method: schisanhenol, schisandrin, deoxyschisandrin, schisantherin B, angeloylgomisin P, tigloylgomisin P, gomisin M_1, M_2, O and J as well as specific rubrisandrins A and B. In 2010, Xiao et al. [17] identified the following lignans in fruit extracts based on the ^{1}H and ^{2}H-NMR spectroscopy: Gomisin G and O, angeloylgomisin P, wulignan A_2, epiwulignan A_1 and rubrisandrin C. Mu et al. [16] conducted in 2011 an isolation and structure elucidation of the following lignans from *S. rubriflora* fruit extracts using preparative HPLC and ^{13}C-NMR: Schisandrin, schisandrin A and C, rubschisantherin, angeloygomisin Q, benzoylgomisin Q, gomisin J, Q, C, B, K, N, S, T, isogomisin O, wilsonilignangomisin G, marlignan L and G. Moreover, the latter authors detected for the first time two new lignans in the fruit extract, i.e., rubrilignans A and B. In comparison to those results, the current study identified eight additional new compounds in fruit extracts: Gomisin A, 6-O-benzoylgomisin O, angeloylgomisin H and O, pregomisin, mesodihydroguaiaretic acid and licarin A and B.

Extracts from aerial parts of *S. rubriflora* were studied by Li et al. [25] using ^{1}H- and ^{13}C-NMR and they found the following compounds: gomisin K, M_1 and R, dimethylgomisin J, angeloylgomisin K_3 and R, interiotherin B, schisantherin D, mesodihydroguaiaretic acid, dihydroguaiaretic acid and pregomisin. Li et al. [35] detected rubriflorin A and B in stem extracts. In the present study, additional nineteen compounds were detected both in shoot and leaf extracts: Schisantherin A and B, schisandrin, schisandrin C, gomisin A, D, G, J, N, O, 6-O-benzoylgomisin O, schisandrin A, rubrisandrin A, epigomisin O, schisanhenol, angeloylgomisin H and O, wulignan A_1 and fragransin A_2. Moreover, our study included quantitative estimation and division on female and male plants, which had not been done before (Table 1).

A comparative lignan profiling of extracts from fruits, leaves and shoots of *S. chinensis* using the UHPLC-MS MS method (Tables 2 and 5) was carried out in the present study. The total content of the tested group of compounds in *S. rubriflora* fruit extracts was 1.60-fold lower than in *S. chinensis* fruit extracts. The total lignan contents in *S. rubriflora* fruit extracts of F specimen was: 1.97-, 2.55-fold higher, respectively, than in *S. chinensis* leaf extracts (Table 1). The total lignan content in extracts from F and M shoots of *S. rubriflora* was: 1.78- and 1.23-fold higher, respectively, than in *S. chinensis* shoot extracts (Table 1).

Schisanhenol, schisantherin B, schisandrin A, gomisin O and angeloylgomisin H were predominant in *S. rubriflora* fruit extracts in terms of quantity (Table 3). The contents of schisanhenol, gomisin O and angeloylgomisin H in *S. rubriflora* fruit extracts was: 27.92- and 1.14-fold higher, respectively, than in *S. chinensis* fruit extracts (Table 1). The total content of schisantherin B and schisandrin A in *S. rubriflora* fruit extracts was: 1.57- and 2.04-fold lower, respectively, than in *S. chinensis* fruit extracts (Table 1).

Schisantherin A and schisantherin B, 6-O-benzoylgomisin O and angeloylgomisin H were among the quantitatively predominant compounds in *S. rubriflora* leaf extracts (Table 2). The quantities of these compounds in F and M leaf extracts of *S. rubriflora* were: 58.76- and 27.76-; 2.84- and 1.02-; 12.42- and 52.13-; and 2.13- and 2.73-fold higher, respectively than in *S. chinensis* leaf extracts (Table 5).

Similarly as in leaf extracts, schisantherin A and schisantherin B, 6-O-benzoylgomisin O and angeloylgomisin H were among the quantitatively predominant compounds in *S. rubriflora* shoot extracts (Table 2). The quantities of these compounds in F and M leaf extracts of *S. rubriflora* were: 38.00- and 10.93-; 6.78- and 9.68-; 6.98- and 52.13-; and 2.36- and 1.67-fold higher, respectively, than in *S. chinensis* shoot extracts (Table 5).

Different compounds were proved to be dominant in *S. chinensis* extracts. The quantities of schisandrin, gomisin D, gomisin J, gomisin A, schisandrin A, gomisin N and schisandrin C in

S. chinensis fruit extracts were: 31.37-; 55.46-; 26.36-; 157.25-; 2.04-; 13.49- and 3.47-fold higher, respectively, than in *S. rubriflora* fruit extracts (Table 5). The dominant compounds in *S. chinensis* leaf and shoot extracts included schisandrin, gomisin A, gomisin N and schisandrin C. The amount of these lignans in leaf extracts, in comparison to their quantities in F and M leaf extracts, were: 3.99- and 5.71-; 11.53- and 17.58-; 32.95- and 24.91-; and 84.55- and 195.79-fold higher, respectively. Correspondingly, the quantities of these compounds in shoot extracts, compared to *S. rubriflora* F and M shoot extracts, were: 8.20- and 14.61-; 12.14- and 22.00-; 77.40- and 58.05; and 106.93- and 332.67-fold higher, respectively (Table 5).

Table 5. The comparison of the amounts (mg/100g DW) of the dominant lignans in the studied *S. rubriflora* and *S. chinensis* extracts.

Lignans	Schisandra rubriflora					Schisandra chinensis		
	Fruits	Leaves		Shoots		Fruits	Leaves	Shoots
		F	M	F	M			
Schisandrin	6.57	8.15	5.69	4.01	2.25	206.08	32.51	32.87
Gomisin D	3.52	16.45	116.51	8.25	20.26	195.22	9.62	11.33
Gomisin J	5.40	0.97	0.76	0.46	0.36	142.35	18.06	13.22
Gomisin A	0.75	6.40	4.20	2.99	1.65	177.94	73.82	36.29
Gomisin G	66.39	11.13	8.23	5.25	3.67	44.56	12.18	11.69
Schisantherin A	27.19	226.80	107.17	84.35	24.27	31.32	3.86	2.22
Schisantherin B	118.07	291.47	104.28	239.11	169.04	185.82	102.47	35.27
Schisanhenol	268.02	2.05	2.73	1.13	2.53	9.60	1.00	0.91
Schisandrin A	104.32	0.22	0.38	0.12	0.50	212.50	17.74	13.89
Gomisin N	19.20	1.58	2.21	0.81	1.08	259.05	55.06	62.69
6-O-Benzoylgomisin O	35.28	134.51	564.62	72.38	52.18	33.64	10.83	7.48
Schisandrin C	4.96	0.44	0.19	0.28	0.09	18.54	37.20	29.94
Angeloylgomisin H	185.10	100.83	129.28	105.80	74.73	161.90	47.34	44.84
Total content	1055.65	853.33	1106.80	559.97	384.80	1686.95	433.59	313.83

The present study determined the complex in vitro anti-inflammatory activity of fruit and leaf extracts of *S. rubriflora* and *S. chinensis*. Moreover, the analyses were also performed on individual most abundant lignans (6-O-benzoylgomisin O, schisandrin, gomisin D, gomisin N and schisantherin A) and a synthetic average sample of lignan composition mixture (MIX) (representing mean concentrations of 16 most abundant lignans from fruit and leaf extracts of *S. rubriflora* and *S. chinensis*; see Table S3). In most cases, dose-dependent inhibition of selected enzymes by plant extracts or lignan solutions was observed. However, the most pronounced exceptions were found for 15-LOX inhibition (and partially COXs), especially by lignan solutions. An inverse dose-dependence in case of plant extracts can be contributed to a relatively wide confidence interval between dilutions (overlapping SDs), and thus the lack of significant difference. However, more interesting is an inverse dose-dependent inhibition of 15-LOX by lignan solutions. This is a clear deviation from the competitive inhibition mechanism. The observed dependencies allow us to suggest the occurrence of another mechanism, namely inhibitor acceleration of the enzyme by lignans. Known mechanisms of inhibitor acceleration rely on allostery and multiple active sites [36]. This assumption could be partially supported by the results for plant extracts, providing also some arguments for significant participation of lignans in their anti-inflammatory properties, albeit of low confidence (as mentioned earlier). This hypothesis should be confirmed in further studies. On the other hand, plant secondary metabolism is so rich and complicated that it is difficult to conclude that particular antioxidant properties, or anti-inflammatory in this case, are driven only by one compound group. Further, a parallel occurrence of components inhibiting as well as increasing the activity is possible due to the natural complexity of such a plant extract.

Previously, there were studies involving anti-inflammatory properties of certain schisandra lignans, but they were conducted on different models. Gomisin N, J and schisandrin C were proved to exert anti-inflammatory effect by reducing nitric oxide (NO) production from lipopolysaccharide-stimulated (LPS) RAW 264.7 cells [28]. Schisantherin A was shown to be an anti-inflammatory agent that down-regulated NF-κB and MAPK signaling pathways in LPS-treated RAW 264.7 cells [29]. Another study [30] demonstrated that schisandrin, deoxyschisandrin, schisandrin B and C and schisantherin A reduced LPS-induced NO production in RAW 264.7 cells. In addition, schisandrin was shown to exert a protective effect on LPS-induced sepsis [31]. In vitro studies performed by Guo et al. [31] showed that anti-inflammatory properties of schisandrin resulted from NO production inhibition, prostaglandin E_2 (PGE_2) release, COX-2 and inducible nitric oxide synthase (iNOS) expression, which in turn was caused by the inhibition of nuclear factor kappa B (NF-κB), c-Jun N-terminal kinase (JNK) and p38 mitogen-activated protein kinase (MAPK) activities in the RAW 264.7 macrophage cell line.

Moreover, extracts from *S. chinensis* fruits were tested for the anti-inflammatory activity by Huyke et al. [26]. Non-polar *S. chinensis* fruit extracts showed that the dose-dependent COX-2 inhibition (at 20 µg/mL) catalyzed prostaglandin production [26].

Lim et al. [27] conducted in vitro tests on such representative schisandra lignans as schisandrin, schisandrin A and C, gomisin B, C, G and N, as well as on methanolic extracts of *S. chinensis* fruits, for 5-lipoxygenase (5-LOX) inhibitory activity. The tested compounds inhibited 5-LOX-catalyzed leukotriene production by A23187-treated rat basophilic leukemia (RBL-1) cells at concentrations of 1–100 µM. Compounds, such as schisandrin and gomisins showed moderate inhibitory activity ($IC_{50} < 10$ µM) against 5-LOX-catalyzed leukotriene production, but they were significantly less active against COX-2-catalyzed PGE_2 and inducible NO production [27].

We have also proved in our study the high anti-inflammatory potential of *S. chinensis* fruit extracts against COX-1 and COX-2 enzyme activities (Table 3). Positive results were also obtained for leaf extracts (Table 3). Moreover, the inhibitory activity against 15-LOX and $sPLA_2$ enzymes has been demonstrated for the first time. To the best of our knowledge, the extracts from fruits and leaves of *S. rubriflora* have not yet been studied for anti-inflammatory activity. The obtained results from *S. rubriflora* plant materials showed a higher activity in comparison to *S. chinensis* (Table 3).

We have demonstrated, based on the results of individual lignan analyses, that 6-*O*-benzoylgomisin O, schisandrin, gomisin D, gomisin N and schisantherin A display significant 15-LOX, COX-1 and COX-2 inhibitory activities, and that they are virtually inactive against $sPLA_2$ (Table 4). Our study also analyzed for the first the anti-inflammatory activity of 6-*O*-benzoylgomisin O and gomisin D.

4. Materials and Methods

4.1. Plant Material

Plant material was obtained as part of cooperation with Clematis—Źródło Dobrych Pnączy Spółka z o.o. spółka jawna with a registered office in Pruszków (address: ul. Duchnicka 27, 05-800 Pruszków, Poland) [37]. Plant species were identified by dr. eng. Szczepan Marczyński (owner of the Clematis arboretum). For the purpose of comparative phytochemical analysis, fruits and leaves and shoots (stems with leaves) of about 10 years old female (F) (100 individuals) and male (M) (50 individuals) *S. rubriflora* (Franch.) Rehd. et Wils specimens, and about 10 years old monoecious specimens of *S. chinensis* Turcz. Baill (100 individuals) were collected and dried. Leaves and shoots were harvested in May, fruits in September 2017. The fruits were lyophilized and the leaves and shoots were air-dried (about 25–30 °C). Dry plant material was pulverized in a mixing ball mill (MM 400, Retch, Haan, Germany).

4.2. Plant Sample Extraction

Methanol extracts were prepared from fruits, shoots and leaves of F and M *S. rubriflora* plants. The samples (0.3 g, 5 replicates) were extracted with 3 mL of methanol (grade-HPLC, Merck, Darmstadt,

Germany). The extraction process was carried out twice in an ultrasonic bath (Sonic 2, POLSONIC Palczyński Sp.J., rsaw, Poland) for 20 min. The obtained extracts were centrifuged for 5 min (4000 rpm) in a centrifuge (Centrifuge MPW–223E, MPW Med. Instruments, Warsaw, Poland). The centrifuged extracts were filtered using sterilizing syringe filters (Millex®GP, 0.22 µm, Filter Unit, Millipore, Bedford, MA, USA).

4.3. UHPLC–MS/MS Lignan Targeted Profiling

Lignan-targeted profiling was carried out in methanolic extracts of *S. rubriflora* and *S. chinensis* by means of ultra-high performance liquid chromatography coupled to a tandem mass spectrometer (UHPLC-MS/MS). An external standard addition method was used. Filtered plant extracts were aliquoted in two 45 µL portions. To the first, 5 µL of methanol was added, while to the second 5 µL of the standard lignan solution (all monitored compounds). Samples were analyzed on a UHPLC Infinity 1260 (Agilent, Wolbrom, Germany) coupled to a quadrupole tandem mass spectrometer 6410 QQQ LC/MS (Agilent, Santa Clara, CA, USA). Samples were separated on an analytical column (Kinetex C18 150 × 4.6 mm, 2.7 µm) in a gradient mode of 50% methanol in water (A) versus 100% methanol (B) with 0.1% of formic acid. A linear gradient was applied, 20% to 65% of B in 22 min at 0.5 mL/min at 60 °C; the injection volume was 2 µL. Standard lignan substances were purchased from ChemFaces Biochemical Co. Ltd. (Wuhan, China). The studied lignans and their structures and synonymous names are listed in Table S1. Lignans were analyzed in the MRM mode after ESI ionization (Table S2).

In addition to lignans, whose standards were commercially available (Tables S1 and S2, Figure S1), compounds from the dibenzocyclooctadiene lignan group (angeloylgomisins H and O) were also identified based on the UHPLC-MS/MS result analysis for the tested extracts. The identification was based on analyzing fragmentation ions of these compounds visible in mass spectra. The quantitative analysis of angeloylgomisyn H and O was based on their content conversion, according to schisandrin standard curve (UHPLC-MS/MS)—the main compound from the dibenzocyclooctadiene lignan group; according to pharmacopoeial requirements, the raw material should be standardized based on the content of this compound [3].

4.4. Anti-Inflammatory Activity

Plant material methanolic extracts of fruits and leaves of *S. rubriflora* (F and M) and *S. chinensis* were tested for anti-inflammatory activity. Additionally, the following most abundant lignans present in the plant samples were analyzed for anti-inflammatory activity: 6-*O*-benzoylgomisin O, schisandrin, gomisin D, gomisin N and schisantherin A (No: L10, L1, L16, L14, L6; Tables S1, S2 and Figure S1, respectively); in addition, the mixture of lignans, representing the average plant sample composition (MIX) (mean concentrations of 16 most abundant lignans of fruit and leaf extracts of *S. rubriflora* and *S. chinensis*; see Table S3), underwent analogous analysis.

The plant extracts (concentrations: 175.0 and 17.5 µg/mL, Table 3) and solutions of selected lignans (concentrations: 1.75 and 0.175 µg/mL, Table 4) were serially diluted in methanol. The tests were based on in vitro inhibition of 15-lipooxygenase (15-LOX), phospholipases A_2 (sPLA$_2$), cyclooxygenase-1 (COX-1) and cyclooxygenase-2 (COX-2) enzymes.

4.4.1. Inhibitory Activity against 15-Lipooxygenase (15-LOX)

Samples were tested for their inhibitory activity against 15-LOX using an assay kit (760700, Cayman Chem. Co., Ann Arbor, MI, USA), according to the manufacturer's instructions; arachidonic acid at 0.91 mM was the substrate; nordihydroguaiaretic acid (NDGA) at 100 µM served as a positive control inhibitor. The kit measures the concentration of hydroperoxides produced in the lipooxygenation reaction using purified soy 15-lipooxygenase standard at pH 7.4 in 10 mM Tris-HCl buffer. The reagent's colorimetric composition is vendor proprietary. The measurements were carried out in 96-well plate using a Synergy II reader (Biotek, Winooski, VT, USA) at 490 nm. The end-point absorbance was recorded after 5-min incubation of enzyme and inhibitor followed by 15-min incubation after substrate addition and 5-min incubation after chromogen addition.

4.4.2. Inhibitory Activity against Cyclooxygenase-1 and Cyclooxygenase-2 (COX-1 and COX-2)

Samples were tested for their ability to inhibit COX-1 and COX-2 using the COX-1 (ovine) and COX-2 (human) inhibitor assay kit (701050, Cayman Chem. Co.), according to the manufacturer's instructions; arachidonic acid at 1.1 mM was the substrate; ibuprofen at 10 μM served as a positive control inhibitor. The kit measures the peroxidase component of COXs. The appearance of oxidized *N,N,N',N'*-tetramethyl-*p*-phenylenediamine (TMPD) was monitored kinetically for 5 min in a 96-well plate format at 590 nm using a Synergy II reader.

4.4.3. Inhibitory Activity against Phospholipases A$_2$ (sPLA$_2$)

Inhibition of sPLA$_2$ activity was tested using an assay kit (10004883, Cayman Chem. Co.), according to the manufacturer's instructions; diheptanoyl thio-PC at 1.44 mM was the substrate; thioetheramide-PC at 100 μM served as a positive control inhibitor. Human recombinant Type V sPLA$_2$ was used. Free thiols released by cleavage of the diheptanoyl thio-PC ester bond were measured kinetically using DTNB (5-5'-dithio-bis-(2-nitrobenzoic acid), Ellman's reagent) in a 96-well plate format at 420 nm using a Synergy II reader. The percent of inhibition was calculated according to Equation (1):

$$\%Inh = [(IA - Inhibitor)/IA] \times 100 \tag{1}$$

where: %Inh—percent of inhibition; IA—100% enzyme activity (without inhibitor); Inhibitor—enzyme activity with inhibitor added.

All samples were assayed in triplicate, including 100% enzyme activity, positive control inhibitor and tested extracts and lignan solutions.

4.4.4. Statistical Analysis

Quantitative results are expressed in mg/100 g DW (dry weight) as the mean ± SD (standard deviation) of three or five samples ($n = 3$, $n = 5$) in the experiments that were repeated three times.

5. Conclusions

The present study is the first comparative, complex, qualitative and quantitative analyses of *S. rubriflora* and *S. chinensis* lignan composition derived from different groups. The contents of shoot and fruit extracts of both plant species were determined for the first time using the UHPLC-MS/MS method. The study identified and characterized twenty-four lignans representing four chemical groups: dibenzocyclooctadiene lignans (schisantherin A and B, schisandrin, schisandrin C, gomisin A, D, G, J, N, O, 6-*O*-benzoylgomisin O, schisandrin A, rubrisandrin A, epigomisin O, schisanhenol, rubriflorin A, angeloylgomisin H and O), aryltetralin lignan (wulignan A$_1$), dibenzylbutane lignans (pregomisin, mesodihydroguaiaretic acid), tetrahydrofuran lignan (fragransin A$_2$) and dihydrobenzofuran neolignans (licarin A and B). Qualitative and quantitative differences in lignan composition were recorded depending on the origin of samples (fruit, leaf and shoot) as well as plant species.

Additionally, the current work determined for the first time the anti-inflammatory activity, based on the in vitro inhibition of 15-lipooxygenase (15-LOX), phospholipases A$_2$ (sPLA$_2$), cyclooxygenases 1 and 2 (COX-1; COX-2) enzymes, of fruit and leaf extracts of the analyzed species as well as individual lignans: 6-*O*-benzoylgomisin O, schisandrin, gomisin D, gomisin N and schisantherin A; furthermore, a mixture of lignans representing an average plant sample composition was also tested. The results revealed a high competitiveness of *S. rubriflora* in relation to known, pharmacopoeial plant species—*S. chinensis*.

Based on our research, we suggest to consider the extracts of *S. rubriflora* (fruit, leaf and shoot), as a rich, valuable source of lignans with a promising anti-inflammatory potential. The objects of interest exhibited very interesting differences and showed new research directions involving these compounds, e.g., phenolic composition and other biological activities of *S. rubriflora* would be worth investigating.

Supplementary Materials: The following are available online, Figure S1: Exemplary UHPLC–MS/MS MRM chromatogram of lignan standard mixture at 100 ug/mL. Table S1: The standard lignan substances used in the performer studies. Table S2: The monitored fragmentation reactions (multiple reactions monitoring, MRM) for studied lignans. Table S3: Quantitative composition of "average sample lignan composition" (MIX) at 1.75 µg/mL.

Author Contributions: Conceptualization, A.S.; Methodology, A.S. and M.D.; Investigation, A.S., M.D., A.W., P.K. and M.K.-S.; Data Curation, A.S., M.D. and A.W.; Writing-Original Draft Preparation, A.S. and M.D.; Writing-Review & Editing, A.S., M.D., M.K.-S., H.E.; Project Administration, A.S.; Funding Acquisition, A.S.

Funding: This research was funded by National Science Centre, Poland, rant number: 2016/23/D/NZ7/01316.

Conflicts of Interest: All authors declare that they have no conflict of interest.

References

1. Saunders, R.M.K. Monograph of *Schisandra* (*Schisandraceae*). In *Systematic Botany Monographs*; American Society of Plant Taxonomists: Laramie, WY, USA, 2000; Volume 58, pp. 1–146. ISBN 978-0912861586.

2. Szopa, A.; Barnaś, M.; Ekiert, H. Phytochemical studies and biological activity of three Chinese *Schisandra* species (*Schisandra sphenanthera*, *Schisandra henryi* and *Schisandra rubriflora*): Current findings and future applications. *Phytochem. Rev.* **2018**, 1–20. [CrossRef]

3. European Directorate for the Quality of Medicines. Schisandrae chinensis fructus. In *European Pharmacopoeia 9.0.*; Council of Europe: Strasbourg Cedex, France, 2017.

4. Szopa, A.; Ekiert, R.; Ekiert, H. Current knowledge of *Schisandra chinensis* (Turcz.) Baill. (Chinese magnolia vine) as a medicinal plant species: A review on the bioactive components, pharmacological properties, analytical and biotechnological studies. *Phytochem. Rev.* **2017**, *16*, 195–218. [CrossRef] [PubMed]

5. Szopa, A.; Klimek, M.; Ekiert, H. Chinese magnolia vine (*Schisandra chinensis*)—therapeutic and cosmetic importance. *Polish J. Cosmetol.* **2016**, *19*, 274–284.

6. European Directorate for the Quality of Medicines. Schisandrae chinensis fructus. In *European Pharmacopoeia 6.0.*; Council of Europe: Strasbourg Cedex, France, 2008.

7. Chinese Pharmacopoeia Commission. *Pharmacopoeia of the People's Republic of China*; China Chemical Industry Press: Beijing, China, 2005.

8. Committee of the Japanese Pharmacopoeia. Evaluation and Licensing Division Pharmaceuticals and Food Safety. In *Japanese Pharmacopoeia*; Bureau Ministry of Health, Labour and Welfare: Tokyo, Japan, 2006.

9. Central Pharmaceutical Affairs Council of Korea. *Korean Pharmacopoeia*; Central Pharmaceutical Affairs Council of Korea: Seoul, Korea, 2002.

10. Xu, L.; Grandi, N.; Del Vecchio, C.; Mandas, D.; Corona, A.; Piano, D.; Esposito, F.; Parolin, C.; Tramontano, E. From the traditional Chinese medicine plant *Schisandra chinensis* new scaffolds effective on HIV-1 reverse transcriptase resistant to non-nucleoside inhibitors. *J. Microbiol.* **2015**, *53*, 288–293. [CrossRef] [PubMed]

11. Upton, R.; Graff, A.; Jolliffe, G.; Länger, R.; Williamson, E. *American Herbal Pharmacopoeia: Botanical Pharmacognosy—Microscopic Characterization of Botanical Medicines*; CRC Press: Boca Raton, FL, USA, 2011; ISBN 1420073281.

12. World Health Organization. *Fructus Schisandrae. WHO Monographs on Selected Medicinal Plants*; WHO: Geneva, Switzerland, 2007; Volume 3.

13. Panossian, A.; Wikman, G. Pharmacology of *Schisandra chinensis* Bail.: An overview of Russian research and uses in medicine. *J. Ethnopharmacol.* **2008**, *118*, 183–212. [CrossRef]

14. Hancke, J.L.; Burgos, R.A.; Ahumada, F. *Schisandra chinensis* (Turcz.) Baill. *Fitoterapia* **1999**, *70*, 451–471. [CrossRef]

15. Li, G.; Zhao, J.; Tu, Y.; Yang, X.; Zhang, H.; Li, L. Chemical constituents of *Schisandra rubriflora* Rehd. et Wils. *J. Integr. Plant Biol.* **2005**, *47*, 362–367. [CrossRef]

16. Mu, H.X.; Li, X.S.; Fan, P.; Yang, G.Y.; Pu, J.X.; Sun, H.D.; Hu, Q.F.; Xiao, W.L. Dibenzocyclooctadiene lignans from the fruits of *Schisandra rubriflora* and their anti-HIV-1 activities. *J. Asian Nat. Prod. Res.* **2011**, *13*, 393–399. [CrossRef]

17. Xiao, W.L.; Wang, R.R.; Zhao, W.; Tian, R.R.; Shang, S.Z.; Yang, L.M.; Yang, J.H.; Pu, J.X.; Zheng, Y.T.; Sun, H.D. Anti-HIV-1 activity of lignans from the fruits of *Schisandra rubriflora*. *Arch Pharm. Res.* **2010**, *33*, 697–701. [CrossRef]

18. Chen, M.; Kilgore, N.; Lee, K.H.; Chen, D.F. Rubrisandrins A and B, lignans and related anti-HIV compounds from *Schisandra rubriflora*. *J. Nat. Prod.* **2006**, *69*, 1697–1701. [CrossRef]

19. Xiao, W.L.; Li, X.; Wang, R.R.; Yang, L.M.; Li, M.; Huang, S.X.; Pu, J.X.; Zheng, Y.T.; Li, R.T.; Sun, H.D. Triterpenoids from *Schisandra rubriflora*. *J. Nat. Prod.* **2007**, *70*, 1056–1059. [CrossRef] [PubMed]

20. Lu, H.; Liu, G.T. Anti-oxidant activity of dibenzocyclooctene lignans isolated from *Schisandraceae*. *Planta Med.* **1992**, *58*, 311–313. [CrossRef] [PubMed]

21. Opletal, L.; Sovová, H.; Bártlová, M. Dibenzo [a,c] cyclooctadiene lignans of the genus *Schisandra*: Importance, isolation and determination. *J. Chromatogr. B* **2004**, *812*, 357–371. [CrossRef]

22. Chang, J.; Reiner, J.; Xie, J. Progress on the chemistry of dibenzocyclooctadiene lignans. *Chem. Rev.* **2005**, *105*, 4581–4609. [CrossRef] [PubMed]

23. Hu, D.; Han, N.; Yao, X.; Liu, Z.; Wang, Y.; Yang, J.; Yin, J. Structure-activity relationship study of dibenzocyclooctadiene lignans isolated from *Schisandra chinensis* on lipopolysaccharide-induced microglia activation. *Planta Med.* **2014**, *80*, 671–675. [CrossRef] [PubMed]

24. Chen, M.; Xu, X.; Xu, B.; Yang, P.; Liao, Z.; Morris-Natschke, S.L.; Lee, K.H.; Chen, D. Neglschisandrins E-F: Two new lignans and related cytotoxic lignans from *Schisandra neglecta*. *Molecules* **2013**, *18*, 2297–2306. [CrossRef] [PubMed]

25. Li, H.M.; Luo, Y.M.; Pu, J.X.; Li, X.N.; Lei, C.; Wang, R.R.; Zheng, Y.T.; Sun, H.D.; Li, R.T. Four new dibenzocyclooctadiene lignans from *Schisandra rubriflora*. *Helv. Chim. Acta* **2008**, *91*, 1053–1062. [CrossRef]

26. Huyke, C.; Engel, K.; Simon-Haarhaus, B.; Quirin, K.-W.; Schempp, C. Composition and Biological Activity of Different Extracts from *Schisandra sphenanthera* and *Schisandra chinensis*. *Planta Med.* **2007**, *73*, 1116–1126. [CrossRef]

27. Lim, H.; Son, K.H.; Bae, K.H.; Hung, T.M.; Kim, Y.S.; Kim, H.P. 5-Lipoxygenase-inhibitory constituents from *Schisandra fructus* and *Magnolia flos*. *Phyther. Res.* **2009**, *23*, 1489–1492. [CrossRef]

28. Oh, S.-Y.; Kim, Y.H.; Bae, D.S.; Um, B.H.; Pan, C.-H.; Kim, C.Y.; Lee, H.J.; Lee, J.K. Anti-Inflammatory Effects of Gomisin N, Gomisin J, and Schisandrin C Isolated from the Fruit of *Schisandra chinensis*. *Biosci. Biotechnol. Biochem.* **2010**, *74*, 285–291. [CrossRef]

29. Ci, X.; Ren, R.; Xu, K.; Li, H.; Yu, Q.; Song, Y.; Wang, D.; Li, R.; Deng, X. Schisantherin A exhibits anti-inflammatory properties by down-regulating NF-κB and MAPK signaling pathways in lipopolysaccharide-treated RAW 264.7 cells. *Inflammation* **2010**, *33*, 126–136. [CrossRef] [PubMed]

30. Qiu, H.; Zhao, X.; Li, Z.; Wang, L.; Wang, Y. Study on main pharmacodynamic effects for *Schisandra* lignans based upon network pharmacology. *Chin. J. Chin. Mater. Med.* **2015**, *40*, 522–527.

31. Guo, L.Y.; Hung, T.M.; Bae, K.H.; Shin, E.M.; Zhou, H.Y.; Hong, Y.N.; Kang, S.S.; Kim, H.P.; Kim, Y.S. Anti-inflammatory effects of schisandrin isolated from the fruit of *Schisandra chinensis* Baill. *Eur. J. Pharmacol.* **2008**, *591*, 293–299. [CrossRef] [PubMed]

32. Whiting, D.A. Lignans and neolignans. *Nat. Prod. Rep.* **1985**, *193*, 191–211. [CrossRef]

33. Whiting, D.A. Lignans, neolignans, and related compounds. *Nat. Prod. Rep.* **1990**, *7*, 349–364. [CrossRef]

34. Gottlieb, O.R. Chemosystematics of the Lauraceae. *Phytochemistry* **1972**, *11*, 1537–1570. [CrossRef]

35. Li, L.; Ren, H.Y.; Yang, X.D.; Zhao, J.F.; Li, G.P.; Zhang, H. Bin Rubriflorin A and B, two novel partially saturated dibenzocyclooctene lignans from *Schisandra rubriflora*. *Helv. Chim. Acta* **2004**, *87*, 2943–2947. [CrossRef]

36. Wales, M.E.; Madison, L.L.; Glaser, S.S.; Wild, J.R. Divergent allosteric patterns verify the regulatory paradigm for aspartate transcarbamylase. *J. Mol. Biol.* **1999**, *294*, 1387–1400. [CrossRef]

37. Clematis–Źródło Dobrych Pnączy Spółka z o.o. spółka jawna. Available online: http://www.clematis.com.pl/pl/ (accessed on 3 October 2018).

Sample Availability: Samples of the compounds (all tested lignans) are available from the authors.

molecules

MDPI

Article

Quality Evaluation of Wild and Cultivated *Schisandrae Chinensis* Fructus Based on Simultaneous Determination of Multiple Bioactive Constituents Combined with Multivariate Statistical Analysis

Shuyu Chen [1], Jingjing Shi [1], Lisi Zou [1], Xunhong Liu [1,*], Renmao Tang [2], Jimei Ma [2], Chengcheng Wang [1], Mengxia Tan [1] and Jiali Chen [1]

[1] College of Pharmacy, Nanjing University of Chinese Medicine, Nanjing 210023, China; 18305172513@163.com (S.C.); shijingjingquiet@163.com (J.S.); zlstcm@126.com (L.Z.); ccw199192@163.com (C.W.); 18816250751@163.com (M.T.); 18994986833@163.com (J.C.)
[2] SZYY Group Pharmaceutical Limited, Taizhou 225500, China; ktang@vip.163.com (R.T.); mjm0607@hotmail.com (J.M.)
* Correspondence: liuxunh1959@163.com; Tel./Fax: +86-25-8581-1524

Academic Editor: David Barker
Received: 7 March 2019; Accepted: 3 April 2019; Published: 4 April 2019

Abstract: *Schisandrae Chinensis* Fructus, also called wuweizi in China, was a widely used folk medicine in China, Korea, and Russia. Due to the limited natural resources and huge demand of wuweizi, people tend to cultivate wuweizi to protect this species. However, the quality of wild and cultivated herbs of the same species may change. Little attention has been paid to comparing wild and cultivated wuweizi based on simultaneous determination of its active components, such as lignans and organic acids. An analytical method based on UFLC-QTRAP-MS/MS was used for the simultaneous determination of 15 components, including 11 lignans (schisandrin, gomisin D, gomisin J, schisandrol B, angeloylgomisin H, schizantherin B, schisanhenol, deoxyschizandrin, γ-schisandrin, schizandrin C, and schisantherin) and 4 organic acids (quinic acid, D(−)-tartaric acid, L-(−)-malic acid, and protocatechuic acid) in wuweizi under different ecological environments. Principal components analysis (PCA), partial least squares discrimination analysis (PLS-DA), independent sample t-test, and gray relational analysis (GRA) have been applied to classify and evaluate samples from different ecological environments according to the content of 15 components. The results showed that the differential compounds (i.e., quinic acid, L-(−)-malic acid, protocatechuic acid, schisandrol B) were significantly related to the classification of wild and cultivated wuweizi. GRA results demonstrated that the quality of cultivated wuweizi was not as good as wild wuweizi. The protocol not just provided a new method for the comprehensive evaluation and quality control of wild and cultivated wuweizi, but paved the way to differentiate them at the chemistry level.

Keywords: *Schisandrae Chinensis* Fructus; wild; cultivated; multiple bioactive components; simultaneous quantitation

1. Introduction

Schisandrae Chinensis Fructus (wuweizi) is the dried ripe fruit of the magnolia plant *Schisandra chinensis* (Turcz.) Baillon. Wuweizi was a folk medicine in China, Korea, and Russia, which was used as a sedative and tonic. With the ability to replenish vital energy and promote fluid production, benefitting the kidneys and tranquilizing the mind, wuweizi was used to treat seminal emission, excessive sweating, diarrhea, insomnia, fatigue, and neurasthenia in clinics [1,2]. Existing research studies have shown that lignans with a dibenzocyclooctadiene skeleton and organic acids are major types of phytochemicals in wuweizi [3,4]. It has been demonstrated that lignans have abundant bioactivities, including antihepatotoxic, antioxidant, antitumor, nervous system protection, and anticancer properties [5–7]. Organic acids have the beneficial pharmacological effects of arresting coughs and removing phlegm [2,8,9].

With the deeper understanding on the effectiveness of wuweizi, many industries, such as medicine, prescriptions, health products, and food and beverage, have a wide range of applications for this species. Various kinds of products, such as beverages, fermented wine, jam, tea, jelly, and dye, have aroused the demand for wuweizi. Due to excessive logging of unsustainable wild resources, the wild resources of wuweizi in northeast China have rapidly decreased and cannot meet the increasing demand of the markets. Therefore, planting cultivars has become a trend throughout northeast China. Unlike the wild ecological environment, the cultivated ecological environment is an artificial intervention. When plants face adversity, specialized metabolites accumulate significantly more than usual [10]. It is generally believed that most of the bioactive components isolated from herbs belong to specialised metabolites [11]. Varied ecological environments can lead to plants producing and accumulating various specialized metabolites, which cause uneven qualities of wuweizi. It is the amount or proportion among the medicinal constituents that mainly contribute to the quality difference for herbal medicine of the same species [12,13].

Nowadays, wuweizi is officially documented in Chinese, Japanese, European, Korean, and Russian pharmacopoeias, and other worldwide pharmacopoeial monographs [2]. Schisandrin is the quality indicator in the Chinese and United States pharmacopoeia. In addition, the content of lignans as the sum of the percentages of schisandrin, schisandrol B, deoxyschisandrin, and γ-schisandrin shall be not less than 0.95% by HPLC for quality control, as has been listed in the United States pharmacopeia. The control of several simple components may provide certain guidance, but for the multi-component and multi-target medicinal plants, there are still some limitations [14] because the pharmacological effect of wuweizi was usually a comprehensive effect of various kinds of compounds. For example, lignans and organic acids all have pharmacological effects, but they take different effects. However, most reports of quantitative analysis were focused on each particular class of compounds or certain active ingredients for the quality control of wuweizi [4,15–18]. The analysis of particular lignans alone was not sufficient for further study of the wuweizi quality control. Therefore, it is necessary to study the differences between wild and cultivated wuweizi and comprehensively evaluate the quality of them based on the simultaneous determination of lignans and organic acids.

The aim of this paper is to evaluate the quality of wild and cultivated wuweizi based on the simultaneous determination of multiple bioactive constituents combined with multivariate statistical analysis. A reliable method based on ultra-fast performance liquid chromatography coupled with triple quadrupole-linear ion trap mass spectrometry (UFLC-QTRAP-MS/MS) was used to simultaneously determine the content of 15 constituents, including 11 lignans and 4 organic acids in 12 batches of wuweizi samples from different ecosystems. Furthermore, principal component analysis (PCA) was introduced to get a good overview of the sample distribution. Partial least squares-discriminant analysis (PLS-DA) and t-tests were performed to show the difference of each compound between two types of wuweizi according to the contents of the tested constituents. Gray relational analysis (GRA) was carried out for the comprehensive evaluation. The protocol not just provided a new method for the comprehensive evaluation and quality control of wild and cultivated wuweizi, but allowed for differentiation of them at the chemistry level.

2. Results and Discussion

2.1. Optimization of Extraction Conditions

The orthogonal experiment was carried out to obtain a satisfactory extraction efficiency of major compounds in samples. Extraction solvent, material-solvent ratio, and ultrasonic time were optimized through 9 extraction experiments. The orthogonal table of levels and factors was shown in Table S1. The total peak area was calculated for evaluation of the orthogonal experiment, and the results were shown in Table S2. Finally, the optimum extraction conditions were achieved through ultrasonic extraction with a 40:1 ratio of methanol for 40 min at room temperature.

2.2. Optimization of UFLC Conditions

The peak area of schisandrin, theoretical plate number, resolution, and retention time were taken into consideration to investigate the different chromatographic columns, flow rates, column temperatures and mobile phases. Three kinds of columns, including an Agilent ZORBAX SB-C_{18} column (250 mm × 4.6 mm, 5 μm), a XBridge C_{18} (100 mm × 4.6 mm, 3.5 μm), and a Synergi^{TM} Hydro-RP 100Å (100 mm × 2.0 mm, 2.5 μm), were all compared to test samples. The results showed that the separation effect of three types of chromatographic columns were quite different, and the last one had the desirable resolution, peak shape, and retention time. Consequently, the column of Synergi^{TM} Hydro-RP 100Å (100 mm × 2.0 mm, 2.5 μm) was employed for this analysis. To achieve an efficient and rapid analysis, several UFLC parameters, including different kinds of mobile phases (methanol/water, acetonitrile/water, acetonitrile/0.1% aqueous formic acid, acetonitrile containing 0.1% formic acid solution/water, and acetonitrile containing 0.1% formic acid solution/0.1% aqueous formic acid), flow rates (0.3 mL/min, 0.35 mL/min and 0.4 mL/min), and column temperatures (30, 35, and 40 °C) were examined systematically. When acetonitrile aqueous solution was used for gradient elution, the shape of the peaks were better. When the flow rate was 0.40 mL/min, the resolution was better. The results of resolution under three column temperatures were all satisfactory, however, the column temperature of 40 °C had a higher theoretical plate number. Finally, it was determined that a gradient elution using water as eluent A and acetonitrile as eluent B at a flow rate of 0.4 mL/min under the column temperature of 40 °C resulted in the desired separation in a short analysis time.

2.3. Optimization of MS Conditions

In order to develop a sensitive and accurate quantitative method, individual solutions of all standard compounds (about 100 ng/mL) were examined separately with infusion mode by a full-scan mass spectrometry method in both positive and negative modes. After trial and error inspection, we found that lignans show maximum sensitivity under the positive ion mode. However, organic acids display desirable results in the negative ion mode. ESI, the electrospray ionizationsource of MS, could obtain abundant fragment ions of compounds. MRM (multiple reaction monitoring) is suitable for the quantification of components as a promising technology for the sensitivity and robustness. MRM transition from the MS/MS spectrum was chosen when the most abundant fragment ions appeared. For example, compounds **1**, **2**, **3**, and **4** the showed most intense ion for $[M - H]^-$. All the optimum values, including retention time (t_R), mass data (m/z), precursor and product ions, fragmentor voltage (FV), and collision energy (CE) for each compound are summarized in Table 1.

Table 1. Precursor/product ion pairs and parameters for MRM of the target compounds.

No	Coumpounds	t_R (min)	Mass Data (*m/z*)	Precursor Ion	Product Ion	FV (V)	CE (eV)
1	Quinic acid	0.79	191.1[M − H]−	191.1	85.02	−120	−26
2	D(−)-Tartaric acid	0.82	149.0[M − H]−	149.0	87	−55	−16
3	L-(−)-Malic acid	0.84	133.0 [M − H]−	133.0	114.9	−80	−14
4	Protocatechuic acid	1.11	153.0 [M − H]−	153.0	106	−85	−16
5	Schisandrin	5.65	433.3[M + H]+	433.3	415.34	146	13
6	Gomisin D	5.75	553.3[M + Na]+	553.3	507.32	21	35
7	Gomisin J	5.92	389.3[M + H]+	389.3	287.1	156	27
8	Schisandrol B	6.11	417.3[M + H]+	417.3	399.2	131	15
9	Angeloylgomisin H	6.67	501.3[M + H]+	501.3	401.2	146	11
10	Schizantherin B	7.34	515.3[M + H]+	515.3	415.2	56	11
11	Schisanhenol	7.63	403.2[M + H]+	403.2	340.2	1	27
12	Deoxyschizandrin	8.83	417.3[M + H]+	417.3	316.18	241	31
13	γ-schisandrin	9.50	401.3[M + H]+	401.3	300.15	231	31
14	Schizandrin C	9.85	385.2[M +H]+	385.2	285.16	201	29
15	Schisantherin	11.96	537.4[M + H]+	537.4	282.3	56	15

2.4. UFLC Method Validation

All method validations of quantification were performed using the UFLC-QTRAP-MS/MS technique. Each standard calibration curve was constructed by plotting the peak areas (Y) against the corresponding concentrations (X), which achieved appropriate determination coefficients ($r^2 > 0.9991$), and the test range covered the concentrations of investigated compounds in samples. The limits of detection (LOD) and limits of quantification (LOQ) were in the ranges of 0.49–2.95 ng/mL and 1.38–12.65 ng/mL of 15 analytes, respectively. The relative standard deviation (RSD) values of intra-day and inter-day variations of 15 components ranged from 0.67% to 3.09% and 0.15% to 2.63%, respectively. The RSD of repeatability and stability tests of the 15 analytes were less than 5%, and the overall recoveries varied between 95.62% and 99.97%, with RSDs less than 4.12%, demonstrating that this method was validated for all kinds of analytes. The detailed results of each method validation were presented in Table 2 and Table S3.

Molecules **2019**, *24*, 1335

Table 2. Regression equation, limits of detection, limits of quantification, precision, repeatability, stability, and recovery of 30 investigated compounds.

No.	Compounds	Regression Equation	r^2	Liner Range (ng/mL)	LODs (ng/mL)	LOQs (ng/mL)	Precision (RSD, %) Intra-Day (n = 6)	Precision (RSD, %) Inter-Day (n = 3)	Repeatability (RSD, %) (n = 6)	Stability (RSD, %) (n = 6)	Recovery (%) (n = 3) Low Mean	Low RSD	Medium Mean	Medium RSD	High Mean	High RSD
1	Quinic acid	Y = 1330X + 76100	1.0000	133 – 13300	2.66	6.57	2.40	2.63	3.5	1.1	99.55	0.22	99.65	0.36	99.83	0.05
2	D(−)-Tartaric acid	Y = 1240X + 1190	0.9995	3.75 – 24.9	0.49	1.62	3.09	1.60	2.1	2.3	95.62	2.1	98.07	0.57	97.34	0.74
3	L-(−)-Malic acid	Y = 2040X + 149000	0.9991	148 – 29500	2.95	12.42	1.79	1.45	3.2	1.2	99.92	0.06	99.93	0.03	99.96	0.06
4	Protocatechuic acid	Y = 162X + 3100	0.9997	255 – 12800	1.96	9.03	2.31	1.27	2.5	3.3	99.98	0.01	99.96	0.02	99.93	0.06
5	Schisandrin	Y = 213X + 1550	0.9991	11.6 – 9320	1.36	7.84	3.06	1.86	0.7	3.1	99.36	0.3	99.8	0.1	99.63	0.06
6	Gomisin D	Y = 3770X + 34900	0.9992	10.1 – 506	0.84	6.61	1.11	0.81	1.5	1.2	96.63	2.36	96.96	1.87	96.33	4.12
7	Gomisin J	Y = 790X + 7100	0.9993	23.3 – 931	1.53	6.07	2.03	1.26	3.0	3.0	98.63	0.62	98.59	0.54	97.7	0.45
8	Schisandrol B	Y = 4140X + 260000	0.9997	107 – 5350	1.92	8.15	2.56	0.31	2.4	0.8	99.05	0.38	98.85	0.18	99.13	0.7
9	Angeloylgomisin H	Y = 258000X + 623000	0.9991	5.65 – 2260	1.36	5.38	2.45	0.15	1.6	3.1	99.84	0.22	99.24	0.74	99.68	0.32
10	Schizantherin B	Y = 240X + 3640	0.9998	55.2 – 2210	2.29	12.65	1.00	0.54	1.3	2.3	99.02	0.34	98.98	0.73	99.33	0.69
11	Schisanhenol	Y = 2900X + 10800	0.9999	19.9 – 1194	1.99	5.82	2.59	1.58	4.9	1.5	98.19	1.12	98.63	0.88	96.92	3.3
12	Deoxyschizandrin	Y = 26600X + 217000	1.0000	24 – 2400	2.91	11.21	0.87	1.10	1.6	1.3	97.83	0.74	98.86	0.34	99.42	0.02
13	γ-schisandrin	Y = 157000X + 214000	1.0000	15.2 – 4560	1.79	6.92	1.91	0.43	0.5	1.1	99.63	0.3	99.75	0.09	99.69	0.13
14	Schizandrin C	Y = 6410X + 24300	0.9991	29.3 – 1470	0.52	1.38	2.25	0.66	3.8	2.5	98.57	0.41	97.64	0.61	98.24	0.61
15	Schisantherin	Y = 25X + 1860	0.9998	12 – 7200	1.59	5.62	0.67	1.61	2.2	1.0	99.89	0.04	99.89	0.04	99.97	0.02

2.5. Quantification of Lignans and Organic Acids

Twelve batches of cultivated and wild wuweizi were collected from Heilongjiang, Jilin, and Liaoning and dealt with using the same processing method (sun drying). Sample information is listed in Table 3. The validated analytical method of UFLC-QTRAP-MS/MS was successfully applied to the simultaneous determination of 11 lignans and 4 organic acids in wuweizi. Each sample was determined three times and the quantitative results of 15 compounds are presented in Table 4. Total ion chromatograms of the representative wild and cultivated samples are shown in Figure S1. Typical MRM chromatograms are shown in Figure 1. The histogram (Figure 2) suggests lignans were found in higher concentrations in wild than cultivated-type. However, the content of organic acids was slightly higher in the cultivated-type. It was clearly shown that the total contents of 15 compounds varied from 25,598.77 µg/g to 35,179.73 µg/g. The total contents of each type of constituent was also calculated, and 11 lignans ranged from 14,960.30 µg/g to 22,853.62 µg/g, and in the following order: (highest) S1 (wild) > S6 (wild) > S3 (wild) > S5 (wild) > S4 (wild) > S10 (cultivated) > S2 (wild) > S8 (cultivated) > S7 (cultivated) > S11 (cultivated) > S9 (cultivated) > S12 (cultivated) (lowest). The four organic acids ranged from 9075.64 µg/g to 13,646.81 µg/g. By comparing the amounts, it was found that the compounds of wuweizi from different ecosystems were quite different. In this study, the contents of lignans were similar to previous studies [4,15–18]. However, the contents of organic acids were slightly lower than the reported, which may be related to the harvest period, processing, and origin [19].

Table 3. Detailed information of samples.

Sample No.	Habitats	GPS Records	Harvesting Time	Processing Method
S1	Mulan, Heilongjiang	45°56′54″ N, 128°02′14″ E	10 August 2017	sun drying
S2	Jingyu, Jilin	42°23′11″ N, 126°48′28″ E	14 August 2017	sun drying
S3	Jingyu, Jilin	42°23′11″ N, 126°48′28″ E	17 August 2017	sun drying
S4	Xinbin, Liaoning	41°43′53″ N, 125°02′01″ E	14 August 2017	sun drying
S5	Hengren, Liaoning	41°15′13″ N, 125°22′15″ E	10 August 2017	sun drying
S6	Baoqing, Heilongjiang	46°19′29″ N, 132°11′22″ E	20 August 2017	sun drying
S7	Jingyu, Jilin	42°23′11″ N, 126°48′28″ E	16 August 2017	sun drying
S8	Jingyu, Jilin	42°23′11″ N, 126°48′28″ E	17 August 2017	sun drying
S9	Jingyu, Jilin	42°23′11″ N, 126°48′28″ E	20 August 2017	sun drying
S10	Fengcheng, Liaoning	41°48′19″ N, 123°27′47″ E	10 August 2017	sun drying
S11	Shuangyang, Jilin	43°31′22″ N, 125°39′31″ E	10 August 2017	sun drying
S12	Heihe, Heilongjiang	50°14′37″ N, 127°31′16″ E	16 August 2017	sun drying

Table 4. Contents of 15 compounds in wuweizi (μg/g, mean ± SD, n = 3).

Analyte	Wild						Cultivated					
	S1 [a]	S2 [a]	S3 [a]	S4 [a]	S5 [a]	S6 [a]	S7 [a]	S8 [a]	S9 [a]	S10 [a]	S11 [a]	S12 [a]
1 [b]	5093.16 ± 30.08	4213.46 ± 7.52	4283.63 ± 18.92	5155.81 ± 4.34	3461.58 ± 52.63	4070.60 ± 22.56	5108.20 ± 15.04	5243.53 ± 30.08	5228.50 ± 22.56	6220.98 ± 15.04	6175.86 ± 22.56	6115.71 ± 0.00
2 [b]	5.27 ± 0.04	6.23 ± 0.01	6.45 ± 0.04	6.80 ± 0.03	5.05 ± 0.01	4.73 ± 0.01	8.15 ± 0.48	8.31 ± 0.32	4.60 ± 0.06	7.06 ± 0.01	6.69 ± 0.03	8.31 ± 0.32
3 [b]	657.03 ± 0.57	2050.65 ± 1.98	2687.42 ± 1.13	1564.71 ± 0.85	1245.42 ± 0.28	3781.21 ± 2.26	4196.57 ± 8.49	5413.89 ± 5.66	3585.78 ± 5.09	5379.58 ± 19.81	4673.53 ± 2.55	4884.48 ± 11.32
4 [b]	3382.10 ± 24.69	5061.52 ± 0.71	3680.04 ± 2.85	5562.96 ± 3.21	4363.58 ± 0.00	4469.14 ± 1.07	2470.58 ± 7.13	2981.07 ± 0.36	1426.13 ± 1.43	1716.67 ± 2.14	2408.23 ± 2.49	2290.53 ± 1.78
5 [b]	6171.36 ± 46.95	5847.42 ± 72.88	7110.33 ± 37.56	7287.17 ± 9.77	6457.75 ± 4.69	7482.79 ± 5.42	5847.42 ± 54.14	6451.49 ± 2.71	5093.11 ± 49.31	5988.26 ± 101.02	5847.42 ± 16.93	5510.95 ± 42.34
6 [b]	107.63 ± 0.31	64.04 ± 0.15	61.48 ± 1.6	94.81 ± 0.67	93.75 ± 1.62	119.30 ± 0.93	37.16 ± 0.46	63.69 ± 0.80	42.29 ± 0.77	54.58 ± 0.31	60.86 ± 0.41	54.49 ± 0.55
7 [b]	662.92 ± 2.34	601.58 ± 0.78	562.34 ± 6.1	389.13 ± 2.07	539.33 ± 4.35	617.82 ± 2.82	312.90 ± 3.91	309.29 ± 3.13	260.13 ± 1.56	478.89 ± 4.06	600.68 ± 3.58	382.81 ± 1.35
8 [b]	4800.32 ± 27.89	2537.84 ± 13.95	3528.18 ± 13.95	2859.90 ± 48.31	3560.39 ± 53.14	3850.24 ± 48.31	1512.88 ± 1.39	1762.48 ± 6.08	1802.74 ± 2.79	2029.79 ± 3.69	3278.58 ± 27.89	3061.19 ± 60.79
9 [b]	1814.35 ± 5.92	1216.16 ± 38.76	1518.49 ± 3.88	1344.07 ± 11.63	1446.14 ± 8.07	1690.97 ± 22.35	934.51 ± 13.61	916.42 ± 4.48	878.95 ± 15.50	1014.61 ± 7.75	1182.83 ± 2.05	1147.69 ± 2.24
10 [b]	840.39 ± 2.41	547.33 ± 4.17	877.89 ± 10.49	684.83 ± 15.02	870.94 ± 18.79	714.00 ± 12.50	714.00 ± 12.50	557.06 ± 6.36	458.44 ± 18.79	726.50 ± 8.33	690.39 ± 12.73	1205.67 ± 36.32
11 [b]	626.16 ± 5.27	528.46 ± 10.53	523.86 ± 6.90	485.93 ± 10.34	468.69 ± 3.45	508.92 ± 12.11	457.20 ± 8.68	481.33 ± 1.99	268.92 ± 0.80	341.1 ± 0.00	263.63 ± 0.53	245.59 ± 0.34
12 [b]	1749.99 ± 2.17	978.71 ± 4.34	974.32 ± 2.56	965.35 ± 3.76	849.74 ± 10.85	971.36 ± 41.41	916.18 ± 5.74	1028.87 ± 1.30	602.78 ± 4.24	729.21 ± 3.39	538.68 ± 4.7	521.74 ± 3.57
13 [b]	4046.88 ± 3.18	3149.85 ± 3.68	3165.77 ± 1.84	3158.56 ± 0.37	3504.42 ± 9.73	3774.06 ± 16.03	2635.63 ± 0.74	2647.94 ± 5.15	2208.87 ± 5.74	2495.71 ± 6.05	2489.55 ± 6.37	2374.27 ± 5.44
14 [b]	787.58 ± 1.00	333.70 ± 0.90	689.40 ± 2.38	516.23 ± 3.25	704.79 ± 2.05	737.24 ± 5.40	249.41 ± 0.95	281.70 ± 1.56	247.38 ± 0.00	277.80 ± 0.78	437.55 ± 1.58	455.91 ± 5.01
15 [b]	705.60 ± 12.00	708.27 ± 2.31	1692.27 ± 2.31	-	2046.93 ± 8.33	2386.93 ± 6.11	1889.60 ± 4.00	1421.60 ± 0.00	3490.93 ± 4.62	3093.60 ± 6.93	19.87 ± 1.01	-
Total	31,450.74 ± 55.65	27,844.58 ± 40.31	31,361.47 ± 100.10	30,076.44 ± 28.88	29,619 ± 105.38	35,179.73 ± 61.85	27,289.59 ± 94.08	29,568.49 ± 19.79	25,598.77 ± 42.20	30,553.57 ± 169.43	28,674.25 ± 21.71	28,259.65 ± 12.66

[a] The sample number is same as in Table 1; [b] the analyte number is the same as in Table 2.

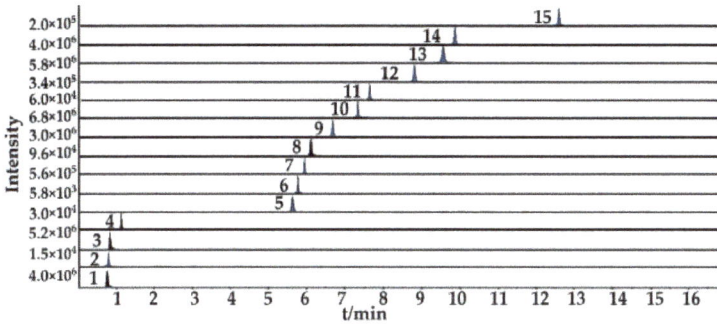

Figure 1. Representative extract ion chromatograms (XIC) of MRM chromatograms of 15 investigated compounds.

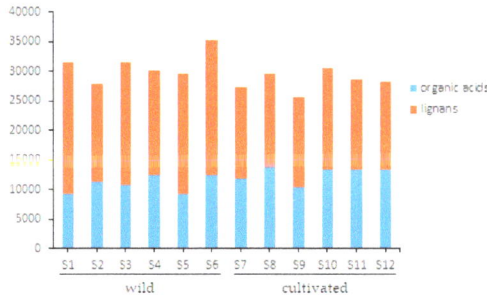

Figure 2. Histogram of the mean accumulative contents of different types of wuweizi from 12 batches.

2.6. PCA of the Samples

To evaluate the differences of components in wuweizi of two types, unsupervised PCA was performed. 12 Samples were set as observations, while the contents of 15 compounds were set as variables. The score scatter plot and the loading plot were displayed in Figure 3. In Figure 3, all samples were separated into two relative clusters, i.e., wild and cultivated wuweizi. This classification indicated that the content and distribution of chemical constituents varied between different types of wuweizi. The first and second principal components described 59.4% and 17.6% of the variability in the original observations, respectively, and the first two principal components accounted for 77.0% of the total variance.

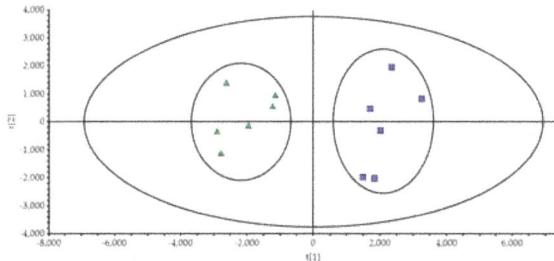

Figure 3. Score scatter plot of PCA processed data acquired from wild and cultivated wuweizi scanned by both positive and negative ion modes. Each of the green triangles represent a batch of wild wuweizi, while the blue squares represent a batch of cultivated ones.

2.7. PLS-DA of the Samples

The selected initial data was further processed by PLS-DA in order to reveal differences in the chemical composition among wild and cultivated wuweizi. The results were shown in Figure 4. Model parameters were set as follows: confidence level was 95%, $R^2Y = 0.967$, and $Q^2 = 0.924$, and the parameters showed that the established PLS-DA model was effective. The PLS-DA score plot displayed that the two clusters representing the wild and cultivated groups were well separated, thereby indicating the remarkable differences between the two types.

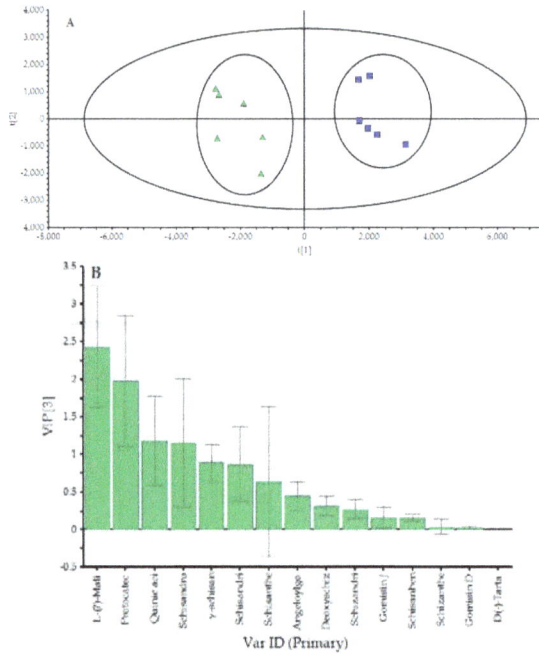

Figure 4. Score scatter plot (**A**) and VIP (**B**) by PLS-DA processed data obtained from wild and cultivated wuweizi scanned by positive and negative ion mode.

The VIP (variable importance for the projection) plot summarizes the importance of the variables both to explain X and to correlate to Y. The VIP plot is arranged from high to low, and the VIP-values greater than 1 indicated important variables, and four potential biological markers of quinic acid, L-(−)-malic acid, protocatechuic acid, and schisandrol B have high contributions to classification.

2.8. T-test

More than two-thirds of bioactive components quantified in this study showed a significant difference between two types of wuweizi, according to the T-test (Figure 5). Quantitation of major compounds, including protocatechuic acid, gomisin D, angeloylgomisin H, schisanhenol, and γ-schisandrin, showed strikingly higher levels ($p < 0.01$) in wild-type, while quinic acid and L-(−)-malic acid showed lower contents ($p < 0.01$) compared with cultivated-type. Secondly, quantitation of schisandrin, gomisin J, schisandrol B, and schizandrin C displayed highly contents ($p < 0.05$) in wild wuweizi than its cultivated-type.

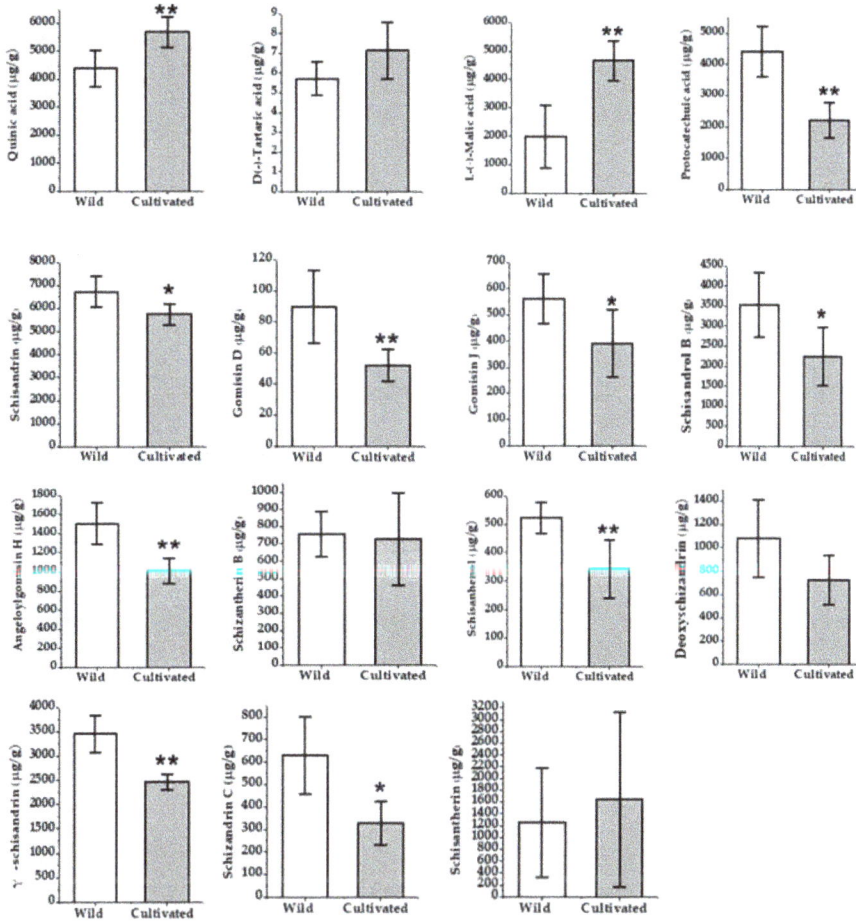

Figure 5. The contents of 15 compounds in wild and cultivated wuweizi (* $p < 0.05$; ** $p < 0.01$).

2.9. Gray Relational Analysis (GRA)

Because the contents of the 15 target components of lignans and organic acids in samples were different, it is difficult to judge the quality of samples intuitively. Therefore, gray relational analysis was carried out for comprehensive evaluation. It could be seen that the quality of wild wuweizi was better according to the results shown in Table S4. The quality sequencing of the samples was in the following order: S1 (wild) > S6 (wild) > S3 (wild) > S4 (wild) > S5 (wild) > S2 (wild) > S10 (cultivated) > S12 (cultivated) > S8 (cultivated) > S11 (cultivated) > S7 (cultivated) > S9 (cultivated).

3. Materials and Methods

3.1. Plant Materials

Twelve batches of cultivated and wild wuweizi dealing with the same processing method were studied in this research. The botanical origin of materials was authenticated by Prof. Xunhong Liu of the Nanjing University of Chinese Medicine. Voucher specimens were deposited at Herbarium in Nanjing University of Chinese Medicine. These samples were collected from Heilongjiang, Jilin, and Liaoning at around August 2017, dealing with sun drying for about 20 days. All batches of wuweizi were ripe when they were collected. Wild wuweizi grew under forests, valleys, and besides streams, places which were shady and moist. Cultivated wuweizi grew in arable land, which had enough sunshine and wind, or low-lying and rainy land. Detailed information is shown in Table 3.

3.2. Chemicals and Reagents

The reference compounds of L-(−)-malic acid (**3**), deoxyschizandrin (**12**), γ-schisandrin (**13**), schisandrin (**5**), schisantherin (**15**) were purchased from National Institutes for Food and Drug Control (Beijing, China). Quinic acid (**1**), D(-)-Tartaric acid (**2**), gomisin D (**6**), gomisin J (**7**), schizandrin C (**14**), schisandrol B (**8**), angeloylgomisin H (**9**), schizantherin B (**10**), and schisanhenol (**11**) were purchased from Nanjing Liangwei biotechnology Co., Ltd. (Nanjing, China). Protocatechuic acid (**4**) was purchased from Nanjing Spring and Autumn Biological Engineering Co., Ltd., China. The purity of all compounds by HPLC analysis was greater than 98%. The structures of the 15 reference compounds are shown in Figure 6. Formic acid of MS grade, acetonitrile, and methanol of HPLC grade were purchased from Merck (Darmstadt, Germany). Ultrapure water was prepared using a Milli-Q water purification system (Millipore, Bedford, MA, USA).

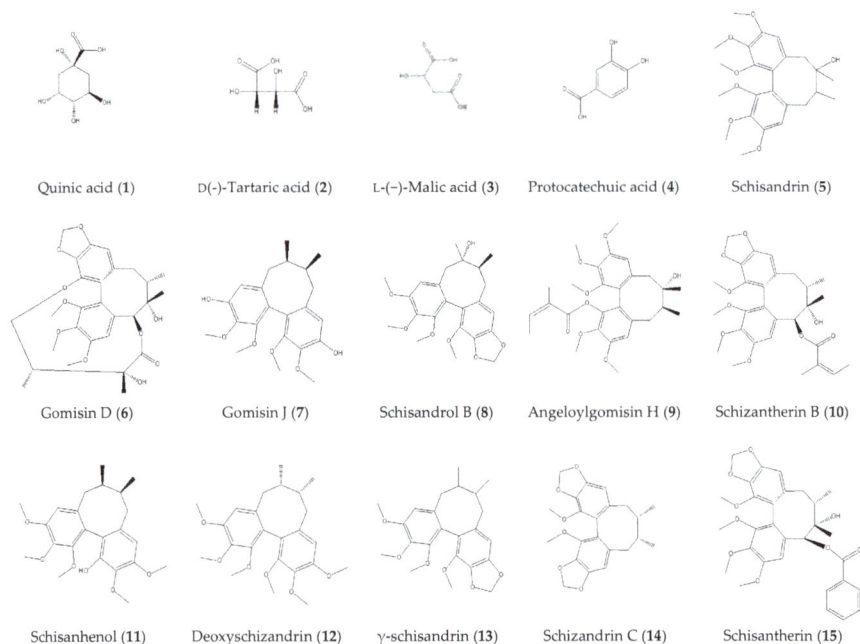

| Quinic acid (1) | D(-)-Tartaric acid (2) | L-(-)-Malic acid (3) | Protocatechuic acid (4) | Schisandrin (5) |

| Gomisin D (6) | Gomisin J (7) | Schisandrol B (8) | Angeloylgomisin H (9) | Schizantherin B (10) |

| Schisanhenol (11) | Deoxyschizandrin (12) | γ-schisandrin (13) | Schizandrin C (14) | Schisantherin (15) |

Figure 6. Chemical structures of 15 reference substances.

3.3. Preparation of Standard Solutions

A mixed standard stock solution containing 15 reference standards was prepared in methanol and their concentrations were as follows: **1**, 26.60 μg/mL; **2**, 0.10 μg/mL; **3**, 29.52 μg/mL; **4**, 12.77 μg/mL; **5**, 9.31 μg/mL; **6**, 1.01 μg/mL; **7**, 0.93 μg/mL; **8**, 5.35 μg/mL; **9**, 2.26 μg/mL; **10**, 2.21 μg/mL; **11**, 1.20 μg/mL; **12**, 2.40 μg/mL; **13**, 4.60 μg/mL; **14**, 1.47 μg/mL; **15**, 7.17 μg/mL. This standard stock solution was then diluted with methanol to a series of appropriate concentrations to generate the calibration curves. The solutions were stored at 4 °C for a day prior to LC-MS analysis.

3.4. Preparation of Sample Solutions

The dried fruits were pulverized into powders and screened through the 50-mesh sieve. Each sample (0.5 g) was accurately weighed and extracted by ultrasonication (500 W, 40 kHz) in 20 mL methanol for 40 min. After cooling down at room temperature, methanol was added to compensate for the weight lost during extraction. After centrifugation (12,000 rpm, 10 min) and filtering (0.22 µm membrane filter), the supernatants were stored in a sample plate at 4 °C prior to LC-MS analysis.

3.5. Chromatographic and Mass Spectrometric Conditions

The mass spectrometry detection was performed using an API5500 triple quadrupole mass (AB SCIEX, Framingham, MA, USA). The MS was equipped with an electrospray ionization (ESI) source operating in MRM and under both positive and negative ion modes. The MS parameters were set as follows: gas temperature 550 °C; pressures of nebulizer of MS, 5500 V (positive) and −4500 V (negative); GSI flow 65 L/min; CUR flow 30 L/min and all MS data were acquired and analyzed using the Analyst 1.5.2 software (AB SCIEX, Framingham, MA, USA). The cone voltage and collision energy parameter of each compound were individually optimized.

The chromatographic analysis was performed on a Shimadzu SIL-20A XR system (Shimadzu, Kyoto, Japan), consisting of a binary solvent delivery system and an automatic sampler. A Synergi™ Hydro-RP 100Å column (100 mm × 2.0 mm, 2.5 µm, (Phenomenex, Los Angeles, CA, USA) was used for eluting samples. The mobile phase was composed of water (A) and acetonitrile (B) using a gradient elution of 30%–52% B at 0–4 min, 52%–75% B at 4–8 min, 75%–90% B at 8–11 min, 90%–30% B at 11–15 min, 30% B at 15–17.10 min. The column temperature was 40 °C, the flow rate kept at 0.4 mL/min, and the sample injection volume was 1 µL.

3.6. Validation of the Method

3.6.1. Linearity, LOD, and LOQ

The linearity of the calibration curves was obtained by plotting the peak areas (Y) against the corresponding concentrations (X) of each analyte. The lowest concentration of standard solution for calibration use was diluted with methanol to a series of appropriate concentrations. The LODs and LOQs of 15 analytes were determined using a series of diluted standard solutions until the signal-to-noise (S/N) ratios were about 3 and 10, respectively.

3.6.2. Precision, Repeatability, Stability, Accuracy

The analysis method developed in this study was validated for precision (the intra- and inter-day), repeatability, stability, and accuracy. The intra-day and inter-day variability tests were determined by measuring the mixed standard solutions in six replicates in a day and once a day during three consecutive days, respectively. To evaluate the repeatability, six different analytical sample solutions prepared from the same sample (sample 1) were parallel processed and analyzed. To confirm the stability, the sample solution mentioned above was stored at room temperature and analyzed at 0, 2, 4, 8, 12, and 24 h, respectively. All the variations were expressed in RSD. A recovery test was used to evaluate the accuracy of this method. A certain amount of the 15 standards with low (80%), medium (100%), and high (120%) levels were added into a known amount of samples (0.25 g), and then extracted and analyzed with the same procedures. To be specific, a recovery test was conducted by standard protocol and calculated by the formula: (%) = (found amount– original amount in sample)/spiked amount × 100%.

3.7. Multivarite Statistical Analysis

Multivariate statistical analysis was performed using the Simca-P 13.0 software (for Windows, Umetrics AB, Umeå, Sweden) by PCA and PLS-DA. PCA and PLS-DA were used to evaluate the variations of the two types of wuweizi according to the contents of the 15 components. PCA is an unsupervised pattern recognition method used for analyzing, classifying, and reducing the dimensionality of numerical datasets in a multivariate problem [20,21], and it has been widely used for the quality control of herbal medicines [22–24]. PLS-DA is good for highlighting the differences between two groups. It is possible to identify and select the important markers in samples via multivariate analysis of LC/MS data, even at low concentration levels [25]. Data of the contents of 15 compounds in wild and cultivated samples were listed. When the contents of investigated components were below the quantitation limit or not detected in the samples, the values of such elements were considered to be 0. All experimental data were statistically analyzed by independent sample t-test (SPSS 16.0 for Windows, IBM, Armonk, NY, USA). The columns were charted by Origin pro 8 (Origin Lab, Northampton, MA, USA), showing the difference of each compound between two types of wuweizi. GRA provides a reliable guarantee for the quality evaluation of traditional Chinese medicine on the basis of the contents of the 15 index constituents [26].

4. Conclusions

An analytical method based on UFLC-QTRAP-MS/MS was used for the simultaneous determination of 15 components, including 11 lignans and 4 organic acids, in wuweizi under different ecosystems (wild and cultivated). Multivariate statistical analyses, such as PCA, PLS-DA, independent sample t-test, and GRA, have been successfully applied to comprehensively analyze and evaluate the wuweizi under different ecosystems according to the contents of the 15 components. The data of content determination showed that lignans had higher contents in wild-type, while cultivated-type contained more organic acids. This phenomenon is probably ascribed to environmental stress (such as water, soil strength, and nutrient status), which can have a great influence on the accumulation of the active compounds of medicinal plants. PCA and PLS-DA results showed that there are great difference between wild and cultivated wuweizi, and 4 different compounds (protocatechuic acid, quinic acid and L-(−)-malic acid, and schisandrol B) were significantly related to sample classification. It could be seen that the quality of the samples collected from the wild environment was better according to the GRA results. Wild wuweizi usually suffered more stress than cultivated-type and produced more specialized metabolites, even though the quality indicator of the two types all conformed to the requirements of Chinese pharmacopoeia. Findings from this research may provide a new method for the comprehensive evaluation and quality control of wuweizi from different ecosystems. It may also provide a basis for differentiating wild and cultivated wuweizi at the chemistry level.

Supplementary Materials: The following are available online. Table S1: Levels and factors of orthogonal table. Table S2: Result of $L_9(3)^3$ orthogonal experiment. Table S3: Recoveries and relative standard deviations (RSD) of fifteen components (%, $n = 3$). Table S4: Quality sequencing of the samples. Figure S1: Total ion chromatogram of the representative wild and cultivated samples.

Author Contributions: S.C. and J.S. carried out the experiments, analyzed data, and composed the draft of the manuscript. C.W., M.T., and J.C. proposed the study and helped to perform the experiments. L.Z., R.T., and J.M. provided the samples of wuweizi. X.L. participated in the supervision of this study and edited the manuscript. All authors read and approved the manuscript.

Funding: This research was supported by the Standardization of Chinese Materia Medica Program (No. ZYBZH-C-JS-32).

Conflicts of Interest: The authors declare no conflict of interest.

References

1. Zhou, Y.; Huang, S.X.; Pu, J.X.; Li, J.R.; Ding, L.S.; Chen, D.F.; Sun, H.D.; Xu, H.X. Ultra performance liquid chromatography coupled with quadrupole time-of-flight mass spectrometric procedure for qualitative and quantitative analyses of nortriterpenoids and lignans in the genus Schisandra. *J. Pharmaceut. Biomed.* **2011**, *56*, 916–927. [CrossRef]
2. Szopa, A.; Ekiert, R.; Ekiert, H. Current knowledge of Schisandra chinensis (Turcz.) Baill. (Chinese magnolia vine) as a medicinal plant species: A review on the bioactive components, pharmacological properties, analytical and biotechnological studies. *Phytochem. Rev.* **2017**, *16*, 195–215. [CrossRef]
3. Li, X.G.; Gao, Q.; Weng, W.; Zhang, P.F.; Xiao, F.; Luo, H.M. Research progress of effective parts and its pharmacological action of *Schisandrae Chinesis* Fructus. *J. Chin. Med. Mater.* **2005**, *28*, 156–159.
4. Hu, J.Y.; Mao, C.Q.; Gong, X.D.; Lu, T.L.; Chen, H.; Huang, Z.J.; Cai, B.C. Simultaneous determination of eleven characteristic lignans in Schisandra chinensis by high-performance liquid chromatography. *Pharmacogn. Mag.* **2013**, *9*, 155–161.
5. Ming, Y.B.; Zhao, H.; Zhou, L.; Tian, Y.H. Reseach progress of *Schisandra Chinesis* (Turcz) Bail. *Pract. Pharm. Clin. Remed.* **2007**, *10*, 365–367.
6. Kormosh, N.; Laktionov, K.; Antoshechkina, M. Effect of a combination of extracts from several plants on cell-mediated and humoral immunity of patients with advanced ovarian cancer. *Phytother. Res.* **2006**, *20*, 424–425. [CrossRef]
7. Liu, K.T.; Cresteil, T.; Columelli, S.; Lesca, P. Pharmacological properties of dibenzo[a,c]cyclooctene derivatives isolated from Fsh, N.; LaII. Induction of phenobarbital-like hepatic monooxygenases. *Chem-Biol. Interact.* **1982**, *39*, 315–330. [CrossRef]
8. Yang, H.J.; Yang, S.M. General situation of pharmacological action of Fructus *Schisandra Chinesis* (Turcz) Bail. *Ginseng Res.* **1999**, *11*, 5–8.
9. Zhong, S.; Nie, Y.C.; Gan, Z.Y.; Liu, X.D.; Fang, Z.F.; Zhong, B.N.; Tian, J.; Huang, C.Q.; Lai, K.F.; Zhong, N.S. Effects of *Schisandra chinensis* extracts on cough and pulmonary inflammation in a cough hypersensitivity guinea pig model induced by cigarette smoke exposure. *J. Ethnopharmacol.* **2015**, *165*, 73–82. [CrossRef]
10. Wang, C.C.; Cai, H.; Zhao, H.; Yan, Y.; Shi, J.J.; Chen, S.Y.; Tan, M.X.; Chen, J.L.; Zou, L.S.; Chen, C.H. Distribution patterns for metabolites in medicinal parts of wild and cultivated licorice. *J. Pharm. Biomed. Anal.* **2018**, *161*, 464–473. [CrossRef]
11. Brunetti, C.; George, R.M.; Tattini, M.; Field, K.; Davey, M.P. Metabolomics in plant environmental physiology. *J. Exp. Bot.* **2013**, *64*, 4011–4020. [CrossRef] [PubMed]
12. Astaf'eva, O.V.; Sukhenko, L.T. Comparative analysis of antibacterial properties and chemical composition of Glycyrrhiza glabra L. from Astrakhan region (Russia) and Calabria region (Italy). *Bull. Exp. Biol. Med.* **2014**, *156*, 829–832. [CrossRef]
13. Wei, S.S.; Yang, M.; Chen, X.; Wang, Q.R.; Cui, Y.J. Simultaneous determination and assignment of 13 major flavonoids and glycyrrhizic acid in licorices by HPLC-DAD and Orbirap mass spectrometry analyses. *Chin. J. Nat. Med.* **2015**, *13*, 232–240. [CrossRef]
14. Ren, M.T.; Chen, J.; Song, Y.; Sheng, L.S.; Li, P.; Qi, L.W. Identification and quantification of 32 bioactive compounds in Lonicera species by high performance liquid chromatography coupled with time-of-flight mass spectrometry. *J. Pharm. Biomed. Anal.* **2008**, *48*, 1351–1360. [CrossRef] [PubMed]
15. Deng, X.X.; Chen, X.H.; Cheng, W.M.; Cheng, W.M.; Shen, Z.D.; Bi, K.S. Simultaneous LC–MS Quantification of 15 Lignans in Schisandra chinensis (Turcz.) Baill. Fruit. *Chromatographia* **2008**, *67*, 559–566. [CrossRef]
16. Zhang, H.; Zhang, G.Q.; Zhu, Z.Y.; Zhao, L.; Fei, Y.; Jing, J.; Chai, Y.F. Determination of six lignans in *Schisandra chinensis* (Turcz.) Baill. Fruits and related Chinese multiherb remedies by HPLC. *Food Chem.* **2009**, *115*, 735–739. [CrossRef]
17. Liu, H.; Lai, H.; Jia, X.; Liu, J.; Zhang, Z.; Qi, Y.; Zhang, J.; Song, J.; Wu, C.; Zhang, B. Comprehensive chemical analysis of Schisandra chinensis by HPLC–DAD–MS combined with chemometrics. *Phytomedicine*. **2013**, *20*, 1135–1143. [CrossRef]
18. An, K.L.; Li, D.K.; Zhou, D.Z.; Ye, Z.L.; Guo, Q.S. Effects of different drying methods on quality of *schisandrae chinensis* fructus. *Chin. J. Chin. Mater. Med.* **2014**, *39*, 2900–2906.
19. Xu, J.H.; Su, L.L.; Wang, Q.H.; Mao, C.Q.; Wang, D.D. Determination of citric acid, L-malic acid and 5-hydroxyl methyl furfural in diffe-rent processed products of Schisandra chinensis by HPLC. *Northwest Pharm. J.* **2017**, *32*, 548–551.

20. Ji-ye, A.; He, J.; Sun, R.B. Multivariate statistical analysis for metabolomic data: the key points in principal component analysis. *Acta Pharm. Sin.* **2018**, *53*, 929–937.

21. Bo, N.; Broberg, P.; Lindberg, C.; Plymoth, A. Analysis and understanding of high-dimensionality data by means of multivariate data analysis. *Chem. Bio-divers.* **2005**, *2*, 1487–1494.

22. Mediani, A.; Abas, F.; Maulidiani, M.; Khatib, A.; Tan, C.P.; Ismail, I.S.; Shaari, K.; Ismail, A. Characterization of Metabolite Profile in Phyllanthus niruri and Correlation with Bioactivity Elucidated by Nuclear Magnetic Resonance Based Metabolomics. *Molecules* **2017**, *22*, 902. [CrossRef]

23. Han, D.Q.; Zhao, J.; Xu, J.; Peng, H.S.; Chen, X.J.; Li, S.P. Quality evaluation of Polygonum multiflorum in China based on HPLC analysis of hydrophilic bioactive compounds and chemometrics. *J. Pharm. Biomed. Anal.* **2013**, *72*, 223–230. [CrossRef] [PubMed]

24. Da, J.; Wu, W.Y.; Hou, J.J.; Long, H.L.; Yao, S.; Yang, Z.; Cai, L.Y.; Yang, M.; Jiang, B.H.; Liu, X.; et al. Comparison of two officinal Chinese pharmacopoeia species of Ganoderma based on chemical research with multiple technologies and chemometrics analysis. *J. Chromatogr. A* **2012**, *1222*, 59–70. [CrossRef] [PubMed]

25. Xu, X.; Xu, S.; Zhang, Y.; Zhang, H.; Liu, M.N.; Liu, H.; Gao, Y.; Xue, X.; Xiong, H.; Lin, R.; et al. Chemical Comparison of Two Drying Methods of Mountain Cultivated Ginseng by UPLC-QTOF-MS/MS and Multivariate Statistical Analysis. *Molecules* **2017**, *22*, 717. [CrossRef] [PubMed]

26. Chen, C.H.; Liu, Z.X.; Zou, L.S.; Liu, X.H.; Chai, C.; Zhao, H.; Yan, Y.; Wang, C.C. Quality evaluation of Apocyni Veneti Folium from different habitats and commercial herbs based on simultaneous determination of multiple bioactive constituents combined with multivariate statistical analysis. *Molecules* **2018**, *23*, 573. [CrossRef] [PubMed]

Sample Availability: Samples of the compounds are available from the authors.

molecules

MDPI

Article
Lignans: A Chemometric Analysis

Lisa I. Pilkington [ID]

School of Chemical Sciences, The University of Auckland, Private Bag 92019, Auckland 1142, New Zealand; lisa.pilkington@auckland.ac.nz; Tel.: +64-9-373-7599 (ext. 86776)

Received: 20 June 2018; Accepted: 7 July 2018; Published: 9 July 2018

Abstract: The physicochemical properties of classical lignans, neolignans, flavonolignans and carbohydrate-lignan conjugates (CLCs) were analysed to assess their ADMET profiles and establish if these compounds are *lead-like/drug-like* and thus have potential to be or act as leads in the development of future therapeutics. It was found that while no studied compounds were *lead-like*, a very large proportion (>75%) fulfilled all the requirements to be deemed as present in *drug-like* space and almost all compounds studied were in the known drug space. Principal component analysis was an effective technique that enabled the investigation of the relationship between the studied molecular descriptors and was able to separate the lignans from their sugar derivatives and flavonolignans, primarily according to the parameters that are considered when defining chemical space (i.e., number of hydrogen bond donors, acceptors, rotatable bonds, polar surface area and molecular weight). These results indicate that while CLCs and flavonolignans are less *drug-like*, lignans show a particularly high level of *drug-likeness*, an observation that coupled with their potent biological activities, demands future pursuit into their potential for use as therapeutics.

Keywords: lignans; chemometrics; neolignans; flavonolignans; chemical space; drug-like

---

1. Introduction

Lignans are a class of secondary metabolites that are derived from the oxidative dimerisation of two or more phenylpropanoid units [1]. Despite their common biosynthetic precursors, lignans show vast structural diversity due to the numerous potential coupling modes of the phenoxy radicals [2]. The nature of the molecular linkage of the phenylpropanoids provides the most fundamental level of classification of lignans into two main subclasses—classical lignans and neolignans—although there exist other smaller subclasses, including flavonolignans and coumarolignans [1,3–6].

Classical lignans are phenylpropane dimers that have a β-β′ linkage and there six main subtypes of classical lignans—dibenzylbutanes, dibenzylbutyrolactones, arylnaphthalenes/aryltetralins, dibenzocyclooctadienes, substituted tetrahydrofurans, and 2,6-diarylfurofurans (Figure 1) [3,6,7]. Neolignan was a classification initially coined by Gottlieb to distinguish phenylpropanoid dimers that did not contain the β-β′ (also referred to as an 8-8′) phenylpropane linkage characteristic of classical lignans [8,9]. Neolignans have more varied structures than classical lignans; there are 15 subtypes designated by the nature and position of the linkage between the phenylpropane units [3,6,10], the most common subtypes being benzofurans, 1,4-benzodioxanes, alkyl aryl ethers, biphenyls, cyclobutanes, 8-1′-bicyclo[3.2.1]octanes, 8-3′-bicyclo[3.2.1]octanes and biphenyl ethers, examples of which are shown in Figure 1.

Lignans have been found in more than 70 plant families and an extensive range of localities within plants, from roots to leaves, seeds and flowers [1,3–5,11]. Most importantly, this class of compound has exhibited several potent, significant, biological activities, including anticancer [3,4,12,13], antimicrobial [4], antiviral [12–15], immunosuppressive [4], anti-inflammatory [4], antioxidant [3,4,16], and hepaprotective [15–17] actions as well as cancer [18,19] and osteoporosis [20] prevention properties; activities that have contributed an ever-increasing interest in lignans and their synthesis [3–5,7,11,15,21–41].

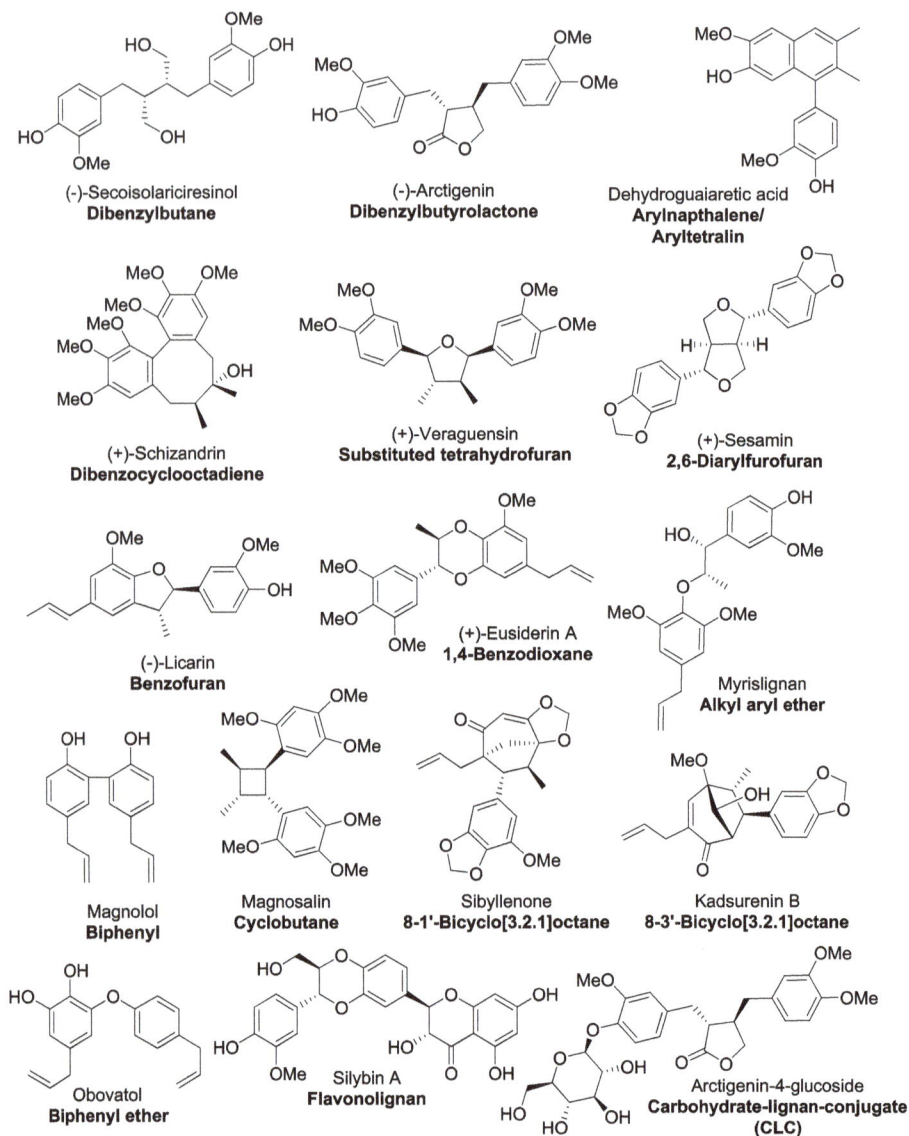

Figure 1. Examples of the six types of classical lignans, the eight types of neolignans investigated in this study, a flavonolignan and carbohydrate-lignan conjugate (CLC).

Throughout human history, plants with a high lignan content have been utilised to treat illnesses and ailments, playing a vital role in traditional folk medicine [5,11]. These lignan-containing plants have been documented in medical pharmacopoeias from a large number of cultures including English, Korean, Native American, Chinese, Japanese, South American and Tibetan. For many, the uptake of modern medicine has supplanted the need and use of traditional medicines, however the continued use of folk medicine exists in a large number of cultures including Ayurvedic, Unani, Siddhi, Kampo, Jamu and in traditional Chinese medicine [4]. Furthermore, traditional medicine has been a critical source of inspiration in the pursuit of modern drug therapies, with a number of presently-used medicines being engendered by compounds of natural origins–circa 40% of commercially-available drugs are either natural products or derivatives thereof [5]. In this capacity, lignans constitute an important class of compounds that provide a starting point for the development of therapeutic agents [1,3,5,11,21,42–44].

Potentially the most well-known example of a lignan as a currently-utilised and lead compound is the aryltetralin lactone, podophyllotoxin (Figure 2) [11,21,45]. It has been known for centuries that the plants of the *Podophyllum* genus possess medicinal properties. These plants have particularly been used by the indigenous peoples of the Himalayas and North America [45]. Podophyllotoxin was first isolated in 1880 from one of these plants [46], and is a cytotoxic compound that binds to tubulin, thereby inhibiting microtubule assembly during mitosis and thus interrupts the cell cycle [47–49]. Podophyllotoxin has a mode of action and level of potency that lends it to be a possible cancer chemotherapeutic, however this possibility was tempered by the discovery that it exhibits high levels of gastrointestinal toxicity [50]. Podophyllotoxin, however, has been approved for use as a topically-administered treatment for genital warts. Additionally, podophyllotoxin was used as a lead compound for antitumour agents [51], resulting in the development of etoposide, its water-soluble phosphate ester prodrug, etopophos and teniposide as anticancer agents that are all in current use to treat a range of cancers, including testicular, lung and ovarian cancer, lymphoma, leukemia, neuroblastoma and various types of brain tumours [15,50,52]. It should be noted that etoposide, etopophos and teniposide exhibit an alternative mode of action to their lead compound, podophyllotoxin, in that they are potent DNA topoisomerase II inhibitors [50–52].

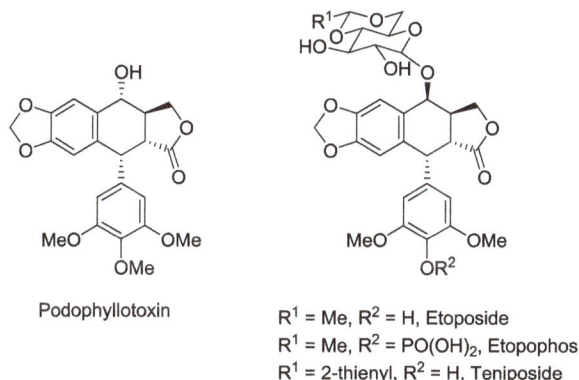

Podophyllotoxin

R^1 = Me, R^2 = H, Etoposide
R^1 = Me, R^2 = PO(OH)$_2$, Etopophos
R^1 = 2-thienyl, R^2 = H, Teniposide

Figure 2. Structure of podophyllotoxin and three of its most notable derivatives; etoposide, etopophos and teniposide.

Podophyllotoxin provides an inspiring example of the potential that lignans possess as a foundation for the development of medicines to target diseases and conditions, many of which that have an unmet need for cures and treatments.

While potent biological activity, which many lignans possess, is the most critical property of a potential drug or lead compound, it is also important to assess the Absorption, Distribution, Metabolism, Excretion and Toxicity (ADMET) profiles of these compounds to evaluate their likelihood of being effective drug leads [53]. To do this, the physicochemical properties of the compounds can be calculated–these molecular descriptors can subsequently be assessed against various existing and verified benchmarks. *Drug-like* chemical space is defined by the Lipinski's rule of five–the most widely used and recognised set of parameters that are used to assess properties of potential therapeutics. Compounds that fall within these boundaries are indicated to be able to be orally absorbed [54,55]. The two other definitions of chemical space are *lead-like* space and known drug space (KDS). Compounds within *lead-like* chemical space are typically compounds that are less complex, hence have low molecular weights and lower lipophilicities (LogP) [56]—*lead-like* compounds have very low limits for these parameters as they generally increase during the optimisation process in medicinal chemistry; *lead-like* compounds are more likely to become real therapeutics once modified [57,58]. KDS is defined by criterion that includes all small organic compounds that have been assessed in human clinical trials and were/are subsequently in medical use [59]. The upper limits for each chemical space referred to in this study are provided (Table 1).

Table 1. Definition of *lead-like*, *drug-like* and known drug space (KDS) in terms of molecular descriptors.

Descriptor	*Lead-Like* Space	*Drug-Like* Space	Known Drug Space
Molecular weight (g mol^{-1})	300	500	800
Lipophilicity (Log P)	3	5	6.5
Hydrogen bond donors	3	5	7
Hydrogen bond acceptors	3	10	15
Polar surface area (Å^2)	60	140	180
Rotatable bonds	3	10	17

We wished to assess the ADMET profile of lignans and related compounds to explore their position in the predefined chemical spaces, as well as if there are notable differences within, and between, these groups. Presented herein is the result of the investigation into the physicochemical properties of traditional lignans, neolignans, flavonolignans and sugar derivatives of lignans (carbohydrate-lignan conjugates; CLCs) to establish if these compounds are *lead-like*/*drug-like* and thus have potential to be or act as leads in the development of future therapeutics.

2. Methodology

Representative compounds for each of the main subclasses of lignan and neolignan compounds were found by doing a substructure search using Scifinder and choosing the ten lignan compounds with the highest number of references. There were six subclasses of classical lignans (dibenzylbutanes, dibenzylbutyrolactones, arylnapthalenes/aryltetralins, dibenzocyclooctadienes, substituted tetrahydrofurans and 2,3-diarylfurans) and eight subclasses of neolignans (benzofurans, 1,4-benzodioxanes, alkyl aryl ethers, biphenyls, cyclobutanes, 8-1′-bicyclo[3.2.1]octanes, 8-3′-bicyclo[3.2.1]octanes and biphenyl ethers) included in this study–examples of these subclasses are given in Figure 1 and the compound details (name, class and CAS number) of each compound included in this study are given in the Supplementary Information. Furthermore, additional lignan-like compounds were also found–flavonolignans and sugar derivatives (carbohydrate-lignan conjugates, CLCs) of classical lignan and neolignan subclasses. In total, 16 different groups of compounds were studied, each group consisting of ten compounds. Hence, 160 compounds were included in this representative study.

The 3D structures of the compounds were drawn using ChemBioDraw as part of the ChemOffice software package [60]. The structures were then optimised using the MM2 [61] force field in Chem3D [60]. The molecular descriptors were calculated using QikProp 4.42 [62], which has been shown to be an accurate and reliable tool for the calculation of the molecular descriptors analysed in this study [63].

Following generation of the molecular descriptors for all the compounds in the study, the mean, median and standard deviation of each descriptor was calculated (see Section 3.1). Graphs of the distributions of these molecular descriptors were generated with R (version 3.2.2) [64] and R Studio (version 0.99.486) [65] using the ggplot2 package [66]. Compounds were categorised as *lead-like*, *drug-like* and in KDS for each of the parameters by comparing the values for the descriptors against those stated in Table 1 and including them in the chemical space if the calculated value was less than or equal to the stipulated benchmark.

Principal Component Analysis (PCA) was carried out using all compounds and parameters included in this study (see Section 3.2) using R (version 3.2.2) [64] and R Studio (version 0.99.486) [65]. PCA analysis was performed using the prcomp function as part of the stats package, by singular value decomposition of the centred and scaled data matrix [64]. Results of this analysis were visualised using the factoextra package (version 1.0.5) [67].

3. Results and Discussion

3.1. Molecular Descriptors

Using the aforementioned methods, ten molecular descriptors were calculated for each of the 160 compounds studied. Molecular weight, lipophilicty (LogP), the number of hydrogen bond donors, hydrogen bond acceptors and rotatable bonds and polar surface area (PSA) have been extensively used in the assessment of a molecules' suitability to be considered as a drug [68]. The other molecular descriptors—dipole moment, polarisability, ionisation potential and water solubility (LogS) have been used less extensively, however their association with desirable characteristics has led them to being increasingly examined in recent times [63,69,70].

To analyse the molecular descriptors, summary statistics–the mean, median and standard deviation for each of these parameters—for each compound type, as well as for all 160 compounds (all classical lignans, neolignans, flavonolignans and CLCs) were calculated and are in the table provided (Table 2).

Table 2. Mean, standard deviation (std dev) and median values of the compound types for the ten molecular descriptors analysed in this study.

Compound Type	Molecular Weight (g mol^{-1})			Lipophilicity (LogP)			Hydrogen Bond Donors		
	Mean	Std Dev	Median	Mean	Std Dev	Median	Mean	Std Dev	Median
Overall	381.5	70.4	372.4	3.0	1.3	3.1	1.8	1.9	2.0
Classical lignans and neolignans	361.2	42.3	364.4	3.3	1.0	3.4	1.4	1.3	1.0
Flavonolignans	478.6	7.0	482.4	1.6	0.6	1.5	4.1	0.7	4.0
CLCs	567.4	67.5	534.6	0.4	1.1	0.5	6.1	2.1	5.5
Dibenzylbutanes	354.8	36.1	346.4	3.2	1.1	3.2	2.5	1.5	2.0
Dibenzylbutyrolactones	369.6	31.4	372.4	2.5	0.7	2.5	1.6	0.8	2.0
Arylnapthalenes/Aryltetralins	348.2	23.6	350.3	3.1	0.8	3.1	0.7	0.9	0.0
Dibenzocyclooctadienes	413.3	17.8	416.5	3.9	0.7	3.7	0.7	0.8	0.5
Substituted tetrahydrofurans	362.2	28.1	350.4	3.6	0.5	3.6	0.7	0.9	0.0
2,6-Diarylfurofurans	377.8	24.2	371.4	2.7	0.6	2.9	1.0	0.9	1.0
Benzofurans	351.2	21.6	352.4	3.4	1.1	3.1	2.1	1.3	2.5
1,4-Benzodioxanes	352.0	37.3	359.4	3.3	1.2	3.5	1.6	1.4	1.5
Alkyl aryl ethers	387.0	15.5	377.4	3.1	1.4	2.7	2.4	1.8	3.0
Biphenyls	313.8	62.6	298.4	3.5	1.6	3.9	2.4	1.5	2.0
Cyclobutanes	371.6	51.1	372.4	3.7	0.9	3.9	1.0	1.4	0.0
8-1'-Bicyclo[3.2.1]octanes	377.8	15.3	373.4	3.3	0.6	3.2	0.4	0.8	0.0
8-3'-Bicyclo[3.2.1]octanes	381.4	30.0	386.4	3.5	0.5	3.6	0.8	0.4	1.0
Biphenyl ethers	296.5	27.4	287.3	3.1	1.1	2.6	1.3	0.8	1.5

Compound Type	Hydrogen Bond Acceptors			Polar Surface Area (Å^2)			Rotatable Bonds		
	Mean	Std Dev	Median	Mean	Std Dev	Median	Mean	Std Dev	Median
Overall	6.3	3.4	6.0	79.5	37.1	70.2	7.0	3.8	6.0
Classical lignans and neolignans	5.4	1.7	5.8	68.3	22.5	66.9	6.4	3.1	6.0
Flavonolignans	9.0	0.8	9.7	158.5	11.4	159.2	7.1	0.7	7.0
CLCs	16.6	3.8	14.9	156.9	29.3	142.1	15.1	5.0	14.5
Dibenzylbutanes	4.9	1.5	5.0	71.2	22.2	69.1	11.0	1.9	11.0
Dibenzylbutyrolactones	6.5	1.2	6.0	90.8	13.6	87.9	7.6	1.5	8.0
Arylnapthalenes/Aryltetralins	4.7	1.5	5.3	61.6	15.4	69.4	3.4	1.9	3.0
Dibenzocyclooctadienes	5.1	0.8	4.9	51.2	11.9	49.4	5.0	1.6	5.0
Substituted tetrahydrofurans	4.9	0.5	4.7	50.6	12.6	47.8	3.4	1.6	4.0
2,6-Diarylfurofurans	6.8	0.5	6.4	66.3	14.0	66.6	3.3	2.2	4.0
Benzofurans	5.1	1.8	6.4	69.2	25.2	80.7	7.1	2.5	8.5
1,4-Benzodioxanes	5.0	1.4	4.5	70.7	23.8	59.3	5.8	1.7	6.0
Alkyl aryl ethers	6.7	1.6	7.2	76.9	27.8	80.9	12.1	1.3	12.5
Biphenyls	4.1	3.1	2.6	63.2	28.4	56.2	8.4	2.3	7.0
Cyclobutanes	4.5	1.5	4.3	66.8	36.4	59.6	4.6	1.6	5.0
8-1'-Bicyclo[3.2.1]octanes	6.3	0.7	6.0	69.5	8.5	66.5	4.7	1.3	5.0
8-3'-Bicyclo[3.2.1]octanes	6.7	0.8	6.5	71.8	7.6	72.9	5.8	1.4	6.0
Biphenyl ethers	4.1	1.5	4.5	76.1	26.2	84.4	7.1	1.8	7.5

Table 2. *Cont.*

	Dipole Moment (D)			Water Solubility (LogS)			Ionisation Potential (eV)		
	Mean	Std Dev	Median	Mean	Std Dev	Median	Mean	Std Dev	Median
Overall	4.1	1.9	3.9	−4.1	1.4	−4.1	9.0	0.4	9.0
Classical lignans and neolignans	4.1	1.9	4.0	−4.2	1.4	−4.2	9.0	0.4	9.0
Flavonolignans	4.0	2.5	3.2	−4.4	0.8	−4.4	9.1	0.2	9.1
CLCs	4.7	1.9	3.9	−2.5	1.3	−2.4	9.0	0.2	9.0
Dibenzylbutanes	4.1	1.2	3.9	−3.4	1.2	−3.0	9.1	0.2	9.0
Dibenzylbutyrolactones	5.5	2.2	4.7	−2.6	0.6	−2.6	9.2	0.2	9.2
Arylnapthalenes/Aryltetralins	5.1	2.8	4.9	−3.5	1.3	−3.5	8.4	0.2	8.4
Dibenzocyclooctadienes	2.6	1.0	2.5	−5.6	0.9	−5.4	8.6	0.3	8.5
Substituted tetrahydrofurans	3.5	1.6	3.8	−5.4	1.1	−5.0	9.0	0.3	9.0
2,6-Diarylfurofurans	2.7	1.3	2.8	−3.9	1.0	−4.4	8.9	0.3	9.0
Benzofurans	3.0	0.9	3.1	−4.7	1.1	−4.4	8.7	0.2	8.7
1,4-Benzodioxanes	3.9	2.0	3.8	−4.9	1.2	−4.7	9.0	0.2	9.1
Alkyl aryl ethers	4.0	1.5	4.4	−4.2	1.9	−3.2	9.1	0.1	9.1
Biphenyls	3.6	1.7	3.5	−3.5	0.9	−4.0	8.7	0.2	8.8
Cyclobutanes	3.7	1.8	3.9	−5.8	1.5	−5.4	9.2	0.5	9.0
8-1'-Bicyclo[3.2.1]octanes	4.9	1.2	5.2	−3.5	0.5	−3.5	8.9	0.5	8.8
8-3'-Bicyclo[3.2.1]octanes	4.5	1.8	4.5	−4.1	0.7	−4.3	9.0	0.6	8.9
Biphenyl ethers	5.3	2.1	4.9	−3.6	0.6	−3.7	9.6	0.3	9.6

	Polarisability (Å^3)		
	Mean	Std Dev	Median
Overall	36.3	5.1	36.0
Classical lignans and neolignans	35.1	4.2	35.1
Flavonolignans	42.7	1.8	43.2
CLCs	45.6	4.6	46.5
Dibenzylbutanes	32.2	4.1	30.4
Dibenzylbutyrolactones	32.0	2.8	30.9
Arylnapthalenes/Aryltetralins	32.2	1.7	31.9
Dibenzocyclooctadienes	51.2	1.6	49.4
Substituted tetrahydrofurans	38.7	3.5	37.1
2,6-Diarylfurofurans	37.3	2.8	37.1
Benzofurans	35.3	1.8	35.1
1,4-Benzodioxanes	35.7	3.8	35.7
Alkyl aryl ethers	35.7	3.5	34.0
Biphenyls	31.0	3.7	29.9
Cyclobutanes	38.7	4.6	38.1
8-1'-Bicyclo[3.2.1]octanes	36.3	2.8	36.1
8-3'-Bicyclo[3.2.1]octanes	37.5	3.0	37.0
Biphenyl ethers	29.9	1.9	29.6

3.1.1. Molecular Weight

The molecular weights of the compounds in this study are approximately normally distributed (Figure 3), with an overall mean of 381.5 g mol^{-1} and standard deviation of 70.4 g mol^{-1} (Table 2).

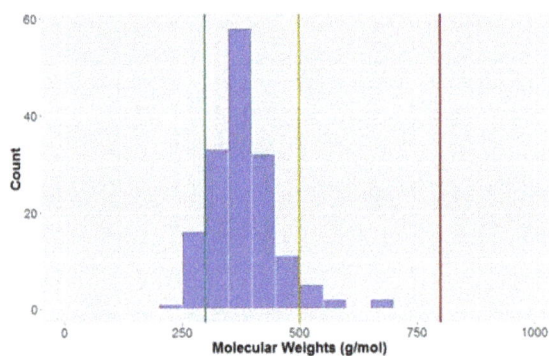

Figure 3. The statistical distribution of the molecular weight of all analysed compounds (green = 300 g mol^{-1}, compounds < 300 g mol^{-1} are in the *lead-like* space; yellow = 500 g mol^{-1}, compounds < 500 g mol^{-1} are in the *drug-like* space; red = 800 g mol^{-1}, compounds < 800 g mol^{-1} are in the KDS). Total number of compounds = 160.

Unsurprisingly, the categories with the highest average molecular weights were CLCs and flavonolignans, with mean molecular weights of 567.4 $\pm$ 67.5 and 478.6 $\pm$ 7.0 g mol^{-1}, respectively. By definition, flavonolignans are the result of a dimerisation of a phenyl propanoid unit and flavone nucleus, a flavone moiety having a higher molecular weight than another phenyl propanoid unit that forms the basis of a classical lignan/neolignan. The CLCs in this study are classical lignans/neolignans with a least one additional saccharide unit attached. Of all of the sub-classes, flavonolignans had the lowest standard deviation for molecular weight, indicative that the compounds of this type have very similar molecular composition. Of the classical lignans and neolignans, dibenzocyclooctadienes had a significantly higher average than other classical lignans and neolignans (413.3 g mol^{-1} vs. 361.2 g mol^{-1}). Conversely, biphenyls (313.8 $\pm$ 62.6 g mol^{-1}) and biphenyl ethers (296.5 $\pm$ 27.4 g mol^{-1}) had the lowest molecular weights, on average. Looking at these compounds, they generally have lower numbers of substituents on the aromatic ring and less elaboration of the sidechains, which could account for this observation. Compounds in the KDS have molecular weights lower than 800 g mol^{-1} (red line in Figure 3); as can be seen, all of the compounds studied exist in KDS for this parameter. Almost all (94.5%) of the compounds would be considered to be *drug-like* when considering molecular weight, however only ~10% of compounds are also considered *lead-like* (<300 g mol^{-1}).

3.1.2. The Octanol–Water Partition Coefficient (LogP)

Like the molecular weights, the lipophilicites (LogP values)—the octanol-water partition coefficient of the molecules—are approximately normally distributed (mean = 3.0, standard deviation = 1.0, Table 2, Figure 4). All compounds studied have a calculated LogP less than the benchmark for KDS (LogP = 6.5), and all but one can be considered *drug-like* (LogP < 5) for this parameter. Approximately half of the compounds had a calculated lipophilicity allowing it to be in *lead-like* space. The compound classes that were calculated to exhibit the highest degree of lipophilicity were dibenzocyclooctadienes and cyclobutanes (LogP = 3.9 $\pm$ 0.7 and 3.7 $\pm$ 0.9, respectively). Contrastingly, CLCs have the lowest average calculated LogP (0.4 $\pm$ 1.1), thereby demonstrating a low affinity for non-aqueous systems and the highest degree of hydrophilicity. Flavonolignans also had low LogP values (mean = 1.6) which was notably lower than the classical lignans and neolignans studied (mean = 3.3).

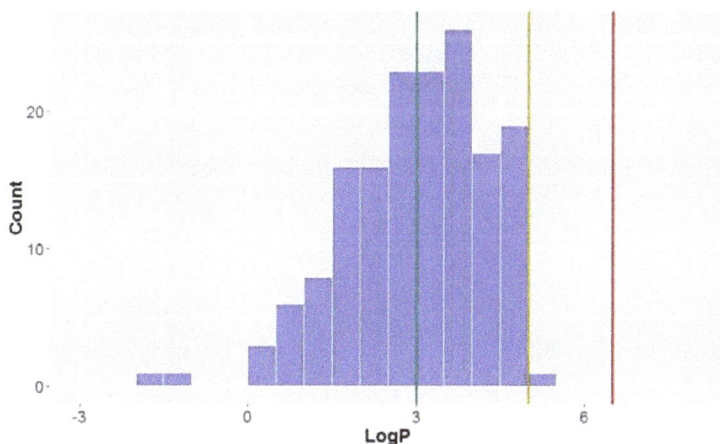

Figure 4. The statistical distribution of the octanol–water partition coefficient (LogP) of all analysed compounds (green = 3, compounds < 3 are in the *lead-like* space; yellow = 5, compounds < 5 are in the *drug-like* space; red = 6.5, compounds < 6.5 are in the KDS). Total number of compounds = 160.

3.1.3. Hydrogen Bond Donors and Acceptors

Ideally, compounds should not have too many hydrogen bond donors and acceptors; the number of hydrogen bond donors should be lower than seven, five and three to be considered to be in KDS, *drug-like* space and *lead-like* space, respectively. On average, the compounds in this study conform reasonably well with the three aforementioned definitions used for the chemical spaces (mean = 3.0, standard deviation = 1.3, Figure 5, Table 2) for hydrogen bond donors. As can be seen, most compounds have three or less hydrogen bond donors (81.3%), allowing them to be classified in *lead-like* space and the majority of compounds have less than two. There are a proportion of compounds that do have more than three hydrogen bond donors–these compounds were mainly CLCs and flavonolignans, with their mean number of hydrogen bond donors being 6.1 and 4.1, respectively. Dibenzylbutanes and alkyl aryl ethers also had a significant percentage of compounds excluded from *lead-like* space according to this parameter.

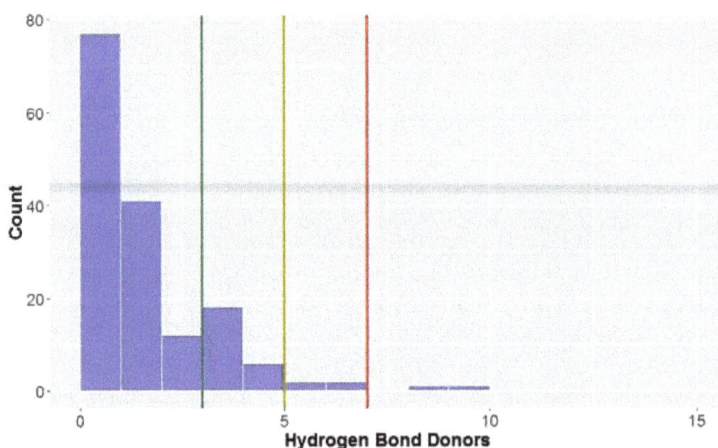

Figure 5. The statistical distribution of the hydrogen bond donors of all analysed compounds (green = 3, compounds < 3 are in the *lead-like* space; yellow = 5, compounds < 5 are in the *drug-like* space; red = 7, compounds < 7 are in the KDS). Total number of compounds = 160.

Only 15% of compounds were classified as being *lead-like* in terms of the number of hydrogen bond acceptors ($\leq$3 hydrogen bond acceptors)–much lower than that observed for the hydrogen bond donors, although greater than 90% of compounds had $\leq$10 hydrogen bond acceptors, classifying them as *drug-like*.

The number of hydrogen bond acceptors displayed a slightly-left skewed normal distribution (Figure 6)–far different to the strongly-skewed distribution seen for the aforementioned number of hydrogen bond donors (Figure 5). The overall mean number of hydrogen bond acceptors was 6.3, although this was largely inflated due to the CLCs (hydrogen bond donors = 16.6 $\pm$ 3.8) and to a lesser degree, flavonolignans (hydrogen bond donors = 9.0 $\pm$ 0.8); without these two compound types included in the analysis, the mean decreased to 5.4 hydrogen bond acceptors. The only compounds studied with greater than 15 hydrogen bond acceptors, thus not in KDS, were CLCs.

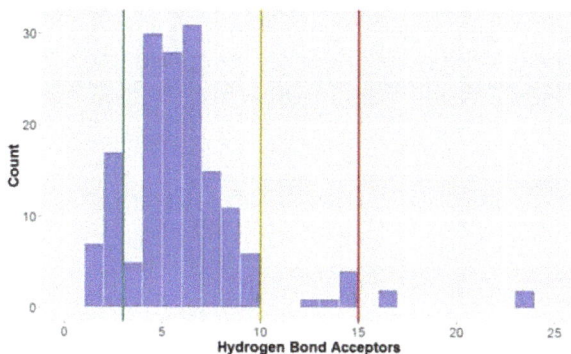

Figure 6. The statistical distribution of the hydrogen bond acceptors of all analysed compounds (green = 3, compounds < 3 are in the *lead-like* space; yellow = 5, compounds < 5 are in the *drug-like* space; red = 15, compounds < 15 are in the KDS). Total number of compounds = 160.

3.1.4. Polar Surface Area (PSA)

The polar surface area (PSA) of all the compounds in the study was found to be 79.5 ± 37.1 Å^2 (Figure 7, Table 2). This parameter is inherently-linked to the number of hydrogen bond acceptors and donors, thus it is no surprise that CLCs had a much higher average polar surface area than the overall average (PSA = 156.9 ± 29.3 Å^2). CLCs, however, did not have the highest mean PSA–flavonolignans had a marginally higher average PSA (158.5 ± 11.4 Å^2). Of the classical lignans/neolignans, the subclass with the highest PSA were dibenzylbutyrolactones (PSA = 90.8 ± 13.6 Å^2); the compounds with the lowest PSAs were dibenzylcyclooctadienes and substituted THF's (mean = 51.2 Å^2 and 50.6 Å^2, respectively). The highest PSA at which oral absorption is able to occur has been reported to be 140 Å^2 – this value thereby benchmarks the upper PSA limit for *drug-like* space [71,72]. A large proportion of studied compounds (88.1%) are in *drug-like* space when considering PSA, while nearly all compounds are in KDS (PSA $\leq$ 180 Å^2), while only a third of compounds were within the strict bounds of *lead-like* space.

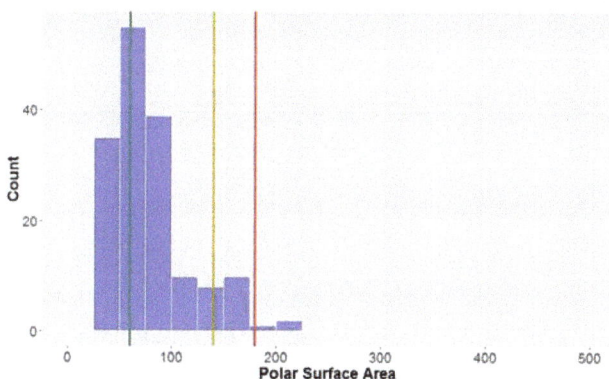

Figure 7. The statistical distribution of the polar surface area (PSA) of all analysed compounds (green = 60, compounds < 60 Å^2 are in the *lead-like* space; yellow = 140, compounds < 140 Å^2 are in the *drug-like* space; red = 180, compounds < 180 Å^2 are in the KDS). Total number of compounds = 160.

3.1.5. Rotatable Bonds

Looking at the general structures of classical lignan/neolignan types, it is apparent that most, in addition to the aryl rings present, have additional ring cycles present in the structures (Figure 1). The exceptions to this general rule are the biphenyl structures, dibenzylbutanes and alkyl aryl ethers; this is reflected in the number of rotatable bonds (Table 2, Figure 8), where alkyl aryl ethers and dibenzylbutanes have the largest average number of rotatable bonds of all the classical lignans/neolignans (mean = 12.1 and 11.0, respectively). CLCs also had high counts for number of rotatable bonds, whereas highly-constricted structures, with a more fused-ring scaffold had far lower averages for this parameter, i.e., 3.3 ± 2.2 for 2,6-diarylfurofurans, and mean = 3.4 ± 1.6 for substituted tetrahydrofurans and arylnapthalenes/aryltetralins.

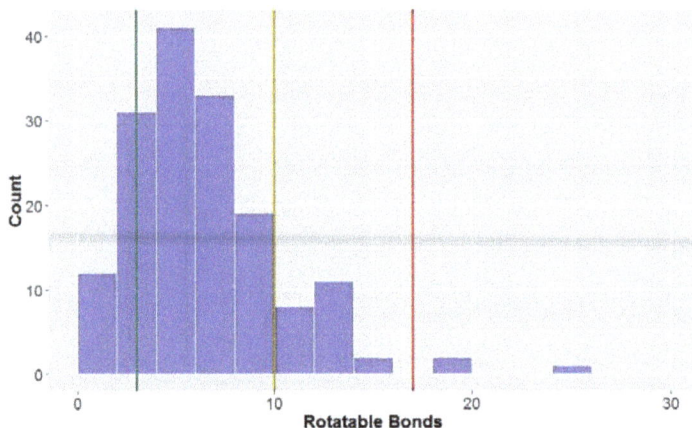

Figure 8. The statistical distribution of the rotatable bonds of all analysed compounds (green = 3, compounds < 3 are in the lead-like space; yellow = 10, compounds < 10 are in the drug-like space; red = 17, compounds < 17 are in the known drug space). Total number of compounds = 160.

The number of rotatable bonds had the second-lowest proportion of compounds classified as *lead-like*, (12.5%) indicating that this parameter (along with molecular weight with 10.6% in *lead-like* space) is one of the more effective descriptors for eliminating potential drug candidates. Lu et al. found that compounds were considered to be in *privileged property space* if they had $\leq$10 rotatable bonds–hence, this is the benchmark used to define *drug-like* space for the number of rotatable bonds [72]. The number of rotatable bonds was the most discerning factor for inclusion of compounds in *drug-like* space (85.0% of compounds met the criteria of $\leq$10 rotatable bonds, Table 3). However, 98.1% of the compounds tested were within the bounds of KDS, with $\leq$17 rotatable bonds.

3.1.6. Other Molecular Descriptors

The calculated dipole moments of the compounds are approximately normally distributed, with a mean of 4.1 ± 1.9 D (Table 2, Figure S1). The compounds with the lowest dipole moments were dibenzocyclooctadienes (mean = 2.6 ± 1.0 D) and 2,6-diarylfurofurans (mean = 2.7 ± 1.3 D), while dibenzylbutyrolactones and biphenyl ethers were the types that had the highest average dipole moments (mean = 5.5 D and 5.3 D, respectively). Density functional theory (DFT) has previously been applied to dipole moment measurements of compounds in KDS, a study which found that compounds within KDS have dipole moments $\leq$10 [70]. Furthermore, it has been reported that to be orally available, a drug should have a dipole moment <13 D. All the compounds in this study had dipole moments below 10 D, indicating that all of the compounds lie within KDS for this parameter and would be orally-available, using dipole moments as a measure of this desirable characteristic in drug therapeutics.

Intrinsically linked to lipophilicity (LogP), the water solubility (LogS) of a compound is an important property to consider [73]. Akin to the dipole moment, the LogS of a compound can be a signifier of oral-availability; it has been shown that the majority of orally available drugs have a LogS between 0 and −7, centring between −4 and −3 [67]. The distribution of the calculated hydrophilicity of the compounds in this study has an approximately normal distribution in those ranges (Figure S2, Table 2). Using the above-mentioned range as the criterion for this parameter as a yardstick of oral-availability, 97% are likely to be orally-available. The mean LogS was −4.1 ± 1.4 for all compounds and it can be noted that CLCs and dibenzylbutyrolactones (mean = −2.5 and −2.6, respectively) have the highest LogS values and thus greater aqueous solubility. In contrast, cyclobutanes (mean = −5.8 ± 1.5), dibenzylcyclooctadienes (mean = −5.6 ± 0.9), and substituted tetrahydrofurans (mean = −5.4 ± 1.1), had lower mean hydrophilicity values, indicating a low affinity for aqueous media.

The ionisation potential of the compounds were normally distributed but had very low variability (mean = 0.9 and standard deviation = 0.4, for all compounds) and no significant differences between compound groups (Table 2, Figure S3). Analogous to the findings for dipole moment and water solubility, the studied compounds have a high degree of compliance with the benchmarks set for ionisation potentials as an indicator of oral availability. All but two of the 160 compounds investigated had an ionisation potential between 8 and 10 eV—it has been shown that orally administered medicinal drugs have ionisation potentials in this range [67]. Of importance to note is that these three descriptors that have been particularly associated with oral availability; ionisation potential, LogS and dipole moment, are very closely correlated in the PCA analysis (see following section for further discussion, Figure 9), with these vectors all having similar bearings. Ionisation potential has also been shown to predicate the redox stability of compounds and thus their ease of metabolism in the body [70].

Polarisability is defined as the ability of a compound to form instantaneous dipoles and can be associated with the ability of the drug to permeate the cell [70]. It has previously been shown that there is a high correlation between the polarisability and molecular weight of a compound (r^2 = 0.90)–an observation that accounts for why molecular weight is such a crucial parameter in chemical space definition [74]. Evidently, this was also the case this study–the principal component analysis shows very close alignment of the polarisability and molecular weight vectors, signifying a high correlation between these parameters (see following section for further discussion, Figure 9). The mean calculated polarisability of all the compounds was found to be 36.3 ± 5.1 Å^3 with an approximately normal distribution, with a slight right-skew (Table 2, Figure S4). CLCs and flavonolignans have high mean polarisability values (45.6 ± 4.6 Å^3 and 42.7 ± 1.8 Å^3, respectively), although dibenzylcclooctadienes had the highest average polarisability (51.2 ± 1.6 Å^3). Biphenyls and biphenyl ethers had the lowest average polarisabilities. Convention dictates that polarisability values should be ≤68 Å^3 to be classified as being in KDS [70] all of the compounds in this study met this criterion.

3.2. PCA (Principal Component Analysis)

After looking at the individual molecular descriptors separately, it was decided to conduct an overall analysis using PCA (Principal Component Analysis). PCA is a statistical technique that transforms the data into a series of new, uncorrelated, dimensions called principal components. These principal components are made up by a combination of the variables studied (in this case, the molecular descriptors). Conducting PCA and analysing these principal components allows for the in-depth, simultaneous investigation of all the descriptors, their interactions and interrelationships. Furthermore, PCA can be used to discover groupings of samples–in this study, compounds—and identify the variables that distinguish these clusters of compounds.

A way to view and analyse the results of a PCA is to plot the principal components against each other, producing a biplot (see Figure 9). In the biplot, the vectors/arrows indicate the direction of influence for each molecular descriptor studied and the data points represent each molecule studied. The biplot given shows the first two principal components–the first principal component is displayed on the x-axis and is the principal component that explains the highest amount of variability of the data, while the second principal component is the dimension in which the data is second-most variable.

In the PCA for this study, nearly half of the variability in the data set is explained by principal component one while principal component two accounts for ~20% of the variability. Thus, these first two principal components account for ~70% of the variability seen in the data which means that the PCA is an effective technique for this investigation, that captures much of the differences and similarities of the compounds in terms of their molecular descriptors and can provide many useful conclusions and observations.

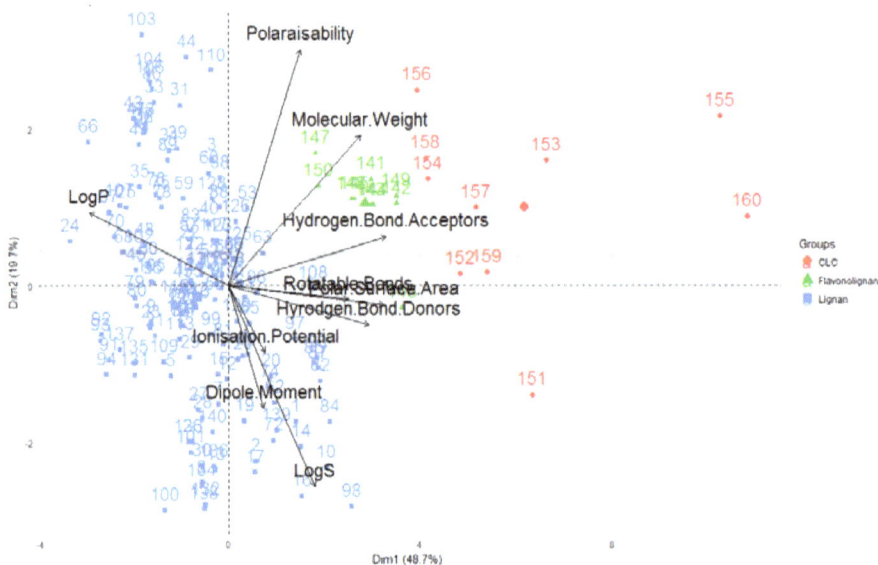

Figure 9. Biplot representing the PCA analysis on the studied compounds and molecular descriptors. The arrows represent molecular descriptors and the direction in which they hold influence. Each point represents a molecule in this study (blue = classical lignans and neolignans, green = flavonolignans, red = CLCs).

Looking at the molecular descriptors, it can be seen that dimension one (corresponding to principal component one–the dimension in which data has the greatest variability; x-axis) is largely influenced by the LogP values, number of hydrogen bond donors, acceptors, rotatable bonds and polar surface area—Figure 9 shows these variables having a significant horizontal component to their direction, with Figure 10 quantifying this contribution. The molecular weight of the compound is a significant contributor to dimension one, but also strongly influences principal component two (y axis, as it also has a large vertical component to its direction), along with the polarisability, ionisation potential, dipole moment and LogS.

It can also be seen from this PCA, that the greater the LogP value (i.e., the greater the lipophilicity), the lower the number of hydrogen bond donors, acceptors and polar surface area for these compounds, as these vectors are in opposite direction to the LogP, signifying an inverse relationship. This coincides with what one would anticipate—compounds with greater hydrogen bond donors, acceptors and polar surface area are expected to be less lipophillic and have lower LogP values. Furthermore, the number of hydrogen bond donors, acceptors and polar surface area are highly correlated, as is signified through them acting in very similar directions in the biplot. This again concurs with what one would expect–a compound with the more hydrogen bond donors and acceptors would be envisaged to have a greater polar surface area. As mentioned when these variables were discussed, ionisation potential, dipole moment and LogS of a compound are shown to be highly correlated. Additionally, LogS is aligned in the opposite orientation to the LogP vector–expected for these two inversely-related measures.

It should also be noted that the variables that are the highest contributors to the first principal component (number of hydrogen donors, acceptors, rotatable bonds, LogP, molecular weight and polar surface area) are those that are considered when defining chemical spaces (Figure 10, Table 1).

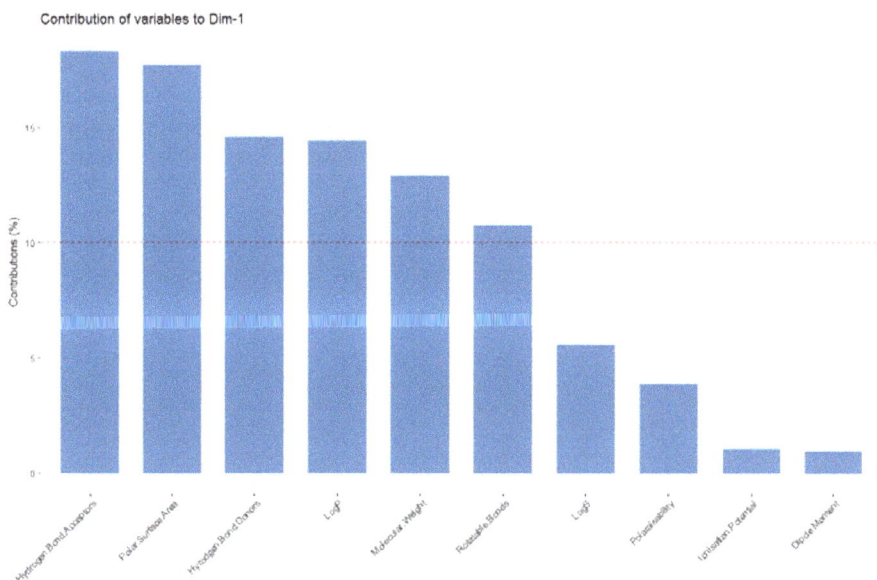

Figure 10. Representation of the contributors to the first principal component.

Where the points are situated on a PCA plot, their relative positions, are a culmination of their values for the various parameters and any observed groupings indicate similarities in the physicochemical properties between compounds in that group. It is apparent that PCA with the ten aforementioned descriptors are able to separate the compounds into groups–classical lignans and neolignans (blue), flavonolignans (green) and CLCs (red). These compounds types are separated on the x-axis–PC1–indicating that these groups can be separated by the variables that contribute to this principal component. Therefore, this analysis can give general indications about the compound classes. In general, it appears that CLCs have higher molecular weights, hydrogen bond acceptors, hydrogen bond donors, rotatable bonds and polar surface areas and lower LogP values (lower lipophilicities). Conversely, classical lignans and neolignans appear to be typified by higher lipophilicities and lower numbers of hydrogen bond acceptors, donors and polar surface areas. In terms of these descriptors, flavonolignans appear to be characterised between these two groups. Additionally, all of the flavonolignans seems to be less variable, while both classical lignans/neolignans and CLCs are significantly more variable. As the CLC group is composed of classical lignan/neolignan types with additional saccharide unit(s), one can see the influence of the inclusion of a sugar moiety, on the molecular descriptors. Furthermore, clemastanin B and secoisolariciresinol diglucoside represented by 155 and 160 in Figure 9, were the two compounds in the study that lie furthest to the right in the biplot shown and were the two compounds in this study that contained two sugar units–this indicates that the number of saccharide units in a CLC can also be differentiated using PCA.

3.3. Lignans in Chemical Space

The proportion of all compounds studied in this investigation that lie within the benchmarks for *lead-like*, *drug-like* and KDS as specified in Table 1 are shown, for each molecular descriptor and when all molecular descriptors are taken into account (Table 3). Immediately apparent is that no compound in this study fits in *lead-like* chemical space, when all of the molecular descriptors that define *lead-likeness* are taken into account. While most compounds have ≤3 hydrogen bond donors and fulfil the requirements for this parameter, the molecular weight, the number of hydrogen bond acceptors and rotatable bonds are the parameters that limit the inclusion of lignans into *lead-like* space. To be *lead-like*, lead structures generally possess low molecular complexity (i.e., low molecular weighs, along with minimal numbers of hydrogen bond acceptors, hydrogen bond donors and rotatable bonds) [56]. Furthermore, these structures are more hydrophyllic and less *drug-like*, hence the strictest criterion for the chemical spaces are those that define *lead-like* space. The purpose of lead structures is to offer a simple scaffold, upon which further complexity can then be added, to provide *drug-like* compounds. Rather than being *lead-like*, the majority of compounds–approximately $\frac{3}{4}$ of those studied, are *drug-like* in that they already have structures with greater complexity than one would expect from a lead compound. The majority of the studied compounds fulfil the requirements to be considered *drug-like*, thus by definition possess properties and characteristics that indicate they would be appropriate for use as therapeutics. An even higher proportion of the compounds studied are in KDS, thus are in the chemical space that is defined by known drugs.

Table 3. All compounds in this study and their inclusion within the defined chemical spaces.

Overall	*Lead-Like* Space	*Drug-Like* Space	Known Drug Space
Molecular weight (g mol^{-1})	10.6%	94.4%	100%
Lipophilicity (Log P)	46.3%	99.4%	100%
Hydrogen bond donors	81.3%	96.3%	98.8%
Hydrogen bond acceptors	15.0%	93.8%	97.5%
Polar surface area (Å^2)	33.1%	88.1%	98.8%
Rotatable bonds	12.5%	85.0%	98.1%
All criteria	0.0%	75.6%	97.5%

The success of podophyllotoxin as both a lead compound and as a drug were discussed previously–podophyllotoxin itself is an approved therapeutic for genital warts and associated ailments, while its structurally-related derivatives are clinically approved cancer treatments. Podophyllotoxin is an aryltetralin lactone and as such can be classified as a dibenylbutryolactone or an aryltetralin. For the purposes of this study, it was classified as a dibenzylbutyrolactone as it was apparent that its molecular descriptors and structural scaffold were most similar to compounds of this type. Looking at its molecular parameters, podophyllotoxin meets all the requirements of the Lipinski's rule of five, with a molecular weight =414.4 gmol^{-1}, one hydrogen bond donor, eight hydrogen bond acceptors, LogP = 2.31, PSA = 98.3 Å^2 and only four rotatable bonds, hence meets all the requirements of being *drug-like*, and therefore it is no surprise that it is an effective medicine. Furthermore, podophyllotoxin has values for the other measures of oral availability, namely LogS, ionisation potential and dipole moment, that signify it to be readily orally-available–a positive and desirable feature for therapeutics.

Podophyllotoxin is an excellent example of the potential of *drug-like* lignans for use a medicines. Another example of a lignan currently in use is a dibenzylbutane, masoprocol, a form of dihydroguaiaretic acid (Figure 1). Masoprocol is a lipoxygenase inhibitior and is an antineoplastic medicine that is indicated to treat skin growths that result from exposure to the sun [21,75,76]. Like podophyllotoxin, masoprocol fulfils all the criterion that dictate the requirements of a *drug-like* compound.

Over $\frac{3}{4}$ of the compounds included in this study, including podophyllotoxin and masoprocol, exist in *drug-like* chemical space, exhibiting properties that allow them to be considered *drug-like* compounds and more likely to be successful therapeutics. There are numerous other lignans that also exhibit potent biological activities and meet all the stipulated benchmarks to be considered *drug-like*, such examples include arctigenin [77], matairesinol [78], sesamin [79] and schizandrin A (Figure 1) [80]. These compounds are just a few of the many hundreds of compounds yet to be fully explored, highlight the vast prospects that lignans provide in medicinal chemistry.

3.4. Classical Lignans and Neolignans

There were 140 classical lignans and neolignans in this study which included the ten most well-known (based on number of references for each compound) of each type. Summary statistics for each of the parameters for the classical lignans/neolignans grouped together, and separate are given in Table 2. Statistical distributions of each of the molecular parameters studied for the classical lignans and neolignans are given (Figure 11, Figures S5–S13) and analysis of their positions in chemical spaces are provided (Table 4, Tables S1–S15).

While some of the compound sub-classes are very similar in the characteristics studied, there are some compound types that are notably different for various parameters. The compound class that most frequently had markedly higher/lower averages than other lignans/neolignans for the molecular descriptors was the dibenzocyclooctadienes (e.g., (+)-Schizandrin A, Figure 1). Dibenzocyclooctadienes appear to have relatively high molecular weights, polarisability, polar surface area and lipophilicity (LogP), while conversely having relatively low dipole moments and water solubility (LogS). Cyclobutanes and substituted tetrahydrofurans also had high lipophilicities (LogP) and low hydrophillicity measures (LogS), while substituted tetrehydrofurans also had high polar surface areas and lower numbers of rotatable bonds. Arylnapthalenes and 2,6-diarylfurofurans are highly fused scaffolds and this is reflected in their lower rotatable bonds count, compared to other classical lignans and neolignans. Conversely, alkyl aryl ethers and dibenzylbutanes have a less rigid/cyclic structural motif, hence have more rotatable bonds. As well as having a lower amount of rotatable bonds, 2,6-diarylfurofurans also exhibit lower lipophilicities and a lower average dipole moment. Dibenzylbutyrolactones also have lower LogP values and higher water solubilities (LogS), along with higher polar surface areas and dipole moments. Biphenyl ethers are another type that have higher relative calculated dipole moments, and along with biphenyls have lower average molecular weights and polar surface areas. Benzofurans, 1,4-benzodioxanes, 8-1′-bicyclo-[3.2.1]octanes and 8-3′-bicyclo[3.2.1]octanes were compound types that were not notably higher/lower than other classical lignans and neolignans for the parameters investigated.

Overall, as stated previously, none of the classical lignans/neolignans in this study fulfil all the requirements to be considered *lead-like*, although almost all (86.4%) are within the limits that define *drug-like* space and all lignans are in KDS (Table 4).

The *lead-like* benchmarks that classical lignans and neolignans are least-frequently able to realise are those concerning molecular weight ($\leq$300 g mol^{-1}; Figure S5), number of hydrogen bond acceptors and the number of rotatable bonds (both $\leq$3; Figure S8 and Figure 11). In view of the biosynthesis and definition of lignan structures, it is not surprising that very few of the compounds meet the individual requirements for these parameters, and no compounds are able to meet them all collectively. By definition, lignans and neolignans are the product of an oxidative dimerisation of two or more phenyl propanoid units, which alone would have a weight of at least 240 g mol^{-1}. These phenyl propanoid units are almost always oxygenated, often containing several oxygen-containing substituents; the inclusion of more than three oxygen atoms, as is frequently the case with naturally-occurring lignans, would not only increase the molecular weight above the cut-off for *lead-likeness*, but would also exceed the number of allowable hydrogen bond acceptors.

Table 4. All classical lignans and neolignans (not CLCs or flavonolignans) studied within the defined chemical spaces.

Overall	Lead-Like Space	Drug-Like Space	Known Drug Space
Molecular weight (g mol^{-1})	12.1%	100%	100%
Lipophilicity (Log P)	38.6%	99.3%	100%
Hydrogen bond donors	91.4%	99.3%	100%
Hydrogen bond acceptors	17.1%	100%	100%
Polar surface area (Å^2)	37.9%	99.3%	100%
Rotatable bonds	14.3%	87.9%	100%
All criteria	0.0%	86.4%	100%

Conversely, it is apparent that lignans and neolignans are generally very *drug-like* and all are within KDS (Table 4). The most discerning *drug-like* space parameter that ~13.5% of the compounds violated were the number of rotatable bonds (criteria: ≤10 rotatable bonds, Figure 11). Related to this, from this study, it can be stated that dibenzylbutanes (e.g., phyllanthin, Figure 12) are the least *drug-like* of all the lignan sub-classes, largely owing to their high number of rotatable bonds–of the ten dibenzylbutanes in this study, six are considered *undrug-like*. In contrast, compounds with similar functional groups but having a more-fused ring scaffold (e.g., (−)-grandisin, Figure 12) are more likely to be drug-like. In all other aspects lignans, in general, almost always fulfil every other requirement that defines *drug-like* space and in several groups, namely dibenzylbutyrolactones, arylnapthalenes/aryltetralins, substituted THFs, 2,6-diarylfurofurans, benzofurans, 1,4-benzodioxanes, 8-1′-bicyclo[3.2.1]octanes, 8-3′-bicyclo[3.2.1]octanes and biphenyl ethers, all members were considered *drug-like*. The high-proportion of *drug-likeness* of classical lignans and neolignans, particularly these aforementioned sub-classes, is a very notable and promising observation that promotes the justifiability and importance of investigating lignans as drugs.

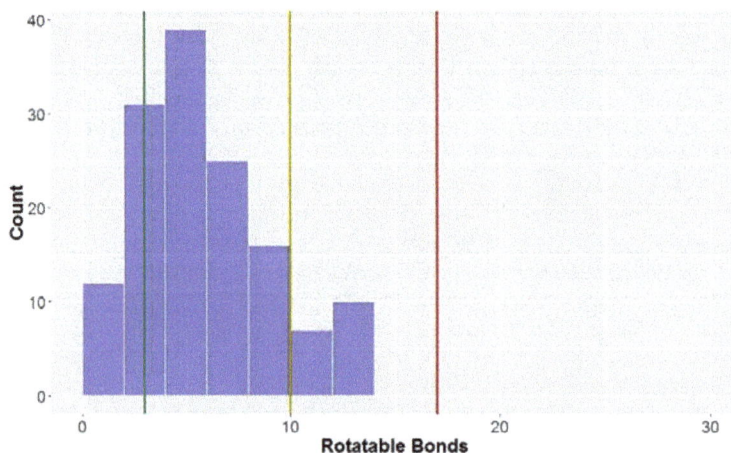

Figure 11. The statistical distribution of the rotatable bonds of the classical lignans and neolignans (green = 3, compounds < 3 are in the lead-like space; yellow = 10, compounds < 10 are in the drug-like space; red = 17, compounds < 17 are in the known drug space). Total number of compounds = 140.

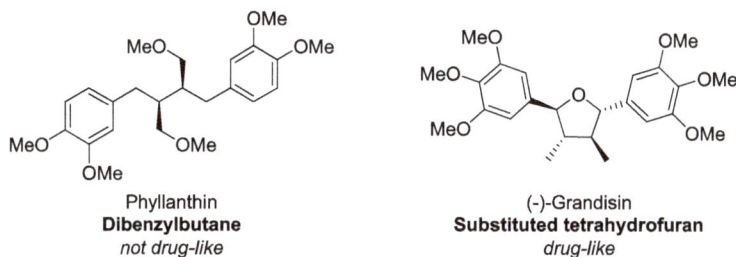

Figure 12. Structures of phyllanthin and (−)-grandisin.

Considering substituents on these core lignan scaffolds—while for most of the studied descriptors, the most *drug-like* compounds would be those with no/few substituents on the core structures which results in lower values for almost all of the parameters–this, however, is a rarity amongst naturally-occurring lignans and it is very likely that the lipophilicity would increase, possibly beyond acceptable levels. The addition of large polar groups (i.e., sugar moieties as for the CLC's, see Section 3.6) do have a significant effect on many of the properties of lignans, that largely result in their exclusion from *drug-like* chemical space. In contrast, it can be seen that many of the commonly-occurring, smaller, oxygenated substituents that feature in naturally-occurring lignans (i.e., hydroxy, methoxy and methylenedioxy groups) are well-tolerated within *drug-like* chemical space. This is evidenced by the fact that the parameters (i.e., molecular weight, lipophilicity, hydrogen bond donors/acceptors and polar surface area) that would be most affected by the inclusion of these moieties are very rarely (<1% of all lignans in the study) exceeded. It can therefore be said, that naturally-occurring lignans do have an excellent balance of parameters and structures closely analogous to these, with similar levels of substitution would be of interest.

Furthermore, as noted when discussing the individual molecular parameters, oral availability can be linked to the dipole moment (ideally < 13 D), LogS (ideally between 0 and −7) and ionisation potential (ideally between 8 and 10 eV). All classical lignans and neolignans had dipole moments under the threshold of 13 D (Figure S10), while all but two and five of the 140 classical lignans and neolignans had an ionisation potential and LogS within the above ranges, respectively (Figures S11 and S12). This is predicates lignans and neolignans to have excellent oral bioavailability–an extremely desirable trait of drugs.

3.5. Flavonolignans

As their name suggests, flavonolignans are a structurally very similar to classical lignans and neolignans, however while lignans are formed through the oxidative dimerisation of two or more phenyl propanoid units, the biosynthetic precursors of flavonolignans are a phenyl propanoid unit and a flavone [81]. Flavonolignans are of particular interest to many, owing to their potent biological activities that have been utilised worldwide, for millennia, particularly in the form of silymarin. Silymarin (commonly known as milk thistle extract) is isolated from the seeds of milk thistle, *Silybum marianum*, and is a complex mixture of, predominantly flavonolignan, compounds [82]. Silymarin is a popular liver protectant that has been used in traditional medicine for centuries and is commonly available and used in present-day society [83,84].

Studying the earlier PCA, it is apparent through the close proximity of all flavonolignans in the biplot, that they are all very structurally similar (Figure 9). Furthermore, it can be seen that while flavonolignans are structurally similar to lignans, they are able to be separated on the basis of their molecular descriptors. Flavonolignans are clustered to the right of almost all classical lignans/neolignans and located higher on the y-axis on the PCA than many. One can use the knowledge in which direction the molecular descriptors hold influence to discuss general trends of this compound type. It can be surmised from the PCA, that flavonolignans generally have higher molecular weights, polarisability, number of hydrogen bond donors and acceptors. They also appear to have lower lipophilicities (LogP). These observations are further corroborated by the results in Table 2 (see Figure 13, Figures S14–S22 for analysis of each of the studied molecular descriptors for flavonolignans alone).

The ten studied flavonolignans were also assessed in relation to the various criterion that define the *lead-like*, *drug-like* and known drug spaces (Table 5). It is notable that flavonolignans are less *drug-like* than classical lignans and neolignans, with no flavonolignans fulfilling all the requirements for *drug-likeness*. The only constraint that flavonolignans exceeded was the polar surface area–no flavonolignans had a PSA ≤ 140 Å^2 and the mean PSA was 158.5 Å^2 (Figure 13). It should be noted, however, that all of the compounds had a PSA within the realm of KDS, and the flavonolignans were all in KDS when considering all molecular descriptors.

Table 5. Flavonolignans studied within the defined chemical spaces.

Overall	*Lead-Like* Space	*Drug-Like* Space	Known Drug Space
Molecular weight (g mol^{-1})	0%	100%	100%
Lipophilicity (Log P)	100%	100%	100%
Hydrogen bond donors	20%	100%	100%
Hydrogen bond acceptors	0%	100%	100%
Polar surface area (Å^2)	0%	0%	100%
Rotatable bonds	0%	100%	100%
All criteria	0%	0%	100%

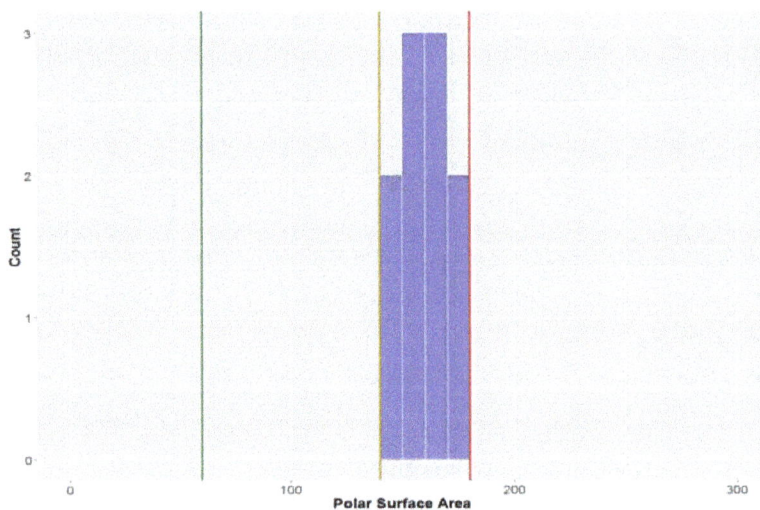

Figure 13. The statistical distribution of the polar surface area (PSA) of the flavonolignans (green = 60, compounds < 60 Å^2 are in the *lead-like* space; yellow = 140, compounds < 140 Å^2 are in the *drug-like* space; red = 180, compounds < 180 Å^2 are in the KDS). Total number of compounds = 10.

Reviewing the additional gauges of oral availability; all flavonolignans meet the benchmarks set for the dipole moment, LogS and ionisation potential, thus there is strong indication that flavonolignans are orally available.

3.6. CLCs; Carbohydrate-Lignan Conjugates

As the CLCs included in this study are saccharide-containing representatives of various types of classical lignans and neolignans, the differences seen in the CLCs from lignans are due to the sugar moiety. Through the principal component analysis, it was shown that CLCs can be differentiated from both classical lignans/neolignans and flavonolignans on the basis of the molecular descriptors that were included in this study (Figure 9). Analysis of the PCA suggests that the inclusion of a saccharide unit to a lignan increases its mass, number of hydrogen bond donors, acceptors and number of rotatable bonds, as well as its polarisability. They are also indicated to have lower lipophilicities (LogP, Figure 14), and as shown through the comparison of means, a higher affinity for water and other aqueous systems (Table 2, see Figure 14, Figures S23–S31 for the distributions of each of the molecular descriptors).

The low calculated lipophilicities of CLCs are an asset in defining this compound type in chemical space, with all of the CLCs studied having sufficiently low LogP values to be considered *lead-like* (Figure 14), although interestingly this is the only parameter that any CLCs do not exceed for *lead-likeness*—CLCs meet none of the other *lead-like* criteria (Table 6). The molecular descriptor that was the most discriminating for CLCs was they number of hydrogen bond acceptors—no CLCs had sufficiently low enough number of hydrogen bond acceptors to be considered either *lead-like* or *drug-like* (Figure S25). Furthermore, only six out of ten CLCs in this study were in KDS for this parameter. For the other molecular descriptors, very few compounds fell within the discerning bounds of *drug-like* space, less than half of the CLCs within the limits for *drug-like* space for molecular weight (Figure S23), number of hydrogen bond donors (Figure S24), number of rotatable bonds (Figure S27) and polar surface area (Figure S26).

Table 6. CLCs studied within the defined chemical spaces.

Overall	*Lead-Like* Space	*Drug-Like* Space	Known Drug Space
Molecular weight (g mol^{-1})	0%	10%	100%
Lipophilicity (Log P)	100%	100%	100%
Hydrogen bond donors	0%	50%	80%
Hydrogen bond acceptors	0%	0%	60%
Polar surface area (Å^2)	0%	20%	80%
Rotatable bonds	0%	30%	70%
All criteria	0%	0%	50%

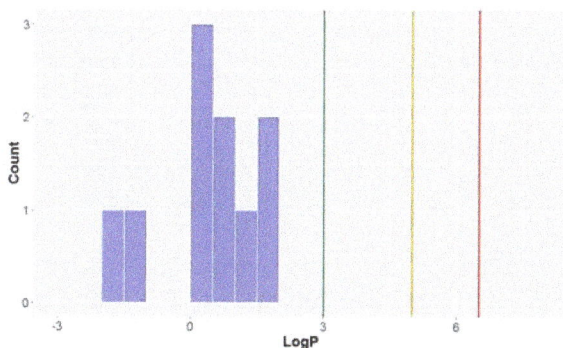

Figure 14. The statistical distribution of the octanol–water partition coefficient (LogP) of the CLCs (green = 3, compounds < 3 are in the *lead-like* space; yellow = 5, compounds < 5 are in the *drug-like* space; red = 6.5, compounds < 6.5 are in the KDS). Total number of compounds = 10.

As a rule, CLCs are distinguished as having many alcohol moieties, which entails that they have low lipophilicities and high molecular weights, number of hydrogen bond acceptors, number of rotatable bonds and polar surface area. One must, however, mention that there is supposition that *drug-likeness* is not a suitable measure of saccharides, which are adsorbed through active transport [85]. Saccharide-based drugs are atypical in KDS [59], and this is reflected in only half of the CLCs studied being entirely in KDS.

4. Summary

In this study, 160 lignans and related compounds were analysed to study their physicochemical properties, their general trends and the variability in these parameters between and within compound types. Furthermore, these molecular descriptors allowed for the defining of these compounds in various chemical spaces, particularly to highlight their *drug-likeness*. It was found, that while no compounds in this study fulfilled all the requirements for six key molecular descriptors to be considered to be *lead-like*, over 3/4 of the compounds were deemed to be in the *drug-like* space and nearly all (~97.5%) were in the KDS. These results strongly advocate for the *drug-likeness* of the majority of lignan compounds that should be further investigated as potential therapeutics. Notably, all compounds from dibenzylbutyrolactones, arylnapthalenes/aryltetralins, substituted THFs, 2,6-diarylfurofurans, benzofurans, 1,4-benzodioxanes, 8-1'-bicyclo[3.2.1]octanes, 8-3'-bicyclo[3.2.1]-octanes and biphenyl ethers sub-classes were shown to *drug-like*, indicating that these sub-classes, in particular, should be further studied for their potential as therapeutic agents.

A PCA analysis of the molecular descriptors particularly highlighted the complex inter-relationships between these physicochemical properties–while the number of hydrogen bond donors, acceptors, rotatable bonds and polar surface area appear to be strong, positive relationship, they collectively have an inverse relationship with the lipophilicity (LogP) of compounds. The first principal component is the dimension that accounts for the greatest variability in the data and it was found that the largest contributors to this principal component were the variables that are considered when defining the chemical spaces. PCA was also able to separate the groups of lignans (classical lignans and neolignans), flavonolignans and CLCs–it was shown that flavnonolignans are more similar to lignans, than their sugar-derivatives, with this separation being distinctive along the first principal component. The differences seen between the different groups in the PCA were also reflected in the differing proportions of lignans, flavonolignans and CLCs that were included in the chemical spaces, particularly in the *drug-like* chemical space. Lignans were almost all (86.4%) *drug-like* and all were in KDS, whereas no flavonolignans or CLCs were *drug-like*. All flavonolignans were in KDS and only half of the CLCs were in KDS. This suggests that lignans, in general, have an excellent balance of the often-paradoxical molecular properties, allowing them to be considered to be *drug-like*, whereas flavonolignans and CLCs have larger values for the variables other than lipophilicity (i.e., polar surface area, number of rotational bonds, hydrogen bond donors and acceptors, as they are further to the right of the PCA plot) that exclude them from *drug-like* chemical space. This is particularly evident for the CLCs, where LogP was the only parameter that all CLCs met the requirements for, to be classified as present in *drug-like* chemical space.

Within the lignans, there were marked differences between different compound types, with dibenzylcyclooctadienes proving to be the most distinctive compound type, exhibiting relatively high molecular weights, polarisability, polar surface area and lipophilicity (LogP), while conversely having relatively low dipole moments and water solubility (LogS). Overall, the results presented here demonstrate that lignans are very *drug-like*. Coupled with their potent biological activities, their physicochemical properties indicate there is significant value in their study as promising future drug leads.

Supplementary Materials: The supplementary materials are available online.

Funding: The author would like to acknowledge the Gavin and Ann Kellaway Medical Research Fellowship for funding that contributed to this research.

Acknowledgments: The author would also like to thank David Barker for his helpful discussions.

Conflicts of Interest: The author declared no conflict of interest.

References

1. Ayres, D.C.; Loike, J.D. *Lignans Chemical, Biological and Clinical Properties*; Cambridge University Press: Cambridge, UK, 1990.
2. Ward, R.S. The synthesis of lignans and neolignans. *Chem. Soc. Rev.* **1982**, *11*, 75–125. [CrossRef]
3. Pan, J.-Y.; Chen, S.-L.; Yang, M.-H.; Wu, J.; Sinkkonen, J.; Zou, K. An update on lignans: Natural products and synthesis. *Nat. Prod. Rep.* **2009**, *26*, 1251–1292. [CrossRef] [PubMed]
4. Saleem, M.; Kim, H.J.; Ali, M.S.; Lee, Y.S. An update on bioactive plant lignans. *Nat. Prod. Rep.* **2005**, *22*, 696–716. [CrossRef] [PubMed]
5. Zhang, J.; Chen, J.; Liang, Z.; Zhao, C. New lignans and their biological activities. *Chem. Biodivers.* **2014**, *11*, 1–54. [CrossRef] [PubMed]
6. Whiting, D.A. Lignans and neolignans. *Nat. Prod. Rep.* **1985**, *2*, 191–211. [CrossRef]
7. Hu, C.-Q. *Introduction to Natural Products Chemistry*; Zhao, W., Ed.; CRC Press: Boca Raton, FL, USA, 2011; pp. 225–245.
8. Gottlieb, O.R. Chemosystematics of the lauraceae. *Phytochemistry* **1972**, *11*, 1537–1570. [CrossRef]
9. Gottlieb, O.R. Neolignans. *Fortschr. Chem. Org. Naturst.* **1978**, *35*, 1–72.
10. Teponno, R.B.; Kusari, S.; Spiteller, M. Recent advances in research on lignans and neolignans. *Nat. Prod. Rep.* **2016**, *33*, 1044–1092. [CrossRef] [PubMed]
11. Gordaliza, M.; García, P.A.; Miguel del Corral, J.M.; Castro, M.A.; Gómez-Zurita, M.A. Podophyllotoxin: Distribution, sources, applications and new cytotoxic derivatives. *Toxicon* **2004**, *44*, 441–459. [CrossRef] [PubMed]
12. Yousefzadi, M.; Sharifi, M.; Behmanesh, M.; Moyano, E.; Bonfill, M.; Cusido, R.M.; Palazon, J. Podophyllotoxin: Current approaches to its biotechnological production and future challenges. *Eng. Life Sci.* **2010**, *10*, 281–292. [CrossRef]
13. MacRae, W.D.; Towers, G.H.N. Biological activities of lignans. *Phytochemistry* **1984**, *23*, 1207–1220. [CrossRef]
14. Cos, P.; Maes, L.; Vlietinck, A.; Pieters, L. Plant-derived leading compounds for chemotherapy of human immunodeficiency virus (HIV) infection—An update (1998–2007). *Planta Med.* **2008**, *74*, 1323–1337. [CrossRef] [PubMed]
15. Pilkington, L.I.; Wagoner, J.; Polyak, S.J.; Barker, D. Enantioselective synthesis, stereochemical correction, and biological investigation of the rodgersinine family of 1,4-benzodioxane neolignans. *Org. Lett.* **2015**, *17*, 1046–1049. [CrossRef] [PubMed]
16. Fauré, M.; Lissi, E.; Torres, R.; Videla, L.A. Antioxidant activities of lignans and flavonoids. *Phytochemistry* **1990**, *29*, 3773–3775. [CrossRef]
17. Negi, A.S.; Kumar, J.K.; Luqman, S.; Shanker, K.; Gupta, M.M.; Khanuja, S.P.S. Recent advances in plant hepatoprotectives: A chemical and biological profile of some important leads. *Med. Res. Rev.* **2008**, *28*, 746–772. [CrossRef] [PubMed]
18. Huang, W.-Y.; Cai, Y.-Z.; Zhang, Y. Natural phenolic compounds from medicinal herbs and dietary plants: Potential use for cancer prevention. *Nutr. Cancer* **2009**, *62*, 1–20. [CrossRef] [PubMed]
19. Webb, A.L.; McCullough, M.L. Dietary lignans: Potential role in cancer prevention. *Nutr. Cancer* **2005**, *51*, 117–131. [CrossRef] [PubMed]
20. Habauzit, V.; Horcajada, M.-N. Phenolic phytochemicals and bone. *Phytochem. Rev.* **2008**, *7*, 313–344. [CrossRef]
21. Apers, S.; Vlietinck, A.; Pieters, L. Lignans and neolignans as lead compounds. *Phytochem. Rev.* **2003**, *2*, 201–217. [CrossRef]

22. Cunha, W.R.; e Silva, M.L.A.; Sola, R.C.; Veneziani, S.R.A.; Bastos, J.K. Lignans: Chemical and biological properties. In *Phytochemicals—A global Perspective of Their Role in Nutrition and Health*; Venketeshwer, R., Ed.; In Tech: Rijeka, Croatia, 2012; pp. 213–234.

23. Pilkington, L.I.; Barker, D. Synthesis and biology of 1,4-benzodioxane lignan natural products. *Nat. Prod. Rep.* **2015**, *32*, 1369–1388. [CrossRef] [PubMed]

24. Pilkington, L.I.; Barker, D. Asymmetric synthesis and CD investigation of the 1,4-benzodioxane lignans eusiderins A, B, C, G, L, and M. *J. Org. Chem.* **2012**, *77*, 8156–8166. [CrossRef] [PubMed]

25. Pilkington, L.I.; Barker, D. Total synthesis of (−)-isoamericanin A and (+)-isoamericanol A. *Eur. J. Org. Chem.* **2014**, *2014*, 1037–1046. [CrossRef]

26. Jung, E.; Pilkington, L.I.; Barker, D. Enantioselective synthesis of 2,3-disubstituted benzomorpholines: Analogues of lignan natural products. *J. Org. Chem.* **2016**, *81*, 12012–12022. [CrossRef] [PubMed]

27. Jung, E.; Dittrich, N.; Pilkington, L.I.; Rye, C.E.; Leung, E.; Barker, D. Synthesis of aza-derivatives of tetrahydrofuran lignan natural products. *Tetrahedron* **2015**, *71*, 9439–9456. [CrossRef]

28. Rye, C.; Barker, D. An acyl-Claisen approach to tetrasubstituted tetrahydrofuran lignans: Synthesis of fragransin A2, talaumidin, and lignan analogues. *Synlett* **2009**, 3315–3319.

29. Barker, D.; Dickson, B.; Dittrich, N.; Rye, C.E. An acyl-Claisen approach to the synthesis of lignans and substituted pyrroles. *Pure Appl. Chem.* **2012**, *84*, 1557–1565. [CrossRef]

30. Dickson, B.D.; Dittrich, N.; Barker, D. Synthesis of 2,3-syn-diarylpent-4-enamides via acyl-Claisen rearrangements of substituted cinnamyl morpholines: Application to the synthesis of magnosalicin. *Tetrahedron Lett.* **2012**, *53*, 4464–4468. [CrossRef]

31. Duhamel, N.; Rye, C.E.; Barker, D. Total Synthesis of ent-hyperione A and ent-hyperione B. *Asian J. Org. Chem.* **2013**, *2*, 491–493. [CrossRef]

32. Paterson, D.L.; Barker, D. Synthesis of the furo[2,3-b]chromene ring system of hyperaspindols A and B. *Beilstein J. Org. Chem.* **2015**, *11*, 265–270. [CrossRef] [PubMed]

33. Davidson, S.J.; Barker, D. Synthesis of various lignans via the rearrangements of 1,4-diarylbutane-1,4-diols. *Tetrahedron Lett.* **2015**, *56*, 4549–4553. [CrossRef]

34. Pilkington, L.I.; Barker, D. Synthesis of 3-methyllobovatol. *Synlett* **2015**, *26*, 2425–2428.

35. Rye, C.E.; Barker, D. Asymmetric synthesis and anti-protozoal activity of the 8,4′-oxyneolignans virolin, surinamensin and analogues. *Eur. J. Med. Chem.* **2013**, *60*, 240–248. [CrossRef] [PubMed]

36. Tran, H.; Dickson, B.; Barker, D. Unexpected O-alkylation and ester migration in phenolic 2,3-diaryl-2,3-dihydrobenzo[b]furans. *Tetrahedron Lett.* **2013**, *54*, 2093–2096. [CrossRef]

37. Rye, C.E.; Barker, D. Asymmetric synthesis of (+)-galbelgin, (−)-kadangustin J, (−)-cyclogalgravin and (−)-pycnanthulignenes A and B, three structurally distinct lignan classes, using a common chiral precursor. *J. Org. Chem.* **2011**, *76*, 6636–6648. [CrossRef] [PubMed]

38. Pilkington, L.I.; Song, S.M.; Fedrizzi, B.; Barker, D. Efficient total synthesis of (±)-isoguaiacin and (±)-isogalbulin. *Synlett* **2017**, *28*, 1449–1452.

39. Davidson, S.J.; Barker, D. Total synthesis of ovafolinins A and B: Unique polycyclic benzoxepin lignans through a cascade cyclization. *Angew. Chem. Int. Ed.* **2017**, *56*, 9483–9486. [CrossRef] [PubMed]

40. Davidson, S.J.; Pearce, A.N.; Copp, B.R.; Barker, D. Total synthesis of (−)-bicubebin A, B, (+)-bicubebin C and structural reassignment of (−)-cis-cubebin. *Org. Lett.* **2017**, *19*, 5368–5371. [CrossRef] [PubMed]

41. Davidson, S.J.; Rye, C.E.; Barker, D. Using NMR to determine the relative stereochemistry of 7,7′-diaryl-8,8′-dimethylbutan-1-ol lignans. *Phytochem. Lett.* **2015**, *14*, 138–142. [CrossRef]

42. Hanessian, S.; Reddy, G.J.; Chahal, N. Total synthesis and stereochemical confirmation of manassantin A, B, and B1. *Org. Lett.* **2006**, *8*, 5477–5480. [CrossRef] [PubMed]

43. Cui, Y.; Wang, Q.; Shi, X.; Zhang, X.; Sheng, X.; Zhang, L. Simultaneous quantification of 14 bioactive constituents in Forsythia suspensa by liquid chromatography-electrospray ionisation-mass spectrometry. *Phytochem. Anal.* **2010**, *21*, 253–260. [CrossRef] [PubMed]

44. Pan, L.; Zhang, X.F.; Deng, Y.; Zhou, Y.; Wang, H.; Ding, L.S. Chemical constituents investigation of Daphne tangutica. *Fitoterapia* **2010**, *81*, 38–41. [CrossRef] [PubMed]

45. Lee, K.-H.; Xiao, Z. *Anticancer Agents from Natural Products*; CRC Press: Boca Raton, FL, USA, 2011; pp. 95–122.

46. Podwyssotzki, D.V. Pharmakologische studien uber *Podophvllum peltatum*. *Arch. Exp. Pathol. Pharmakol.* **1880**, *13*, 29–52. [CrossRef]

47. Hamel, E. Antimitotic natural products and their interactions with tubulin. *Med. Res. Rev.* **1996**, *16*, 207–231. [CrossRef]

48. Sullivan, B.J.; Wechsler, H.I. The cytological effects of podophyllin. *Science* **1947**, *105*, 433. [CrossRef] [PubMed]

49. Hartwell, J.L.; Shear, M.J. Chemotherapy of cancer: Classes of compounds under investigation, and active components of podophyllin. *Cancer Res.* **1947**, *7*, 716–717.

50. Canel, C.; Moraes, R.M.; Dayan, F.E.; Ferreira, D. Podophyllotoxin. *Phytochemistry* **2000**, *54*, 115–120. [CrossRef]

51. Stähelin, H.F.; von Wartburg, A. The chemical and biological route from podophyllotoxin glucoside to etoposide: Ninth Cain memorial Award lecture. *Cancer Res.* **1991**, *51*, 5–15. [PubMed]

52. Hainsworth, J.D.; Greco, F.A. Etoposide: Twenty years later. *Ann. Oncol.* **1995**, *6*, 325–341. [CrossRef] [PubMed]

53. Baurin, N.; Baker, R.; Richardson, C.; Chen, I.; Foloppe, N.; Potter, A.; Jordan, A.; Roughley, S.; Parratt, M.; Greaney, P.; et al. Drug-like Annotation and Duplicate Analysis of a 23-Supplier Chemical Database Totalling 2.7 Million Compounds. *J. Chem. Inf. Comput. Sci.* **2004**, *44*, 643–651. [CrossRef] [PubMed]

54. Lipinski, C.A.; Lombardo, F.; Dominy, B.W.; Feeney, P.J. Experimental and computational approaches to estimate solubility and permeability in drug discovery and development settings. *Adv. Drug Deliv. Rev.* **1997**, *23*, 3–25. [CrossRef]

55. Lipinski, C.A. Lead- and drug-like compounds: The rule-of-five revolution. *Drug Discov. Today Technol.* **2004**, *1*, 337–341. [CrossRef] [PubMed]

56. Oprea, T.I.; Davis, A.M.; Teague, S.J.; Leeson, P.D. Is there a difference between leads and drugs? A historical perspective. *J. Chem. Inf. Comput. Sci.* **2001**, *41*, 1308–1315. [CrossRef] [PubMed]

57. Oprea, T.I. Current trends in lead discovery: Are we looking for the appropriate properties? *Mol. Divers.* **2002**, *5*, 199–208. [CrossRef] [PubMed]

58. Lipinski, C.A. *Practice of Medicinal Chemistry*, 2nd ed.; Wermuth, C.G., Ed.; Academic Press: San Diego, CA, USA, 2003; pp. 341–349.

59. Bade, R.; Chan, H.-F.; Reynisson, J. Characteristics of known drug space. Natural products, their derivatives and synthetic drugs. *Eur. J. Med. Chem.* **2010**, *45*, 5646–5652. [CrossRef] [PubMed]

60. *ChemOffice Professional 16.0*; CambridgeSoft: Akron, OH, USA, 1986–2016.

61. Allinger, N.L. Conformational analysis. 130. MM2. A hydrocarbon force field utilizing V1 and V2 torsional terms. *J. Am. Chem. Soc.* **1977**, *99*, 8127–8134. [CrossRef]

62. *Schrödinger Release 2018-2: QikProp, Schrödinger*; LLC: New York, NY, USA, 2018.

63. Ioakimidis, L.; Thoukydidis, L.; Naeem, S.; Mirza, A.; Reynisson, J. Benchmarking the reliability of QikProp. Correlation between experimental and predicted values. *QSAR Comb. Sci.* **2008**, *27*, 445–456. [CrossRef]

64. R Core Team. *R: A Language and Environment for Statistical Computing*; R Foundation for Statistical Computing: Vienna, Austria, 2015.

65. RStudio Team. *RStudio: Integrated Development for R*; RStudio, Inc.: Boston, MA, USA, 2015.

66. Wickham, H. *ggplot2: Elegant Graphics for Data Analysis*; Springer-Verlag: New York, NY, USA, 2009.

67. Kassambara, A.; Mundt, F. *Factoextra: Extract and Visualize the Results of Multivariate Data Analyses*; R Package Version 1.0.3; 2016. Available online: https://cran.r-project.org/web/packages/factoextra/index.html (accessed on 2 July 2018).

68. Muchmore, S.W.; Edmunds, J.J.; Stewart, K.D.; Hajduk, P.J. Cheminformatic tools for medicinal chemists. *J. Med. Chem.* **2010**, *53*, 4830–4841. [CrossRef] [PubMed]

69. Zhu, F.; Logan, G.; Reynisson, J. Wine compounds as a source for HTS screening collections. A feasibility study. *Mol. Inf.* **2012**, *31*, 847–855. [CrossRef] [PubMed]

70. Matuszek, A.; Reynisson, J. Defining known drug space using DFT. *Mol. Inf.* **2016**, *35*, 46–53. [CrossRef] [PubMed]

71. Veber, D.F.; Johnson, S.R.; Cheng, H.-Y.; Smith, B.R.; Ward, K.W.; Kopple, K.D. Molecular properties that influence the oral bioavailability of drug candidates. *J. Med. Chem.* **2002**, *45*, 2615–2623. [CrossRef] [PubMed]

72. Lu, J.J.; Crimin, K.; Goodwin, J.T.; Crivori, P.; Orrenius, C.; Xing, L.; Tandler, P.J.; Vidmar, T.J.; Amore, B.M.; Wilson, A.G.E.; et al. Influence of molecular flexibility and polars area metrics on oral bioavailability in the rat. *J. Med. Chem.* **2004**, *47*, 6104–6107. [CrossRef] [PubMed]

73. Jain, N.; Yalkowsky, S.H. Estimation of the aqueous solubility I: Application to organic nonelectrolytes. *J. Pharm. Sci.* **2000**, *90*, 234–252. [CrossRef]

74. Hann, M.M.; Keserì, G.M. Finding the sweet spot: The role of nature and nurture in medicinal chemistry. *Nat. Rev. Drug Dis.* **2012**, *11*, 355–365. [CrossRef] [PubMed]

75. Olsen, E.A.; Abernethy, M.L.; Kulp-Shorten, C.; Callen, J.P.; Glaaer, S.D.; Huntley, A.; McCray, M.; Monroe, A.B.; Tschen, E.; Wolf, F.E., Jr. A double-blind, vehicle-controlled study evaluating masoprocol cream in the treatment of actinic keratoses on the head and neck. *J. Am. Acad. Dermatol.* **1991**, *24*, 738–743. [CrossRef]

76. Luo, J.; Chuang, T.; Cheung, J.; Quan, J.; Tsai, J.; Sullivan, C.; Hector, R.F.; Reed, M.J.; Meszaros, K.; King, S.R.; et al. Masoprocol (nordihydroguaiaretic acid): A new antihyperglycemic agent isolated from the creosote bush (*Larrea tridentata*). *Eur. J. Pharmacol.* **1998**, *346*, 77–79. [CrossRef]

77. Wu, X.; Tong, B.; Yan, Y.; Luo, J.; Yuan, X.; Wei, Z.; Yue, M.; Xia, Y.; Dai, Y. Arctigenin functions as a selective agonist of estrogen receptor β to restrict mTORC1 activation and consequent Th17 differentiation. *Oncotarget* **2016**, *7*, 83893–83906. [CrossRef] [PubMed]

78. Xu, P.; Huang, M.-W.; Xiao, C.-X.; Long, F.; Wang, Y.; Liu, S.-Y.; Jia, W.-W.; Wu, W.-J.; Yang, D.; Hu, J.-F.; et al. Matairesinol Suppresses Neuroinflammation and Migration Associated with Src and ERK1/2-NF-κB Pathway in Activating BV2 Microglia. *Neurochem. Res.* **2017**, *42*, 2850–2860. [CrossRef] [PubMed]

79. Majdalawieh, A.F.; Massri, M.; Nasrallah, G.K. A comprehensive review on the anti-cancer properties and mechanisms of action of sesamin, a lignan in sesame seeds (Sesamum indicum). *Eur. J. Pharmacol.* **2017**, *815*, 512–521. [CrossRef] [PubMed]

80. Song, F.; Zeng, K.; Liao, L.; Yu, Q.; Tu, P.; Wang, X. Schizandrin A inhibits microglia-mediated neuroninflammation through inhibiting TRAF6-NF-κB and Jak2-Stat3 signaling pathways. *PLoS ONE* **2016**, *11*, e0149991. [CrossRef] [PubMed]

81. Begum, S.A.; Sahai, M.; Ray, A.B. Non-conventional lignans: Coumarinolignans, flavonolignans, and stilbenolignans. *Fortschr. Chem. Org. Naturst.* **2010**, *93*, 1–70. [PubMed]

82. Biedermann, D.; Vavrikova, E.; Cvak, L.; Kren, V. Chemistry of silybin. *Nat. Prod. Rep.* **2014**, *31*, 1138–1157. [CrossRef] [PubMed]

83. Morishima, C.; Shuhart, M.C.; Wang, C.C.; Paschal, D.M.; Apodaca, M.C.; Liu, Y.; Sloan, D.D.; Graf, T.N.; Oberlies, N.H.; Lee, D.Y.; et al. Silymarin inhibits in vitro T-cell proliferation and cytokine production in hepatitis C virus infection. *Gastroenterology* **2010**, *138*, 671–681. [CrossRef] [PubMed]

84. Polyak, S.J.; Ferenci, P.; Pawlotsky, J.-M. Hepatoprotective and antiviral functions of silymarin components in hepatitis C virus infection. *Hepatology* **2013**, *57*, 1262–1271. [CrossRef] [PubMed]

85. Cundy, K.C.; Branch, R.; Chernov-Rogan, T.; Dias, T.; Estrada, T.; Hold, K.; Koller, K.; Liu, X.; Mann, A.; Panuwat, M.; et al. XP13512 [(±)-1-([(α-Isobutanoyloxyethoxy)carbonyl] aminomethyl)-1-cyclohexane Acetic Acid], A novel gabapentin prodrug: I. Design, synthesis, enzymatic conversion to gabapentin, and transport by intestinal solute transporters. *J. Pharm. Exp. Therap.* **2004**, *311*, 315–323. [CrossRef] [PubMed]

Sample Availability: Not available.

molecules
MDPI

Review

Naturally Lignan-Rich Foods: A Dietary Tool for Health Promotion?

Carmen Rodríguez-García [1,2], **Cristina Sánchez-Quesada** [1,2,3], **Estefanía Toledo** [4,5,6],
Miguel Delgado-Rodríguez [1,2,7] and **José J. Gaforio** [1,2,3,7,*]

[1] Center for Advanced Studies in Olive Grove and Olive Oils, University of Jaen, Campus las Lagunillas s/n, 23071 Jaén, Spain; crgarcia@ujaen.es (C.R.-G.); csquesad@ujaen.es (C.S.-Q.); mdelgado@ujaen.es (M.D.-R.)
[2] Department of Health Sciences, Faculty of Experimental Sciences, University of Jaén, 23071 Jaén, Spain
[3] Agri-food Campus of International Excellence (ceiA3), 14071 Córdoba, Spain
[4] Department of Preventive Medicine and Public Health, University of Navarra, 31008 Pamplona, Spain; etoledo@unav.es
[5] CIBER Fisiopatología de la Obesidad y Nutrición (CIBERObn), Instituto de Salud Carlos III, 28029 Madrid, Spain
[6] IdiSNA, Navarra Institute for Health Research, 31008 Pamplona, Spain
[7] CIBER Epidemiología y Salud Pública (CIBER-ESP), Instituto de Salud Carlos III, 28029 Madrid, Spain
* Correspondence: jgaforio@ujaen.es; Tel.: +34-953-212-002

Academic Editor: David Barker
Received: 28 January 2019; Accepted: 4 March 2019; Published: 6 March 2019

Abstract: Dietary guidelines universally advise adherence to plant-based diets. Plant-based foods confer considerable health benefits, partly attributable to their abundant micronutrient (e.g., polyphenol) content. Interest in polyphenols is largely focused on the contribution of their antioxidant activity to the prevention of various disorders, including cardiovascular disease and cancer. Polyphenols are classified into groups, such as stilbenes, flavonoids, phenolic acids, lignans and others. Lignans, which possess a steroid-like chemical structure and are defined as phytoestrogens, are of particular interest to researchers. Traditionally, health benefits attributed to lignans have included a lowered risk of heart disease, menopausal symptoms, osteoporosis and breast cancer. However, the intake of naturally lignan-rich foods varies with the type of diet. Consequently, based on the latest humans' findings and gathered information on lignan-rich foods collected from Phenol Explorer database this review focuses on the potential health benefits attributable to the consumption of different diets containing naturally lignan-rich foods. Current evidence highlight the bioactive properties of lignans as human health-promoting molecules. Thus, dietary intake of lignan-rich foods could be a useful way to bolster the prevention of chronic illness, such as certain types of cancers and cardiovascular disease.

Keywords: lignans; diet; antioxidants; health promotion; chronic diseases

1. Introduction

Polyphenol-rich diets are suggested to possess health benefits. Polyphenols are micronutrients found in plants, and include flavonoids, stilbenes, phenolic acids, lignans and others [1]. They are secondary plant metabolites implicated in protection against pathogens and ultraviolet radiation [2]. Given their diverse chemical structures, different polyphenol classes likely possess differing health benefits [3]. It is therefore important to elucidate the specific potential benefits of each polyphenolic compound. Significant interest has been elicited by lignans, due to their steroid-analogous chemical structure. Accordingly, they are considered to be phytoestrogens. Lignans are bioactive compounds exhibiting various biological properties, including anti-inflammatory, antioxidant and antitumor activities [4]. Additionally, some epidemiological studies have proposed that lignans decrease the risk of cardiovascular disease, but their effects on other chronic diseases (e.g., breast cancer) remain controversial [5].

Lignans are found in relatively low concentrations in various seeds, grains, fruits and vegetables, and in higher concentrations in sesame and flax seeds [6]. Therefore, the level of lignan ingestion—and, thus, lignan bioavailability, depends on the type of diet consumed [7,8] and can be highly variable. The present review attempts to describe the potential beneficial effects of lignan intake on human chronic disease, depending on the dietary source.

2. Biosynthesis, Classification and Presence of Lignans in Foods

Lignans are a type of secondary plant metabolite exhibiting diverse structures [9]. Plants derive a complex array of secondary metabolites from only a handful of relatively simple propenyl phenols [10]. Biosynthesis of lignans is characterized by a remarkable increase in molecular complexity [10].

Lignans share common biosynthetic pathways, consist of two propyl-benzene units coupled by a β,β'-bond [11], and thus belong to the group of diphenolic compounds [12].

Lignans may be organized into eight structural subgroups (according to the manner in which oxygen is incorporated and the pattern of cyclization): Dibenzylbutyrolactol, dibenzocyclooctadiene, dibenzylbutyrolactone, dibenzylbutane, arylnaphthalene, aryltetralin, furan and furofuran (Figure 1). Each subgroup can be further subdivided according to lignan molecule oxidation level and identities of non-propyl aromatic rings present on side chains [13,14].

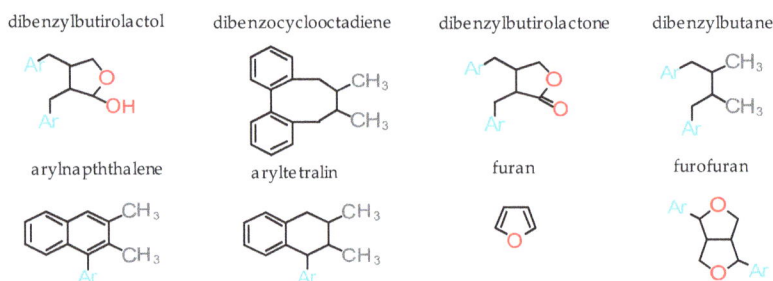

Figure 1. Structural subgroups of lignans (Ar=Aryl).

Of the eight lignan subclasses, synthesis of furofurans—which exhibit a 2,6-diaryl-3,7-dioxabicyclooctane skeleton—is initiated by the enantioselective dimerization of two coniferyl alcohol units derived from the shikimate biosynthetic pathway (Figure 2) [14]. To date, 53 species of furofuran lignans have been reported in 41 genera of 27 plant families, including Thymelaeaceae, Styracaceae, Scrophulariaceae, Saururaceae, Rutaceae, Rhizophoraceae, Piperaceae, Pedaliaceae, Orobanchaceae, Myristicaceae, Magnoliaceae, Lauraceae, Lamiaceae, Geraniaceae, Dioscoreaceae, Cyperaceae, Cupressaceae, Compositae, Combretaceae, Cactaceae, Aristolochiaceae, Arecaceae, Araliaceae, Aquifoliaceae, Apocynaceae, Acoraceae and Acanthaceae. Furofuran lignans are present in the bark, bulbs, leaves, seeds, stems and roots of these plants [14].

However, depending on the enzyme that catalyzes modification of the precursor metabolite, a variety of lignans can be synthesized (Figure 2). The major lignans—which possess numerous pharmacological properties—are artigenin, enterodiol, enterolactone, sesamin, syringaresinol, medioresinol, (−)-matairesinol, (−)-secoisolariciresinol, (+)-lariciresinol and (+)-pinoresinol, among others [15].

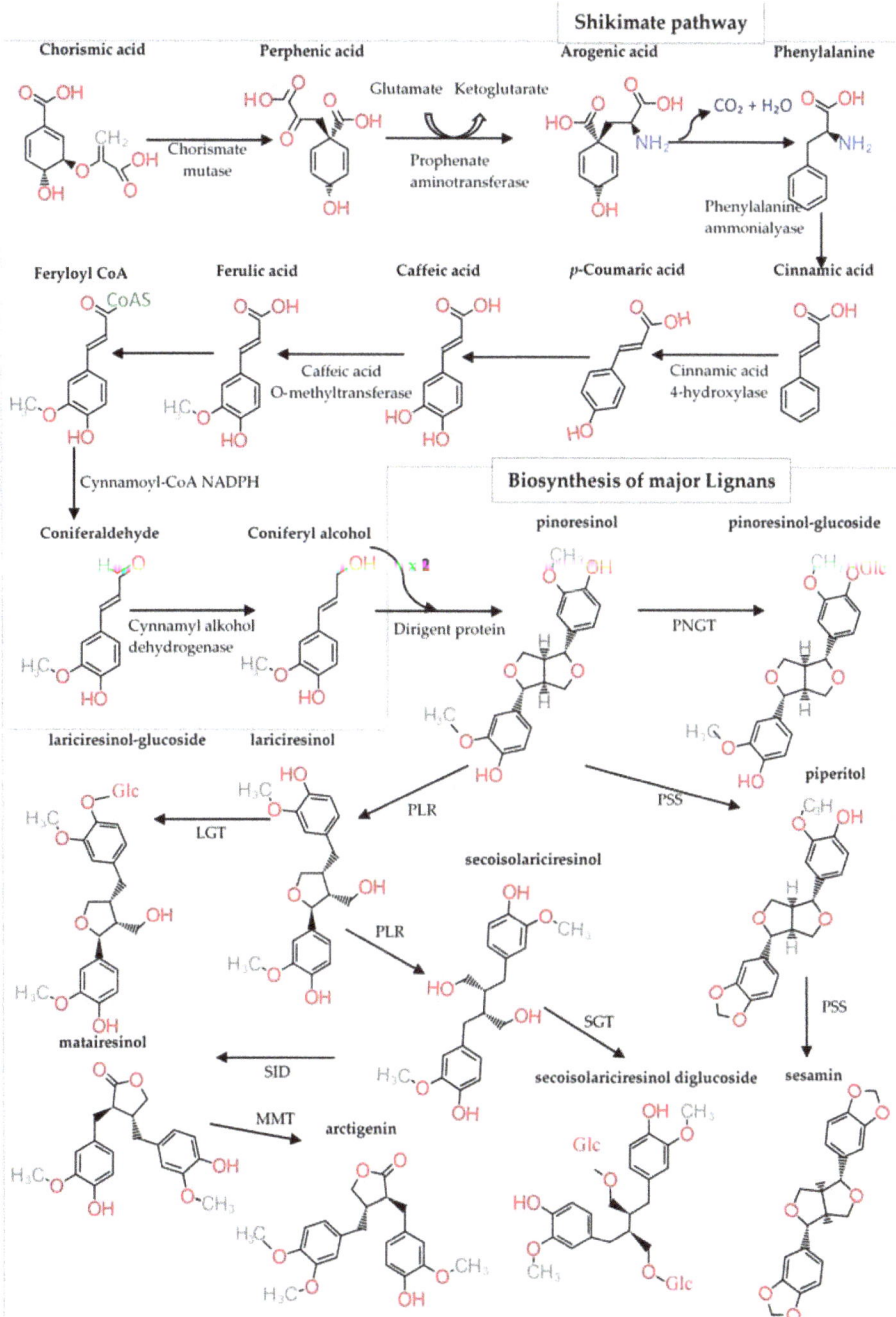

Figure 2. Biosynthetic pathway of lignans. NGT (pinoresinol glucosyltransferase), PSS (piperitol/ sesamin synthase), PLR (pinoresinol/lariciresinol reductase), LGT (lariciresinol glycosyltransferase), SGT (secoisolariciresinol glycosyltransferase), SID (matairesinol O-methyltransferase), MMT (matairesinol O-methyltransferase), Glc (Glucoside).

Currently, there is a growing interest in the presence of lignans in foodstuffs, given the potentially beneficial bioactive properties of the former (anti-estrogenic, antioxidant and anti-carcinogenic activities) [16]. The chief sources of dietary lignans are various vegetables and fruits, legumes, whole grain cereals and oilseeds [16,17]. Among edible plant components, the most concentrated lignan sources are sesame and flax seeds (Tables 1 and 2) [6]. Specifically, flax seeds contain approximately 294.21 mg/100 g lignan, at present the maximal known content of any foodstuff. Sesame seeds exhibit the second-highest lignan concentration, with sesaminol as the major constituent, at 538.08 mg/100 g [6]. Flaxseed and cashew nuts are also relatively rich in lignans (containing 257.6 and 56.33 mg/100 g, respectively) [6].

Table 1. Lignan content of sesame seed (mg/100g food). Data collected from phenol explorer [18].

Seeds	HMA	HSE	OXO	ARC	CYC	CON	DIM
Sesame seed	7.2	0.01	0.7	0.01	1.77	0.75	0.39
	ISO	LAR	LAS	MAT	MED	NOR	SEC
	1.61	10.37	0.08	29.79	4.15	0.08	0.1
	SECS	SES	SEI	SEN	SYR	TOD	Total
	0.01	538.08	102.86	133.94	0.2	2.47	834.57

Lignans: 7-Hydroxymatairesinol (HMA), 7-Hydroxysecoisolariciresinol (HSE), 7-Oxomatairesinol (OXO), Arctigenin (ARC), Conidendrin (CON), Cyclolariciresinol (CYC), Dimethylmatairesinol (DIM), Isohydroxymatairesinol (IHM), Isolariciresinol (ISO), Lariciresinol (LAR), Lariciresinol-sesquilignan (LAS), Matairesinol (MAT), Medioresinol (MED), Nortrachelogenin (NOR), Secoisolariciresinol (SEC), Secoisolariciresinol-sesquilignan (SECS), Sesamin (SES), Sesaminol (SEI), Sesamolin (SEN), Syringaresinol (SYR), Todolactol A (TOD).

Table 2. Lignan content of seeds (mg/100g food) [18].

	LAR	MAT	MED	SEC	SYR	Total
		Other Seeds				
Flaxseed	11.46	6.68	-	257.6	-	257.6
Sunflower seed	0.67	0.67	-	0.18	-	1.52
		Nuts				
Almond	0.03	3×10^{-4}	-	0.07	-	0.10
Brazil nut	-	0.01	-	0.77	-	0.78
Cashew nut	49.6	2.5×10^{-3}	-	6.73	-	56.33
Chesnut	7.8×10^{-3}	8.42×10^{-3}	-	0.2	-	0.21
Hazelnut	0.01	3.3×10^{-3}	-	0.05	-	0.06
Peanut	4.1	2.5×10^{-3}	-	2.7	-	6.8
Pecan nut	8.4×10^{-3}	3.15×10^{-3}	-	0.01	-	0.02
Pistachio	0.12	1×10^{-4}	-	0.04	-	0.16
Walnut	7.2×10^{-3}	3.8×10^{-3}	-	0.12	-	0.13
		Pulses-Beans				
Common bean white	0.12	1×10^{-3}	-	0.08	8×10^{-3}	0.2
Broad bean seed whole	-	8.9×10^{-4}	-	0.09	-	0.09
Mung bean	-	-	-	0.18	-	0.18
		Soy and soy products				
Soy paste, miso	0.02	3.6×10^{-3}	-	0.01	-	0.03
Soy flour	-	7.5×10^{-3}	-	0.3	-	0.3
Soy tempe	0.01	5×10^{-4}	-	0.01	-	0.02
Soy tofu	0.04	7.27×10^{-5}	8.5×10^{-3}	9.91×10^{-3}	0.04	0.09
Soy yogurt	0.01	3×10^{-3}	-	0.02	-	0.03
Soyben edamame	0.07	-	0.02	0.07	0.2	0.3
Soybean sprout	0.03	5×10^{-4}	0.01	0.03	0.05	0.12

Regarding cereal grains (Table 3), lignans are largely concentrated in their outer layers [19,20]. In cereal grains, the highest lignan concentration is found in the fiber-rich outer layers (seed coat and pericarp), as well as the aleurone layer, whereas the lowest concentration is found in the inner endosperm [21,22].

Table 3. Lignan content of cereals (mg/100g food) [18].

	LAR	MAT	MED	SEC	SYR	Total
		Cereal products				
Bread (whole grain flour)	0.05	3.1×10^{-4}	-	8.68×10^{-3}	-	0.05
Bread (refined flour)	0.01	1.23×10^{-3}	-	7.19×10^{-3}	0.04	0.05
Bread, rye, whole grain flour	0.01	0.02	-	0.14	-	0.17
Breakfast cereals, bran	0.01	4.87×10^{-3}	-	0.03	-	0.04
Breakfast cereals, corn	-	1.67×10^{-3}	-	5.5×10^{-3}	-	0.007
Breakfast cereals, muesli	0.14	5.6×10^{-3}	-	0.08	-	0.22
Breakfast cereal, oat	-	0.06	-	0.02	-	0.08
Pasta	-	1.85×10^{-3}	-	2.3×10^{-3}	-	0.004
Pasta Whole Grain	-	1.5×10^{-3}	-	5×10^{-3}	-	0.006
		Cereals				
Barley, whole grain flour	0.08	3×10^{-3}	0.01	0.03	0.16	0.28
Buckwheat, whole grain flour	0.36	1×10^{-3}	0.03	0.13	0.24	0.76
Common wheat, germ	-	9×10^{-3}	-	0.02	-	0.02
Common wheat, refined flour	0.18	2.14×10^{-4}	-	0.02	-	0.2
Common wheat, whole grain flour	0.1	9×10^{-4}	0.03	0.01	0.37	0.52
Hard wheat, semolin	-	-	-	2×10^{-3}	-	0.002
Maize, whole grain	0.12	6.55×10^{-5}	-	0.14	0.07	0.33
Oat, whole grain flour	0.18	0.07	0.04	0.01	0.35	0.65
Rye, whole grain flour	0.32	0.01	0.14	0.02	0.97	1.46

Ordering species by lignan content produces the following list: Dhurra < brown rice < red rice < quinoa < millet < corn < amaranth < barley < buckwheat < wild rice < Japanese rice < spelt < oat < triticale < wheat < rye [6]. Regarding vegetables (Table 4), the brassica family may contain between 185 and 2.321 mg /100 g of lignan, mainly pinoresinol. Peppers, French beans, carrots and courgettes also exhibit a relatively high lignan content, ranging from 0.113 to 0.273 mg/100 g. Other foods, such as spinach, white potatoes and mushrooms—contain below 0.1 mg/100 g of lignan. Fruits exhibit a lower lignan content than seeds or vegetables (Tables 5 and 6), ranging from 11.57 mg/100 g for apricots to 0 mg/100 g for banana, with green grapes and kiwi fruit falling somewhere between these extremes [6].

Table 4. Lignan contents of vegetables (mg/100g food) [18].

	LAR	MAT	MED	SEC	SYR	Total
		Cabbages				
Broccoli	97.2	2.44×10^{-5}	-	1.31	-	98.51
Brussel sprouts	49.3	4×10^{-5}	-	1.06	-	50.36
Cauliflower	9.31	2.4×10^{-5}	0.02	0.13	0.02	9.48
Collards	0.06	4×10^{-4}	-	5.9×10^{-3}	-	0.06
Green cabbage	0.03	3.5×10^{-5}	-	9.2×10^{-3}	-	0.03
Red cabbage	17.8	4.44×10^{-5}	-	0.3	-	18.1
White cabbage	21.2	-	-	0.31	-	21.51
Kale	59.9	1.2	-	1.9	-	63
Sauerkraut	11.6	-	-	6.7	-	18.3

Table 4. *Cont.*

	LAR	MAT	MED	SEC	SYR	Total
			Fruit vegetales			
Avocado	0.03	7.67×10^{-3}	0.24	0.02	0.44	0.73
Eggplant purple	0.05	-	7×10^{-3}	7.79×10^{-3}	6×10^{-3}	0.07
Black olive	0.03	5.62×10^{-3}	-	5.75×10^{-3}	-	0.04
Green olive	3.9×10^{-3}	3.34×10^{-3}	-	0.02	-	0.02
Green sweet pepper	12.32	-	1×10^{-3}	0.22	4×10^{-3}	12.54
Red sweet pepper	7.97	-	-	0.24	-	8.21
Yellow sweet pepper	0.07	-	-	5.5×10^{-3}	-	0.07
Tomato (Cherry)	0.03	-	3×10^{-3}	0.01	4.5×10^{-3}	0.04
Tomato (Whole)	2.1	8.33×10^{-6}	3.5×10^{-3}	0.05	4.5×10^{-3}	2.15
			Gourds			
Cucumber	3.55	-	-	0.25	-	3.8
Pumpkin	0.01	2.5×10^{-5}	-	0.1	-	0.11
Squash	-	-	-	9×10^{-3}	-	0.009
Zucchini	6.4	-	-	0.62	-	7.02
			Leaf vegetables			
Arugula	-	2×10^{-4}	-	0.1	-	0.1
Chicory (green)	0.6	1.24×10^{-4}	-	0.57	-	1.17
Lettuce (green)	0.3	2.24×10^{-4}	-	0.18	-	0.48
Spinach	0.06	2.37×10^{-5}	-	4.85×10^{-3}	-	0.06
Broad bean pod	-	-	-	0.02	-	0.02
			Pod vegetables			
Green bean	22	-	-	0.67	-	22.67
			Pulse vegetables			
Fresh pea	0.05	-	3.5×10^{-3}	7.56×10^{-4}	-	0.0542
			Root vegetables			
Carrot	4.5	3.89×10^{-3}	-	3.16	-	7.66
Celeriac	-	3×10^{-5}	-	0.02	-	0.02
Parsnip	-	0.02	-	0.03	-	0.05
Radish	0.01	1.25×10^{-4}	5.5×10^{-3}	6.57×10^{-3}	0.02	0.04
Swede	-	7.43×10^{-5}	-	4.93×10^{-3}	-	0.005
Turnip root	0.1	-	4×10^{-3}	9.83×10^{-3}	0.03	0.14
			Shoot vegetables			
Asparagus	0.07	3.97×10^{-3}	4×10^{-3}	0.25	0.05	0.37
Fennel	-	0.01	-	0.05	-	0.06
			Stalks vegetables			
Celery stalks	-	-	-	5.99×10^{-3}	-	0.005
			Tubers			
Potato	2.8	7.69×10^{-4}	-	0.09	-	2.89
Sweet potato	0.07	0.1	-	0.12	-	0.29

Table 5. Lignan contents of fruits berries (mg/100g food) [18].

	HMA	OXO	CON	CYC	LAR	LAS
			Fruit Berries			
Bilberry	-	-	-	6.24×10^{-3}	0.04	0.09
Blackberry	-	-	-	7.96×10^{-3}	0.15	0.15
Blackcurrant	-	-	-	0.01	7.3×10^{-3}	0.01
Cloudberry	-	-	-	-	0.65	0.25
Black grape	-	-	-	-	5.2	-
Green grape	-	-	-	-	1.88	-
Lingonberry	-	-	1.04×10^{-3}	0.03	0.03	0.01
Strawberry	8.55×10^{-4}	4.59×10^{-4}	9.45×10^{-3}	0.01	5.87	0.1

	MAT	MED	SEC	SECS	SYR	Total
Bilberry	-	0.08	0.06	0.01	0.12	0.4
Blackberry	9.07×10^{-4}	0.05	0.1	0.13	0.19	0.77
Blackcurrant	1.47×10^{-3}	0.01	0.09	0.03	-	0.15
Cloudberry	-	0.48	0.05	0.01	0.41	1.85
Black grape	0.11	-	0.09	-	-	5.4
Green grape	0.09	-	0.28	-	-	2.25
Lingonberry	-	0.23	0.37	0.02	0.14	0.83
Strawberry	1.58×10^{-5}	0.03	0.14	0.01	0.03	6.2

Table 6. Lignan contents of fruits (mg/100g food) [18].

	LAR	MAT	MED	SEC	SYR	Total
			Fruits Citrus			
Grapefruit	7.13	0.05	-	0.26	-	7.44
Lemon	-	-	-	0.02	-	0.02
Orange	2.4	0.05	9.5×10^{-3}	0.14	0.12	2.71
Tangerine	5.7	0.02	-	0.08	-	5.8
			Fruits Drupes			
Apricot	10.5	3.11×10^{-5}	-	1.07	-	11.57
Nectarine	4.1	-	-	0.61	-	4.71
Peach	6	1.71×10^{-4}	-	0.83	-	6.83
Plum	0.31	2.22×10^{-4}	1×10^{-3}	0.09	-	0.4
			Fruits-Gourds			
Cantaloupe	1.8×10^{-3}	-	-	4.7×10^{-3}	-	0.006
Melon	4.4	1.05×10^{-5}	-	0.09	-	4.49
Watermelon	0.04	-	1×10^{-3}	0.02	0.02	0.08
			Fruits-Pomes			
Apple	0.1	2.71×10^{-5}	-	1.79×10^{-3}	-	0.1
Pear	15.5	4.3×10^{-5}	-	0.06	-	15.56
			Fruits-Tropical			
Banana	2.2×10^{-3}	5.45×10^{-5}	-	7.73×10^{-5}	0.01	0.01
Kiwi	1.03	1.93×10^{-3}	4.5×10^{-3}	3.13	4×10^{-3}	4.17
Mango	-	1.06×10^{-3}	-	0.01	-	0.01
Passion fruit	-	-	-	0.02	-	0.02
Papaya	-	2×10^{-3}	-	-	-	0.002
Persimmon	-	-	-	4×10^{-3}	-	0.004
Pineapple	0.2	0.16	2×10^{-3}	0.21	0.09	0.66
Pomegranate	-	9×10^{-3}	-	0.29	-	0.29

The highest lignan content is observed in non-alcoholic beverages, such as tea (0.0392–0.0771 mg/100 g), which also contains other polyphenols (Table 7). Coffee is another important source of lignans, although concentration varies by type of coffee, ranging from 0.0187 to 0.0313 mg/100 g. Regarding alcoholic beverages, red wine contains an average of 0.080 mg/100 mL, whereas white wine contains only approximately 0.022 mg/100 g [23].

Table 7. Lignan content of beverages (mg/100g drink and mg/100 mL wine) [18].

	ISO	LAR	MAT	SEC	SYR	Total
			Alcoholic Beverages			
Red Wine	0.07	7.56×10^{-3}	5.51×10^{-3}	0.04	3.43×10^{-3}	0.12
White Wine	0.03	6.65×10^{-3}	2.68×10^{-3}	7.45×10^{-3}	1.45×10^{-3}	0.04
Dark Beer	-	-	-	0.04	-	0.04
Beer	-	-	-	0.03	-	0.03
Cider	-	-	-	0.04	-	0.04
Scotch whisky	-	-	-	4×10^{-3}	-	0.004
Sherry	-	-	-	0.02	-	0.02
			Non-alcoholic Beverages			
Cocoa	-	-	-	0.03	-	0.03
Coffee	-	9×10^{-4}	4×10^{-4}	8.67×10^{-3}	-	0.009
Decaffeinated Coffe	-	1.1×10^{-3}	4.25×10^{-4}	8.35×10^{-3}	-	0.009
Roman camomile	-	-	5×10^{-4}	1×10^{-3}	-	0.001
Lemon juice	-	-	-	2×10^{-3}	-	0.002
Orange juice	-	2×10^{-4}	-	8×10^{-3}	-	0.008
Soy milk	-	6.17×10^{-3}	5×10^{-5}	2.25×10^{-3}	-	0.008
Black Tea	-	2×10^{-4}	2.65×10^{-3}	0.03	-	0.03
Green Tea	-	1×10^{-4}	3.38×10^{-3}	0.03	-	0.03
Oolong Tea	-	-	1.8×10^{-3}	0.02	-	0.02

Furthermore, the chief source of dietary fat in Mediterranean countries—extra virgin olive oil (EVOO)—has garnered much interest regarding its beneficial properties, largely attributable to its polyphenol profile (Table 8). Lignans are the second most abundant polyphenolic class present in EVOO; of these, the most abundant across different EVOO types are pinoresinol (1.17–4.12 mg/ 100 g) and 1-acetoxypinoresinol (0.27–6.69 mg/ 100 g) [7,24,25].

Table 8. Lignan content of oils (mg/100 g food) [18].

Fruit oils	ACE	LAR	MAT	PIN	SEC	Total
Extra virgin Olive Oil	0.66	3.43×10^{-3}	7.5×10^{-5}	0.42	2.5×10^{-4}	1.08
			Nut oils			
Peanut, butter	-	8.8×10^{-3}	7.52×10^{-3}	-	0.05	0.06

Other seed oils	EPI	EPL	SES	SEI	SEO	SEN	SEL	Total
Sesame seed oil	192.6	51.97	420.99	305.43	24.92	243.13	55.71	1294.75
Sesame seed black oil	-	-	644.5	226.92	21.55	287.33	43	1223.3

1-Acetoxypinoresinol (ACE), Episesamin (EPI), Episesaminol (EPL), Pinoresinol (PIN), Sesamol (SEO), Sesamolinol (SEL).

Thus, given the presence of lignan in many common foodstuffs and beverages, its intake occurs frequently, on a near-daily basis. For example, in a Dutch population, the major dietary sources of lignan were fruits (7%), bread (9%), seeds and nuts (14%), vegetables (24%), and beverages (37%) [6]. Similarly, in a cohort of French women, the major dietary sources of lignan were vegetables and fruits (0.2% from legumes, 0.6% from potatoes, 30% from vegetables, and 35% from fruits), followed by alcoholic beverages (5%), coffee (5%), cereals (7%) and tea (11%) [6,26,27].

3. Bioavailability

Only a handful of studies exist regarding post-consumption lignan bioavailability, including only very limited human pharmacokinetic studies. After ingestion, plant lignans are metabolized by intestinal bacteria, undergoing transformation to mammalian lignans (enterolactones and enterodiols (Figure 3)) prior to absorption [16,28]. This apparently considerably decreases the risk of diverse types of cancer, particularly of the colon, prostate and breast [16,29].

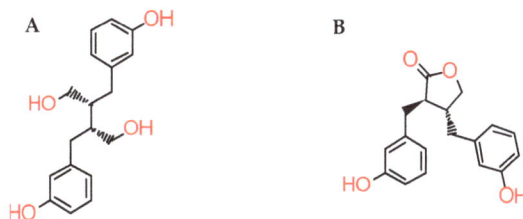

Figure 3. Chemical structure of enterodiol (**A**) and enterolactone (**B**).

Many studies demonstrate a positive correlation between plant lignan intake and plasma enterolignan levels [30]. After lignan ingestion, enterolactone and enterodiol are the first lignans to become detectable in human biological fluids [28]. The half-lives of these compounds in plasma are approximately 13 and 5 h, respectively [31], and they remain detectable even up to 8–10 h after plant lignan consumption [32]. Furthermore, their intestinal metabolism into mammalian forms appears indispensable for colonic absorption, and the colonic barrier is capable of conjugating enterolignans [28,33].

The concentration of enterodiol and enterolactone in biological fluids varies significantly by geographic region [28]. A study examining mammalian lignan pharmacokinetics in both men and women after lignan solution intake found that enterodiol and enterolactone, respectively, exhibit absorption half-lives of 3.4 and 8.4 h, reach maximum plasma concentrations of 65 and 42 mmol/L [28], exhibit elimination half-lives of 4.6 and 15.1 h, and exhibit maximum retention times of 23.9 and 43.2 h [28,34]. Thus, while enterolactone is more rapidly absorbed than enterodiol, the former attains a lower maximum plasma concentration [28].

During lignan metabolism, the initial (cytochrome P450-mediated) step involves conjugation to glucuronic acid and sulfate, followed by enterohepatic recirculation [35]. Chaojie et al. (2013) that glucuronidation of flax seed lignans significantly involves liver and intestinal microsomes [36]. Some studies demonstrate that flax seed-derived lignan metabolites distribute mainly to the intestine (largely to the caecum), kidneys, uterus, prostate and liver [37]. Of these locations, the highest concentration of lignan metabolites is observed in the liver [37].

Human breast cyst, prostatic, and seminal fluid (as well as prostate tissue) lignan concentration has been determined [38,39]. As in circulation, the common mammary form of lignan is enterolignan, while urinary forms are essentially monoglucuronides [28]. Furthermore, inter-individual variations in gut microbiota and hepatic enzymes may modulate mammalian lignan metabolism and bioactivity [33].

Moreover, lignan bioavailability also depends on diet. For example, diets rich in flax seed increase production of gut microbiota-derived enterolignans in a murine model, and lead to high tissue and plasma concentrations of sulfate and glucuronide conjugates (the major flax-derived lignan metabolites) [8,40].

Other studies have demonstrated that plant lignans, such as sesamin are quickly absorbed, apparently from the small intestine and become detectable in systemic circulation within a few hours after ingestion [22,41]. For example, lignans have been observed in porcine plasma 3 h after cereal intake [42]. On the one hand, it has been empirically demonstrated that plant lignans are rapidly absorbed from the small intestine after intake of a diet rich in cereals [22]. On the other hand, various factors—e.g., the use of oral antibiotics and inter-individual variations in gut microflora, as well as diet—impact lignan pharmacokinetics [43]. For example, seed maturation state can alter oral lignan bioavailability [44].

4. Lignan Content of Various Regional Diets

Dietary lignan consumption varies mainly with geographic location, but diet patterns are also subject to cultural and ethnic group influences.

4.1. Mediterranean Diet

The traditional Mediterranean diet is predominantly plant-based, characterized by a low intake of sweets; low meat products and red meat; a moderate intake of fish, poultry and fermented dairy products; a high intake of unprocessed cereals, legumes, nuts, fruits and vegetables [45]; the use of EVOO as the principal source of added fat; and moderate consumption of red wine [45]. Health benefits of this diet are essentially attributable to increased consumption of fiber and bioactive compounds (including antioxidants and functional fatty acids and lipids), as well as to a low intake of saturated fats [45,46].

Lignan sources in the diet of a Mediterranean population included garlic, onions, vegetables, including leafy greens, grains and seasonal fruits, including citrus, with each accounting for diverse proportions (11–70%) and subtypes of total polyphenols consumed [47].

Indeed, many typical Mediterranean diet foods (e.g., cereals) exhibit a high concentration of both lignans and other phenolic compounds [48].

Recently, the role of whole grain cereal intake in chronic disease prevention has been evaluated. Numerous studies propose a connection between lignan intake—as part of a wholegrain-based diet—and decreased incidence of chronic diseases, including cardiovascular disease, cancer and diabetes [5].

Thus, the major dietary lignan sources in the Mediterranean diet are vegetables and fruits, legumes, wholegrain cereals and oilseeds [3]. Additionally, another component of the Mediterranean diet, the chestnut, represents an excellent source of calcium, antioxidants and phenolic compounds [16,49]. Furthermore, EVOO consumption is an essential part of the Mediterranean diet. In fact, regular EVOO consumption is associated with a lower incidence of atherosclerosis, cardiovascular disease and some types of cancer [50–52]. This effect may be attributable to the high concentrations of (+)-1-acetoxypinoresinol and (+)-pinoresinol present in EVOO [53,54].

4.2. Northern Hemisphere Diet

This diet is observed in Northern and Nordic European regions, and is characterized by a high level of consumption of seaweed, shellfish, fatty fish (such as mackerel, herring and salmon), lean meats, rapeseed oil, legumes, nuts (such as almonds), vegetables, fruits (such as berries), whole grains (such as oats), low-fat dairy, and restricted salt and sugar intake [55,56]. In Nordic countries, the major dietary sources of plant lignans are vegetables, fruits and wholegrain cereals [57].

Among the many frequently-consumed plant species exhibiting a high lignan content, some species occur mainly in the Northern Hemisphere (e.g., *Cirsium spp.* of the family Asteraceae) [58]. The vegetative structures of these plants contain triterpenes, polyacetylenes, phenolic acids, flavonoids and alkaloids [58]. The most recent phytochemical studies of European *Cirsium spp.* demonstrate that their seeds are rich sources of neolignans and lignans [58,59].

4.3. Indian Diet

Various categories of food products make up a significant portion of the typical Indian diet, including fish, grapes, chocolate, oils, coffee, tea, biscuits and bread [60].

The fruit of *Morinda citrifolia* (Indian mulberry) has been extensively traditionally utilized in the treatment of cancer, diabetes, high blood pressure, diarrhea, headache and inflammation, largely due to its high lignan content [61,62].

Sesame is a typical component of the Indian diet, and both sesame seeds and oil are rich in lignans [63]. Sesame oil is recognized for both its notable resistance to oxidation and its nutritional value [64–66]. Despite lignans comprising only a small proportion (0.5 to 1.0%) of total sesame seed mass, the main sesame lignans—such as (+)-sesaminol, (+)-sesamolin and (+)-sesamin glucosides—have garnered attention for their notable health-promoting properties (demonstrated both in vitro and in vivo), including anti-inflammatory, antioxidant and anti-hypertensive activities [63].

Long-term intake of (+)-sesaminol has been proposed to inhibit the pathogenic extracellular β-amyloid aggregation observed in Alzheimer's Disease [67]. Similarly, (+)-sesamin exhibits protective activity against prostate and breast cancers [68], and is a precursor to enterodiol and enterolactone (which have been shown to possess anti-cancer, antidiabetic and anti-ageing properties [64]).

4.4. Asian Diet

The Asian diet is characterized by an elevated consumption of rice, noodles, spices and vegetables, sesame seeds and oil [69]. Additionally, seafood, tofu and other soy products are commonly consumed [70]. Many major plant sources of lignans occur in Asia; these are habitually included in the diet, and in China are also used as medicinal plants. Such plants include *Articum lappa*, whose fruit extracts and seeds are a rich source of bioactive lignans [70], including arctiin and arctigenin. These two lignans exhibit anti-inflammatory activities (e.g., inhibition of lipopolysaccharide-induced nitric oxide production and release of pro-inflammatory cytokines in murine macrophages in vivo) [70,71]. In addition, when tested on diverse cancer cell lines, arctigenin possesses potent apoptotic and anti-proliferative activities [70,72].

Certain medicinal herbs are usually used as an aqueous infusion. Among them, *Isodon spp.* and *Tripterygium spp.*

The genus Isodon comprises nearly 150 species found in the subtropical and tropical regions of Asia and represents an excellent lignan source [73]. Some species, such as *Isodon japonica*, have been used in traditional Chinese medicine to treat (for example) arthralgia, stomach-ache, mastitis, gastritis and hepatitis [73,74]. *Isodon rubescens* has also been used in traditional medicine for its hypotensive, antioxidant, immunological, antimicrobial, antitumor and anti-inflammatory properties [73].

Tripterygium wilfordii Hook f., a traditional medicinal herb, may ameliorate symptoms of rheumatoid arthritis and other autoimmune diseases [75]. Several phytochemical research studies have isolated hundreds of bioactive compounds—including lignans—from the root of this plant [75,76].

Chinese traditional medicine has long made use of *Schisandra chinensis Baill.* fruit as a sedative and antitussive tonic [77]. This fruit is additionally used in other countries in the production of functional foods, jam and beverages. Dibenzocyclooctadiene lignans isolated from *S. chinensis* exhibit anti-inflammatory and antioxidant properties, as well as improving cognitive functions (e.g., memory) [77]. In addition, prior studies have reported that *S. chinensis* fruit extracts—in which the major bioactive constituents are lignans—exert a neuroprotective effect and possess bioactivity which may help prevent Alzheimer's Disease [78]. Furthermore, *S. chinensis* fruit may have positive effects on the liver, as well as on the gastrointestinal, immune, sympathetic and central nervous systems [79,80]. Lignan extracts have been shown to successfully suppress hepatocellular carcinoma cell proliferation and to prevent chemical toxin-induced hepatic injury [79]. However, only 2% of the total *S. chinensis* fruit is made up of lignans, and most of these are present in the seeds, which are usually removed during manufacture of fruit-derived products [79].

The *Schisandra glaucescens Diels* vine is extensively distributed across the Southeastern Sichuan and Western Hubei regions of China [81]. The stem of this vine has been used as an analgesic in diverse conditions, including arthritis, rheumatism, and contusions. As yet, one sesquiterpenoid, 25 lignans and 43 triterpenoids have been isolated from *S. glaucescens* [81]. In addition, *S. glaucescens* berries are thought to exert beneficial effects on the kidneys and lungs, relieving the symptoms of asthma for example [82].

Crataegus pinnatifida has been employed by the functional foods industry. Some studies have reported that it has the ability to protect against low-density lipoprotein (LDL) oxidation, to scavenge free radicals, and to exert an anti-inflammatory effect [83,84]. *C. pinnatifida* is mostly consumed as fresh fruit, processed juice or jam. Juice and jam manufacture results in a significant quantity of by-products, including seeds and leaves [84].

Schisandra sphenanthera is mainly located in Southwest China. A diversity of triterpenoids and lignans has been isolated from its leaves, stems, and fruit [85].

The roots, stems, fruit, and leaves of *Kadsura coccinea* are used medicinally, and its fruit, particularly, exhibits significant medicinal and nutritional properties [86]. Its bioactive triterpenoids and lignans have garnered interest for their reported bioactivities, including anti-inflammatory and anti-tumor effects [86–88].

Zanthoxylum schinifolium has been employed to stimulate blood circulation, as well as in the treatment of various diseases [89,90]. Due to its exceptional taste and characteristic aroma (usually described as green, spicy, floral, and fresh), *Z. schinifolium* fruit is used as a spice in many traditional Asiatic cuisines [89]. Prior pharmacological studies have demonstrated that the leaves and fruit of this plant possess medicinal properties, including antitumor, anti-inflammatory, and antioxidant activities, as well as inhibition of both platelet aggregation and monoamine oxidase production [89,91].

4.5. Latin-American Diet

The basis of the Latin-American diet consists of maize (corn), potatoes, peanuts and beans. This diet also includes flax seed. As mentioned above, *Linum usitatissimum* L. (flax seed) represents one of the best dietary sources of lignans, exhibiting a higher lignan content than legumes or grains [8]. Diets rich in flax seed are associated with a reduced risk of various diseases, including cardiovascular disease, osteoporosis, diabetes, and prostate and breast cancers [8,92]. Likely mechanisms include the ability to decrease circulating glucose, LDL and total cholesterol levels [93,94]. Furthermore, *L. usitatissimum* has significant commercial applications, in the manufacture of linen fiber for example [94]. In terms of lignans, flax seed contains mainly secoisolariciresinol and secoisolariciresinol diglucoside, but matairesinol is also present in small quantities [95]. Indeed, >95% of total flax seed mass consists of secoisolariciresinol diglucoside, which is predominantly localized in the seed's fibrous hull [96] rather than its interior [97].

Asian diet appears to facilitate the highest intake of lignans, in forms which also result in higher bioavailability. This is due largely to a high level of vegetable consumption, as well as the use of lignan-rich plant infusions in traditional medicine.

5. Human Studies Concerning Lignan Bioactivity

Recently, interest in identifying new sources of health-promoting natural compounds has increased. However, there are few human epidemiological studies that evaluate lignans bioactivity. Laboratory research, carried out on cell and animal models, concluded that lignans possess antimicrobial, anti-inflammatory and anti-oxidant activities, among others.

About antimicrobial activity, various lignans have exhibited antiviral and antibacterial activity, e.g., against Gram-positive bacteria through alteration of biofilm formation, bacteria metabolites, membrane receptors and ion channels [98]. For instance, pinoresinol has demonstrated activity against some virus [99].

Concerning anti-inflammatory activity, some lignans have the capacity to inhibit NF-kB activity (transcription factor involves on the expression of inflammatory cytokines) on human mast cells (HMC-1). Thus, reduced pro-inflammatory cytokines production. Furthermore, lignans are able to suppress nitric oxide (NO) generation and decrease inflammatory cell infiltration [100–102].

Regarding anti-oxidant activity, various bioactive natural compounds—including phenols from grains, vegetables and fruits—are rich dietary sources of phytochemicals and vitamins, both of which guard against oxidative stress [84,103]. A free radical formation is an inevitable byproduct of cellular metabolism, and cells also *require* a certain level of reactive oxygen species (ROS) to carry out a normal cellular process [70]. Nevertheless, accumulation and/or overproduction of ROS can damage cellular constituents, including DNA [70], and play an important role in the pathogenesis of various severe disorders, including chronic inflammation, cancer, neurodegeneration and atherogenesis [84].

Many studies have demonstrated the strong antioxidant activity of plant extracts, attributable to several highly-effective antioxidants, including lignans (e.g., lariciresinol, matairesinol, secoisolariciresinol, pinoresinol, and nortrachelogenin) [104]. Among the natural antioxidants, lignans exhibit particularly high antioxidant efficiency and thus have potential as preventive and/or therapeutic clinical tools [105].

In recent years, a significant effort has been devoted to analyzing the lignan consumption of various populations (Table 9). Most studies have focused on post-menopausal women, due to lignans being phytoestrogens that ameliorate menopausal symptoms and consequences (e.g., climacteric symptoms, osteoporosis and estrogen-dependent cancers) [106].

5.1. Cancer

Various cohort studies have investigated dietary lignan anticancer bioactivity. As McCann et al. (2010) describe in the "Western New York Exposures and Breast Cancer" study, lignan intake among post-menopausal women with breast cancer significantly reduced the risk of mortality from breast cancer (Hazard Ratio (HR) 0.29, 95% Confidence interval (CI) 0.11–0.76), as well as significantly reducing the risk of all-cause mortality (HR 0.49, 95% CI 0.26–0.91) [107]. Other research based on the Swedish Mammography Cohort (SMC) also detected a statistically significant inverse association between breast cancer risk and lignan consumption among post-menopausal breast cancer patients [108]. Interestingly, the "Ontario Women's Diet and Health Study" reported that neither lignan nor isoflavone consumption by a Canadian cohort correlated with a significant reduction in breast cancer risk [109]. Nonetheless, some studies do propose that isoflavone consumption correlates with a minor reduction in breast cancer risk in both pre- and post-menopausal women [109,110]. In addition, a cohort study examining the association between flax seed and flax bread intake and breast cancer risk demonstrated that flax seed intake was associated with a significant reduction in breast cancer risk (Odds Ratio (OR) 0.82, 95% CI 0.69–0.97) [111]. Furthermore, Buck et al. (2011) demonstrated that high serum enterolactone levels in post-menopausal breast cancer patients are associated with improved overall survival rates [109,112].

Another study, based on data from the United States Cancer Center Support Grant, investigated the association between individual breast cancer estrogen receptor (ER) status and lignan intake [113]. Higher lignan consumption was inversely correlated with the risk of ER⁻ breast cancer among premenopausal women (OR 0.16, 95% CI 0.03–0.44) and with the risk of ER⁺ breast cancer among post-menopausal women (OR 0.64, 95% CI 0.42–1.00) [113]. Although this effect was largely independent of specific lignan class, it predominantly correlated with matairesinol and lariciresinol intake levels [113]. In addition, this study examined associations between breast tumor subtype and dietary lignan intake, demonstrating that a reduction in premenopausal triple-negative (HER2⁻PR⁻ER⁻) breast cancer risk (OR 0.16, 95% CI 0.04–0.62) was associated with higher lariciresinol and pinoresinol intake [113]. This finding agrees with that of a German case-control study that demonstrated a correlation between high intake of pumpkin and sunflower seeds (rich sources of lariciresinol and pinoresinol) and a statistically significant reduction in post-menopausal ER⁺ breast cancer risk (OR = 0.88, 95% CI = 0.77–0.99, p for trend = 0.02) [109,114].

Two recent meta-analyses have corroborated that high levels of plant lignan consumption correlate with a modest reduction in post-menopausal breast cancer risk (13 studies; Risk Estimated (RE) 0.86, 95% CI 0.78–0.94) [115,116].

Dietary lignan intake is also associated with a reduced risk for other cancer types (e.g., esophageal and gastric adenocarcinoma, as well as colon cancer), but very few human studies have been conducted.

A Swedish study indicates that dietary lignan intake correlates with decreased risk of gastroesophageal junction adenocarcinoma [117]. However, another Swedish study examining the Swedish Cancer Registry database did not find a clear association between dietary lignan consumption and development of gastric or esophageal adenocarcinoma [118]. Yet another (case-control) study indicated that a diet rich in resveratrol, quercetin and lignans (characterized by low intake of milk, but high intake of wholegrain bread, vegetables, wine and tea) may decrease the risk of developing such cancers [103].

Regarding colorectal cancer, Zamora-Ros et al. (2015) evaluated the association of lignan and flavonoid consumption with overall survival time and risk of recurrence in Barcelona (Spain) [119]. After a mean of 8.6 years' follow-up, 77 of the 319 (24.1%) patients in the cohort had experienced recurrence (excluding cases with metastasis that could not be resected), 133 of 409 (32.5%) patients had died, and no association was noted between consumption of any flavonoid subclass or total lignans and colorectal cancer risk [119].

Concerning prostate cancer risk, it has been studied its association with plasma enterolactone concentrations. Wallström et al. (2018) evaluated a population of Swedish men with 1010 cases and 1817 controls. After a mean follow-up of 14.6 years; there were no significant associations between the incidence of prostate cancer and plasma enterolactone (OR 0.99, 95% CI 0.77–1.280) [120]. Other study carried out at Danish men, neither found an association between prostate cancer mortality and plasma enterolactone [121]. However, two other pieces of research on humans, from 2003 and 2006, obtained positive results based on dietary phytoestrogen intake [122,123]. A Swedish case-control study indicated that lower prostate cancer risk is related to certain phytoestrogen-rich foods [123].

Given such mixed results, additional studies examining the effect of human lignan intake on cancer risk are necessary. Specifically, most existing studies have not examined the relevance of the specific dietary lignan source.

5.2. Cardiovascular Disease

Neolignans and flax lignans are reportedly relevant in diabetes, hypercholesterolemia and cardiovascular disorders [124]. In addition, the anti-aging role of lignans has recently been described [125]. Such lignan characteristics may be relevant to the reduction of cardiovascular disease risk in post-menopausal women. Indeed, an inverse association exists between high lignan consumption and the development of hypertension and cardiovascular disease [126]. Furthermore, prospective and cross-sectional epidemiological evidence suggests that dietary lignan intake reduces cardiovascular disease risk in post-menopausal women and elderly men by modifying traditional risk factors [127].

Jacobs et al. (2000) demonstrated that the risk of mortality is inversely associated with whole grain consumption in post-menopausal women [128]. Another study described how four weeks' consumption of a whole grain cereal-rich diet exerted a reasonable cholesterol-lowering effect in healthy post-menopausal women [17].

Table 9. Association between naturally lignan-rich foods and health promotion.

Author, Year	Methods	Results
Breast Cancer		
Lowcock, E.C. et al. (2013) [111]	Case-control study (2999 cases and 3370 controls) FFQ	Consumption of flaxseed and flax bread was associated with a significant reduction in breast cancer risk (OR 0.82, 95% CI 0.69–0.97; and OR 0.77, 95% CI 0.67–0.89), respectively.
McCann et al. (2012) [113]	Case-control study (638 cases and 611 controls) BioRepository at Roswell Park Cancer Institute FFQ	Lignan intakes were inversely associated with risk of ER (−) breast cancer among premenopausal women (OR 0.16, 95% CI 0.03–0.44) and particularly triple negative tumors (OR 0.16, 95% CI 0.04–0.62).
Zaineddin AK et al. (2012) [114]	Case-control study (2884 cases and 5509 controls) FFQ	High and low consumption of soybeans, as well as of sunflower and pumpkin seeds were associated with significantly reduced breast cancer risk compared to no consumption (OR 0.83, 95% CI 0.70–0.97; and OR 0.66, 95% CI 0.77–0.97, respectively).
Buck K et al. (2011) [112]	1140 postmenopausal patients (age 50 to 74 years) FFQ Serum Enterolactone	Serum enterolactone was associated with a significantly reduced risk of death only for estrogen receptor-negative tumors (HR 0.27; 95% CI 0.08 to 0.87)
Buck K et al. (2010) [116]	Meta-analysis. Medline, to identify epidemiologic studies published between 1997 and August 2009	Lignan exposure was not associated with overall breast cancer risk (RE 0.92; 95% CI 0.81, 1.02).
McCann, S.E et al. (2010) [107]	Breast cancer patients; National Death Index Food frequency questionnaire (FFQ), DietSys (3.7)	Lignan intake among post-menopausal women with breast cancer significantly reduced risk of mortality from breast cancer (HR 0.29, 95% CI, 0.11–0.76), as well as significantly reducing risk of all-cause mortality (HR 0.49, 95% CI 0.26–0.91).
Velentzis LS et al. (2009) [115]	Meta-analy sesMedline, BIOSIS and EMBASE databases publications up to 30 September 2008	Overall, there was little association between high plant lignan intake and breast cancer risk (11 studies, OR 0.93, 95% CI 0.83–1.03).
Cotterchio, M et al. (2008) [109]	Ontario Cancer Registry; Controls: Age-stratified random sample of women FFQ	Total phytoestrogen intake in pre-menopausal women was associated with a significant reduction in breast cancer risk among overweight women (OR 0.51, 95% CI 0.30, 0.87).
Suzuki, R. et al. (2008) [108]	Swedish Mammography Cohort FFQ and Swedish National Food database Serum Enterolactone: Fluoroimmunoassay Receptor status of tumors: Immunohistochemical	A significant 17% risk reduction for breast cancer overall in high lignan intake was observed, but no heterogeneity across Estrogen Receptor/Progesterone Receptor subtypes.
Trock BJ et al. (2006) [110]	Meta-analysis of 18 epidemiologic studies published from 1978 through 2004	High soy intake was discreetly associated with reduction of breast cancer risk (OR 0.86, 95% CI: 0.75 to 0.99); association was not statistically significant among women in Asian countries (OR 0.89, 95% CI 0.71 to 1.12).

Molecules **2019**, *24*, 917

Table 9. *Cont.*

Author, Year	Methods	Results
	Gastroesophageal Cancer	
Lin Y et al. (2012) [117]	Case-control study (1995–1997); 806 controls, 181 cases of esophageal adenocarcinoma, 255 cases of gastroesophageal junctional adenocarcinoma, and 158 cases of esophageal squamous cell carcinoma. Interviews and questionnaires; FFQ	No clear associations were found between risk of esophageal carcinoma and lignan intake.
Lin Y et al. (2012) [118]	Cohort study in Sweden, 81,670 (followed up 1998 to 2009). Cancer cases: Swedish Cancer Register FFQ	There was no statistically significant association between dietary intake of lignans and any of the studied adenocarcinomas.
	Colon Cancer	
Zamora-Ros, R. et al. (2015) [119]	409 CRC cases in Barcelona (Spain). FFQ; Phenol-Explorer database.	No associations were also observed with either total lignans or any flavonoid subclass intake.
	Prostate Cancer	
Wallström P et al. (2018) [120]	Case-control study (1010 cases and 1817 controls) National registers and hospital records FFQ Plasma Enterolactone: Fluoroimmunoassay	There were no significant associations between plasma enterolactone and incidence of prostate cancer (OR 0.99, 95% CI 0.77–1.280)
Eriksen AK et al. (2017) [121]	1390 men diagnosed with prostate cancer from the Danish Diet, Cancer and Health cohort Plasma Enterolactone: Fluoroimmunoassay	No associations between plasma enterolactone concentrations and prostate cancer aggressiveness.
Hedelin M et al. (2006) [123]	Swedish case-control study (1499 prostate cancer cases and 1130 controls) FFQ	No association was found between dietary intake of total or individual lignans or isoflavonoids and risk of prostate cancer.
Bylund A. et al. (2003) [122]	10 men with prostate cancer were randomized to a daily supplement of rye bran bread and 8 men of wheat bread Blood and urine samples. Ultrasound-guided core biopsies of the prostate.	In the rye group, there was a significant increase in plasma enterolactone. However, only small changes were observed in plasma concentrations of prostate specific antigen (PSA).
	Cardiovascular disease	
Witkowska AM et al. (2018) [126]	2599 postmenopausal women, participants of the Multi-center National Population Health Examination Surveys. 24-h Dietary recall and food databases.	In postmenopausal women, total and individual lignan intakes (secoisolariciresinol, pinoresinol, matairesinol) were not associated with the prevalence of CVD and its risk factors.
Pellegrini N et al. (2010) [127]	Cross-sectional study in 151 men and 91 post-menopausal women. Anthropometric characteristics. Soluble intercellular adhesion molecule-1 (sICAM-1), CRP, insulin, glucose, total cholesterol, HDL-cholesterol and triacylglycerols. Three-day weighed food record	No relationship between intake of pinoresinol, lariciresinol or total lignans and sICAM-1 values was observed.

Table 9. *Cont.*

Author, Year	Methods	Results
Jacobs DR. et al. (2000) [128]	11,040 postmenopausal women enrolled in the Iowa Women's Health Study Followed from baseline 1986–997.	Women who consumed on average 1.9 g refined grain fiber/2000 kcal and 4.7 g whole grain fiber/2000 kcal had a 17% lower mortality rate (RR = 0.83, 95% CI = 0.73–0.94) than women who consumed predominantly refined grain fiber.
Vanharanta M. et al. (2003) [129]	A prospective study of Finnish men. 1889 men aged 42 to 60 years. Followed up 12.2 years.	Multivariate analyses showed significant associations between elevated serum enterolactone concentration and reduced risk of CVD-related mortality.
Other diseases		
Franco OH. et al. (2005) [130]	Community-based survey among 394 postmenopausal women. FFQ; Cognitive function:Mini-Mental Examination	Increasing dietary lignans intake was associated with better performance on the MMSE (OR 1.49, 95% CI 0.94–2.38). Results were most pronounced in women who were 20–30 years.
Eichholzer M. et al. (2014) [131]	2028 participants of NHANES 2005-2008 and 2628 participants of NHANES 1999-2004 (aged ≥18 years) Inflammatory marker CRP	Statistically significant inverse associations of urinary lignan, enterodiol, and enterolactone concentrations with circulating CRP counts were observed in the multivariate-adjusted models.

FFQ: Food Frequency Questionnaire; CI: Confidence Interval; HR: Hazard Ratio; OR: Odds Ratio; CVD: Cardiovascular Disease; MMSE: Cognitive function Mini-Mental Examination; CRP: C-Reactive Protein.

However, a Warsaw population-based cross-sectional study conducted by the National Institute of Cardiology demonstrated that total dietary lignan consumption does not correlate with the occurrence of cardiovascular diseases, nor with cardiovascular risk factors (including central obesity, hypercholesterolemia and hypertension) in post-menopausal women [126]. Nevertheless, this study attributed a potentially-beneficial effect of lignan intake on hypercholesterolemia specifically to lariciresinol [126].

In a Finnish population, the highest serum enterolactone concentrations correlated with a lower risk of all-cause mortality, including from cardiovascular disease [129]. Enterolactone is a metabolite of lariciresinol, pinoresinol, secoisolariciresinol and matairesinol, and very low matairesinol intake does demonstrate an inverse relationship with endothelial dysfunction and vascular inflammation [127].

5.3. Other Diseases

Most studies have focused on the effects of lignan-rich food consumption in the prevention of cancer and cardiovascular disease. However, some observational studies have investigated the relationship between regular consumption of plant lignans and the risk of developing other lifestyle-related diseases. A study based on the European Prospective Investigation into Cancer and Nutrition cohort proposed that improved cognitive performance in post-menopausal women is associated with higher dietary phytoestrogen consumption (predominantly lignans in Western diets) [130]. Thus, it has been suggested that low-grade chronic inflammation contributes to the prevalence of chronic lifestyle-related diseases. The relationship between lignan consumption and inflammatory markers (e.g., C-reactive protein (CRP)) was studied in a United States cohort, demonstrating that a beneficial inflammatory marker profile is associated with adult lignan consumption [131].

6. Conclusions

Taken together, reviewed data support the recently increased interest in lignan health-promoting properties. Due to their various bioactive properties, dietary intake of lignan-rich foods may prevent certain types of cancers (e.g., breast cancer in post-menopausal women and colon cancer). Regarding chronic lifestyle-related diseases, some pieces of evidence indicate that lignan intake is associated with a lower risk of developing cardiovascular disease. Nonetheless, further human studies are warranted to evaluate lignan bioavailability resulting from different traditional dietary patterns, in order to influence the rational promotion of healthy lignan-rich diets.

Author Contributions: All authors have participated actively in the design and conception of this review. All authors have assessed the present form of the review and have approved it for publication.

Funding: Carmen Rodríguez-García received a pre-doctoral research grant from the University of Jaén (Ayudas predoctorales para la formación del personal investigador, Acción 4).

Conflicts of Interest: The authors declare no conflict of interest.

References

1. Marilena, V.; Olga, V.; Maria, M.; Enzo, B.; Stefano, D.P.; Carlo, B.; Giorda, A.N.; Sebastiano, S.; Stefania, A.; Anna, C.; et al. Rivellese, Polyphenol intake, cardiovascular risk factors in a population with type 2 diabetes: The TOSCA.IT study. *Clin. Nutr.* **2017**, *36*, 1686–1692. [CrossRef]
2. Rocha, L.; Monteiro, M.; Anderson, T. Anticancer Properties of Hydroxycinnamic Acids-A Review. *Cancer Clin. Oncol.* **2012**, *1*, 109–121. [CrossRef]
3. Adlercreutz, H. Lignans, human health. *Crit. Rev. Clin. Lab. Sci.* **2007**, *44*, 483–525. [CrossRef] [PubMed]
4. Ionkova, I. Anticancer lignans—From discovery to biotechnology. *Mini Rev. Med. Chem.* **2011**, *10*, 843–856. [CrossRef]
5. Peterson, J.; Dwyer, J.; Adlercreutz, H.; Scalbert, A.; Jacques, P.; McCullough, M.L. Dietary lignans: Physiology, potential for cardiovascular disease risk reduction. *Nutr. Rev.* **2010**, *10*, 571–603. [CrossRef] [PubMed]
6. Landete, J.M. Plant, mammalian lignans: A review of source, intake, metabolism, intestinal bacteria, health. *Food Res. Int.* **2012**, *46*, 410–424. [CrossRef]
7. Touré, A.; Xu, X. Flaxseed Lignans: Source, Biosynthesis, Metabolism, Antioxidant Activity, Bio-Active Components, Health Benefits. *Compr. Rev. Food Sci. Food Saf.* **2010**, *9*, 261–269. [CrossRef]
8. Marcotullio, M.C.; Curini, M.; Becerra, J.X. An Ethnopharmacological, Phytochemical, Pharmacological Review on Lignans from *Mexican Bursera* spp. *Molecules* **2018**, *23*, 1976. [CrossRef] [PubMed]
9. Magoulas, G.E.; Papaioannou, D. Bioinspired syntheses of dimeric hydroxycinnamic acids (lignans), hybrids, using phenol oxidative coupling as key reaction, medicinal significance thereof. *Molecules* **2014**, *19*, 19769–19835. [CrossRef] [PubMed]
10. Li, Y.; Xie, S.; Ying, J.; Wei, W.; Gao, K. Chemical Structures of Lignans, Neolignans Isolated from Lauraceae. *Molecules* **2018**, *23*, 3164. [CrossRef] [PubMed]
11. Pan, J.Y.; Chen, S.L.; Yang, M.H.; Wu, J.; Sinkkonen, J.; Zou, K. An update on lignans: Natural products, synthesis. *Nat. Prod. Rep.* **2009**, *26*, 1251–1292. [CrossRef] [PubMed]
12. Suzuki, S.; Umezawa, T. Biosynthesis of lignans, norlignans. *J. Wood Sci.* **2007**, *53*, 273–284. [CrossRef]
13. Solyomvary, A.; Beni, S.; Boldizsar, I. Dibenzylbutyrolactone Lignans-A Review of Their Structural Diversity, Biosynthesis, Occurrence, Identification, Importance. *Mini Rev. Med. Chem.* **2017**, *17*, 1053–1074. [CrossRef] [PubMed]
14. Xu, W.-H.; Zhao, P.; Wang, M.; Liang, Q. Naturally occurring furofuran lignans: Structural diversity, biological activities. *Nat. Prod. Res.* **2018**, *16*, 1–17. [CrossRef] [PubMed]
15. Zhang, J.; Chen, J.; Liang, Z.; Zhao, C. New lignans, their biological activities. *Chem. Biodivers* **2014**, *11*, 1–54. [CrossRef] [PubMed]
16. Durazzo, A.; Zaccaria, M.; Polito, A.; Maiani, G.; Carcea, M. Lignan Content in Cereals, Buckwheat, Derived Foods. *Foods* **2013**, *2*, 53–63. [CrossRef] [PubMed]

17. Durazzo, A.; Turfani, V.; Azzini, E.; Maiani, G.; Carcea, M. Phenols, lignans, antioxidant properties of legume, sweet chestnutflours. *Food Chem.* **2013**, *140*, 666–671. [CrossRef] [PubMed]
18. Rothwell, J.A.; Pérez-Jiménez, J.; Neveu, V.; Medina-Ramon, A.; M'Hiri, N.; Garcia Lobato, P.; Manach, C.; Knox, K.; Eisner, R.; Wishart, D.; et al. Phenol-Explorer 3.0: A major update of the Phenol-Explorer database to incorporate data on the effects of food processing on polyphenol content. *Database* **2013**, *2013*. [CrossRef] [PubMed]
19. Smeds, A.I.; Jauhiainen, L.; Tuomola, E. Peltonen-Sainio, P. Characterization of variation in the lignan content, composition of winter rye, spring wheat, spring oat. *J. Agric. Food Chem.* **2009**, *57*, 5837–5842. [CrossRef] [PubMed]
20. Esposito, F.; Arlotti, G.; Maria Bonifati, A.; Napolitano, A.; Vitale, D.; Fogliano, V. Antioxidant activity, dietary fibre in durum wheat bran by-products. *Food Res. Int.* **2005**, *38*, 1167–1173. [CrossRef]
21. Fardet, A. New hypotheses for the health-protective mechanisms of whole-grain cereals: What is beyond fibre? *Nutr. Res. Rev.* **2010**, *23*, 65–134. [CrossRef] [PubMed]
22. Bolvig, A.K.; Adlercreutz, H.; Theil, P.K.; Jorgensen, H.; Bach Knudsen, K.E. Absorption of plant lignans from cereals in an experimental pig model. *Br. J. Nutr.* **2016**, *115*, 1711–1720. [CrossRef] [PubMed]
23. Milder, I.E.; Arts, I.C.; van de Putte, B.; Venema, D.P.; Hollman, P.C. Lignan contents of Dutch plant foods: A database including lariciresinol, pinoresinol, secoisolariciresinol, matairesinol. *Br. J. Nutr.* **2005**, *93*, 393–402. [CrossRef] [PubMed]
24. Ruiz-Aracama, A.; Goicoechea, E.; Guillén, M.D. Direct study of minor extra-virgin olive oil components without any sample modification. 1H NMR multisupression experiment: A powerful tool. *Food Chem.* **2017**, *228*, 301–314. [CrossRef] [PubMed]
25. Ricciutelli, M.; Marconi, S.; Boarelli, M.C.; Caprioli, G.; Sagratini, G.; Ballini, D.; Fiorini, R. Olive oil polyphenols: A quantitative method by high-performance liquid-chromatography-diode-array detection for their determination, the assessment of the related health claim. *J. Chromatogr. A* **2017**, *1481*, 53–63. [CrossRef] [PubMed]
26. Milder, I.E.; Feskens, E.J.; Arts, I.C.; Bueno de Mesquita, H.B.; Hollman, P.C.; Kromhout, D. Intake of the plant lignans secoisolariciresinol, matairesinol, lariciresinol, pinoresinol in Dutch men, women. *J. Nutr.* **2005**, *135*, 1202–1207. [CrossRef] [PubMed]
27. Sun, Q.; Wedick, N.M.; Pan, A.; Townsend, M.K.; Cassidy, A.; Franke, A.A.; Rimm, E.B.; Hu, F.B.; van Dam, R.M. Gut microbiota metabolites of dietary lignans, risk of type 2 diabetes: A prospective investigation in two cohorts of U.S. women. *Diabetes Care* **2014**, *37*, 1287–1295. [CrossRef] [PubMed]
28. McCann, M.J.; Gill, C.I.; McGlynn, H.; Rowland, I.R. Role of mammalian lignans in the prevention, treatment of prostate cancer. *Nutr. Cancer* **2005**, *52*, 1–14. [CrossRef] [PubMed]
29. Szewczyk, M.; Abarzua, S.; Schlichting, A.E.; Nebe, B.; Piechulla, B.; Volker, B.; Dagmar-Ulrike, R. Effects of extracts from Linum usitatissimum on cell vitality, proliferation, cytotoxicity in human breast cancer cell lines. *J. Med. Plant Res.* **2014**, *8*, 237–245. [CrossRef]
30. Björck, I.; Östman, E.; Kristensen, M.; Mateo Anson, N.; Price, R.K.; Haenen, G.R.M.M.; Havenaar, R.; Bach Knudsen, K.E.; Frid, A.; Mykkänen, H.; et al. Cereal grains for nutrition, health benefits: Overview of results from in vitro, animal, human studies in the HEALTHGRAIN project. *Trends Food Sci. Technol.* **2012**, *25*, 87–100. [CrossRef]
31. Kuijsten, A.; Arts, I.C.; Vree, T.B.; Hollman, P.C. Pharmacokinetics of enterolignans in healthy men, women consuming a single dose of secoisolariciresinol diglucoside. *J. Nutr.* **2005**, *135*, 795–801. [CrossRef] [PubMed]
32. Tetens, I.; Turrini, A.; Tapanainen, H.; Christensen, T.; Lampe, J.W.; Fagt, S.; Håkansson, N.; Lundquist, A.; Hallund, J.; Valsta, L.M.; et al. Dietary intake, main sources of plant lignans in five European countries. *Food Nutr. Res.* **2013**, *57*. [CrossRef] [PubMed]
33. Heinonen, S.; Nurmi, T.; Liukkonen, K.; Poutanen, K.; Wahala, K.; Deyama, T.; Nishibe, S.; Adlercreutz, H. In vitro metabolism of plant lignans: New precursors of mammalian lignans enterolactone, enterodiol. *J. Agric. Food Chem.* **2001**, *49*, 3178–3186. [CrossRef] [PubMed]
34. Saarinen, N.M.; Thompson, L.U. Prolonged administration of secoisolariciresinol diglycoside increases lignan excretion, alters lignan tissue distribution in adult male, female rats. *Br. J. Nutr.* **2010**, *104*, 833–841. [CrossRef] [PubMed]

35. Mukker, J.K.; Singh, R.S.; Muir, A.D.; Krol, E.S.; Alcorn, J. Comparative pharmacokinetics of purified flaxseed, associated mammalian lignans in male Wistar rats. *Br. J. Nutr.* **2015**, *113*, 749–757. [CrossRef] [PubMed]
36. Chaojie, L.; Ed, S.K.; Jane, A. The Comparison of Rat, Human Intestinal, Hepatic Glucuronidation of Enterolactone Derived from Flaxseed Lignans. *Nat. Prod. J.* **2013**, *3*, 159–171. [CrossRef]
37. Murray, T.; Kang, J.; Astheimer, L.; Price, W.E. Tissue distribution of lignans in rats in response to diet, dose-response, competition with isoflavones. *J. Agric. Food Chem.* **2007**, *55*, 4907–4912. [CrossRef] [PubMed]
38. Thompson, L.U.; Chen, J.M.; Li, T.; Strasser-Weippl, K.; Goss, P.E. Dietary flaxseed alters tumor biological markers in postmenopausal breast cancer. *Clin. Cancer Res.* **2005**, *11*, 3828–3835. [CrossRef] [PubMed]
39. Clavel, T.; Dore, J.; Blaut, M. Bioavailability of lignans in human subjects. *Nutr. Res. Rev.* **2006**, *19*, 187–196. [CrossRef] [PubMed]
40. Adlercreutz, H. Phyto-oestrogens, cancer. *Lancet. Oncol.* **2002**, *3*, 364–373. [CrossRef]
41. Kuijsten, A.; Arts, I.C.; van't Veer, P.; Hollman, P.C. The relative bioavailability of enterolignans in humans is enhanced by milling, crushing of flaxseed. *J. Nutr.* **2005**, *135*, 2812–2816. [CrossRef] [PubMed]
42. Lærke, H.N.; Mortensen, M.A.; Hedemann, M.S.; Bach Knudsen, K.E.; Penalvo, J.L.; Adlercreutz, H. Quantitative aspects of the metabolism of lignans in pigs fed fibre-enriched rye, wheat bread. *Br. J. Nutr.* **2009**, *102*, 985–994. [CrossRef] [PubMed]
43. Johnson, T.W.; Dress, K.R.; Edwards, M. Using the Golden Triangle to optimize clearance, oral absorption. *Bioorganic. Med. Chem. Lett.* **2009**, *19*, 5560–5564. [CrossRef] [PubMed]
44. Li, J.J.; Cheng, L.; Shen, G.; Qiu, L.; Shen, C.Y.; Zheng, J.; Xu, R.; Yuan, H.L. Improved stability, oral bioavailability of Ganneng dropping pills following transforming lignans of herpetospermum caudigerum into nanosuspensions. *Chin. J. Nat. Med.* **2018**, *16*, 70–80. [CrossRef]
45. Tierney, A.C.; Zabetakis, I. Changing the Irish dietary guidelines to incorporate the principles of the Mediterranean diet: Proposing the MedEire diet. *Public Health Nutr.* **2018**, 1–7. [CrossRef] [PubMed]
46. Trichopoulou, A.; Costacou, T.; Bamia, C.; Trichopoulos, D. Adherence to a Mediterranean Diet, Survival in a Greek Population. *N Engl. J. Med.* **2003**, *348*, 2599–2608. [CrossRef] [PubMed]
47. Pounis, G.; Di Castelnuovo, A.; Bonaccio, M.; Costanzo, S.; Persichillo, M.; Krogh, V.; Donati, M.B.; de Gaetano, G.; Iacoviello, L. Flavonoid, lignan intake in a Mediterranean population: Proposal for a holistic approach in polyphenol dietary analysis, the Moli-sani Study. *Eur. J. Clin. Nutr.* **2016**, *70*, 338–345. [CrossRef] [PubMed]
48. Bolvig, A.K.; Norskov, N.P.; van Vliet, S.; Foldager, L.; Curtasu, M.V.; Hedemann, M.S.; Sorensen, J.F.; Laerke, H.N.; Bach Knudsen, K.E. Rye Bran Modified with Cell Wall-Degrading Enzymes Influences the Kinetics of Plant Lignans but Not of Enterolignans in Multicatheterized Pigs. *J. Nutr.* **2017**, *147*, 2220–2227. [CrossRef] [PubMed]
49. Bolling, B.W.; Chen, C.Y.; McKay, D.L.; Blumberg, J.B. Tree nut phytochemicals: Composition, antioxidant capacity, bioactivity, impact factors. A systematic review of almonds, Brazils, cashews, hazelnuts, macadamias, pecans, pine nuts, pistachios, walnuts. *Nutr. Res. Rev.* **2011**, *24*, 244–275. [CrossRef] [PubMed]
50. Guasch-Ferré, M.; Hu, F.B.; Martínez-González, M.A.; Fitó, M.; Bulló, M.; Estruch, R.; Ros, E.; Corella, D.; Recondo, J.; Gómez-Gracia, E.; et al. Olive oil intake, risk of cardiovascular disease, mortality in the PREDIMED Study. *BMC Med.* **2014**, *12*, 78. [CrossRef] [PubMed]
51. Toledo, E.; Salas-Salvado, J.; Donat-Vargas, C.; Buil-Cosiales, P.; Estruch, R.; Ros, E.; Corella, D.; Fito, M.; Hu, F.B.; Aros, F.E.; et al. Mediterranean Diet, Invasive Breast Cancer Risk Among Women at High Cardiovascular Risk in the PREDIMED Trial: A Randomized Clinical Trial. *JAMA Int. Med.* **2015**, *175*, 1752–1760. [CrossRef] [PubMed]
52. Medina-Remón, A.; Casas, R.; Tressserra-Rimbau, A.; Ros, E.; Martínez-González, M.A.; Fitó, M.; Corella, D.; Salas-Salvadó, J.; Lamuela-Raventos, R.M.; Estruch, R.; et al. Polyphenol intake from a Mediterranean diet decreases inflammatory biomarkers related to atherosclerosis: A substudy of the PREDIMED trial. *Br. J. Clin. Pharmacol.* **2017**, *83*, 114–128. [CrossRef] [PubMed]
53. Lopez-Biedma, A.; Sanchez-Quesada, C.; Beltran, G.; Delgado-Rodriguez, M.; Gaforio, J.J. Phytoestrogen (+)-pinoresinol exerts antitumor activity in breast cancer cells with different oestrogen receptor statuses. *BMC Compl. Altern. Med.* **2016**, *16*, 350. [CrossRef] [PubMed]
54. Antonini, E.; Farina, A.; Scarpa, E.S.; Frati, A.; Ninfali, P. Quantity, quality of secoiridoids, lignans in extra virgin olive oils: The effect of two-, three-way decanters on Leccino, Raggiola olive cultivars. *Int. J. Food Sci. Nutr.* **2016**, *67*, 9–15. [CrossRef] [PubMed]

55. Ramezani-Jolfaie, N.; Mohammadi, M.; Salehi-Abargouei, A. The effect of healthy Nordic diet on cardio-metabolic markers: A systematic review, meta-analysis of randomized controlled clinical trials. *Eur. J. Nutr.* **2018**, *57*, 1–16. [CrossRef] [PubMed]

56. Galbete, C.; Kröger, J.; Jannasch, F.; Iqbal, K.; Schwingshackl, L.; Schwedhelm, C.; Weikert, C.; Boeing, H.; Schulze, M.B. Nordic diet, Mediterranean diet, the risk of chronic diseases: The EPIC-Potsdam study. *BMC Med.* **2018**, *16*, 99. [CrossRef] [PubMed]

57. Smeds, A.I.; Eklund, P.C.; Sjoholm, R.E.; Willfor, S.M.; Nishibe, S.; Deyama, T.; Holmbom, B.R. Quantification of a broad spectrum of lignans in cereals, oilseeds, nuts. *J. Agric. Food Chem.* **2007**, *55*, 1337–1346. [CrossRef] [PubMed]

58. Konye, R.; Toth, G.; Solyomvary, A.; Mervai, Z.; Zurn, M.; Baghy, K.; Kovalszky, I.; Horvath, P.; Molnar-Perl, I.; Noszal, B.; et al. Chemodiversity of Cirsium fruits: Antiproliferative lignans, neolignans, sesquineolignans as chemotaxonomic markers. *Fitoterapia* **2018**, *127*, 413–419. [CrossRef] [PubMed]

59. Boldizsar, I.; Kraszni, M.; Toth, F.; Noszal, B.; Molnar-Perl, I. Complementary fragmentation pattern analysis by gas chromatography-mass spectrometry, liquid chromatography tandem mass spectrometry confirmed the precious lignan content of Cirsium weeds. *J. Chromatogr. A* **2010**, *1217*, 6281–6289. [CrossRef] [PubMed]

60. Singh, L.; Agarwal, T. PAHs in Indian diet: Assessing the cancer risk. *Chemosphere* **2018**, *202*, 366–376. [CrossRef] [PubMed]

61. Liu, W.J.; Chen, Y.; Chen, D.; Wu, Y.; Gao, Y.J.; Li, J.; Zhong, W.J.; Jiang, L. A new pair of enantiomeric lignans from the fruits of Morinda citrifolia, their absolute configuration. *Nat. Prod. Res.* **2018**, *32*, 933–938. [CrossRef] [PubMed]

62. Nguyen, P.H.; Yang, J.L.; Uddin, M.N.; Park, S.L.; Lim, S.I.; Jung, D.W.; Williams, D.R.; Oh, W.K. Protein tyrosine phosphatase 1B (PTP1B) inhibitors from Morinda citrifolia (Noni), their insulin mimetic activity. *J. Nat. Prod.* **2013**, *76*, 2080–2087. [CrossRef] [PubMed]

63. Chen, J.; Chen, Y.; Tian, J.; Ge, H.; Liang, X.; Xiao, J.; Lin, H. Simultaneous determination of four sesame lignans, conversion in Monascus aged vinegar using HPLC method. *Food Chem.* **2018**, *256*, 133–139. [CrossRef] [PubMed]

64. Yashaswini, P.S.; Sadashivaiah, B.; Ramaprasad, T.R.; Singh, S.A. In vivo modulation of LPS induced leukotrienes generation, oxidative stress by sesame lignans. *J. Nutr. Biochem.* **2017**, *41*, 151–157. [CrossRef] [PubMed]

65. Namiki, M. Nutraceutical functions of sesame: A review. *Crit. Rev. Food Sci. Nutr.* **2007**, *47*, 651–673. [CrossRef] [PubMed]

66. Dar, A.A.; Arumugam, N. Lignans of sesame: Purification methods, biological activities, biosynthesis—A review. *Bioorganic Chem.* **2013**, *50*, 1–10. [CrossRef] [PubMed]

67. Katayama, S.; Sugiyama, H.; Kushimoto, S.; Uchiyama, Y.; Hirano, M.; Nakamura, S. Effects of Sesaminol Feeding on Brain Aβ Accumulation in a Senescence-Accelerated Mouse-Prone 8. *J. Agric. Food Chem.* **2016**, *64*, 4908–4913. [CrossRef]

68. Liu, Z.; Saarinen, N.M.; Thompson, L.U. Sesamin Is One of the Major Precursors of Mammalian Lignans in Sesame Seed (Sesamum indicum) as Observed In Vitro, in Rats. *J. Nutr.* **2006**, *136*, 906–912. [CrossRef] [PubMed]

69. Hsu, W.C.; Lau, K.H.K.; Matsumoto, M.; Moghazy, D.; Keenan, H.; King, G.L. Improvement of Insulin Sensitivity by Isoenergy High Carbohydrate Traditional Asian Diet: A Randomized Controlled Pilot Feasibility Study. *PLoS ONE* **2014**, *9*, e106851. [CrossRef] [PubMed]

70. Su, S.; Wink, M. Natural lignans from Arctium lappa as antiaging agents in Caenorhabditis elegans. *Phytochemistry* **2015**, *117*, 340–350. [CrossRef] [PubMed]

71. Kou, X.; Qi, S.; Dai, W.; Luo, L.; Yin, Z. Arctigenin inhibits lipopolysaccharide-induced iNOS expression in RAW264.7 cells through suppressing JAK-STAT signal pathway. *Int. Immunopharmacol.* **2011**, *11*, 1095–1102. [CrossRef] [PubMed]

72. Susanti, S.; Iwasaki, H.; Inafuku, M.; Taira, N.; Oku, H. Mechanism of arctigenin-mediated specific cytotoxicity against human lung adenocarcinoma cell lines. *Phytomedicine* **2013**, *21*, 39–46. [CrossRef] [PubMed]

73. Zhang, Y.; Wang, K.; Chen, H.; He, R.; Cai, R.; Li, J.; Zhou, D.; Liu, W.; Huang, X.; Yang, R.; et al. Anti-inflammatory lignans, phenylethanoid glycosides from the root of Isodon ternifolius (D.Don) Kudô. *Phytochemistry* **2018**, *153*, 36–47. [CrossRef] [PubMed]

74. Chen, Y.; Tang, Y.M.; Yu, S.L.; Han, Y.W.; Kou, J.P.; Liu, B.L.; Yu, B.Y. Advances in the pharmacological activities, mechanisms of diosgenin. *Chin. J. Nat. Med.* **2015**, *13*, 578–587. [CrossRef]

75. Chen, F.; Li, C.; Ma, J.; Ni, L.; Huang, J.; Li, L.; Lin, M.; Hou, Q.; Zhang, D. Diterpenoids, lignans from the leaves of Tripterygium wilfordii. *Fitoterapia* **2018**, *129*, 133–137. [CrossRef] [PubMed]

76. Xu, J.; Lu, J.; Sun, F.; Zhu, H.; Wang, L.; Zhang, X.; Ma, Z. Terpenoids from Tripterygium wilfordii. *Phytochemistry* **2011**, *72*, 1482–1487. [CrossRef] [PubMed]

77. Hu, D.; Yang, Z.; Yao, X.; Wang, H.; Han, N.; Liu, Z.; Wang, Y.; Yang, J.; Yin, J. Dibenzocyclooctadiene lignans from Schisandra chinensis, their inhibitory activity on NO production in lipopolysaccharide-activated microglia cells. *Phytochemistry* **2014**, *104*, 72–78. [CrossRef] [PubMed]

78. Yang, B.Y.; Han, W.; Han, H.; Liu, Y.; Guan, W.; Li, X.M.; Kuang, H.X. Effects of Lignans from Schisandra chinensis Rattan Stems against Abeta1-42-Induced Memory Impairment in Rats, Neurotoxicity in Primary Neuronal Cells. *Molecules* **2018**, *23*, e870. [CrossRef] [PubMed]

79. Wang, O.; Cheng, Q.; Liu, J.; Wang, Y.; Zhao, L.; Zhou, F.; Ji, B. Hepatoprotective effect of Schisandra chinensis (Turcz.) Baill. lignans, its formula with Rubus idaeus on chronic alcohol-induced liver injury in mice. *Food Funct.* **2014**, *5*, 3018–3025. [CrossRef] [PubMed]

80. Panossian, A.; Wikman, G. Pharmacology of Schisandra chinensis Bail.: An overview of Russian research, uses in medicine. *J. Ethnopharmacol.* **2008**, *118*, 183–212. [CrossRef] [PubMed]

81. Wu, W.; Ruan, H. Triterpenoids, lignans from the stems of Schisandra glaucescens. *Nat. Prod. Res.* **2018**, *32*, 1–7. [CrossRef] [PubMed]

82. Yu, H.Y.; Chen, Z.Y.; Sun, B.; Liu, J.; Meng, F.Y.; Liu, Y.; Tian, T.; Jin, A.; Ruan, H.L. Lignans from the fruit of Schisandra glaucescens with antioxidant, neuroprotective properties. *J. Nat. Prod.* **2014**, *77*, 1311–1320. [CrossRef] [PubMed]

83. Liu, P.; Kallio, H.; Yang, B. Phenolic Compounds in Hawthorn (*Crataegus grayana*) Fruits, Leaves, Changes during Fruit Ripening. *J. Agric. Food Chem.* **2011**, *59*, 11141–11149. [CrossRef] [PubMed]

84. Huang, X.-X.; Bai, M.; Zhou, L.; Lou, L.-L.; Liu, Q.-B.; Zhang, Y.; Li, L.-Z.; Song, S.-J. Food Byproducts as a New, Cheap Source of Bioactive Compounds: Lignans with Antioxidant, Anti-inflammatory Properties from Crataegus pinnatifida Seeds. *J. Agric. Food Chem.* **2015**, *63*, 7252–7260. [CrossRef] [PubMed]

85. Jiang, K.; Song, Q.Y.; Peng, S.J.; Zhao, Q.Q.; Li, G.D.; Li, Y.; Gao, K. New lignans from the roots of Schisandra sphenanthera. *Fitoterapia* **2015**, *103*, 63–70. [CrossRef] [PubMed]

86. Liu, Y.; Yang, Y.; Tasneem, S.; Hussain, N.; Daniyal, M. Lignans from Tujia Ethnomedicine Heilaohu: Chemical Characterization, Evaluation of Their Cytotoxicity, Antioxidant Activities. *Molecules* **2018**, *23*, e2147. [CrossRef] [PubMed]

87. Sun, J.; Yao, J.; Huang, S.; Long, X.; Wang, J.; García-García, E. Antioxidant activity of polyphenol, anthocyanin extracts from fruits of *Kadsura coccinea* (Lem.) A.C. Smith. *Food Chem.* **2009**, *117*, 276–281. [CrossRef]

88. Kim, K.H.; Choi, J.W.; Ha, S.K.; Kim, S.Y.; Lee, K.R. Neolignans from Piper kadsura, their anti-neuroinflammatory activity. *Bioorg. Med. Chem. Lett.* **2010**, *20*, 409–412. [CrossRef] [PubMed]

89. Li, W.; Sun, Y.N.; Yan, X.T.; Yang, S.Y.; Kim, E.J.; Kang, H.K.; Kim, Y.H. Coumarins, lignans from Zanthoxylum schinifolium, their anticancer activities. *J. Agric. Food Chem.* **2013**, *61*, 10730–10740. [CrossRef] [PubMed]

90. Cui, H.Z.; Choi, H.R.; Choi, D.H.; Cho, K.W.; Kang, D.G.; Lee, H.S. Aqueous extract of Zanthoxylum schinifolium elicits contractile, secretory responses via beta1-adrenoceptor activation in beating rabbit atria. *J. Ethnopharmacol.* **2009**, *126*, 300–307. [CrossRef] [PubMed]

91. Min, B.K.; Hyun, D.G.; Jeong, S.Y.; Kim, Y.H.; Ma, E.S.; Woo, M.H. A new cytotoxic coumarin, 7-[(E)-3′,7′-dimethyl-6′-oxo-2′,7′-octadienyl] oxy coumarin, from the leaves of Zanthoxylum schinifolium. *Arch. Pharm. Res.* **2011**, *34*, 723–726. [CrossRef] [PubMed]

92. Teponno, R.B.; Kusari, S.; Spiteller, M. Recent advances in research on lignans, neolignans. *Nat. Prod. Rep.* **2016**, *33*, 1044–1092. [CrossRef] [PubMed]

93. Fuentealba, C.; Figuerola, F.; Estevez, A.M.; Bastias, J.M.; Munoz, O. Bioaccessibility of lignans from flaxseed (Linum usitatissimum L.) determined by single-batch in vitro simulation of the digestive process. *J. Sci. Food Agric.* **2014**, *94*, 1729–1738. [CrossRef] [PubMed]

94. Zahir, A.; Ahmad, W.; Nadeem, M.; Giglioli-Guivarc'h, N.; Hano, C.; Abbasi, B.H. In vitro cultures of Linum usitatissimum L.: Synergistic effects of mineral nutrients, photoperiod regimes on growth, biosynthesis of lignans, neolignans. *J. Photochem. Photobiol. B* **2018**, *187*, 141–150. [CrossRef] [PubMed]

95. Gabr, A.M.M.; Mabrok, H.B.; Abdel-Rahim, E.A.; El-Bahr, M.K.; Smetanska, I. Determination of lignans, phenolic acids, antioxidant capacity in transformed hairy root culture of Linum usitatissimum. *Nat. Prod. Res.* **2017**, *32*, 1867–1871. [CrossRef] [PubMed]

96. Schogor, A.L.B.; Huws, S.A.; Santos, G.T.D.; Scollan, N.D.; Hauck, B.D.; Winters, A.L.; Kim, E.J.; Petit, H.V. Ruminal Prevotella spp. May Play an Important Role in the Conversion of Plant Lignans into Human Health Beneficial Antioxidants. *PLoS ONE* **2014**, *9*, e87949. [CrossRef] [PubMed]

97. Côrtes, C.; Gagnon, N.; Benchaar, C.; Da Silva, D.; Santos, G.T.D.; Petit, H.V. In vitro metabolism of flax lignans by ruminal, faecal microbiota of dairy cows. *J. Appl. Microbiol.* **2008**, *105*, 1585–1594. [CrossRef] [PubMed]

98. Alvarez-Martinez, F.J.; Barrajon-Catalan, E.; Encinar, J.A.; Rodriguez-Diaz, J.C.; Micol, V. Antimicrobial Capacity of Plant Polyphenols against Gram-positive Bacteria: A Comprehensive Review. *Curr. Med. Chem.* **2018**. [CrossRef] [PubMed]

99. Nor Azman, N.S.; Hossan, M.S.; Nissapatorn, V.; Uthaipibull, C.; Prommana, P.; Jin, K.T.; Rahmatullah, M.; Mahboob, T.; Raju, C.S.; Jindal, H.M.; et al. Anti-infective activities of 11 plants species used in traditional medicine in Malaysia. *Exp. Parasitol.* **2018**, *194*, 67–78. [CrossRef] [PubMed]

100. Chen, P.; Pang, S.; Yang, N.; Meng, H.; Liu, J.; Zhou, N.; Zhang, M.; Xu, Z.; Gao, W.; Chen, B.; et al. Beneficial effects of schisandrin B on the cardiac function in mice model of myocardial infarction. *PLoS ONE* **2013**, *8*, e79418. [CrossRef] [PubMed]

101. Chun, J.N.; Cho, M.; So, I.; Jeon, J.H. The protective effects of Schisandra chinensis fruit extract, its lignans against cardiovascular disease: A review of the molecular mechanisms. *Fitoterapia* **2014**, *97*, 224–233. [CrossRef] [PubMed]

102. Olaru, O.T.; Nițulescu, G.M.; Orțan, A.; Dinu-Pîrvu, C.E. Ethnomedicinal, Phytochemical, Pharmacological Profile of Anthriscus sylvestris as an Alternative Source for Anticancer Lignans. *Molecules* **2015**, *8*, 15003–15022. [CrossRef] [PubMed]

103. Lin, Y.; Yngve, A.; Lagergren, J.; Lu, Y. A dietary pattern rich in lignans quercetin resveratrol decreases the risk of oesophageal cancer. *Br. J. Nutr.* **2014**, *112*, 2002–2009. [CrossRef] [PubMed]

104. Kyselka, J.; Rabiej, D.; Dragoun, M.; Kreps, F.; Burčová, Z.; Němečková, I.; Smolová, J.; Bjelková, M.; Szydłowska-Czerniak, A.; Schmidt, S.; et al. Antioxidant, antimicrobial activity of linseed lignans, phenolic acids. *Eur. Food Res. Technol.* **2017**, *243*, 1633–1644. [CrossRef]

105. Vo, Q.V.; Nam, P.C.; Bay, M.V.; Thong, N.M.; Cuong, N.D.; Mechler, A. Density functional theory study of the role of benzylic hydrogen atoms in the antioxidant properties of lignans. *Sci. Rep.* **2018**, *8*, 12361. [CrossRef] [PubMed]

106. Sammartino, A.; Tommaselli, G.A.; Gargano, V.; di Carlo, C.; Attianese, W.; Nappi, C. Short-term effects of a combination of isoflavones, lignans, Cimicifuga racemosa on climacteric-related symptoms in postmenopausal women: A double-blind, randomized, placebo-controlled trial. *Gynecol. Endocrinol.* **2006**, *22*, 646–650. [CrossRef] [PubMed]

107. McCann, S.E.; Thompson, L.U.; Nie, J.; Dorn, J.; Trevisan, M.; Shields, P.G.; Ambrosone, C.B.; Edge, S.B.; Li, H.-F.; Kasprzak, C.; et al. Dietary lignan intakes in relation to survival among women with breast cancer: The Western New York Exposures, Breast Cancer (WEB) Study. *Breast Cancer Res. Treat.* **2010**, *122*, 229–235. [CrossRef] [PubMed]

108. Suzuki, R.; Rylander-Rudqvist, T.; Saji, S.; Bergkvist, L.; Adlercreutz, H.; Wolk, A. Dietary lignans, postmenopausal breast cancer risk by oestrogen receptor status: A prospective cohort study of Swedish women. *Br. J. Cancer* **2008**, *98*, 636–640. [CrossRef] [PubMed]

109. Cotterchio, M.; Boucher, B.A.; Kreiger, N.; Mills, C.A.; Thompson, L.U. Dietary phytoestrogen intake—lignans, isoflavones—and breast cancer risk (Canada). *Cancer Causes Control.* **2008**, *19*, 259–272. [CrossRef] [PubMed]

110. Trock, B.J.; Hilakivi-Clarke, L.; Clarke, R. Meta-analysis of soy intake, breast cancer risk. *J. Natl. Cancer Inst.* **2006**, *98*, 459–471. [CrossRef] [PubMed]

111. Lowcock, E.C.; Cotterchio, M.; Boucher, B.A. Consumption of flaxseed, a rich source of lignans, is associated with reduced breast cancer risk. *Cancer Causes Control.* **2013**, *24*, 813–816. [CrossRef] [PubMed]

112. Buck, K.; Vrieling, A.; Zaineddin, A.K.; Becker, S.; Husing, A.; Kaaks, R.; Linseisen, J.; Flesch-Janys, D.; Chang-Claude, J. Serum enterolactone, prognosis of postmenopausal breast cancer. *J. Clin. Oncol.* **2011**, *29*, 3730–3738. [CrossRef] [PubMed]

113. McCann, S.E.; Hootman, K.C.; Weaver, A.M.; Thompson, L.U.; Morrison, C.; Hwang, H.; Edge, S.B.; Ambrosone, C.B.; Horvath, P.J.; Kulkarni, S.A. Dietary intakes of total, specific lignans are associated with clinical breast tumor characteristics. *J. Nutr.* **2012**, *142*, 91–98. [CrossRef] [PubMed]

114. Zaineddin, A.K.; Buck, K.; Vrieling, A.; Heinz, J.; Flesch-Janys, D.; Linseisen, J.; Chang-Claude, J. The association between dietary lignans, phytoestrogen-rich foods, fiber intake, postmenopausal breast cancer risk: A German case-control study. *Nutr. Cancer* **2012**, *64*, 652–665. [CrossRef] [PubMed]

115. Velentzis, L.S.; Cantwell, M.M.; Cardwell, C.; Keshtgar, M.R.; Leathem, A.J.; Woodside, J.V. Lignans, breast cancer risk in pre-, post-menopausal women: Meta-analyses of observational studies. *Br. J. Cancer* **2009**, *100*, 1492–1498. [CrossRef] [PubMed]

116. Buck, K.; Zaineddin, A.K.; Vrieling, A.; Linseisen, J.; Chang-Claude, J. Meta-analyses of lignans, enterolignans in relation to breast cancer risk. *Am. J. Clin. Nutr.* **2010**, *92*, 141–153. [CrossRef] [PubMed]

117. Lin, Y.; Yngve, A.; Lagergren, J.; Lu, Y. Dietary intake of lignans, risk of adenocarcinoma of the esophagus, gastroesophageal junction. *Cancer Causes Control.* **2012**, *23*, 837–844. [CrossRef] [PubMed]

118. Lin, Y.; Wolk, A.; Hakansson, N.; Lagergren, J.; Lu, Y. Dietary intake of lignans, risk of esophageal, gastric adenocarcinoma: A cohort study in Sweden. *Cancer Epidemiol. Biomarkers Prev.* **2013**, *22*, 308–312. [CrossRef] [PubMed]

119. Zamora-Ros, R.; Guinó, E.; Alonso, M.H.; Vidal, C.; Barenys, M.; Soriano, A.; Moreno, V. Dietary flavonoids, lignans, colorectal cancer prognosis. *Sci. Rep.* **2015**, *5*, 14148. [CrossRef] [PubMed]

120. Wallstrom, P.; Drake, I.; Sonestedt, E.; Gullberg, B.; Bjartell, A.; Olsson, H.; Adlercreutz, H.; Tikkanen, M.J.; Wirfält, E. Plasma enterolactone, risk of prostate cancer in middle-aged Swedish men. *Eur J. Nutr.* **2018**, *57*, 2595–2606. [CrossRef] [PubMed]

121. Eriksen, A.K.; Kyrø, C.; Nørskov, N.; Bolvig, A.K.; Christensen, J.; Tjønneland, A.; Overvad, K.; Landberg, R.; Olsen, A. Prediagnostic enterolactone concentrations, mortality among Danish men diagnosed with prostate cancer. *Eur. J. Clin. Nutr.* **2017**, *71*, 1235–1240. [CrossRef] [PubMed]

122. Bylund, A.; Lundin, E.; Zhang, J.X.; Nordin, A.; Kaaks, R.; Stenman, U.H.; Aman, P.; Adlercreutz, H.; Nilsson, T.K.; Hallmans, G.; et al. Randomised controlled short-term intervention pilot study on rye bran bread in prostate cancer. *Eur. J. Cancer Prev.* **2003**, *12*, 407–415. [CrossRef] [PubMed]

123. Hedelin, M.; Klint, A.; Chang, E.T.; Bellocco, R.; Johansson, J.E.; Andersson, S.O.; Heinonen, S.M.; Adlercreutz, H.; Adami, H.O.; Grönberg, H.; et al. Dietary phytoestrogen, serum enterolactone, risk of prostate cancer: The cancer prostate Sweden study (Sweden). *Cancer Causes Control.* **2006**, *17*, 169–180. [CrossRef] [PubMed]

124. Anjum, S.; Abbasi, B.H.; Doussot, J.; Favre-Réguillon, A.; Hano, C. Effects of photoperiod regimes, ultraviolet-C radiations on biosynthesis of industrially important lignans, neolignans in cell cultures of *Linum usitatissimum* L. (Flax). *J. Photochem. Photobiol. B* **2017**, *167*, 216–227. [CrossRef] [PubMed]

125. Correa, R.C.G.; Peralta, R.M.; Haminiuk, C.W.I.; Maciel, G.M.; Bracht, A.; Ferreira, I. New phytochemicals as potential human anti-aging compounds: Reality, promise, challenges. *Crit. Rev. Food Sci. Nutr.* **2018**, *58*, 942–957. [CrossRef] [PubMed]

126. Witkowska, A.M.; Waśkiewicz, A.; Zujko, M.E.; Szcześniewska, D.; Stepaniak, U.; Pająk, A.; Drygas, W. Are Total, Individual Dietary Lignans Related to Cardiovascular Disease, Its Risk Factors in Postmenopausal Women? A Nationwide Study. *Nutrients* **2018**, *10*, 865. [CrossRef] [PubMed]

127. Pellegrini, N.; Valtuena, S.; Ardigo, D.; Brighenti, F.; Franzini, L.; Del Rio, D.; Scazzina, F.; Piatti, P.M.; Zavaroni, I. Intake of the plant lignans matairesinol, secoisolariciresinol, pinoresinol, lariciresinol in relation to vascular inflammation, endothelial dysfunction in middle age-elderly men, post-menopausal women living in Northern Italy. *Nutr. Metab. Cardiovasc. Dis.* **2010**, *20*, 64–71. [CrossRef] [PubMed]

128. Jacobs, D.R.; Pereira, M.A.; Meyer, K.A.; Kushi, L.H. Fiber from whole grains, but not refined grains, is inversely associated with all-cause mortality in older women: The Iowa women's health study. *J. Am. Coll. Nutr.* **2000**, *19*, 326–330. [CrossRef]

129. Vanharanta, M.; Voutilainen, S.; Rissanen, T.; Adlercreutz, H.; Salonen, J.T. Risk of cardiovascular disease-related, all-cause death according to serum concentrations of enterolactone: Kuopio Ischaemic Heart Disease Risk Factor Study. *Arch. Intern. Med.* **2003**, *163*, 1099–1104. [CrossRef] [PubMed]

130. Franco, O.H.; Burger, H.; Lebrun, C.E.; Peeters, P.H.; Lamberts, S.; Grobbee, D.E.; Van Der Schouw, Y.T. Higher dietary intake of lignans is associated with better cognitive performance in postmenopausal women. *J. Nutr.* **2005**, *135*, 1190–1195. [CrossRef] [PubMed]

131. Eichholzer, M.; Richard, A.; Nicastro, H.L.; Platz, E.A.; Linseisen, J.; Rohrmann, S. Urinary lignans, inflammatory markers in the US National Health, Nutrition Examination Survey (NHANES) 1999–2004, 2005–2008. *Cancer Causes Control.* **2014**, *25*, 395–403. [CrossRef] [PubMed]

molecules

MDPI

Article

Insight into the Influence of Cultivar Type, Cultivation Year, and Site on the Lignans and Related Phenolic Profiles, and the Health-Promoting Antioxidant Potential of Flax (*Linum usitatissimum* L.) Seeds

Laurine Garros [1,2,3,†] , Samantha Drouet [1,2,†], Cyrielle Corbin [1,2], Cédric Decourtil [1,2], Thibaud Fidel [1,2], Julie Lebas de Lacour [1,2], Emilie A. Leclerc [1,2], Sullivan Renouard [1,2], Duangjai Tungmunnithum [1,2,4], Joël Doussot [1,2,5], Bilal Haider Abassi [1,2,6], Benoit Maunit [2,3], Éric Lainé [1,2], Ophélie Fliniaux [7], François Mesnard [7] and Christophe Hano [1,2,*]

[1] Laboratoire de Biologie des Ligneux et des Grandes Cultures (LBLGC) EA1207 INRA USC1328, Plant LIGNANS Team, Université d'Orléans, 28000 Chartres, France; laurine.garros@univ-orleans.fr (L.G.); samantha.drouet@univ-orleans.fr (S.D.); cyrielle.corbin@univ-orleans.fr (C.C.); cedric.decourtil@univ-orleans.fr (C.D.); thibaud.fidel@univ-orleans.fr (T.F.); julie.lebas-de-lacour@univ-orleans.fr (J.L.d.L.); emilie.leclerc@univ-orleans.fr (E.A.L.); sullivan.renouard@univ-orleans.fr (S.R.); duangjai.tun@mahidol.ac.th (D.T.); joel.doussot@lecnam.net (J.D.); bhabbasi@qau.edu.pk (B.H.A.); eric.laine@univ-orleans.fr (É.L.)
[2] COSM'ACTIFS, Bioactifs et Cosmétiques, CNRS GDR3711, 45067 Orléans Cedex 2, France; benoit.maunit@univ-orleans.fr
[3] Institut de Chimie Organique et Analytique (ICOA) UMR7311, Université d'Orléans-CNRS, 45067 Orléans CEDEX 2, France
[4] Department of Pharmaceutical Botany, Faculty of Pharmacy, Mahidol University, 447 Sri-Ayuthaya Road, Rajathevi, Bangkok 10400, Thailand
[5] Le CNAM, Ecole Sciences Industrielles et Technologies de l'Information (SITI), Chimie Alimentation Santé Environnement Risque (CASER), 75141 Paris Cedex 3, France
[6] Department of Biotechnology, Quaid-i-Azam University, 45320 Islamabad, Pakistan
[7] Biologie des Plantes et Innovation (BIOPI) EA 3900, Université de Picardie Jules Verne, 80000 Amiens, France; ophelie.fliniaux@u-picardie.fr (O.F.); francois.mesnard@u-picardie.fr (F.M.)
[*] Correspondence: hano@univ-orleans.fr; Tel.: +33-237-309-753
[†] These two authors contributed equally to this work and should be considered both as first authors.

Received: 23 September 2018; Accepted: 11 October 2018; Published: 14 October 2018

Abstract: Flaxseeds are a functional food representing, by far, the richest natural grain source of lignans, and accumulate substantial amounts of other health beneficial phenolic compounds (i.e., flavonols, hydroxycinnamic acids). This specific accumulation pattern is related to their numerous beneficial effects on human health. However, to date, little data is available concerning the relative impact of genetic and geographic parameters on the phytochemical yield and composition. Here, the major influence of the cultivar over geographic parameters on the flaxseed phytochemical accumulation yield and composition is evidenced. The importance of genetic parameters on the lignan accumulation was further confirmed by gene expression analysis monitored by RT-qPCR. The corresponding antioxidant activity of these flaxseed extracts was evaluated, both in vitro, using ferric reducing antioxidant power (FRAP), oxygen radical absorbance capacity (ORAC), and iron chelating assays, as well as in vivo, by monitoring the impact of UV-induced oxidative stress on the lipid membrane peroxidation of yeast cells. Our results, both the in vitro and in vivo studies, confirm that flaxseed extracts are an effective protector against oxidative stress. The results point out that secoisolariciresinol diglucoside, caffeic acid glucoside, and *p*-coumaric acid glucoside are the main contributors to the antioxidant capacity. Considering the health benefits of these compounds, the present study demonstrates that the flaxseed cultivar type could greatly influence the phytochemical intakes and, therefore, the associated biological activities. We recommend that this crucial parameter be considered in epidemiological studies dealing with flaxseeds.

Keywords: cultivar; environment; flax; flavonol; genetic; hydroxycinnamic acid; lignan; seed

1. Introduction

The consumption of fruit, vegetables, and grains has been associated with lower risks of chronic and degeneration diseases [1]. Considering their numerous beneficial effects on human health, during the last decades, there has been an increasing interest in their uses, and flaxseeds are, therefore, considered as functional food [2]. Flaxseeds are the richest natural grain source of lignan and accumulate a substantial amount of other phenolic compounds (e.g., flavonols, hydroxycinnamic acids). In flaxseed, the foremost part of these phytochemicals is accumulated under the form of a macromolecular complex (also known as lignan macromolecule) composed of the lignan secoisolariciresinol diglucoside (SDG, Figure 1A) as the main component, and of flavonol herbacetin diglucoside (HDG, Figure 1A), as well as hydroxycinnamic acid derivatives: *p*-coumaric acid glucoside (CouG, Figure 1A), caffeic acid glucoside (CafG, Figure 1A), and ferulic acid glucoside (FerG, Figure 1A), ester-linked together to hydroxymethylglutaryl spacers (Figure 1B) [3,4].

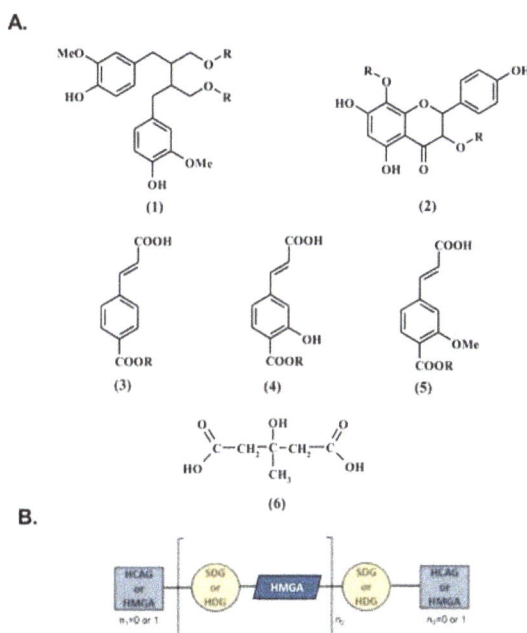

Figure 1. Structure of phenolic compounds involved in the lignan macromolecular complex. (**A**) Structure of the complex components: (**1**) secoisolariciresinol (R = H), or secoisolariciresinol diglucoside (SDG, R = β-D-glucose), (**2**) herbacetin (R = H) or herbacetin diglucoside (HDG, R = β-D-glucose), (**3**) *p*-coumaric acid (R = H), or *p*-coumaric acid glucoside (CouG, R = β-D-glucose), (**4**) caffeic acid (R = H) or caffeic acid glucoside (CafG, R = β-D-glucose), (**5**) ferulic acid (R = H) or ferulic acid glucoside (FerG, R = β-D-glucose), (**6**) hydroxymethylglutaric acid (HMGA). (**B**) Schematic representation of lignan macromolecule, where a unit of SDG or HDG is ester-linked to another unit, thanks to HMGA, which can be replaced by one hydroxycinnamic acid glucoside (HCAG) unit (CouG, CafG, or FerG) in terminal position of the chain.

The beneficial effects of lignans on human health are well recognized [5,6]. Particularly, the chemopreventive actions of SDG toward cancer, diabetes mellitus, and cardiovascular diseases have been largely described [5,7,8]. The pharmacological activity of this compound is thought to be due to its high antioxidant capacity [9–11] and to its phytoestrogenic activity [12]. Flavonols and hydroxycinnamic acids, the other constituents of the flaxseed lignan macromolecule, also display a wide range of health-promoting effects. The favorable actions on cardiovascular health of vegetable-rich diets have been ascribed to flavonols, and hydroxycinnamic acids have revealed powerful antioxidant properties and might be of particular interest for dermatologic applications [13,14].

Although both in vivo and in vitro data are globally in favor of a chemopreventive effect of lignans, epidemiological studies are much less conclusive, and the mechanism by which phytoestrogenic lignans prevent cancers still remains unclear [7] and requires further elucidation. This could be explained by the fact that our current knowledge concerning the genetic and environmental factors affecting productivity and yield stability of these phenolic compounds, in flaxseeds, remains partial, and little is known about the variation in antioxidant capacities of different flaxseed cultivars. Moreover, no study has put efforts toward linking the lignan content of different cultivars and the expression of genes involved in their biosynthetic pathway.

Herein we present a complete dataset concerning the relative impact of cultivar, edaphic, and climatic parameters on productivity of the main constituents of the lignan macromolecule of flaxseeds, in relation to their antioxidant capacities determined using both in vitro and in vivo systems. Such data could be useful to predict, more precisely, the accumulation and, therefore, the nutritional intakes of these compounds, with health benefits for pharmaceutical, nutraceutical, and/or cosmetic applications.

2. Materials and Methods

2.1. Chemicals

All chemicals were of analytical grade quality and purchased from Thermo (Illkirch, France). The deionized water was produced using a milli-Q water purification system (Merck Millipore, Molsheim, France). SDG and HDG standard were purchased from LGC Standards (Molsheim, France). The hydroxycinnamic acid glucosides—*p*-coumaric acid glucoside, caffeic acid glucoside, ferulic acid glucoside—were synthesized according to Beejmohun et al. (2004) [15] and Beejmohun et al. (2006) [16]. Prior to their use for HPLC or LC-MS analysis, all solutions were filtered through 0.45 μm nylon syringe membranes (Merck Millipore, Molsheim, France).

2.2. Plant Materials and Cultivation

Flax cultivars Astral, Baïkal, Baladin, Barbara, and Oliver were provided by Laboulet Semences (Airaines, France), Coopérative Linière Terre de Lin (Saint-Pierre-le-Viger, France) and Arvalis-Institut Technique du Lin (Boigneville, France). Flax was grown up to seed at the following locations in France: Eure (Gamaches-en-Vexin, GAM, 49°16′14″N/1°37′02″E/89 m), Somme (Airaines, AIR, 49°57′57″N/1°56′39″/70 m), and Eure-et-Loir (Chartres, CHA, 48°27′21.05″N/1°29′3.06″E/141 m). Sowings were performed on March 30th of each year, with 450 seeds per m². Fields were fertilized, immediately after sowing, with 80 units of nitrogen, 60 units of potassium, and 60 units of phosphorus per hectare (Figure S1). The soils of these sites were of clay loam type balanced, well-structured with a granulometry of ca. 25% 2000–63 μm, 50% 63–2 μm, and 25% <2 μm particles, and a pH around 7.8. The final harvest took place on August 15th of each year at the same ripening stage for each cultivar. Throughout the experiments, no visible disease or insect attack occurred at either location. During the growing period, the experimental stations received 182.8 mm (year 2003), 305.0 mm (year 2004), 269.8 mm (year 2005) for GAM, 387.4 mm for AIR (year 2005), and 308.6 mm for CHA (year 2005) of rainfall over the growing period. The day temperatures at an elevation of 2 m averaged 15.72 °C (year 2003), 13.67 °C (year 2004), 13.96 °C (year 2005) for GAM, 13.48 °C for AIR (year 2005), and 14.45 °C for CHA (year 2005) over the growing period. All these meteorological characteristics are displayed in Table S1 and Figure S2.

2.3. Gene Expression Analysis by RT-qPCR

Total RNA was extracted from 100 mg of frozen plant material in liquid nitrogen as described by Hano et al. (2006) [17]. Expression patterns of *LuDIR5*, *LuPLR1*, and *LuUGT74S1* were analyzed using RT-qPCR, using specific primers described by Dalisay et al. (2015) [18]. For reverse transcription, 50 ng of total RNA was incubated for 60 min at 50 °C with 1× RT buffer, 0.5 mM of each dNTP, 1 μM of oligo-dT primers, 1 unit of RiboLock, and 4 units of Omniscript Reverse Transcriptase in a total volume of 20 μL (Qiagen, Hilden, Germany). qPCR was performed with a PikoReal™ Real-Time PCR System (Thermo Fisher Scientific, Villebon-sur-Yvette, France) using DyNAmo ColorFlash SYBR Green qPCR (ThermoScientific) and specific primers. Two reference genes (*CYC* and *ETIF5A*) were used for data normalization [19]. The qPCR parameters were as follows: an initial denaturation at 95 °C for 5 min, then 40 three-step cycles of 94 °C for 10 s, primer annealing at 65 °C for 10 s, and extension at 72 °C for 30 s. After 40 cycles, an additional extension step was performed at 72 °C for 90 s. The presence of a single amplicon was confirmed by the observation of a single peak in the melting curve obtained after amplification. Expression levels were calculated and normalized using $2^{-\Delta\Delta Ct}$ method [20]. Reactions were performed in three biological and two technical replicates.

2.4. Extraction, HPLC, and LC-ESI-MS Analysis

Extractions (4 biological and 2 technical replicates), quantification of compounds was carried out on a Varian liquid chromatographic system (Agilent Technology, Les Ulis, France), as well as LC-ESI-MS analyses using a Waters 2695 Alliance coupled with a single quadrupole mass spectrometer ZQ (Waters-Micromass, Manchester, UK), equipped with an electrospray ion source (ESI-MS), were performed as described in Corbin et al. (2015) [21].

2.5. Determination of the Ferric-Reducing Antioxidant Power (FRAP)

Ferric-reducing antioxidant power (FRAP) was measured as described by Benzie & Strain, (1996) [22] with little modification. Briefly, 10 μL of the extracted sample was mixed with 190 μL of FRAP (10 mM TPTZ; 20 mM $FeCl_3 \cdot 6H_2O$, and 300 mM acetate buffer pH 3.6; ratio 1:1:10 (v/v/v)). Incubation lasted 15 min at room temperature. Absorbance of the reaction mixture was measured at 630 nm with a BioTek ELX800 Absorbance Microplate Reader (Thermo Fisher Scientific, Villebon-sur-Yvette, France). Assays were made in triplicate and antioxidant capacity was expressed as Trolox C equivalent antioxidant capacity (TAEC).

2.6. Determination of Oxygen Radical Absorbance Capacity (ORAC)

Oxygen radical absorbance capacity (ORAC) assay was performed as described by Prior et al. (2003) [23]. Briefly, 10 μL of the extracted sample was mixed with 190 μL of fluorescein (0.96 μM) in 75 mM phosphate buffer pH 7.4, and incubated for at least 20 minutes at 37 °C with intermittent shaking. Then, 20 μL of 119.4 mM 2,2′-azobis-amidinopropane (ABAP, Sigma Aldrich, Saint-Quentin Fallavier, France) was added and the fluorescence intensity was measured every 5 min for 2.5 h at 37 °C using a fluorescence spectrophotometer (Bio-Rad, Marnes-la-Coquette, France) set with an excitation at 485 nm and emission at 535 nm. Assays were made in triplicate, and antioxidant capacity was expressed as Trolox C equivalent antioxidant capacity (TAEC).

2.7. Determination of the Iron-Chelating Capacity

The iron-chelating capacity was determined as described by Mladenka et al. (2011) [24]. Briefly, 10 μL of extract sample were mixed with ferrous iron at a final concentration of 50 μM in HEPES (pH 6.8) buffer and 50 μL ferrozine (5 mM aqueous solution). All experiments were performed in

96-well microplates. Each sample was measured with and without (blank) the addition of ferrozine. Absorbance was measured at 550 nm immediately after addition of ferrozine, and 5 min later with a BioTek ELX800 Absorbance Microplate Reader (Thermo Fisher Scientific, Villebon-sur-Yvette, France). Chelating activity values were expressed in μM of fixed iron.

2.8. Yeast Cells Cultivation and Treatments

Yeast (*Saccharomyces cerevisiae*) strain MAV203 (Invitrogen, Thermo Fisher Scientific Villebon-sur-Yvette, France) were used. Cells were grown aerobically at 30 °C in an orbital shaker (150 rpm) in complete 2.0% (w/v) glucose YPD medium (Sigma Aldrich, Saint-Quentin Fallavier, France). All extracts evaporated under nitrogen flow, dissolved in DMSO at 50 μg/mL, and added to the cells 6 h before oxidative stress induction at a final concentration of 1 mg/mL. The final concentration of DMSO applied on the cell was 1 % (v/v). For the control sample, DMSO to 0.1% of the final volume, was added. Cells were irradiated with 106.5 J/m^2 UV-C (254 nm) under a Vilber VL-6.C filtered lamp (Thermo Fisher Scientific, Villebon-sur-Yvette, France), as described by Bisquert et al. (2018) [25], and then incubated overnight at 30 °C before membrane lipid peroxidation determination.

2.9. Determination of Membrane Lipid Peroxidation Using Thiobarbituric Acid-Reactive Substances (TBARS) Assay

Measurement of membrane lipid peroxide was carried out with the thiobarbituric acid (TBA; Sigma Aldrich, Saint-Quentin Fallavier, France) method described by Hano et al. (2008) [26]. Briefly, ca. 10^7 cells were ground using a mortar and pestle in distilled water, and centrifuged at $10,000 \times g$ for 10 min. Supernatant fractions (75 μL) were mixed with 25 μL of 3% (w/v) SDS, 50 μL of 3% TBA (w/v) in 50 mM NaOH, and 50 μL of 23% (v/v) of HCl throughout mixing between each addition. The mixture was heated at 80 °C for 20 min. After cooling on ice, the absorbance at 532 nm (A532) was measured, and non-specific absorbance at 600 nm (A600) was subtracted.

2.10. Statistical Treatment of Data

All data presented in this study are the means and the standard deviations of at least three independent replicates. ANOVAs and Pearson correlations were performed using R software version 3.0.2. PCA was performed with XL-STAT2017 software (Addinsoft, Paris, France), with each parameter considered as a discrete variable; the initial dataset was then converted into principal components (PCs), and it was possible to graphically display the relationships among the considered parameters. Gene expression and SDG content were represented using MeV4 software. All statistical tests were considered significant at $p < 0.05$.

3. Results and Discussion

3.1. Influence of Genetic Variations on the Accumulation of the Main Constituents of the Lignan Macromolecule

The flax cultivars, herein studied, showed an SDG content ranging from 8.23 to 21.85 mg/g of dry weight (DW) (Table 1). Barbara and Oliver are high SDG-producing cultivars, Baladin presents an intermediate content, whereas Astral and Baïkal are poor in SDG, as compared to the other cultivars. A similar range of variation in SDG content has been reported in a flax germplasm collection by Diederichsen and Fu (2008) [27]. Lower SDG content was reported by Zimmermann et al. (2007, 2006) [28,29] for cultivars grown in Spain and Germany. Nevertheless, it should be noted that these authors employed an extraction method based on acid hydrolysis, which is known to be potentially destructive for SDG [30], leading to a possible underestimation in the actual contents. SDG is the main component of the lignan macromolecule accumulated in flaxseed, but other compounds, such as hydroxycinnamic acid glucosides (caffeic acid glucoside (CafG), *p*-coumaric acid glucoside (CouG), and ferulic acid glucoside (FerG), as well as the flavonol herbacetin diglucoside (HDG), are also incorporated in substantial amounts in this macromolecule [4,31,32]. Here, the whole set of these compounds was assayed. In our hands, CouG contents ranged from 4.78 to 10.48 mg/g DW, and FerG content from 1.03 to 2.28 mg/g DW (Table 1). These results sound consistent with those described by Westcott and Muir (1996), Jonhson et al. (2000), and Eliasson et al. (2003) [33–35] for cultivars grown respectively in Canada, Denmark, and Sweden. To date, only semi-quantitative evaluation of the HDG variations in flax cultivars have been studied through NMR [36], therefore, to the best of our knowledge, the present work is the first study focusing on the quantitative variations in HDG contents in linseed cultivars. Concerning the quantitative variations in CafG, only Wang et al. (2017) [37] reported very low contents ranging from 2.40 to 8.70 µg/g DW for Chinese cultivars. Here, HDG content ranged from 0.75 to 1.18 mg/g DW, and CafG contents from 0.80 to 1.90 mg/g DW (Table 1).

Table 1. Influence of the cultivar (C), cultivation site (L), and year (Y) on the accumulation of the main constituents of the lignan macromolecule in flaxseeds.

Cultivar	Location_Year	SDG [a]	HDG [a]	FerG [a]	CouG [a]	CafG [a]
Astral	AIR_05	12.85 ± 0.14	0.98 ± 0.06	1.65 ± 0.06	5.88 ± 0.09	0.85 ± 0.04
	CHA_05	12.53 ± 0.11	1.05 ± 0.03	1.90 ± 0.08	6.05 ± 0.08	0.98 ± 0.06
	GAM_03	11.73 ± 0.06	0.88 ± 0.04	1.58 ± 0.07	4.80 ± 0.08	1.23 ± 0.06
	GAM_04	11.68 ± 0.09	1.10 ± 0.05	1.58 ± 0.04	4.78 ± 0.05	1.25 ± 0.05
	GAM_05	13.48 ± 0.13	0.85 ± 0.01	1.57 ± 0.02	6.07 ± 0.22	0.87 ± 0.02
Barbara	AIR_05	21.68 ± 0.17	0.93 ± 0.06	1.95 ± 0.03	10.48 ± 0.12	1.83 ± 0.04
	CHA_05	20.88 ± 0.07	1.05 ± 0.03	2.18 ± 0.07	8.63 ± 0.65	1.63 ± 0.04
	GAM_03	20.03 ± 0.10	0.95 ± 0.03	1.78 ± 0.03	9.95 ± 0.11	1.33 ± 0.06
	GAM_04	21.85 ± 0.34	0.85 ± 0.03	1.78 ± 0.02	9.85 ± 0.26	1.53 ± 0.06
	GAM_05	21.85 ± 0.81	0.88 ± 0.01	1.66 ± 0.01	9.84 ± 0.07	1.53 ± 0.01
Baladin	AIR_05	16.03 ± 0.11	1.15 ± 0.03	2.10 ± 0.05	7.88 ± 0.07	1.48 ± 0.07
	CHA_05	15.68 ± 0.19	1.18 ± 0.03	2.28 ± 0.07	7.58 ± 0.10	1.43 ± 0.09
	GAM_03	18.20 ± 0.38	1.03 ± 0.03	2.03 ± 0.04	7.88 ± 0.12	1.33 ± 0.02
	GAM_04	16.45 ± 0.13	1.08 ± 0.07	2.08 ± 0.02	7.33 ± 0.21	1.40 ± 0.08
	GAM_05	16.20 ± 0.38	0.95 ± 0.01	1.92 ± 0.01	6.91 ± 0.04	1.19 ± 0.01
Baïkal	AIR_05	8.33 ± 0.13	1.03 ± 0.04	1.75 ± 0.07	4.85 ± 0.07	0.85 ± 0.01
	CHA_05	8.23 ± 0.07	1.18 ± 0.02	1.75 ± 0.05	4.85 ± 0.06	1.05 ± 0.03
	GAM_03	8.23 ± 0.17	1.15 ± 0.05	1.88 ± 0.07	4.90 ± 0.07	0.80 ± 0.05
	GAM_04	8.45 ± 0.10	1.05 ± 0.08	1.98 ± 0.03	4.98 ± 0.08	0.98 ± 0.04
	GAM_05	8.40 ± 0.04	0.98 ± 0.01	1.79 ± 0.02	4.88 ± 0.05	0.83 ± 0.01
Oliver	AIR_05	21.00 ± 0.25	0.75 ± 0.03	1.55 ± 0.03	10.15 ± 0.09	1.90 ± 0.05
	CHA_05	19.50 ± 0.19	0.85 ± 0.03	1.33 ± 0.06	9.03 ± 0.26	1.75 ± 0.06
	GAM_03	19.95 ± 0.11	0.98 ± 0.04	1.18 ± 0.06	9.38 ± 0.07	1.33 ± 0.06
	GAM_04	19.93 ± 0.11	1.03 ± 0.07	1.03 ± 0.03	9.35 ± 0.09	1.45 ± 0.05
	GAM_05	21.55 ± 0.37	0.83 ± 0.01	1.06 ± 0.01	9.12 ± 0.08	1.37 ± 0.01
F values	Cultivar (C)	284.62 ***	4.06 *	19.18 ***	95.33 ***	14.76 ***
	Location (L)	0.02	1.21	1.04	0.13	0.61
	Year (Y)	0.03	2.09	0.68	0.02	0.48
	C × L	194.69 ***	3.57 *	21.90 ***	75.96 ***	11.94 ***
	C × Y	198.99 ***	4.58 *	17.25 ***	59.81 ***	11.19 ***
	Y × L	0.02	1.484	0.536	0.062	0.441
	C × L × Y	148.70 ***	4.23 *	16.18 ***	51.26 ***	9.62 ***

[a] All contents are given in mg/g DW. Values are mean ± SD of 4 independent replicates. ANOVA, F represents the effect. Significance level: * $p < 0.05$; ** $p < 0.01$; *** $p < 0.001$.

In flaxseed, the lignan biosynthesis involves the dirigent protein (*LuDIR5*; Figure 2A)-mediated stereoselective coupling of two *E*-coniferyl alcohol moieties, resulting in the formation of (−)-pinoresinol [18,38]. The two following reaction steps leading to the conversion of (−)-pinoresinol to (−)-lariciresinol, and (−)-lariciresinol to (+)-secoisolariciresinol, are catalyzed by the same bifunctional enzyme pinoresinol–lariciresinol reductase (*LuPLR1*, Figure 2A) [17,39,40]. Secoisolariciresinol is then glycosylated into SDG under the control of UDP-glycosyltransferase (*LuUGT74S1*, Figure 2A) glycosylating the C-9 and C-9′ hydroxyl positions [41,42]. SDG is stored as a

3-hydroxy-3-methylglutaryl ester-linked complex (HMG-SDG), as shown in Figure 1. Formation of the HMG–SDG ester-linked oligomers, occurs by linking hydroxylmethylglutaryl (HMG) to C-6a and C-6a′ position, via action of HMG CoA-transferase [43].

A.

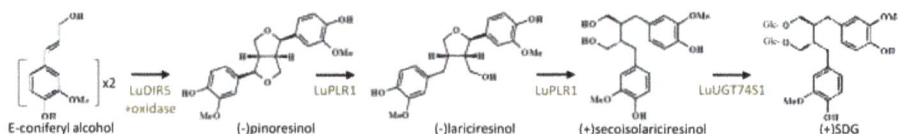

B.

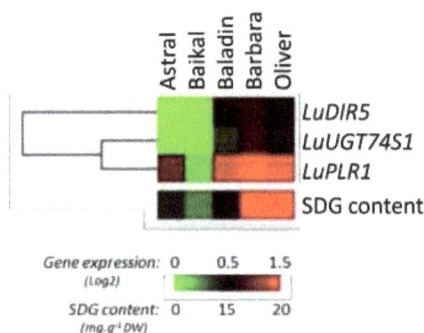

C.

	DIRS	PLR1	UGT74S1
SDG	0.817***	0.957***	0.833***

Gene expression: 0   0.5   1.5
(Log2)

SDG content: 0   15   20
(mg·g⁻¹ DW)

Figure 2. Expression profile of flax lignan biosynthetic gene and SDG accumulation in five flaxseed cultivars. (**A**) Biosynthetic pathway leading to the formation of (+)-SDG in flaxseed. (**B**) Expression of *LuDIR5*, *LuPLR1*, and *LuUGT74S1* determined by RT-qPCR (normalized with *CYC* and *ETIF5A* reference genes) visualized using MeV4 (*n* = 3) and SDG content measured by HPLC and visualized using MeV4 (*n* = 3). (**C**) Pearson correlation matrix between (+)-SDG accumulation and the corresponding biosynthetic gene expression. Significance level: * *p* < 0.05; ** *p* < 0.01; *** *p* < 0.001.

Correlation analysis between the different constituents of the flax lignan macromolecule revealed significant positive correlations between the CafG, CouG, and SDG contents, on the one hand, and between HDG and FerG, on the other hand (Table 2). This correlation was in agreement with our previous results [36]. On the contrary, significant negative correlations were noted between the SDG vs HDG and FerG yields (Table 2), which confirmed our previous observations [36]. From a metabolic point of view, *p*-coumaric acid (Figure 1A) is a branch point leading to the biosynthesis of either flavonoids or lignans [44]. Therefore, caffeic acid and *p*-coumaric acid (Figure 1A) could be considered as more direct precursors for the HDG biosynthesis, whereas ferulic acid (Figure 1A) constitutes a precursor for SDG biosynthesis. These biosynthetic links could explain, in part, the observed correlations. Studying the possible metabolic channel regulation of the carbon allocation between these two branches, during flaxseed development, could be of particular interest.

As a step forward, lignan biosynthetic gene expression analysis performed on immature flaxseed (developmental stage 2; [17]) by RT-qPCR using the 3 specific genes involved in SDG biosynthesis (*LuDIR5*, *LuPLR1*, and *LuUGT74S1*; Figure 2A,B) appeared in good agreement with the HPLC quantification (Figure 2C). High expression of *LuPLR1* was detected in high SDG-producing cultivars, Barbara and Oliver, whereas Astral and Baïkal cultivars, accumulating lower SDG content, showed a lower expression of these biosynthetic genes (Table 1). The steady state levels of the key *LuPLR1* transcripts [40], and the two other biosynthetic genes (*LuDIR5* and *LuUGT74S1*) are correlated with the SDG content measured in the corresponding mature seeds (Figure 2C), confirming the great influence of genetic parameters (i.e., the cultivar), and indicated that most of the regulation occurred at transcriptional level.

Table 2. Correlation analysis using Pearson correlation coefficient (PCC).

Variables	SDG	HDG	FerG	CouG	CafG	FRAP	ORAC	Iron Chelation	MDA inhibition
SDG									
HDG	−0.515 **								
FerG	−0.231 ns	0.510 **							
CouG	0.966 ***	−0.476 *	−0.215 ns						
CafG	0.835 ***	−0.360 ns	−0.063 ns	0.832 ***					
FRAP	0.676 ***	−0.624 ***	−0.545 **	0.639 ***	0.671 ***				
ORAC	0.669 ***	−0.325 ns	−0.170 ns	0.676 ***	0.717 ***	0.573 **			
Iron Chelation	0.758 ***	−0.627 ***	−0.692 ***	0.768 ***	0.661 ***	0.817 ***	0.665 ***		
MDA inhibition	0.867 ***	−0.666 ***	−0.482 *	0.806 ***	0.721 ***	0.774 ***	0.617 ***	0.875 ***	

Significance level: * $p < 0.05$; ** $p < 0.01$; *** $p < 0.001$; ns: not significant.

3.2. Influence of Geographic Parameters on the Accumulation of the Main Constituents of Lignan Macromolecule

It is well accepted that environmental conditions, such as the climate of the culture year and the location (soil conditions), could also greatly affect the accumulation of phenolic compounds, as previously observed by Oomah et al. (1996) [45] for the accumulation of total flavonoids in flaxseeds. Here, three different locations have been selected to provide access to the potential influence of edaphic condition on lignan accumulation in flaxseed. Bordered by four different seas, three mountain ranges, and the edge of the central European lowlands, France is known to be a country with very diverse climatic conditions, resulting in very different weather patterns. Here, the three selected experimental sites are representative of the major flax-growing areas in France, i.e., the western part, and the present contrasting weather patterns. The CHA site is characterized by the highest temperatures and the lowest rainfall during the seed maturation phase. On the contrary, AIR location presents the lowest temperatures and the highest rainfall observed during the same period. The last location, GAM, is considered as an intermediate in terms of climate. The impact of these different conditions, on the composition and amount of the main constituents of the lignan macromolecule accumulated in the seed of the five selected cultivars, are presented in Table 1. Analysis of the variance revealed that cultivar was the main contributor for the observed variability (cultivars, C, Table 1). Edaphic factor (location L, Table 1) has no significant effect on the accumulation of these phytochemicals, whereas significant interactions with genetic factor were noted, but evidenced the prominent effect of genetic background at a particular location according to F values (Table 1).

Nonetheless, the location constitutes a complex variable, differing by both climatic and edaphic parameters, thus, to evaluate the sole contribution of climate, we decided to compare flaxseeds grown at the same site, GAM (i.e., the same edaphic parameters) but in different cultivation years (i.e., different climatic parameters). Here, we chose to consider three consecutive years with very contrasting weather patterns and, for this reason, the 2003–2005 period was selected. Indeed, it must be noted that the summer of 2003 was the hottest and driest in recent decades, and must be regarded as extremely unusual. The 2003–2005 period was also the warmest period recorded in France since 1950, whereas the low rainfall observed from June 2004 to December 2005 led to a dramatic soil water deficit for 2005, with a soil humidity index close to 0.25 for GAM region (considering that a soil humidity index of 1 is for water-saturated soil whereas 0 is for water-depleted soil; see Table S1, Figure S2 for complete meteorological condition descriptions). As flax is known to be a water-demanding crop during its flowering period (i.e., June), we therefore decided to evaluate how these climate changes, leading to water deficiency during this period, have affected the flaxseed metabolism. The results are reported in

Table 1, and the analysis of the variance evidenced the genetic background (cultivars, C, Table 1) as the sole significant factor influencing the SDG, FerG, CouG, and CafG content (Table 1). The climatic parameters considered here (cultivation year, Y, Table 1) did not influence the accumulation of any molecules in the analyzed cultivars, whereas significant interaction between genetic and climatic parameters (C × Y, Table 1) was noted, but with lower *F* values as compared to genetic parameter alone (C, Table 1), indicating that the main contributions have to be attributed to this latter parameter. Our results are in good agreement with the results of Saastamoinen et al. (2013) [46], who also reported a lower impact of the cultivation year compared to the cultivar parameter on SDG accumulation. On the contrary, Wescott et al. (2002) [47] reported that the cultivation year could also influence SDG yield. This apparently contradictory result can be due to the complexity of the climatic variable, that could also be influenced by the nature of the soil considered (edaphic parameters). The nature of the soil could greatly affect the influence of the drought period as its ability to retain water greatly relied on its composition and granulometry. Indeed, a high soil ability to retain water could alleviate the effect of temporary drought, and differences in this feature could explain such apparent discrepancies, moreover, the rainfall regime differs between Scandinavian and Canadian summers, making it more probable that drought occurs during the latter.

All these phytochemical profiles were subjected to principal component analysis. The resulting biplot representation accounts for 99.43% (F1 + F2) of the initial variability of the data (Figure 3). Discrimination occurs mainly in the first dimension, and SDG content was the main contributor for this F1 axis that explains 89.43% of the initial variability. The concentrations of hydroxycinnamic acid glucosides (particularly CouG) were the main contributors for the second dimension (F2 axis), accounting for only 10% of the initial variability (Figure 3). PCA showed a significant grouping of samples as a function of their SDG content. Using this analysis, the different cultivars could also be easily discriminated. This PCA confirmed the prominence of the genetic background over the environmental (edaphic and climatic) factors studied here.

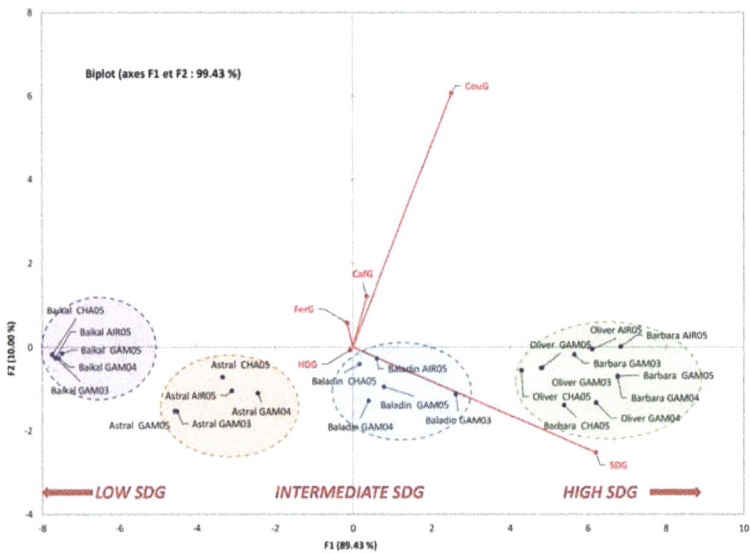

Figure 3. Correlation circle for principal component analysis. The SDG, HDG, CafG, CouG, and FerG contents for 5 cultivars (Astral, Baïkal, Baladin, Barbara, and Oliver) growing at 3 different locations (GAM, AIR, and CHA) and over 3 different years (03 (2003), 04 (2004), or 05 (2005)) were submitted for analysis by the PCA algorithm in Excel-XLSTAT software, using the Pearson correlation matrix (at a significance level of *p* < 0.05).

3.3. Evaluation and Comparison of In Vitro and In Vivo Antioxidant Capacities

To evaluate the influence of genetic and edaphic variables on the health benefit potential of these flaxseeds, the antioxidant capacity of the corresponding extracts was then evaluated using both in vitro and in vivo assays. On the basis of the chemical reaction involved, the major antioxidant capacity assays can be roughly divided into two categories: i) hydrogen atom transfer (HAT) reaction-based assay, such as ORAC assay, or ii) electron transfer (ET) reaction-based assay, such as FRAP assay (Table 3).

In our hands, the antioxidant capacity of our flaxseed extracts revealed by these two different assays ranged from 217.35 (Baladin, AIR_05) to 355.75 (Oliver, GAM_05) µM of Trolox C equivalent antioxidant capacity (TEAC) using FRAP assay, and from 269.97 (Baïkal, GAM_04) to 375.76 (Oliver, GAM_05) µM TEAC using ORAC assay (Table 3). The antioxidant capacity of polyphenolic compounds, such as lignans, has been previously attributed to their capacity for HAT, from their OH groups to the free radicals [48]. However, the radical scavenging capacity of these extracts occurring through an ET-based mechanism cannot be excluded, according to the high antioxidant values calculated from the FRAP assay (Table 3). Here these two in vitro antioxidant assays were significantly correlated with the presence of SDG, CouG and CafG (Table 2).

Besides these two mechanisms involved in the scavenging of reactive oxygen species, transient metal ion chelation is also considered as an antioxidant mechanism, since the Fenton reaction, responsible for the hydroxyl radical formation and, subsequently, radical chain reaction propagation, could be inhibited through this chelating mechanism [49,50]. Here, we evidenced that flaxseed extracts displayed an efficient iron (Fe^{2+})-chelating activity, ranging from 7.18 µM (Baïkal, GAM_03) to 14.72 µM (Oliver, GAM_04) of fixed iron (Table 3), that could also contribute to their antioxidant activity, largely described in the literature [9,10]. In good agreement with the recent rationalization of the iron-chelating capacity of SDG and its aglycone form secoisolariciresinol, high SDG quantities associated with elevated contents of CouG and CafG, appeared to significantly contribute to the development of a high iron-chelating capacity of the corresponding flaxseed extracts (Table 2).

It is necessary to emphasize that the assays described herein are strictly predictive results based on the chemical reaction in vitro, however, they not necessary bear a great similarity to biological systems. The validity of these data has to be, therefore, considered as limited to a strict chemical sense with context interpretation. For this reason, in order to better reflect the in vivo situation, the antioxidant activity of these extracts was further investigated for their capacity to inhibit membrane lipid peroxidation induced by UV-C in yeast cells. Yeast cells have been proven to be an excellent model to evaluate in vivo antioxidant capacity in a cellular oxidative stress context [51]. Indeed, baker's yeast (*Saccharomyces cerevisiae*) is an attractive and reliable model. This organism is a true eukaryote, and the mechanisms of defense and adaptation to oxidative stress are well understood [25,52]. The in vivo anti-lipoperoxidation activity (inhibition of malondialdehyde (MDA) formation), determined using the TBARS assay, ranged from 21.68% (Baïkal, GAM_04) to 47.02% (Oliver, AIR_05) (Table 3). Interestingly, a strong and significant correlation was observed between this cellular antioxidant capacity and the SDG (PCC = 0.867), CouG (PCC = 0.806), and CafG (PCC = 0.721) contents (Table 2). However, we can note that, since the contents of SDG, CouG, and CafG are highly correlated, these parameters are not independent, and it is, therefore, difficult to definitely judge their respective contribution to this biological activity (cellular antioxidant capacity) by single correlation analysis. In yeast models, a similar protective effect against oxidative stress was previously observed on yeast cells treated with thiamine [52] and melatonin [25]. To the best of our knowledge, this the first time that this system is applied to characterize a flax extract. Our results are in agreement with those obtained using in vitro assays, and highlighted the great in vivo antioxidant potential of flaxseed extracts as already proposed by Wang et al. (2017) [37], using another cellular antioxidant assay in HepG2 cells.

Table 3. Influence of the cultivar (C), cultivation site (L), and year (Y) on the in vitro and in vivo antioxidant activities of flaxseed extracts.

Cultivar	Location_Year	FRAP [a]	ORAC [a]	Iron Chelation [b]	MDA Inhibition [c]
Astral	AIR_05	252.55 ± 2.45	281.55 ± 11.16	9.57 ± 0.25	37.10 ± 0.28
	CHA_05	264.41 ± 10.28	317.87 ± 8.00	10.11 ± 0.41	36.95 ± 0.43
	GAM_03	322.81 ± 1.41	263.13 ± 11.91	9.66 ± 0.12	38.47 ± 0.75
	GAM_04	276.68 ± 4.33	312.34 ± 1.12	9.31 ± 0.53	37.71 ± 1.18
	GAM_05	252.55 ± 5.23	263.92 ± 4.56	9.57 ± 0.22	38.17 ± 1.07
Barbara	AIR_05	332.55 ± 4.57	339.45 ± 1.95	11.97 ± 0.19	45.95 ± 2.26
	CHA_05	317.61 ± 9.33	341.29 ± 16.09	11.35 ± 0.72	44.58 ± 1.08
	GAM_03	292.81 ± 5.84	329.45 ± 8.65	11.79 ± 0.06	43.05 ± 3.23
	GAM_04	296.41 ± 8.20	336.55 ± 2.14	12.32 ± 0.53	46.41 ± 0.43
	GAM_05	331.48 ± 5.23	309.45 ± 3.72	10.99 ± 0.28	45.65 ± 1.94
Baladin	AIR_05	217.35 ± 4.67	286.82 ± 2.79	8.69 ± 0.31	33.44 ± 2.16
	CHA_05	240.55 ± 6.31	334.18 ± 5.95	8.24 ± 0.44	33.28 ± 0.97
	GAM_03	254.01 ± 2.30	307.61 ± 4.18	8.16 ± 0.43	32.06 ± 0.64
	GAM_04	288.41 ± 11.27	321.82 ± 2.60	8.87 ± 0.06	28.85 ± 0.43
	GAM_05	255.08 ± 14.24	281.03 ± 2.70	7.80 ± 0.59	34.50 ± 2.16
Baïkal	AIR_05	262.01 ± 3.21	278.66 ± 5.76	8.42 ± 0.37	25.34 ± 1.18
	CHA_05	274.95 ± 3.49	281.03 ± 2.79	8.16 ± 0.40	24.89 ± 1.19
	GAM_03	227.61 ± 1.41	305.24 ± 2.32	7.18 ± 0.56	23.66 ± 1.62
	GAM_04	243.75 ± 5.42	269.97 ± 4.09	7.62 ± 0.55	21.68 ± 3.13
	GAM_05	243.78 ± 1.27	286.29 ± 2.51	8.07 ± 0.12	26.11 ± 0.86
Oliver	AIR_05	350.68 ± 0.81	368.66 ± 21.39	14.45 ± 0.25	47.02 ± 2.16
	CHA_05	306.28 ± 5.27	374.45 ± 6.42	14.54 ± 0.44	46.87 ± 1.19
	GAM_03	302.41 ± 6.17	278.92 ± 5.39	13.74 ± 0.38	43.21 ± 0.75
	GAM_04	355.75 ± 8.20	326.55 ± 2.70	14.72 ± 0.47	45.34 ± 2.37
	GAM_05	343.08 ± 5.28	375.76 ± 7.91	14.10 ± 0.38	46.87 ± 1.40
F values	Genetic (C)	11.91 ***	5.37 ***	188.00 ***	161.33 ***
	Location (L)	0.02	1.05	0.04	0.03
	Year (Y)	0.06	0.15	0.04	0.08
	C × L	7.20 **	4.59 **	133.03 ***	105.31 ***
	C × Y	7.34 **	3.40 *	133.16 ***	130.60 ***
	Y × L	0.03	0.64	0.04	0.06
	C × L × Y	4.91 *	3.41 *	114.80 ***	109.27 ***

[a] expressed in mM of Trolox C equivalent antioxidant capacity (TEAC); [b] expressed in µM of fixed Fe^{2+}; [c] expressed in % inhibition of MDA formation relative to control cells; values are mean ± SD of 4 independent replicates. ANOVA, F represents the effect. Significance level: * $p < 0.05$; ** $p < 0.01$; *** $p < 0.001$.

4. Conclusions

During the last decades, flaxseeds have emerged as one of the key sources of antioxidant phytochemicals. Knowledge about the variation in the accumulation of these valuable constituents is, hence, of particular interest. This study constitutes the first work devoted to the influence of genetic, edaphic, and climatic parameters on the main compounds constituting the so-called lignan macromolecule of flaxseeds, and the antioxidant activities of the obtained extracts. Our results evidenced the predominant influence of genetic factors (cultivar) on the accumulation of the constituents of the lignan macromolecule in flaxseeds. The results of gene expression suggest a transcriptional regulation of this accumulation, knowledge of which would help to manipulate the phenolic contents of flax. Elucidating the complete transcription regulation of lignan biosynthesis in flax would, therefore, help to better control their accumulation. In our hands, other environmental parameters, such as geographic and climatic variables, did not result in significant changes in the lignan macromolecule accumulation. Both in vitro and in vivo antioxidant activity relied on SDG, CafG, and CouG accumulations. Future works using purified compounds will be conducted to further elucidate their respective contribution to the cellular antioxidant capacity observed with flaxseed extracts. Considering the health benefits of these compounds, the present study evidenced the importance of a better knowledge of the flax cultivar type that could greatly influence the phytochemical intakes and the associated biological activities. Therefore, we recommend that this crucial parameter be considered in epidemiological studies dealing with flaxseeds.

Supplementary Materials: The following are available online, Table S1: Meteorological characteristics of the cultivation site; Figure S1: Scheme describing cultivation conditions; Figure S2: Climatic data for the trial sites Airaines (AIR), Gamaches-en-Vexin (GAM) and Chartres (CHA) for the years 2003 (03), 2004 (04), and 2005 (05). Precipitations are expressed as cumulative monthly rainfall in mm, and temperatures are an average of daily temperature in °C.

Author Contributions: Conceptualization, C.H., F.M. and E.L.; Methodology, L.G., S.D., C.C., C.D., J.L.L., E.A.L.; Software, T.F.; Validation, C.H., E.A.L., S.R., E.L. and F.M.; Formal Analysis, C.H., D.T., J.D., B.H.A., B.M. and O.F.; Investigation L.G., S.D., and C.C.; Resources, C.H. and E.L.; Data Curation, C.H., E.L. and S.R.; Writing—Original Draft Preparation, C.H.; Writing—Review & Editing, C.H., O.F., D.T., S.R., E.L., B.H.A and F.M.; Visualization, L.G., S.D. and C.H.; Supervision, C.H., B.M. and E.L.; Project Administration, C.H.; Funding Acquisition, C.H., E.L. and B.M.

Funding: This research was supported by Cosmetosciences, a global training and research program dedicated to the cosmetic industry. Located in the heart of the Cosmetic Valley, this program led by University of Orleans is funded by the Region Centre-Val de Loire. This research was also supported by the Conseil Departemental d'Eure et Loir. BHA acknowledges research fellowship of Le Studium-Institute for Advanced Studies, Loire Valley, Orléans, France. DT gratefully acknowledges the support of French government via the French Embassy in Thailand in the form of Junior Research Fellowship Program 2018.

Acknowledgments: The authors wish to thank Laboulet Semences, Coopérative Linière Terre de Lin, Arvalis Institut Technique du Lin, and Graines de Lin 28 for the donation of flaxseed used in this study and agronomical data on cultivars and access to experimental fields.

Conflicts of Interest: The authors declare no conflict of interest.

References

1. Nayak, B.; Liu, R.H.; Tang, J. Effect of Processing on Phenolic Antioxidants of Fruits, Vegetables, and Grains—A Review. *Crit. Rev. Food Sci. Nutr.* **2015**, *55*, 887–919. [CrossRef] [PubMed]

2. Oomah, B.D. Flaxseed as a functional food source. *J. Sci. Food Agric.* **2001**, *81*, 889–894. [CrossRef]

3. Kamal-Eldin, A.; Peerlkamp, N.; Johnsson, P.; Andersson, R.; Andersson, R.E.; Lundgren, L.N.; Åman, P. An oligomer from flaxseed composed of secoisolariciresinoldiglucoside and 3-hydroxy-3-methyl glutaric acid residues. *Phytochemistry* **2001**, *58*, 587–590. [CrossRef]

4. Struijs, K.; Vincken, J.-P.; Doeswijk, T.G.; Voragen, A.G.J.; Gruppen, H. The chain length of lignan macromolecule from flaxseed hulls is determined by the incorporation of coumaric acid glucosides and ferulic acid glucosides. *Phytochemistry* **2009**, *70*, 262–269. [CrossRef] [PubMed]

5. Westcott, N.; Muir, A. Flax seed lignan in disease prevention and health promotion. *Phytochem. Rev.* **2003**, *2*, 401–417. [CrossRef]

6. McCann, M.J.; Gill, C.I.R.; McGlynn, H.; Rowland, I.R. Role of Mammalian Lignans in the Prevention and Treatment of Prostate Cancer Mark. *Nutr. Cancer* **2005**, *52*, 1–14. [CrossRef] [PubMed]

7. Lainé, E.; Hano, C.; Lamblin, F.F. *Phytoestrogens: Lignans*; Knasmüller, S., DeMarini, D.M., Johnson, I., Gerhäuser, C., Eds.; WILEY-VCH: Weinheim, Germany, 2009; ISBN 9783527320585.

8. Hano, C.; Renouard, S.; Molinié, R.; Corbin, C.; Barakzoy, E.; Doussot, J.; Lamblin, F.; Lainé, E. Flaxseed (*Linum usitatissimum* L.) extract as well as (+)-secoisolariciresinol diglucoside and its mammalian derivatives are potent inhibitors of α-amylase activity. *Bioorg. Med. Chem. Lett.* **2013**, *23*, 3007–3012. [CrossRef] [PubMed]

9. Prasad, K. Hydroxyl radical-scavenging property of secoisolariciresinol diglucoside (SDG) isolated from flax-seed. *Mol. Cell. Biochem.* **1997**, *168*, 117–123. [CrossRef] [PubMed]

10. Kitts, D.D.; Yuan, Y.V.; Wijewickreme, A.N.; Thompson, L.U. Antioxidant activity of the flaxseed lignan secoisolariciresinol diglycoside and its mammalian lignan metabolites enterodiol and enterolactone. *Mol. Cell. Biochem.* **1999**, *202*, 91–100. [CrossRef] [PubMed]

11. Hano, C.; Corbin, C.; Drouet, S.; Quéro, A.; Rombaut, N.; Savoire, R.; Molinié, R.; Thomasset, B.; Mesnard, F.; Lainé, E. The lignan (+)-secoisolariciresinol extracted from flax hulls is an effective protectant of linseed oil and its emulsion against oxidative damage. *Eur. J. Lipid Sci. Technol.* **2017**, *119*. [CrossRef]

12. Adlercreutz, H.; Mousavi, Y.; Clark, J.; Höckerstedt, K.; Hämäläinen, E.; Wähälä, K.; Mäkela, T.; Hase, T. Dietary phytoestrogen and cancer: In vitro and In vivo studies. *J. Steroid Biochem. Mol. Biol.* **1992**, *41*, 8012–8020. [CrossRef]

13. Schoenrock, U.; Untiedt, S.; Kux, U.; Inoue, K. Application of Ferulic Acid Glucosides as Anti-Irritants in Cosmetic and Topical Dermatological Preparations. German Patent AN 1997:708582, 1997.

14. Kosinska, A.; Penkacik, K.; Wiczkowski, W.; Amarowicz, R. Presence of caffeic acid in flaxseed lignan macromolecule. *Plant Foods Hum. Nutr.* **2011**, *66*, 270–274. [CrossRef] [PubMed]

15. Beejmohun, V.; Grand, E.; Mesnard, F.; Fliniaux, M.A.; Kovensky, J. First synthesis of (1,2-13C2)-monolignol glucosides. *Tetrahedron Lett.* **2004**, *45*, 8745–8747. [CrossRef]

16. Beejmohun, V.; Grand, E.; Lesur, D.; Mesnard, F.; Fliniaux, M.A.; Kovensky, J. Synthesis and purification of [1,2-13C2]coniferin. *J. Label. Compd. Radiopharm.* **2006**, *49*, 463–470. [CrossRef]

17. Hano, C.; Martin, I.; Fliniaux, O.; Legrand, B.; Gutierrez, L.; Arroo, R.R.J.; Mesnard, F.; Lamblin, F.; Lainé, E. Pinoresinol-lariciresinol reductase gene expression and secoisolariciresinol diglucoside accumulation in developing flax (*Linum usitatissimum*) seeds. *Planta* **2006**, *224*, 1291–1301. [CrossRef] [PubMed]

18. Dalisay, D.S.; Kim, K.W.; Lee, C.; Yang, H.; Rübel, O.; Bowen, B.P.; Davin, L.B.; Lewis, N.G. Dirigent Protein-Mediated Lignan and Cyanogenic Glucoside Formation in Flax Seed: Integrated Omics and MALDI Mass Spectrometry Imaging. *J. Nat. Prod.* **2015**, *78*, 1231–1242. [CrossRef] [PubMed]

19. Huis, R.; Hawkins, S.; Neutelings, G. Selection of reference genes for quantitative gene expression normalization in flax (*Linum usitatissimum* L.). *BMC Plant Biol.* **2010**, *10*, 71. [CrossRef] [PubMed]

20. Livak, K.J.; Schmittgen, T.D. Analysis of relative gene expression data using real-time quantitative PCR and the 2(-Delta Delta C(T)) Method. *Methods* **2001**, *25*, 402–408. [CrossRef] [PubMed]

21. Corbin, C.; Fidel, T.; Leclerc, E.A.; Barakzoy, E.; Sagot, N.; Falguiéres, A.; Renouard, S.; Blondeau, J.; Ferroud, C.; Doussot, J.; et al. Development and validation of an efficient ultrasound assisted extraction of phenolic compounds from flax (*Linum usitatissimum* L.) seeds. *Ultrason. Sonochem.* **2015**, *26*, 176–185. [CrossRef] [PubMed]

22. Benzie, I.; Strain, J. The ferric reducing ability of plasma (FRAP) as a measure of "antioxidant power": The FRAP assay. *Anal. Biochem.* **1996**, *239*, 70–76. [CrossRef] [PubMed]

23. Prior, R.L.; Hoang, H.; Gu, L.; Wu, X.; Bacchiocca, M.; Howard, L.; Hampsch-Woodill, M.; Huang, D.; Ou, B.; Jacob, R. Assays for hydrophilic and lipophilic antioxidant capacity (oxygen radical absorbance capacity (ORACFL)) of plasma and other biological and food samples. *J. Agric. Food Chem.* **2003**, *51*, 3273–3279. [CrossRef] [PubMed]

24. Mladěnka, P.; MacÁková, K.; Filipský, T.; Zatloukalová, L.; Jahodář, L.; Bovicelli, P.; Silvestri, I.P.; Hrdina, R.; Saso, L. In vitro analysis of iron chelating activity of flavonoids. *J. Inorg. Biochem.* **2011**, *105*, 693–701. [CrossRef] [PubMed]

25. Bisquert, R.; Muñiz-Calvo, S.; Guillamón, J.M. Protective role of intracellular Melatonin against oxidative stress and UV radiation in Saccharomyces cerevisiae. *Front. Microbiol.* **2018**, *9*, 1–11. [CrossRef] [PubMed]

26. Hano, C.; Addi, M.; Fliniaux, O.; Bensaddek, L.; Duverger, E.; Mesnard, F.; Lamblin, F.; Lainé, E. Molecular characterization of cell death induced by a compatible interaction between Fusarium oxysporum f. sp. linii and flax (*Linum usitatissimum*) cells. *Plant Physiol. Biochem.* **2008**, *46*, 590–600. [CrossRef] [PubMed]

27. Diederichsen, A.; Fu, Y.B. Flax Genetic Diversity as the Raw Material for Future Success. In Proceedings of the International Conference on Flax and Other Bast Plants, Saskatoon, SK, Canada, 21–23 July 2008; pp. 270–279.

28. Zimmermann, R.; Bauermann, U.; Morales, F. Effects of growing site and nitrogen fertilization on biomass production and lignan content of linseed (*Linum usitatissimum* L.). *J. Sci. Food Agric.* **2006**, *86*, 415–419. [CrossRef]

29. Zimmermann, R.; Bauermann, U.; Spedding, C. Effects of nitrogen fertilisation and two growing sites on biomass production and lignan content of linseed (*Linum usitatissimum* L.): Second year. *Acta Agron. Hungarica* **2007**, *55*, 173–181. [CrossRef]

30. Li, H.B.; Wong, C.C.; Cheng, K.W.; Chen, F. Antioxidant properties in vitro and total phenolic contents in methanol extracts from medicinal plants. *LWT-Food Sci. Technol.* **2008**, *41*, 385–390. [CrossRef]

31. Struijs, K.; Vincken, J.P.; Verhoef, R.; van Oostveen-van Casteren, W.H.M.; Voragen, A.G.J.; Gruppen, H. The flavonoid herbacetin diglucoside as a constituent of the lignan macromolecule from flaxseed hulls. *Phytochemistry* **2007**, *68*, 1227–1235. [CrossRef] [PubMed]

32. Struijs, K.; Vincken, J.P.; Verhoef, R.; Voragen, A.G.J.; Gruppen, H. Hydroxycinnamic acids are ester-linked directly to glucosyl moieties within the lignan macromolecule from flaxseed hulls. *Phytochemistry* **2008**, *69*, 1250–1260. [CrossRef] [PubMed]

33. Westcott, N.D.; Muir, A.D. Variation in the concentration of the flax seed lignan concentration with variety, location and year. In Proceedings of the 56th Flax Institute of the United States Conference, Fargo, ND, USA, 20–22 March 1996; pp. 77–80.

34. Johnsson, P.; Kamal-Eldin, A.; Lundgren, L.N.; Aman, P. HPLC method for analysis of secoisolariciresinol diglucoside in flaxseeds. *J. Agric. Food Chem.* **2000**, *48*, 5216–5219. [CrossRef] [PubMed]

35. Eliasson, C.; Kamal-Eldin, A.; Andersson, R.; Aman, P. High-performance liquid chromatographic analysis of secoisolariciresinol diglucoside and hydroxycinnamic acid glucosides in flaxseed by alkaline extraction. *J. Chromatogr. A* **2003**, *1012*, 151–159. [CrossRef]

36. Ramsay, A.; Fliniaux, O.; Fang, J.; Molinie, R.; Roscher, A.; Grand, E.; Guillot, X.; Kovensky, J.; Fliniaux, M.A.; Schneider, B.; et al. Development of an NMR metabolomics-based tool for selection of flaxseed varieties. *Metabolomics* **2014**, *10*, 1258–1267. [CrossRef]

37. Wang, H.; Wang, J.; Qiu, C.; Ye, Y.; Guo, X.; Chen, G.; Li, T.; Wang, Y.; Fu, X.; Liu, R.H. Comparison of phytochemical profiles and health benefits in fiber and oil flaxseeds (*Linum usitatissimum* L.). *Food Chem.* **2017**, *214*, 227–233. [CrossRef] [PubMed]

38. Corbin, C.; Drouet, S.; Markulin, L.; Auguin, D.; Lainé, É.; Davin, L.B.; Cort, J.R.; Lewis, N.G.; Hano, C. A genome-wide analysis of the flax (*Linum usitatissimum* L.) dirigent protein family: From gene identification and evolution to differential regulation. *Plant Mol. Biol.* **2018**, *97*, 73–101. [CrossRef] [PubMed]

39. von Heimendahl, C.B.I.; Schäfer, K.M.; Eklund, P.; Sjöholm, R.; Schmidt, T.J.; Fuss, E. Pinoresinol–lariciresinol reductases with different stereospecificity from Linum album and Linum usitatissimum. *Phytochemistry* **2005**, *66*, 1254–1263. [CrossRef] [PubMed]

40. Renouard, S.; Tribalatc, M.; Lamblin, F.; Mongelard, G.; Fliniaux, O.; Corbin, C.; Marosevic, D.; Pilard, S.; Demailly, H.; Gutierrez, L.; et al. RNAi-mediated pinoresinol lariciresinol reductase gene silencing in flax (*Linum usitatissimum* L.) seed coat: Consequences on lignans and neolignans accumulation. *J. Plant Physiol.* **2014**, *171*, 1372–1377. [CrossRef] [PubMed]

41. Ghose, K.; Selvaraj, K.; McCallum, J.; Kirby, C.W.; Sweeney-Nixon, M.; Cloutier, S.J.; Deyholos, M.; Datla, R.; Fofana, B. Identification and functional characterization of a flax UDP-glycosyltransferase glucosylating secoisolariciresinol (SECO) into secoisolariciresinol monoglucoside (SMG) and diglucoside (SDG). *BMC Plant Biol.* **2014**, *14*, 82. [CrossRef] [PubMed]

42. Fofana, B.; Ghose, K.; McCallum, J.; You, F.M.; Cloutier, S. UGT74S1 is the key player in controlling secoisolariciresinol diglucoside (SDG) formation in flax. *BMC Plant Biol.* **2017**, *17*, 1–13. [CrossRef] [PubMed]

43. Ford, J.D.; Huang, K.; Wang, H.; Davin, L.B.; Lewis, N.G. Biosynthetic Pathway to the Cancer Chemopreventive Secoisolariciresinol Diglucoside-Hydroxymethyl Glutaryl Ester-Linked Lignan Oligomers in Flax (*Linum usitatissimum*) Seed. *J. Nat. Prod.* **2001**, *2*, 1388–1397. [CrossRef]

44. Żuk, M.; Kulma, A.; Dymińska, L.; Szołtysek, K.; Prescha, A.; Hanuza, J.; Szopa, J. Flavonoid engineering of flax potentiate its biotechnological application. *BMC Biotechnol.* **2011**, *11*, 10. [CrossRef] [PubMed]

45. Dave Oomah, B.; Mazza, G.; Kenaschuk, E.O. Flavonoid content of flaxseed. Influence of cultivar and environment. *Euphytica* **1996**, *90*, 163–167. [CrossRef]

46. Saastamoinen, M.; Pihlava, J.M.; Eurola, M.; Klemola, A.; Jauhiainen, L.; Hietaniemi, V. Yield, SDG lignan, cadmium, lead, oil and protein contents of linseed (*Linum usitatissimum* L.) cultivated in trials and at different farm conditions in the south-western part of Finland. *Agric. Food Sci.* **2013**, *22*, 296–306. [CrossRef]

47. Westcott, N.D.; Muir, A.D.; Lafond, G.; McAndrew, D.W.; May, W.; Irvine, B.; Grant, C.; Shirtliffe, S.; Bruulsema, T.W. Factors Affecting the Concentration of a Nutraceutical Lignan in Flaxseed. In Proceedings of the Symposium on Fertilizing Crops for Functional Food, Indianapolis, IN, USA, 11 November 2002; pp. 1–3.

48. Podloucká, P.; Berka, K.; Fabre, G.; Paloncýová, M.; Duroux, J.L.; Otyepka, M.; Trouillas, P. Lipid bilayer membrane affinity rationalizes inhibition of lipid peroxidation by a natural lignan antioxidant. *J. Phys. Chem. B* **2013**, *117*, 5043–5049. [CrossRef] [PubMed]

49. Donoso-Fierro, C.; Becerra, J.; Bustos-Concha, E.; Silva, M. Chelating and antioxidant activity of lignans from Chilean woods (Cupressaceae). *Holzforschung* **2009**, *63*, 559–563. [CrossRef]

50. Fucassi, F.; Heikal, A.; Mikhalovska, L.I.; Standen, G.; Allan, I.U.; Mikhalovsky, S.V.; Cragg, P.J. Metal chelation by a plant lignan, secoisolariciresinol diglucoside. *J. Incl. Phenom. Macrocycl. Chem.* **2014**, *80*, 345–351. [CrossRef]

51. Steels, E.L.; Learmonth, R.P.; Watson, K. Stress tolerance and membrane lipid unsaturation in *Saccharomyces cerevisiae* grown aerobically or anaerobically. *Microbiology* **1994**, *140*, 569–576. [CrossRef] [PubMed]

52. Wolak, N.; Kowalska, E.; Kozik, A.; Rapala-Kozik, M. Thiamine increases the resistance of baker's yeast Saccharomyces cerevisiae against oxidative, osmotic and thermal stress, through mechanisms partly independent of thiamine diphosphate-bound enzymes. *FEMS Yeast Res.* **2014**, *14*, 1249–1262. [CrossRef] [PubMed]

Review

A Review of Lignan Metabolism, Milk Enterolactone Concentration, and Antioxidant Status of Dairy Cows Fed Flaxseed

André F. Brito * and Yu Zang

Department of Agriculture, Nutrition, and Food Systems, University of New Hampshire, Durham, NH 03824, USA; yz1040@wildcats.unh.edu
* Correspondence: andre.brito@unh.edu; Tel.: +1-603-862-1341

Academic Editor: David Barker
Received: 26 November 2018; Accepted: 18 December 2018; Published: 22 December 2018

Abstract: Lignans are polyphenolic compounds with a wide spectrum of biological functions including antioxidant, anti-inflammatory, and anticarcinogenic activities, therefore, there is an increasing interest in promoting the inclusion of lignan-rich foods in humans' diets. Flaxseed is the richest source of the lignan secoisolariciresinol diglucoside—a compound found in the outer fibrous-containing layers of flax. The rumen appears to be the major site for the conversion of secoisolariciresinol diglucoside to the enterolignans enterodiol and enterolactone, but only enterolactone has been detected in milk of dairy cows fed flaxseed products (whole seeds, hulls, meal). However, there is limited information regarding the ruminal microbiota species involved in the metabolism of secoisolariciresinol diglucoside. Likewise, little is known about how dietary manipulation such as varying the nonstructural carbohydrate profile of rations affects milk enterolactone in dairy cows. Our review covers the gastrointestinal tract metabolism of lignans in humans and animals and presents an in-depth assessment of research that have investigated the impacts of flaxseed products on milk enterolactone concentration and animal health. It also addresses the pharmacokinetics of enterolactone consumed through milk, which may have implications to ruminants and humans' health.

Keywords: animal health; cattle; enterolignan; human health; pharmacokinetic; ruminant; secoisolariciresinol diglucoside

---

1. Introduction

Lignans are polyphenolic, phytoestrogenic compounds known to display a wide range of biological functions, including weak estrogenic and cardioprotective activities, as well as antiestrogenic, antioxidant, anti-inflammatory, and anticarcinogenic properties [1–3]. The weak and antiestrogenic effects of lignans are caused by distinct transactivation activities of estrogen receptors between the enterolignans enterodiol (ED) and enterolactone (EL) [4]. There is a growing interest in promoting the consumption of lignan-rich foods because of the potential benefits to human health. The outer fibrous-containing layers of flaxseed (*Linum usitatissimum* L.) is the richest source of the lignan secoisolariciresinol diglucoside (SDG) [5], which accounts for over 95% of the total lignans found in flax [6]. In ruminants, the rumen appears to be the main site for conversion of SDG into the mammalian lignans ED and EL [7–10]. However, only EL was detected in milk of dairy cows fed flaxseed meal (FM) [11] possibly because of ruminal dehydrogenation reactions converting ED to EL like those occurring in humans [12]. This suggests that EL-enriched milk can be used as a source of lignans for humans due to the following reasons: (1) milk is consumed by a large part of the world population despite regional differences in per capita consumption [13], (2) global consumption of milk is projected to increase by 60% between 2005/2007 and 2050, particularly in regions where the population traditionally consumes less milk such as East and North Africa, sub-Saharan Africa, and South and East Asia [14], and (3) a poor and variable consumption of plant lignans worldwide [15].

Hulls, meal, and whole seeds are flaxseed products that have been used as sources of the lignan SDG to improve the concentration of EL in milk of dairy cows [11,16–19]. It is important to note that other ingredients (e.g., forages, cereal grains, protein supplements) used in diets of dairy cows also provide lignans. Therefore, comparison of milk EL concentrations across experiments should consider the contribution of lignans from non-flaxseed feedstuffs. Diets containing sources of nonstructural carbohydrates (NSC) with different ruminal degradability (e.g., ground corn vs. liquid molasses) also have been shown to affect the EL concentration in milk of dairy cows fed FM [18]. Despite the growing knowledge regarding the impact of flaxseed supplementation on milk EL concentration in the last 10 years, little is known about how dietary manipulation affects the ruminal microbiome and EL production in dairy cows. Research in this area is needed to unravel dietary strategies suitable to modulate the concentration of EL in dairy cows' milk.

In addition to human health benefits, flaxseed lignans can be also used as natural antioxidants to improve animal health via upregulation of antioxidant enzymes. Newborn dairy calves and periparturient dairy cows are prone to oxidative stress and immune depression [20,21]. Previous research revealed that the antioxidant activity of plant enterolignans is stronger than that of vitamin E [22]. Furthermore, weanling albino rats receiving 10% flaxseed (1.5 g/kg of body weight) during 14 d followed by a challenge with a toxin (i.e., carbon tetrachloride) known to downregulate the hepatic expression of antioxidant enzymes were able to restore the activities of superoxide dismutase (SOD), catalase (CAT), and glutathione peroxidase (GPx) by 95, 182, and 136%, respectively, compared with the control treatment [23]. Altogether, these results are encouraging and open new opportunities to explore the use of flaxseed products or flaxseed-derived lignans as bioactive sources to mitigate oxidative stress in newborn, growing, and adult dairy cattle.

The primary objective of this review is to present an in-depth summary and evaluation of research that have investigated the impacts of flaxseed hulls (FH), FM, and whole seeds flaxseed (WF) on milk EL concentration and animal health. We also covered the metabolism of lignans in the gastrointestinal tract of humans and animals and the pharmacokinetics of milk EL consumed by newborn dairy calves, which may have implications to ruminants and humans' health.

2. Metabolism of Lignans in the Gastrointestinal Tract

The flaxseed lignans SDG, secoisolariciresinol (SECO), pinoresinol, lariciresinol, and matairesinol are converted by the gut microbiota of humans [6,24] and ruminants [7–10] to the enterolignans ED and EL. In contrast, the lignan isolariciresinol, also derived from flaxseed, is not converted to ED and EL [25]. Enterodiol and EL are named mammalian lignans or enterolignans because they are produced in the gut of humans and other mammals and not found in plant tissues [26]. A simplified pathway highlighting the conversion of plant lignans to enterolignans in humans is presented in Figure 1. Consortia of gut microorganisms appear to be involved in the sequential catalytic reactions reported in Figure 1, including 28 bacterial species belonging to 12 different genera such as *Bacteroides*, *Clostridium*, *Bifidobacterium*, and *Ruminococcus* among others according to previous research [12,27–33]. After conversion of lignans into ED and EL, these enterolignans are absorbed in the large intestine followed by conjugation as glucuronides and sulfates based on in vitro work using human colon epithelial cells [34]. Conjugated EL and ED undergo extensive first-pass metabolism and enterohepatic recirculation [34,35], as well as deconjugation by colonic bacterial β-glucuronidases and sulfatases followed by reabsorption [36]. It has also been shown that conjugation of EL takes place not only in the colon, but also in the small intestine and liver microsomes of humans and rats according to in vitro enzymatic kinetic analysis of EL glucuronidation [35].

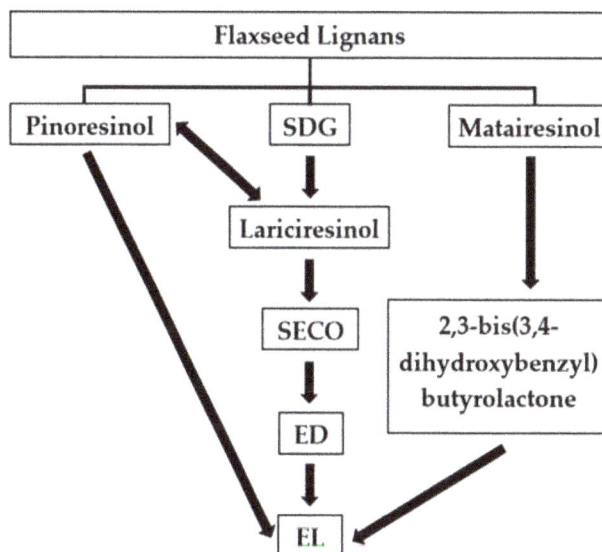

Figure 1. Metabolic pathways for enterolignans production from flaxseed lignans by the gut microbiota of humans. SDG = secoisolariciresinol diglucoside; SECO = secoisolariciresinol; ED = enterodiol; EL = enterolactone. Adapted from [29].

An investigation of the relationship among the gut microbial community, urinary EL excretion, and diet from a 3-d food record of 115 premenopausal American women (40–45 years old) revealed a significant positive association between EL excretion and either the gut microbial community or its diversity [37]. They also demonstrated that the gut microbial community associated with high EL production was distinct and enriched in *Moryella*, *Acetanaerobacterium*, *Fastidiosipila* spp., and *Streptobacillus* spp. [37]. Interestingly, these 4 bacterial genera were not part of those typically related to the sequential pathway of lignans catabolism [12,27–33]. However, despite these genera not being previously linked to EL production, they are closely related to those involved in the metabolism of lignans [37]. Recently, the complete metabolic pathway of pinoresinol and lariciresinol was unraveled using comparative genomics and transcriptional profiling (RNAseq) prepared from stool samples, thus indicating that the conversion of dietary lignans to bioactive enterolignans is a common route adopted by the gut microbiota of humans [38]. These results are an important step for advancing the molecular genetic understanding of the gut bioactivation of lignans and other plant secondary compounds to downstream metabolites relevant to humans' health [38].

In ruminants, it is conceivable that deglycosylation, demethylation, dehydroxylation, and dehydrogenation reactions like those reported in humans (Figure 1) are also involved in the metabolism of lignans, but little is known about which ruminal bacteria species or consortia participate in these reactions. Lignans present in FH and WF were both converted to mammalian lignans by ruminal and fecal microbiota of dairy cows during in vitro incubations [7]. While EL was the major enterolignan produced by the ruminal microbiota, the fecal counterpart yielded primarily ED [7]. In a study conducted using ruminally-cannulated goats, the concentrations of SDG, ED, and EL increased significantly in both rumen and serum following ruminal infusion of SDG (1 mg/kg of body weight) [9]. These authors also observed that the ruminal and serum concentrations of EL were approximately 2-fold greater than those of ED [9], indicating that EL is the predominant enterolignan in the rumen, which agree with results from another study [7]. The role of the ruminal microbiota and the effects of flaxseed oil (FO) in the metabolism of flaxseed-lignans and concentrations of EL in biological fluids have been also investigated [8]. Flaxseed oil is a rich source of polyunsaturated fatty

Molecules **2019**, *24*, 41

acids (PUFA) [39], which are known to be toxic for certain species of ruminal microorganisms [40,41]. Therefore, feeding sources rich in PUFA may interfere with the ruminal metabolism of flaxseed-lignans and ultimately affect the concentrations of EL in biological fluids. The concentrations of EL increased by an average of 1,755% in urine, 238% in plasma, and 925% in milk of cows administered with FH in the rumen compared with FO and FH infused in the abomasum [8]. However, no significant differences in the concentrations of EL in urine, plasma, and milk were observed when FO was administered in the rumen and FH infused in the abomasum [8], which confirm that rumen is the major site for conversion of SDG to EL. In addition, the ruminal concentration of EL increased linearly and a strong correlation (r = 0.76) between EL concentrations in ruminal fluid and milk was observed in dairy cows fed incremental amounts of FM [0, 5, 10, and 15% of the diet dry matter (DM)] [10,42], further reinforcing the key role of the ruminal microbiota in the metabolism of flaxseed-SDG.

It appears that in ruminants, ED and EL are absorbed in the rumen and intestines [10,43,44], possibly as conjugated forms like other phytoestrogens including formononetin, daidzein, and equol [43]. Interestingly, sheep had a greater conjugative activity than cattle in most parts of the gastrointestinal tract evaluated (i.e., rumen, reticulum, omasum) except in the small intestine [43]. In humans, deconjugation performed by gut microbial β-glucuronidases and sulfatases is known to enhance the reabsorption of ED and EL [36,45,46]. Studies conducted with lactating dairy cows showed no relationship between flaxseed supplementation (FH or FM) and activity of microbial β-glucuronidase in the rumen [8,10,47], thus suggesting that this enzyme has little or no involvement in the ruminal absorption of EL, possibly because conjugation occurs during or after cell uptake of enterolignans [43]. In fact, when the ruminal activity of microbial β-glucuronidase decreased in dairy cows fed FH [48], the concentrations of EL in rumen, plasma, urine, and milk increased compared with the control diet. However, additional research is needed to elucidate the actual mechanisms involved in the absorption of enterolignans in ruminant animals. Likewise, research investigating the potential effects of intestinal β-glucuronidases on deconjugation of enterolignans before reabsorption in the large intestine of ruminants is warranted.

Studies in which oil (FO or sunflower) was administered in the rumen or infused in the abomasum also helped to shed light on the gastrointestinal tract metabolism of lignans in dairy cows. Oil sources rich in n-3 PUFA such as FO are known to inhibit the growth of ruminal microorganisms involved in fiber degradation (e.g., *Butyrivibrio*, *Ruminococcus*) and methanogenesis (e.g., *Methanobrevibacter*) [40,41]. β-glucuronidase activity in humans has been attributed to colonic bacteria belonging to the genera *Ruminococcus*, *Bacteroides*, *Bifidobacterium*, and *Eubacterium* [49], which are also found in the rumen [50,51]. Thus, it is conceivable that FO may inhibit ruminal bacteria with β-glucuronidase activity. In fact, FO reduced microbial β-glucuronidase activity when it was administered in the rumen, but not during abomasal infusion in lactating dairy cows [8]. These results [8] imply that ruminal bacteria with predominant β-glucuronidase activity may be more susceptible to the toxic effects of FO than those primarily involved in the conversion of SDG to EL as the concentration of EL in the rumen was not affected by the site of FO supplementation (rumen or abomasum). Compared with the control treatment, fecal β-glucuronidase activity tended to increase in dairy cows fed FH and no change was detected with abomasal infusion of FO in another experiment [48]. In contrast, it was found that feeding FM and infusing sunflower oil (n-6 PUFA source) in the abomasum of lactating dairy cows decreased fecal β-glucuronidase activity relative to the control treatment [47]. It has been shown that the ruminal microbiota can be modulated by modifying the dietary PUFA profile and similar processes may also take place in the large intestine of ruminants, which may explain to a certain extent these inconsistent results in fecal β-glucuronidase activity [8,47,48]. Changes (increase or decrease) in 16S rRNA copy numbers of ruminal microorganisms such as *Butyrivibrio*, ciliate protozoa, methanogens, *Selenomonas ruminantium*, and *Streptococcus bovis* were detected during an in vitro rumen simulation technique study in which fermenters were dosed with diets rich in n-6 PUFA (i.e., sunflower oil) or a n-6/n-3 PUFA mix (i.e., sunflower oil plus fish and algae oil) [52]. Overall, ruminal or fecal microbiota β-glucuronidase activity appears to have limited biological importance for the absorption of EL in lactating dairy cows fed different flaxseed products or abomasally-infused with n-3 or n-6 PUFA oil sources.

As mentioned earlier, there is scarce information about the role of ruminal microbiota species in the metabolism of plant-derived lignans. Ruminal supplementation of SDG stimulated the growth of the bacterium *Ruminococcus gnavus*, which is likely involved with glucuronidase activity in the rumen [9]. In fact, *R. gnavus* E1, an anaerobic bacterium belonging to the dominant human gut microbiota, expresses the gene *gnus* known to encode for the β-glucuronidase enzyme [49]. In a more recent study, the concentration of total ruminal bacteria 16S rRNA obtained using qPCR did not differ in cows fed incremental amounts of FM [42]. However, additional PCR-DGGE and DNA extraction analyses using bands from cows fed 15% FM showed that several genera contributed to the metabolism of lignans, particularly *Prevotella* spp. [42]. Moreover, a follow-up in vitro pure culture assay revealed that 11 ruminal bacteria species were able to metabolize SDG to SECO, with bacteria from the genus *Prevotella* being the most efficient followed by *Butyrivibrio fibrisolvens* and *Peptostreptococcus anaerobius*, whereas *Ruminococcus albus*, *Eubacterium ruminantium*, *Butyrivibrio proteoclasticus*, and *Ruminococcus flavefaciens* showed the least conversion efficiency [42]. Their data also suggested that intermediate compounds between the SDG to EL pathway were formed during in vitro pure culture incubations due to the presence of unidentified peaks in the chromatograms [42]. Overall, the genus *Prevotella* appears to be the most relevant in the metabolism of plant lignans to enterolignans in ruminants. However, the current knowledge regarding ruminal microbiota diversity and function in young and adult ruminants fed different sources of flaxseed is limited and warrants further research.

3. Effects of Flaxseed Products on Milk EL Concentration

Table 1 summarizes results from 15 studies in which milk EL concentration was measured in dairy cows fed different flaxseed products (i.e., FH, FM, WF) and NSC sources. The nutritional profile of flaxseed products used in studies summarized in Table 1 are presented in Table 2.

3.1. Dose-Response Studies and Milk EL Concentration

Four dose-response studies using FH (1 experiment) [17], FM (2 experiments) [10,11], and WF (1 experiment) [53] have been conducted to date (Table 1). In three out of four experiments, the milk concentration of EL increased linearly in response to incremental amounts (diet DM basis) of FH (0, 5, 10, and 20%) or FM (0, 5, 10, and 15%). Compared with the control diet, feeding 20% FH increased the concentration of milk EL by approximately 250% [17]. The milk concentrations of EL increased by approximately 110% [11] and 330% [10] relative to control treatments when cows were fed the greatest amount of FM (i.e., 15%). In contrast, only a positive linear trend in milk EL concentration was observed in response to increasing amounts of WF (0, 5, 10, and 15%) [53]. Flaxseed hulls (mean = 1% SDG) and FM (mean = 1.6% SDG) contain greater concentrations of SDG than WF (mean = 0.6% SDG; see Table 2), thus consistent with a more pronounced response in milk EL concentration with feeding FH or FM versus WF. No curvilinear responses were detected in these four dose-response studies, indicating that a theoretical maximum concentration of milk EL was not achieved in diets containing up to 15% FM, 15% WF, or 20% FH. These results also suggest that ruminal and intestinal absorptive mechanisms were not saturated by increased concentrations of EL. However, there are limitations regarding the amount of flaxseed products that can be included in dairy diets due to environmental and milk production concerns associated with excess intake of crude protein or crude fat depending on the flax source used. As shown in Table 2, FM is a protein supplement (mean = 37.2% crude protein), while FH can be used as both lipid (mean = 28.4% crude fat) and protein sources (mean = 22.4% crude protein); likewise, WF contains high concentration of lipids (mean = 34.9% crude fat) and moderate crude protein content (mean = 23.5%).

Table 1. Milk enterolactone concentration in dairy cows fed different flaxseed products.

References	No. of Cows	DIM [1]	Experimental Design [2]	Treatments [3]	Milk Enterolactone Concentration [4]
[16]	24	119	RCB	CON, 10% FM, 10% WF	10% FM = 10% WF > CON
[8]	4	92	4 × 4 Latin square	FO & FH at ABO/ABO, RUM/ABO, RUM/RUM, ABO/RUM	ABO/RUM = RUM/RUM > RUM/ABO = ABO/ABO
[11]	32	231	RCB	0%, 5%, 10%, 15% FM	Linear increase
[54]	12	61	RCB	CON, 20% FM	20% FM > CON
[53]	32	175	RCB	0%, 5%, 10%, 15% WF	Tendency for linear increase
[55]	4	190	4 × 4 Latin square	CON, 20% FH, MON, 20% FH + MON	20% FH = 20% FH + MON > CON = MON
[17]	45	140	RCB	0%, 5%, 10%, 15%, 20% FH	Linear increase
[56]	8	163	4 × 4 Latin square	CON, 9.88% FH, 500 g/d FO at ABO, 9.88% FH + 500 g/d FO at ABO	9.88% FH = 9.88% FH + 500 g/d < CON > 500 g/d FO
[48]	6	95	6 × 6 Latin square	2 × 3 factorial: FH (0%, 15.9%) × FO (0, 250, 500 g/d)	15.9% FH diets > 0% FH diets
[18]	16	135	4 × 4 Latin square	2 × 2 factorial: GRC + 16% SBM-SFM mix, GRC + 16% FM, LM + 16% SBM-SFM mix, LM + 16% FM	16% FM diets > 16% SBM-SFM mix diets & LM diets > GRC diets
[47]	8	56	4 × 4 Latin square	2 × 2 factorial: CON, 13.7% FM, 250 g/d SO at ABO, 13.7% FM + 250 g/d SO at ABO	No treatment differences
[57]	8	108	4 × 4 Latin square	2 × 2 factorial: CON, 12.4% FM, 250 g/d FO at ABO, 12.4% FM + 250 g/d FO at ABO	12.4% FM = 12.4% FM + 250 g/d FO > CON = 250 g/d FO
[10]	8	112	4 × 4 Latin square	0%, 5%, 10%, 15% FM	Linear increase
[58]	16	95	4 × 4 Latin square	CON, 15% FM + 5% sucrose, 15% FM + 3% FO, 15% FM + 5% sucrose + 3% FO	15% FM + 3% FO = 15% FM + 5% sucrose + 3% FO > CON
[19]	16	101	4 × 4 Latin square	Different GRC to LM ratios (12:0, 8:4, 4:8, and 0:12) + 15% FM	Tendency for cubic effect

[1] DIM = days in milk; [2] RCB = randomized complete block design; [3] CON = control, FM = flaxseed meal, WF = whole flaxseed, FO = flaxseed oil, FH = flaxseed hulls, ABO = abomasum, RUM = rumen, MON = monensin, GRC = ground corn, LM = liquid molasses, SBM = soybean meal, SFM = sunflower meal, SO = sunflower oil; [4] Significant differences in the cited references were declared at $p \le 0.05$ and trends at $0.05 < p \le 0.10$; no treatment differences ($p > 0.10$).

Table 2. Nutritional profile (% of dry matter) of flaxseed products used in studies listed in Table 1 [1].

Item	Flax Products		
	Flaxseed Hulls [2] (*n* = 5)	Flaxseed Meal [3] (*n* = 6)	Whole Flaxseed [4] (*n* = 1)
Crude protein	22.4 ± 2.41	37.2 ± 1.35	23.5
Neutral detergent fiber	20.6 ± 2.64	30.6 ± 4.61	20.7
Acid detergent fiber	15.8 ± 3.44	15.9 ± 1.39	13.7
Crude fat	28.4 ± 3.09	3.70 ± 4.11	34.9
SDG	1.00 ± 0.08	1.60 ± 0.21	0.60

[1] Values are presented as mean ± standard deviation, unless otherwise noted. [2] Values were calculated using data reported by [8,17,48,55,56]; 4 studies including [8,48,55,56] reported the same nutritional composition for flaxseed hulls except for a different secoisolariciresinol diglucoside (SDG) concentration value reported by [55]; no SDG concentration for flaxseed hulls was reported by [17]. [3] Values were calculated using data from [11,18,19,47,57,58]; SDG concentrations were not reported by [47] and [57]. [4] Values were calculated using data from [53].

High intake of crude protein can lead to excess N excretion to the environment and poor N use efficiency in lactating dairy cows [59,60]. Excess consumption of fat (>5% of the diet DM) has been associated with depressed DM intake, milk production, and ruminal fiber digestibility [59].

3.2. Comparison of Flaxseed Products and Animal Variation in Milk EL Concentration

We are aware of only one publication that compared, in the same experiment, the effect of flaxseed products on milk EL concentration in dairy cows (i.e., [16]; see Table 1). In this study [16], 24 lactating dairy cows were used in a randomized complete block design in which animals were assigned to a control diet without flaxseed supplementation or 10% of the diet DM as FM or WF. It was observed that relative to the control treatment, the milk concentration of EL increased by an average of 178% in cows fed FM or WF. However, no differences in the concentration of milk EL was found between FM and WF. Even though milk EL yield (mg/d) did not differ with feeding FM versus WF, only cows supplemented with FM had a significant increase in milk EL output (+216%) compared with the control animals. For the remaining studies summarized in Table 1, including the dose-response experiments (discussed above) and the feeding trials that evaluated different NSC sources and FM supplementation (discussed next section), milk EL concentration improved in all except one study (i.e., [47]). In their experiment [47], 8 ruminally-cannulated dairy cows were used in a replicated 4 × 4 Latin square design with a 2 × 2 factorial arrangement of treatments. The concentrations of milk EL averaged 75 and 122 n*M* in cows fed diets without and with FM supplementation, respectively. Despite an average increase of 63% in milk EL concentration comparing FM- versus non-FM diets [47], this difference did not reach statistical significance possibly because of the low number of animals used and the large cow-to-cow variability in milk EL content. For instance, the 95% confidence interval for milk EL concentration ranged from 32 to 161 n*M* (control), 35 to 175 n*M* (250 g/d abomasal infusion of sunflower oil), 46 to 221 m*M* (13.7% FM), and 63 to 312 n*M* (13.7% FM plus 250 g/d abomasal infusion of sunflower oil) [47].

A large interindividual variation in the concentration of the phytoestrogen equol in milk of dairy cows has been reported, with values ranging from 400 to 2,600 μg/kg across treatments in two experiments [61]. Similarly, we [19] observed a large interindividual variation in milk EL yield in dairy cows fed varying levels of NSC sources and 15% FM (see Figure 2), which is consistent with previous research [61]. This large cow-to-cow variability cannot be entirely explained by differences in dietary composition or phytoestrogens intake so that other factors such as ruminal microbiota profile, digesta passage rate, and dairy cattle genetics may be also involved [61]; however, the actual biological mechanisms underpinning this wide interindividual variability are not well understood. Previous researchers reported that EL is a transported substrate and likely a competitive inhibitor of the ATP-binding cassette subfamily G2 (ABCG2) protein [62], which is known to transport phytoestrogens and their conjugated metabolites [63–65]. It was further demonstrated that the milk-to-plasma ratio of EL decreased significantly in the Abcg2$^{(-/-)}$ knockout female mice phenotype compared with the wild-type group (0.4 vs. 6.4) [62]. A subsequent study showed that EL was used as substrate to the bovine ABCG2 variant Y in vitro and was also actively secreted in milk resulting in a 2-fold increase in its milk-to-plasma ratio in Y/S heterozygous versus Y/Y homozygous cows [66]. The bovine ABCG2 Y581S variant has been described as a gain-of-function polymorphism that increases milk secretion and decreases plasma levels of its substrates [67–69]. Taken together, the ABCG2 protein and its variant Y581S appear to contribute to the interindividual variation of EL secretion in milk of dairy cows opening the possibility for controlling, through genetic selection or other management tools, the amount of enterolignans consumed by the population [61]. Improved knowledge of lignans metabolism in ruminants is needed because high intake of phytoestrogens may result in adverse health effects, particularly in critical stages of infant development [70,71] and during lactation and pregnancy [72]. Therefore, timing of exposure to phytoestrogens is key for capitalizing on health benefits while minimizing undesirable health outcomes [73]. In a recent literature review, the authors stated that current evidences regarding the potential health benefits of phytoestrogens are not so convincing that clearly outweigh the possible health risks (e.g., decreased fertility, increased risk of cancer in estrogen-sensitive tissues) [74]. They concluded that data currently available are not sufficient to support a more refined (semi) quantitative risk–benefit analysis, implying that a definite conclusion on potential health benefit outcomes of phytoestrogens cannot be made [74].

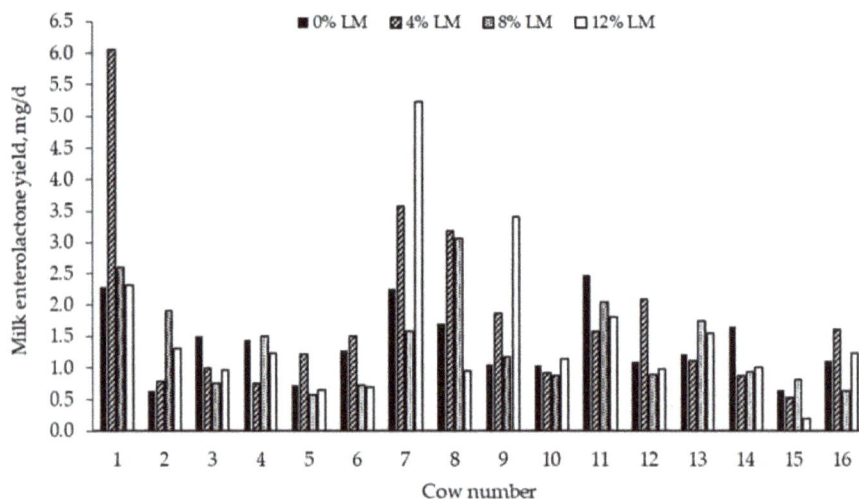

Figure 2. Interindividual variation in milk enterolactone yield in dairy cows fed (% of diet dry matter) diets in which ground corn was replaced by incremental amounts of liquid molasses (LM) (see [18] for study details).

3.3. Impact of NSC Sources and FM on Milk EL Concentration

To the best of our knowledge, only three studies have investigated the impact of different NSC sources on milk EL concentration in dairy cows fed FM (see Table 1). It is well established that in relation to starch, sugars are more rapidly fermented in the rumen [75], implying that NSC sources with different degradability in the rumen may change ruminal fermentation processes, digesta passage rate, and microbiota growth and species composition. Compared with ground corn, liquid or dried molasses has greater concentration of sucrose [18,76]. The effects of supplemental NSC (ground corn vs. liquid molasses) and rumen-degradable protein (soybean meal-sunflower meal mix vs. FM) on milk EL concentration have been evaluated in dairy cows fed grass hay-based diets [18]. No significant rumen-degradable protein by NSC source interaction was observed for milk EL concentration. However, significant rumen-degradable protein and NSC source effects were detected;

cows fed diets containing (DM basis) 16% FM and 12% liquid molasses had 288 and 53% more EL in milk than those fed rations consisting of 16% soybean meal-sunflower meal mix and 12% ground corn, respectively. Therefore, liquid molasses may select for ruminal microorganisms with better capacity to convert FM-SDG to EL than ground corn [18]. A follow-up study evaluated the effects of replacing ground corn with incremental amounts of liquid molasses (0, 4, 8, and 12% of the diet DM) on milk EL concentration in dairy cows fed 15% FM [19]. It was hypothesized that the concentration of EL in milk would be modulated by possible changes in DM intake (also affecting SDG intake) when varying the dietary proportions of liquid molasses and ground corn. Only a cubic trend was observed for milk EL concentration despite the linear decrease in SDG intake with replacing ground corn by liquid molasses [19]. Although this cubic trend is difficult to explain biologically, the lack of a precursor-product relationship suggests that the ruminal output of EL seems to be more affected by the microbiota metabolism of SDG than by SDG supply. Milk EL yield did not differ and averaged 1.38, 1.61, 1.36, and 1.52 mg/d in diets containing 0, 4, 8, or 12% liquid molasses, respectively [19]. *Prevotella* spp. have been reported to be one of the main converters of SDG to SECO, a lignan-derived metabolite that is further metabolized to ED and EL, presumably by additional ruminal microbiota species [42]. *Prevotella* species are also capable of utilizing starch, other non-cellulosic polysaccharides, and simple sugars as energy sources, yielding succinate as the major end-product of ruminal fermentation [77]. Therefore, it was not entirely surprising to obtain a curvilinear response for milk EL concentration with feeding various dietary levels of liquid molasses [19] because *Prevotella* spp. can utilize both starch and sugars [77].

Our laboratory conducted a third study evaluating the effect of sucrose and FO on milk EL concentration of dairy cows fed 15% FM [58]. Specifically, 16 lactating dairy cows were used in a replicated 4 × 4 Latin square design with the following arrangement of treatments (% of diet DM): (1) 8% soybean meal (control); (2) 5% sucrose + 15% FM; (3) 3% FO + 15% FM; and (4) 5% sucrose + 3% FO + 15% FM. As discussed above, *Prevotella* spp. have been shown to be involved in the metabolism of SDG [42] and NSC [77]. In addition, the genus *Prevotella* dominated the ruminal bacterial community when steers were fed diets containing 4% FO, suggesting that *Prevotella* species are possibly involved in the metabolism of PUFA [78]. We hypothesized [58] that sucrose and FO could synergistically interact to increase the concentration of EL in milk as sugars [77] and FO [78] have been shown to promote growth of *Prevotella* spp. Compared with the control diet (mean = 76.8 nM of milk EL), the average concentration of EL in milk increased 4-fold in cows fed 15% FM (mean = 321 nM). However, no differences in milk EL concentration was observed among the treatments containing FM supplemented with sucrose or FO or both [58]. Overall, our data [18,19,58] indicate that the use of NSC sources with different ruminal degradability did not consistently improve milk EL concentration. Differences in DM intake, milk production, type of forage, and forage-to-concentrate ratio may have contributed to the inconsistent results in milk EL content across our studies.

3.4. Dairy Breed and Milk EL Concentration

Holstein cows were used in all studies presented in Table 1 except in two experiments where Jerseys were selected [18,19]. A large interindividual variation for the milk concentration of equol has been reported, but this variability was more pronounced in Swedish Red than Norwegian Red dairy cows [61]. These results suggest that dairy cattle genetics may influence the output of phytoestrogens in milk. It is well known that Jersey cows produce milk with greater concentrations of fat and protein than Holsteins (e.g., [79]). However, we are not aware of any publication that has simultaneously compared Holstein versus Jersey cows in terms of milk EL concentration and yield. Therefore, data from [10,11,19,58] were used to make inferences regarding the concentration of milk EL between breeds. In these four studies cows received 15% FM in at least one dietary treatment (see Table 1 for details). The concentration of milk EL averaged 259 nM in Jerseys [19], and 78 nM [11], 650 nM [10], and 321 nM [58] in Holsteins. Compared with one study using Holsteins [11], the concentration and yield of milk EL in Jerseys increased by 3.3- and 2.8-fold, respectively [19]. Contrarily, the concentration and yield of milk EL were greater in two other studies with Holsteins [10,58] than Jerseys [19], suggesting that no conclusive relationship between dairy breed and milk EL could be established. It is important to note that this exercise is a gross evaluation of the potential effect of dairy breed on milk EL concentration so that any association between breed and milk EL should be done cautiously. Nevertheless, the mean concentration of milk EL ranged from 78 to 650 nM implying that genetics, dietary composition, and even analytical methods may be involved in this variation in milk EL across experiments [10,11,19,58]. For instance, a chromatographic method (i.e., HPLC) was used in one (i.e., [11]) of the four studies resulting in the lowest milk EL content (i.e., 78 nM). The concentration of EL in the remaining three studies [10,19,58] were analyzed colorimetrically using a commercial competitive enzymatic immunoassay, which led to an average milk EL concentration 425% greater than that obtained with HPLC [11]. Moreover, the ingredient composition of the basal diet, forage-to-concentrate ratio, and forage source may have changed the ruminal environment among these four studies ultimately impacting the concentration of EL in milk. Plant lignans such as matairesinol, pinoresinol, and lariciresinol are also converted to enterolignans [6,24,80], with pinoresinol and lariciresinol present in greater concentrations than SDG and matairesinol in several plant species [81]. Thus, it is conceivable that dietary ingredients other than flaxseed may also supply lignans to the ruminal microbiota, which can contribute to variation in milk EL concentration reported in the literature.

4. Pharmacokinetics of Milk EL and Potential Implications on Animal and Human Health

Elevated blood concentrations of ED and EL have been associated with reduced risk of coronary diseases and colorectal adenoma in humans [82–84]. A dose-response relationship between flaxseed intake and serum concentrations of ED or EL was observed in a study conducted with healthy young women [85]. Moreover, a 5-fold increase in the urinary excretion of EL was found in rats fed pure EL compared with those fed plant lignans [86]. These authors [86] hypothesized that EL may be passively absorbed along the intestinal tract, while plant lignans must be first converted to EL by colonic microorganisms followed by absorption in a limited segment of the gut. A large interindividual variation in the blood concentration of enterolignans has been observed in humans, thus revealing differences in the capacity of the colonic microbiota in converting plant lignans to ED and EL [46,85,87]. Therefore, EL-enriched milk has potential to be used as an enterolignan source for improving human health, particularly because EL appears to be more bioavailable than plant lignans [86]. Periparturient dairy cows, as well as newborn and nursing dairy calves could also benefit from the antioxidant properties of EL due to their susceptibility to oxidative stress and depressed immune system [20,21]. However, there is limited information regarding the pharmacokinetics of EL derived from milk and we are not aware of any published research that have instigated the effects on EL-enriched milk on human or animal health.

Recently, we investigated the pharmacokinetics of EL in newborn dairy calves fed milk replacer or EL-enriched milk [58]. In newborn calves, suckling stimulates the reflex closure of the esophageal groove so that ingested milk or milk replacer bypass the reticulo-rumen down to the abomasum [50]. Thus, calves may be used as a translational model to make inferences about the pharmacokinetics of EL in simple-stomach mammals including humans. We hypothesized that the area under the curve and plasma concentration of EL would be greater in Holstein calves fed a single bolus of EL-enriched milk versus milk replacer [58]. The EL-enriched milk was collected from a Jersey cow fed 15% FM. On d 5 of life, 20 calves (10 males and 10 females) were administered 2 L of milk replacer (low-EL treatment: 123 nM of EL) or 2 L of EL-enriched milk (high-EL treatment: 481 nM of EL) during the morning feeding. The area under the curve for the plasma concentration of EL, which was determined using the trapezoidal rule between 0 and 12 h after treatment administration was greater in high- (26 nM × h) than low-EL calves (4.30 nM × h). Similarly, the maximum concentration of EL in plasma was greater in high- (5.06 nM) versus low-EL calves (1.95 nM). Furthermore, the time after treatment administration to reach maximum plasma concentration of EL was faster in the high- (4.31 h) compared with the low-EL (4.44 h) treatment. Our results showed that newborn calves were able to absorb

EL, suggesting that EL-enriched milk can potentially be used as a natural source of antioxidants to pre-weaned ruminants. We also calculated the apparent efficiency of EL absorption between 0 and 12 h after the oral administration of treatments; calves fed EL-enriched milk tended to have lower apparent efficiency of EL absorption than those fed milk replacer (1.31 vs. 1.80%, respectively). In a study in which 12 healthy volunteers (6 men and 6 women) ingested a single dose of purified SDG (1.31 µmol/kg of body weight), ED and EL reached their maximum plasma concentrations at 14.8 and 19.7 h after intake of SDG, respectively [87]. In addition, the area under the curve of EL (mean = 1762 nM × h) increased by 2-fold compared with that of ED (mean = 966 nM × h), indicating a greater systemic exposure to EL than ED [87]. Although our study shed some light in the metabolism of milk EL in vivo [58], future research using humans or animal models that better represent the anatomy and physiology of humans' gastrointestinal tract is warranted to provide further insights about the pharmacokinetics of EL consumed through milk.

An association between serum EL concentration $\geq$ 10 nM and decreased mortality risk (i.e., all-causes and breast cancer-specific) after breast cancer surgery has been reported in women [88]. Milk concentration of EL averaged 395 nM in two studies in which Jersey cows received 15–16% FM [18,19]. Thus, 1 daily serving (250 mL) of EL-enriched milk with a concentration of 395 nM of EL would result in 1.3 nM of EL in plasma assuming an apparent efficiency of absorption of 1.31% based on our previous work [58]. These results imply that EL-enriched milk needs to be consumed in combination with other lignan-rich foods to reach EL concentration in blood that has been linked to decreased mortality and positive health outcomes in humans [88]. However, our inferences should be interpreted cautiously because calves were fed milk as the sole dietary source [58], which may have increased digesta passage rate ultimately limiting the intestinal absorption of EL.

5. Antioxidant Activity of Flaxseed Products and Dairy Cow Health

Periparturient dairy cows mobilize triacylglycerols from the adipose tissue to support elevated energy demand during early lactation [59,89]. As lactation advances, dairy cows also experience extensive metabolic adaptations for maintenance and high milk production [90]. This increased metabolic activity requires more oxygen consumption, which stimulates production of reactive oxygen species (ROS) [91]. When ROS generation exceeds the endogenous antioxidant defense capacity, animals are susceptible to oxidative damage to DNA, lipids, protein, and other cellular components [92]. Oxidative stress may also impair the immune system of dairy cows [91,93] so that they are likely more vulnerable to a variety of metabolic disorders, including udder edema, milk fever, retained placenta, mastitis, and reproductive issues [90,91]. It has been shown that newborn calves had greater blood concentration of free radicals than pregnant cows, suggesting that they undergo a more severe oxidative stress [20]. Therefore, mitigation of oxidative stress has great potential to improve dairy cattle health and profitability of dairy enterprises. In recent years, several studies were conducted to investigate the effects of flaxseed products on the activity of antioxidant enzymes in plasma and erythrocytes, and their gene expression in mammary and hepatic tissues and results are discussed below.

Superoxide dismutase, CAT, and GP$_X$ are antioxidant enzymes commonly involved in combating free radicals in animals' blood and tissues. Superoxide dismutase catalyzes the reaction of highly reactive superoxides to form less reactive peroxides [94]. Peroxides can then be converted to water and oxygen under the catalyzation of CAT [95]. Glutathione peroxidase is an enzyme that facilitates reduction reactions of hydroperoxides such as organic hydroperoxides and peroxides [94]. According to previous work [96], CAT mainly works against free radicals when animals experience severe oxidative stress, whereas GP$_X$ protects those with less oxidative stress pressure.

The activity of antioxidant enzymes in lactating dairy cows fed different flaxseed products are summarized in Table 3. Overall, the activities of SOD, CAT, and GPx in plasma, erythrocytes, and mammary and hepatic tissues were not affected by supplementation of FH, FM, WF, and whole linola (see Table 3). Linola is a cultivar of flaxseed containing approximately 70% linoleic acid [97]. A potential explanation for the inability of flaxseed products to modify the activity of antioxidant enzymes in most studies listed in Table 3 may be due to the use of mid-lactation dairy cows experiencing low oxidative stress. Contrarily, a study [98] reported that inclusion of 12.4% FM lowered plasma CAT activity and tended to elevate that of erythrocytes. Likewise, a tendency for increased activity of SOD in mammary tissues was observed with feeding 9.88% FH [56]. It is also important to note that significant treatment by sampling time interactions were found for plasma CAT and GPx activity with FM supplementation [99]. Plasma CAT and GPx activity responded quadratically and cubically to increasing amounts of FM (0, 5, 10, 15%) when blood samples were collected before feeding, but no treatments effect was observed with sampling 3 h post-feeding [99]. These interactions were probably caused by a longer-lasting supply of antioxidants from the diet with the greatest intake of SDG (i.e., 15% FM) compared with the lower levels [99].

The effect of flaxseed products on mRNA abundance of antioxidant enzymes genes in the mammary gland of lactating dairy cows are summarized in Table 4. Feeding 9.88% FH [56] and incremental amounts of FM (0, 5, 10, and 15%) [99] increased mRNA abundance of CAT gene, whereas no changes were observed with inclusion of 13.7% FM [100]. Additionally, GPx1 and GPx3, two isoforms of GPx, were not impacted with feeding varying amounts of FM [98–100]. However, GPx1 and GPx3 were up- and downregulated, respectively, in dairy cows fed 9.88% FH compared with those fed the control diet [56]. These contradictory effects on GPx1 and GPx3 mRNA abundance with feeding 9.88% FH may be associated with different functions of GPx genes [101]. In addition to CAT and GPx, the mRNA abundance of three isoforms of SOD genes including SOD1, SOD2, and SOD3 were quantified. Both De Marchi et al. [100] and Schogor et al. [99] showed that the mRNA abundance of SOD genes was not modified by FM supplementation to lactating dairy cows. In contrast, an increase in the mRNA abundance of SOD1 and decreases in that of SOD2 and SOD3 were detected in dairy cows fed 9.88% FH [56]. The promoter region of SOD1 contains an antioxidant response element not found in SOD2 and SOD3, thereby consistent with the variable responses of SOD genes to FH supplementation [102]. Collectively, the effects of flaxseed products on modifying antioxidant enzymes or their expression in mammary or hepatic tissues were limited.

The nuclear factor (erythroid-derived 2)-like 2 (*NFE2L2*) relative mRNA abundance in mammary tissues increased linearly in cows fed incremental amounts of FM [99] (see Table 4). The *NFE2L2* gene encodes for a transcription factor involved in activating the expression of a series of genes that are transcribed and translated into antioxidant proteins [103,104]. It is noteworthy that increased *NFE2L2* [99] did not coincide with changes in mRNA abundance of most antioxidant enzymes as discussed above. A trend was observed for decreased relative mRNA abundance of the nuclear factor kappa-light-chain-enhancer of activated B cells subunit 1 (NF-κB1) gene with feeding 13.7% FM to lactating dairy cows [100]; however, two other studies [98,99] did not detect changes in mRNA abundances of NF-κB and NF-κB1, respectively, when similar amounts of FM were fed. The NF-κB1 gene is one of the five members of the NF-κB family, which regulates numerous genes involved in inflammatory and immune responses, apoptosis, and tumor progression [105–107]. The polyphenolic compound quercetin protected interstitial Leydig cells against atrazine-induced toxicity by decreasing the expression of NF-κB and preventing oxidative stress [107]. As shown in Table 2, FM is the richest source of the lignan SDG, a polyphenolic compound like quercetin, thus in line with the reduced expression of NF-κB1 gene [100]. These results suggest that FM supplementation has potential to decrease inflammation and cell death in mammary tissues [100]. Interestingly, decreased NF-κB1 was not associated with changes in the relative mRNA abundance of antioxidant enzymes [100], possibly because FM supplementation did not affect the nuclear factor erythroid 2–related factor 2 (*NRF2*) mRNA abundance, which agrees with previous work [98]. As known, *NRF2* is a transcription factor that activates the expression of multiple genes holding an antioxidant response element in their promoters for codifying antioxidant proteins and phase 2 detoxifying enzymes [105]. Future research is needed to better understand how the relationship between flaxseed supplementation and expression of antioxidant enzyme genes may interact to modulate inflammatory, immunological, and health responses in dairy cows experiencing oxidative stress.

Table 3. Activity of antioxidant enzymes in plasma, erythrocytes, and mammary and hepatic tissues in lactating dairy cows fed different flaxseed products [1].

Item [3]	Treatments and References [2]					
	Non-FH vs. 9.88% FH Diets [56]	0%, 5%, 10%, 15% FM [99]	Non-FM vs. 16% FM Diets [18]	Non-FM vs. 13.7% FM Diets [100]	Non-FM vs. 12.4% FM Diets [98]	CON vs. 7.7% WF, 7.7% WL [108]
Plasma [4]						
CAT	NS	NS [8]	–	NS	↓	–
GPx	NS	NS [9]	NS	NS	NS	–
SOD	NS	NS	NS	NS	NS	–
Erythrocytes [5]						
CAT	NS	NS	–	NS	↑, †	–
GPx	NS	NS	–	NS	NS	–
SOD	NS	NS	–	NS	NS	–
Mammary tissue [6]						
CAT	NS	NS	–	NS	NS	–
GPx	NS	NS	–	NS	NS	–
SOD	↑, †	NS	–	NS	NS	–
Hepatic tissue [7]						
CAT	–	–	–	–	–	NS
GPx	–	–	–	–	–	NS
SOD	–	–	–	–	–	NS

[1] Significant differences in the cited references were declared at $p \leq 0.05$ and trends at $0.05 < p \leq 0.10$; NS = not significant ($p > 0.10$). [2] FH = flaxseed meal; WL = whole flaxseed (linola is a cultivar of flaxseed containing approximately 70% linoleic acid [97]). [3] CAT = catalase; GPx = glutathione peroxidase; SOD = superoxide dismutase; ↑ = positive effect; ↓ = negative effect; † = tendency for significance; – = no measurement. [4] CAT and GPx units were reported as μmol/min per mg of protein, nmol/min per mg of protein, or nmol/min per mL; SOD units were reported as U/mg of protein, U/g of protein, U/mg of protein, or nmol/min per g of protein; GPx units were reported as nmol/min per mg of protein or nmol/min per g of protein. [5] CAT units were reported as μmol/min per mg of protein; SOD units were reported as U/g of protein, U/mg of protein, or nmol/min per mg of protein; GPx units were reported as μmol/min per mg of protein, or nmol/min per g of protein. [6] CAT units were reported as μmol/min per mg of protein, nmol/min per mg of protein, or nmol/min per g of protein; SOD units were reported as U/g of protein, U/mg of protein, or nmol/min per mg of protein. [7] CAT, GPx, and SOD units were reported as U/mg of protein. [8] no overall treatment effect, but a significant treatment by sampling time interaction was reported [quadratic and cubic effects before feeding (0 h) and no effect at 3 h post-feeding]. [9] no overall treatment effect, but a significant treatment by sampling time interaction was observed [quadratic and cubic effects before feeding (0 h) and no effect at 3 h post-feeding].

Table 4. Relative mRNA abundance of oxidative stress-related genes in mammary tissues of lactating dairy cows fed flaxseed products [1].

Items [3]	Treatments [2] and References			
	Non-FH vs. 9.88% FH diets [56]	0%, 5%, 10%, 15% FM [99]	Non-FM vs. 13.7% FM diets [100]	Non-FM vs. 12.4% FM diets [98]
CAT	↑	linear increase†	NS	–
GP$_{X1}$	↑	NS	NS	NS
GP$_{X3}$	↓	NS	NS	–
SOD1	↑	NS	NS	–
SOD2	↓	NS	NS	–
SOD3	↓	NS	NS	–
NFE2L2	–	linear increase	–	–
NF-κB	–	NS	–	–
NF-κB1	–	–	↓, †	NS
NRF2	–	–	NS	NS

[1] Significant differences in the cited references were declared at $p \leq 0.05$ and trends at $0.05 < p \leq 0.10$; NS = not significant ($p > 0.10$). [2] FH = flaxseed hulls; FM = flaxseed meal. [3] CAT = catalase; GP$_X$ = glutathione peroxidase; SOD = superoxide dismutase; NFE2L2 = nuclear factor (erythroid-derived 2)-like 2; NF-κB1 = nuclear factor Kappa-B1; NRF2 = nuclear factor (erythroid-derived 2)-like 2; ↑ = positive effect; ↓ = negative effect; † = tendency for significance; – = no measurement.

Thiobarbituric acid-reactive substances (TBARS) are markers of oxidative status and mainly used to estimate oxidative damage to lipids or lipoperoxidation [109]. Lipoperoxidation can cause damages to cell membranes and membrane-bound enzymes [110]. The impact of flaxseed products supplementation on TBARS concentration in milk, plasma, and ruminal fluid are summarized in Table 5. Quadratic and cubic responses for milk TBARS production were observed in cows fed incremental amounts of FM, with 5% FM and 10% FM resulting in the lowest values [99]. They [99] also reported a significant treatment × sampling time interaction for ruminal TBARS concentration; a linear decrease in TBARS was found with increasing FM supplementation at 2 h after feeding, but no changes were detected at 0 (pre-feeding), 4, and 6 h post-feeding. It was hypothesized that the defense of FM-lignans against oxidation in the rumen is a time-dependent process, with protection being more effective within the first hours after feeding and weakening over time [99]. However, another study [111] reported no changes in ruminal TBARS concentration at 0 (pre-feeding) and 2 h post-feeding but decreased thereafter (4 and 6 h) with feeding 12.4% FM. A third experiment [112] showed a significant decrease in ruminal TBARS concentration in dairy cows fed 13.7% FM despite no treatment × sampling time interaction effect. None of the studies listed in Table 5 (i.e., [99,111,112]) reported effects of FM on plasma TBARS concentration. Similarly, no effects of FM supplementation were observed for the plasma peroxidizability index and total antioxidant capacity [111,112]. As pointed out earlier, research using dairy cattle during stages of life (e.g., transition period, neonatal phase, weaning) more conducive of oxidative stress is needed to better assess the role of flaxseed lignans on animal oxidative status and overall health.

Table 5. Indicators of oxidative stress in lactating dairy cows fed flaxseed products [1].

Items [3]	Treatments [2] and References		
	0%, 5%, 10%, 15% FM [99]	Non-FM vs. 12.4% FM Diets [111]	Non-FM vs. 13.7% FM Diets [112]
Milk TBARS	Q,C [4]	NS	NS
Plasma TBARS	NS	NS	NS
Ruminal TBARS	NS [5]	NS [6]	↑ [7]
Plasma peroxidizability index	–	NS	NS
Plasma total antioxidant capacity	–	NS	NS

[1] Significant differences in the cited references were declared at $p \leq 0.05$ and trends at $0.05 < p \leq 0.10$; NS = not significant ($p > 0.10$). [2] FM = flaxseed meal. [3] TBARS = thiobarbituric acid-reactive substances (nmol of malondialdehyde equivalent/mL); plasma peroxidizability index = (% dienoic fatty acid × 1) + (% trienoic fatty acid × 2) + (% tetraenoic fatty acid × 3) + (% pentaenoic fatty acid × 4) + (% hexaenoic fatty acid × 5) [113]; plasma total antioxidant capacity expressed in mM. [4] Quadratic and cubic effects were observed. [5] no overall treatment effect, but a significant treatment by sampling time interaction was reported [linear decrease at 2 h post-feeding; no changes at 0 (pre-feeding), 4, and 6 h post-feeding]. [6] no overall treatment effect, but a significant treatment by sampling time interaction was reported [no effects at 0 (pre-feeding) and 2 h post-feeding but decreased with FM at 4 and 6 h post-feeding]. [7] ↑ = positive effect.

6. Conclusions

Our review showed that flaxseed products, particularly FM and FH were effective in enhancing the concentration of EL in milk. The metabolism of SDG to EL by the ruminal microbiota possibly involves deglycosylation, demethylation, dehydroxylation, and dehydrogenation reactions. In vitro work showed that ruminal bacteria from the genus *Prevotella* were the most efficient converters of SDG to SECO. The large interindividual variation in milk EL yield suggests that the ruminal microbiota vary in their effectiveness for metabolizing SDG to EL. This opens the possibility for controlling, through genetic selection or other management tools, the amount of EL consumed by the population. Scientific information related to the pharmacokinetics of EL consumed via milk is limited. Recent research showed that EL is absorbed by newborn dairy calves, indicating that EL-enriched milk has potential to be used as a natural source of antioxidants for pre-weaned ruminants.

We specifically call for future research to assess how the relationship between flaxseed supplementation and expression of antioxidant enzyme genes may interact to modulate inflammatory, immunological, and health responses in dairy cattle experiencing oxidative stress. Microbiome work is also needed to elucidate the profile and function of the ruminal microbiota species and genomes involved in the metabolism of lignans in ruminants. The impact of forage types (e.g., low- vs. high-lignan legumes), forage conservation methods, and different sources of NSC and fibrous by-products (e.g., soyhulls, beet pulp, citrus pulp) on ruminal microbiome and milk EL concentration in cows fed FM deserves specific attention. These complex research questions should be tackled through collaborative efforts of laboratories with complementary expertise so that an in-depth understanding of the opportunities and challenges of lignans research in dairy cattle can be successfully implemented. To do so, the scientific community, dairy processors, and the population need to be informed and engaged concerning the implications of phytoestrogens to animals and humans' health.

Author Contributions: A.F.B. conceptualized the manuscript. Y.Z. conducted the literature review and prepared all tables and Figure 1. Y.Z. also wrote Section 5, prepared the reference list, and organized references in text and figures. A.F.B. wrote the abstract and Section 1, Section 2, Section 3, Section 4, and Section 6, and prepared Figure 2. Both authors edited the manuscript and approved the final version of the document.

Acknowledgments: Research conducted at the University of New Hampshire and presented in this review was funded by New Hampshire Agricultural Experiment Station (Durham, NH), USDA-National Institute of Food and Agriculture-Organic Agriculture Research and Extension Initiative (Project Number NHW-2011-01950; Project Accession Number 226410, Washington, DC), and USDA-National Institute of Food and Agriculture Hatch Multistate NC-2042 (Project Number NH00616-R; Project Accession Number 1001855). This is New Hampshire Agricultural Experiment Station Scientific Contribution Number 2801.

Conflicts of Interest: The authors declare no conflicts of interest.

Abbreviations

ABCG2	ATP-binding cassette subfamily G2
CAT	catalase
DM	dry matter
ED	enterodiol
EL	enterolactone
FH	flaxseed hulls
FM	flaxseed meal
FO	flaxseed oil
GPx	glutathione peroxidase
NFE2L2	nuclear factor (erythroid-derived 2)-like 2
NF-κB	nuclear factor kappa-light-chain-enhancer of activated B cells
NF-κB1	nuclear factor kappa-light-chain-enhancer of activated B cells subunit 1
NFR2	nuclear factor erythroid 2–related factor 2
NSC	nonstructural carbohydrate
PUFA	polyunsaturated fatty acids

ROS   reactive oxygen substances
SECO   secoisolariciresinol
SDG   secoisolariciresinol diglucoside
SOD   superoxide dismutase
TBARS thiobarbituric acid-reactive substances
WF   whole flaxseed

References

1. Adolphe, J.L.; Whiting, S.J.; Juurlink, B.H.J.; Thorpe, L.U.; Alcorn, J. Health effects with consumption of the flax lignan secoisolariciresinol diglucoside. *Br. J. Nutr.* **2010**, *103*, 929–938. [CrossRef] [PubMed]
2. Högger, P. Nutrition-derived bioactive metabolites produced by gut microbiota and their potential impact on human health. *Nutr. Med.* **2013**, *1*, 1–32.
3. Imran, M.; Ahmad, N.; Anjum, F.M.; Khan, M.K.; Mushtaq, Z.; Nadeem, M.; Hussain, S. Potential protective properties of flax lignan secoisolariciresinol diglucoside. *Nutr. J.* **2015**, *14*, 71–77. [CrossRef] [PubMed]
4. Carreau, C.; Flouriot, G.; Bennetau-Pelissero, C.; Potier, M. Enterodiol and enterolactone, two major diet-derived polyphenol metabolites have different impact on ERα transcriptional activation in human breast cancer cells. *J. Steroid Biochem. Mol. Biol.* **2008**, *110*, 176–185. [CrossRef] [PubMed]
5. Adlercreutz, H.; Mazur, W. Phyto-oestrogens and western diseases. *Ann. Med.* **1997**, *29*, 95–120. [CrossRef]
6. Thompson, L.U.; Robb, P.; Serraino, M.; Cheung, F. Mammalian lignan production from various foods. *Nutr. Cancer* **1991**, *16*, 43–52. [CrossRef]
7. Côrtes, C.; Gagnon, N.; Benchaar, C.; Da Silva, D.; Santos, G.T.D.; Petit, H.V. *In vitro* metabolism of flax lignans by ruminal and faecal microbiota of dairy cows. *J. Appl. Microbiol.* **2008**, *105*, 1585–1594. [CrossRef]
8. Gagnon, N.; Côrtes, C.; da Silva, D.; Kazama, R.; Benchaar, C.; dos Santos, G.; Zeoula, L.; Petit, H.V. Ruminal metabolism of flaxseed (*Linum usitatissimum*) lignans to the mammalian lignan enterolactone and its concentration in ruminal fluid, plasma, urine and milk of dairy cows. *Br. J. Nutr.* **2009**, *102*, 1015–1023. [CrossRef]
9. Zhou, W.; Wang, G.; Han, Z.; Yao, W.; Zhu, W. Metabolism of flaxseed lignans in the rumen and its impact on ruminal metabolism and flora. *Anim. Feed Sci. Technol.* **2009**, *150*, 18–26. [CrossRef]
10. Schogor, A.L.B.; Palin, M.F.; Santos, G.T.; Benchaar, C.; Petit, H.V. β-glucuronidase activity and enterolactone concentration in ruminal fluid, plasma, urine, and milk of Holstein cows fed increased levels of flax (*Linum usitatissimum*) meal. *Anim. Feed Sci. Technol.* **2017**, *223*, 23–29. [CrossRef]
11. Petit, H.V.; Gagnon, N. Milk concentrations of the mammalian lignans enterolactone and enterodiol, milk production, and whole tract digestibility of dairy cows fed diets containing different concentrations of flaxseed meal. *Anim. Feed Sci. Technol.* **2009**, *152*, 103–111. [CrossRef]
12. Jin, J.S.; Kakiuchi, N.; Hattori, M. Enantioselective oxidation of enterodiol to enterolactone by human intestinal bacteria. *Biol. Pharm. Bull.* **2007**, *30*, 2204–2206. [CrossRef] [PubMed]
13. Muehlhoff, E.; Bennett, A.; McMahon, D. *Milk and Dairy Products in Human Nutrition*; Food and Agriculture Organization of the United Nations (FAO): Rome, Itay, 2013; pp. 11–30.
14. Alexandratos, N.; Bruinsma, J. *World Agriculture Towards 2030/2050: The 2012 Revision*; Food and Agriculture Organization of the United Nations (FAO): Rome, Itay, 2012; pp. 4–5.
15. De Kleijn, M.J.J.; van der Schouw, Y.T.; Wilson, P.W.F.; Adlercreutz, H.; Mazur, W.; Grobbee, D.E.; Jacques, P.F. Intake of dietary phytoestrogens is low in postmenopausal women in the United States: The framingham study. *J. Nutr.* **2001**, *131*, 1826–1832. [CrossRef] [PubMed]
16. Petit, H.V.; Gagnon, N.; Mir, P.S.; Cao, R.; Cui, S. Milk concentration of the mammalian lignan enterolactone, milk production, milk fatty acid profile, and digestibility in dairy cows fed diets containing whole flaxseed or flaxseed meal. *J. Dairy Res.* **2009**, *76*, 257–264. [CrossRef] [PubMed]
17. Petit, H.V.; Gagnon, N. Production performance and milk composition of dairy cows fed different concentrations of flax hulls. *Anim. Feed Sci. Technol.* **2011**, *169*, 46–52. [CrossRef]
18. Brito, A.F.; Petit, H.V.; Pereira, A.B.D.; Soder, K.J.; Ross, S. Interactions of corn meal or molasses with a soybean-sunflower meal mix or flaxseed meal on production, milk fatty acid composition, and nutrient utilization in dairy cows fed grass hay-based diets. *J. Dairy Sci.* **2015**, *98*, 443–457. [CrossRef]

19. Ghedini, C.P.; Moura, D.C.; Santana, R.A.V.; Oliveira, A.S.; Brito, A.F. Replacing ground corn with incremental amounts of liquid molasses does not change milk enterolactone but decreases production in dairy cows fed flaxseed meal. *J. Dairy Sci.* **2018**, *101*, 2096–2109. [CrossRef]

20. Gaál, T.; Ribiczeyné-Szabó, P.; Stadler, K.; Jakus, J.; Reiczigel, J.; Kövér, P.; Mezes, M.; Sümeghy, L. Free radicals, lipid peroxidation and the antioxidant system in the blood of cows and newborn calves around calving. *Comp. Biochem. Physiol. B Biochem. Mol. Biol.* **2006**, *143*, 391–396. [CrossRef]

21. Abuelo, A.; Hernandez, J.; Benedito, J.L.; Castillo, C. A pilot study to compare oxidative status between organically and conventionally managed dairy cattle during the transition period. *Reprod. Domest. Anim.* **2015**, *50*, 538–544. [CrossRef]

22. Prasad, K. Antioxidant activity of secoisolariciresinol diglucoside-derived metabolites, secoisolariciresinol, enterodiol, and enterolactone. *Int. J. Angiol.* **2000**, *9*, 220–225. [CrossRef]

23. Rajesha, J.; Murthy, K.N.C.; Kumar, M.K.; Madhusudhan, B.; Ravishankar, G.A. Antioxidant potentials of flaxseed by *in vivo* model. *J. Agric. Food Chem.* **2006**, *54*, 3794–3799. [CrossRef]

24. Gaya, P.; Medina, M.; Sánchez-Jiménez, A.; Landete, J.M. Phytoestrogen metabolism by adult human gut microbiota. *Molecules* **2016**, *21*, 1034. [CrossRef] [PubMed]

25. Thompson, L.U. Flaxseed, lignans, and cancer. In *Flaxseed in Human Nutrition*; Thompson, L.U., Cunnane, S.C., Eds.; AOCS Press: Champaign, IL, USA, 2003; pp. 199–227.

26. Morris, D.H. *Flax: A Health and Nutrition Primer*; Flax Council of Canada: Winnipeg, MB, Canada, 2007.

27. Xie, L.; Akao, T.; Hamasaki, K.; Deyama, T.; Hattori, M. Biotransformation of pinoresinol diglucoside to mammalian lignans by human intestinal microflora, and isolation of Enterococcus faecalis strain PDG-1 responsible for the transformation of (#)-pinoresinol to (+)-lariciresinol. *Chem. Pharm. Bull.* **2003**, *51*, 508–515. [CrossRef]

28. Clavel, T.; Borrmann, D.; Braune, A.; Doré, J.; Blaut, M. Occurrence and activity of human intestinal bacteria involved in the conversion of dietary lignans. *Anaerobe* **2006**, *12*, 140–147. [CrossRef] [PubMed]

29. Clavel, T.; Doré, J.; Blaut, M. Bioavailability of lignans in human subjects. *Nutr. Res. Rev.* **2006**, *19*, 187–196. [CrossRef]

30. Clavel, T.; Lippman, R.; Gavini, F.; Doré, J.; Blaut, M. *Clostridium saccharogumia* sp. nov. and *Lactonifactor longoviformis* gen. nov., sp. nov., two novel human faecal bacteria involved in the conversion of the dietary phytoestrogen secoisolariciresinol diglucoside. *Syst. Appl. Microbiol.* **2007**, *30*, 16–26. [CrossRef] [PubMed]

31. Jin, J.; Zhao, Y.; Nakamura, N.; Akao, T.; Kakiuchi, N.; Min, B.; Hattori, M. Enantioselective dehydroxylation of enterodiol and enterolactone precursors by human intestinal bacteria. *Biol. Pharm. Bull.* **2007**, *30*, 2113–2119. [CrossRef] [PubMed]

32. Jin, J.; Hattori, M. Human intestinal bacterium, strain END-2 is responsible for demethylation as well as lactonization during plant lignan metabolism. *Biol. Pharm. Bull.* **2010**, *33*, 1443–1447. [CrossRef]

33. Roncaglia, L.; Amaretti, A.; Raimondi, S.; Leonardi, A.; Rossi, M. Role of bifidobacteria in the activation of the lignan secoisolariciresinol diglucoside. *Appl. Microbiol. Biotechnol.* **2011**, *92*, 159–168. [CrossRef] [PubMed]

34. Jansen, G.H.E.; Arts, I.C.W.; Nielen, M.W.F.; Müller, M.; Hollman, P.C.H.; Keijer, J. Uptake and metabolism of enterolactone and enterodiol by human colon epithelial cells. *Arch. Biochem. Biophys.* **2005**, *435*, 74–82. [CrossRef]

35. Lin, C.; S Krol, E.; Alcorn, J. The comparison of rat and human intestinal and hepatic glucuronidation of enterolactone derived from flaxseed lignans. *J. Nat. Prod.* **2013**, *3*, 159–171. [CrossRef]

36. Setchell, K.D.R.; Adlercreutz, H. Mammalian lignans and phytooestrogens recent studies on their formation, metabolism and biological role in health and disease. In *Role of the Gut Flora in Toxicity and Cancer*; Rowland, I.R., Ed.; Elsevier Inc.: Carshalton, UK, 1988.

37. Hullar, M.A.J.; Lancaster, S.M.; Li, F.; Tseng, E.; Beer, K.; Atkinson, C.; Wähälä, K.; Copeland, W.K.; Randolph, T.W.; Newton, K.M. Enterolignan-producing phenotypes are associated with increased gut microbial diversity and altered composition in premenopausal women in the United States. *Cancer Epidemiol. Biomarkers Prev.* **2015**, *24*, 546–554. [CrossRef] [PubMed]

38. Bess, E.N.; Bisanz, J.E.; Spanogiannopoulos, P.; Ang, Q.Y.; Bustion, A.; Kitamura, S.; Alba, D.L.; Wolan, D.W.; Koliwad, S.K.; Turnbaugh, P.J. The genetic basis for the cooperative bioactivation of plant lignans by a human gut bacterial consortium. *bioRxiv* **2018**, 357640. [CrossRef]

39. Brossillon, V.; Reis, S.F.; Moura, D.C.; Galvão, J.G.B., Jr.; Oliveira, A.S.; Côrtes, C.; Brito, A.F. Production, milk and plasma fatty acid profile, and nutrient utilization in Jersey cows fed flaxseed oil and corn grain with different particle size. *J. Dairy Sci.* **2018**, *101*, 2127–2143. [CrossRef] [PubMed]

40. Henderson, C. The effects of fatty acids on pure cultures of rumen bacteria. *J. Agric. Sci.* **1973**, *81*, 107–112. [CrossRef]

41. Maia, F.J.; Branco, A.F.; Mouro, G.F.; Coneglian, S.M.; Santos, G.T.d.; Minella, T.F.; Guimarães, K.C. Feeding vegetable oil to lactating goats: Milk production and composition and milk fatty acids profile. *R. Bras. Zootec.* **2006**, *35*, 1504–1513. [CrossRef]

42. Schogor, A.L.B.; Huws, S.A.; Santos, G.T.D.; Scollan, N.D.; Hauck, B.D.; Winters, A.L.; Kim, E.J.; Petit, H.V. Ruminal *Prevotella* spp. may play an important role in the conversion of plant lignans into human health beneficial antioxidants. *PLoS ONE* **2014**, *9*, e87949. [CrossRef]

43. Lundh, T.J.O. Conjugation of the plant estrogens formononetin and daidzein and their metabolite equol by gastrointestinal epithelium from cattle and sheep. *J. Agric. Food Chem.* **1990**, *38*, 1012–1016. [CrossRef]

44. Njåstad, K.M.; Adler, S.A.; Hansen-Møller, J.; Thuen, E.; Gustavsson, A.M.; Steinshamn, H. Gastrointestinal metabolism of phytoestrogens in lactating dairy cows fed silages with different botanical composition. *J. Dairy Sci.* **2014**, *97*, 7735–7750. [CrossRef]

45. Jenab, M.; Thompson, L.U. The influence of flaxseed and lignans on colon carcinogenesis and β-glucuronidase activity. *Carcinogenesis* **1996**, *17*, 1343–1348. [CrossRef]

46. Raffaelli, B.; Hoikkala, A.; Leppälä, E.; Wähälä, K. Enterolignans. *J. Chromatogr. B* **2002**, *777*, 29–43. [CrossRef]

47. De Marchi, F.E.; Palin, M.F.; Santos, G.T.; Benchaar, C.; Petit, H.V. Effects of duodenal infusion of sunflower oil on β-glucuronidase activity and enterolactone concentration in dairy cows fed flax meal. *Anim. Feed Sci. Technol.* **2016**, *220*, 143–150. [CrossRef]

48. Côrtes, C.; da Silva-Kazama, D.; Kazama, R.; Benchaar, C.; dos Santos, G.; Zeoula, L.M.; Gagnon, N.; Petit, H.V. Effects of abomasal infusion of flaxseed (*Linum usitatissimum*) oil on microbial β-glucuronidase activity and concentration of the mammalian lignan enterolactone in ruminal fluid, plasma, urine and milk of dairy cows. *Br. J. Nutr.* **2013**, *109*, 433–440. [CrossRef] [PubMed]

49. Beaud, D.; Tailliez, P.; Anba-Mondoloni, J. Genetic characterization of the β-glucuronidase enzyme from a human intestinal bacterium, *Ruminococcus gnavus*. *Microbiology* **2005**, *151*, 2323–2330. [CrossRef] [PubMed]

50. Hofmann, R.R. Anatomy of the gastro-intestinal tract. In *The Ruminant Animal: Digestive Phisiology and Nutrition*; Church, D.C., Ed.; Waveland Press: Long Grove, IL, USA, 1993; pp. 14–43.

51. Dehority, B.A. *Rumen Microbiology*; Nottingham University Press: Nottingham, UK, 2003.

52. Vargas, J.E.; Andrés, S.; Snelling, T.J.; López-Ferreras, L.; Yáñez-Ruíz, D.R.; García-Estrada, C.; López, S. Effect of sunflower and marine oils on ruminal microbiota, *in vitro* fermentation and digesta fatty acid profile. *Front. Microbiol.* **2017**, *8*, 1124. [CrossRef] [PubMed]

53. Petit, H.V.; Gagnon, N. Concentration of the mammalian lignans enterolactone and enterodiol in milk of cows fed diets containing different concentrations of whole flaxseed. *Animal* **2009**, *3*, 1428–1435. [CrossRef] [PubMed]

54. Gagnon, N.; Côrtes, C.; Petit, H.V. Weekly excretion of the mammalian lignan enterolactone in milk of dairy cows fed flaxseed meal. *J. Dairy Res.* **2009**, *76*, 455–458. [CrossRef]

55. Petit, H.V.; Côrtes, C.; da Silva, D.; Kazama, R.; Gagnon, N.; Benchaar, C.; dos Santos, G.T.; Zeoula, L.M. The interaction of monensin and flaxseed hulls on ruminal and milk concentration of the mammalian lignan enterolactone in late-lactating dairy cows. *J. Dairy Res.* **2009**, *76*, 475–482. [CrossRef]

56. Côrtes, C.; Palin, M.F.; Gagnon, N.; Benchaar, C.; Lacasse, P.; Petit, H.V. Mammary gene expression and activity of antioxidant enzymes and concentration of the mammalian lignan enterolactone in milk and plasma of dairy cows fed flax lignans and infused with flax oil in the abomasum. *Br. J. Nutr.* **2012**, *108*, 1390–1398. [CrossRef]

57. Lima, L.S.; Palin, M.F.; Santos, G.T.; Benchaar, C.; Petit, H.V. Dietary flax meal and abomasal infusion of flax oil on microbial β-glucuronidase activity and concentration of enterolactone in ruminal fluid, plasma, urine and milk of dairy cows. *Anim. Feed Sci. Technol.* **2016**, *215*, 85–91. [CrossRef]

58. Ghedini, C.P. Improving the Understanding of Different Diets on the Concentration and Metabolism of the Mammalian Lignan Enterolactone in Dairy Cattle. Ph.D. Thesis, University of New Hampshire, Durham, NH, USA, 2017.

59. NRC. *Nutrient Requirements of Dairy Cattle*, 7th ed.; NRC: Washington, DC, USA, 2001.

60. Olmos Colmenero, J.O.; Broderick, G. Effect of dietary crude protein concentration on milk production and nitrogen utilization in lactating dairy cows. *J. Dairy Sci.* **2006**, *89*, 1704–1712. [CrossRef]

61. Höjer, A.; Adler, S.; Purup, S.; Hansen-Møller, J.; Martinsson, K.; Steinshamn, H.; Gustavsson, A.M. Effects of feeding dairy cows different legume-grass silages on milk phytoestrogen concentration. *J. Dairy Sci.* **2012**, *95*, 4526–4540. [CrossRef] [PubMed]

62. Miguel, V.; Otero, J.A.; Garcia-Villalba, R.; Tomás-Barberán, F.; Espin, J.C.; Merino, G.; Álvarez, A.I. Role of ABCG2 in transport of the mammalian lignan enterolactone and its secretion into milk in Abcg2 knockout mice. *Drug Metab. Dispos.* **2014**, *42*, 943–946. [CrossRef]

63. Zhu, W.; Xu, H.; Wang, S.W.; Hu, M. Breast cancer resistance protein (BCRP) and sulfotransferases contribute significantly to the disposition of genistein in mouse intestine. *AAPS J.* **2010**, *12*, 525–536. [CrossRef] [PubMed]

64. Álvarez, A.I.; Vallejo, F.; Barrera, B.; Merino, G.; Prieto, J.G.; Tomás-Barberán, F.; Espín, J.C. Bioavailability of the glucuronide and sulfate conjugates of genistein and daidzein in Bcrp1 knockout mice. *Drug Metab. Dispos.* **2011**, *39*, 2008–2012. [CrossRef] [PubMed]

65. Tan, K.W.; Li, Y.; Paxton, J.W.; Birch, N.P.; Scheepens, A. Identification of novel dietary phytochemicals inhibiting the efflux transporter breast cancer resistance protein (BCRP/ABCG2). *Food Chem.* **2013**, *138*, 2267–2274. [CrossRef] [PubMed]

66. Otero, J.A.; Miguel, V.; González-Lobato, L.; García-Villalba, R.; Espín, J.C.; Prieto, J.G.; Merino, G.; Álvarez, A.I. Effect of bovine ABCG2 polymorphism Y581S SNP on secretion into milk of enterolactone, riboflavin and uric acid. *Animal* **2016**, *10*, 238–247. [CrossRef]

67. Real, R.; González-Lobato, L.; Baro, M.F.; Valbuena, S.; de la Fuente, A.; Prieto, J.G.; Alvarez, A.I.; Marques, M.M.; Merino, G. Analysis of the effect of the bovine adenosine triphosphate-binding cassette transporter G2 single nucleotide polymorphism Y581S on transcellular transport of veterinary drugs using new cell culture models. *J. Anim. Sci.* **2011**, *89*, 4325–4338. [CrossRef]

68. Otero, J.A.; Real, R.; de la Fuente, Á.; Prieto, J.G.; Marqués, M.; Álvarez, A.I.; Merino, G. The bovine ATP-binding cassette transporter ABCG2 Tyr581Ser single-nucleotide polymorphism increases milk secretion of the fluoroquinolone danofloxacin. *Drug Metab. Dispos.* **2013**, *41*, 546–549. [CrossRef]

69. Otero, J.A.; Barrera, B.; de la Fuente, A.; Prieto, J.G.; Marqués, M.; Alvarez, A.I.; Merino, G. The gain-of-function Y581S polymorphism of the ABCG2 transporter increases secretion into milk of danofloxacin at the therapeutic dose for mastitis treatment. *J. Dairy Sci.* **2015**, *98*, 312–317. [CrossRef]

70. Mendez, M.A.; Anthony, M.S.; Arab, L. Soy-based formulae and infant growth and development: A review. *J. Nutr.* **2002**, *132*, 2127–2130. [CrossRef] [PubMed]

71. Tuohy, P. Soy infant formula and phytoestrogens. *J. Paediatr. Child Health* **2003**, *39*, 401–405. [CrossRef] [PubMed]

72. Troina, A.A.; Figueiredo, M.S.; Passos, M.C.F.; Reis, A.M.; Oliveira, E.; Lisboa, P.C.; Moura, E.G. Flaxseed bioactive compounds change milk, hormonal and biochemical parameters of dams and offspring during lactation. *Food Chem. Toxicol.* **2012**, *50*, 2388–2396. [CrossRef] [PubMed]

73. Steinshamn, H.; Purup, S.; Thuen, E.; Hansen-Møller, J. Effects of clover-grass silages and concentrate supplementation on the content of phytoestrogens in dairy cow milk. *J. Dairy Sci.* **2008**, *91*, 2715–2725. [CrossRef] [PubMed]

74. Rietjens, I.M.C.M.; Louisse, J.; Beekmann, K. The potential health effects of dietary phytoestrogens. *Br. J. Pharmacol.* **2017**, *174*, 1263–1280. [CrossRef] [PubMed]

75. Chamberlain, D.G.; Robertson, S.; Choung, J.J. Sugars versus starch as supplements to grass silage: Effects on ruminal fermentation and the supply of microbial protein to the small intestine, estimated from the urinary excretion of purine derivatives, in sheep. *J. Sci. Food Agric.* **1993**, *63*, 189–194. [CrossRef]

76. Broderick, G.A.; Radloff, W.J. Effect of molasses supplementation on the production of lactating dairy cows fed diets based on alfalfa and corn silage. *J. Dairy Sci.* **2004**, *87*, 2997–3009. [CrossRef]

77. Purushe, J.; Fouts, D.E.; Morrison, M.; White, B.A.; Mackie, R.I.; Coutinho, P.M.; Henrissat, B.; Nelson, K.E.; Bacteria, N.A.C.f.R. Comparative genome analysis of *Prevotella ruminicola* and *Prevotella bryantii*: Insights into their environmental niche. *Microb. Ecol.* **2010**, *60*, 721–729. [CrossRef]

78. Li, X.; Park, B.K.; Shin, J.S.; Choi, S.H.; Smith, S.B.; Yan, C.G. Effects of dietary linseed oil and propionate precursors on ruminal microbial community, composition, and diversity in Yanbian yellow cattle. *PLoS ONE* **2015**, *10*, e0126473. [CrossRef]

79. Aikman, P.C.; Reynolds, C.K.; Beever, D.E. Diet digestibility, rate of passage, and eating and rumination behavior of Jersey and Holstein cows. *J. Dairy Sci.* **2008**, *91*, 1103–1114. [CrossRef]

80. Heinonen, S.; Nurmi, T.; Liukkonen, K.; Poutanen, K.; Wähälä, K.; Deyama, T.; Nishibe, S.; Adlercreutz, H. *In vitro* metabolism of plant lignans: New precursors of mammalian lignans enterolactone and enterodiol. *J. Agric. Food Chem.* **2001**, *49*, 3178–3186. [CrossRef] [PubMed]

81. Milder, I.E.J.; Feskens, E.J.M.; Arts, I.C.W.; de Mesquita, H.B.B.; Hollman, P.C.H.; Kromhout, D. Intake of the plant lignans secoisolariciresinol, matairesinol, lariciresinol, and pinoresinol in Dutch men and women. *J. Nutr.* **2005**, *135*, 1202–1207. [CrossRef] [PubMed]

82. Kuijsten, A.; Arts, I.C.W.; Hollman, P.C.H.; van't Veer, P.; Kampman, E. Plasma enterolignans are associated with lower colorectal adenoma risk. *Cancer Epidemiol. Biomarkers Prev.* **2006**, *15*, 1132–1136. [CrossRef] [PubMed]

83. Vanharanta, M.; Voutilainen, S.; Lakka, T.A.; van der Lee, M.; Adlercreutz, H.; Salonen, J.T. Risk of acute coronary events according to serum concentrations of enterolactone: A prospective population-based case-control study. *Lancet* **1999**, *354*, 2112–2115. [CrossRef]

84. Vanharanta, M.; Voutilainen, S.; Rissanen, T.H.; Adlercreutz, H.; Salonen, J.T. Risk of cardiovascular disease–related and all-cause death according to serum concentrations of enterolactone: Kuopio ischaemic heat disease risk factor study. *Arch. Intern. Med.* **2003**, *163*, 1099–1104. [CrossRef] [PubMed]

85. Nesbitt, P.D.; Lam, Y.; Thompson, L.U. Human metabolism of mammalian lignan precursors in raw and processed flaxseed. *Am. J. Clin. Nutr.* **1999**, *69*, 549–555. [CrossRef] [PubMed]

86. Saarinen, N.M.; Smeds, A.; Mäkelä, S.I.; Ämmälä, J.; Hakala, K.; Pihlava, J.M.; Ryhänen, E.L.; Sjöholm, R.; Santti, R. Structural determinants of plant lignans for the formation of enterolactone *in vivo*. *J. Chromatogr. B* **2002**, *777*, 311–319. [CrossRef]

87. Kuijsten, A.; Arts, I.C.W.; Vree, T.B.; Hollman, P.C.H. Pharmacokinetics of enterolignans in healthy men and women consuming a single dose of secoisolariciresinol diglucoside. *J. Nutr.* **2005**, *135*, 795–801. [CrossRef]

88. Guglielmini, P.; Rubagotti, A.; Boccardo, F. Serum enterolactone levels and mortality outcome in women with early breast cancer: A retrospective cohort study. *Breast Cancer Res. Treat.* **2012**, *132*, 661–668. [CrossRef]

89. Grummer, R.R. Impact of changes in organic nutrient metabolism on feeding the transition dairy cow. *J. Anim. Sci.* **1995**, *73*, 2820–2833. [CrossRef]

90. Miller, J.K.; Brzezinska-Slebodzinska, E.; Madsen, F.C. Oxidative stress, antioxidants, and animal function. *J. Dairy Sci.* **1993**, *76*, 2812–2823. [CrossRef]

91. Sordillo, L.M.; Aitken, S.L. Impact of oxidative stress on the health and immune function of dairy cattle. *Vet. Immunol. Immunopathol.* **2009**, *128*, 104–109. [CrossRef]

92. Brenneisen, P.; Steinbrenner, H.; Sies, H. Selenium, oxidative stress, and health aspects. *Mol. Aspects Med.* **2005**, *26*, 256–267. [CrossRef]

93. Finch, J.M.; Turner, R.J. Effects of selenium and vitamin E on the immune responses of domestic animals. *Res. Vet. Sci.* **1996**, *60*, 97–106. [CrossRef]

94. Matés, J.M.; Sánchez-Jiménez, F. Antioxidant enzymes and their implications in pathophysiologic processes. *Front. Biosci.* **1999**, *4*, 0339–0345. [CrossRef]

95. Lledías, F.; Rangel, P.; Hansberg, W. Oxidation of catalase by singlet oxygen. *J. Biol. Chem.* **1998**, *273*, 10630–10637. [CrossRef] [PubMed]

96. Hong, Y.; Harding, J.J. Glycation-induced inactivation and loss of antigenicity of catalase and superoxide dismutase. *Biochem. J.* **1997**, *328*, 599–605. [CrossRef]

97. Dribnenki, J.C.P.; McEachern, S.F.; Green, A.G.; Kenaschuk, E.O.; Rashid, K.Y. Linola™'1084'low-linolenic acid flax. *Can. J. Plant Sci.* **1999**, *79*, 607–609. [CrossRef]

98. Lima, L.S.; Palin, M.F.; Santos, G.T.; Benchaar, C.; Petit, H.V. Effects of supplementation of flax meal and flax oil on mammary gene expression and activity of antioxidant enzymes in mammary tissue, plasma and erythrocytes of dairy cows. *Livest. Sci.* **2015**, *176*, 196–204. [CrossRef]

99. Schogor, A.L.B.; Palin, M.-F.; dos Santos, G.T.; Benchaar, C.; Lacasse, P.; Petit, H.V. Mammary gene expression and activity of antioxidant enzymes and oxidative indicators in the blood, milk, mammary tissue and ruminal fluid of dairy cows fed flax meal. *Br. J. Nutr.* **2013**, *110*, 1743–1750. [CrossRef]

100. De Marchi, F.E.; Palin, M.F.; Dos Santos, G.T.; Lima, L.S.; Benchaar, C.; Petit, H.V. Flax meal supplementation on the activity of antioxidant enzymes and the expression of oxidative stress- and lipogenic-related genes in dairy cows infused with sunflower oil in the abomasum. *Anim. Feed Sci. Technol.* **2015**, *199*, 41–50. [CrossRef]

101. Brigelius-Flohé, R.; Maiorino, M. Glutathione peroxidases. *Biochim. Biophys. Acta* **2013**, *1830*, 3289–3303. [CrossRef] [PubMed]

102. Miao, L.; Clair, D.K.S. Regulation of superoxide dismutase genes: Implications in disease. *Free Radic. Biol. Med.* **2009**, *47*, 344–356. [CrossRef]

103. Kang, K.W.; Lee, S.J.; Kim, S.G. Molecular mechanism of nrf2 activation by oxidative stress. *Antioxid. Redox Signal.* **2005**, *7*, 1664–1673. [CrossRef] [PubMed]

104. Nguyen, T.; Nioi, P.; Pickett, C.B. The Nrf2-antioxidant response element signaling pathway and its activation by oxidative stress. *J. Biol. Chem.* **2009**, *284*, 13291–13295. [CrossRef]

105. Zhu, H.; Itoh, K.; Yamamoto, M.; Zweier, J.L.; Li, Y. Role of Nrf2 signaling in regulation of antioxidants and phase 2 enzymes in cardiac fibroblasts: Protection against reactive oxygen and nitrogen species-induced cell injury. *FEBS Lett.* **2005**, *579*, 3029–3036. [CrossRef] [PubMed]

106. Lawrence, T. Cold Spring Harbor Perspectives in Biology. Available online: https://cshperspectives.cshlp.org/content/early/2009/10/04/cshperspect.a001651.abstract (accessed on 24 November 2018).

107. Abarikwu, S.O.; Pant, A.B.; Farombi, E.O. Quercetin decreases steroidogenic enzyme activity, NF-κB expression, and oxidative stress in cultured Leydig cells exposed to atrazine. *Mol. Cell. Biochem.* **2013**, *373*, 19–28. [CrossRef] [PubMed]

108. Do Prado, R.M.; Palin, M.F.; Do Prado, I.N.; Dos Santos, G.T.; Benchaar, C.; Petit, H.V. Milk yield, milk composition, and hepatic lipid metabolism in transition dairy cows fed flaxseed or linola. *J. Dairy Sci.* **2016**, *99*, 8831–8846. [CrossRef] [PubMed]

109. Esterbauer, H. Estimation of peroxidative damage: A critical review. *Pathol. Biol.* **1996**, *44*, 25–28.

110. Chang, Y.; Chang, W.; Tsai, N.; Huang, C.; Kung, C.; Su, Y.; Lin, W.; Cheng, B.; Su, C.; Chiang, Y. The roles of biomarkers of oxidative stress and antioxidant in Alzheimer's disease: A systematic review. *BioMed Res. Int.* **2014**, *2014*. [CrossRef] [PubMed]

111. Lima, L.S.; Palin, M.F.; Santos, G.T.; Benchaar, C.; Lima, L.C.R.; Chouinard, P.Y.; Petit, H.V. Effect of flax meal on the production performance and oxidative status of dairy cows infused with flax oil in the abomasum. *Livest. Sci.* **2014**, *170*, 53–62. [CrossRef]

112. De Marchi, F.E.; Santos, G.T.; Petit, H.V.; Benchaar, C. Oxidative status of dairy cows fed flax meal and infused with sunflower oil in the abomasum. *Anim. Feed Sci. Technol.* **2017**, *228*, 115–122. [CrossRef]

113. Scislowski, V.; Bauchart, D.; Gruffat, D.; Laplaud, P.M.; Durand, D. Effects of dietary n-6 or n-3 polyunsaturated fatty acids protected or not against ruminal hydrogenation on plasma lipids and their susceptibility to peroxidation in fattening steers. *J. Anim. Sci.* **2005**, *83*, 2162–2174. [CrossRef] [PubMed]

Perspective

Dietary Lignans: Definition, Description and Research Trends in Databases Development

Alessandra Durazzo [1,*], Massimo Lucarini [1], Emanuela Camilli [1], Stefania Marconi [1], Paolo Gabrielli [1], Silvia Lisciani [1], Loretta Gambelli [1], Altero Aguzzi [1], Ettore Novellino [2], Antonello Santini [2], Aida Turrini [1] and Luisa Marletta [1]

[1] CREA Research Centre for Food and Nutrition, Via Ardeatina 546, 00178 Rome, Italy;
massimo.lucarini@crea.gov.it (M.L.); emanuela.camilli@crea.gov.it (E.C.); stefania.marconi@crea.gov.it (S.M.);
paolo.gabrielli@crea.gov.it (P.G.); silvia.lisciani@crea.gov.it (S.L.); loretta.gambelli@crea.gov.it (L.G.);
altero.aguzzi@crea.gov.it (A.A.); aida.turrini@crea.gov.it (A.T.); luisa.marletta@crea.gov.it (L.M.)
[2] Department of Pharmacy, University of Napoli Federico II, Via D. Montesano 49, 80131 Napoli, Italy;
ettore.novellino@unina.it (E.N.); asantini@unina.it (A.S.)
* Correspondence: alessandra.durazzo@crea.gov.it; Tel.: +30-065-149-4430

Received: 9 November 2018; Accepted: 6 December 2018; Published: 8 December 2018

Abstract: The study aims to communicate the current status regarding the development and management of the databases on dietary lignans; within the phytochemicals, the class of the lignan compounds is of increasing interest because of their potential beneficial properties, i.e., anticancerogenic, antioxidant, estrogenic, and antiestrogenic activities. Furthermore, an introductory overview of the main characteristics of the lignans is described here. In addition to the importance of the general databases, the role and function of a food composition database is explained. The occurrence of lignans in food groups is described; the initial construction of the first lignan databases and their inclusion in harmonized databases at national and/or European level is presented. In this context, some examples of utilization of specific databases to evaluate the intake of lignans are reported and described.

Keywords: dietary lignans; national databases; food groups; dietary intake; harmonized databases

---

1. Introduction

Within phytochemicals, phenolic compounds called lignans have attracted the interest of food chemists and nutrition researchers over the years. Lignans are vascular plant secondary metabolites, with widespread occurrence in the plant kingdom, and which are ascribed a wide range of physiological functions, positively affecting human health [1]. They are a class of secondary plant metabolites that belong to the group of diphenolic compounds derived from the combination of two phenylpropanoid C6–C3 units at the β and β′ carbon, and can be linked to additional ether, lactone, or carbon bonds; they have a chemical structure like the 1,4-diarylbutan [2]. The range of their structures and biological activities is broad. They are derived from the shikimic acid biosynthetic pathway [3]. The range relative to structurally different forms of lignans and biological activities is broad [4,5]. The main commonly studied and reported compounds are secoisolariciresinol, lariciresinol, matairesinol, pinoresinol, medioresinol, and syringaresinol (shown in Figure 1), while, recently, the isolation and structure elucidation of new lignan compounds have been carried out [6–8] and the spectrum of their attributing properties has been widened [9–11].

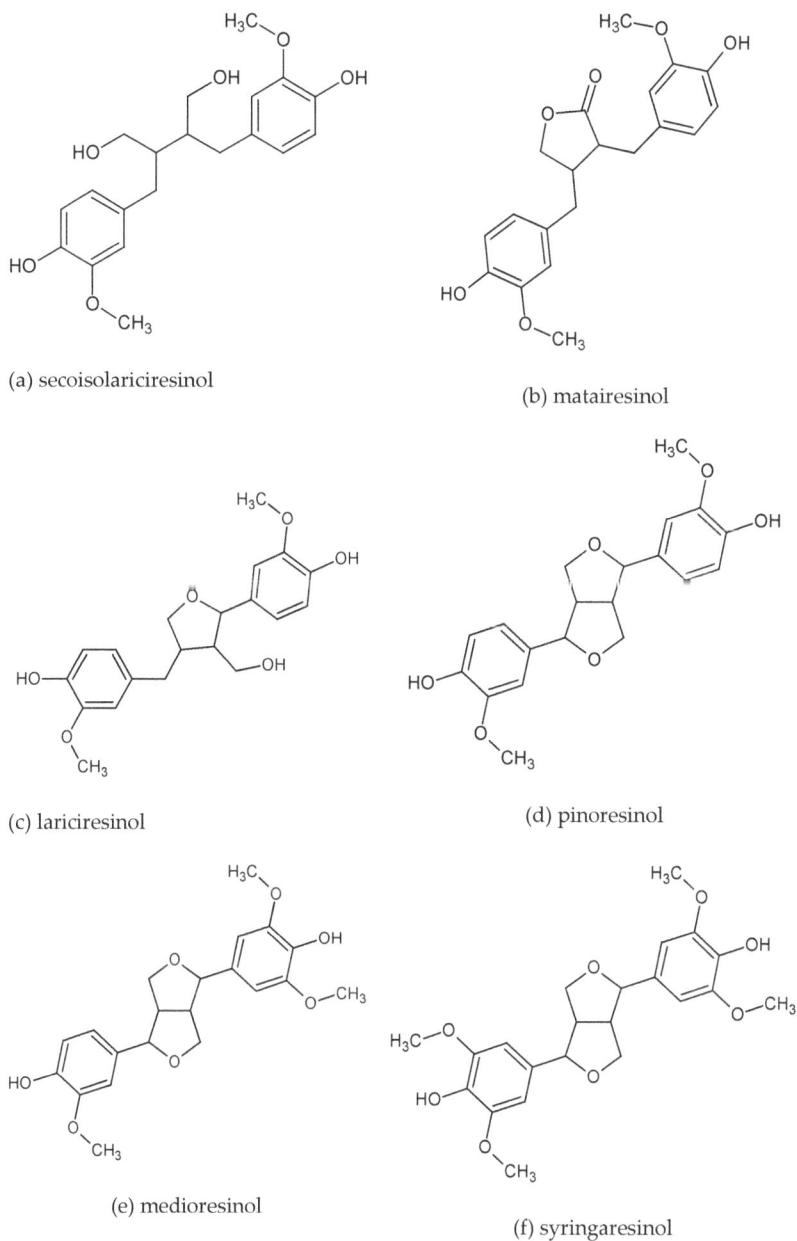

Figure 1. The chemical structure of main dietary lignans, (**a**) secoisolariciresinol, (**b**) matairesinol, (**c**) lariciresinol, (**d**) pinoresinol, (**e**) medioresinol, and (**f**) syringaresinol.

Plant lignans give rise to metabolites, enterodiol, and enterolactone [12], generally called enterolignans due to their colonic origin (named also mammalian lignans) (shown in Figure 2).

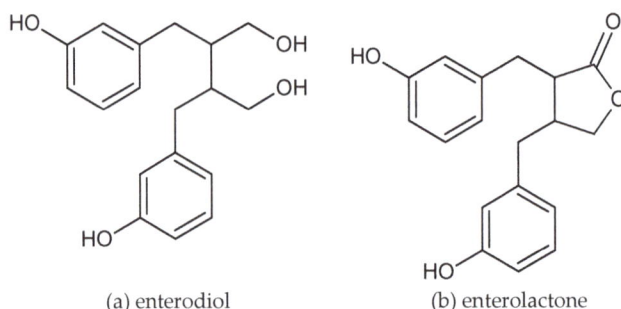

(a) enterodiol   (b) enterolactone

Figure 2. The chemical structure of enterolignans, (**a**) enterodiol and (**b**) enterolactone.

Enterolignans, and some of their plant precursors, are reported to have several biological activities—antitumorigenic [13], anticarcinogenic [14], estrogenic or anti-estrogenic [15,16], as well as antioxidant properties [17].

Lignans, in line with other natural compounds, contribute in disease prevention and health promotion [18,19]; several studies have showed the potential of lignan-rich diets against the development of various diseases, particularly hormone-dependent cancer, cardiovascular diseases, and diabetes [20–27].

Lignans are the basis for novel perspectives for health promotion and disease prevention as nutraceuticals and functional foods [28–32]. Currently, Pilkington, [33], by using a chemometric approach, have analyzed the physicochemical properties of classical lignans, neolignans, flavonolignans, and carbohydrate–lignan conjugates to assess their absorption, distribution, metabolism, excretion and toxic (ADMET) profiles, and establish if these compounds are lead-like/drug-like and, thus, have potential to be, or act as, a driver in the development of future therapeutics; the results showed how carbohydrate–lignan conjugates and flavonolignans are less drug-like, while lignans showed a particularly high level of drug-likeness [33].

Nowadays, lignan species and their quantity in food products are determined. Different methodologies have been defined for the extraction and identification of lignans [34–40]. The extraction procedure from the food matrix represents a key issue and, in particular, the type of hydrolysis step (alkaline, acid hydrolysis, enzymatic hydrolysis, or a mixture of them). The expanding demand for lignans are stimulating the interest in identification of new sources and in improvement of analytical and purification procedures. Analytical values using HPLC, as well as either gas or liquid chromatography–mass spectrometry, were developed and carried out [41,42]. The development and the assessment of methodologies for the extraction, identification, and determination of lignans are achieved [17,43,44]. Also, the "new" emerging lignans, due to LC combined with HR-MS/MS, have been, and will continue, broadening the view regarding dietary lignans [45]; simultaneously, the synthesis [46,47] and the design [48] of new compounds are being carried out.

The complex relationship between food, nutrition, and health [49] is explored via nutrients and bioactive compounds, i.e., beneficial food components [50], and via non-beneficial food components [51]. In this direction, a directory of information about bioactive component databases, specialized, at a national and European level, is being developed, and will be useful for the planning and evaluation of clinical and epidemiological research studies on bioactive components. Databases of lignans in food products are being creating in several countries (Finland, Netherlands, United States, Canada, United Kingdom, Japan, and Spain), and represent the first step for establishing comprehensive and harmonized dietary databases, including all or nearly all bioactive compounds [1]. Reliable methods of exposure measurement are essential for understanding the potential benefits of lignans [52].

2. Databases: Significance, Principles and Common Criteria/Measures

Databases, also called electronic databases, represent a system to generate and collect any data, information, and documentation specially organized for rapid search and retrieval by a computer [53]. Databases are tools constructed to facilitate the storage, retrieval, modification, and deletion of data in conjunction with various data-processing operations [54].

A comprehensive food composition database (FCDB) should be a repository of all numeric, descriptive, and graphical information on the nutrient characteristics of foods [55]; the term food composition data indicates all information referring to the description and identification of foods and their food components (nutrient values, number of sample collections and analyses, analytical methods, descriptive coding, photos, data source, value documentation, etc.) and include various steps in the production, generation, compilation, and publication of data [55].

The EuroFIR project (European Food Information Resource Network of Excellence) was born to develop and integrate a comprehensive, coherent, and validated network of databanks providing a single, authoritative source of food composition data for Europe [56,57]. In this project, efforts in developing procedures for defining and establishing a standardized approach of study have been carried out from the various European partners within their FCDB [56,57].

The establishment of the "Project Committee—Food composition data" (CEN/TC 387, 2008–2013) [58] was an important milestone for the EuroFIR Network of Excellence to reach this objective. A common European standard, established within the CEN-European Committee for Standardization framework, represents a key tool enabling unambiguous identification and description of food composition data and its quality in e.g. databases, for dissemination and interchange [58].

Generally, the use of database management system allows the administration of large volumes of information and data by providing epidemiological research to store large varieties of food consumed for each individual subject and the comparability of data, representing a basic tool for obtaining reliable information on the relationship between nutrients and foods [59,60].

The utilization made by different users requires that FCDBs follow very specific compilation criteria, such as representativeness, accuracy in the production and selection of analytical values, traceability of data taken from other sources at the nutrient level, and clarity in the designation and description of the food [60].

In this context, the food grouping systems in food composition databases represent a key tool. Currently, Durazzo et al. [60] summarized and discussed how the food grouping systems of the various international food composition databases (FCDBs), in terms of number, type and class of consumed foods (e.g., ingredients, commercial products, cooked food, recipes, mixed dishes, etc.) vary between different countries (usually, 10 and 25 food groups), and are constantly evolving according to their changes and updates; the authors marked how these groupings are structured according to the convenience of using the nutritional composition of specific foods and, therefore, there is not an internationally standardized approach.

3. Distribution of Lignans in Food: Occurrence

Lignans are in a wide variety of plants from different origins, including the major edible plants. Amongst the latter, flaxseed and sesame seeds represent rich sources of lignans [40,61–65], whereas wood knots in coniferous trees, particularly Norway spruce, are identified as the most concentrated lignan sources known so far [66].

The main sources of dietary lignans are oilseeds (i.e., flax, soy, rapeseed, and sesame), whole-grain cereals (i.e., wheat, oats, rye, and barley), legumes, various vegetables and fruit (particularly berries), as well as beverages, such as coffee, tea, and wine, and, recently, lignans are also reported in dairy products, meat, and fish [64,65,67–84]. The types and amounts vary from one source to another. The content of some lignans, as well as the degree of esterification of their glycosides, could vary with different growing conditions, geographic location, climate, and genetic characteristics. Some examples of profile and distribution of lignans in common food groups are here reported, from research in the literature applying different methodological approaches. As concluded by Durazzo et al. [17], in a systematized assessment of lignans in cereals and cereal-based products for grains studied in [65,73,76], the total average values in grains ranged between 23 and 401 μg/100 g dry weight, with lariciresinol the most representative. As, for instance, for vegetables, Milder et al. [64] reported a content of total lignans (as the sum of secoisolariciresinol, matairesinol, lariciresinol, and pinoresinol, and expressed as μg/100 g fresh edible weight) of 1325 for broccoli, 185 for cauliflower, 787 for white cabbage, 171 for carrot, 58 for tomato, and 48 for chicory. Another example was given by Penalvo et al. [65] that described, for asparagus, a following profile of lignan concentrations: secoisolariciresinol 183 μg/100 g wet basis, syringaresinol 58 μg/100 g wet basis, pinoresinol 49 μg/100 g wet basis, lariciresinol 47 μg/100 g wet basis, medioresinol 5 μg/100 g wet basis, matairesinol 2 μg/100 g wet basis whereas, for eggplant, tomato, and radish, the most representative was lariciresinol [65]. For the fruit group, as reported in a work of Kuhnle [75] secoisolariciresinol and matairesinol were identified, respectively, in orange (peel and pith removed, 21 and <1 μg/100 g wet weight), nectarine (stoned, 24 and <1 μg/100 g wet weight), apricot (stoned, 51 and <1 μg/100 g wet weight), mango (skinned and stoned, 17 and 1 μg/100 g wet weight), melon (cantaloupe, skin and seeds removed, 16 and <1 μg/100 g wet weight), and others [75]. Moreover, Penalvo et al. [70] showed for avocado, a profile of decreasing concentration of lignans, syringaresinol > pinoresinol > medioresinol > secoisolariciresinol > lariciresinol > matairesinol and for pineapple, syringaresinol > lariciresinol > matairesinol > secoisolariciresinol > pinoresinol > medioresinol, whereas, the most representative lignan for navel orange was lariciresinol, and secoisolariciresinol for kiwifruit. In berries, as reported by Smeds et al. [78], the most representative lignans among those studied were lariciresinol for cloudberries (5008 μg/100 g dry weight); secoisolariciresinol for blackberries (2902 μg/100 g dry weight), lingoberries (2319 μg/100 g dry weight), blackcurrants (446 μg/100 g dry weight); syringaresinol for cranberries (2578 μg/100 g dry weight), sea buckthorns (1177 μg/100 g dry weight), bilberries (801 μg/100 g dry weight), and red gooseberries (498 μg/100 g dry weight); and pinoresinol for strawberries (1403 μg/100 g dry weight); for raspberries the most representatives were lariciresinol (406 μg/100 g dry weight), syringaresinol (388 μg/100 g dry weight) and pinoresinol (377 μg/100 g dry weight).

Within the beverage group, a recent work of Angeloni et al. [84] reported, for coffee samples from different Countries, secoisolariciresinol from 27.9 to 52.0 μg L^{-1} and lariciresinol from 5.3 to 27.8 μg L^{-1} respectively, contrary to matairesinol, that was not possible to detect it in each type of coffee.

For foods of animal origin, Kuhnle et al. [72] reported the content of lignans for the first time; in milk and its derived products, the content of dietary lignans was reported (as the sum of secoisolariciresinol, matairesinol, and shonanin) as follows: about 1 μg/100 g wet weight for skimmed, semi-skimmed, or whole milk; in the cheese group, from <1 μg/100 g wet weight for feta cheese derived from ewe's and goat's milk, to 4 μg/100 g wet weight for mascarpone, 5 μg/100 g wet weight for parmesan, 6 μg/100 g wet weight for mozzarella (derived from buffalo milk), 13 μg/100 g wet weight for soft Philadelphia cheese (full fat), and to 25 μg/100 g wet weight for Wensleydale cheese. Moreover, cow milk, also condensed and evaporated, showed a content of enterolactone in a range of 3–9 μg/100 g wet weight, and cheese in a range of 3–23 μg/100 g wet weight.

The same authors [72] reported a dietary lignan content for meat (including different meat cuts and offal) at various cooking of 1–2 μg/100 g wet weight in chicken, 3–9 μg/100 g wet weight in pork, 4–16 μg/100 g wet weight in beef, 4–17 μg/100 g wet weight in lamb; whereas, for eggs, 2–3 μg/100 g wet weight for egg whites and 6–10 μg/100 g wet weight for egg yolks. Small quantities of enterolignans (<6 μg/100 g wet weight) were detected in some type of eggs and meat cuts.

Most of the foods are consumed after cooking or processing, depending on the type of food matrices and the eating habits of the consumers, indeed, researches are moving in this direction [72,85,86]; indeed, the evaluation of the effects of all type of factors on lignan content in different food matrices increase the reliability of lignan intake estimations.

At the same time, procedures to improve the content of lignans such as milling, parboiling, or supplementation diet in animals [86–88] were optimized.

Nowadays, attention is paid to less common species and agro-industrial side streams [89–91], in order to continually explore new sources of lignans.

4. Lignans and Databases: The Current Workflow

Studies that examine the relationship between diet and health have led to increased interest in all biologically active constituents that are present together with nutrients in food, and data on these, as well as other compounds, are increasingly required in the database system.

A complete and comprehensive harmonized databases on the content of lignans in foods are useful in dietary assessment and in the evaluation of formulated diet, in order to be used in observational studies as key elements for healthy nutritional patterns [92]. Knowledge of the dietary intake of lignans is needed for understanding the relationship between a lignan-rich diet and the potential lower risk of development of various diseases, that is, hormone-related cancers, heart diseases, menopausal symptoms, and osteoporosis.

Detailed and accurate information on the lignans in foods is crucial in determining exposure and to investigate health effects in vivo.

To reach this objective, limitations were given by numerous existing factors—from one side, the diversity of the chemical features of compounds, the great number of dietary sources, and the large variability in content from a specified source, to the other side, the different extraction procedures and analytical techniques and methodologies [93]. Additional factors, in some cases, are given by the fact that several studies have been focused only on few compounds within a class, and by the lack of appropriate analytical methods.

In the last decade, researchers are addressing the identification and determination of lignan profiles in main food groups and in food chain products; when a new dataset for nutritional values is used, it is very important to evaluate the quality of the analytical information [55]. New experimental and analytical data on lignan content are now available for updating and expanding food composition databases [64,65,67–84]. In Table 1 the main national databases of lignans are described.

Table 1. National databases of lignans.

Country	Type of Database	Main/Common Lignan Compounds	N° Total Foods	Food Groups and Subgroups	References
Finland	Phytoestrogen Database including lignans	Secoisolariciresinol Matairesinol	180	Vegetables, Herbs and spices, Mushrooms, Fruits, Miscellaneous	[67]
Netherland	Lignan Database	Secoisolariciresinol Matairesinol Lariciresinol Pinoresinol	109	Oilseeds and nuts, Grain products, Vegetables and legumes, Fruits, Vegetable oils and fats, Other solid foods, Alcoholic beverages, Non-alcoholic beverages, Juices, Other beverages,	[64]

Table 1. *Cont.*

Country	Type of Database	Main/Common Lignan Compounds	N° Total Foods	Food Groups and Subgroups	References
Canada	Phytoestrogen Database including lignans	Secoisolariciresinol Matairesinol Lariciresinol Pinoresinol	121	Soy products. Legumes. Nuts and oil seeds. Vegetables. Fruits. Cereals and bread. Meat products and other processed foods. Non-alcoholic beverages. Alcoholic beverages	[69]
Japan	Lignan Database	Secoisolariciresinol Matairesinol Lariciresinol Pinoresinol Syringaresinol Medioresinol	86	Vegetables. Tubers and roots. Mushrooms. Fruits. Legumes. Soybean-based products. Cereal-based products. Animal-derived products	[70]
Spain	Alkylresorcinols and Lignans Database	Secoisolariciresinol Matairesinol Lariciresinol Pinoresinol Syringaresinol Medioresinol	593	Vegetables. Grains. Animal. Fats. Drinks	[77]
United Kingdom	Phytoestrogen Database including lignans	Secoisolariciresinol Matairesinol (and Shonanin)	496	Cereal and cereal-based foods, Fresh and processed fruit and vegetables including soya-based foods and legumes, Nuts and seeds, Oils. Alcoholic beverages. Tea and coffee. Dairy products, Eggs, Meat, Fish and seafood	[71,72,74,75]

The first examples of databases including lignans were movements toward the development of phytoestrogen databases [67,94]. Valsta et al. [67] reported on expansion of the Finnish National Food Composition Database (Fineli®), compiling values for plant lignans, matairesinol, and secoisolariciresinol (shown in Figure 1), and the isoflavones, daidzein and genistein.

Further, Milder et al. [64] developed a lignan database for 83 solid foods and 26 beverages commonly consumed in the Netherlands: the amount of lignans in plant foods varied widely, from 0 to 301,129 µg/100 g fresh weight; in detail, the lignan values varied from 10 to 30,129 µg/100 g fresh edible weight of oilseeds and nuts, from 7 to 12,474 µg/100 g fresh edible weight of grain products, from 0 to 2321 µg/100 g fresh edible weight of vegetables, from 0 to 450 µg/100 g fresh edible weight of fruits, from 26 to 37 µg/100 g fresh edible weight of legumes, and in beverages ranged from 0 to 91 µg/100 mL. Only five of the studied foods did not contain a measurable amount of lignans and, in most cases, the amount of lariciresinol and pinoresinol was larger than that of secoisolariciresinol and matairesinol.

On the basis of above mentioned lignan databases, in another work, Milder et al. [68] have assessed the lignan intake in a representative sample of 4660 Dutch adults (Dutch Food Consumption Survey, carried out in 1997–1998), reporting the following contribution percentages to lignan intake: lariciresinol and pinoresinol contributed 75%, whereas secoisolariciresinol and matairesinol contributed 25%; and the major food sources of lignans were beverages (37%), followed by vegetables (24%), nuts and seeds (14%), bread (9%), and fruits (7%) [68].

Thompson et al. [69] developed a lignan database of foods consumed in Canada: nine phytoestrogens were identified in 121 food products of Canada by GC–MS, including lignans; decreasing amounts (on wet weight, µg per 100 g) of total lignans are reported in the following order: nuts and oilseeds (25–379012), cereals and breads (2.0–7239.3), legumes (1.8–979.4), fruits (0.3–61.8), vegetables (1.2–583.2), soy products (2.2–269.2), meat products and other processed foods (0.2–415.1), alcoholic beverages (1.1–37.3), and non-alcoholic beverages (0.9–12). Matairesinol was the least-concentrated lignan in most studied foods, whereas secoisolariciresinol reached the highest concentration in 63 foods, lariciresinol in 44 foods, and pinoresinol in 14 foods [69].

Peñalvo et al. [70] have reported the content of six plant lignans (shown in Figure 1) in 86 food items commonly consumed in Japan: the amount of plant lignans ranged from 0 to 1724 µg/100 g (wet basis); in details, as for instance, considering the food group of vegetables, most of the lignans were in the stems and leaves of Japanese parsley, asparagus, Japanese spinach, bitter oranges, and Chinese citrus, and related concentrations in vegetables ranged from 19 to 1724 µg/100 g wet basis.

Moreno-Franco et al. [77] have developed the Aligna databases, by collecting data from scientific publications for alkylresorcinols and lignans in common foods and beverages, and by analyzing foods particularly consumed in Spain; moreover, the assess of lignans intake in Spain was evaluated and reported as follows: 0.76 mg/day, with the major contributors, i.e. oils and fats (33 percent), fruits and vegetables (30 percent), bread (14 percent), and wine and beer (10 percent) [77].

In several works, Kuhnle et al. [71,72,74,75] reported the content of secoisolariciresinol and matairesinol in 115 foods of animal origin, 240 different foods based on fresh and processed fruit and vegetables, 101 cereal and cereal-based foods including bread, breakfast cereals, biscuits, pasta, and rice, and about 40 beverages, nuts, seeds, and oils. The study of Mulligan et al. [81] estimates the average intakes of isoflavones, lignans, enterolignans, and coumestrol in the Norfolk arm of the European Prospective Investigation into Cancer and Nutrition (EPIC-Norfolk) from 7-days food diaries, and provides data on total isoflavone, lignan, and phytoestrogen consumption by food group—the mean daily total lignan intake was 361 (SD 230) µg in soya-consuming men, and 311 (SD 178) µg in non-soya-consuming men; the mean daily total lignans intake was 318 (SD 212) µg in soya-consuming women and 251 (SD 141) µg in non-soya-consuming women [81].

It is worth mentioning the work of Tetens et al. [95] which estimated and evaluated the scale of consumption and the main food sources of lignans in five European countries using the Finnish databases [67], including lignans and Dutch lignan databases [64], respectively; in detail, 42 food groups known to contribute to the total lignan intake were selected and a value attributed for secoisolariciresinol and matairesinol from the Finnish lignan database (Fineli®) or for secoisolariciresinol, matairesinol, lariciresinol, and pinoresinol from the Dutch database. The total intake of lignans was estimated from food consumption data for adult men and women (19–79 years) from Denmark, Finland, Italy, Sweden, United Kingdom, and the contribution of aggregated food groups calculated using the Dutch lignan database [75]. The authors showed that, compared to the total lignan intakes among Dutch men and women, the total lignan intakes were higher in Denmark and Sweden, and within similar range in Finland, Italy, and United Kingdom [75].

Here, also, are some examples of utilization of lignan databases to investigate the association between lignan intake and prevention of some chronic pathologies.

A recent study was undertaken by Witkowska et al. [96] that examined the total and individual lignan intakes and their dietary sources in postmenopausal Polish women: for lignan content, the Dutch lignan database was used [64]; for beverages, nuts, seeds, and oils, data from Kuhnle et al. [71] were taken, and when data on lignan content were missing, values were taken from Thompson et al. [69]; in women with cardiovascular disease (CVD), secoisolariciresinol accounted for 50.15% lignan intake from plant foods, as compared to 44.8% in the control. Pinoresinol, lariciresinol, and matairesinol contributed to the total lignan intakes of CVD and non-CVD women in 24.0% vs. 26.1%, 22.7% vs. 26.1%, and 3.1% vs. 2.9%, respectively [96].

Nowadays, the major core public databases that gather extensive data on the polyphenol content of foods and beverages include lignans—Phenol-Explorer [97], the first comprehensive database on polyphenol content in foods [98] and eBASIS (Bioactive Substances in Food Information Systems) [99–101], published through the EuroFIR project.

Phenol-Explorer was the first comprehensive web-based database on polyphenol content in foods and an open-access database and, now, throughout several updates [102,103], includes new data on pharmacokinetic and metabolites, the effect of food processing and cooking and, in the last update (version 3.6), 1451 new content values for lignans have been added (to the database). The development of the Phenol-Explorer database included five main steps: literature search, data compilation, data evaluation, data aggregation, and final data exportation to the MySQL database which is used by the web interface. Composition data were collected from peer-reviewed scientific publications, and evaluated before they were aggregated to produce final representative mean content values.

The eBASIS database contains composition data and biological effects of over 300 major European plant foods of 24 compound classes, such as glucosinolates, phytosterols, polyphenols, isoflavones, glycoalkaloids, and xanthine alkaloids in 15 EU languages. EuroFIR eBASIS resource is a compilation of expert critically evaluated data extracted from peer-reviewed literature as raw data. This could be seen and considered as the first effort to establish a harmonized food composition information system in EU. Indeed, eBASIS should be defined as the first EU harmonized food composition database. Currently, 2695 data points for lignans were inserted in eBASIS, in detail, 658 values for secoisolariciresinol, 550 values for matairesinol, 313 values for lariciresinol, 276 values for pinoresinol, 93 values for medioresinol, and 86 values for syringaresinol [99,101].

Indeed, considering the importance of metabolic pathways and the benefits of bioactive compounds in humans, it is worth mentioning the Human Metabolome Database or HMDB 4.0 [104], a web metabolomic database on human metabolites including lignans and their metabolites [105], as well as PhytoHub [106], a freely electronic database containing detailed information about all phytochemicals and their metabolites commonly ingested in diets [107].

Author Contributions: A.D., M.L., A.S. and L.M. have conceived the work and wrote the manuscript. All authors have made a substantial contribution to revise the work, and approved it for publication.

Funding: This research did not receive any specific grant from funding agencies in the public, commercial, or not-profit sectors.

Conflicts of Interest: The authors declare no conflict of interest.

References

1. Durazzo, A. Lignans. In *Phenolic Compounds in Food: Characterization and Analysis (Food Analysis and Properties)*; Leo, M.L.N., Janet, A.G.-U., Eds.; CRC Press: Boca Raton, FL, USA, 2018; Chapter 11.

2. Lewis, N.G.; Davin, L.B. Lignans: Biosynthesis and function. In *Comprehensive Natural Products Chemistry*; Barton, D., Nakanishi, K., Meth-Cohn, O., Eds.; Elsevier: Amsterdam, the Netherlands, 1999; pp. 639–712.

3. Imai, T.; Nomura, M.; Fukushima, K. Evidence for involvement of the phenylpropanoid pathway in the biosynthesis of the norlignan agatharesinol. *J. Plant Physiol.* **2006**, *163*, 483–487. [CrossRef] [PubMed]

4. Pan, J.-Y.; Chen, S.-L.; Yang, M.-H.; Wu, J.; Sinkkonen, J.; Zou, K. An update on lignans: Natural products and synthesis. *Nat. Prod. Rep.* **2009**, *26*, 1251–1292. [CrossRef] [PubMed]

5. Teponno, R.B.; Kusari, S.; Spiteller, M. Recent advances in research on lignans and neolignans. *Nat. Prod. Rep.* **2016**, *33*, 1044–1092. [CrossRef] [PubMed]

6. Huang, X.Y.; Feng, Z.M.; Yang, Y.N.; Jiang, J.S.; Zhang, P.C. Four new neolignan glucosides from the fruits of Arctium lappa. *J. Asian Nat. Prod. Res.* **2015**, *17*, 504–511. [CrossRef] [PubMed]

7. Duarte, S.L.F.; Nascimento, Y.M.; Madeiro, S.A.L.; Costa, V.C.O.; Agra, M.F.A.; Sobrala, M.V.; Braz-Filho, R.; Carvalho, M.G.; Carvalho, J.E.; Ruiz, A.L.T.G.; et al. Luclaricin, a new lignan Phyllanthus acuminatus. *Quim. Nova* **2018**, *41*, 880–883. [CrossRef]

8. Mo, X.; Chen, Y.; Han, Y.; Hao, H.; Huang, R. A New Benzylbutane Lignan from the Stems of Schisandra bicolor. *Chem. Nat. Comp.* **2018**, *54*, 872–874. [CrossRef]

9. Gnabre, J.; Bates, R.; Huang, R.C. Creosote bush lignans for human disease treatment and prevention: Perspectives on combination therapy. *J. Trad. Complem. Med.* **2015**, *5*, 119–126. [CrossRef] [PubMed]

10. Su, S.; Wink, M. Natural lignans from Arctium lappa as antiaging agents in *Caenorhabditis elegans*. *Phytochemistry* **2015**, *117*, 340–350. [CrossRef] [PubMed]

11. Zhu, Y.; Huang, R.Z.; Wang, C.G.; Ouyang, X.L.; Jing, X.T.; Liang, D.; Wang, H.S. New inhibitors of matrix metalloproteinases 9 (MMP-9): Lignans from Selaginella moellendorffii. *Fitoterapia* **2018**, *130*, 281–289. [CrossRef]

12. Landete, J.M. Plant and mammalian lignans: A review of source, intake, metabolism, intestinal bacteria and health. *Food Res. Int.* **2012**, *46*, 410–424. [CrossRef]

13. Saarinen, N.M.; Tuominen, J.; Pylkkänen, L.; Santti, R. Assessment of information to substantiate a health claim on the prevention of prostate cancer by lignans. *Nutrients* **2010**, *2*, 99–115. [CrossRef] [PubMed]

14. Velentzis, L.S.; Cantwell, M.M.; Cardwell, C.; Keshtgar, M.R.; Leathem, A.J.; Woodside, J.V. Lignans and breast cancer risk in pre- and post-menopausal women: Meta-analyses of observational studies. *Br. J. Canc.* **2009**, *100*, 1492–1498. [CrossRef] [PubMed]

15. Aehle, E.; Müller, U.; Eklund, P.C.; Willför, S.M.; Sippl, W.; Dräger, B. Lignans as food constituents with estrogen and antiestrogen activity. *Phytochemistry* **2011**, *72*, 2396–2405. [CrossRef] [PubMed]

16. Kiyama, R. Biological effects induced by estrogenic activity of lignans. *Trends Food Sci. Technol.* **2016**, *54*, 186–196. [CrossRef]

17. Durazzo, A.; Turfani, V.; Azzini, E.; Maiani, G.; Carcea, M. Phenols, lignans and antioxidant properties of legume and sweet chestnut flours. *Food Chem.* **2013**, *140*, 666–671. [CrossRef] [PubMed]

18. Muir, A.D. Flax lignans: New opportunities for functional foods. *Food Sci. Technol. Bull.* **2010**, *6*, 61–79. [CrossRef]

19. Cunha, W.R.; e Silva, M.L.A.; Sola, R.C.; Veneziani, S.R.A.; Bastos, J.K. Lignans: Chemical and biological properties. In *Phytochemicals—A global Perspective of Their Role in Nutrition and Health*; Venketeshwer, R., Ed.; In Tech: Rijeka, Croatia, 2012; pp. 213–234.

20. Peterson, J.; Dwyer, J.; Adlercreutz, H.; Scalbert, A.; Jacques, P.; McCullough, M.L. Dietary lignans: Physiology and potential for cardiovascular disease risk reduction. *Nutr. Rev.* **2010**, *68*, 571–603. [CrossRef]

21. Peñalvo, J.L.; Lopez-Romero, P. Urinary enterolignan concentrations are positively associated with serum HDL cholesterol and negatively associated with serum triglycerides in U.S. adults. *J. Nutr.* **2012**, *142*, 751–756. [CrossRef]

22. Zamora-Ros, R.; Agudo, A.; Lujan-Barroso, L.; Isabelle, R.; Pietro, F.; Viktoria, K.; Bueno-de-Mesquita, H.B.; Max, L.; Ruth, C.T.; Carmen, N.; et al. Dietary flavonoid and lignan intake and gastric adenocarcinoma risk in the European Prospective Investigation into Cancer and Nutrition (EPIC) study. *Am. J. Clin. Nutr.* **2012**, *96*, 1398–1408. [CrossRef]

23. Zamora-Ros, R.; Touillaud, M.; Rothwell, J.A.; Romieu, I.; Scalbert, A. Measuring exposure to the polyphenol metabolome in observational epidemiologic studies: Current tools and applications and their limits. *Am. J. Clin. Nutr.* **2014**, *100*, 11–26. [CrossRef]

24. Durazzo, A.; Carcea, M.; Adlercreutz, H.; Azzini, E.; Polito, A.; Olivieri, L.; Zaccaria, M.; Meneghini, C.; Maiani, F.; Bausano, G.; et al. Effects of consumption of whole grain foods rich in lignans in healthy postmenopausal women with moderate serum cholesterol: A. pilot study. *Int. J. Food Sci. Nutr.* **2014**, *65*, 637–645. [CrossRef] [PubMed]

25. Sun, Q.; Wedick, N.M.; Pan, A.; Townsend, M.K.; Cassidy, K.; Franke, A.A.; Rimm, E.B.; Hu, F.B.; van Dam, R.B. Gut Microbiota Metabolites of Dietary Lignans and Risk of Type 2 Diabetes: A Prospective Investigation in Two Cohorts of U.S. Women. *Diabetes Care* **2014**, *37*, 1287–1295. [CrossRef] [PubMed]

26. Chang, V.C.; Cotterchio, M.; Boucher, B.A.; Jenkins, D.J.A.; Mirea, L.; McCann, S.E.; Thompson, L.U. Effect of Dietary Flaxseed Intake on Circulating Sex Hormone Levels among Postmenopausal Women: A Randomized Controlled Intervention Trial. *Nutr. Cancer.* **2018**, *30*, 1–14. [CrossRef] [PubMed]

27. Barre, D.E.; Mizier-Barre, K.A. Lignans' potential in pre- and post-onset type 2 diabetes management. *Curr. Diabetes Rev.* **2018**, in press. [CrossRef] [PubMed]

28. Santini, A.; Novellino, E.; Armini, V.; Ritieni, A. State of the art of Ready-to-Use Therapeutic Food: A tool for nutraceuticals addition to foodstuff. *Food Chem.* **2013**, *140*, 843–849. [CrossRef] [PubMed]

29. Santini, A.; Tenore, G.C.; Novellino, E. Nutraceuticals: A paradigm of proactive medicine. *Eur. J. Pharm. Sci.* **2017**, *96*, 53–61. [CrossRef]

30. Adefegha, S.A. Functional foods and nutraceuticals as dietary Intervention in chronic diseases; novel perspectives for health promotion and disease prevention. *J. Diet. Suppl.* **2018**, *15*, 977–1009. [CrossRef]

31. Durazzo, A.; D'Addezio, L.; Camilli, E.; Piccinelli, R.; Turrini, A.; Marletta, L.; Marconi, S.; Lucarini, M.; Lisciani, S.; Gabrielli, P.; et al. From plant compounds to botanicals and back: A current snapshot. *Molecules* **2018**, *23*, 1844. [CrossRef]

32. Santini, A.; Novellino, E. Nutraceuticals—shedding light on the grey area between pharmaceuticals and food. *Expert. Rev. Clin. Pharmacol.* **2018**, *11*, 545–547. [CrossRef]

33. Pilkington, L.I. Lignans: A Chemometric Analysis. *Molecules* **2018**, *23*, 1666. [CrossRef]

34. Obermeyer, W.R.; Musser, S.M.; Betz, J.M.; Casey, R.E.; Pohland, A.E.; Page, S.W. Chemical studies of phytoestrogens and related compounds in dietary supplements: Flax and chaparral. *Proc. Soc. Exp. Biol. Med.* **1995**, *208*, 6–12. [CrossRef]

35. Mazur, W.; Fotsis, T.; Wahala, K.; Ojala, S.; Salakka, A.; Adlercreutz, H. Isotope dilution gas chromatographic-mass spectrometric method for the determination of isoflavonoids, coumestrol, and lignans in food samples. *Anal. Biochem.* **1996**, *233*, 169–180. [CrossRef] [PubMed]

36. Nilsson, M.; Åman, P.; Härkönen, H.; Hallmans, G.; Knudsen, K.E.B.; Mazur, W.; Adlercreutz, H. Content of nutrients and lignans in roller milled fractions of rye. *J. Sci. Food Agric.* **1997**, *73*, 143–148. [CrossRef]

37. Meagher, L.P.; Beecher, G.R.; Flanagan, V.P.; Li, B.W. Isolation and characterization of the lignan, isolariciresinol and pinoresinol in flaxseed meal. *J. Agric. Food Chem.* **1999**, *47*, 3173–3180. [CrossRef] [PubMed]

38. Johnsson, P.; Kamal Eldin, A.; Lundgren, L.N.; Aman, P. HPLC method for analysis of secoisolariciresinol diglucoside in flaxseed. *J. Agric. Food Chem.* **2000**, *48*, 5216–5219. [CrossRef] [PubMed]

39. Kraushofer, T.; Sontag, G. Determination of matairesinol in flaxseed by HPLC with coulometric electrode array detection. *J. Chrom. B* **2002**, *777*, 61–66. [CrossRef]

40. Muir, A.D.; Westcott, N.D. Flaxseed constituents and human health. In *Flax: The Genus Linum*; Muir, A.D., Westcott, N.D., Eds.; Taylor & Francis: London, UK, 2003; pp. 243–251.

41. Willför, S.M.; Smeds, A.I.; Holmbom, B.R. Chromatographic analysis of lignans. *J. Chromatogr. A* **2006**, *1112*, 64–77. [CrossRef]

42. Smeds, A.I.; Eklund, P.C.; Sjöholm, R.E.; Willför, S.M.; Nishibe, S.; Deyama, T.; Holmbom, B.R. Quantification of a broad spectrum of lignans in cereals, oilseeds, and nuts. *J. Agric. Food Chem.* **2007**, *55*, 1337–1346. [CrossRef]

43. Schwartz, H.; Sontag, G. Analysis of lignans in food samples-impact of sample preparation. *Curr. Bioact. Compd.* **2011**, *7*, 156–171. [CrossRef]

44. Nørskov, N.P.; Knudsen, K.E.B. Validated LC-MS/MS Method for the Quantification of free and bound lignans in cereal-based diets and feces. *J. Agric. Food Chem.* **2016**, *64*, 8343–8351. [CrossRef]

45. Hanhineva, K.; Rogachev, I.; Aura, A.M.; Aharoni, A.; Poutanen, K.; Mykkänen, H. Identification of novel lignans in the whole grain rye bran by non–targeted LC–MS metabolite profiling. *Metabolomics* **2012**, *8*, 399–409. [CrossRef]

46. Linder, T.; Schnürch, M.; Mihovilovic, M.D. Construction of heterocyclic lignans in natural product synthesis and medicinal chemistry. In *Targets in Heterocyclic Systems (Reviews and Accounts on Heterocyclic Chemistry)*; Attanasi, O.A., Merino, P., Spinelli, D., Eds.; Società Chimica Italiana: Rome, Italy, 2015; volume 19.

47. Soorukram, D.; Pohmakotr, M.; Kuhakarn, C.; Reutrakul, V. Stereoselective synthesis of tetrahydrofuran lignans. *Synthesis* **2018**, in press. [CrossRef]

48. Vo, Q.V.; Nam, P.C.; Bay, M.V.; Thong, N.M.; Cuong, N.D.; Mechler, A. Density functional theory study of the role of benzylic hydrogen atoms in the antioxidant properties of lignans. *Sci. Rep.* **2018**, *8*, 12361. [CrossRef] [PubMed]

49. Waltner-Towes, D.; Lang, T. A new conceptual base for food and agricultural policy: The emerging model of links between agriculture, food, health, environment and society. *Glob. Chang. Hum. Health* **2000**, *1*, 116–130. [CrossRef]

50. Astley, S.; Finglas, P. Nutrition and Health. *Ref. Module Food Sci.* **2016**. Available online: https://doi.org/10.1016/B978-0-08-100596-5.03425-9 (accessed on 7 May 2016).

51. World Health Organization, Food Safety. 2017. Available online: http://www.who.int/mediacentre/factsheets/fs399/en/ (accessed on 31 October 2017).

52. Lampe, J.W.; Atkinson, C.; Hullar, M.A. Assessing exposure to lignans and their metabolites in humans. *J. AOAC Int.* **2006**, *89*, 1174–1181.

53. Encyclopaedia Britannica. Available online: https://www.britannica.com/technology/database (accessed on 20 November 2018).

54. Sofroniou, A. *Relational Databases and Distributed Systems*; lulu.com: Morrisville, CA, USA, 2018.

55. Greenfield, H.; Southgate, D.A.T. *Food Composition Data. Production, Management, and Use*, 2nd ed.; Food and Agriculture Organization of The United Nations: Rome, Italy, 2003.

56. Pakkala, H.; Christensen, T.; de Victoria, I.M.; Presser, K.; Kadvan, A. Harmonised information exchange between decentralised food composition database systems. *Eur. J. Clin. Nutr.* **2010**, *64*, S58–S63. [CrossRef] [PubMed]

57. Finglas, P.M.; Berry, R.; Astley, S. Assessing and improving the quality of food composition databases for nutrition and health applications in Europe: The contribution of EuroFIR. *Adv. Nutr.* **2014**, *5*, 608–614. [CrossRef]

58. Becker, W. CEN/TC387 Food Data. Towards a CEN Standard on food data. *Eur. J. Clin. Nutr.* **2010**, *64*, S49–S52. [CrossRef]

59. Finglas, P.; Roe, M.; Pinchen, H.; Astley, S. The contribution of food composition resources to nutrition science methodology. *Nutr. Bull.* **2017**, *42*, 198–206. [CrossRef]

60. Durazzo, A.; Camilli, E.; D'Addezio, L.; Le Donne, C.; Ferrari, M.; Marconi, S.; Marletta, L.; Mistura, L.; Piccinelli, R.; Scalvedi, M.L.; et al. Food Groups and Individual Foods: Nutritional Attributes and Dietary Importance. *Ref. Module Food Sci.* **2018**, 1–13. [CrossRef]

61. Thompson, L.U.; Richard, S.E.; Cheung, F.; Kenaschuk, E.O.; Obermeyer, W.R. Variability in anticancer lignan levels in flaxseed. *Nutr. Cancer* **1997**, *27*, 26–30. [CrossRef] [PubMed]

62. Mazur, W.M.; Adlercreutz, H. Natural and anthropogenic environmental estrogens: The scientific basis for risk assessment; naturally occurring estrogens in food. *Pure. Appl. Chem.* **1998**, *70*, 1759–1776. [CrossRef]

63. Coulman, K.D.; Liu, Z.; Hum, W.Q.; Michaelides, J.; Thompson, L.U. Whole sesame seed is as rich a source of mammalian lignan precursors as whole flaxseed. *Nutr. Cancer* **2005**, *52*, 156–165. [CrossRef] [PubMed]

64. Milder, I.E.; Arts, I.C.; van de Putte, B.; Venema, D.P.; Hollman, P.C. Lignan contents of Dutch plant foods: A database including lariciresinol, pinoresinol, secoisolariciresinol and matairesinol. *Brit. J. Nutr.* **2005**, *93*, 393–402. [CrossRef] [PubMed]

65. Peñalvo, J.L.; Haajanen, K.M.; Botting, N.; Adlercreutz, H. Quantification of lignans in food using isotope dilution gas chromatography/mass spectrometry. *J. Agric. Food Chem.* **2005**, *53*, 9342–9347. [CrossRef] [PubMed]

66. Holmbom, B.; Eckerman, C.; Eklund, P.; Hemming, J.; Nisula, L.; Reunanen, M.; Sjöholm, R.; Sundberg, A.; Sundberg, K.; Willför, S. Knots in trees—A new rich source of lignans. *Phytochem. Rev.* **2003**, *2*, 331–340. [CrossRef]

67. Valsta, L.M.; Kilkkinen, A.; Mazur, W.; Nurmi, T.; Lampi, A.M.; Ovaskainen, M.L.; Korhonen, T.; Adlercreutz, H.; Pietinen, P. Phyto-oestrogen database of foods and average intake in Finland. *Br. J. Nutr.* **2003**, *89*, S31–S38. [CrossRef]

68. Milder, I.E.J.; Feskens, E.J.M.; Arts, I.C.W.; Bueno-de-Mesquita, H.B.; Hollman, P.C.H.; Kromhout, D. Intake of the plant lignans secoisolariciresinol, matairesinol, lariciresinol and pinoresinol in Dutch men and women. *J. Nutr.* **2005**, *135*, 1202–1207. [CrossRef]

69. Thompson, L.U.; Boucher, B.A.; Liu, Z.; Cotterchio, M.; Kreiger, N. Phytoestrogen content of foods consumed in Canada, including isoflavones, lignans and coumestan. *Nutr. Cancer* **2006**, *54*, 184–201. [CrossRef]

70. Peñalvo, J.L.; Adlercreutz, H.; Uehara, M.; Ristimaki, A.; Watanabe, S. Lignan content of selected foods from Japan. *J. Agric. Food Chem.* **2008**, *56*, 401–409. [CrossRef]

71. Kuhnle, G.G.C.; Dell'Aquila, C.; Aspinall, S.M.; Runswick, S.A.; Mulligan, A.A.; Bingham, S.A. Phytoestrogen content of beverages, nuts, seeds, and oils. *J. Agric. Food Chem.* **2008**, *56*, 7311–7315. [CrossRef] [PubMed]

72. Kuhnle, G.G.C.; Dell'Aquila, C.; Aspinall, S.M.; Runswick, S.A.; Mullingan, A.A.; Bingham, S.A. Phytoestrogen content of foods of animal origin: Dairy products, eggs, meat, fish, and seafood. *J. Agric. Food Chem.* **2008**, *56*, 10099–10104. [CrossRef] [PubMed]

73. Durazzo, A.; Raguzzini, A.; Azzini, E.; Foddai, M.S.; Narducci, V.; Maiani, G.; Carcea, M. Bioactive molecules in cereals. *Tecnica Molitoria Int.* **2009**, *60*, 150–162.

74. 68Kuhnle, G.G.C.; Dell'Aquila, C.; Aspinall, S.M.; Runswick, S.A.; Mulligan, A.A.; Bingham, S.A. Phytoestrogen content of cereals and cereal based foods consumed in the UK. *Nutr. Cancer* **2009**, *61*, 302–309.

75. Kuhnle, G.G.C.; Dell'Aquila, C.; Sue, M.A.; Runswick, S.A.; Joosen, A.M.C.P.; Mulligan, A.A.; Bingham, S.A. Phytoestrogen content of fruits and vegetables commonly consumed in the UK based on LC-MS and 13C-labelled standards. *Food Chem.* **2009**, *116*, 542–554. [CrossRef]

76. Smeds, A.I.; Jauhiainen, L.; Tuomola, E.; Peltonen-Sainio, P. Characterization of variation in the lignan content and composition of winter rye, spring wheat and spring oat. *J. Agric. Food Chem.* **2009**, *57*, 5837–5842. [CrossRef] [PubMed]

77. Moreno-Franco, B.; Garcia-Gonzalez, A.; Montero-Bravo, A.M.; Iglesias-Gitierrez, E.; Ubeda, N.; Maroto-Nunez, L.; Adlercreutz, H.; Penãlvo, J. Dietary alkylresorcinols and lignans in the Spanish diet: Development of the Alignia database. *J. Agric. Food Chem.* **2011**, *59*, 9827–9834. [CrossRef]

78. Smeds, A.I.; Eklund, P.C.; Willför, S.M. Content, composition, and stereochemical characterisation of lignans in berries and seeds. *Food Chem.* **2012**, *134*, 1991–1998. [CrossRef]

79. Durazzo, A.; Azzini, E.; Turfani, V.; Polito, A.; Maiani, G.; Carcea, M. Effect of cooking on lignans content in wholegrain pasta made with different cereals and other seeds. *Cereal Chem.* **2013**, *90*, 169–171. [CrossRef]

80. Durazzo, A.; Zaccaria, M.; Polito, A.; Maiani, G.; Carcea, M. Lignan content in cereals, buckwheat and derived foods. *Foods* **2013**, *2*, 53–63. [CrossRef]

81. Mulligan, A.A.; Kuhnle, G.G.; Lentjes, M.A.; van Scheltinga, V.; Powell, N.A.; McTaggart, A.; Bhaniani, A.; Khaw, K.T. Intakes and sources of isoflavones, lignans, enterolignans, coumestrol and soya-containing foods in the Norfolk arm of the European Prospective Investigation into Cancer and Nutrition (EPIC-Norfolk), from 7 d food diaries, using a newly updated database. *Public Health Nutr.* **2013**, *16*, 1454–1462. [CrossRef] [PubMed]

82. Durazzo, A.; Turfani, V.; Narducci, V.; Azzini, E.; Maiani, G.; Carcea, M. Nutritional characterization and bioactive components of commercial carobs flours. *Food Chem.* **2014**, *153*, 109–113. [CrossRef] [PubMed]

83. Turfani, V.; Narducci, V.; Durazzo, A.; Galli, V.; Carcea, M. Technological, nutritional and functional properties of wheat bread enriched with lentil or carob flours. *LWT Food Sci. Technol.* **2017**, *78*, 361–366. [CrossRef]

84. Angeloni, A.; Navarini, L.; Sagratini, L.; Torregiani, E.; Vittori, S.; Caprioli, G. Development of an extraction method for the quantification of lignans in espresso coffee by using HPLC-MS/MS triple quadrupole. *J. Mass Spectrom.* **2018**, *53*, 842–848. [CrossRef] [PubMed]

85. Gerstenmeyer, E.; Reimer, S.; Berghofer, E.; Schwartz, H.; Sontag, G. Effect of thermal heating on some lignans in flax seed, sesame seeds and rye. *Food Chem.* **2013**, *138*, 1847–1855. [CrossRef] [PubMed]

86. Pihlava, J.M.; Nordlund, E.; Heinio, R.L.; Hietaniemi, V.; Lehtinen, P.; Poutanen, K. Phenolic compounds in wholegrain rye and its fractions. *J. Food Comp. Anal.* **2015**, *38*, 89–97. [CrossRef]

87. Durazzo, A.; Azzini, E.; Raguzzini, A.; Maiani, G.; Finocchiaro, F.; Ferrari, B.; Gianinetti, A.; Carcea, M. Influence of processing on the lignans content of cereal based foods. *Tecnica Molitoria Int.* **2009**, *60*, 163–173.

88. Mattioli, S.; Ruggeri, S.; Sebastiani, B.; Brecchia, G.; Dal Bosco, A.; Mancinelli, A.C.; Castellini, C. Performance and egg quality of laying hens fed flaxseed: Highlights on n-3 fatty acids, cholesterol, lignans and isoflavones. *Animal* **2017**, *11*, 705–712. [CrossRef]

89. Inostroza, J.P.; Troncoso, J.; Mardones, C.; Vergara, C. Lignans in olive stones discarded from the oil industry. Comparison of three extraction Methods followed by HPLC-DAD-MS/MS and antioxidant capacity determination. *J. Chil. Chem. Soc.* **2018**, *63*, 4001–4005. [CrossRef]

90. Jablonský, M.; Škulcová, A.; Malvis, A.; Šima, J. Extraction of value-added components from food industry based and agro-forest biowastes by deep eutectic solvents. *J. Biotechnol.* **2018**, *282*, 46–66. [CrossRef]

91. Lucarini, M.; Durazzo, A.; Romani, A.; Campo, M.; Lombardi-Boccia, G.; Cecchini, F. Bio-based vompounds from grape seeds: A biorefinery approach. *Molecules* **2018**, *23*, 1888. [CrossRef] [PubMed]

92. Blitz, C.L.; Murphy, S.P.; Au, D.L.M. Adding lignan values to a food composition database. *J. Food Compos. Anal.* **2007**, *20*, 99–105. [CrossRef]

93. Scalbert, A.; Andres-Lacueva, C.; Arita, M.; Kroon, P.; Manach, C.; Urpi-Sarda, M.; Wishart, D. Databases on food phytochemicals and their health-promoting effects. *J. Agric. Food Chem.* **2011**, *59*, 4331–4348. [CrossRef] [PubMed]

94. Horn-Ross, P.L.; Barnes, S.; Lee, M.; Coward, L.; Mandel, J.E.; Koo, J.; John, E.M.; Smith, M. Assessing phytoestrogen exposure in epidemiologic studies: Development of a database (United States). *Cancer Causes Control.* **2000**, *11*, 289–298. [CrossRef] [PubMed]

95. Tetens, I.; Turrini, A.; Tapanainen, H.; Christensen, T.; Lampe, J.W.; Fagt, S.; Hakansson, N.; Lundquist, A.; Hallund, J.; Valsta, L.M.; et al. Dietary intake and main sources of plant lignans in five European countries. *Food Nutr. Res.* **2013**, *57*, 1. [CrossRef] [PubMed]

96. Witkowska, A.M.; Waskiewicz, A.; Zujko, M.E.; Szczesniewska, D.; Stepaniak, U.; Pajak, A.; Drygas, W. Are total and individual dietary lignans related to cardiovascular disease and its risk factors in postmenopausal women? A. Nationwide Study. *Nutrients* **2018**, *10*, 865. [CrossRef] [PubMed]

97. Phenol-Explorer—Database on Polyphenol Content in Foods. Available online: http://phenol-explorer.eu/ (accessed on 29 November 2018).

98. Neveu, V.; Perez-Jiménez, J.; Vos, F.; Crespy, V.; du Chaffaut, L.; Mennen, L.; Knox, C.; Eisner, R.; Cruz, J.; Wishart, D.; Scalbert, A. Phenol-Explorer: An online comprehensive database on polyphenol contents in foods. *Database* **2010**, *2010*, bap024. [CrossRef]

99. eBASIS—Bioactive Substances in Food Information System. Available online: http://ebasis.eurofir.org/Default.asp (accessed on 29 October 2018).

100. Kiely, M.; Black, L.J.; Plumb, J.; Kroon, P.A.; Hollman, P.C.; Larsen, J.C.; Speijers, G.J.; Kapsokefalou, M.; Sheehan, D.; Gry, J.; et al. EuroFIR consortium. EuroFIR eBASIS: Application for health claims submissions and evaluations. *Eur. J. Clin. Nutr.* **2010**, *3*, S101. [CrossRef]

101. Plumb, J.; Pigat, S.; Bompola, F.; Cushen, M.; Pinchen, H.; Nørby, E.; Astley, S.; Lyons, J.; Kiely, M.; Finglas, P. eBASIS (Bioactive Substances in Food Information Systems) and Bioactive Intakes: Major Updates of the Bioactive Compound Composition and Beneficial Bioeffects Database and the Development of a Probabilistic Model to Assess Intakes in Europe. *Nutrients* **2017**, *9*, 320. [CrossRef]

102. Rothwell, J.A.; Urpi-Sarda, M.; Boto-Ordonez, M.; Knox, C.; Llorach, R.; Eisner, R.; Cruz, J.; Neveu, V.; Wishart, D.; Manach, C.; et al. Phenol-Explorer 2.0: A major update of the Phenol-Explorer database integrating data on polyphenol metabolism and pharmacokinetics in humans and experimental animals. *Database* **2012**, *2012*, bas031. [CrossRef]

103. Rothwell, J.A.; Perez-Jimenez, J.; Neveu, V.; Medina-Remon, A.; M'Hiri, N.; Garcia-Lobato, P.; Manach, C.; Knox, C.; Eisner, R.; Wishart, D.S.; et al. Phenol-Explorer 3.0: A major update of the Phenol-Explorer database to incorporate data on the effects of food processing on polyphenol content. *Database* **2013**, *2013*, bat070.

104. HMDB—Human Metabolome Database. Available online: www.hmdb.ca (accessed on 29 October 2018).

105. Wishart, D.S.; Feunang, Y.D.; Marcu, A.; Guo, A.C.; Liang, K.; Vázquez-Fresno, R.; Sajed, T.; Johnson, D.; Li, C.; Karu, N.; et al. HMDB 4.0—The Human Metabolome Database for 2018. *Nucleic Acids Res.* **2018**, *46*, D608–D617. [CrossRef] [PubMed]

106. PhytoHub Database. Available online: www.phytohub.eu (accessed on 8 June 2018).

107. Bento da Silva, A.; Giacomoni, F.; Pavot, B.; Fillâtre, Y.; Rothwell, J.A.; Sualdea, B.B.; Veyrat, C.; Garcia-Villalba, R.; Gladine, C.; Kopec, R.; et al. PhytoHub V1.4: A new release for the online database dedicated to food phytochemicals and their human metabolites. In Proceedings of the 1st International Conference on Food Bioactivities & Health, Norwich, UK, 13–15 September 2016.

Sample Availability: Not available.

molecules

MDPI

Article

Lignans in Spirits: Chemical Diversity, Quantification, and Sensory Impact of (±)-Lyoniresinol

Delphine Winstel and Axel Marchal *

Unité de recherche Œnologie, EA 4577, USC 1366 INRA, Université de Bordeaux, ISVV, 33882 Villenave d'Ornon, France; delphine.winstel@u-bordeaux.fr
* Correspondence: axel.marchal@u-bordeaux.fr; Tel.: +33-557-575-867

Academic Editor: David Barker
Received: 15 November 2018; Accepted: 25 December 2018; Published: 30 December 2018

Abstract: During barrel aging, spirits undergo organoleptic changes caused by the release of aroma and taste compounds. Recently, studies have revealed the bitter properties of oak wood lignans, such as (±)-lyoniresinol, and their contribution to wine taste. To evaluate the impact of lignans in spirits, a targeted screening of 11 compounds was set up and served to validate their presence in this matrix, implying their release by oak wood during aging. After development and validation of a quantification method, the most abundant and the bitterest lignan, (±)-lyoniresinol, was assayed by liquid chromatography–high resolution mass spectrometry (LC-HRMS) in spirits. Its gustatory detection threshold was established at 2.6 mg/L in spirits. A large number of samples quantified were above this detection threshold, which suggests its effect of increased bitterness in spirit taste. Significant variations were observed in commercial spirits, with concentrations ranging from 0.2 to 11.8 mg/L, which could be related to differences in barrel aging processes. In "eaux-de-vie" of cognac, concentrations of (±)-lyoniresinol were observed in the range from 1.6 mg/L to 12 mg/L. Lower concentrations were measured for older vintages.

Keywords: Lignan; bitterness; taste-active compound; quantification; oak ageing

1. Introduction

Spirits are alcoholic beverages, traditionally consumed for human enjoyment. Their sensory quality is strongly influenced by their different production stages [1]. First, the elaboration of spirits needs a raw material with a high content of sugars naturally present in fruits (apple, pear, or grape) for calvados, cognac, armagnac, and brandy; of carbohydrates and starch (corn, barley, or rye) for bourbon and whiskey; or of sugarcane for rum. Thereafter, yeast fermentation, distillation techniques in pots or column stills, maturation in oak barrels, and bottling play their own special part in aroma and taste formation [2,3]. From a chemical point of view, spirits are a highly complex matrix characterized by an ethanol concentration that is usually between 36% to 55% v/v and a high content of volatile and nonvolatile compounds.

With the development of sensitive and resolutive analytical techniques, such as GC-olfactometry and GC-MS, hundreds of volatile components have been identified in spirits over the last few decades [3–7]. These compounds are mainly esters, alcohols, aldehydes, and volatile acids coming from grapes or formed during fermentation.

Contrary to volatile compounds, nonvolatile molecules present in spirits are exclusively acquired after distillation. Most of them are released from oak wood during aging. Indeed, barrel aging is a crucial step during which the organoleptic properties of spirits are fine-tuned [8]. The color, structure, aroma, and taste of spirits are modified during oak aging [9–12], and it is commonly acknowledged that the overall quality improves. While the key aromatic compounds released from oak wood in spirits are well known [13,14], modifications to gustatory properties during this process have been only partially explained. Recent studies have demonstrated the influence of oak wood on the softening of wine and spirits with the discovery of natural sweet compounds called quercotriterpenosides [15,16]. Furthermore, the astringency and bitterness of ellagitannins, a major class of oak wood compounds, have also been studied. Glabasnia and Hofmann showed that the detection thresholds (DTs) of the main hydrolysable tannins were significantly higher than their concentrations in wines, suggesting their low influence on wine bitterness [17]. So far, no correlation between bitterness and oak tannins has been clearly established [18]. Moreover, a large number of studies have examined the phenolic composition of spirits. Mainly nonflavonoids have been identified and quantified in aged spirits, most of them of low molecular weight [19,20]. Phenolic acids such as gallic acid are the most abundant phenolic compounds in spirits, followed by phenolic aldehydes (vanillin, sinapaldehyde, syringaldehyde), lignans, phenyl ketones, and coumarins such as scopoletin [21–25].

Among these nonvolatile compounds present in spirits, lignans appear to be particularly interesting, since previous studies have demonstrated the bitter properties of lyoniresinol [16] and various derivatives. Lyoniresinol has been described as the most abundant lignin in sessile oak, and its detection threshold has been established at 1.5 mg/L in wine [26]. A complementary study showed that only (+)-lyoniresinol was bitter, with a detection threshold established at 0.5 mg/L in wines [27]. This enantiomer was quantified in various wines at concentrations higher than this value, demonstrating its contribution to wine bitterness [28]. Yet the sensory properties of a molecule, and in particular its detection threshold, can be strongly influenced by the nature of the matrix from an olfactory or gustatory point of view [29–32]. Consequently, the contribution of lignans to spirit taste cannot be presumed from the results obtained in wine. A better knowledge on oak wood and spirits composition appears particularly interesting to improve the quality of these products with high economic interest.

Based on these observations, the present work aimed at studying the occurrence of oak wood lignans in spirits and their taste contribution. First, samples of commercial spirits were screened by liquid chromatography–high resolution mass spectrometry (LC-HRMS) to search for the presence of targeted oak wood lignans. Considering the strong bitterness induced by (±)-lyoniresinol, the detection threshold of this lignan was established in a spirit matrix. Finally, after development and validation of an LC-HRMS method, a racemic mixture of lyoniresinol was quantified in various samples of commercial and experimental spirits in order to investigate its sensory impact along with the influence of oenological parameters on its content.

2. Results and Discussion

2.1. Chemical Diversity of Oak Lignans in Spirits

Previous studies have focused on the diversity and gustatory importance of lignans present in oak wood [33]. Using an LC-HRMS guided purification protocol, 11 lignans (Compounds **1–11**, Figure 1), natural derivatives of lyoniresinol, were isolated and identified from an oak wood extract [26,34]. In addition, it has been proven that all these molecules are released in wine aged in oak barrels. In this matrix, some of the galloyl, glucosyl, and xylosyl derivatives have been described as bitter but with a lower intensity than lyoniresinol [26].

Their presence has not been reported in spirits until now, even though maturation in barrels is crucial for fine-tuning their sensory properties. To determine whether these lignans were likely to impact the taste of spirits, their presence was first investigated by screening a cognac aged for 23 years. Based on mass measurement accuracy, LC-HRMS is a powerful technique in screening samples by targeting the characteristic *m/z* ions of specific empirical formulas. To this end, the chromatographic and spectrometric conditions described by Marchal et al. [26] were applied. Extracted ion chromatograms (XICs) were obtained in an oak wood extract (Figure 2a) and in a spirit (cognac, Figure 2b) by considering the *m/z* ratios specific to the deprotonated ions ([M − H]⁻) of (±)-lyoniresinol and lignans **1** to **11** with a 3-ppm tolerance window (Figure 2).

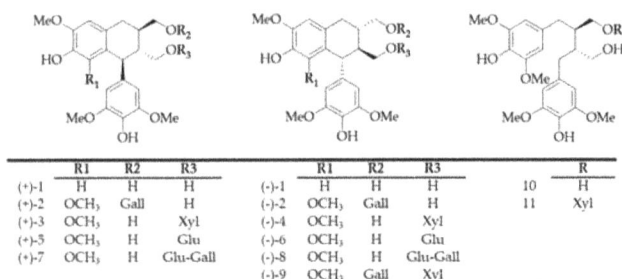

Figure 1. Chemical structures of lignans **1–11**. Xyl, Glu, and Gall correspond, respectively, to β-xylopyranose, β-glucopyranose, and galloyl.

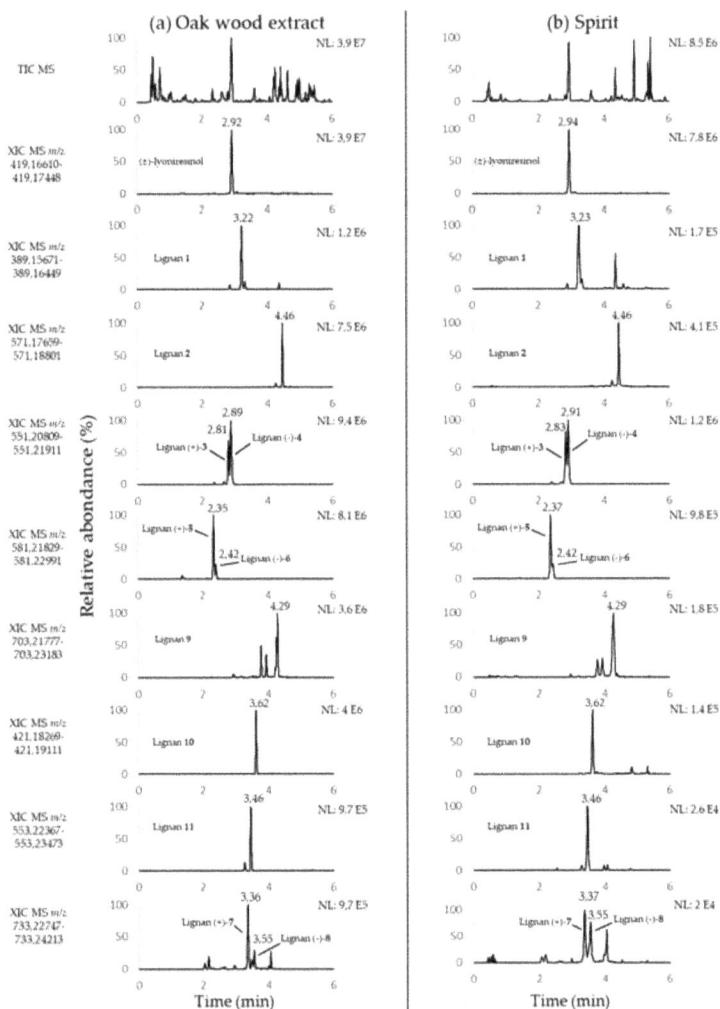

Figure 2. Negative LC-ESI-FTMS XIC of an (**a**) oak wood extract and (**b**) a spirit corresponding to [M − H]⁻ ions of lyoniresinol and lignans **1–11** (from top to bottom).

Similar signals were detected in XICs of both samples. Moreover, analysis in higher energy collision dissociation (HCD) fragmentation mode revealed the same main fragment ions in the two matrices. Concomitantly with the specificity of mass measurement (<3 ppm) and retention time similarity (<0.02 min), these results demonstrated that lignans **1–11** were present in the analyzed spirits. The presence of lyoniresinol had already been established in oak wood extracts [35,36], wines [21], and spirits [8,22,35,37]. Lignans **7** and **8** had also been identified in these three matrices [34]. However, lignans **1–6** and **9–11** had been described in wines [26,34], but never in spirits. A comparison of the signal intensity of all the lignans suggested that (±)-lyoniresinol might be the most abundant of them in spirits. These results confirmed recent studies, in which the combination of sensory analysis and quantitative studies established that lyoniresinol is both the most abundant lignan released from oak wood in wine and one of the bitterest.

The detection threshold of a compound is the concentration beyond which this molecule is perceived by one half of a panel. In oenology, detection thresholds have been measured, mainly on olfactory compounds, in different matrices such as water, wine, or spirits [29,30,32]. These studies have shown that the nature of the matrix can have a significant impact on the olfactory properties of a compound. In the same way, taste-active compounds can also be strongly affected by the matrix. For instance, the detection threshold was calculated for a bitter compound, caffeine, in water and liquid food at 94 and 184 mg/kg, respectively [31]. These results underline the importance of calculating a detection threshold of (±)-lyoniresinol in a spirit matrix to determine its effective impact on spirit taste.

2.2. Sensory Impact and Quantification of (±)-Lyoniresinol in Spirits

Previous studies have demonstrated that lyoniresinol significantly contributes to the bitterness of oaked wines [26]. However, depending on the studied matrix, the sensory properties of lyoniresinol can change. For this reason, the detection threshold must be evaluated in "eau-de-vie" of cognac and compared to quantitative values measured in spirits by LC-HRMS.

2.2.1. Determination of (±)-Lyoniresinol Detection Threshold in Spirits

The gustatory impact of (±)-lyoniresinol had been previously established in a white wine by Marchal et al. [26]. The group taste threshold was calculated to be 1.5 mg/L, with a wide range of individual detection thresholds from 0.125 to 11.3 mg/L. Furthermore, Cretin et al. demonstrated that only (+)-lyoniresinol exhibited a bitter taste compared to its enantiomer, described as tasteless or slightly sweet [27,28]. As not enough (+)-lyoniresinol was available in our laboratory to determine its detection threshold, the sensory studies presented in this work were carried out with a racemic mixture, which is naturally present in oak wood.

The detection threshold of lyoniresinol was determined using the 3AFC (three-alternative forced choice) method [38,39]. Solutions of (±)-lyoniresinol at various concentrations were prepared according to a geometric progression with a ratio of 2, and presented in 3AFC tests. Two sessions were organized to avoid excessive tiredness among the tasters. The (±)-lyoniresinol group threshold was established at 2.6 mg/L with strong inter-individual variability. Indeed, individual detection thresholds covered a range from 0.35 mg/L to 32 mg/L. The same trends had been described in wine, but the gustatory threshold values were significantly higher in "eau-de-vie". These variations could be partly due to the higher level of ethanol in the matrix. They confirmed the results of a preliminary study on "eau-de-vie" of armagnac [35].

2.2.2. Development of an LC-HRMS Method to Quantitate Lyoniresinol in Spirits

Previous studies have shown the relevance of using LC-HRMS to quantify (±)-lyoniresinol in spirits, wines, and oak wood macerates [26,28]. The same chromatographic conditions were used for (±)-lyoniresinol quantification in spirits.

The LC-HRMS method previously described for quantification in wine involved an ionization in negative mode [26]. However, preliminary tests were carried out in the same conditions and showed insufficient results regarding linearity and accuracy with the spirit samples. Consequently, a quantification method based on a positive ionization mode was developed for this matrix. LC-HRMS quantification was performed in full scan mode, the selectivity of the detection being ensured by the mass accuracy measurement of the Orbitrap analyzer (<3 ppm) and the repeatability of the retention time.

The full scan HRMS spectrum of (±)-lyoniresinol ($C_{22}H_{28}O_8$) presented several ions corresponding to adducts ([M + Na]$^+$, [M + K]$^+$, [M + NH$_4$]$^+$); pseudodimers ([2M + H]$^+$, [2M + Na]$^+$); and fragments ($C_{14}H_{17}O_4{}^+$, $C_{14}H_{19}O_5{}^+$) (Figure 3), but the signal corresponding to protonated ion ($C_{22}H_{29}O_8{}^+$) was very low, whatever the ionization parameters. Preliminary tests showed that quantification was more reliable by using the fragment ion at *m/z* 249.11214 ($C_{14}H_{17}O_4{}^+$), which was the most intense signal. This chemical species might result from a loss of a dimethoxyphenol group ($C_8H_{10}O_3$) jointly with dehydration. Such fragmentation reactions are well known for lignin derivatives [40] and can be caused by high temperature [41]. Spectrometry parameters were tuned to optimize the response of the *m/z* 249.11214 ion.

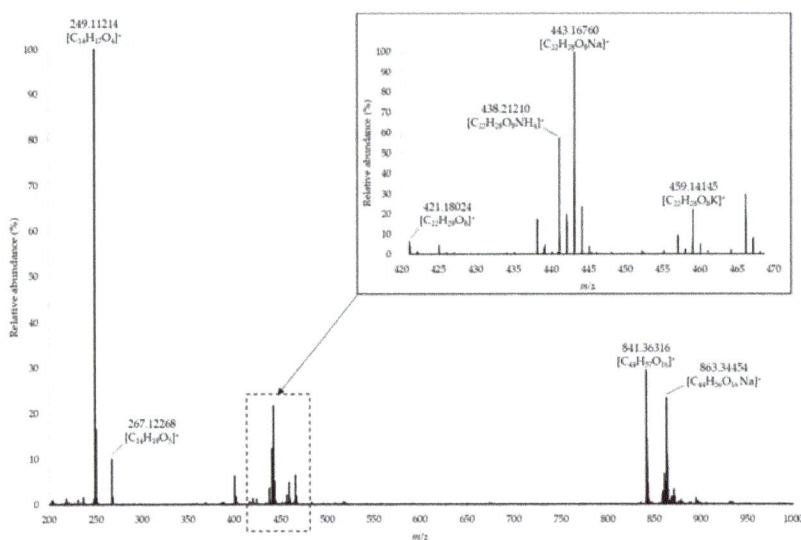

Figure 3. High resolution mass spectrometry (HRMS) spectrum of (±)-lyoniresinol in positive ionization mode.

Absolute quantification was carried out by preparing calibration solutions of pure lyoniresinol in a model solution at 8% *v/v*. Orbitrap analysis afforded high accuracy of mass measurement, so extracted ion chromatograms were built with a 3-ppm window around the theoretical *m/z* of the $C_{14}H_{17}O_4{}^+$ ion. Injection of pure lyoniresinol indicated a characteristic retention time of 2.90 min. This t_R was considered for automatic integration of the XIC.

Sensitivity

Given the high selectivity of the mass measurement, the notion of signal-to-noise is not suitable for this technique. The detection limit of a molecule is defined as the lowest concentration of this molecule for which a reliable and reproducible signal is observed. In addition, the signal must be different from a blank made under the same conditions. In this study, the method described by De Paepe et al. [42] was used. The lowest levels of the calibration curve (from 2 to 50 µg/L) were injected into five replicates. Precision (relative standard deviation (RSD)%) and accuracy (recovery of back-calculated concentrations) were obtained for each concentration. The instrumental detection limit (IDL) was defined as the standard deviation at the lowest concentration that could be measured with a precision lower than, e.g., 10%, and an accuracy higher than, e.g., 90%. With this method, it was evaluated at 5 µg/L in spirits. The instrumental quantification limit (IQL) was defined as twice the corresponding IDL (10 µg/L). Limits of detection (LOD) and quantification (LOQ), reassessed using the dilution factor, were calculated at 25 µg/L and 50 µg/L, respectively.

Linearity and Accuracy

The working range was chosen taking into account the IQL previously determined. A quadratic calibration curve ($1/x$ statistical weight) was obtained with a good correlation coefficient (R^2 of 0.9999) for a range from 10 µg/L to 10 mg/L. The recovery of back-calculated concentrations was higher than 90% at each method calibration level, establishing the accuracy.

Specificity

Specificity was assessed by mass accuracy and the repeatability of retention times. These parameters were checked for each injection of calibration solutions and samples. Low variations in retention time (<0.04 min) and a mass deviation lower than 2.2 ppm between experimental and theoretical values were observed for lyoniresinol, guaranteeing the specificity of the method. Moreover, no signal was detected in non-oaked "eau-de-vie".

Repeatability and Trueness

To determine intraday repeatability (RSD%), five replicates of two concentrations (100 µg/L and 1 mg/L) of the calibration curve were successively injected. Values lower than 4% were obtained for both concentrations, guaranteeing the repeatability of the method.

Trueness was determined by calculating the recovery ratios of three different samples of cognac spiked with stock solutions for additions of 100 µg/L and 1 and 2 mg/L. These recovery ratios ranged from 85% to 98%. These slight variations could be explained by the highly complex matrix of spirits. However, the results remained in accordance with common specifications [43] and established the trueness of the method. Interday repeatability was estimated by injections of the same standard solutions for five successive days. As usually observed for LC-ESI-MS analysis, the RSD values were quite high. To overcome this issue, all the calibration solutions were injected for each quantitative analysis of an unknown sample.

All of the results proved the ability of the LC-HRMS method to quantify lyoniresinol in spirits (Table 1).

Table 1. Validation parameters for HRMS quantitation of (±)-lyoniresinol in spirits.

Parameters	Matrix/Spirits			
Sensitivity	IDL (µg/L)	IQL (µg/L)	LOD (µg/L)	LOQ (µg/L)
	5	10	25	50
Linearity and Accuracy	Working Range		R^2	
	10 µg/L–10 mg/L		0.9999	
Specificity	t_R variation		Mass Accuracy	
	<0.04 min		<2.2 ppm	
Repeatability and Trueness	Intraday Repeatability			
	100 µg/L 3.35%		1 mg/L 3.43%	
	Recovery			
	Samples	100 µg/L	1 mg/L	2 mg/L
	EDV-C7	94%	91%	98%
	EDV-C8	87%	85%	97%
	EDV-1995	88%	86%	87%

IDL: Instrumental detection limit; IQL: Instrumental quantification limit; LOD: Limit of detection; LOQ: Limit of quantification; EDV: eau-de-vie.

2.2.3. Application of the Method to Quantitate Lyoniresinol in Spirits

Content of Lyoniresinol in Various Commercial Spirits

Twenty-four commercial spirits were analyzed to assess the range of lyoniresinol amounts in some cognacs, but also whiskies, rums, and other brandies using the LC-HRMS method previously validated. Lyoniresinol was detected in all samples, at concentrations ranging from 0.2 to 11.8 mg/L with a mean value of 3.3 mg/L. The results are illustrated in Figure 4. There were not enough samples to carry out a statistical study, but among the spirits richest in lyoniresinol, there was a majority of cognacs (C-2 and C-5 to C-11), as well as two brandies (B-13 and B-14), one rum (R-19), and one whiskey (W-22). A comparison to sensory data highlighted that the lyoniresinol concentration was above the detection threshold in these 12 spirits, establishing the sensory relevance of this compound, which is likely to contribute to the bitterness of oaked spirits.

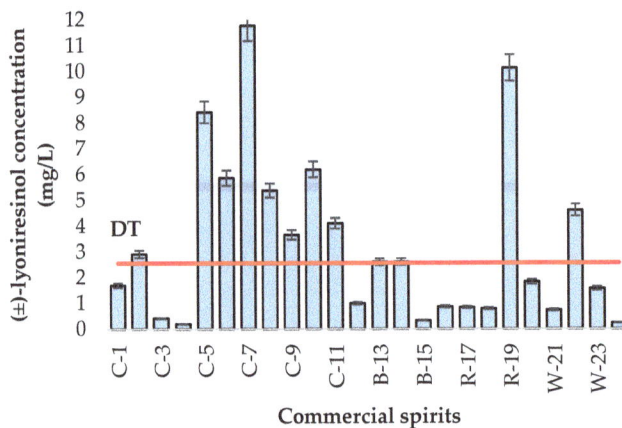

Figure 4. Variations in (±)-lyoniresinol content in 24 commercial spirits. The red line represents the level of the gustatory detection threshold (DT).

Furthermore, the significant variations in lyoniresinol observed between these commercial spirits could have been due to various factors, and some hypotheses can be evoked. First, the analyzed commercial spirits were aged in contact with oak wood, so these observations could have been related to aging conditions. The influence of the aging container on the lyoniresinol content of a white wine has already been established, confirming that lyoniresinol is released from oak wood to wine [26]. Additionally, this study showed that new oak barrels contained more lyoniresinol. Thus, the variations observed in the analyzed spirits could have been due to the proportion of new oak barrels used during the aging of the spirits. Previous studies have also demonstrated that the concentrations of molecules vary according to the use of new or used barrels, barrels previously used in a maturation cycle [9]. Indeed, Piggott et al. noted differences in the levels of phenolic compounds depending on three types of casks. The study showed an increase in concentrations of nonvolatile compounds, such as vanillin, syringaldehyde, and syringic acid, as well as a reduction in coniferaldehyde and sinapaldehyde concentrations in 36-month-old whiskey distillates [10]. As spirits are aged for a longer period than wine [8], aging time and aging container could explain the significant variations in lyoniresinol observed in these samples, but such information was not available for the commercial spirits analyzed in this study. The potential influence of aging conditions on lyoniresinol concentrations in spirits needs to be clarified in further studies.

Second, various cooperage parameters could have affected the chemical composition of oak wood. Previous works have shown that the concentration of oak molecules released in wine or spirits could vary according to the botanical species or geographical origin of the oak [44–46]. However, Cretin et al. demonstrated that there was no significant effect of oak species on lyoniresinol content [28].

In addition, during barrel making, the wood undergoes a series of stages that influence its oenological quality, most notably seasoning and toasting of the staves. These technological features affect the structure and the chemical composition of the wood, but also the future matrix with which it will be in contact [47–51]. Indeed, Cretin et al. studied the influence of wood toasting temperature on lyoniresinol and showed that this compound was slightly degraded at around 250 °C [28], in line with another study [52].

Content of Lyoniresinol in Various Vintages of the Same Spirits

Lyoniresinol was quantified in a series of "eaux-de-vie" of cognac of 10 different vintages from the same distillery and using similar aging conditions. The samples were not commercial cognac, but "eaux-de-vie" still undergoes the aging process in barrels. They were matured in used barrels (a 350-L coarse grain oak barrel). For each vintage, a sample was collected from five different barrels and analyzed. The concentrations presented in Figure 5 correspond to the mean values of these five replicates. The measured values ranged from 1.6 mg/L (2015) to 12 mg/L (1995). For each vintage, the coefficient of variation between the five replicates was relatively low (from 5.7% to 33.2%), revealing a good homogeneity between barrels. From 2015 to 1995, the results showed that the older the "eau-de-vie", the higher the level of lyoniresinol, confirming previous observations [27,35,37]. Conversely, for older vintages, lower concentrations were measured. This could suggest a degradation of lyoniresinol in long-time barrel storage. Furthermore, aging in used barrels implies a lower extractable potential of the compounds than in new oak barrels and can influence the lyoniresinol content in these vintages. However, this hypothesis needs to be studied more deeply, since the results could also have been due to modifications to aging practices in the distillery or changes in barrel supplies.

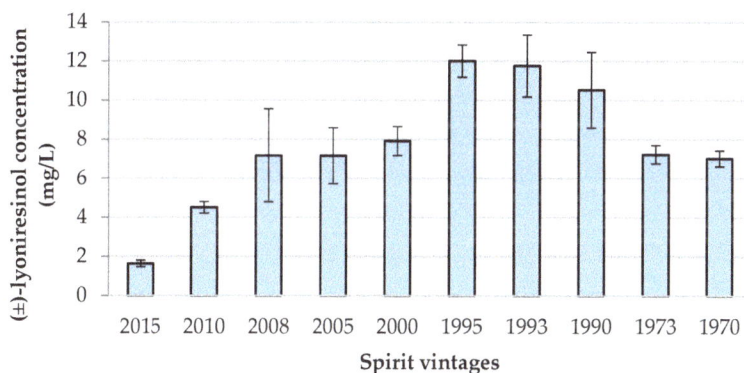

Figure 5. Concentrations of (±)-lyoniresinol in 10 vintages of cognac "eaux-de-vie" coming from the same distillery.

Despite the release of lyoniresinol, a bitter compound, spirits are known to improve during oak wood aging. Research has highlighted the impact of other taste-active compounds such as quercotriterpenosides [16], which could be released at the same time from oak wood and might modulate the effect of lyoniresinol on the taste balance in spirits.

3. Materials and Methods

3.1. Chemicals

D-(+)-glucose, D-(−)-fructose, and quinine sulfate were purchased from Sigma-Aldrich (Saint-Quentin-Fallavier, France). Ultrapure water (Milli-Q purification system, Millipore, France) and HPLC grade solvent (acetonitrile, ethanol, ethyl acetate, *n*-heptane, methanol, and propan-2-ol; VWR International, Pessac, France) were used for sample preparation and lignan purification. Acetonitrile (ACN) and water used for chromatographic separation were LC-MS grade and were purchased from Fisher Chemical (Illkirch, France). Lignans were isolated as previously described by Marchal et al. [26].

3.2. LC Analysis

The HPLC appliance consisted of an HTC PAL autosampler (CTC Analytics AG, Zwingen, Switzerland) and an Accela U-HPLC system with quaternary pumps. For (±)-lyoniresinol quantitation, a C18 column (Hypersil Gold 2.1 × 100 mm, 1.9-μm particle size, Thermo Fisher Scientific) was used with water (Eluent A) and ACN (Eluent B) as mobile phases. The flow rate was set at 600 μL/min, and the injection volume was 5 μL. Eluent B varied as follows: 0 min, 14%; 0.5 min, 14%; 1.5 min, 19%; 2 min, 19%; 4.5 min, 38%; 4.6 min, 98%; 6.9 min, 98%; 7 min, 14%; 8.6 min, 14%.

3.3. HRMS

An Exactive Orbitrap mass spectrometer equipped with a heated electrospray ionization (HESI II) probe (both from Thermo Fisher Scientific, Les Ulis, France) was used. The mass analyzer was calibrated each week using Pierce®ESI Negative and Positive Ion Calibration solutions (Thermo Fisher Scientific).

3.3.1. Screening

To perform targeted screening, the ionization and spectrometric parameters, optimized in negative mode, were previously described by Marchal et al. [26]. Table 2 summarizes all the data.

Table 2. Ionization and spectrometric conditions for HRMS analyses.

Mass Spectrometer	Exactive	
Use	LC-MS Screening	LC-MS Quantification
Ionization Mode	Negative	Positive
Sheath gas flow [a]	75	75
Auxiliary gas flow [a]	18	20
HESI probe temperature	320 °C	320 °C
Capillary temperature	350 °C	350 °C
Electrospray voltage	−3 kV	3.5 kV
Capillary voltage	−60 V	35 V
Tube lens voltage offset	−135 V	120 V
Skimmer voltage	−26 V	18 V
Mass range (in Th)	200–800	200–800
Resolution [b]	25 000	25 000
Automatic gain control value	10^6	10^6

[a] Sheath gas and auxiliary gas flows (both nitrogen) are expressed in arbitrary units. [b] Resolution $m/\Delta m$, fwhm at m/z 200 Th.

3.3.2. Quantification

For quantitation of (±)-lyoniresinol in spirits, mass acquisitions were performed and optimized in positive Fourier transform (FTMS) ionization mode. The ionization and spectrometric parameters are described in Table 3. All data were processed using the Qual Browser and Quan Browser applications of Xcalibur version 3.0 (Thermo Fisher Scientific) [45].

Table 3. Features of commercial spirits.

Samples	Brands	Type	Origin	ABV (Alcohol by Volume)
C-1	Brand A	Cognac	France	40
C-2	Brand A	Cognac	France	40
C-3	Brand A	Cognac	France	40
C-4	Brand A	Cognac	France	40
C-5	Brand B	Cognac	France	40
C-6	Brand B	Cognac	France	40
C-7	Brand C	Cognac	France	40
C-8	Brand C	Cognac	France	40
C-9	Brand C	Cognac	France	40
C-10	Brand D	Cognac	France	40
C-11	Brand D	Cognac	France	40
C-12	Brand E	Cognac	France	42
B-13	Brand F	Brandy	South Africa	38
B-14	Brand G	Brandy	France	40
B-15	Brand H	Brandy	Germany	38
B-16	Brand I	Brandy	France	40
R-17	Brand J	Rum	Jamaica	43
R-18	Brand K	Rum	Guyana	40
R-19	Brand L	Rum	Barbados	43
W-20	Brand M	Whisky	Ireland	40
W-21	Brand N	Whisky	Scotland	40
W-22	Brand O	Whisky	Scotland	40
W-23	Brand O	Whisky	Scotland	40
Bo-24	Brand P	Bourbon	United States	50

3.4. Spirits and Sample Preparation

Two series of spirits were used in this study. Lyoniresinol quantitation was assessed in 24 commercial spirits (including 12 cognacs, 4 grape brandies, 3 rums, 4 whiskies, and 1 bourbon). All of these were aged in oak wood. Table 3 summarizes all the features of these spirits, which were randomly chosen among spirits commercially available and well distributed in France.

The second set of spirits, supplied by Rémy-Martin, consisted of 10 vintages from 1970 to 2015, with five replicates for each year. The samples came from the same distillery and used similar aging conditions.

All concentrations were expressed in mg/L of spirits.

A spirit is a matrix with a high alcohol content, so a dilution is necessary before any injection. This prevents deterioration of the chromatographic separation and allows all concentrations to be included in the working range. For quantitative analysis, the percentage of alcohol ranged from 71 to 38% v/v, and the spirit samples were reduced to 8% alcohol with water and 0.45 μm filtered. The diluted spirits were injected directly into LC-HRMS using the chromatographic and spectrometric parameters described above.

3.5. Preparation of Calibration Solution for Lignan Quantitation

A stock solution of (±)-lyoniresinol (1 g/L) was prepared in ethanol. One range of calibration was prepared by successive dilution of this solution in hydroethanolic solution (8%, v/v) in order to supply calibration samples (10 mg/L, 5 mg/L, 2 mg/L, 1 mg/L, 500 μg/L, 200 μg/L, 100 μg/L, 50 μg/L, 20 μg/L, 10 μg/L, 5 μg/L, 2 μg/L). Detection of (±)-lyoniresinol was based on the theoretical exact mass of the most intense ion, the fragment ion at m/z 249.11214, and its retention time at 2.90 min. Peak areas were determined by automatic integration of extracted ion chromatograms built in a 3-ppm window around the exact mass of the $C_{14}H_{17}O_4^+$ ion.

3.6. Method Validation for Quantitation of Lyoniresinol on C18 Column

The quantitation method of (±)-lyoniresinol in spirits was validated by studying sensitivity, linearity, specificity, intraday repeatability, and trueness. A calibration curve was established by plotting the areas for each concentration level versus the nominal concentration. Quadratic regression was chosen with a $1/x$ statistical weight.

The sensitivity of the LC-HRMS method was determined following the approach described by De Paepe et al. [42]. Linearity was evaluated by correlation coefficient (R^2) and by deviations of each back-calculated standard concentration from the nominal value. To evaluate repeatability, the intraday precision was determined by injecting five replicates of two intermediate calibration solutions (100 and 1000 µg/L), and the relative standard deviation (RSD%) was calculated. Trueness was checked by calculating the recovery ratio (between measured and expected areas) from three samples of spirits (C-7, C-8, EDV-1995). They were chosen among the analyzed samples and were spiked with calibration solution corresponding to an addition of 100 µg/L and 1 and 2 mg/L of (±)-lyoniresinol. Specificity was assessed by evaluating the mass accuracy and retention time repeatability. These parameters were determined concomitantly with the precision and trueness analysis described above.

3.7. Sensory Analysis

Tasting sessions took place in a specific air-conditioned room at 20 °C equipped with individual booths and normalized glasses. The "eau-de-vie" used for sensory analysis was a non-oaked spirit adjusted to 40% v/v of ethanol with pure and demineralized water (eau de source de Montagne, Laqueuille, France). The absence of (±)-lyoniresinol in this matrix was checked by LC-HRMS analysis.

Results obtained from the sensory tests were statistically interpreted following the norms published by the International Organization for Standardization (ISO) [38].

3.7.1. Panel Training

The panel consisted of 24 wine tasters, 7 men and 17 women, aged from 20 to 45 years. The aim of this training session was, first, to accustom the panel to sweet and bitter perceptions (glucose, fructose, and quinine sulfate aqueous solutions), and in the second session to familiarize all panelists with a new matrix with a higher alcohol concentration (40%, v/v).

First, two aqueous solutions containing, respectively, glucose and fructose at 10 g/L, and quinine sulfate at 20 mg/L, were presented to the tasters to illustrate sweet and bitter tastes. These solutions were made with demineralized water (eau de source de Montagne, Laqueuille, France).

In a second session, solutions containing these compounds at the same concentrations, but in a non-oaked "eau-de-vie" (40%, v/v), were presented to the tasters. They were asked to rate the sweetness and bitterness intensities of each solution on a 0–10 scale. The results were interpreted by a one-way analysis of variance (ANOVA). For each parameter, the homogeneity of the variance was assessed using the Levene test. ANOVA's statistics were considerably below the p-value of 0.05. Indeed, results were significant with a p-value of 0.0001 for the sweet solution and a p-value of 0.003 for the bitter solution. Training showed the reliability of the panel to distinguish sweet and bitter tastes even in a complex matrix.

3.7.2. Determination of Lyoniresinol Taste Threshold in Spirits

The taste threshold of (±)-lyoniresinol was evaluated in a non-oaked "eau-de-vie" adjusted at 40% v/v. Due to the higher alcohol concentration present in this matrix and the remanence of the bitter taste, two different sessions were planned to avoid tiredness among the panelists. In the first session, three concentrations (2, 4, and 8 mg/L) were presented in ascending order. Each concentration was displayed according to the 3AFC (three-alternative forced choice) described by ISO 4120:2007 [38]. Concentrations presented in the second session depended on the results from the first session for each taster. If the panelist had given correct answers, three lower concentrations (0.5, 1, and 2 mg/L) were presented to him or her, following a geometric progression of ratio 2, starting with the lowest. Conversely, tasters who did not give any correct answers during the first session received three higher concentrations (8, 16, and 32 mg/L) in the other session.

Individual thresholds were estimated as the geometrical mean between the lowest concentration of a continuous series of three correct answers and the concentration just below this level. The group threshold was estimated as the geometrical mean between all the individual thresholds.

4. Conclusions

This study focused on the use of analytical and sensory techniques to highlight the presence of lignans in spirits and the importance of lyoniresinol. First, an LC-HRMS targeted screening of spirits aged in barrels revealed the presence of 11 lignans. Next, the study focused on the most abundant and the bitterest lignan, (±)-lyoniresinol, to assess its sensory role in spirits. After development and validation of the LC-HRMS method, a racemic mixture of lyoniresinol was quantified in 24 commercial spirits and 10 different vintages of an "eau-de-vie" of cognac. Results showed that in various spirits, (±)-lyoniresinol was above its detection threshold, which was estimated at 2.6 mg/L. This work revealed that this lignan has a significant impact on the taste balance of spirits, as it increases its bitterness. Additionally, high concentrations of (±)-lyoniresinol (up to 12 mg/L) were observed in an oaked "eau-de-vie" of cognac, and its level was detected above the detection threshold for the considered samples, covering a period of almost 50 years. Furthermore, in the analyzed commercial spirits, significant variations were observed, ranging from 0.2 to 11.8 mg/L, and could be partly explained by differences in aging modalities. A similar analytical strategy could be developed to determine the importance of the dextrorotatory enantiomer of lyoniresinol, which exhibits a strong bitter taste, in brandies. This work brings new insight into and a better understanding of the molecular origin of spirit taste. From a practical point of view, studying the parameters likely to affect the level of (+)-lyoniresinol in oak wood and spirits would open up interesting perspectives for better monitoring of the organoleptic properties of spirits.

Author Contributions: Conceptualization, D.W. and A.M.; Methodology, D.W. and A.M.; Software, D.W.; Validation, D.W. and A.M.; Formal Analysis, D.W.; Investigation, D.W. and A.M.; Resources, D.W.; Data Curation, D.W.; Writing—Original Draft Preparation, D.W. and A.M..; Writing—Review & Editing, D.W. and A.M.; Supervision, A.M.; Project Administration, A.M.; Funding Acquisition, A.M.

Funding: Delphine Winstel's grant is supported by Remy-Martin and Seguin-Moreau. Analytical techniques were funded by Biolaffort.

Acknowledgments: The authors acknowledge François Clavero for technical assistance. The authors are also very grateful to J.-C. Mathurin and L. Urruty for providing samples.

Conflicts of Interest: The authors declare no conflicts of interest.

References

1. Cantagrel, R.; Mazerolles, G.; Vidal, J.; Galy, B. *Élaboration et Connaissance des Spiritueux: Recherche de la qualité, tradition & innovation*; Lavoisier—Tec & Doc.: Paris, France, 1992; ISBN 2-87777-357-4.
2. MacNamara, K.; van Wyk, C.; Brunerie, P.; Augustyn, O.P.; Rapp, A. Flavour components of whiskey. III. ageing changes in the low-volatility fraction. *South Afr. J. Enol. Vitic.* **2001**, *22*, 82–92. [CrossRef]
3. Piggott, J.R. *Flavour of Distilled Beverages: Origin and Development*; E. Horwood Limited: London, UK, 1983; ISBN 978-0-89573-131-9.
4. Lafon, J.; Couillud, P.; Gaybellile, F. *Le Cognac*; J.B. Baillière: Paris, France, 1973.
5. Ledauphin, J.; Le Milbeau, C.; Barillier, D.; Hennequin, D. Differences in the Volatile Compositions of French Labeled Brandies (Armagnac, Calvados, Cognac, and Mirabelle) Using GC-MS and PLS-DA. *J. Agric. Food Chem.* **2010**, *58*, 7782–7793. [CrossRef] [PubMed]
6. Ledauphin, J.; Basset, B.; Cohen, S.; Payot, T.; Barillier, D. Identification of trace volatile compounds in freshly distilled Calvados and Cognac: Carbonyl and sulphur compounds. *J. Food Compos. Anal.* **2006**, *19*, 28–40. [CrossRef]

7. Puentes, C.; Joulia, X.; Vidal, J.-P.; Esteban-Decloux, M. Simulation of spirits distillation for a better understanding of volatile aroma compounds behavior: Application to Armagnac production. *Food Bioprod. Proc.* **2018**, *112*, 31–62. [CrossRef]

8. MacNamara, K.; Dabrowska, D.; Baden, M.; Helle, N. Advances in the ageing chemistry of distilled spirits matured in oak barrels. **2011**, *6*, 448–466.

9. Mosedale, J.R.; Puech, J.-L. Wood maturation of distilled beverages. *Trends Food Sci. Technol.* **1998**, *9*, 95–101. [CrossRef]

10. Piggott, J.R.; Conner, J.M.; Paterson, A.; Clyne, J. Effects on Scotch whisky composition and flavour of maturation in oak casks with varying histories. *Int. J. Food Sci. Technol.* **1993**, *28*, 303–318. [CrossRef]

11. Puech, J.-L. Extraction of Phenolic Compounds from Oak Wood in Model Solutions and Evolution of Aromatic Aldehydes in Wines Aged in Oak Barrels. *Am. J. Enol. Vitic.* **1987**, *38*, 236–238.

12. Puech, J.-L. Characteristics of Oak Wood and Biochemical Aspects of Armagnac Ageing. *Am. J. Enol. Vitic.* **1984**, *35*, 77–81.

13. Chatonnet, P. Influence des procédés de tonnellerie et des conditions d'élevage sur la composition et la qualité des vins élevés en fûts de chêne. 1995. Available online: https://www.theses.fr/1995BOR20338 (accessed on 31 October 2018).

14. Tominaga, T.; Blanchard, L.; Darriet, P.; Dubourdieu, D. A Powerful Aromatic Volatile Thiol, 2-Furanmethanethiol, Exhibiting Roast Coffee Aroma in Wines Made from Several *Vitis v inifera* Grape Varieties. *J. Agric. Food Chem.* **2000**, *48*, 1799–1802. [CrossRef]

15. Marchal, A.; Génin, E.; Waffo-Téguo, P.; Bibès, A.; Da Costa, G.; Mérillon, J.-M.; Dubourdieu, D. Development of an analytical methodology using Fourier transform mass spectrometry to discover new structural analogs of wine natural sweeteners. *Anal. Chem. Acta* **2015**, *853*, 425–434. [CrossRef] [PubMed]

16. Marchal, A.; Waffo-Téguo, P.; Génin, E.; Mérillon, J.-M.; Dubourdieu, D. Identification of New Natural Sweet Compounds in Wine Using Centrifugal Partition Chromatography–Gustatometry and Fourier Transform Mass Spectrometry. *Anal. Chem.* **2011**, *83*, 9629–9637. [CrossRef] [PubMed]

17. Glabasnia, A.; Hofmann, T. Identification and Sensory Evaluation of Dehydro- and Deoxy-ellagitannins Formed upon Toasting of Oak Wood (*Quercus alba* L.). *J. Agric. Food Chem.* **2007**, *55*, 4109–4118. [CrossRef] [PubMed]

18. Hufnagel, J.C.; Hofmann, T. Orosensory-Directed Identification of Astringent Mouthfeel and Bitter-Tasting Compounds in Red Wine. *J. Agric. Food Chem.* **2008**, *56*, 1376–1386. [CrossRef] [PubMed]

19. Puech, J.-L. Phenolic compounds in oak wood extracts used in the ageing of brandies. *J. Sci. Food Agric.* **1988**, *42*, 165–172. [CrossRef]

20. Puech, J.-L. Extraction and evolution of lignin products in armagnac matured in oak. **1981**, *32*, 111–114.

21. Moutounet, M.; Puech, J.-L.; Rabier, P. Analysis by HPLC of Extractable Substances in Oak Wood. Available online: http://agris.fao.org/agris-search/search.do?recordID=FR19890112212 (accessed on 8 October 2018).

22. Nabeta, K.; Yonekubo, J.; Miyake, M.; Obihiro, U.A.V.M. Phenolic Compounds from the Heartwood of European Oak (*Quercus robur* L.) and Brandy. Available online: http://agris.fao.org/agris-search/search.do?recordID=JP880129288 (accessed on 3 October 2018).

23. Puech, J.L.; Moutonnet, M. Phenolic Compounds in an Ethanol-Water Extract of Oak Wood and in a Brandy. Available online: http://agris.fao.org/agris-search/search.do?recordID=US201301769211 (accessed on 8 October 2018).

24. Ribéreau-Gayon, P.; Glories, Y.; Maujean, A.; Dubourdieu, D. *Traité d'oenologie-Tome 2-6e éd.-Chimie du Vin. Stabilisation et Traitements*; Dunod: Buchanan, NY, USA, 2012; ISBN 978-2-10-058875-6.

25. Salagoity-Auguste, M.-H.; Tricard, C.; Sudraud, P. Dosage simultané des aldéhydes aromatiques et des coumarines par chromatographie liquide haute performance: Application aux vins et eaux-de-vie vieillis en fût de chêne. *J. Chromatogr. A* **1987**, *392*, 379–387. [CrossRef]

26. Marchal, A.; Cretin, B.N.; Sindt, L.; Waffo-Téguo, P.; Dubourdieu, D. Contribution of oak lignans to wine taste: Chemical identification, sensory characterization and quantification. *Tetrahedron* **2015**, *71*, 3148–3156. [CrossRef]

27. Cretin, B.N.; Sallembien, Q.; Sindt, L.; Daugey, N.; Buffeteau, T.; Waffo-Teguo, P.; Dubourdieu, D.; Marchal, A. How stereochemistry influences the taste of wine: Isolation, characterization and sensory evaluation of lyoniresinol stereoisomers. *Anal. Chem. Acta* **2015**, *888*, 191–198. [CrossRef]

28. Cretin, B.N.; Dubourdieu, D.; Marchal, A. Development of a quantitation method to assay both lyoniresinol enantiomers in wines, spirits, and oak wood by liquid chromatography-high resolution mass spectrometry. *Anal. Bioanal. Chem.* **2016**, *408*, 3789–3799. [CrossRef]

29. Chatonnet, P.; Dubourdieu, D.; Boidron, J.; Pons, M. The origin of ethylphenols in wines. *J. Sci. Food Agric.* **1992**, *60*, 165–178. [CrossRef]

30. Gammacurta, M.; Tempere, S.; Marchand, S.; Moine, V.; De Revel, G. Ethyl 2-hydroxy-3-methylbutanoate enantiomers: Quantitation and sensory evaluation in wine. *OENO One* **2018**, *52*, 57–65. [CrossRef]

31. Mackey, A.; Valassi, K. The discernement of primary tastes in the presence of different food textures. *Food Technol.* **1956**, *10*, 238–240.

32. Poisson, L.; Schieberle, P. Characterization of the Key Aroma Compounds in an American Bourbon Whisky by Quantitative Measurements, Aroma Recombination, and Omission Studies. *J. Agric. Food Chem.* **2008**, *56*, 5820–5826. [CrossRef] [PubMed]

33. Nonier, M.-F.; Vivas, N.; Vivas de Gaulejac, N.; Fouquet, E. Origin of brown discoloration in the staves of oak used in cooperage–Characterization of two new lignans in oak wood barrels. *Comptes Rendus Chim.* **2009**, *12*, 291–296. [CrossRef]

34. Sindt, L.; Gammacurta, M.; Waffo-Teguo, P.; Dubourdieu, D.; Marchal, A. Taste-Guided Isolation of Bitter Lignans from *Quercus petraea* and Their Identification in Wine. *J. Nat. Prod.* **2016**, *79*, 2432–2438. [CrossRef] [PubMed]

35. Arramon, G. Les triterpènes et lignanes des bois de chêne européen *Quercus robur* L. Et Quercus Petraea Liebl.: Quantification et apports qualitatifs aux eaux de vie d'Armagnac. Available online: https://www. theses.fr/2001BOR2085.2 (accessed on 6 October 2018).

36. Seikel, M.K.; Hostettler, F.D.; Niemann, G.J. Phenolics of Quercus rubra wood. *Phytochemistry* **1971**, *10*, 2249–2251. [CrossRef]

37. Koga, K.; Taguchi, A.; Koshimizu, S.; Suwa, Y.; Yamada, Y.; Shirasaka, N.; Yoshizumi, H. Reactive Oxygen Scavenging Activity of Matured Whiskey and Its Active Polyphenols. *J. Food Sci.* **2007**, *72*, S212–S217. [CrossRef]

38. ISO 4120:2007 Sensory analysis-Methodology-Triangle test. In *Analyse Sensorielle: Recueil, Normes, Agroalimentaire*; AFNOR: Geneva, Switzerland, 2007.

39. García, J.; Prieto, L.; Guevara, A.; Malagon, D.; Osorio, C. Chemical Studies of Yellow Tamarillo (Solanum betaceum Cav.) Fruit Flavor by Using a Molecular Sensory Approach. *Molecules* **2016**, *21*, 1729. [CrossRef]

40. Li, M.; Liu, D.; Cong, X.-S.; Wu, J.-H.; Wu, C.-C.; Xia, W. Identification of Lignin-Derived p-Bis(2,6-dimethoxyphenol)yl Compounds in Bio-oil with Mass Spectrometry. *Energy Fuels* **2016**, *30*, 10950–10953. [CrossRef]

41. Jenkins, R.W.; Sutton, A.D.; Robichaud, D.J. Chapter 8-Pyrolysis of Biomass for Aviation Fuel. In *Biofuels for Aviation*; Chuck, C.J., Ed.; Academic Press: Cambridge, MA, USA, 2016; pp. 191–215. ISBN 978-0-12-804568-8.

42. De Paepe, D.; Servaes, K.; Noten, B.; Diels, L.; De Loose, M.; Van Droogenbroeck, B.; Voorspoels, S. An improved mass spectrometric method for identification and quantification of phenolic compounds in apple fruits. *Food Chem.* **2013**, *136*, 368–375. [CrossRef] [PubMed]

43. Center for Veterinary Medicine (CVM). *Guidance for Industry Bioanalytical Method Validation*; Center for Drug Evaluation and Research (CDER), Food and Drug Administration, U.S. Department of Health and Human Services: Washington, DC, USA, 2018.

44. De Rosso, M.; Cancian, D.; Panighel, A.; Dalla Vedova, A.; Flamini, R. Chemical compounds released from five different woods used to make barrels for ageing wines and spirits: Volatile compounds and polyphenols. *Wood Sci. Technol.* **2009**, *43*, 375–385. [CrossRef]

45. Marchal, A.; Prida, A.; Dubourdieu, D. New Approach for Differentiating Sessile and Pedunculate Oak: Development of a LC-HRMS Method To Quantitate Triterpenoids in Wood. *J. Agric. Food Chem.* **2016**, *64*, 618–626. [CrossRef] [PubMed]

46. Prida, A.; Boulet, J.-C.; Ducousso, A.; Nepveu, G.; Puech, J.-L. Effect of species and ecological conditions on ellagitannin content in oak wood from an even-aged and mixed stand of *Quercus robur* L. and *Quercus petraea* Liebl. *Ann. For. Sci.* **2006**, *63*, 415–424. [CrossRef]

47. Canas, S.; Caldeira, I.; Mateus, A.M.; Belchior, A.P.; Clímaco, M.C.; Bruno-de, R. Effect of natural seasoning on the chemical composition of chestnut wood used for barrel making. *Ciência e técnica vitivinícola* **2006**, *21*, 1–16.

48. Chatonnet, P.; Boidron, J.-N.; Dubourdieu, D.; Pons, M. Evolution de certains composés volatils du bois de chêne au cours de son séchage premiers résultats. *OENO One* **1994**, *28*, 359. [CrossRef]

49. Fernández de Simón, B.; Cadahía, E.; del Álamo, M.; Nevares, I. Effect of size, seasoning and toasting in the volatile compounds in toasted oak wood and in a red wine treated with them. *Anal. Chem. Acta* **2010**, *660*, 211–220. [CrossRef] [PubMed]

50. Mosedale, J.R.; Puech, J.-L.; Feuillat, F. The Influence on Wine Flavor of the Oak Species and Natural Variation of Heartwood Components. *Am. J. Enol. Vitic.* **1999**, *50*, 503–512.

51. Snakkers, G.; Boulesteix, J.-M.; Estréguil, S.; Gaschet, J.; Lablanquie, O.; Faure, A.; Cantagrel, R. Effect of oak wood heating on cognac spirit matured in new barrel: A pilot study. *OENO One* **2003**, *37*, 243. [CrossRef]

52. Sarni, F.; Moutounet, M.; Puech, J.L.; Rabier, P. Effect of heat treatment of oak wood on extractable compounds. *Holzforsch.-Int. J. Biol. Chem. Phys. Technol. Wood* **1990**, *44*, 461–466. [CrossRef]

Sample Availability: Samples of the compounds **1**, **2**, **3**, **5**, **6**, **10** and lyoniresinol are available from the authors.

molecules

MDPI

Article

Molecular Dynamics on Wood-Derived Lignans Analyzed by Intermolecular Network Theory

Thomas Olof Sandberg [1,*] **, Christian Weinberger** [2] **and Jan-Henrik Smått** [1]

[1] Centre of Excellence for Functional Materials, Laboratory for Physical Chemistry, Åbo Akademi University, Porthansgatan 3–5, FI-20500 Åbo, Finland; jsmatt@abo.fi

[2] Department of Chemistry–Inorganic Functional Materials, Paderborn University, 33098 Paderborn, Germany; christian.weinberger@upb.de

* Correspondence: tsandber@abo.fi; Tel.: +358-40-539-4636

Received: 3 July 2018; Accepted: 8 August 2018; Published: 10 August 2018

Abstract: The dynamics of interactions to a solvent is a key factor in the proper characterization of new molecular structures. In molecular dynamics simulations, the solvent molecules are explicitly present, thereby defining a more accurate description on how the solvent molecules affect the molecular conformation. Intermolecular interactions in chemical systems, e.g., hydrogen bonds, can be considered as networks or graphs. Graph theoretical analyses can be an outstanding tool in analyzing the changes in interactions between solvated and solid. In this study, the software *ChemNetworks* is applied to interaction studies between TIP4P solvent molecules and organic solutes, i.e., wood-derived lignan-based ligands called LIGNOLs, thereby supporting the research of interaction networks between organic molecules and solvents. This new approach is established by careful comparisons to studies using previously available tools. In the hydration studies, tetramethyl 1,4-diol is found to be the LIGNOL which was most likely to form hydrogen bonds to the TIP4P solvent.

Keywords: lignan; molecular dynamics; intermolecular interactions; graph theory

---

1. Introduction

Computational chemistry is of utmost importance in scientific research today. Theoretical methods are frequently being used in most branches to explain and predict various chemical phenomena. A shortcoming, especially in organic synthesis, is the proper characterization of new, molecular structures at the atomic level. The key factor often affecting the molecular minimum energy configuration is the solvent and especially the dynamics of hydrogen bonding to the solvent [1].

Traditional quantum chemical (QC) calculations usually do not account for the solvent and are best suited for studies of single molecules. With continuum models [2], the solvent can be taken into account as a bulk medium. However, in molecular dynamics (MD) simulations [3] the solvent molecules are explicitly present, thereby defining a more accurate description of how the solvent molecules affect the molecular conformation.

Intermolecular interactions in chemical systems can be considered as networks or graphs. Graph theory is commonly used in mathematics and computer science to model pairwise relations between objects making up a graph, consisting of vertices connected by edges. In chemistry, the structure of the intermolecular interaction can be treated as a graph, wherein the molecules are regarded as vertices, and the interactions are regarded as edges. Considering intramolecular interactions, the atoms serve as vertices; the edges represent chemical bonds between different atoms, and the number of edges associated with a given vertex is the valence of the atom. Graph theoretical analyses can dissect complex local and global changes occurring within the chemical network over multiple time and length scales, thereby making it an outstanding tool in analyzing the changes in interactions, e.g., hydrogen bond topologies [4], between solvent and solute in MD simulations. Graph theory's similarities to statistical mechanical simulations make it a unique new tool for analysis of large chemical systems, as has recently been shown [5–7]. The implementation of graph theory in chemistry belongs to a suite of approaches called Intermolecular Network Theory.

ChemNetworks [8] is a software originating from the previous edition, *moleculaRnetworks* [5] for analysis of statistical mechanical data on the hierarchical structure and dynamics of water. The purpose of this software is to process coordinates of chemical systems into a graph formalism and apply topological network analyses that include network neighborhood, determination of geodesic pathways (the shortest edge path between two vertices, i.e., the shortest contiguous hydrogen bond pathway in the systems), the vertex degree census (the average distribution of hydrogen bonds), direct structural searches, and distribution of defect states of the network. The package enables a more comprehensive overview of different conformers compared to classical approaches, as explained in more detail within this study.

In this work, the *ChemNetworks* software has been applied to interaction studies between TIP4P solvent molecules and organic solutes, which thus provides a new way of contemplating large chemical systems. The software has been utilized to support the research of interaction networks between organic molecules and solvents and applied to fit the trajectory data format of common MD simulation programs, e.g., *Gromacs*. By utilizing *ChemNetworks*, the objective in this first study was to establish the new approach by carefully comparing results of the analyses by *ChemNetworks* with previously described analyses performed with *Gromacs*, and to investigate a more detailed treatment of the solvent organization around the solutes than has been possible with previously available tools.

In this study, the *ChemNetworks* software has been applied to TADDOL-like lignan-based chiral ligands (LIGNOLs) as shown in Figure 1. TADDOLs [9] have hindered structures containing two adjacent stereocenters, resulting in a fixed angle between the metal-complexing hydroxyl groups, and are often used as ligands for transition metal catalyzed asymmetric synthesis. The LIGNOL structures have been studied extensively [10–14] and are known on a molecular level to a great extent. LIGNOLs are a class of 1,4-diols based on the natural product group lignans, with the same catalytic functionality as TADDOLs [9].

Figure 1. General structure and numbering of atoms of the LIGNOLs (R, R′ = phenyl, methyl or hydrogen, which is explained in more detail in Table 1).

Table 1. Description of the investigated structures.

	R	R′
2Ph	Phenyl	H
3PhR	Phenyl	Phenyl, H
3PhS	Phenyl	H, Phenyl
4Met	Methyl	Methyl
4Ph	Phenyl	Phenyl

2. Results and Discussion

2.1. Hydration

In the LIGNOLs, there are four hydrogen bonding acceptor oxygen atoms in methoxy groups. However, the most interesting oxygens from the reaction point of view are those in the hydroxyl groups, O9–H and O9′–H, see Figure 1. In those groups, there are also two hydrogen bonding donors, which can be seen in Figure 2.

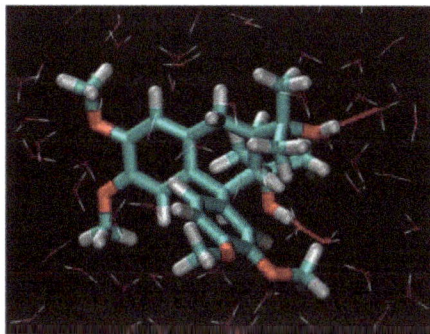

Figure 2. Hydrogen bonding of tetramethyl 1,4-diol with TIP4P water (hydrogen atoms white, carbon atoms cyan, oxygen atoms red; red dashed lines indicate hydrogen bonds).

In order to understand the hydration effect more properly the g_hbond analyzing program [15–19] implemented in *Gromacs* was used in a previous study [11] to get the number of hydrogen bonds for the oxygen atoms O9 and O9′, and totally for each LIGNOL conformer, as well as the average lifetime of the uninterrupted hydrogen bonds. The g_hbond routine calculates the hydrogen bond correlation function (CHB$_{(t)}$) by accounting for the hydrogen bonds between a specific donor–acceptor pair at different times (t_1, t_2), even if the hydrogen bond is absent in the interval between t_1 and t_2. By integrating the resulting hydrogen bond correlation function, the average lifetime of a hydrogen bond (the term is explained in detail in Section 2.2) can be calculated [15–19].

These are shown for comparison in the uneven columns in Table 2. The g_hbond program computes and analyses hydrogen bonds between all possible donors (D) and acceptors (A), but cannot perform analyses in graph formalism, e.g., geodesics, nor identify water-mediated hydrogen bond bridges.

In Table 2, the average number of hydrogen bonds per timeframe between the LIGNOLs and the TIP4P solvent used in ref. [11], are compared to the average hydrogen bond degree between TIP4P and the sites of interest in the LIGNOLs. For each property, a mean value was calculated for the three conformers of each LIGNOL structure chosen from previous studies [11] to avoid individual deviations. The notation for the LIGNOLs is taken from ref. [10], i.e., **2Ph** meaning diphenyl 1,4-diol, **3Ph** meaning triphenyl 1,4-diol, **4Ph** meaning tetraphenyl 1,4-diol, and **4Met** meaning tetramethyl 1,4-diol, and are summarized in Table 2. The numbers originally referred to the three quantum chemically most stable conformers [10] of each of the LIGNOLs, but when used in simulations, they refer to different starting points to cover more phase space.

The columns with a "g_" before contain the number of hydrogen bonds per timeframe calculated in ref. [11] with *Gromacs* tools. "Deg" is a weighted average hydrogen bond degree, as the output of *ChemNetworks* gives the number of observations (time frames) separately for each degree of a specific site. The analysis takes both hydrogens of water into account separately, so if a water molecule arranges itself to an interaction site such that both of the two H atoms are close to that site within the hydrogen bond cutoff, then the degree will be counted as two. The actual number is divided by two to get the degree concerning the number of water molecules interacting with the LIGNOL, which is comparable to the g_hbond analysis. The indication that the same water is counted twice (instead of a bifurcated hydrogen bond) can also be seen in the bond distance section.

In general, Table 2 (and the visualized data in Figure 3) shows that the results from *ChemNetworks* compared to the *Gromacs* analysis tool g_hbond are coherent, although slightly different hydrogen bond definitions are employed, as described in the experimental section. It can be seen that O9 exhibits a slightly higher affinity to form hydrogen bonds to the solvent molecules compared to O9′ [11]. The steric hindrance within the molecule strongly depends on the kind and number of substituents. Phenyl fragments show a much higher impact on the configuration of the molecule and towards restrictions to form hydrogen bonds.

Table 2. The average number of hydrogen bonds per timeframe (Gromacs) compared to the average hydrogen bond degree between TIP4P water and the sites of interest in the LIGNOLs (*ChemNetworks* = CNw).

Software	Gromacs	CNw	Gromacs	CNw	Gromacs	CNw
Conformation	g_num$_{O9}$	Deg$_{O9}$	g_num$_{O9'}$	Deg$_{O9'}$	g_num$_{tot}$	Deg$_{tot}$
2Ph1	1.24	1.00	0.98	0.75	6.23	5.66
2Ph2	1.23	1.06	0.98	0.80	6.26	5.93
2Ph9	1.41	1.22	0.80	0.74	6.08	5.92
Mean	1.29 ± 0.08	1.09 ± 0.09	0.92 ± 0.08	0.76 ± 0.03	6.19 ± 0.08	5.83 ± 0.12
3PhR3	1.16	1.02	0.71	0.70	5.54	5.64
3PhR4	1.11	0.99	0.78	0.76	5.53	5.64
3PhR5	1.27	0.94	1.15	0.91	6.08	5.79
Mean	1.18 ± 0.07	0.98 ± 0.03	0.88 ± 0.19	0.79 ± 0.09	5.72 ± 0.26	5.69 ± 0.07
3PhS3	1.25	1.08	0.65	0.69	5.58	5.71
3PhS7	1.12	1.01	0.80	0.77	5.60	5.67
3PhS10	1.10	0.95	1.14	0.95	6.51	5.89
Mean	1.16 ±0.07	1.01 ± 0.05	0.86 ± 0.20	0.80 ± 0.11	5.90 ± 0.43	5.75 ± 0.10
4Met2	1.20	0.97	1.33	0.97	6.35	5.93
4Met3	1.37	1.13	1.24	0.99	7.05	6.13
4Met6	1.37	1.16	1.24	1.02	7.04	6.28
Mean	1.31 ± 0.08	1.08 ± 0.08	1.27 ± 0.04	0.99 ± 0.02	6.81 ± 0.33	6.11 ± 0.14
4Ph3	0.62	0.67	1.05	0.90	5.31	5.51
4Ph4	0.68	0.67	0.95	0.84	5.23	5.41
4Ph8	0.79	0.77	0.85	0.80	5.26	5.50
Mean	0.70 ± 0.07	0.70 ± 0.05	0.95 ± 0.08	0.84 ± 0.04	5.27 ± 0.03	5.47 ± 0.04

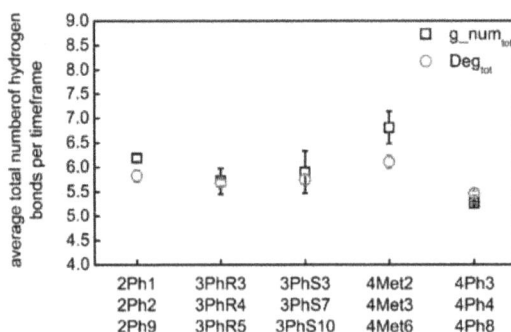

Figure 3. Comparision of the average total number of hydrogen bonds per timeframe derived from *ChemNetworks* (Deg$_{tot}$) and *Gromacs* (g_num$_{tot}$) analysis for different molecules.

A fact which is even more obvious in the *ChemNetworks* analysis is that tetramethyl 1,4-diol is more likely to form hydrogen bonds to TIP4P, and tetraphenyl less, mainly due to the small tendency of O9 to form hydrogen bonds to TIP4P. Figure 2 shows hydrogen bonding of tetramethyl 1,4-diol with TIP4P water and visualizes possible steric hindrance attributed to exchanging of the substituents.

2.2. Lifetime of Hydrogen Bonds

In Table 3, the average lifetime (in ps) of the uninterrupted hydrogen bonds between the LIGNOLs and the TIP4P solvent used in ref. [11], are compared to the mean lifetimes of the hydrogen bonds between water and the sites of interest in the LIGNOLs. The determined timeframe is usually referred to as resident time in case of a molecule or conformation. In the case of a hydrogen bond, it is rather called lifetime. Throughout the study, residence times of molecules were determined and interpreted as lifetimes of hydrogen bonds. For each property, a mean value was calculated for the three conformers of each LIGNOL structure to avoid separate deviations. The lifetimes calculated by *ChemNetworks* are doubled due to the same reason that the degrees were divided in two in the previous section. The water molecule stays in the vicinity of the referred site for a certain time, while, on average, half of the time the first hydrogen of water is interacting, and the other half of the time, the other hydrogen is interacting with the solute, meaning that the water molecule is dynamically mobile when it locates near the LIGNOL. This was also confirmed by changing the analysis code such that only the water molecule ID was traced instead of water hydrogen labels.

Table 3 summarizes the lifetimes of the uninterrupted hydrogen bonds between the LIGNOLs and the TIP4P solvent molecules. The results are visualized in Figure 4a for the O9 and in Figure 4b for the O9' atom. The unexpectedly long lifetime observed for O9' in ref. [11] cannot be found with *ChemNetworks*, although the general trend is the same. However, a correlation can be seen to the number of hydrogen bonds, as the lifetimes are longer for tetramethyl 1,4-diol and shorter for tetraphenyl. Shorter lifetimes for a large average number of hydrogen bonds may imply that they are slightly weak, meaning that the hydrogen bonds from O9' in tetramethyl 1,4-diol might be stronger compared to O9. These findings could be important for the application of these LIGNOLs as metal-binding agents, as their bonding to a metal-atom catalyst is comparable to the hydrogen bonding of the diol to TIP4P water. In other words, by changing the substituents of the LIGNOL molecules it is easily possible to design molecules with a certain strength of coordinating bonds. A comprehensive prediction by chemical simulations might then be advantageous over synthesizing different compounds in the lab without knowing about their coordination properties.

Diphenyl 1,4-diol was the only LIGNOL with phenyls at C9' and not at C9, thus the reason for this phenomenon was concluded to be the electronic effects of the phenyl rings at C9'.

Table 3. Lifetimes (in ps) of the uninterrupted hydrogen bonds between the LIGNOLs and the TIP4P solvent.

Software	Gromacs	CNw	Gromacs	CNw
Conformation	g_life_{O9}	$life_{O9}$	$g\_life_{O9'}$	$life_{O9'}$
2Ph1	3.35	3.30	6.18	3.68
2Ph2	3.26	3.12	5.82	3.52
2Ph9	2.29	3.02	2.24	2.92
Mean	2.97 ± 0.48	3.15 ± 0.12	4.75 ± 1.78	3.37 ± 0.33
3PhR3	2.03	3.02	1.67	2.82
3PhR4	1.97	2.96	1.72	2.96
3PhR5	2.88	3.82	3.15	3.40
Mean	2.29 ± 0.42	3.27 ± 0.39	2.18 ± 0.69	3.06 ± 0.25
3PhS3	2.08	3.08	1.58	2.70
3PhS7	1.96	2.92	1.78	3.04
3PhS10	3.53	3.18	3.35	3.16
Mean	2.52 ± 0.71	3.06 ± 0.11	2.24 ± 0.79	2.97 ± 0.19
4Met2	3.95	3.38	3.63	4.10
4Met3	3.57	3.40	3.83	3.40
4Met6	3.56	3.36	3.86	3.44
Mean	3.69 ± 0.18	3.38 ± 0.02	3.77 ± 0.10	3.65 ± 0.32
4Ph3	1.57	2.56	2.07	3.58
4Ph4	1.82	2.82	2.00	3.34
4Ph8	1.88	2.98	1.77	3.14
Mean	1.76 ± 0.13	2.79 ± 0.17	1.95 ± 0.13	3.35 ± 0.18

The results from *ChemNetworks* shows very clearly the effect of steric hindrance in the activity of the hydroxyl groups of interest. Tetramethyl 1,4-diol exhibits the least sterically demanding substituents, with a lot of space for several water molecules to move close to O9 and O9'.

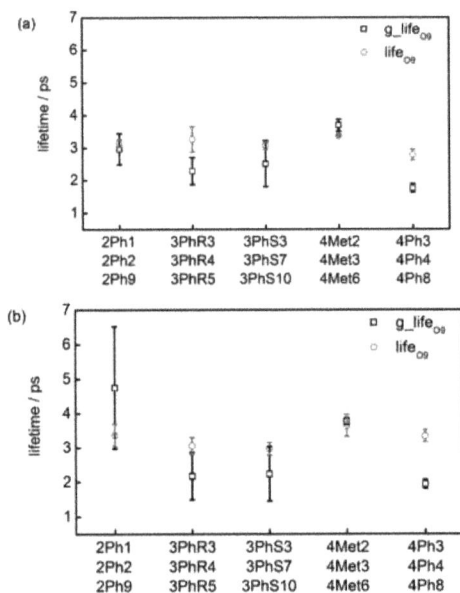

Figure 4. Comparison of the average lifetimes (in ps) of the uninterrupted hydrogen bonds of O9 (**a**) and O9' (**b**) derived from *ChemNetworks* (life) and *Gromacs* (g_life) analysis for different molecules.

The steric hindrance of all molecules is most pronounced for the tetraphenyl 1,4-diol, which is by far the least capable of forming hydrogen bonds combined with the shortest lifetimes especially in case of O9. The lifetimes of the four hydrogen bonding acceptor oxygen atoms in methoxy groups were also analyzed, and they varied between 1.04 and 1.18 ps in most cases with a few exceptions reaching up to 1.29 ps for O4 in triphenyl(S) 1,4-diol.

2.3. Distance Distributions

In the previous study with g_hbond [11], the mean value of the hydrogen bond lengths was calculated as 0.28 nm, which corresponds well with the general approximation of O···H hydrogen bond length of 0.18 nm in addition to the O–H bond length of 0.10 nm. Figure 5 shows the distribution of distances between tetramethyl 1,4-diol and TIP4P water for the whole solute and oxygen atom O9′ calculated by *ChemNetworks*.

As previously stated, in *ChemNetworks*, each interaction is treated separately for each site in every molecule for every frame in the trajectory. The distance distribution shown in Figure 5 shows one peak at around 0.18 nm corresponding to the water molecule that directly interacts with the LIGNOL site (the first hydration shell), while the second partial peak at 0.3 nm may have contribution from the first hydration shell (another hydrogen atom) and the second hydration shell (those water molecules that are not directly interacting with the LIGNOL, but interacting with the first hydration shell water). The second coordination shell usually gives broader distribution, and the reason for the illusionary sharpness of the secondary peak is that it is truncated in this case. *ChemNetworks* does not include the O–H bond length (approximately 0.1 nm) in the value of the hydrogen bond, so the average distance to the closer hydrogen is approximately 0.18 nm. A hydrogen bond distance cutoff of 0.25 nm would be enough for these systems since it is the minimum of this distribution which can exclude the second shell water in the analysis. Besides, this supports the double-interaction from the two hydrogen atoms of a water molecule, which can explain the difference in the degree and probably in the hydrogen bond lifetime.

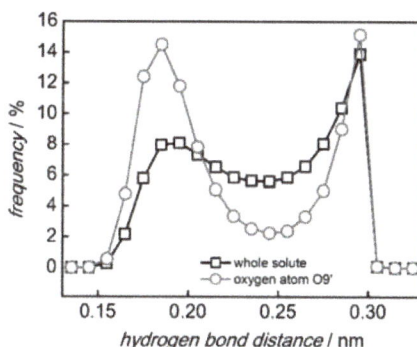

Figure 5. Distribution of distances between tetramethyl 1,4-diol and TIP4P water for the whole solute (black squares) and oxygen atom O9′ (grey circles).

3. Materials and Methods

In the LIGNOLs, there are four hydrogen bonding acceptor oxygen atoms in methoxy groups. However, the most interesting oxygens from the reaction point of view are those in the hydroxyl groups, O9–H and O9′–H, see Figure 1. In those groups, there are also two hydrogen bonding donors, which can be seen in Figure 2.

In order to understand the hydration effect more properly, the g_hbond analyzing program [15–19] implemented in *Gromacs* (version 4.5.3) was used in a previous study [11] to get the number of hydrogen bonds for the oxygen atoms O9 and O9′, and totally for each LIGNOL conformer, as well as the average lifetime of the uninterrupted hydrogen bonds. These are shown for comparison in the uneven columns in Table 3. The g_hbond program computes and analyses hydrogen bonds between all possible donors (D) and acceptors (A), but cannot perform analyses in graph formalism, e.g., geodesics, nor identify water-mediated hydrogen bond bridges. The existence of a hydrogen bond was determined by a geometrical criterion, i.e., $r(DA) \leq 0.35$ nm and $\alpha(HDA) \leq 30°$ [15–19].

The purpose of the *ChemNetworks* (version 2.2) software is to process Cartesian coordinates of the chemical systems into a graph formalism and apply topological network analyses, thereby describing intermolecular chemical networks of entire systems quantitatively at both the local and global levels and as a function of time. In Scheme 1, the analysis steps are described schematically.

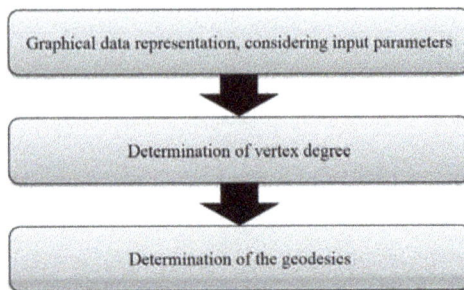

Scheme 1. Schematic flowchart for the *ChemNetworks* run (graph theoretical analysis).

Before the actual graph theoretical analysis, the Cartesian (.xyz) simulation data needs to be converted into a graph (an output file called *.graph). This procedure is accomplished by defining the system topology in an input file, where all atoms for each molecule as well as all internal bonds are listed. In this study, the interactions were calculated with the geometric criteria of the nonbonded distance $r(O \ldots H) < 0.30$ nm and the hydrogen bond angle $\alpha(HO \ldots H)$ unspecified, which means that all the possible angles are accepted as intermolecular interactions to get a proper calculation of the surrounding water molecules. Periodic boundary conditions (PBC) were applied in the original simulations, and these can be accounted for in *ChemNetworks* so that the graph can cross the PBC boundaries.

The first analysis includes determination of each vertex degree and the network neighborhood, which gives the edge distributions. In the .graph file containing the intermolecular interactions, all pairs of vertices sharing an edge are listed. In graph theory, the degree of a vertex is defined as a count of the number of connections to that vertex. In this first-degree analysis step, the numbers of edges per vertex are collected for a histogram of the edge distribution. The edge distribution is essentially the same as the integrated pair distribution function but split into its intrinsic components.

The second analysis involves the determination of the geodesics, in this study meaning the shortest contiguous hydrogen bond paths between every pair of vertices, and their lifetimes. This can be done separately for each site specified in the system; in this case, different oxygen atoms. In chemical systems, all possible paths for the intermolecular interaction graph are not that trivial to analyze mathematically. However, the adjacency matrix can be converted to a geodesic distance (gd) matrix via the Floyd–Warshall algorithm [20,21], which is not possible by *Gromacs* analysis tools. The gd matrix is a square matrix with the dimension N, i.e., the total number of vertices. The entries of the matrix are the geodesics [22] (geodesic distances), i.e., the number of edges in the shortest path connecting two vertices (atoms) in a graph. The output of the geodesics analysis (.geopath) is a list for each graph of all paths between all sets of vertices connected by edges.

The last part of the analysis in this study comprises the residence times of the solute–solvent interactions, which are calculated as the average of all hydrogen bond durations weighted by the relative concentration of each species with a specific residence time in solution. Considering all individual edges for the residence time determination might be too memory intensive for simple spreadsheet programs. Thus, the *geodesic-statistics.c* code of *ChemNetworks* is strongly preferred.

To conclude the description of the analysis algorithms, the solute configurations with a certain number of interactions (*i*) with solvent molecules are identified for every snapshot, and the corresponding number of observed occurrences is recorded as $N(i)$. The degree census is obtained as the statistic histogram of $N(i)$. The solute–solvent interaction is monitored concerning individual sites from the solute molecule such that the degree per site can be resolved. Thus, the degree census depicts the structure of solute–solvent interactions around each site. On the other hand, the solute–solvent residence time is representative of the dynamic feature of the solute–solvent interaction. The continuous persistence (t_i) of every solute–solvent interaction is traced along the trajectory. Finally, the residence time of solute–solvent interactions is calculated as the average interaction persistence weighted by its relative occurrence probability $P(t_i)$. The residence time of the solute–solvent interaction is also labeled concerning the interacting sites so that the residence time of each site can be obtained.

Data Format and Working Procedure

The trajectory data format of common MD simulation programs, e.g., *Gromacs*, is binary, and one of the important practical tasks is to make a recipe for how to treat the trajectory data to be readable by *ChemNetworks*. In this study, it has been done by applying the compressed and portable trajectory format (.xtc) in combination with the structure format (.gro) and saving it as Cartesian coordinates (.xyz) in VMD [23]. Although Gromacs tools (trj_conv) can convert trajectories to human readable formats, this was found to be the most convenient way to make them *ChemNetworks* readable.

For each structure, a multi-level deterministic structural optimization was conducted in earlier studies, including complementary QC calculations [10] (step 1 in Scheme 2), and MD simulations [11] (step 2 in Scheme 2). The optimizations were performed using DFT [24] with the B3LYP hybrid exchange-correlation functional [25–27] in combination with the MARI-J approximation [28–30] and the TZVP basis set [31] for all atoms, as implemented in the *Turbomole* program package. The MD simulations were performed using the *Gromacs* version 4.5.3 software [15–19]. Water was described using the TIP4P model [32], and the LIGNOLs were modeled with the OPLS-AA force field [33] implemented in *Gromacs*. The topologies of the LIGNOLs were constructed manually and they comprised between 533 (**4Ph**) and 369 (**4Met**) internal coordinates, respectively. To get reasonable atomic charges to help choose suitable atom types with the hand-tuned charges available in the force field, electrostatic potential fit (ESP) charges were studied with GAMESS at HF/6-31G* level. For O9 and O9′ (shown in Figure 1), the OPLS atom type opls_154 with the atomic charge −0.683 was found to be the most suitable, and for the other four oxygens (O3, O4, O4′, and O5′), the atom type opls_179 with the atomic charge −0.285 was chosen. The parametrization is crucial information when explaining differences in persistence by electronic effects which are not intrinsically described in a force field. An important detail to consider is also that the sum of the atomic charges in a charge group should be an integer or equal to zero. Each conformation was placed at the center of a cubic box with the dimension between 5.2 and 5.6 nm (volume = 144–174 nm^3) and solvated by 4802–5795 water molecules. The original simulations in ref. [11] were run for ten ns with a one fs time step at 298 K and 1 atm. The whole trajectory was used for the analysis. A cutoff of 0.9 nm was applied to short-range nonbonded interactions, and for long-range electrostatic interactions, the particle mesh Ewald (PME) method [34,35] was used with a grid spacing of 0.12 nm and fourth-order interpolation. In all simulations, system snapshots were collected every 500 steps, i.e., every 0.5 ps, for subsequent analysis. In this time, only electronic excitations and bonding vibrations will occur, but those can be ignored when studying the conformational preferences of the system. It is important to get a comprehensive view of all the studied systems to locate possible shortcomings and develop the software to handle those. The flowchart for the working procedure is shown in Scheme 2.

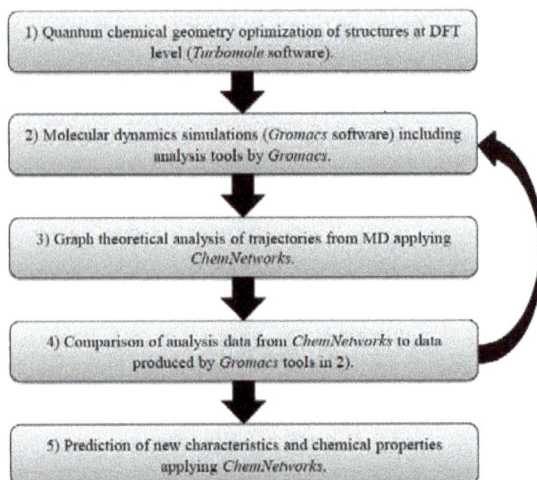

Scheme 2. Flowchart for the working procedure.

4. Conclusions

ChemNetworks gives a more detailed description of solvent organization around the solutes as the output gives the number of observations separately for each degree of a specific site and both hydrogens of the TIP4P solvent are taken into account separately. This result was confirmed in all three parts of the performed analysis. The general objective of this study, i.e., to establish the new approach by careful comparison of the analyses by *ChemNetworks* with previously described analyses performed with *Gromacs*, was accomplished. A successful recipe was proposed for applying the graph theoretical analysis software to molecular dynamics trajectories by *Gromacs*. In the hydration studies, tetramethyl 1,4-diol was found to be most likely to form hydrogen bonds, which could be important for application of the LIGNOL as a metal-binding agent. Also, the residence times for tetramethyl 1,4-diol were found to be longer, meaning that the hydrogen bonds might be strong. The results from *ChemNetworks* show very clearly the effect of a steric hindrance in the activity of the hydroxyl groups of interest. The hydration studies of the MD simulations confirm that several of these LIGNOLs, produced from a renewable source, have great potential as chiral catalysts.

Author Contributions: T.O.S. performed the experiments. T.O.S., C.W. and J.-H.S. interpreted the results and wrote the manuscript.

Funding: This research received no external funding.

Acknowledgments: Svenska tekniska vetenskapsakademien i Finland and the Alfred Kordelin Foundation are kindly acknowledged for mobility grants. T.O.S. is also very grateful to Clark for the opportunity of working in the Clark group at Washington State University as well as utilizing their computational resources.

Conflicts of Interest: The authors declare no conflict of interest.

References

1. Van der Spoel, D.; van Maaren, P.J.; Larsson, P.; Tîmneanu, N. Thermodynamics of hydrogen bonding in hydrophilic and hydrophobic media. *J. Phys. Chem. B* **2006**, *110*, 4393–4398. [CrossRef] [PubMed]
2. Smith, P.E.; Pettitt, B.M. Modeling solvent in biomolecular systems. *J. Phys. Chem.* **1994**, *98*, 9700–9711. [CrossRef]
3. Jorgensen, W.L.; Tirado-Rives, J. Potential energy functions for atomic-level simulations of water and organic and biomolecular systems. *Proc. Natl. Acad. Sci. USA* **2005**, *102*, 6665–6670. [CrossRef] [PubMed]

4.	Kuo, J.L.; Coe, J.V.; Singer, S.J.; Band, Y.B.; Ojamäe, L. On the use of graph invariants for efficiently generating hydrogen bond topologies and predicting physical properties of water clusters and ice. *J. Chem. Phys.* **2001**, *114*, 2527–2540. [CrossRef]

5.	Mooney, B.L.; Corrales, L.R.; Clark, A.E. MoleculaRnetworks: An integrated graph theoretic and data mining tool to explore solvent organization in molecular simulation. *J. Comput. Chem.* **2012**, *33*, 853–860. [CrossRef] [PubMed]

6.	Mooney, B.L.; Corrales, L.R.; Clark, A.E. Novel analysis of cation solvation using a graph theoretic approach. *J. Phys. Chem. B* **2012**, *116*, 4263–4275. [CrossRef] [PubMed]

7.	Hudelson, M.; Mooney, B.L.; Clark, A.E. Determining polyhedral arrangements of atoms using PageRank. *J. Math. Chem.* **2012**, *50*, 2342–2350. [CrossRef]

8.	Ozkanlar, A.; Clark, A.E. ChemNetworks: A complex network analysis tool for chemical systems. *J. Comput. Chem.* **2013**, *35*, 495–505. [CrossRef] [PubMed]

9.	Seebach, D.; Beck, A.K.; Schiess, M.; Widler, L.; Wonnacott, A. Some recent advances in the use of titanium reagents for organic synthesis. *Pure Appl. Chem.* **1983**, *55*, 1807–1822. [CrossRef]

10.	Sandberg, T.; Brusentsev, Y.; Eklund, P.; Hotokka, M. Structural analysis of sterically hindered 1,4-diols from the naturally occurring lignan hydroxymatairesinol a quantum chemical study. *Int. J. Quantum Chem.* **2011**, *111*, 4309–4317. [CrossRef]

11.	Sandberg, T.; Eklund, P.; Hotokka, M. Conformational solvation studies of LIGNOLs with molecular dynamics and conductor-like screening model. *Int. J. Mol. Sci.* **2012**, *13*, 9845–9863. [CrossRef] [PubMed]

12.	Brusentsev, Y.; Sandberg, T.; Hotokka, M.; Sjöholm, R.; Eklund, P. Synthesis and structural analysis of sterically hindered chiral 1,4-diol ligands derived from the lignan hydroxymatairesinol. *Tetrahedron Lett.* **2013**, *54*, 1112–1115. [CrossRef]

13.	Sandberg, T.; Eklund, P. The effect of density functional dispersion correction (DFT-D3) on lignans. *Comput. Theor. Chem.* **2015**, *1067*, 60–63. [CrossRef]

14.	Sandberg, T. *Computational Chemistry Studies of Wood-Derived Lignans*; Åbo Akademi University: Turku, Finland, 2013.

15.	Bekker, H.; Berendsen, H.J.C.; Dijkstra, E.J.; Achterop, S.; Vondrumen, R.; Vanderspoel, D.; Sijbers, A.; Keegstra, H.; Renardus, M.K.R. Gromacs—A parallel computer for molecular-dynamics simulations. *Phys. Comput.* **1993**, *92*, 252–256.

16.	Berendsen, H.J.C.; van der Spoel, D.; van Drunen, R. GROMACS: A message-passing parallel molecular dynamics implementation. *Comput. Phys. Commun.* **1995**, *91*, 43–56. [CrossRef]

17.	Lindahl, E.; Hess, B.; van der Spoel, D. GROMACS 3.0: A package for molecular simulation and trajectory analysis. *J. Mol. Model* **2001**, *7*, 306–317. [CrossRef]

18.	Van der Spoel, D.; Lindahl, E.; Hess, B.; Groenhof, G.; Mark, A.E.; Berendsen Herman, J.C. GROMACS: Fast, flexible, and free. *J. Comput. Chem.* **2005**, *26*, 1701–1718. [CrossRef] [PubMed]

19.	Hess, B.; Kutzner, C.; van der Spoel, D.; Lindahl, E. GROMACS 4: Algorithms for highly efficient, load-balanced, and scalable molecular Simulation. *J. Chem. Theory Comput.* **2008**, *4*, 435–447. [CrossRef] [PubMed]

20.	Floyd, R.W. Shortest path. *J. ACM* **1962**, *5*, 345. [CrossRef]

21.	Warshall, S. A theorem on boolean matrices. *J. ACM* **1962**, *9*, 11–12. [CrossRef]

22.	Bouttier, J.; Di Francesco, P.; Guitter, E. Geodesic distance in planar graphs. *Nucl. Phys. B* **2003**, *663*, 535–567. [CrossRef]

23.	Humphrey, W.; Dalke, A.; Schulten, K. VMD: Visual molecular dynamics. *J. Mol. Graph. Model.* **1996**, *14*, 33–38. [CrossRef]

24.	Treutler, O.; Ahlrichs, R. Efficient molecular numerical integration schemes. *J. Chem. Phys.* **1995**, *102*, 346–354. [CrossRef]

25.	Becke, A.D. Density-functional exchange-energy approximation with correct asymptotic behavior. *Phys. Rev. A* **1988**, *38*, 3098–3100. [CrossRef]

26.	Lee, C.; Yang, W.; Parr, R.G. Development of the colle-salvetti correlation-energy formula into a functional of the electron density. *Phys. Rev. B* **1988**, *37*, 785–789. [CrossRef]

27.	Becke, A.D. Density-functional thermochemistry. III. The role of exact exchange. *J. Chem. Phys.* **1993**, *98*, 5648–5652. [CrossRef]

28. Eichkorn, K.; Treutler, O.; Öhm, H.; Häser, M.; Ahlrichs, R. Auxiliary basis sets to approximate coulomb potentials. *Chem. Phys. Lett.* **1995**, *242*, 652–660. [CrossRef]

29. Eichkorn, K.; Weigend, F.; Treutler, O.; Ahlrichs, R. Auxiliary basis sets for main row atoms and transition metals and their use to approximate coulomb potentials. *Theor. Chem. Acc.* **1997**, *97*, 119–124. [CrossRef]

30. Sierka, M.; Hogekamp, A.; Ahlrichs, R. Fast evaluation of the coulomb potential for electron densities using multipole accelerated resolution of identity approximation. *J. Chem. Phys.* **2003**, *118*, 9136–9148. [CrossRef]

31. Schäfer, A.; Huber, C.; Ahlrichs, R. Fully optimized contracted gaussian basis sets of triple zeta valence quality for atoms Li to Kr. *J. Chem. Phys.* **1994**, *100*, 5829–5835. [CrossRef]

32. Jorgensen, W.L.; Chandrasekhar, J.; Madura, J.D.; Impey, R.W.; Klein, M.L. Comparison of simple potential functions for simulating liquid water. *J. Chem. Phys.* **1983**, *79*, 926–935. [CrossRef]

33. Jorgensen, W.L.; Maxwell, D.S.; Tirado-Rives, J. Development and testing of the OPLS all-atom force field on conformational energetics and properties of organic liquids. *J. Am. Chem. Soc.* **1996**, *118*, 11225–11236. [CrossRef]

34. Darden, T.; York, D.; Pedersen, L. Particle mesh ewald: An N-log (N) method for Ewald sums in large systems. *J. Chem. Phys.* **1993**, *98*, 10089–10092. [CrossRef]

35. Essmann, U.; Perera, L.; Berkowitz, M.L.; Darden, T.; Lee, H.; Pedersen, L.G. A smooth particle mesh Ewald method. *J. Chem. Phys.* **1995**, *103*, 8577–8593. [CrossRef]

Sample Availability: Samples of the compounds are not available from the authors.

MDPI

St. Alban-Anlage 66

4052 Basel

Switzerland

Tel. +41 61 683 77 34

Fax +41 61 302 89 18

www.mdpi.com

Molecules Editorial Office

E-mail: molecules@mdpi.com

www.mdpi.com/journal/molecules

www.ingramcontent.com/pod-product-compliance
Lightning Source LLC
Chambersburg PA
CBHW051708210326
41597CB00032B/5412